GEOSTATISTICAL RESERVOIR MODELING

SECOND EDITION

GEOSTATISTICAL RESERVOIR MODELING

MICHAEL J. PYRCZ

CLAYTON V. DEUTSCH

Oxford University Press is a department of the University of Oxford. It furthers the University's objective of excellence in research, scholarship, and education by publishing worldwide.

Oxford New York
Auckland Cape Town Dar es Salaam Hong Kong Karachi
Kuala Lumpur Madrid Melbourne Mexico City Nairobi
New Delhi Shanghai Taipei Toronto

With offices in
Argentina Austria Brazil Chile Czech Republic France Greece
Guatemala Hungary Italy Japan Poland Portugal Singapore
South Korea Switzerland Thailand Turkey Ukraine Vietnam

Published in the United States of America by
Oxford University Press
198 Madison Avenue, New York, NY 10016

Library of Congress Cataloging-in-Publication Data
Deutsch, Clayton V.
Geostatistical reservoir modeling / Clayton Deutsch, Michael Pyrcz. — 2nd edition.
p. cm.
Includes bibliographical references and index.
ISBN 978-0-19-973144-2 (alk. paper)
1. Hydrocarbon reservoirs—Mathematical models. 2. Petroleum—Geology—Statistical methods. I. Pyrcz, Michael. II. Title.
TN870.53.D48 2014
553.2'82015195—dc23
2013031615
9780199731442

3 5 7 9 8 6 4 2

Printed in the United States of America
on acid-free paper

CONTENTS

PREFACE

This second edition builds on the first with an expanded and more comprehensive treatment, consistent reservoir examples, expanded prerequisites, concepts, algorithms, and applications. The preparation of this much expanded text provided an opportunity to modernize the presentation with the exciting new developments from the past decade and improve the linkages to the geology. Descriptions of the data and thought process associated with problem formulation are much expanded.

This book presents geostatistical tools for building numerical geological models of petroleum reservoirs. The use of geostatistical tools for modeling reservoir heterogeneities and assessing uncertainty in reservoir forecasting has increased significantly since the 1990s. There are many technical papers that track this growth in theory and applications; however, few collected works exist to bring together the practice of petroleum geostatistics into a coherent framework.

Building a suitable reservoir model calls upon the skills of many people. Geologists provide critical input on the sedimentology and stratigraphy of the subsurface. Geophysicists provide valuable information on the geometry of the reservoir and the internal distribution of reservoir properties in the interwell region. Engineers, with their knowledge of flow processes and production data, provide critical information on connectivity and major heterogeneities. We do not address any single discipline; the emphasis is on the interdisciplinary interaction necessary to build numerical geological models consistent with all available information sources.

This book is aimed at the *practice* of geostatistical reservoir modeling; it is not intended to be a theoretical reference textbook. We focus on tools, techniques, examples, tradecraft, and providing guidance on the practice of petroleum reservoir modeling. *Geostatistical Reservoir Modeling* would serve as a reference text for the practitioner and be appropriate for an advanced undergraduate or graduate class on reservoir characterization.

Michael J. Pyrcz
Clayton V. Deutsch

ACKNOWLEDGMENTS

Many people and organizations contributed to the preparation of this book. We would like to extend special thanks to Chevron for permitting Michael to undertake the significant effort required for this project and specifically recognize Dr. Sebastien Strebelle, Dr. Jean-Baptiste Clavaud, and Carlos Hanze for their work in reviewing the book on behalf of Chevron. We would like to acknowledge the industrial affiliates and researchers of the Center for Computational Geostatistics (CCG) at the University of Alberta. We would like to especially thank Saina Lajevardi for assistance in assembling the book and for helping with so many details. Also, Michele Tomlinson from the Society for Sedimentary Research (SEPM) was extremely helpful in providing images and permission to reproduce them in the book. Dr. Steve Bachtel, Dr. Jake Covault, Alte Folkestad, Dr. Tao Sun, and Dr. Brian Willis provided exceptional critical review of sections related to their fields of expertise. We appreciate the willingness of Dr. Olena Babak, Prof. Jeff Boisvert, Prof. Octavian Catuneanu, Jared Deutsch, Dr. Andrea Fildani, Prof. Stephen Hubbard, Prof. Chris Kendall, Dr. Oy Leuangthong, Dr. John Manchuk, Prof. Chris Paola, Dr. Ted Playton, Dr. Henry Posamentier, Prof. David Pyles, Prof. Kyle Straub, Dr. Morgan Sullivan, Brandon Wilde, and Larry Zarra, along with many others, to share their work and expertise.

Michael would like to recognize his family. His young children (Aidan, Clayton, and Emily) have provided comic relief and perspective. His wife Tobi has significantly shared this challenge and provided constant support. During the preparation of this book, both of Michael's parents passed away. Michael wishes to acknowledge their infectious love for natural sciences.

Clayton would also acknowledge his family. His loving wife Pauline has been there through many years of geostatistical adventures. The children (Jared, Rebecca, and Matthew) are far from home, but not far from his thoughts.

M. J. P.
C. V. D.

GEOSTATISTICAL RESERVOIR MODELING

1

Introduction

Geostatistical tools are commonly being used for the modeling of petroleum reservoirs and yet the theoretical descriptions of those tools, case studies, and tradecraft details are widely dispersed in a variety of technical journals, books, and software not all related to petroleum geostatistics. This work aims at collecting that dispersed information into a handbook for production geologists, reservoir geophysicists, and reservoir engineers involved in building and using numerical reservoir models.

No software is included in this book. Algorithms are described with flow charts that illuminate the methodology. Source code for most algorithms can be found in the public domain GSLIB software Deutsch and Journel (1998) or other cited sources. There are a number of commercial software packages available: Halliburton (2012), Schlumberger (2012), and Paradigm (2012), among others.

This book is written for undergraduate to graduate students and the practicing geologist, geophysicist, and engineer desiring to use geostatistical tools. The essential theory will be developed in the context of modeling petroleum reservoirs. Most theoretical sections could be skipped by readers with a "high-level" understanding of geostatistics. Those who seek to major in petroleum geostatistics, however, should master the modest statistics and mathematics presented here. Original work and papers presenting greater detail are cited for those seeking a solid background.

This book is not designed around the modeling of a single reservoir. Instead, a number of different reservoir data sets will be used to illustrate specific techniques. This makes it easier to present the diversity of useful techniques and to demonstrate the relative strengths and weaknesses of alternative approaches.

Various methods are presented. Guidance is given on when to use each particular method. A fascinating, but often frustrating, aspect of geostatistics is that there is a large combination of tools that can be brought to bear on any problem. No attempt has been made to present an exhaustive list of tools; however, citations to methods not developed have been included. Omissions are the responsibility of the authors.

1.1 COMMENTS ON SECOND EDITION

Many new developments have emerged in geostatistics since the release of the first edition in 2002. There have been dramatic changes in the practice of nonstationary statistical characterization, facies modeling, and uncertainty models and management, to name a few. This second edition provides an opportunity to integrate these new topics. In addition, we recognized that the first edition did not thoroughly cover some important topics of geostatistical reservoir modeling. This is an opportunity to produce a more comprehensive treatment, with coverage of the geological concepts and geological story of the reservoir, data inventory, conceptual model, problem formulation, large-scale modeling, reporting and documentation, and uncertainty management.

An attempt has been made to strengthen the linkages to the geological roots of geostatistical reservoir modeling. Geological practice has become more quantitative, and geostatistical methods have developed to improve the integration of geologic information in the models. We capitalize on this to close the gap by describing these efforts in geological sciences and the geostatistical parallels. Consider the recent work by various geological experts to characterize hierarchical architectures for various depositional settings. Some of these have included various statistical descriptions of their associated heterogeneities. By seeking deeper geological understanding, integration of concepts related to geologic process and resulting architectures, and uncovering the geologic story behind the reservoir, improved reservoir models are possible with improved integrations of geologic insights, model credibility, and ultimately decision quality.

A critical feature of geostatistics is the ability to integrate multiple, disparate data sources. A section has been added to cover the aspects of each data source, including typical coverage, resolution, assumptions, and limitations. In addition, other information sources not considered in the first edition, such as analog outcrops, flume experiments, and numerical process-based models, have been added. With the aforementioned increasing quantitative nature of geology, these additional data sources are now providing rich insights that may aid in the statistical inference and decisions required in geostatistical reservoir modeling. A general formalism is presented to represent data in a more complete manner as *data events*. In addition, further statistical details are provided on methodologies that allow for the combination of multiple, redundant data sources.

With the expansion of geostatistical methods, there are more choices on the specific methods to apply. While there are common work flow steps, there is a need to choose between competing techniques and to develop tailored work flows that are fit-for-purpose. This has necessitated additional comparison and contrasting of the available methods with implementation details, strengths, and limitations. This has been formalized in sections dedicated to conceptual models and problem formalization.

With increased rigor in uncertainty modeling, efforts have been made to integrate large-scale modeling and associated uncertainty into the geostatistical work flow. This includes modeling the reservoir container, trends within the reservoir, separation of unique regions, and methods for combining multiple attributes to distill local information. Coverage of these topics allows for improved control on this first-order constraint on reservoir volumetrics.

In addition, with regard to the general practice of uncertainty modeling, there have been important changes. Greater rigor is now typical in seeking out and describing all sources of uncertainty. Each model choice, input parameter, and statistic is a potential source of uncertainty. As mentioned above, the model container, layers, correlation, regions, and trends are now considered potentially uncertain. In addition, new expressions are available to communicate local and global model uncertainty and the important related topic of risk. Additional methods for exploring uncertainty are covered.

Discussion is included on project documentation and reporting. With increasing work flow complexity, number and size of input and analog data sets, and an ever-present reliance on expert judgment, rigorous project documentation and reporting are essential. This is especially important, given that many of today's complicated reservoir projects are executed over decades; the geostatistical reservoir modeling often outlives the project teams. The tendency to document in PowerPoint and not in detailed reports should be questioned. Without the effort and good practice required in this area, significant professional time is wasted and mistakes may be made and even repeated.

While an effort has been made to expand the material in the first edition, this edition preserves the focus on the practice of geostatistics with an effort to utilize accessible language to a competent earth science or engineering professional and the absence of complete theoretical underpinnings. It represents the amalgamation of practical information and tradecraft with reference to theory to those seeking deeper understanding. This book should also prove to be a valuable textbook for advanced undergraduate or graduate-level university classes.

Even with this expanded second edition, it is not possible to cover all geostatistical methods. At times, choices were made to omit or to only include cursory coverage and references for specific topics. This decision was made based on the acceptance and availability of specific methods in practice. Yet, as with the first edition, this edition focuses on methods and work flows that are currently in practice. Current and alternative research will only be briefly discussed in the final chapter and referenced to provide context for the present and a vision of the future state of practice.

1.2 PLAN FOR THE BOOK

There are six chapters addressing the application of geostatistics to reservoir modeling. A petroleum-specific glossary of geostatistical terms and a list of geostatistics notation is given at the end. The bibliography provides a resource for newcomers to the field. Acknowledging that few readers can pick up a book and read it cover to cover, a comprehensive index has been prepared. This book attempts to answer five questions:

1. What is reservoir modeling and what is the role of geostatistics? (Chapter 1)
2. What essential background do we need before we engage in geostatistical reservoir modeling? (Chapters 2 and 3)
3. What are the main steps in reservoir modeling and where/how does geostatistics intervene? (Chapter 4)
4. What do we do with the models once we have them? (Chapter 5)
5. What promising less mature or new developments might we foresee in the future? (Chapter 6)

A more specific description of each chapter follows.

This chapter presents an overview of the book, the steps and methodologies involved in typical reservoir modeling, the information and data that are integrated to accomplish this, and how the different disciplines of geology, geophysics, geostatistics, and petroleum engineering work together. Flow charts are presented throughout the book to illustrate the flow of data, the geostatistical operations, and the decisions to be taken during reservoir modeling. High-level flow charts are presented in this chapter; detailed flow charts for specific tasks are presented in later chapters.

Modeling prerequisites related to reservoir modeling are covered in Chapter 2. These principles are covered in the Preliminary Geological Concepts, Preliminary Statistical Concepts, Quantifying Spatial Continuity, and Preliminary Mapping Concepts sections.

The first step in any geostatistical reservoir model is to uncover the geological story behind the formation of the reservoir as described in the Preliminary Geological Concepts section. This will aid in the integration of all available information and result in consistent, plausible, and credible reservoir models. Understanding of the various geological controls and the geological models that describe their resulting architectures is necessary to communicate with geoscientists in the project team and to best utilize their expert knowledge. Examples are provided of three distinct geologic settings: the McMurray Formation in Northern Alberta, Canada, the Lower Wilcox Formation in present-day deepwater Gulf of Mexico, and the Tengiz Platform in Kazakhstan. Basic geological stories are told for each setting, with associated linkages to reservoir modeling decisions.

These geological concepts must be mapped into statistical concepts. Each core plug is unique, and yet we must pool samples together before calculating any statistics or building a reservoir model. Considerations for choosing geological populations are presented. There are a number of classical statistical tools that can help with the selection and description of geological populations. Introductory geostatistical tools, such as declustering and correcting for nonrepresentative data, are also presented in the Preliminary Statistical Concepts section.

To move away from the data, we need to account for spatial correlation. The Quantifying Spatial Continuity section provides the additional tools to statistically analyze and quantify spatial data central to reservoir modeling. Variogram calculation, interpretation, and modeling are covered with numerous examples from real reservoirs. Indicator variograms, multiple point statistics, and geological rules are presented as more sophisticated measures of spatial correlation.

The Preliminary Mapping Concepts section builds on these previous sections. This section presents the background for kriging-based geostatistical techniques. The use of kriging and cokriging is widespread in geostatistical modeling. In addition, the framework for sequential simulation methods is presented as a preparation for subsequent discussion on sequential Gaussian, indicator, and multiple-point simulation methods. The background of such techniques and their practical application are discussed. The p-field simulation technique is presented. The use of locally varying directions of continuity and trend models is presented. This chapter collects essential geostatistics theory together with small illustrative examples. More extensive case studies are presented in Chapter 4.

Chapter 3 discusses the modeling prerequisites. The sections in this chapter include Data Inventory, Conceptual Model, and Problem Formulation. With the geological, statistical, and geostatistical prerequisites in place, these are the considerations for compiling the data, making decisions on the best work flow and methods to complete the models, and determining how to formulate the problem.

The Data Inventory section provides discussion on the typical data available for geostatistical reservoir models, scale, accuracy, coverage, and data checking. In addition, details are provided on methods for data transformation, calibration, and combination. This is often more challenging than anticipated, as there are often gaps in data coverage, issues with data quality, and multiple, redundant sources of information.

Geostatistics is a toolbox of various spatial modeling tools. It is important to choose the best tool or set of tools given the modeling objectives, modeling constraints, and data available. The Conceptual Model section discusses details on data integration and algorithm selection, along with a general discussion on fit-for-purpose modeling. That is the choice of tools and their sequences to meet the objectives of the reservoir study. Common work flow steps and comparison of the available tools are presented to aid in these modeling choices.

In the Problem Formulation section, details are provided on modeling goals, modeling constraints, and problem scoping. These are essential for fit-for-purpose modeling and form the basis for all model choices. In all reservoir modeling studies, resources, project and professional time, and money are limited. Good problem formulation can maximize use of these resources and maximize the ability of the resulting model to add value to the project.

Chapter 4 provides details on the specific modeling methods and examples of their applications. This includes sections on Large-Scale Modeling, Cell-Based Facies Modeling, Multiple-Point Facies Modeling, Object-based Facies Modeling, Process Mimicking Facies Modeling, Porosity and Permeability Modeling, and Optimization for Model Construction. These are the modeling tools in the geostatistical tool box to produce models of reservoir heterogeneity that honor available data and expert knowledge. The typical hierarchical approach is composed of first a model of the large-scale container, then a model of the facies within the container, and finally a model of reservoir properties within the distinct facies. The distribution of facies must be established prior to modeling of continuous properties such as porosity and permeability. In fact, it is common for reservoir porosity and permeability heterogeneity to be dominated by the facies heterogeneity, as most of the porosity and permeability variability and continuity may be captured in the facies model. This is the rationale behind four facies modeling sections as compared to one porosity and permeability model section in this chapter.

Large-scale reservoir modeling decisions have a major impact on volumetrics and large-scale connectivity. Large-Scale Modeling covers methodologies for modeling the reservoir container, trends within the reservoir, separation of unique regions, methods for combining multiple attributes to distill local information, and subsequent summarization and visualization. Geostatistical tools are particularly well suited to work in convenient rectangular Cartesian coordinate systems; however, most reservoirs are complicated by structural deformation, erosion, and faulting. This chapter reviews coordinate transformations to represent or grid complex real-world reservoirs in Cartesian coordinates. When there is significant uncertainty in these constraints, quantification and integration of uncertainty with respect to large-scale modeling is necessary.

The Cell-Based Facies Modeling section presents cell-based facies modeling techniques—that is, methods that assign facies on a cell-by-cell basis with variogram statistical control. Sequential indicator simulation and truncated Gaussian simulation, with their variants, are the important cell-based techniques. Procedures to remove unwanted short-scale variations (noise) from the resulting realizations are also discussed.

Since the first edition of this book, multiple-point simulation has become part of common practice. The Multiple-Point Facies Modeling section provides details on implementation of this technology and gives examples demonstrating the unique strengths of this method. Building from previous discussion on quantification of multiple-point statistics and the sequential simulation framework, implementation details and examples are provided for this method.

Hierarchical, object-based modeling schemes are considered appropriate in specific geological environments where the facies are organized in clear geometric units. Fluvial channels, crevasse splays, levees, and turbidite lobes are some examples of facies units suitable for object-based modeling. These are discussed in the Object-Based Facies Modeling section along with important conditioning limitations and methods to improve conditioning.

Finally, the Process-Mimicking Facies Modeling section provides methods to improve geologic realism of geostatistical models by integrating geological rules typically based on a sequence of depositional and erosional events. This method has been demonstrated in a variety of settings to produce models with high-resolution geologic details beyond traditional geostatistical methods. Yet, questions remain concerning the flow relevance of these details and the ability of these models to condition to all available reservoir data.

Details of building 3-D models of continuous variables such as porosity and permeability are presented in the Porosity and Permeability Modeling section. The alternative approaches are presented together with practical aspects of uncertainty assessment and accounting for secondary data such as seismic data. Techniques to capture the continuity of extreme permeability values and correlation with porosity are discussed in this section.

Chapter 5, Model Applications, covers what to do with geostatistical models after they are constructed. This includes the Model Checking, Model Post-processing, and Uncertainty Management sections. This chapter provides general concepts central to model quality control, summarization, scaling, ranking, and decision making in the face of uncertainty. We should be cognizant that our models are almost always an intermediate step and that optimum decision making after a transfer function is the usual product of geostatistical reservoir modeling. This chapter does not cover the actual transfer functions that are typically applied to geostatistical models, such as volumetric calculation, flow simulation, and forward seismic.

How do we judge the suitability of our geostatistical reservoir models? What do we do with multiple realizations? These questions and others are related to model checking. The important subjects of model verification and assessing the fairness of uncertainty models are covered in the Model Checking section. With increasing work flow and data integration complexity, model checking is even more essential. Issues may develop due to misunderstanding of the modeling methods and associated assumptions and limitations, data handling blunders, data contradictions, and the omission of important concepts such as data and model scale. The first minimum acceptance criterion is that input statistics must be sufficiently reproduced by the models. Additional checks such as model performance for local prediction and associated uncertainty models are presented to quality check geostatistical reservoir models.

At times the models require model post-processing in preparation for a transfer function or to provide useful model summarization. This may include scale up or scale down and local refinement to provide models at a suitable resolution. Smoothing may be applied to remove noise deemed an artifact of the model-generation process or to test the influence of short-scale heterogeneity on the transfer function. Statistical analysis of the model, or jointly of a suite of models that represent uncertainty, may be a useful exercise to aid in decision making. For example, measures of local uncertainty may aid in decisions related to model interpretation and future data collection.

A critical strength of geostatistical reservoir modeling is the ability to represent reservoir uncertainty. The Uncertainty Management section discusses methods to jointly represent all sources of model uncertainty, formats to communicate local and global model uncertainty, methods to rank realizations, and how to make decisions in the face of uncertainty.

Finally, Chapter 6 reviews a number of special topics in geostatistical reservoir modeling. These topics deserve more complete treatment, but are beyond the scope of this book. Some of the special topics include: (1) other data sources such as microseismic, 4-D seismic, and production monitoring, (2) more advanced descriptions of continuous property heterogeneity, (3) spectral methods for model simulation, (4) ensemble Kalman filtering for data integration, (5) advanced statistical descriptions of geology, and (6) other emerging techniques.

The jargon of geostatistics and reservoir modeling can be overwhelming. Different words are often used to describe the same activity. For example, *variography* and *structural analysis* both refer to the calculation, interpretation, and model fitting of measures of spatial correlation. Moreover, the same words are often used to describe different activities: *simulation* can refer to the stochastic simulation of petrophysical properties or to the subsequent flow simulation. The *Geostatistical Glossary and Multilingual Dictionary* (Olea, 1991) has been used as a starting point for most definitions. The notations developed in GSLIB (Deutsch and Journel, 1998) and expanded upon in the book by Goovaerts (1997) have been used as much as possible. For completeness a glossary and list of notation are included at the end of this book.

1.3 KEY CONCEPTS

Predicting rock properties at unsampled locations and forecasting the future flow behavior of complex geological and engineering systems is difficult. A number of concepts and assumptions make it possible to address this difficult problem.

Petrophysical Properties

Techniques presented in later chapters are concerned with constructing high-resolution 3-D models of facies types, porosity, and permeability. The definition of these three variables is often ambiguous and the related modeling choice subject to debate. *Facies* are typically based on an assessment of geological variables such as grain size or mineralization—for example, limestone and dolomite in a carbonate setting and channel sandstone and shale in a siliciclastic setting. Facies are modeled first because they narrow the range of possible porosity and permeability (see definition below) and because multiphase fluid flow properties can vary significantly between facies types. There are different definitions and types of facies, lithofacies, electrofacies, flow facies, depositional facies, and so on. The choice of facies or rock types is usually evident for a particular reservoir modeling problem. The issue of facies definition is addressed in greater detail in Section 4.2.

Porosity ϕ is the fraction of void space in the rock that may contain fluid. We are interested in the "effective" porosity that contributes to fluid flow rather than the "total" porosity, which includes small isolated pores. The spatial distribution of porosity and the total volume of in-place hydrocarbon is of significant interest to reservoir modelers.

Permeability K is a measure of the ease with which the rock allows fluids to flow through it. Permeability depends not only on the direction of flow but on the local pressure conditions, also called boundary conditions. The commonly used "no flow" boundary conditions are particular to laboratory measurement devices and many scale-up algorithms; however, actual boundary conditions in the reservoir will certainly be different. Moreover, permeability is a tensor; a pressure gradient in the horizontal direction can induce vertical flow through K_{XZ} permeability terms.

It is necessary to have *hard* truth measurements at some scale. Most often, these *hard* data are the facies assignments, porosity, and permeability observations taken from core measurements. In the absence of direct core measurements, well log data may be the hard data. All other data types including well logs and seismic are called *soft* data and must be calibrated to the hard data.

Although this book focuses on facies, porosity, and permeability in the context of reservoir modeling, these petrophysical properties are also characteristic of many earth science-related spatial variables (see Table 1.1).

Modeling Scale

It is neither possible nor optimal to build models of the reservoir properties at the resolution of the *hard* core data. In common work flows, the core data must be scaled up (averaged) to some intermediate resolution. Models are generated at that intermediate geological modeling scale and are then usually scaled to an even coarser resolution for flow simulation. An important case-specific issue is to determine the appropriate intermediate geological modeling scale. In other work flows, the model is constructed at the data scale, with simulated nodes discretizing the model space. Irrespective of the method, a decision must be made on the scale or level of descretization. A too-small choice leads to large and inefficient computer use, which restricts the number of alternative scenarios and sensitivity runs that can be considered.

TABLE 1.1. VARIABLES CONSIDERED IN THIS BOOK, GENERALIZATIONS, AND ANALOGOUS VARIABLES IN OTHER DISCIPLINES

Variable	Generalization	Analogous Variables
Facies	Categorical variable	Rock or soil type
		Predominant biological species
Porosity	Volumetric concentration	Mineral grades
		Contaminant concentration
		Fracture density
		Pest concentration
		Cumulative annual rainfall
Permeability	Dynamic property	Conductivity
		Dispersivity

A too-large choice could lead to incorrect flow results due to inadequate representation of important subgrid geological heterogeneities.

Geostatistical reservoir models are built for specific goals, and the level of detail should be suitable for the stated goals. In practice, the goals tend to change and the models are used for more than originally intended. For this reason, it is appropriate to err on the side of building in too much detail even when not explicitly called for by the present modeling objectives.

In common work flows, all data must be related to each other at the intermediate geological modeling scale. Core and well-log-derived measurements, which represent a smaller scale, must be scaled up. The larger scale of seismic and production-derived data must be accounted for (essentially scaled down to the modeling resolution) during the construction of the geostatistical model. Dealing with data of different scales that measure petrophysical properties with varying levels of precision is a central problem addressed in subsequent chapters.

Numerical Modeling

At any point in geological time, there is a single true distribution of petrophysical properties in each reservoir. This true distribution is the result of a complex succession of physical, chemical, and biological processes. Although the physics of these depositional and diagenetic processes may be understood quite well, we do not completely understand all of the processes and their interaction, and we could never have access to the initial and boundary conditions in sufficient detail to provide the unique true distribution of dynamic properties within the reservoir. We can only hope to create numerical models that mimic the physically significant features.

In general, the large-scale features are the most critical for prediction of reservoir flow performance. Small-scale geological details are summarized by effective properties at a large scale.

We will strive to make our numerical models consistent with the available data, including historical flow results. It is understood that the true distribution of petrophysical properties will not follow *any* of our relatively simplistic mathematical models. It is also understood that visual appearance alone is inadequate to judge the acceptability of a numerical model. The suitability of a numerical model is judged by its ability to accurately predict future flow performance under different boundary conditions.

Uncertainty

It is not possible to establish the unique true distribution of facies, porosity, and permeability between widely spaced wells. All numerical models would be found in error if we were to excavate that interwell volume and take exhaustive measurements: There is uncertainty.

This uncertainty exists because of our ignorance/lack of knowledge. It is not an inherent feature of the reservoir. Notwithstanding the elusiveness of the concept of uncertainty, models of uncertainty will be constructed and care will be taken to ensure that these models honestly represent our state of incomplete knowledge. These assessments of uncertainty are nothing more than models, however, and it will not be possible to rigorously validate them.

Geostatistical techniques allow alternative numerical models, also called realizations, to be generated. The response of these realizations, say time to water breakthrough, could be combined in a histogram as a model of uncertainty. The parameters of the geostatistical modeling technique are also uncertain and they could be made variable, that is, "randomized," to lead to a larger and possibly more realistic assessment of uncertainty. The modeling approach itself could be considered uncertain and alternative approaches or geological scenarios considered. There is uncertainty in uncertainty, but at some point this quest for a realistic assessment of uncertainty must be stopped (Journel, 1996).

Uniqueness and Smoothing

Conventional mapping algorithms were devised to create smooth maps to reveal large-scale geological trends; they are low-pass filters that remove high-frequency property variations. The goal of conventional mapping algorithms such as kriging, splines, inverse distance, and contouring algorithms is *not* to show the full spectrum of patterns or variability of the property being mapped. For fluid flow problems, however, the spatial patterns of extreme high and low values of permeability often have a large effect on the flow response.

Geostatistical simulation techniques, conversely, are devised with the goal of introducing the full variability, creating maps or realizations that are neither unique nor smooth. Although the small-scale variability of these realizations may mask large-scale trends, geostatistical simulation is more appropriate for predicting flow performance and modeling uncertainty.

Analog Data

There are rarely enough data to provide reliable statistics, especially horizontal measures of continuity. For this reason, data from analog outcrops and similar, more densely drilled reservoirs are used to help infer spatial statistics that are impossible to calculate from the present subsurface reservoir data.

We acknowledge that there are general features of certain geological settings that can be transported to other reservoirs, provided that they originate from similar geological processes. Although the use of analog data is essential in reservoir modeling, it should be critically evaluated and adapted to fit any hard data from the reservoir being studied.

Data Integration

The goal of geostatistical reservoir modeling is the creation of detailed numerical 3-D geological models that *simultaneously* account for (honor) a wide range of relevant geological, geophysical, and engineering data of varying degrees of resolution, quality, and certainty. This data integration must be accomplished by *construction* rather than by selection. The probability of finding a realization that happens to match data not used in model construction is essentially zero.

A two-step trial-and-error approach could be taken to account for production data: (1) create many realizations that do not explicitly account for historical production data, and then (2) forward simulate the historical production to identify realizations that just happen to reproduce the production data. This approach is not recommended since no realization will match all of the production data. A priori data integration must be attempted since a posteriori screening of realizations will eliminate all possibilities.

Dynamic Reservoir Changes

Geostatistical modeling provides static descriptions of petrophysical properties. In general, it is not recommended to add time as a fourth dimension in the geostatistical model itself and predict future pressure and saturation changes. Such predictions will not necessarily honor important physics such as the conservation of mass and energy. Time-dependent changes in pressure and fluid saturations are best modeled with a flow simulator that encodes these physical laws [see, for example, Aziz and Settari (1979)].

The Place of Geostatistics

Geostatistics was not developed as a theory in search of practical problems. On the contrary, the discipline was gradually developed by engineers and geologists faced with real problems and searching for a consistent set of numerical tools that would help them address those real problems. Reasons for seeking such comprehensive technology included (1) an increasing number of data to deal with, (2) a greater diversity of available data at different scales and levels of precision, (3) a need to address problems with consistent and reproducible methods, (4) an availability of computational and mathematical developments in related scientific disciplines, and (5) an understanding that more responsible and profitable decisions would be made with improved numerical models.

This book describes a set of geostatistical tools that have proven useful in practical reservoir modeling. The application of geostatistical (or any geological modeling approach for that matter) depends on the aforementioned principles.

The Toolbox

These geostatistical tools can be compared to a toolbox and the practice of geostatistics to carpentry. Skilled carpenters will familiarize themselves with a well maintained set of tools. To perfect their trade, they will practice and learn the capabilities and limits of each tool. The expert must understand subtle details of the tool's performance under a wide variety of applications. Each tool will have a unique utility and may be applied in sequence or combination with other tools. At times, a choice will need to be made between tools, yet this choice will be based on expert judgment of the circumstances and the goals of the project.

A decision will be made on the specific tools to keep readily assessable in the toolbox. It is not reasonable to assume that one tool would be capable of addressing all situations, nor would it be practical to overfill the toolbox with obscure tools that will likely not be needed. The former will likely lead to inefficiency and poor quality in the final product, and the latter may result in mistakes as unfamiliar tools are applied.

As with the carpenter, a skilled geostatistician must acquire a deep understanding of the inner workings, limitations, and utility of the established tools through practice. A black box approach will likely lead to poor results, as would the unskilled application of an unfamiliar woodworking tool.

Non-Geostatistical Mapping Methods

The focus of this book is on geostatistical techniques. There are other mapping and modeling techniques including spline interpolation, triangulation, and inverse distance weighting schemes. These non-geostatistical techniques may work for certain purposes—for example, smooth surface interpolation—but they are not adapted to capture the site-specific details of the geological features, nor adapted to account for the scale of the model, and provide no measure of the inevitable uncertainty arising from the combination of sparse data and heterogeneity. There is increasing acceptance that geostatistical modeling tools provide practical and relevant techniques for numerical geological modeling.

Yet, non-geostatistical mapping methods often play a valuable role in geostatistical work flows. We cannot understate the value of sound geological mapping, and we recognize the need of such expert methods to inform geostatistical methods. For example, interpreted bounding surfaces are required to constrain the reservoir container or consider a hand-contoured map of vertically averaged reservoir properties that may be applied to correct the biased reservoir property distributions and to locally constrain facies proportions and property means. While non-geostatistical mapping methods are not within the scope of this book, we recognize their importance in geostatistical modeling.

1.4 MOTIVATION FOR RESERVOIR MODELS

A high-resolution 3-D reservoir model that is consistent with the available data is often a sufficient reason to consider geostatistical reservoir modeling; reconciling all available hard and soft data in a numerical model has many intangible benefits such as (1) transfer of data between disciplines, (2) a tool to focus attention on critical unknowns, and (3) a vehicle to present spatial variations that may enhance or work against a particular production strategy. In addition to these intangible benefits, there are specific reasons for constructing high-resolution 3-D models:

- There is a need for reliable estimates of the original volume of hydrocarbon in the reservoir. These in situ volumes are important to (1) determine the economic viability of producing a given reservoir, (2) allocate equity among multiple owners, (3) compare the relative economic merits of alternative reservoirs, and (4) determine the appropriate size of production facilities. Geostatistical reservoir modeling provides the numerical models for such volumetric estimates.
- Well locations must be selected to be economically optimal and robust with respect to uncertainty in the reservoir description. Multiple geostatistical realizations make it possible to evaluate the uncertainty underlying well locations and permit selection of "good" well locations. These realizations also make it possible to address related questions such as (1) what type of wells—for example, horizontal, vertical, multilateral—should be considered and (2) how many wells are required?
- There is often a need to reconcile an abundance of soft data (say, from a 3-D seismic survey or years of historical production data) with a limited amount of hard well data. Geostatistical models allow different types of data to be represented in a common format, for example, seismic data can be represented at the scale and with the units of the hard data.
- The 3-D static connectivity of a reservoir can be assessed through simple visualization and various connectivity tools to assess the potential for bypassed oil and the value of infill wells, prior to performing flow simulation.
- Flow simulation allows the prediction of reservoir performance with different production scenarios. The optimal number of wells and operating conditions can be established by comparing the economics of various alternatives. The use of flow simulation was initially hampered by the limited resolution of flow models (primarily a computer hardware constraint) and oversimplistic geological input (layercake models). Geostatistical methods now provide the required numerical models of rock properties such as facies, porosity ϕ, and permeability K.
- Major decisions must be made in the presence of significant uncertainty: how many wells? well locations? timing of injection and infill

production wells? The goal of many energy-producing companies is to make such decisions in a way that maximizes profitable production of the resource. From a geostatistical perspective, these decisions must be robust with respect to the inherent uncertainty in the spatial distribution of reservoir properties. This is illustrated in Section 5.3.

The Benefit of Geostatistics

Ideally, there would be clear documentation of the quantifiable benefits of geostatistics to motivate its use. There are a number of case studies available in the literature from the early days of geostatistical reservoir modeling that illustrate the utility of geostatistical reservoir models (Alabert and Corre, 1991; Alabert and Massonnat, 1990; Cox et al., 1995; Damsleth et al., 1992b; Delfiner and Haas, 2005; Omre, 1992; Rossini et al., 1994; Tyler et al., 1995; Yang et al., 1995). In some of these case studies the combined use of geostatistics and flow simulation lead to a clearly different reservoir management decision than would be obtained with a conventional geological model. One can also argue that the geostatistical models are better since they honor more reservoir-specific information. Notwithstanding the previous arguments, it is difficult to quantify the *worth* or *value* of geostatistics except in simplistic synthetic examples. In some cases, the demonstrated benefit of geostatistical reservoir modeling is in part due to the decision to spend additional effort on reservoir characterization; even conventional methods would do better under such effort.

1.5 DATA FOR RESERVOIR MODELING

The following is an introduction to the data available for reservoir modeling. Section 3.1 provides more detailed discussion on data for reservoir modeling and methods to check, transform, and so on. Figure 1.1 schematically illustrates the different types of relevant data for reservoir modeling. Data for reservoir modeling includes reservoir-specific data:

- Core data (ϕ and K by facies) available from a limited number of wells. Core may be sampled by relatively small core plugs, or entire sections of core may be measured for whole core data.
- Well log data provide precise information on stratigraphic surfaces and faults, as well as measurements of petrophysical properties

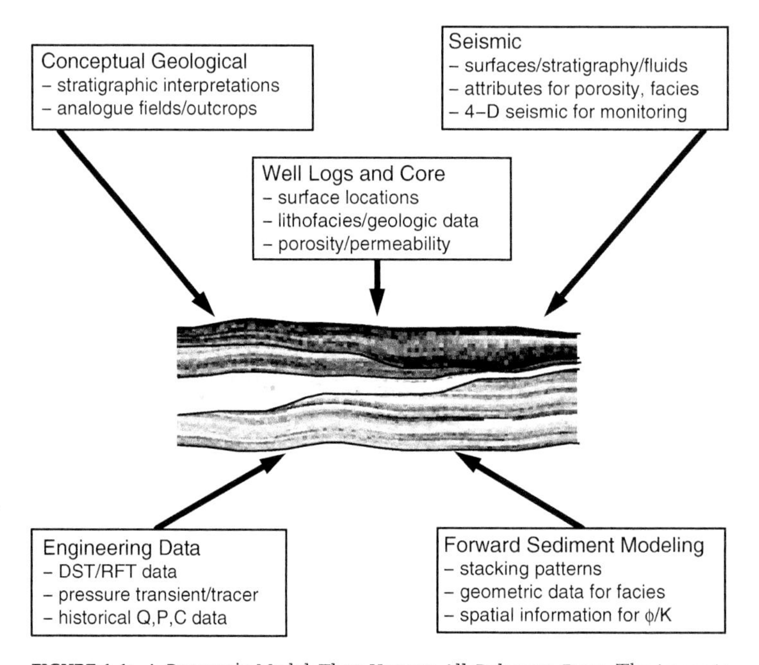

FIGURE 1.1: A Reservoir Model That Honors All Relevant Data. The image in the center schematically illustrates a complex 3-D reservoir. The different types of relevant data are described in the surrounding boxes.

such as facies types, ϕ, and, perhaps, soft K measurements.

- Well image log data provide precise information on stratigraphic surfaces and faults, as well as small-scale heterogeneity that may aid in assessment of beds dips and facies types.
- Seismic-derived structural interpretation, that is, surface grids and fault locations that define large-scale reservoir geometry.
- Seismic-derived attributes provide information on large-scale variations in facies proportions and porosity.
- 4-D seismic surveys that provide local information on the connectivity of a reservoir by monitoring the flow of fluids.
- Well test and production data yield interpreted permeability thickness, channel widths, connected flow paths, and barriers. The use of inversion methods provide coarse scale ϕ, K models with an associated measure of uncertainty.
- A sedimentologic and sequence stratigraphic interpretation provides information on layering and the continuity and trends within each layer of the reservoir.

In addition to these reservoir-specific data, there is more general knowledge available from analog data:

- Trends and stacking patterns available from a regional geological interpretation. For example, it may be known that the porosity distribution is fining upward or coarsening upward. Also, the position of the reservoir in the overall regional geological setting may provide soft information on areal trends in the distribution of facies proportions.
- Outcrops or more densely drilled similar fields may provide size distributions, variogram measures of lateral continuity, and other spatial statistics considered general.
- Knowledge of geological processes/principles established through widely accepted theories (forward geologic modeling and flume experiments) provides general geological information of the type described above.

Although there is a large diversity in the data available for reservoir modeling, those data are sparse relative to the vast interwell region that must be modeled.

There can be problems of consistency between data types that must be resolved. For example, core permeability data may imply an average permeability of 50 mD, and well test data may imply an average of 300 mD. Perhaps there are fractures or other high-permeability features causing the large-scale effective permeability to be higher than the measured core data; it may be necessary to correct or complete the core data before modeling. Alternatively, an inappropriate interpretive model may have been used in the well test analysis program; reinterpretation may be necessary before using that data. Geostatistical tools, at best, identify inconsistent data. They do not help to resolve any discrepancies or inconsistencies; the reconciliation of the different data types must be done by the various professionals involved in the reservoir modeling. Most often, inconsistencies are reduced by interpreting each source of raw data in light of all other data types.

It is relatively easy to construct a numerical reservoir model that honors only some of the available data; for example, (1) a hand-contoured map of average porosity would honor averaged well data and geologic trends, but would represent neither the details of the well data nor the flow properties from well test and historical production data, (2) a statistical model would honor all local well data and certain statistical measures of heterogeneity but may not represent important geologic features that could impact flow predictions, and (3) a large-scale "inversion" from production or seismic data would honor the data from which it was generated but would not represent small-scale heterogeneities and other possibly important geological constraints.

The challenge addressed by geostatistical reservoir modeling and by this book is to simultaneously account for all relevant geological, geophysical, and engineering data of varying degrees of resolution, quality, and certainty. At times there may be a wide variety of data with the potential for redundant and/or contradictory information. An additional challenge is to account for this redundancy and detect and correct for contradiction. This may include statistical methods to identify significant variables and to merge variables into super variables that predict the reservoir property of interest. The idea of a creating a "shared earth model" that represents all relevant data is not new (Gawith et al., 1995).

1.6 THE COMMON WORK FLOW

There are three important model types that are common targets for geostatistical reservoir modeling; (1) large-scale multivariate mapping for resource/reserve assessment, (2) reservoir-scale 3-D modeling for reservoir development planning, and (3) high-resolution modeling to understand effective flow parameters at the flow simulation scale. All of these applications will be covered in this book. The following discussion is adapted from Deutsch (2011).

Although, the scale and other implementation details vary between these approaches, the fundamental approach is the same (Chilès and Delfiner, 2012; Deutsch and Tran, 2002). The core geostatistical approach could be summarized by six steps.

The detailed implementation of these steps will depend on the purpose of the study, and not all steps may be required to meet the project goals. Greater detail on these steps is provided in Section 3.3.

1. **Specify** the goals of the study and take inventory of the available resources, measurements, and conceptual data. All models need to be fit-for-purpose—that is, constructed to improve decisions that add value to the reservoir project. These factors will determine all the model choices, such as the model scale, level of effort, methods and algorithms to apply, the components that should be treated as uncertain, and so on.
2. **Develop** a conceptual model or multiple conceptual models to account for uncertainty of the reservoir integrating all available measurements, analogs, and expert knowledge. This step includes the establishment of stratigraphic hierarchy, layering/coordinates defining the geometry, and stratigraphy of the reservoir interval being modeled. This step also involves the development of a conceptual model for the major architecture and continuity of faces, porosity, and permeability within each layer. Critical to this step is the *choice* to divide of the area/volume of interest into subsets with distinct heterogeneities and to model how the continuous mean or categorical proportions of each variable depends on location within each chosen subset.
   In many geostatistical reservoir modeling problems these subsets include regions with distinct facies and reservoir property heterogeneities, then facies with unique reservoir heterogeneities, and then correlated reservoir properties such as porosity, permeability, and saturations in each facies.
   Facies are modeled by variogram-based, multiple-point, object-based or process-mimicking techniques within each stratigraphic layer. *Porosity* is modeled on a by-facies basis before permeability because there are more porosity data (reliable well log data and, often, an abundance of porosity-related large-scale seismic data). The models of *permeability* are constrained to the porosity, facies, and layering previously established. At times, there may be a conflict between the facies and porosity already established and well test or production derived permeability data. In this case, such permeability data should be expressed as constraints on the porosity and permeability models.
3. **Infer** all required statistical parameters for creating spatial models of each variable within each subset. This typically includes histograms, spatial continuity models, and correlations between variables. Efforts are made to ensure that these statistics are representative of the entire region or facies. This is often challenging, as data are typically very limited and collected preferentially in high-quality reservoirs and not for statistically representative sample sets. This step may include data mining and resulting new lessons that result in a recycle of the conceptual model.
4. **Estimate** the value of each variable at each unsampled location. It is always a good idea to begin with estimation models because they provide an unimpeded visualization of the impact of data, trends, and multivariate relations on the most likely value of each variable at each location. This is a good step for discovery of important features and data and implementation issues.
5. **Simulate** multiple realizations to assess joint uncertainty at different scales. Multiple, equally likely realizations are created by repeating the entire process. Each realization is "equally likely to be drawn"; however,

some realizations are more similar to others, hence their class has higher probability; therefore, the term "equiprobable" is thus avoided by some. We use the term equiprobable realizations in this book, recognizing that while the individual realizations are equiprobable, any assessment of likelihood of classes requires investigation of all realizations.

6. **Post-process** the statistics, estimated models, and simulated realizations to provide decision support information. This post-processing is generally a summarization of the suite of simulated realizations. This may include (a) simple statistical summarizations such as probability density function, specified local percentile value, and mean or (b) more complicated summarizations such as volume, connectivity, and flow performance calculations.

1.7 AN INTRODUCTORY EXAMPLE

An illustrative introductory example is presented to demonstrate the application of the common steps in reservoir modeling listed above to solve various typical reservoir problems. We focus on the value of integrating various data sources into geostatistical analysis and modeling to meet the specified project goals. There is no effort here to discuss the implementation details of the methods utilized in this work flow because they are covered in detail through the rest the book.

Goals and Data Inventory

In this illustrative example, the goal is to predict the flow performance of a simple reservoir and the associated uncertainty. In order to predict flow performance and uncertainty, intermediate steps include inference of reservoir heterogeneity (facies and reservoir properties), understanding the potential reservoir connectivity, sensitivity of flow to that heterogeneity, and characterizing the uncertainty model in the geological heterogeneity. A secondary goal of this study is to produce reservoir heterogeneity uncertainty models that integrate all available information and may be utilized to make future drilling decisions.

Seismic data are available with sufficient resolution to identify a reservoir zone (basal surface interpreted from seismic is shown in Figure 1.2 along with two sections) and to constrain possible reservoir architectures, but it does not provide sufficient information to inform reservoir properties. Well log information from the horizontal and the neighboring vertical well provide significantly biased reservoir property distributions and local conditioning data for the geostatistical models. Basin analysis, geologic analog, and inferred depositional setting and

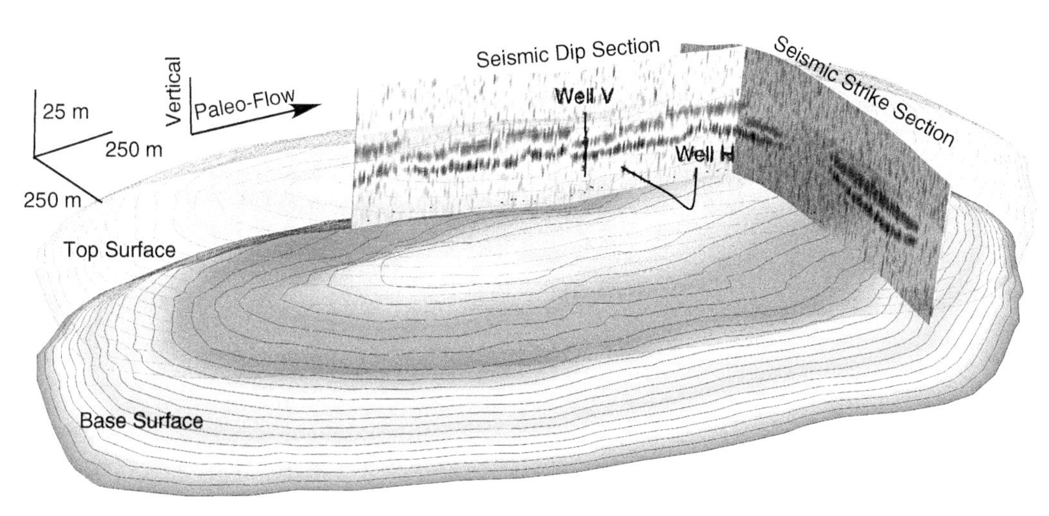

FIGURE 1.2: Data Available for the Introductory Geostatistical Reservoir Model; Two Seismic Lines, Basal Structure Inferred, and Well Data.

associated process concepts provide geological models on potential reservoir architecture.

Conceptual Model

Basin analysis and adjacent reservoir analogs indicate that this reservoir is likely deepwater turbidite deposits within a slope valley complex. The base of the early low stand includes nonreservoir mass transport complexes that result from failures of the slope valley walls. Immediately above the mass transport complexes are various turbidite deposits that are sourced from failure of sands deposited on the continental shelf.

Given the strong seismic signature and limited aerial extent of the reflectors and analysis of the well logs that has identified potential channel fill and overbank facies, it is determined that the reservoir is most likely a set of stacked turbidite channels with good reservoir quality embedded in fine-grained overbank material. Although, it is possible that the reservoir may be composed of a set of stacked distributary lobes with axial, off-axial and marginal fills.

The stacked channels result in the focusing of the best reservoir properties along the channel axis and have limited connectivity between the channels. Distributary lobes are characterized as horizontally continuous reservoir deposits with good horizontal connectivity and local vertical connectivity along the flow axes. This has significant implications on flow response; therefore, some models with this less likely scenario will be included in the uncertainty model. Note that if the seismic data had sufficient resolution, it may be possible to map some or all of the architecture deterministically. In this example, it is assumed that this is not possible.

The late low stand includes isolated channels with decreasing reservoir quality as the system energy wanes due to sediment storage on the continental shelf. The channels tend to be sand-rich good-quality reservoirs encased in fine-grained overbank. These thick sands and the associated acoustic contrast with the fine-grained overbank and hemipelagic muds result in strong seismic signatures. In addition, these channels and sheet fills have a variety of facies, porosity, and permeability trends inferred from reservoir analogs and process knowledge that may impact fluid flow. These trends include permeability streaks caused by focusing of course-grained sands along the channel axis and baffles due to poorer quality marginal facies.

The reservoir zone is deposited on salt and hemipelagic sediments, and salt withdrawal is responsible for the formation of the structural trap. The top seal is composed of hemipelagic sediments from the high-stand system tract. This conceptual model for the reservoir and adjacent stratigraphic units is illustrated in Figure 1.3. This provides a framework for our geostatistical reservoir model. If there was significant uncertainty with respect to this framework, then multiple representations of this framework could be included to account for this uncertainty. For this example, we will that assume that this is not the case and utilize a single framework.

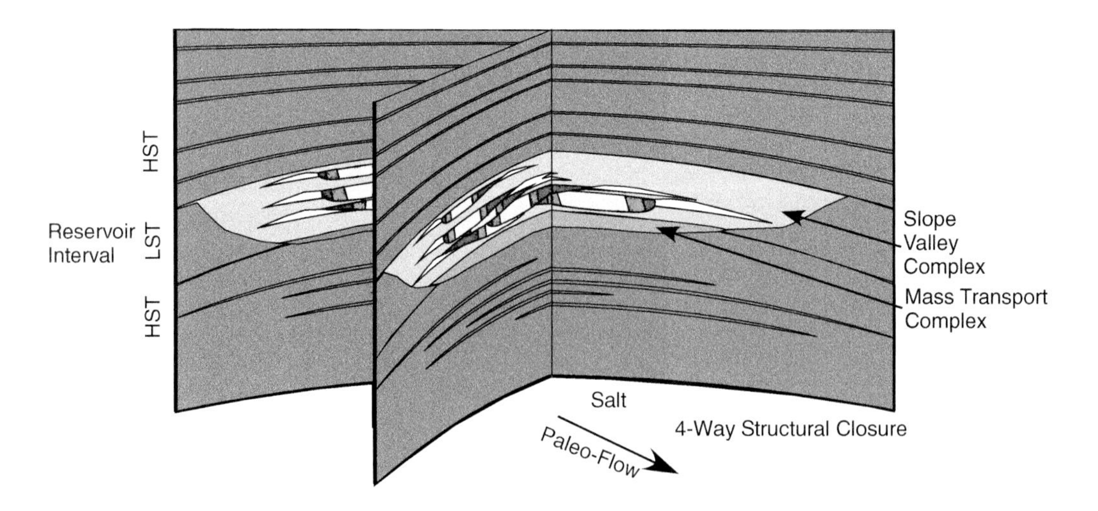

FIGURE 1.3: Conceptual Model of a Slope Valley Complex 4-Way Closure Structural Trap.

Statistical Inference

The horizontal well tracks along a mainly channel axis facies, while the vertical well penetrates multiple stacked channel off-axis or marginal facies (see Figure 1.4). It is observed that facies have significant differences in reservoir properties and that facies provide mappable reservoir architectures that will likely have a significant impact on flow responses; therefore, facies should be integrated into the geostatistical model. Conversely, an attempt to ignore facies and build reservoir property models directly would mix these populations together (for demonstration this is attempted later), and important heterogeneity information may be ignored in the model.

As a first step, representative proportions for facies are inferred utilizing the conceptual model and seismic lines to correct for the bias due to selective well locations. In addition, vertical trends in facies are inferred from the conceptual model. This includes more channels and associated fills at the base of the reservoir and less at the top, resulting in more isolated and less connected channels. These global proportions and associated trends are then applied as inputs to constrain any geostatistical facies modeling method. For this demonstration, only one set of facies inputs is utilized. Once again, if they were considered uncertain, a distribution of possible global proportions and trends may be utilized.

Given our conceptual model, we require statistics to represent the heterogeneity of channel or lobe facies and porosity and permeability within each of these facies. Statistical inference is challenged in this example, given the presence of only two wells and the preferential placement of these wells in a high-quality reservoir. In addition, the long horizontal reach potentially results in overrepresentation of a single stratigraphic layer (see Figure 1.4). In this example the limited information from seismic and analog information from neighboring reservoirs are applied to correct the facies distributions. The naïve facies proportions from wells and facies proportions corrected are shown in Figure 1.5. The two wells

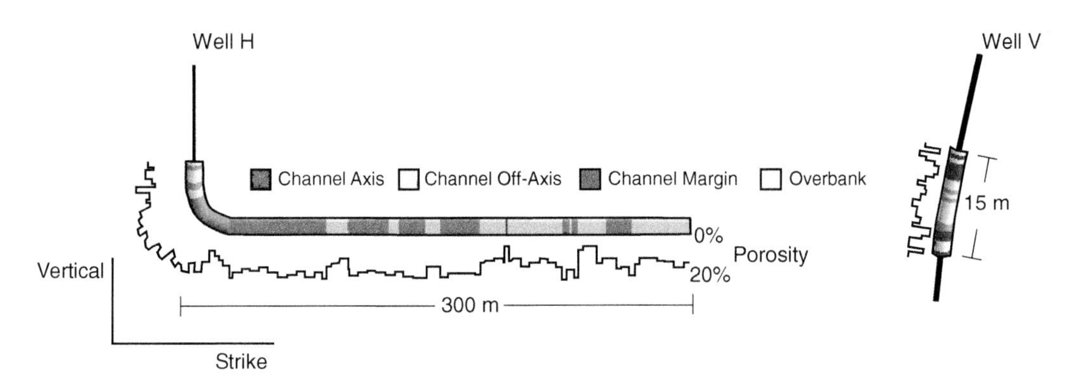

FIGURE 1.4: Horizontal and Near Vertical Wells with Associated Facies and Porosity Samples Indicated.

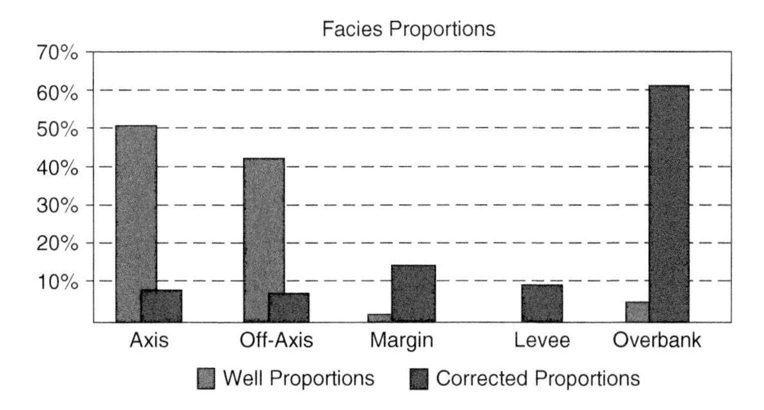

FIGURE 1.5: Naïve and Representative Distributions for Facies. The preferentially placed wells have oversampled good-quality facies and undersampled overbank facies.

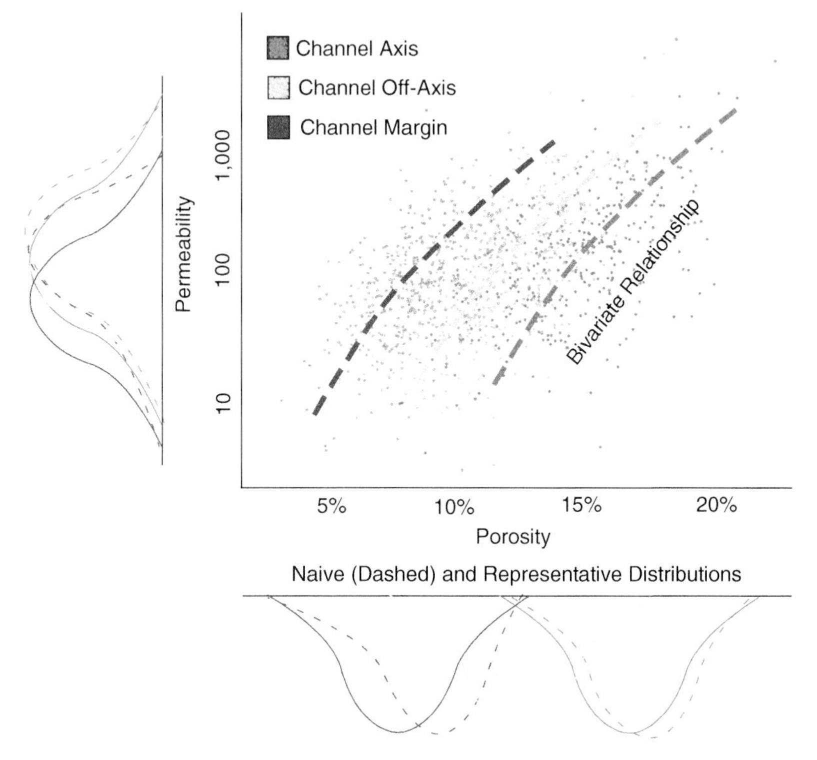

FIGURE 1.6: Naïve and Representative Distributions and Bivariate Relationship for Scaled-up Porosity and Permeability Distributions for Axis, Off-Axis, and Margin Facies.

have significantly oversampled axis and off-axis facies while having undersampling margin and overbank facies, and they even fail to sample any levee facies.

Representative distributions of porosity and permeability must be inferred from the available data for each continuous reservoir property to be distributed within the facies model. The well-based porosity and permeability distributions are likely be biased high due to preferential sampling of the best part of the reservoir. These distributions should be corrected to remove this bias, and scale up must be conducted to account for the change in support size from well sampling to reservoir grid to provide appropriate input distributions for continuous property simulation. In addition, the relationships between these reservoir properties are required for each facies. Figure 1.6 illustrates this concept with univariate naïve and representative distributions of porosity and permeability for axis, off-axis, and margin facies and curvilinear trend lines, indicating the relationship between porosity and permeability for each facies. These statistics, along with spatial statistics discussed in this book, provide the statistical constraints to construct the required geostatistical reservoir models.

Estimation

The calculation of estimation models provides an opportunity to visualize the combination of the porosity well data and porosity trends, without the stochastic features that may cloud issues in simulated realizations. This is an important check to ensure that the channel objects, channel facies framework, and within-facies trends are consistent with the wells and result in reasonable porosity estimates away from the well data. An example of an estimated model with various slices is shown in Figure 1.7. Note that one slice is aligned with the well data, by design, to allow for ease of comparison between data and estimation model. The result seems reasonable; if any issues were found, then this may result in recycling of the well facies assignments, the channel object model or the facies model, including trends and proportions. During the simulation step (covered next), the simulation models exhibit stochastic features that tend to prevent this type of visualization and checking.

Simulation

With geostatistical simulation, multiple equiprobable models are generated to jointly represent model

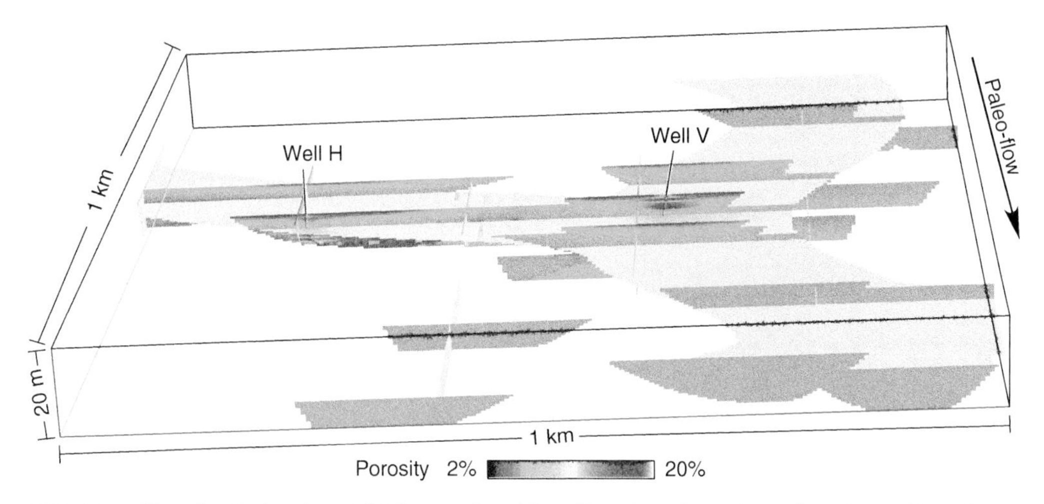

FIGURE 1.7: Slice of an Estimation Model for Porosity with Wells Trajectories and Porosity Values Indicated. Areas outside the channel fills are set transparent to simplify the visualization.

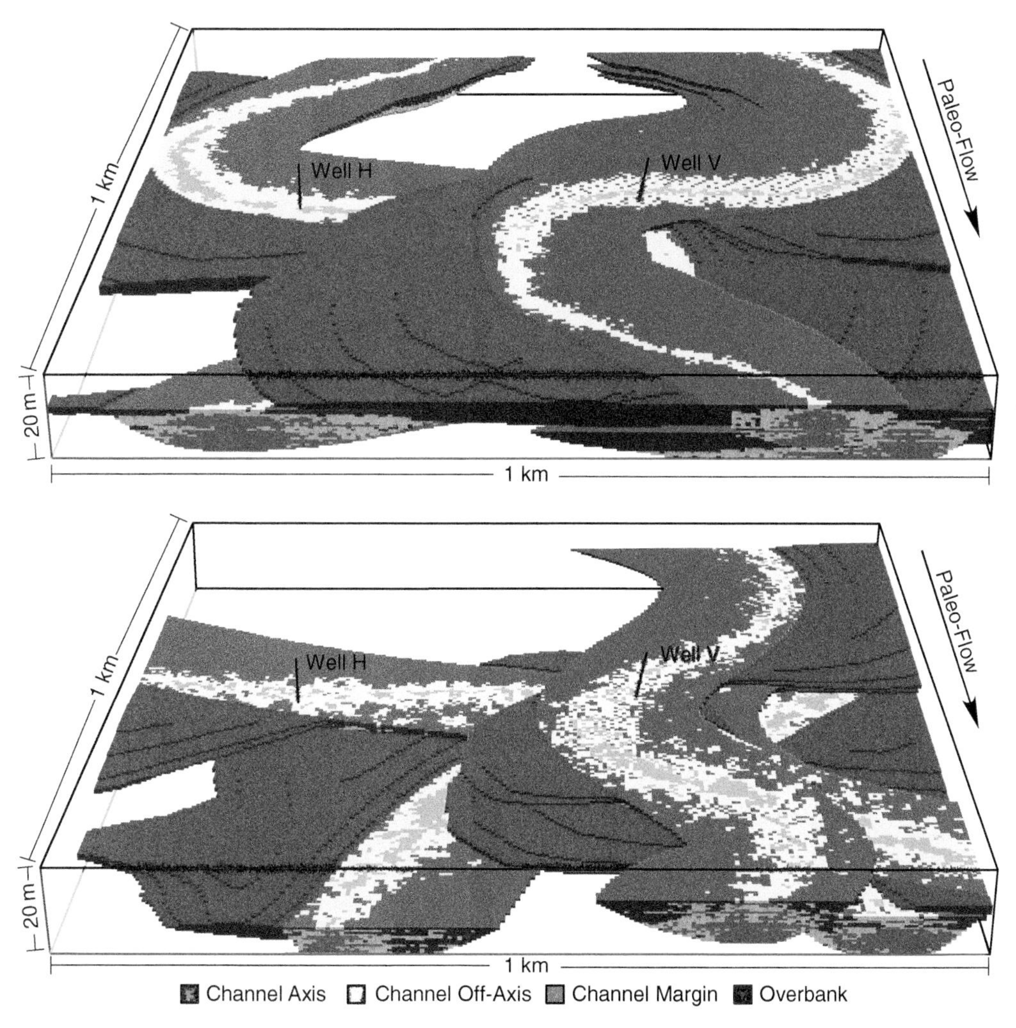

FIGURE 1.8: Two Realizations of Object-Based Channels with Facies, Conditional to the Available Well Data and Input Statistics.

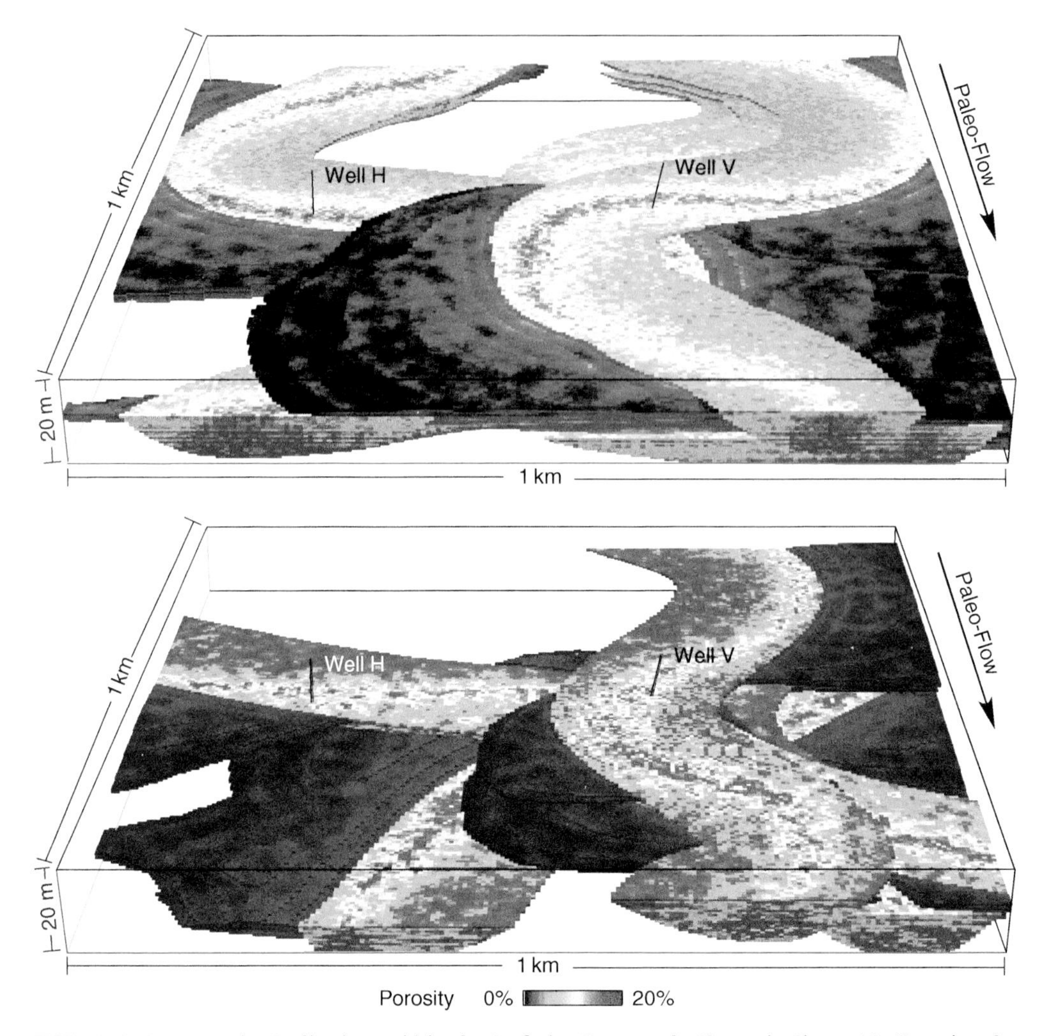

FIGURE 1.9: Two Porosity Realizations within the By-facies Framework, Shown in Figure 1.8, Honoring the Wells and Input Distributions. Each realization has distinct global distributions and trend models to account for their associated uncertainty.

uncertainty. Each realization reproduces the specified representative input statistics, trends, and well data, but varies away from well control and under the constraints from spatial continuity, trend, and secondary information. When the input statistics, trends, or any other model choices are considered uncertain, this may be integrated into the uncertainty model by varying these between scenarios.

Multiple object-based geostatistical realizations and scenarios of stacked channels are simulated honoring the conceptual model of stacked turbidite channels, multiple realizations of facies within these channels are simulated, and finally correlated reservoir properties are simulated within the facies. An oblique view of two realizations of channels and associated facies fills is shown in Figure 1.8.

Given this channel and facies architecture, reservoir properties such as porosity may be simulated on a by-facies basis. The simulation work flow reproduces the representative porosity distributions along with their spatial continuity and associated uncertainty. An oblique view of two porosity realizations is shown in Figure 1.9. The porosity realizations illustrate changes in reservoir quality across facies along with various degrees of spatial continuity and trends, including a high-porosity streak along the axis of the channel fill and discontinuous low porosity in the levees. To account for uncertainty in the global porosity distributions and trends,

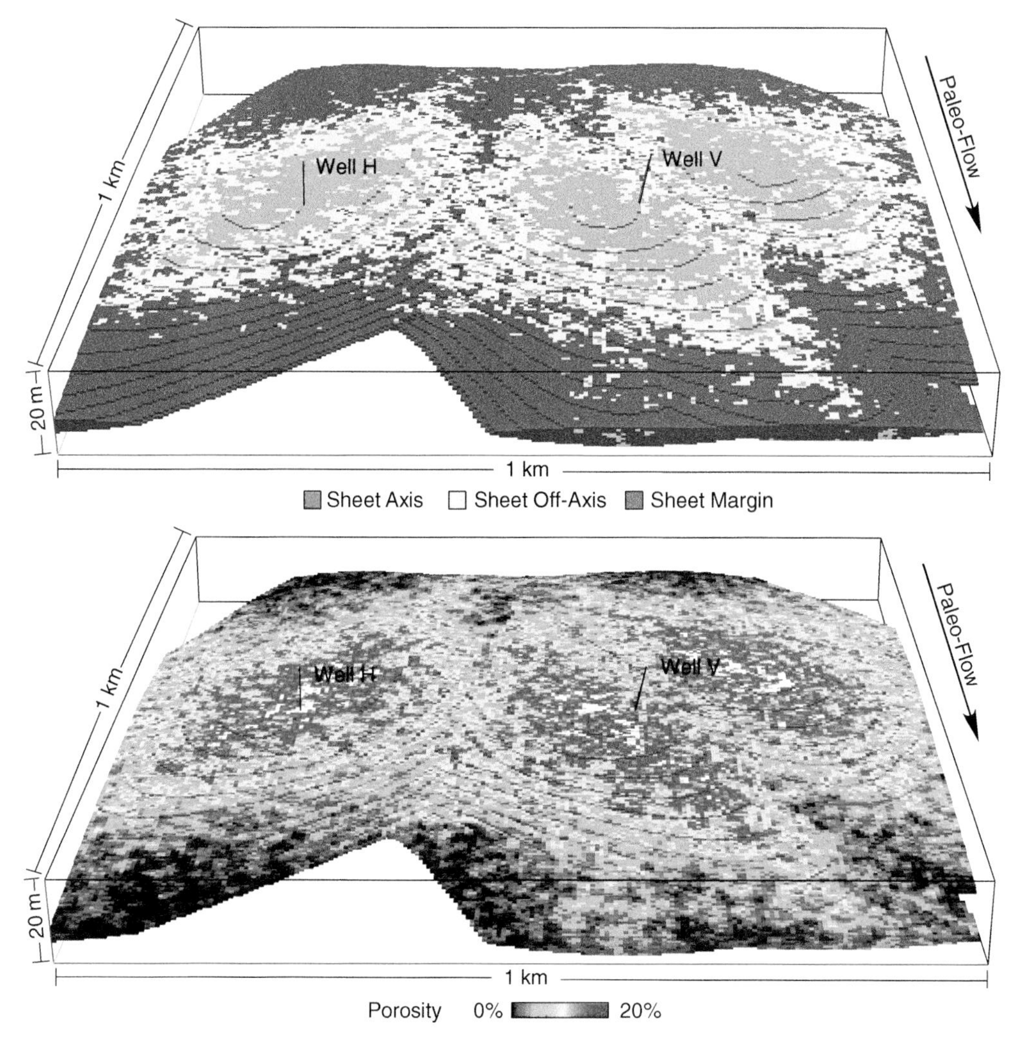

FIGURE 1.10: A Single Facies and Porosity Realization for a Distributary Lobe Reservoir Conditional to Well Data.

these were varied between scenario sets of realizations. One can observe the overall change in porosity distribution and within channel trends for each porosity realization.

It is often important to capture the heterogeneity associated with low and high reservoir quality. These are the critical reservoir flow components including flow conduits, barriers, and baffles. This type of object-based work flow with clean geometries and within object trends is well suited to reproducing these specific features. Of course, this will depend on the purpose of the reservoir study. In our case we are concerned with flow response and so we anticipate that these conduits will have a significant impact. If only volumetrics are required, then these features are not likely critical.

As mentioned previously, there is some uncertainty with regard to the architectural concepts. To account for this uncertainty, we may include various conceptual models. For example, another conceptual model could be distributary lobes. In the case of distributary lobes the reservoir elements are lobes that stack on top of each other in a distributary pattern. Within the lobes, the best-quality reservoir may be focused along the lobe axis and may diminish toward the lobe margins. Also, the reservoir quality may diminish toward the proximal and distal extents of the lobes. The resulting heterogeneity model is more homogeneous than the stacked channels. Distributary lobes will likely result in significantly different flow simulation results than with stacked channels. While the stacked channels result in piping

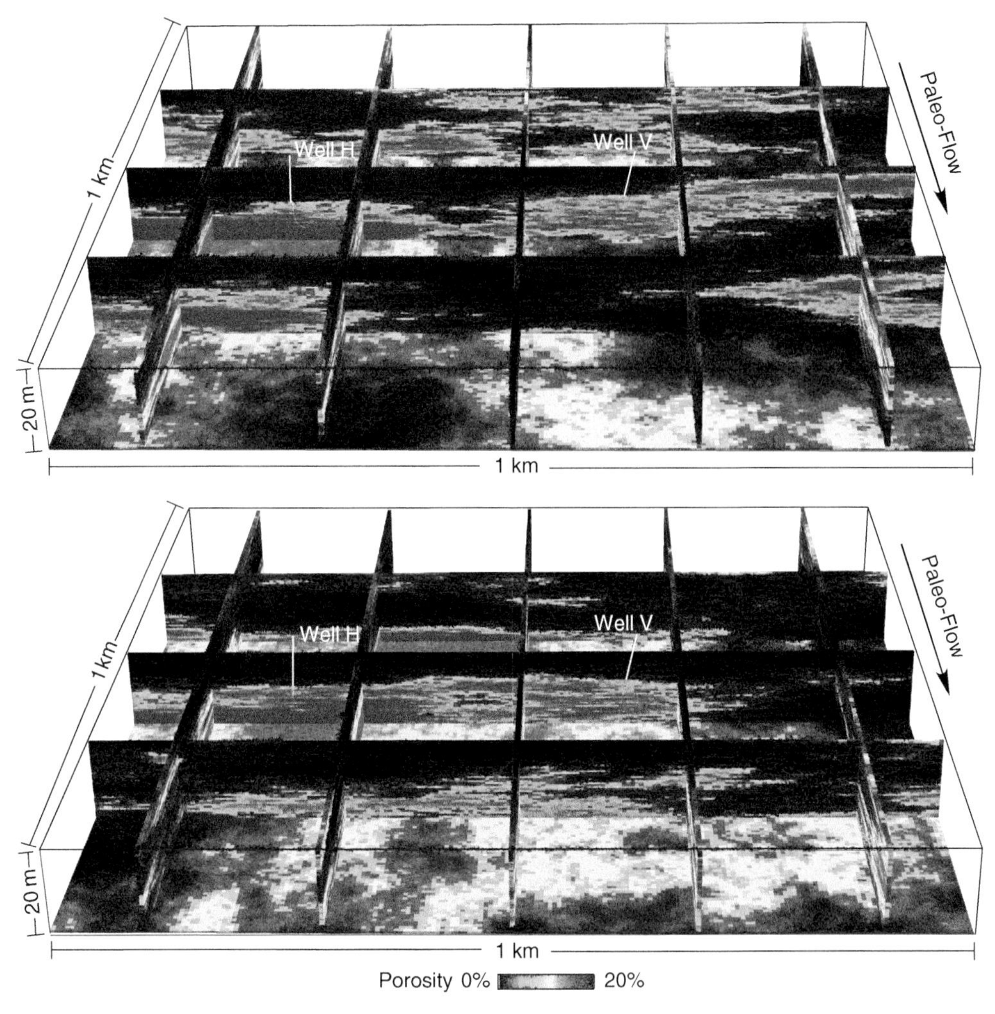

FIGURE 1.11: Two Porosity Realizations Simulated Directly without an Associated Facies Model to Constrain the Spatial Distribution.

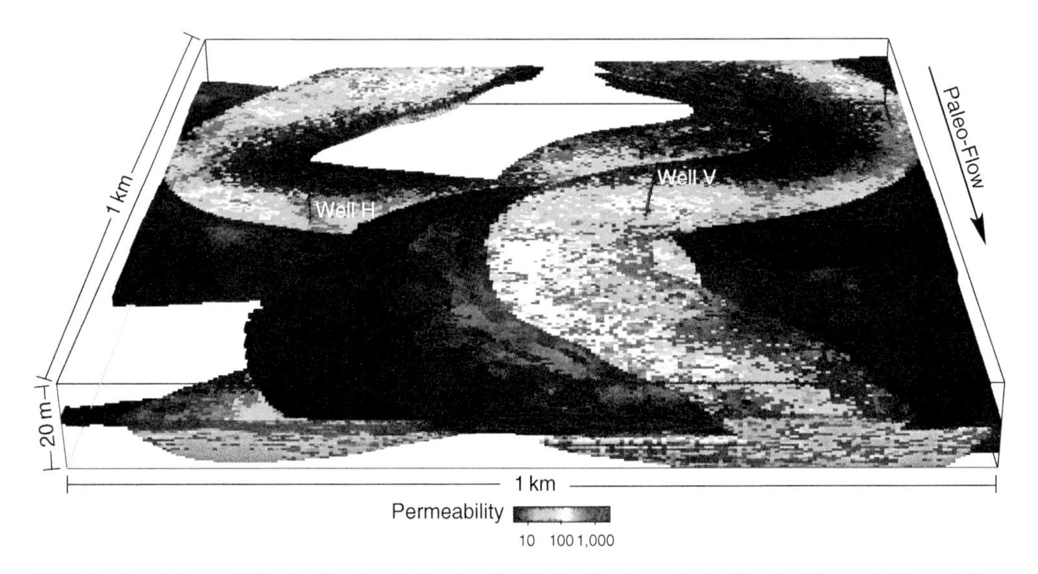

FIGURE 1.12: A Permeability Realization Cosimulated with the First Porosity Realization Shown in Figure 1.9.

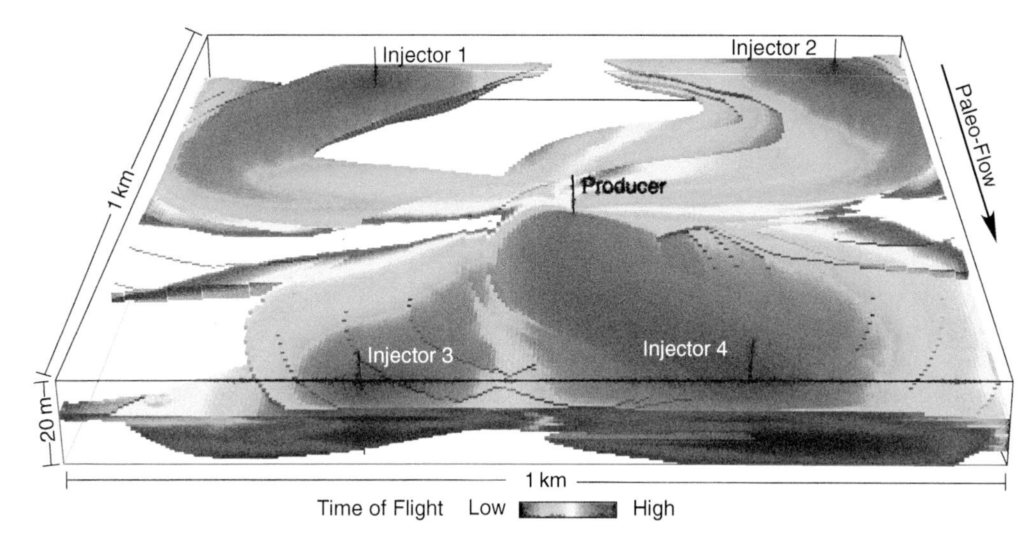

FIGURE 1.13: An Oblique View of the First Channel Architecture Realization with Time of Flight Indicating the Degree of Connectivity of the Reservoir. Note the four injectors on the model edge and single producer in the center of the model.

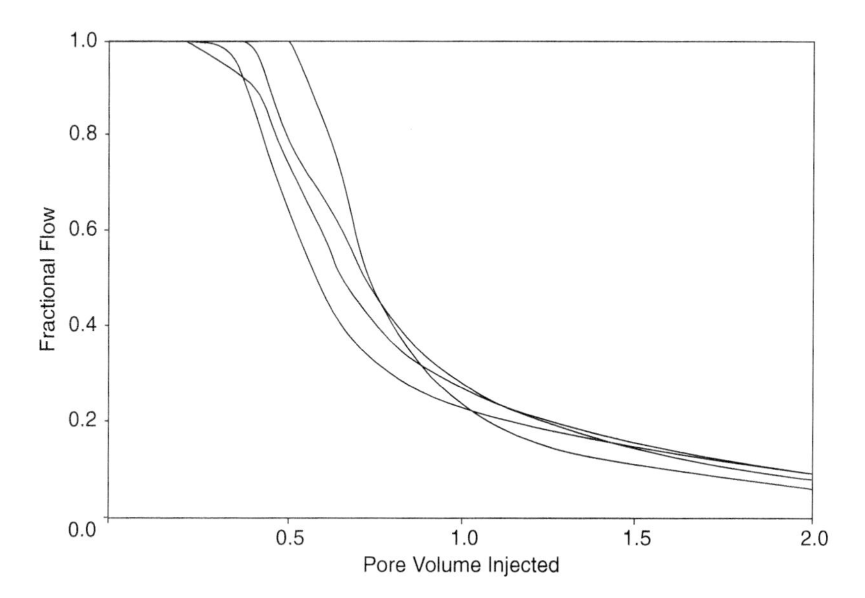

FIGURE 1.14: Fractional Flow or Fraction of Produced Fluids That Are Oil Are Shown Relative to Pore Volumes Injected for Several Realizations. This is useful for quantifying reservoir connectivity and flow response to the specified exploitation plan given the modeled reservoir heterogeneity.

of flow along channel axes, distributary lobes likely result in a more dispersive flow front. The resulting model reservoir property simulation model is more heterogeneous with broad lobe axis to margin trends, but does not have long-range flow conduits (see a single facies and porosity realization in Figure 1.10).

It is illustrative to consider a work flow without facies modeling. In this case, reservoir properties are not constrained within simulated facies frameworks. As discussed later, facies may not always be required. For demonstration, two porosity realizations, unconstrained by facies, are shown in Figure 1.11. Note that beyond data constraints there is no constraint on porosity and the full continuum of porosity values mix over the model. Of course, one could impose trends on the porosity realizations to further

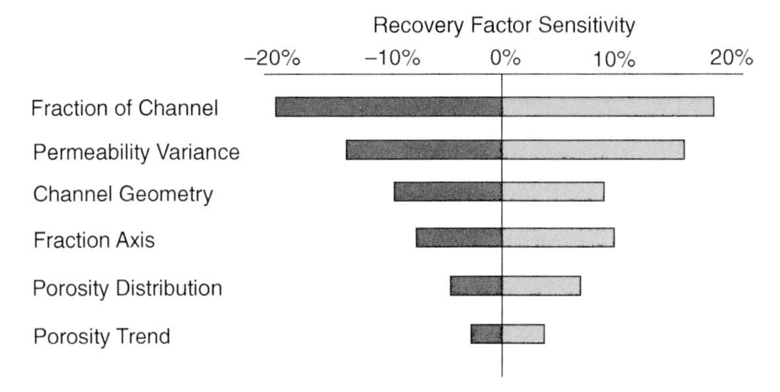

FIGURE 1.15: Tornado Diagram to Illustrate Recovery Factor Sensitivity to Various Modeling Parameters and Choices.

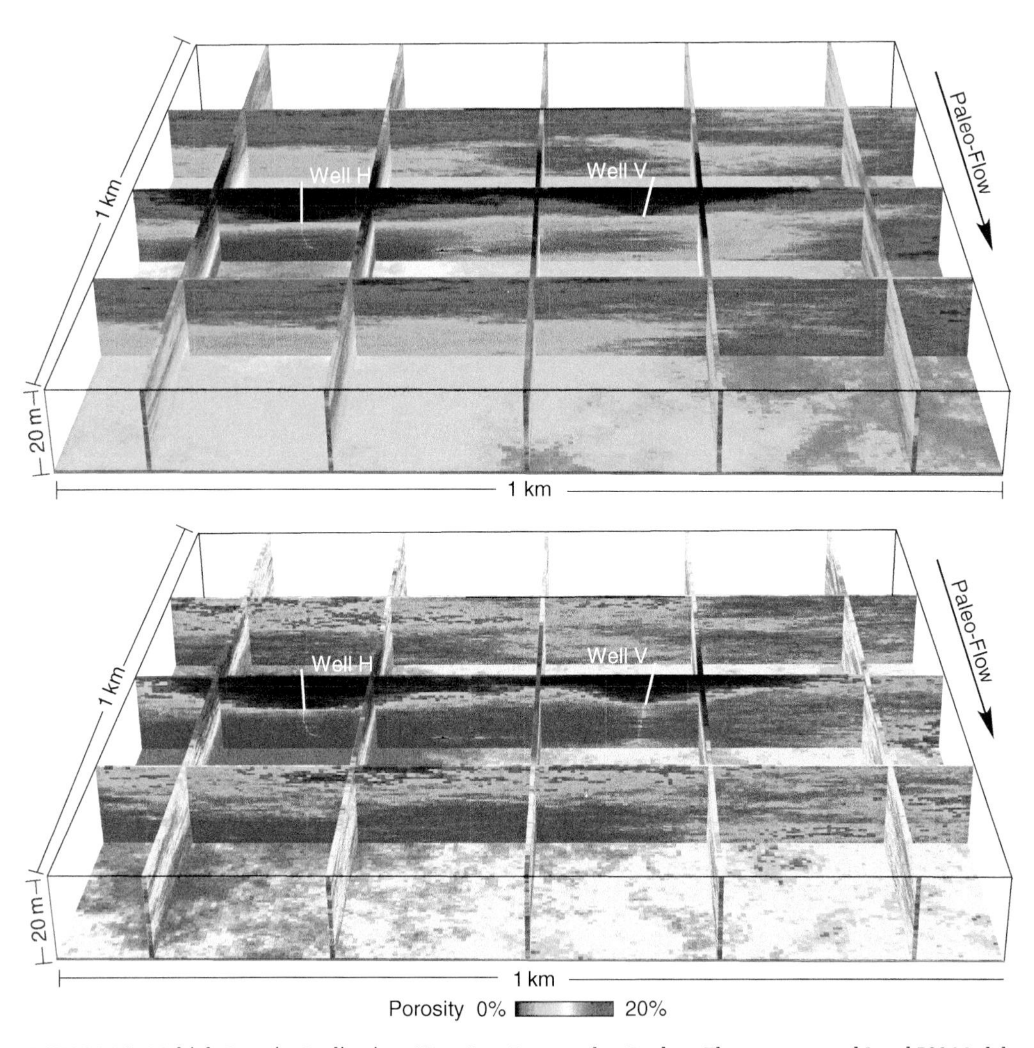

FIGURE 1.16: Multiple Porosity Realizations Were Post-Processed to Produce These e-type and Local P90 Models (Above and Below, Respectively). Post-processing is useful to summarize multiple realizations and the associated uncertainty model. For example, the e-type models illustrate the influence of data, trends, and secondary data on the models, and the P90 models provide an indication of locations that are surely low.

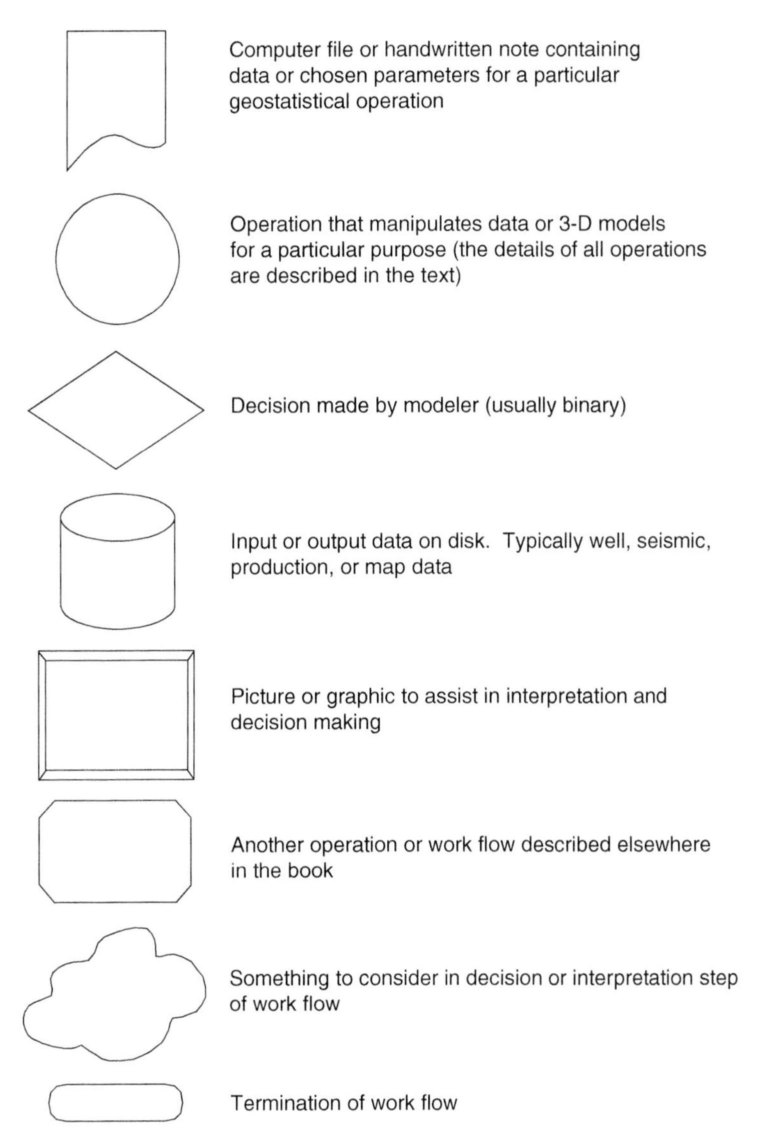

FIGURE 1.17: Work Flow Symbols Used in Subsequent Chapters to Clarify the Operations and Considerations Involved in Geostatistical Reservoir Modeling.

constrain the porosity distribution, but in this case the absence of a facies model removes a potentially important component of the reservoir heterogeneity based on position within channels or lobes.

Permeability may be simulated, given its relationship to porosity within each facies and conditional to any permeability information available along the wells. An example realization for permeability is shown in Figure 1.12. Similarly, water saturation is simulated, given any relationship to facies, porosity, and permeability. The result of this simulation work flow is a suite of realizations each with facies, porosity, permeability, and saturation with the appropriate relationships.

Post-processing

Once these realizations have been compiled, they may be subjected to flow simulation to compute flow variables of interest. In this case, flow simulation for each existing model is practical because the grid is well-behaved and there are few enough cells for reasonable flow simulation run times. In some cases it is necessary to first scale up the reservoir

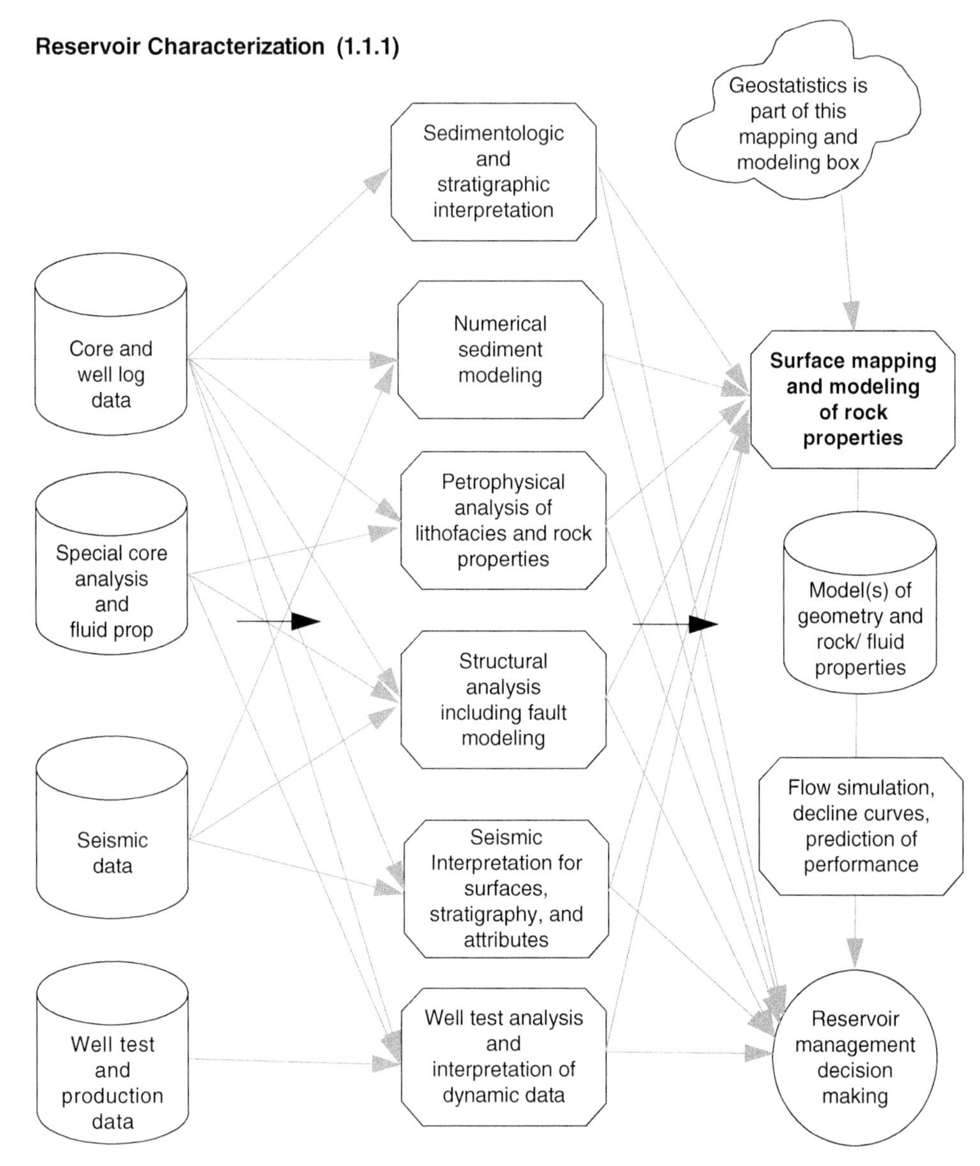

FIGURE 1.18: An Illustration of Some of the Major Operations That Enter into Reservoir Characterization. In this large picture, "Geostatistical Reservoir Modeling" constitutes part of the work flow to build numerical models of reservoir geometry and internal rock property heterogeneity.

model prior to flow simulation, utilizing a scale-up routine that attempts to preserve important heterogeneities or to rank the models with a fast-flow proxy and flow-select models. Nevertheless, it is important to realize during reservoir model construction that there may be limits to the resolution that may be included and limits to features preserved in scale-up if required. Care must be taken not to spend an effort on modeling fine-scale architectures that may not readily be included in the flow simulation. For this demonstration, a simple streamline-based calculation was applied to assess flow within the reservoir. Four injection wells and a single producer were placed in the model for this analysis. The resulting time-of-flight model is shown for the first channel reservoir realization in Figure 1.13. One can observe the impact of heterogeneity near the injectors and producer. Fractional flow for several realizations is shown in Figure 1.14. This provides an indication of the flow response of the reservoir realizations to the specified exploitation plan. Realizations provide the associated uncertainty in reservoir response given

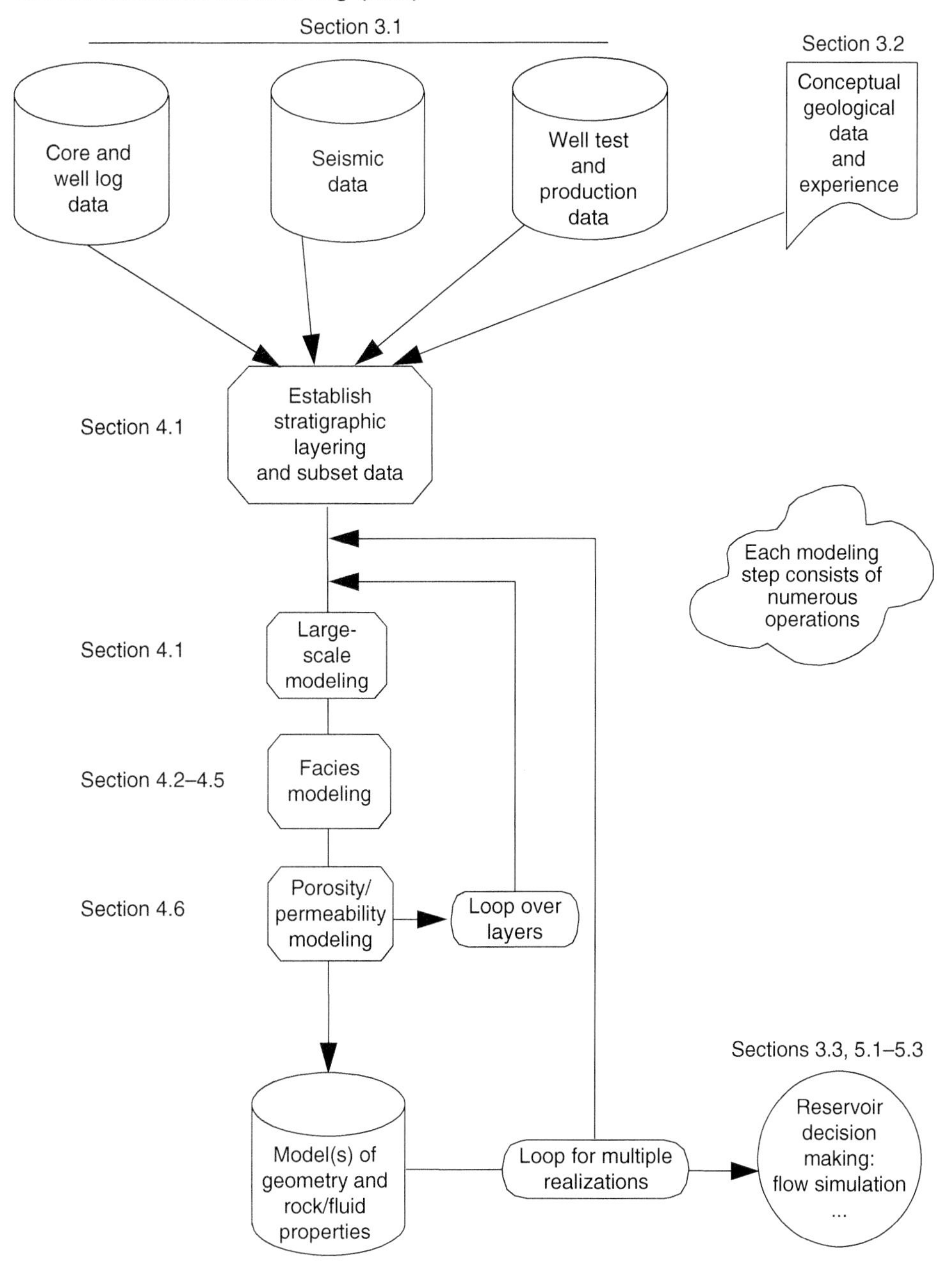

FIGURE 1.19: A High-Level View of Geostatistical Reservoir Modeling. Chapter 2 presents preliminary concepts. Section 3.1 introduces various data types and details on data integration. Section 3.2 discusses the integration of conceptual information, while Section 3.3 discusses problem formulation. Chapter 4 presents critical operations in detail. Chapter 5 presents model applications such as modeling checking, model post-processing, and uncertainty modeling.

each well configuration. Iterations of the exploitation plan allows for the opportunity to optimize well placement in the presence of this uncertainty.

If production data are available, the flow simulation results may be compared to these data to further refine the architectural model. This step may be helpful to reduce architectural uncertainty. For example, if a specific architectural scenario or parameter set does not produce flow results similar to production data, then the decision may be made to remove these combinations from set of realizations.

Another common product of post-processing is a sensitivity analysis to assess the impact of various model choices and parameters on the transfer function of interest. We accomplished this by comparing the recovery factor over multiple realizations with varying a single parameter or modeling choice at a time. The result can be visualized in a tornado diagram. These are useful for assigning the most important components in reservoir architectural uncertainty. This may be utilized to focus future modeling efforts on these important components or to even direct future data collection to reduce uncertainty associated with these components (see Figure 1.15).

Other useful results may include summaries of the multiple reservoir realizations. For example, an "e-type" model is the local average of all realizations at each location in the model. This model provides an indication of the most likely value of each reservoir property at each location in the model that can be applied to assist in well planning. Another useful summarization is local percentile models. For these models the local P value is displayed. These provide a good indication of areas within the model that are surely high or surely low. An example e-type and P90 model based on many porosity realizations of the reservoir are shown in Figure 1.16.

The geostatistical realizations are not the final product of this exercise. The final product is an uncertainty model for reservoir response, given the exploitation method. In general, this production uncertainty model will be combined with economic information to aid in making optimum decisions given reservoir uncertainty. In another potential application the geostatistical realizations are utilized to determine the need for and plan for further drilling. The critical point is that most often the value of geostatistics is realized not in the geostatistical realization themselves, but when the realizations improve decision quality and subsequently adds value to a project.

1.8 WORK FLOW DIAGRAMS

Many sections will close with work flow diagrams. The intent of the work flow diagrams is to help newcomers understand some of the linkages and required inputs to different modeling steps. The goal is neither to document all possible work flows nor to document each minor decision that must be made; the work flow diagrams are conceptual in nature. The diagrams are meant to be a summary of the procedures and material presented in the chapter. Readers should refer to the text in all cases for details about the data, operations, decisions, and interpretation guidelines.

The standard flow chart symbols shown in Figure 1.17 will be used. A brief description of how the symbols will be used is given on the right-hand side of Figure 1.17.

Figure 1.18 shows an illustration of some of the steps in reservoir characterization. The topic is diverse and involves many different disciplines. Geostatistical techniques are involved in a number of operations; however, the main focus is numerical algorithms to aid in surface mapping and modeling of rock properties (upper right of Figure 1.18). Any single book on petrophysics, seismic, geology, well test analysis, flow simulation, or geostatistics could not hope to cover "reservoir characterization." The aim here is to describe the geostatistical tools that enter into the critical problem of mapping surface and rock properties on the basis of sparse and imprecise data.

Geostatistical tools are most widely used to construct models of the geometry and rock properties for decision making and, perhaps, input to flow simulation. Figure 1.19 shows a work flow for this purpose.

2

Modeling Principles

This chapter covers the fundamental prerequisites required for geostatistical reservoir modeling. The *Preliminary Geological Modeling Concepts* section discusses the importance of understanding the geology story behind the reservoir. This understanding allows for consistent integration of all available information into a model that best represents reality. In addition, the modeling concepts used by geologists can be readily translated into the geostatistical modeling concepts discussed in the next three sections.

The *Preliminary Statistical Concepts* section covers the basic statistical concept. These are the input statistics that must be inferred to constrain the geostatistical models. Inference of representative statistics, parameter uncertainty, and methods for combining sources of information are discussed to provide good inference and to account for all available information and associated uncertainty.

The *Quantifying Spatial Correlation* section extends the statistical discussion to spatial statistics. These allow us to model and reproduce spatial features that are essential to the reservoir heterogeneity model. Variogram, and multiple point statistics are introduced along with concepts of volume variance that are important to accounting for data and model grid scale.

The *Preliminary Mapping Concepts* section takes the statistical concepts and introduces statistically driven methods for estimation and simulation. These include a variety of kriging-based methods, multiple point simulation, and object-based and optimization algorithms.

2.1 PRELIMINARY GEOLOGICAL MODELING CONCEPTS

This section covers preliminary geological modeling concepts including a brief outline of some of the geologic principles that in combination tell the story of the genesis of a petroleum reservoir. As a prerequisite for reservoir modeling, these principles are utilized to formulate a consistent and rational story of the reservoir or a set of stories if uncertainty is a main goal of the study. Our conceptual geological model for a reservoir has a large influence on the selected geostatistical tools and parameters.

The Story subsection discusses the importance of formulating this conceptual model or story in contrast with a naïve purely data-driven statistical approach to reservoir modeling. The available data for reservoir modeling are often inadequate to constrain the resulting reservoir model. There is a need to supplement sparse and ambiguous data with an understanding of the expected geological features.

In the *Geological Models* subsection the models utilized by geologists and the models utilized by geostatisticians to understand petroleum reservoirs are compared and contrasted. While there are important distinctions, there are many commonalities in concepts because geostatistics has always aimed to provide a practical framework for describing geology. In addition, a short summary of the considerations in reservoir geology is presented. This includes a brief discussion on reservoir formation and the framework of sequence stratigraphy. A review of basin and reservoir scale process provides more details and context for the conceptual models underlying the application of geostatistics.

The *Basin Formation and Filling* subsection presents the large-scale processes and associated terminology related to the formation of sedimentary basins and their subsequent filling. Allogenic processes are introduced as the extrabasinal controls that determine the sediment supply and accommodation available to place these sediments in a basin. In general, these controls determine whether or not a reservoir forms.

The *Reservoir Architecture* subsection presents smaller-scale processes that occur inside a basin and determine the architecture of a reservoir. This includes discussion on autogenic processes that place sediments in the basin and preservation potential that determines the final form of the reservoir in the ancient sedimentary record. In general, these controls determine the heterogeneity and connectivity of the reservoir.

The *Example Stories and Reservoir Modeling Significance* subsection presents three example reservoir settings: the Cretaceous McMurray Formation, Northern Alberta, Canada; the Lower Tertiary Wilcox Formation, Deepwater Offshore Texas, United States of America; and the Tengiz Carbonate Platform, Precaspian Basin, Kazakhstan. Each includes a short description of the story behind the basin formation, associated filling and reservoir architectures, and how this information may be utilized to construct geologically consistent geostatistical reservoir models.

This section motivates the construction of a conceptual geological model. The way this conceptual model impacts the reservoir model through practical modeling decisions such as the reservoir volume, gridding with associated faults and strata correlation styles, and reservoir geometries and heterogeneities is provided in Section 3.2. Statistical constructs and geostatistical algorithms required to represent these geological concepts are presented in the remaining sections of Chapter 2.

> The concepts in this section are intended to be simple and accessible to all levels of modelers. The novice modeler will benefit from an understanding of the benefits of direct integration of geological information and the similarity of geological and geostatistical models. The intermediate modeler will benefit from the examples of geological stories and methods to link this story to modeling choices as discussed later in Section 3.2, Conceptual Modeling. The expert modeler will benefit from the opportunity to dive deeper into the provided geologic references to improve the geologic integration and communication in their modeling projects.

2.1.1 The Story

Numerical reservoir models could be constructed without an in-depth understanding of the geologic story responsible for the genesis of the reservoir. One could approach reservoir modeling as a purely statistical study. In this scenario, the data would be handed over from the geologists to the reservoir modeler(s) working in isolation. Data would be statistically described, as well as pooled into distinct regions, and trends are mapped. Spatial continuity is characterized within the regions, and simulation and post-processing could proceed as described in subsequent chapters of this book. The resulting models could then be validated only by their reproduction of input statistics and local data, described as minimum acceptance criteria (Boisvert, 2010; Leuangthong et al., 2004). These models would seem to be perfectly correct from a statistical perspective, and some may prefer them for their apparent objectivity and lack of bias due to a preconceived conceptual model.

Yet, this approach would ignore important information not apparent from the local data alone and could even violate concepts fundamental to the geology of the reservoir. Potential consequences include a lack of reproduction of geological features at a scale less than the well spacing and a poor assessment of model uncertainty. The geological information beyond the data may provide important information to constrain extrapolation away from the data and uncertainty constraints. This geological information is central to selecting analogs to aid in inference in sparse data settings and to decide on appropriate high-resolution heterogeneity in mature, dense data settings. A more intangible consequence relates to model communication and credibility. Models that violate or lack integration of fundamental geological knowledge may communicate misconceptions to recipients or the flow simulation results may not be trusted, resulting in issues in project team alignment.

Some important caveats on the value of the geological story should be mentioned. Firstly, the importance of the geological story will depend on the amount of data available. It is clear that in exploration and appraisal, with few wells and poor or no seismic information, the geological story forms the foundation for all reservoir model decisions. On the other hand, for a mature project with dense well data and high-resolution measurements from a seismic data set, a more data-driven approach may be utilized with the geological story providing little more than a consistency check. Secondly, the importance of the geological story may shift with changes in project objectives and scale. For example, if enhanced recovery is later applied in the mature field and the results are sensitive to fine-scale intrawell heterogeneity, then the geological story may once again become more important. Thirdly, it is difficult to know what geological features are important prior to modeling; therefore, ignoring geological information even in a mature field may result in significant lost opportunity.

A reservoir modeler, within the integrated earth science and engineering project team, should attempt to discover and piece together a plausible and consistent story for the reservoir. This effort allows for the integration of not only the local hard and soft data, but also geological insight and concepts. Given the previously mentioned data paucity, incompleteness, and inaccuracies typically encountered, finding the story may be difficult. General concepts with regard to addressing uncertainty (see Section 5.3) can be employed. Uncertainty may be characterized by (a) retaining multiple scenarios or parameter distributions and (b) multiple stories with variations in the details. The simplest explanation(s) that matches the data and concepts should be retained. At times, less likely cases may be retained to aid in risk mitigation.

Issues in Telling the Story

Devising a reservoir story involves many decisions and an assessment of the associated uncertainty. While doing this, we should be aware that the heuristics we naturally apply to solve these difficult problems may lead to bias (Tversky and Kahneman, 1974). These include representativeness, availability, and adjustment and anchoring. For example, representativeness results in the assignment of probability of *A* being part of *B* by the similarity of *A* and *B*. Availability results in the assignment of greater probability to the familiar and does not represent less familiar. Finally, adjustment and anchoring is a natural human tendency to seize onto a position and adjust to formulate an estimate. Firstly, often the initial anchor is unreliable, we may anchor to unrelated or unreliable information; so secondly, we tend to under adjust the estimate from the anchor.

Not only are there potential issues in the manner that we estimate, but once an estimate is made, confirmation bias may become an issue. Confirmation bias is the tendency to ignore or undervalue information that contradicts the currently supported view, and group polarization is the tendency of a group to evolve to the extreme views of the group. The result of these cognitive issues is (1) we typically underestimate uncertainty or, more plainly stated, we think we know more than we actually do, and (2) groups of experts may not work together in an optimum manner. When the project team's efforts are moderated, then group wisdom may result in better estimates for even the most difficult estimation problems.

Value of Telling the Story

A project team utilizes individual expertise to carefully formulate a comprehensive story that integrates all the available data, information, and expertise. This story continues to play a role in the project as a reality check, communication tool, and means to make further inferences. It must be pieced together so that each contributing chapter is consistent and defendable and serves as a tool for future data and concept integration.

A critical skill set is the ability to communicate between the disciplines and to integrate all salient information from the geologic perspective into the numerical model. The following sections provide a brief treatment of the immense science of basin and reservoir geology. The focus is on providing a sampling of the concepts and nomenclature that form the basis for the geological model of a petroleum reservoir with references for more intense study. Section 3.2 provides more details on *Conceptual Models* that represent this story. The following subsection discusses the models applied by geologists to represent the story.

2.1.2 Geological Models

Geologists employ their own models and definitions that have similarities or parallels to the fundamental concepts of geostatistics (see Sections 2.2 and 2.3). This is not a coincidence, as geostatistics was developed to meet the practical need of modeling geologic phenomena [with the later development of theoretical underpinnings; see Journel and Huijbregts (1978) and Krige (1951)]. Philosophically speaking, understanding the geological models provides a geostatistician with a deeper understanding of their own related discipline. Like geostatisticians, geologists have models for characterizing geological heterogeneity at various scales, describing transitions and trends, separating samples that represent distinct subsets of the reservoir, and stating degrees uniqueness and certainty.

A general difference between geology and geostatistics is the geometric and descriptive nature of the former and the statistical nature of the latter. Geology does have some qualitative aspects that are not present in geostatistics. While geologists may

TABLE 2.1. RESERVOIR CONCEPTS AND ASSOCIATED GEOLOGICAL AND GEOSTATISTICAL EXPRESSIONS

Concept	Geological Expression	Geostatistical Expression
Major changes in relationships between reservoir bodies	Architectural complexes and complex sets	Regions—separate units and model with unique methods and input statistics
Changes in reservoir properties within reservoir bodies	Basinward and landward stepping Fining/Coarsening up	Nonstationary mean
Stacking patterns of reservoir bodies	Organization, disorganization, compartmentalization, compensation	Attraction, repulsion, minimum and maximum spacing distributions, interaction rules
Major direction of continuity	Paleo-flow direction	Major direction of continuity, locally variable azimuth model
Relationship between vertical and horizontal continuity	Walther's Law	Geometric and zonal anisotropy
Distinct reservoir property groups	Lithofacies, depositional facies, and architectural elements	Reservoir categories, stationary regions
Heterogeneity	Architecture	Spatial continuity model, geometric parameters, training image patterns

[a]Most geostatistical constructs can be directly mapped to geological constructs that describe the reservoir.

describe a rock unit as laterally continuous, a geostatistician will discuss the horizontal range as quantified by a variogram. While geologists may note a persistent fining up trend, geostatisticians will calculate and model a vertical trend in a reservoir property. While geologists may invoke Walther's Law to explain the relationship between vertical and horizontal features, geostatisticians formulate a geometric model with an anisotropy continuity model. See Table 2.1 for some concepts and the associated geological and geostatistical expressions. We recognize the level of quantification and description inherent to geologic analysis. Consider the outcrop interpretation shown in Figure 2.1. Heterogeneity across scales is characterized by high-resolution vertical measured sections with grain size, bedding structures, paleo flow directions, and interpolated correlations with thickness distributions. In the original field guide these features are described and linked to various processes. As will be shown later in this chapter, this type of data is for not only useful inference of the geologic story, but may also provide useful statistical inputs. The following is a brief overview of components of the geologic model. Specifics with regard to basin models and reservoir architecture are provided in the subsequent subsections.

2.1.3 Geological Model Overview

The geological model for the genesis of the reservoir has two important and interrelated components; rock and fluids. It is the story of how the reservoir and associated seal formed and how reservoir fluids formed in a source and migrated to the reservoir. Yet, for the purpose of reservoir modeling, the story may be limited to the genesis of the reservoir units themselves, because issues related to source, transport, or seal are exploration prerequisites to reservoir modeling that are typically addressed within the geological, geochemical, and geophysical disciplines by remote sensing and mapping, fluid sampling, and fluid analysis and not in the geostatistical model.

The geological process is central to geological models. There are excellent resources for the fundamental geologic processes related to reservoir formation. Catuneanu (2006), Catuneanu et al. (2009), Einsele (2000), Reading (1996), and Walker and James (1992) provide excellent, detailed descriptions of the fundamental processes and models related to depositional setting, while Galloway and Hobday (1996) provide a convenient synthesis for resource investigation. Reservoirs are commonly differentiated as siliciclastic and carbonate and are dealt with separately, due to their distinct genesis.

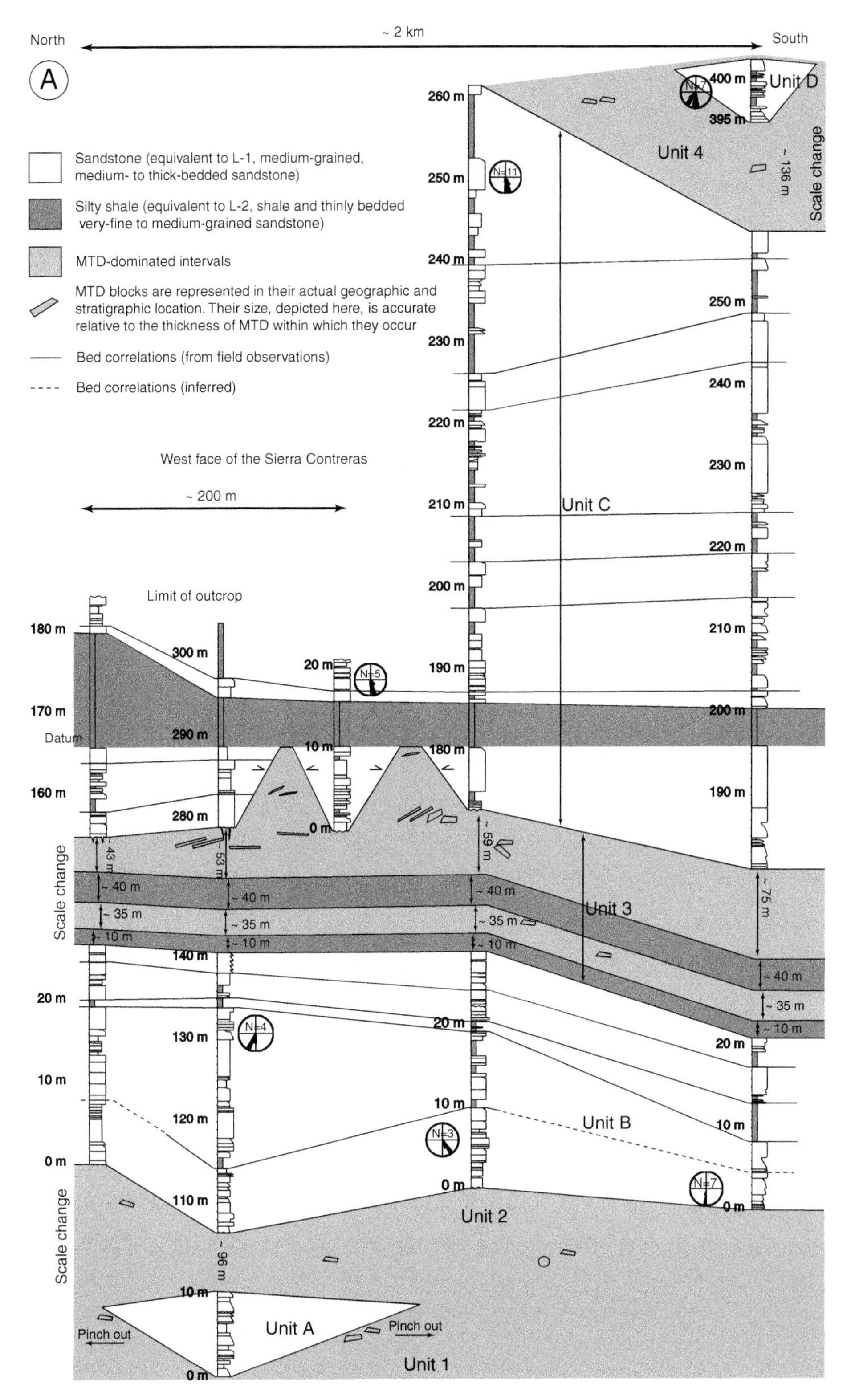

FIGURE 2.1: Measured Vertical Sections with Grain Size and Bedding along with Paleo Flow Directions and Interpolations with Correlation and Bed Thickness Between Sections from the Tres Pasos Formation. Figure reprinted from Armitage et al. (2009), Figure 12.3, with permission from the Society for Sedimentary Research.

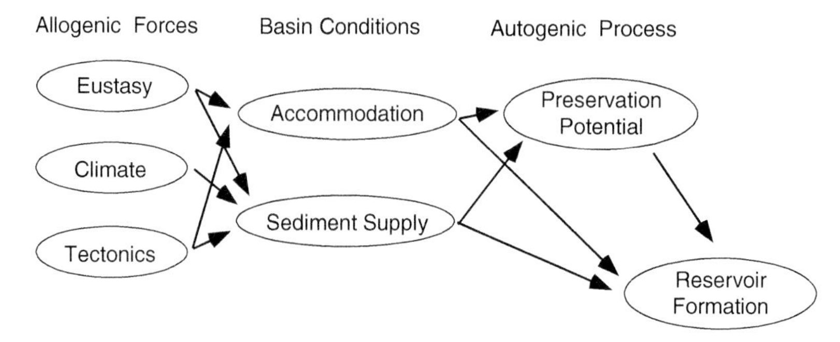

FIGURE 2.2: Gross Simplification of the Complicated Interaction of External to Basin (Allogenic Forces) and within Basin (Autogenic Processes) on Reservoir Formation. The combination of all these processes determines the presence, quality, size, shape, heterogeneity, and so on, of the reservoir.

Siliciclastic reservoirs are dominated by eroded and transported rock detritus, while carbonate reservoirs are dominated by carbonate materials that are grown in place and/or transported to the basin. Note that combinations are possible with carbonate systems receiving terrigenous rock detritus or siliciclastic reservoirs receiving carbonate sourced rock detritus.

Concepts such as accommodation, allogenic forcing, sediment transport and growth, and post-depositional alteration are fundamental to reservoir construction and will be discussed subsequently. All of these factors interact to form a sequence of interconnected *depositional systems,* defined as a volume of the earth influenced by specific conditions and processes. The result is specific depositional settings, such as alluvial, lacustrine, eolian, fluvial, deltaic, and shelf slope. Typically a single reservoir model is a small part of a depositional setting's areal extent, yet a reservoir model may include multiple depositional systems juxtaposition vertically.

The use of *hierarchical analysis* streamlines the communication and comparison of geologic stories. Architectural hierarchy is applied to classify and describe the parts and scales of this system. A key tool in architectural hierarchy is sequence stratigraphy, which is "the subdivision of sedimentary basin fills into genetic packages bounded unconformities" [see Myers and Milton (1996) for quote and detailed discussion]. This is an effort to divide the rock into a sequence of time units and place them in a predictable spatiotemporal framework. These sequences may include multiple genetically linked depositional systems known as *systems tracts.* Each systems track represents a specific time unit significant to basin filling. The result is an understanding of the temporal evolution of the basin fills. This is in contrast with lithostratigraphy where similar lithologies are mapped without consideration of sequence. It should be noted that lithostratigraphy is only considered useful within a sequence stratigraphic framework. This is of interest to the reservoir modeler because naïve interpolation of available data is analogous to lithostratigraphy. Subsequent discussion addresses specific formative processes at basin and reservoir scale; these are shown with their interactions schematically in Figure 2.2.

2.1.4 Basin Formation and Filling

Reservoirs form in *sedimentary basins.* These basins are areas where significant sediments may be deposited and/or grown and preserved (Einsele, 2000). Tectonic controls (movements of the earth's plates) are responsible for the formation of these basins. For example, intracratonic sag may form a basin in the middle of continents, and continental rifting forms new ocean basins. Sediment supply and the linked concepts of base level and accommodation are controls on basin filling. Base level and accommodation are discussed next, followed by sediment supply and then discussion on influence on reservoir framework.

Base level is an imaginary plane below which sediments deposit and above which they erode. Base level grade is determined by the type of depositional setting (i.e., flow discharge and grain size) and is generally hinged to a lake or ocean. The space below the base level is known as *accommodation* and represents the volume available to receive sediments (Myers and Milton, 1996).

Sediment supply is characterized by origin (or provenance), quantity, and type of source materials. The provenance is important because it may constrain the type of sediments (lithology and grain size) that in turn impacts reservoir quality and heterogeneity. The quantity of sediments constrains the volume of a reservoir, specifically the thickness and lateral extent and the connectivity and net to gross as sources with coarser and finer grain size compete to fill the available accommodation. The source type characterizes the location(s) and temporal nature of the sediment source. A sediment source may be a point source such as a river mouth or a line source such as along the front of a delta. The source type may also be continuous in time, such as a sediment-laden river building an alluvial plane, or discontinuous in time, such as the submarine slope failures that result in periodic turbidity flows on the slope and abyssal plane. The source may change across various time and volume scales; for example, a carbonate ramp may continuously grow during deposition of the high-stand systems tract and then shut down with subaerial exposure during the low stand (see discussion on eustasy below).

As discussed, basin filling is constrained by sediment supply and accommodation. Sediment supply and accommodation are, in turn, controlled by *allogenic controls*. Allogenic controls are defined as those factors external to the basin and include tectonics, climate, and eustasy. *Tectonics* refers to the large-scale movements in the earth's crust that alters the shape and depth of the basin and adjacent areas (i.e., notably the source area for the sediment fills). *Climate* refers to long term temperature and moisture regime. *Eustasy* is the rise and fall of the oceans, driven principally by the storage and release of water in continental glaciers. In addition to forming the basin, tectonics influence the availability of sediments to deposit in the basin and the conditions for growing carbonates. For example, mountain building provides a new source of sediments to enter adjacent basins, or a rise in the shelf may expose and kill a reef by exposing it above sea level and remove available accommodation for clastic deposits. Climate influences the rates of erosion, the water discharge rates available to transport sediments into the basin, and the growing conditions and availability of nutrients for carbonates.

Eustasy influences the available accommodation and the erosion rates by influencing base level and the growing conditions for carbonate analogous to the previously mentioned influence of coastal tectonics. For example, many reservoirs form in basins associated with continental margins in margin stratigraphic sequences. In these settings, base level is linked directly to the eustatic cycles. At high stand (or eustatic high level), deposition may occur on the continental shelves, while during low stand (or eustatic low level), deposition moves distal toward the ocean basins. Also, consider that eustatic influence in many basins extend significantly inland, resulting in large amounts of stored sediments (during rise) and erosion (during fall) of river valleys. These deposits may be characterized as high-stand and low-stand systems tracts within a sequence stratigraphy framework.

This model is useful for large scale reservoir framework characterization because it provides information on the external geometry, internal geometry, connectivity, and interrelationships between large-scale reservoir units (discussed in Section 4.1, Large-Scale Modeling). Commonly, this sequence stratigraphy framework is adopted directly into the reservoir model, with reservoir units representing high-order cycles related to allogenic or autogenic controls, gridding schemes based on stratal correlation style characteristic to the systems tracts, architectural concepts based on the depositional system and system track, and major bounding surfaces providing grid data for trends and regions. Even smaller-scale reservoir model choices such facies groups, and associated proportions and continuity are linked back to these large-scale concepts.

These basin-scale concepts are useful for large-scale framework modeling and to constrain smaller-scale reservoir architectural patterns. The actual true reservoir architecture, the position, the geometry, the quality, and the interrelationships of reservoir rock are the product of the preservation of the products of many smaller-scale processes related to transport, deposition, growth, and reworking of sediments. The concept of preservation is needed to translate this evolving landscape into a static reservoir. These autogenic controls (within basin controls) are discussed next.

2.1.5 Reservoir Architecture

The following subsections discuss the varieties of autogenic geologic processes that impact reservoir architecture. We start with siliciclastic reservoirs and then add some of the additional complexities of carbonate reservoirs and alterations; finally we discuss schemes to characterize reservoir architecture that are useful for reservoir modeling.

Siliciclastic Reservoir Architecture

Preservation potential is the likelihood of deposited sediments to remain in place and be subsequently buried and preserved. In other words, preservation potential is the transform required to move from *geomorphology*, the study of the near-surface expression of geology and the processes that are presently forming them, and *sedimentology*, the study of the rock record and the processes that anciently formed and preserved them within the statigraphic framework. While it is tempting to form simple models that directly relate current geomorphology and associated processes to the preserved sediments from observation of the surface expression, this is often wrong. For example, observation of a modern delta may suggest that the related preserved sediments should be composed of individual distributive channels with muddy overbank. Yet, the actual preserved architecture may be composed of cycles of stacked sandy units dipping and finning basinward (prograding clinoforms) (Walker and James, 1992). For another example, consider a meandering river. From the surface it is characterized by a narrow sinuous channel, point bars, and muddy overbank, but in many cases it is preserved as a thick sandy unit with some isolated mud plugs that form as channels are rapidly abandoned due to cutoffs [see coarse-grained meander fluvial style in Miall (1996) and Figure 2.3]. Preservation potential is a warning to geostatisticians to dig deeper into the process of reservoir formation to avoid making major errors in reservoir architecture.

Preservation potential is related to location and scale. In general, for a location where sufficient accommodation exists or is being created, preservation potential is higher than areas with low or decreasing accommodation. For example, deposition in a mountain hanging valley will subsequently be removed and not preserved in place for any significant amount of time. Turbidite sediments in the deep ocean basins will likely be preserved for longer periods because there is abundant accommodation and little to no force to remobilize the deposits, barring subsequent tectonic uplift. On the continental shelf, preservation is more variable because sediments tend to accumulate during rise in sea level (increase in accommodation) and erode during fall in sea level (decrease in accommodation). Preservation potential is also related to scale. For example, a ripple on a beach may be formed and reworked in minutes, while overbank deposits during the flood stage in a river valley may possibly be preserved and remobilized during channel avulsion years later. To reiterate, preservation potential is important in reservoir modeling because it provides us with a model of the reservoir architecture, including the scales and related processes to be concerned with and those to ignore.

FIGURE 2.3: Preservation Potential Demonstrated by an Oblique Block Diagram of a Meandering River. Note that the geomorphology seems to indicate isolated, sinuous channels dominated by overbank, while the true preserved sedimentology is imbricated sandy lateral accretion units forming extensive tabular units.

Once again, the controls on sedimentation within the basin are labeled as *autogenic controls*. These include processes related to the transport and placement of sediments, such as channel avulsion and delta lobe switching. While these cycles tend to be noisier or more difficult to characterize than allogenic controls, the resulting architecture has predictable features at all scales. In general, water is the principal agent in this chain of processes, as mechanical and chemical weather liberate clasts and streams transport the clasts until a loss in energy causes deposition due to a decrease in capacity or competency (the total sediment load and maximum sediment size transported, respectively). For carbonate reservoirs, biological and chemical processes grow the carbonate rock where the water depth, currents, and nutrients are favorable. Carbonate clasts may be eroded and transported or stand in place to form the reservoir. These processes have been extensively studied, and models exist to predict rates of erosion (Traer et al., 2012), carbonate growth potential and rates (Granjeon, 2009), and paleoclimate and associated transportation and deposition rates (see discusion in next section).

While transportation is critical to placing sediments in the reservoir, it also has an important role in the final reservoir architecture at various scales. For example, at the bed scale, tractive flow results in the sorting of grain sizes and the formation of bedding structures such as ripples, cross bedding, and dunes. Extensive discussion on the formation sedimentary structures is available in Boggs (2001). At larger scales, distributary lobes may result in (a) compensatory stacking that increase the areal extent of the reservoir and (b) preserved fine-grained breaks vertically.

Carbonate Reservoir Architecture

Walker and James (1992) provide detailed description of various carbonate reservoir settings. The following provides a short comparison to siliciclastic settings discussed previously. Carbonate reservoirs have many commonalities with siliciclastic reservoirs (Senger et al., 1992). Carbonate settings are sensitive to tectonics and eustasy because they control water depth and siliciclastic sediment supply that may drown or bury the carbonate source or support rapid growth. The majority of deposition is carbonate clasts that have been eroded and redeposited (i.e., not grown in place); therefore, all of the previously discussed concepts related to transport and preservation potential are applicable. Provenance is important, because different types of carbonaceous sources result in different types of sediments. Yet, there are critical differences. The diagenetic alterations (discussed in the next subsection) play an important role and may even override features preserved due to transport and preservation. This is due to the instability of the carbonaceous sediments, relative to typical quartzites in siliciclastic settings, because they may readily dissolve (forming large voids) and reprecipitate (forming cements that destroy porosity). These alterations are a function of biological and geochemical processes. In addition, carbonate sources are different from siliciclastic sources because (1) carbonate sources are very sensitive to environmental factors such as water depth, nutrient supply, water temperature, and turbidity and (2) carbonate sources change over geologic time with shifts between coral, algal, and so on.

Syn- and Post-depositional Alteration

Once sediments have been deposited and preserved in the early reservoir, they may be further altered. This may occur during the formation of the reservoir (syn-depositional) or after the formation of the reservoir (post-depositional). These alterations may include diagenesis, subsidence, and faulting.

Diagenetic alterations refer to changes that occur to sediments after deposition, including cementation and replacement but excluding metamorphism. These changes tend to locally modify reservoir properties and increase heterogeneity. In carbonate systems, diagenesis tends to be more important, with diagenesis enhancing, destroying or even inverting porosity (grains become pores and pores become grains).

Subsidence and faulting refer to large-scale compaction of the reservoir and units below the reservoir and fractures with movement. Both of these complicate the reservoir as units may no longer be horizontal or contiguous as deposited. In reservoir modeling, such features must be modeled and removed because the all geostatistical methods assume continuity within the model grid between adjacent cells and the model to be stratigraphically flattened (see Section 3.2).

Architectural Hierarchy

Reservoir architecture is typically described within a specific architectural hierarchy. The hierarchy represents cycles based on the previously discussed allogenic and autogenic processes. For example, Sprague et al. (2002) and Campion et al. (2005) developed a deepwater architectural hierarchy that is commonly utilized for deepwater reservoirs and Miall (1996) developed a scheme for fluvial settings.

To demonstrate the application of architectural hierarchy to reservoir modeling, the Campion et al. (2005) method is described and linked to reservoir modeling. The basic building block is a reservoir *element*. The architectural element represents a volume of sediment deposited by a single cycle of erosion and deposition separated by an avulsion. For example, a single sand-filled deepwater channel is an element. A set of elements with genetic linkage and predictable organization and depositional trend are classified as a *complex*. For example, a set of deepwater channels that are superimposed with similar geometries and fills are a complex. Two or more complexes that illustrate a genetic linkage are classified as a *complex set*. For example, a disorganized channel complex with latter organized channel complex that fill the same deepwater slope valley (i.e., have the same general trajectory) are a complex set. An illustration of this architectural hierarchy scheme is shown in Figure 2.4. It should be noted that the interpreted architectural hierarchy of a reservoir has some subjectivity due to dependence on interpretation of the local architecture and genetic processes given limited information. Yet, such a scheme does provide a reasonable common model to facilitate

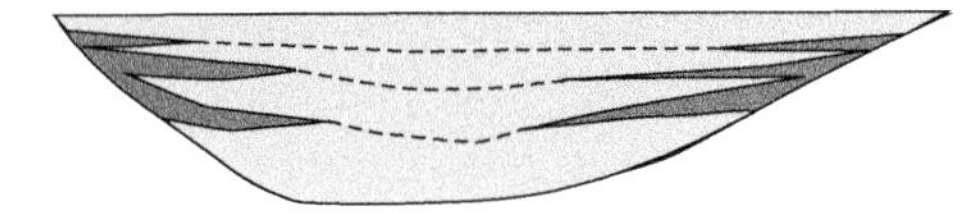

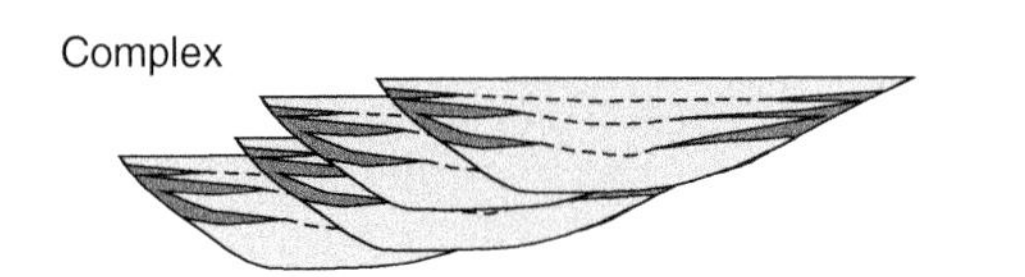

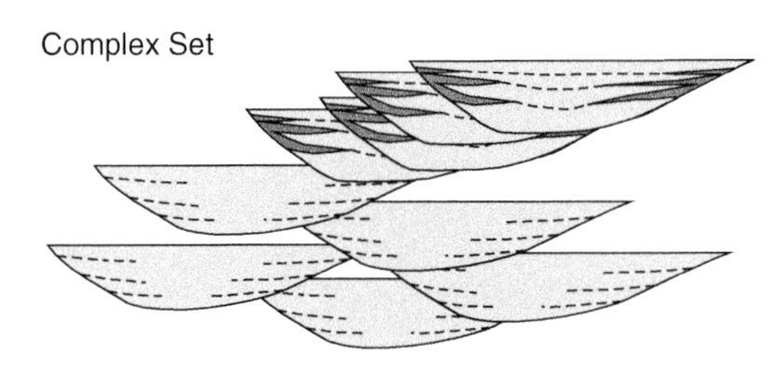

FIGURE 2.4: Architectural Hierarchical Scheme—Elements Are Typically the Smallest Identifiable Reservoir Units. They are defined by their geometries and internal fills. Architectural complex comprises two or more elements with similar geometry, internal fill and consistent stacking pattern. Architectural complex set is a set of related architectural complexes. Modified from Campion et al. (2005) and Sprague et al. (2002).

communication among the project team and/or between teams from analogous projects.

2.1.6 Example Stories and Reservoir Modeling Significance

In the following section, example stories and reservoir significance of three well-known hydrocarbon-bearing formations are discussed. The purpose is to demonstrate the practical linkage between the geologic conceptual model presented in this section and geostatistical models that are the topic of subsequent chapters of this book. Our illustrations and examples are not intended to provide a rigorous geological analysis, but rather are presented simply to provide the examples of diverse geological stories that must be inferred and integrated into geostatistical models (see included references for more detailed geological descriptions).

Reservoir Story for the McMurray Formation, Alberta, Canada

The McMurray Formation, Alberta, Canada is known for world-class oil sands that are a target for both surface mining and in situ steam-assisted drainage oil production. The following is a brief attempt to tell the reservoir story based on the regional studies by Hein and Cotterill (2006) and Fitzgerald (1978). During the Mesozoic Era, the foreland basin subsided, due to crustal loading by uplifted mountains (to the west), and much of current-day Alberta became flooded. Into this subsiding basin fluvial sediments prograded onto older exposed bedrock.

In the location of the current Alberta oil sands, the proximity to the course sediment source (the Canadian shield) and high energy in the river systems resulted in braided channels that broadly swept across their flood plains and preserved few fine-grained overbank deposits. These river systems were generally confined by and filled existing bedrock valleys. This confinement resulted in amalgamated channel reservoir sandstones that were thick along paleo-valley trends and absent over paleo-highs. The preserved valley sandstones are now the sand-rich Lower McMurray Formation, with associated high fraction of sand and good reservoir connectivity.

Accelerated foreland basin subsidence continued to deepen the interior sea during deposition of the Middle and Upper McMurray Formation, which caused gradual transgression and a general landward shift of the depositional environments. The McMurray fluvial systems gradually became lowered-energy meandering rivers that fed expanding estuaries bas-inward. This resulted in channel belts with extensive formation of point bar lateral-accretion elements and increased preservation of fine-grained overbank deposits [see the seismic geomorphology study by Hubbard et al. (2011) and an example of modeling point bars by Willis and Tang (2010)]. As transgression proceeded, very heterolithic lateral accretion units with extensive mud drapes became the dominate reservoir facies [see inclined heterolithic strata described by Thomas et al. (1987)]. Thick mud drapes within the point bar deposits are critical potential barriers and baffles to injected steam during reservoir production of these sandstones. Because paleo-valleys were generally filled by the end of Lower McMurray deposition, Middle and Upper McMurray Formation channel belts were broadly unconfined and these reservoirs are now widely distributed. Although extensive, connectivity is far more complicated than the Lower McMurray due to the complexity of the internal heterogeneities.

As seas continued to transgress, the McMurray fluvial deposits were subsequently buried by the Clearwater Formation, a layer of marine shale

TABLE 2.2. RESERVOIR ARCHITECTURAL CONSIDERATIONS FOR THE BRAIDED, MEANDERING, AND ESTUARINE CHANNEL SETTINGS OF THE MCMURRAY FORMATION

Reservoir Setting	Braided Channels	Meandering Channels	Estuarine Channels
Complex architecture	Locally confined laterally and vertically stacked channels	Extensive imbricated point bars with avulsion and overbank splays	Locally confined tabular sheet sands
Element architecture	Low sinuosity, sand-prone channels, and muddy overbank. Channels may be plugged by local slope failures	Low- to high-sinuosity sand prone channels and some overbank sands	Laterally extensive sands, amalgamated at axis of flow
Barriers	Bank failures	Overbank from adjacent flows	Mud drapes
Net to gross	Moderate to high	High	Moderate to high

and sandstones. Although recently glacial process have eroded the Clearwater and McMurray locally, relative tectonic stability of the region since the Mesozoic era has left the McMurray strata largely un-deformed and little diagenetic alteration has preserved relationships between depositional architecture and reservoir quality. The reservoir considerations for the McMurray Formation, summarized from this story, are shown in Table 2.2.

A Reservoir Story for the Lower Wilcox Formation, Gulf of Mexico

For the Lower Wilcox Formation in modern-day deepwater Gulf of Mexico, its story is dramatically different from the McMurray formation of Northern Alberta, Canada. As will be demonstrated, understanding this story is critical to constructing reservoir models throughout the Lower Wilcox formation. The following story was distilled from the work of Zarra (2007).

The present-day deepwater Lower Wilcox formation has been an important source of hydrocarbons since initial discoveries in at the Baha prospect in 2001. Previously, the deepwater Lower Wilcox Formation had been considered too far from sediment sources and too deep for economic reservoir quality and hydrocarbon presence. Data have been very challenged in this setting due to low seismic quality caused by depth and imaging subsalt and a paucity of wells due to very expensive drilling costs. Yet, the exploitation of the Lower Wilcox has resulted in exploration success with 65% to 70% discovery rates. This has been driven by a good understanding of the geologic story for the genesis of the Lower Wilcox formation, supported by good analog data sets for the onshore equivalent of the deepwater sediments and concepts derived from the study of turbidity currents from outcrops, flume tanks, and numerical experiments.

Tectonic uplift associated with mountain building in Colorado to Montana (Laramide orogeny) provided a significant amount of siliciclastic sediments that were transported, stored on the coast of the Gulf of Mexico, and subsequently transported into deepwater during the Paleocene and Eocene (about 60 million years ago to about 50 million years ago). Eustasy subsequently played an important role. During the deposition of the high-stand systems tract, these sediments were stored on the shelf and inland river valleys. During deposition of the low-stand systems tract, these sediments were eroded from river valleys and added to the sediments stored on and growing the continental shelf. Further falling of sea level exposed the shelf to erosional processes and destabilized the shelf. The result was massive periodic failures of the shelf, thereby generating turbidity flows that traveled through slope valleys down the shelf slope and along the abyssal plane.

This resulted in a continuum of deepwater reservoir environments, each with specific details important to reservoir modeling. During these flow events, course-grained sediments are deposited along channel axes and fine-grained sediments were stripped from the flow and deposited as levees. This resulted in predictable transition in reservoir quality from the axis to the margin of the flows and into the overbank. Also, this resulted in a generally high reservoir quality of channel infills because much of the fine-grained component is removed and deposited outside of channels (contrary to meandering fluvial channels that are often mud-filled). In the slope valley, the instability also resulted in failure of fine-grained materials from the valley walls forming local non-reservoir plugs in the slope valleys, known as mass transport complexes. These may form significant barriers to reservoir fluid flow between channel complexes or may break up the contiguity of antecedent channels.

In the Gulf of Mexico, significant relief was developing coeval to these turbidity current events due to salt tectonics. The density-driven turbidity currents were influenced by this relief. In some places, turbidity events and their associated deposits are concentrated due to confinement of this relief: in other areas, this relief results in depositional shadows and the absence of turbidites, and at times mini basins formed that ponded the turbidites, resulting in sheet-like sand units that onlap the minibasins' margins. Knowledge concerning paleobathymetry and its influence on the turbidites has been essential to determining local exploration sand potential.

Figure 2.5 provides a schematic illustration of the various allogenic and autogenic processes involved in the formation of Lower Wilcox deepwater reservoirs. In addition, several reservoir settings are indicated, including confined channels, weakly confined channels, and ponded turbidites. In Table 2.3, some reservoir considerations derived from the story of the deepwater Lower Wilcox formation are listed.

A Reservoir Story for Tengiz Carbonate Platform, Precaspian Basin, Kazakhstan

The Tengiz carbonate buildup of the Precaspian basin, located in western Kazakhstan, hosts the Tengiz super-giant field. Detailed summaries of Tengiz are available from Weber et al. (2003) and Collins et al. (2006). A brief summary is provided with a focus on salient points of the Tengiz story essential to assembling a reservoir heterogeneity model. We recognize that Tengiz is a large composite platform and have necessarily generalized for illustrative purposes.

The Tengiz isolated platform represents 1500 m of vertical relief with an area of about 580 km^2 with deposits from the late Devonian to the Carboniferous. Weber et al. (2003) divide this deposition into three major stages of growth, all constrained by allogenic controls. Initial carbonate growth was seeded on basin highs and was associated with aggradation

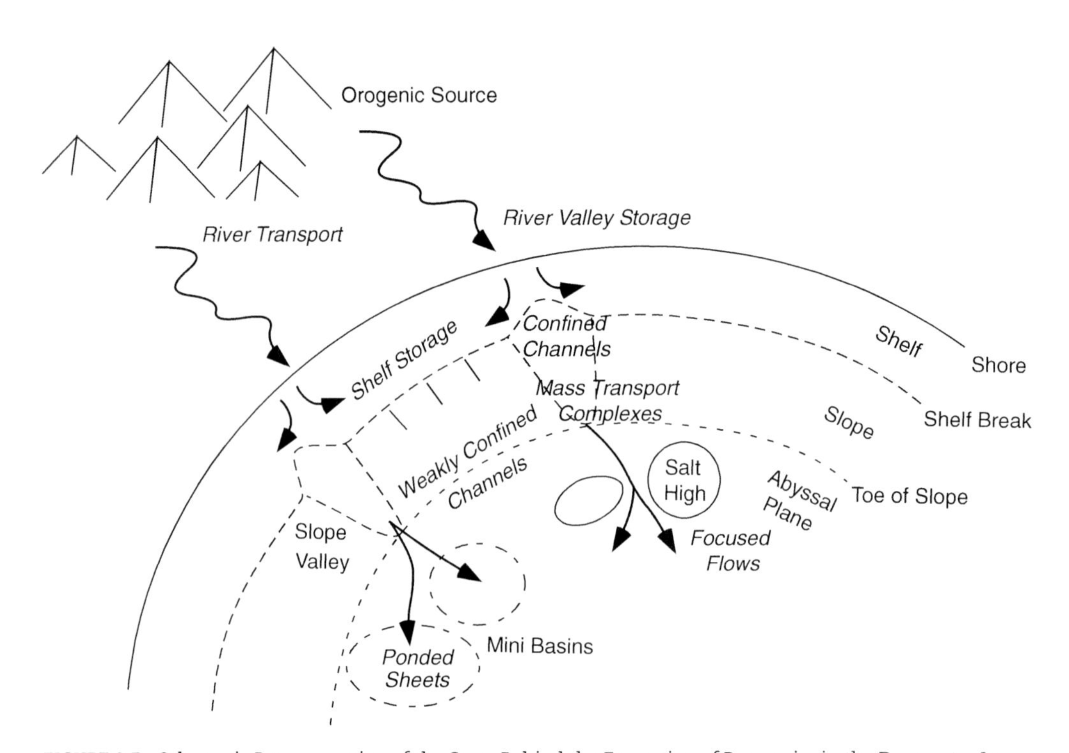

FIGURE 2.5: Schematic Representation of the Story Behind the Formation of Reservoirs in the Deepwater Lower Wilcox Formation in the Gulf of Mexico. The various controls such as sediment supply, transport, and the associated distinct reservoir settings are indicated.

TABLE 2.3. RESERVOIR ARCHITECTURAL CONSIDERATIONS FOR THE CONFINED, CHANNELS, WEAKLY CONFINED CHANNELS, AND PONDED SHEETS OF THE LOWER WILCOX FORMATION

Reservoir Setting	Confined Channels	Weakly Confined Channels	Ponded Sheets
Complex architecture	Laterally and vertical stacked channels	Loosely associated channels with avulsions and overbank splays	Extensive tabular sheet sands onlapping mini-basin confinement
Element architecture	Low sinuosity, sand-prone channels, and muddy overbank; may be plugged by local slope failures	Low to high sinuosity, sand-prone channels, and some overbank sands	Laterally extensive sands, amalgamated at axis flow
Barriers	Mass transport complexes	Overbank from adjacent flows	Mass transport complexes
Net to gross	Moderate to high	Moderate to high	High

and retrogradation, indicating sea level rise outpacing sediment production. This is followed by a stage of platform progradation, indicating that the sediment production outpaced sea level rise. Finally, this was followed by a stage of aggradation, indicating a balance of sediment production and sea level, followed by the drowning of the platform because sea level completely outpaced sediment production and the water depth exceeded that required to support shallow carbonate deposition.

Allogenic controls have a dramatic control on the overall reservoir framework. Large-scale tectonics developed the sag basin that created the needed accommodation, and local tectonics seeded the location of the platform with a local high. Then sea level rise and fall resulted in aggradational, progradational, and retrogradational units each with unique external geometries and internal bounding surfaces. Retrogradation resulted in a narrowing of the platform with backstepping of the margins and with deeper water conditions on the platform resulting in lower energy and increased preservation of mud. Progradation expanded the platform with steeply dipping growth wedge (up to 30° slopes) with significant boundstone component and is generally associated with shallower water conditions on the platform, resulting in higher energy and decreased preservation of mud.

Within this overall framework there are a variety of important autogenic processes that control local reservoir heterogeneity. This includes the local impact of currents, associated nutrient supply, and waves. These processes impact carbonate growth and subsequent transport. In this setting, high-wave energy favors growth of organisms that generate coarse grains and washes the mud component from the deposits. Local highs that accumulate along the rim block wave energy and result in mud deposition on the lee side. Due to asymmetry in the currents and waves, these highs are more extensive along the east and north edges of the platform. During progradation, this rim is dominated by boundstone that is prone to failures as evidenced by allochthonous blocks on the slope, failure scarps on the rim, and fracture sets.

As a result of all of this, Tengiz has an interesting separation between areas dominated by allogenic and autogenic controls. Over much of the platform, heterogeneity is controlled by allogenic sea level cycles with shallowing upward cycles at various scales. Over the rim and slope regions, heterogeneity is controlled mostly by autogenic slope failures that remove sediments from the rim and deliver them to the slope with compensational stacking, resulting in locally variable transition along the slope wedge (rim to toe) from fractured boundstone to breccias and finally to mud.

It would be difficult to cover all the detailed descriptions of facies and diagenetic alterations discussed in the cited papers. Nevertheless, the above discussion provides an indication of the importance of the geologic story for understanding the general geometry and internal heterogeneity of the Tengiz Field. In Table 2.4, some reservoir considerations derived from the story of the Tengiz isolated platform are listed.

TABLE 2.4. RESERVOIR ARCHITECTURAL CONSIDERATIONS FOR THE TENGIZ ISOLATED PLATFORM

Reservoir Setting	Platform	Fairway / Rim	Slope
Complex architecture	Sets of laterally extensive flat layers and channels	Massive, fractured boundstone avulsions and overbank splays	Fining basinward wedge, onlapping mini-basin confinement
Element architecture	Variable cycles	Massive	Imbricated failure aprons
Barriers	Shallow layers	Diagenetic cementation	Toe of slope muds
Net to gross	High	Moderate to high	Low to moderate

2.1.7 Section Summary

These geological model prerequisites are fundamental to constructing geologically realistic reservoir models. A reservoir is the result of the interaction of various processes at various scales of time and space. Knowledge of these processes and their associated conditions is limited, thus resulting in significant uncertainty. Yet, this limited information still provides valuable constraints on the geostatistical models.

The philosophy promoted in this section is to integrate all available information and expertise to reveal the story behind the formation of the reservoir. This will provide a consistent platform for future data integration and provide the best support for decision making. It is difficult to know what geologic information will be significant, so the best practice is to attempt to understand and thoroughly integrate all conceptual geological knowledge.

While some details and illustrative examples were provided in this section, greater depth of detail will be required for each specific reservoir model. Once this story has been characterized in sufficient detail, it can be mapped into geostatistical concepts to constrain the geostatistical reservoir model. The next section provides the statistical prerequisites for this mapping.

2.2 PRELIMINARY STATISTICAL CONCEPTS

The field of statistics is concerned with quantitative methods for collecting, organizing, summarizing, and analyzing data, as well as for drawing conclusions and making reasonable decisions on the basis of such analysis. The field of statistics is extensive; the goal here is a pragmatic presentation of the concepts essential to geostatistical reservoir modeling. Geostatistics is distinct from statistics in three main respects: (1) focus on the geological origin of the data, (2) explicit modeling and treatment of spatial correlation between data, and (3) treatment of data at different volume scales and levels of precision.

In the work flow described above, statistical inference follows the development of a conceptual model. This section and the subsequent one present the statistics that form the basis of geostatistical characterization. Inference of representative statistics is critical, because geostatistical methods aim to reproduce these statistics within the constraints of local conditioning data.

The *Geological Populations and Stationarity* subsection addresses the problem of defining geological and statistical populations for modeling. The concept of *stationarity* is discussed with the implications on subsequent modeling. The *Notation and Definition* subsection covers the fundamentals of probability distributions.

Often, data must be transformed from one distribution to another. The *Data Transformation* subsection describes a widely applicable technique for transformation that preserves the ordering and spatial relationships of the data. This section also introduces Q–Q plots and the concept of despiking.

The important techniques of declustering, debiasing, and histogram smoothing are presented in the *Declustering and Debiasing* subsection. These techniques allow going beyond the available samples and inferring a distribution and summary statistics more representative of the entire volume of interest.

The *Monte Carlo Simulation* subsection introduces Monte Carlo simulation methods for stochastic simulation of one variable. Such Monte Carlo techniques form the foundation of later 3-D geostatistical simulation methods.

It is recognized that rigorous uncertainty modeling requires assessment and integration of *Parameter Uncertainty*. A variant of Monte Carlo methods, known as the *bootstrap*, is a useful tool to assess uncertainty in input statistics under the assumption of independence between the constituent data. Other methods are presented that provide uncertainty models while accounting for the spatial context of the data.

Most of the inferential statistics in this book are derived from a frequentist perspective, based on the concept of pooling sample replicates to construct and model probability distributions. The *Bayesian Statistics* framework provides an alternative that utilizes belief and probability logic to build probability relationships. A short introduction is presented in preparation for topics such as the integration of multiple, redundant data sources covered in Section 3.1, Data Inventory.

This section should be accessible to all readers. The statistics are not advanced. The novice modeler should gain appreciation of the types of statistics required for modeling and methods to improve their inference. The intermediate and expert modelers may gain a deeper understanding of concepts such as stationarity along with concise notation and concepts for communication of the statistics.

2.2.1 Geological Populations and Stationarity

A critical decision in any statistical study is how to pool the data into populations for further analysis. This decision depends on the goals of the study, the specific data available, and the geological setting of that data; there is no universal answer. Often, a reservoir model is built on a by-facies basis one layer at a time. In this case, the data for each facies in each layer should be kept separate.

Populations must be defined, because all sample statistics, including a histogram and associated mean value, refer to a population and not to any sample in particular. Grouping everything together into one population may mask important trends. Excessive subdivision of the data may result in unreliable population statistics due to not enough data. With limited well data, it may be appropriate to combine the data from all of the layers—that is, separate the data on the basis of facies alone. At the time of modeling, the layers will be kept separate to ensure that the geological layering and layer-specific facies proportions are reproduced.

Facies

As mentioned in Section 2.1, Preliminary Geological Modeling Concepts, facies or rock types are a generic subdivision of the reservoir. The main consideration in choosing the number of facies is to balance the level of geological input (tend to consider more) with the statistical and engineering significance of each facies classification (tend to consider less). Ideally, each chosen facies should be geologically significant and yet have enough data to allow reliable inference of the required statistics for reservoir modeling. In practice, it is difficult to support more than four different facies from the data.

The important petrophysical properties used in subsequent modeling are porosity, permeability, and multiphase flow properties such as relative permeability and capillary pressure. The facies must have (1) clearly different petrophysical properties or (2) spatial features that make them easy to model. There is no benefit to separate the data according to facies that do not lead to distinguishable flow properties.

Typically, there are limited wells with cores; most will only have a suite of well logs. The facies are most easily identified by studying core. Once identified from core, it is essential that the facies can also be separated (with reasonable accuracy) from well logs at the wells without core. This is done by some type of calibration procedure (discriminant analysis, neural networks, etc.). If the facies cannot be identified from well logs alone, then “facies associations,” or a broader definition of facies, would have to be considered prior to geostatistical modeling.

Stationarity

A more formal definition of stationarity will be presented later in Section 2.3; however, for all practical purposes, the decision of which data should be pooled together for subsequent statistical analysis and over what volume these statistics are representative *is* the decision of stationarity.

The geological context of reservoir characterization presents unique challenges to the decision on what data to pool for statistical inference. Many statistical problems allow for repeated sampling, resulting in replicates. Consider, for example, products on a production line or water quality at a monitoring site where repeated samples are available over time. In the geological context, once a sample has been taken, measured by a well log or extracted as core, there are no replicates available for inference of statistics; there is only one sample available. The replicates required for inference of population statistics are available only by supplementing samples available from other locations that are considered to be similar from a geological basis.

There is always some temptation to consider each sample as unique and, therefore, not to lump it with other samples. Clearly, without a decision of stationarity, inference between the measured data is not possible. Enough data must be pooled for reliable geostatistical inference. Too few data will result in unreliable statistical descriptions and result in overfitting the local data and underestimating model uncertainty. Conversely, one may be tempted to pool data broadly across subsets of the reservoir with significantly dissimilar properties. Too broad assumption of stationarity may hide important reservoir heterogeneity and overestimate model uncertainty. The decision of stationarity is a critical opportunity to account for the geological conceptual information.

The decision of stationarity may be revisited once data analysis and geostatistical modeling has started. For example, we may notice a bimodal (two peaks) histogram of porosity within one of our chosen facies classifications. This does not mean the data are "nonstationary." It does mean that we should go back to the data and consider separating the data into two classes provided these display distinct statistical and geological properties.

Sometimes trends, such as a systematic fining upward trend or degradation in reservoir quality in the seaward direction, are referred to as nonstationarities. The presence of a trend or gradual areal/vertical change in facies proportions, porosity, or permeability need not be handled by further subdivision of the data. After all, there are no natural subdivisions in the case of a gradual, continuous change in properties. This situation can be handled by modifications to the modeling technique—for example, simple kriging with locally varying means or considering that the average or mean property follows a specified trend (see later chapters).

We are careful to describe stationarity as a decision. Stationarity is not a hypothesis; therefore, it cannot be tested. It is a decision that is made, given the objectives of the study, and the scale of the study, the observed spatial heterogeneity of the variable(s) of interest. While it cannot be tested, it must be defended and it could be changed as more data becomes available.

Input Statistics

A variety of statistical descriptions, such as univariate, bivariate, and spatial statistics that are inferred by pooling samples together under the assumption of stationarity, are now presented. The purpose of this inference is to provide statistical inputs into a variety of geostatistical algorithms for the construction of geostatistical reservoir models. These statistics are inferred from available local data and judiciously selected analogs and with efforts to improve representativity. As mentioned previously, geostatistical methods aim to reproduce these input statistics; geostatistical models have no predictive power with respect to these statistics. In fact, as will be discussed in Section 5.1, a critical minimum acceptance model check is to compare all the input statistics to the statistics of the resulting geostatistical models.

There are limitations in the reproduction of these statistics due to algorithmic limitations, compromise between contradictory inputs, and fluctuations between stochastic realizations. These are discussed below. Most geostatistical algorithms will not perfectly reproduce the specified input statistics. For example, categorical applications of sequential indicator simulation may underrepresent the short-distance spatial continuity (Deutsch and Journel, 1998) and p-field simulation honors data as local extrema (Pyrcz and Deutsch, 2001). The practitioner should be familiar with the limits in statistical reproduction for the algorithms applied to a reservoir modeling problem.

An important strength of geostatistical methods is the ability to integrate a variety of information. Nevertheless, these methods do not explicitly warn the practitioner when the input data are contradictory. Instead they attempt a compromise, based on a hierarchy of data importance built implicitly into their algorithms. For example, if porosity and permeability are cosimulated with a high correlation with each other, yet each has significantly different spatial continuity, the algorithm will not be able to honor these contradictory input statistics and may produce unanticipated results such as poor variogram reproduction and discontinuities at the data locations.

When the model size is large relative to the range of continuity, then the input statistics tend to be reproduced closely. As the model size decreases relative to range of continuity, the statistics over multiple geostatistical realizations tend to fluctuate. These fluctuations are known as ergodic fluctuations and are the consequence of the assumption that all geostatistical algorithms are embedded in an infinite domain. More details will be provided in Section 5.1, Model Checking.

With recognition of the importance and limitations of input statistics to geostatistical analysis, this section and the next cover the variety of statistical inputs for geostatistical modeling.

2.2.2 Notation and Definitions

A *variable* is a measure, such as ϕ, K, i, that can assume any of a prescribed set of values. A variable that can assume a continuum of values between any two given values is called a *continuous variable*; otherwise it is a *discrete* or *categorical* variable. Porosity and permeability are continuous variables often denoted generically by the letter z. Facies classifications are categorical and commonly denoted with the indicator variable i_k, where i_k is 1 if the category k is present and 0 if not.

The basic approach of predictive statistics is to characterize any unsampled (unknown) value z as a random variable (RV) Z. The probability distribution of Z models the uncertainty about the unknown true value. The random variable is traditionally denoted by a capital letter, say Z, while its outcome values are denoted with the corresponding lowercase letter, say z. The RV model Z, and more specifically its probability distribution, is usually location-dependent; hence the notation $Z(\mathbf{u})$, with $\mathbf{u}$ being the location coordinates vector. The RV $Z(\mathbf{u})$ seen as a function of $\mathbf{u}$ is called a random function (RF). The RV $Z(\mathbf{u})$ is also information-dependent in the sense that its probability distribution changes as more data about the unsampled value $z(\mathbf{u})$ become available.

Probability is a central concept in geostatistics. While clearly a vehicle for conveying the chance of a specific outcome, there are distinct differences in how it is assessed. In the discussions in this section a "frequentist" perspective is applied. Probability is calculated by the frequency of observations with the specified outcome as a ratio of all observed outcomes. A "Bayesian" perspective could be equally applied, with a focus on belief and updating in the presence of new information. In the Bayesian statistics subsection, this is introduced in preparation for the various applications of Bayesian methods throughout the book. These applications include the integrating secondary data and inversion of probabilities to move from probability of X event given Y event to a more difficult inversion of probability of Y event given X event.

Histograms and Probability Distributions

The histogram, a bar chart of the frequency of sample values observed, is a familiar statistical display. Figure 2.6 shows a typical histogram created with the GSLIB program histplt (Deutsch and Journel, 1992). This histogram has a fixed bin width. The frequency in each bin is the number of sample values falling into that bin divided by the total number of samples (2993 in this case). The original data values may be assigned unequal weights to correct

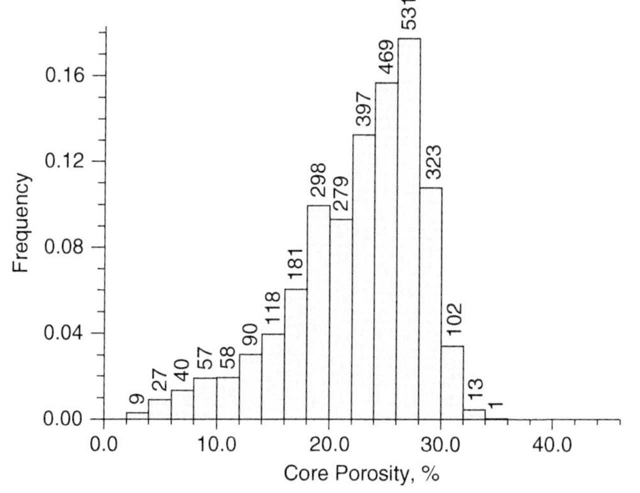

FIGURE 2.6: Histogram of 2993 Core Porosity Measurements. The number of data in each bin is shown above each histogram bar and the relative frequency is shown on the vertical axis.

for preferential clustering (see next section). A too-small bin width would create a noisy histogram with many short scale variations. A too-large bin width creates a too smooth histogram that may mask important features.

Histograms are suitable for gaining a visual appreciation for the distribution of data values; however, cumulative histograms are more extensively used in calculations. The cumulative distribution function (cdf) of the random variable Z is denoted:

$$F(z) = \text{Prob}\{Z \leq z\} \in [0, 1]$$

where the cdf $F(z)$ is a nondecreasing function of z. A plot of $F(z)$ versus z (see Figure 2.7) does not require the data to be classified into bins. Note that $F(z)$ is valued between 0 and 1 and z is valued in the range of data values. We calculate $F(z)$ as the experimental proportion of data values no greater than a threshold z.

Probability Distributions

The *probability density function* (pdf) $f(z)$ is linked to the cdf $F(z)$, provided that $F(z)$ is derivable:

$$f(z) = F'(z) = \lim_{dz \to 0} \frac{F(z + dz) - F(z)}{dz} \quad \text{if it exists}$$

where the pdf $f(z)$ is greater than or equal to zero ($f(z) \geq 0, \forall z$), the integral of $f(z)$ is the cumulative distribution function or cdf ($\int_{-\infty}^{x} f(u)\,du = F(x) = \text{Prob}\{X \leq x\}$), and the integral of $f(z)$ from $-\infty$ to ∞ is one ($\int_{-\infty}^{\infty} f(z)\,dz = 1$). As mentioned above, the cdf $F(z)$ is more extensively used in calculations, although people would rather visualize a histogram or probability density function.

The probability density function or pdf $f(z)$ is related to the histogram; often, a parametric probability density function is overlaid on an experimental histogram. There is, however, an important conceptual difference between the discrete histogram and the continuous pdf curve. The frequency of samples in a histogram class may be interpreted as the probability of a sample value to fall in that class; however, $f(z)$ should not be interpreted as the "probability of z." It is the integral of $f(z)$ between a lower bound z_{lower} and upper bound z_{upper}, $\int_{z_{lower}}^{z_{upper}} f(z)\,dz$, where $z_{upper} > z_{lower}$, which can be interpreted as a discrete probability for z to be in the interval $(z_{lower}, z_{upper}]$.

Parametric Distributions

A *parametric* distribution model is an analytical expression for the pdf or cdf—for example, for the normal or Gaussian density function:

$$f(z) = \frac{1}{\sigma\sqrt{2\pi}} e^{-\frac{1}{2}(z-m)^2/\sigma^2}$$

with mean m and standard deviation σ that control the center and spread of the bell-shaped normal distribution. Parametric models sometimes relate to an underlying theory; for example, the normal distribution is the limit distribution for a number of theorems collectively known as the central limit theorem.

There is no general theory for earth-science-related variables that would predict the parametric form for probability distributions. Nevertheless,

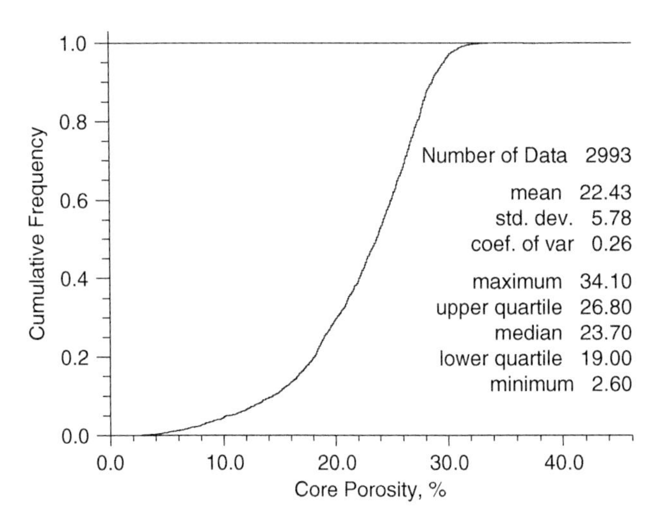

FIGURE 2.7: Cumulative Distribution of 2993 Core Porosity Measurements.

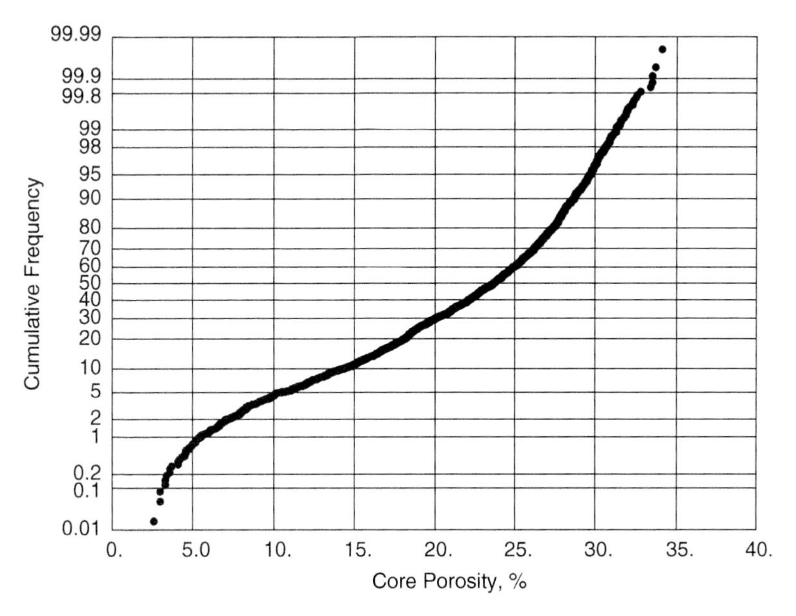

FIGURE 2.8: Probability Plot of 2993 Core Porosity Measurements.

certain distribution shapes are commonly observed. A "lognormal" distribution is one where the logarithm of the variable follows a normal distribution. Often, however, any skewed distribution of a positive variable such as permeability is wrongly called *lognormal*. There are statistical tests to judge whether a set of data values follow a particular parametric distribution. These tests are of little value in reservoir characterization, because they require that the data values all be independent from one another, in practice, taken far away from each other. Probability plots are useful for checking closeness to a normal or lognormal distribution and for identifying outlier data.

Probability plots are cumulative probability plots with the vertical probability axis scaled such that a normal distribution would plot as a straight line. Figure 2.8 shows a probability plot for the 2993 data we have been considering. Note the exaggerated probability scale for the high and low values. Such probability plots were originally used to see how well data fit a normal or lognormal probability distribution. The use of the histogram transformation procedures presented below alleviates the need for the data to follow any particular parametric distribution model; the main application of probability plots is to see all of the data values on a single plot. There are useful interpretations that can be made; for example, a sharp change of slope may indicate a different geologic population. Since we are often interested in the high and low values, the exaggeration of extreme values makes the normal probability scaling more useful than arithmetic probability scaling. The data variable axis may be plotted with logarithmic scale; data falling on a straight line would then indicate a lognormal distribution.

Parametric distributions have three significant advantages: (1) They are amenable to mathematical calculations, (2) the pdf $f(z)$ and cdf $F(z)$ are analytically known for all z values, and (3) they are defined with a few parameters, which are chosen to be easy to infer.[1] The primary disadvantage of parametric distributions is that, in general, real data do not conveniently fit a parametric model. *Nonparametric* distributions (described below) are more flexible in capturing the behavior of real data. *Data transformation* (described below in the next subsection) permits data following any distribution to be transformed to any other distribution, which permits us to capitalize on most of the benefits of parametric distributions.

[1] A general rule is that a minimum of 10 data are needed for a particular "statistic"; for example, 10 data would be needed to compute a mean with any reasonable level of precision. Two hundred data would be needed to compute a 90% probability interval (10 data below the 5% lower limit and 10 above the 95% upper limit). This is, of course, overly simplistic, but useful nonetheless.

Nonparametric Distributions

In the presence of sufficient data (say, more than 200), the data are often not well represented by a parametric distribution model. In this case, the cdf probability distribution may be inferred directly from the data, that is, *nonparametrically*. The cumulative distribution function $F(z)$ is inferred directly as the proportion of data less than or equal to the threshold value z:

$$F(z) = \text{Prob}\{Z \leq z\} \approx \frac{\text{Number of data less than or equal to } z}{\text{Number of data}} \quad (2.1)$$

Thus, a proportion is associated with a probability. The proportion in Eq. (2.1) assumes that the data are equally representative, which may be a poor assumption in the presence of data that are preferentially clustered in specific areas. For that reason, we will develop *declustering* procedures in Section 2.2.6 (below) to calculate a relative weight $\{w_i, i = 1, \ldots, n\}$ to be assigned to each of the n data $\{z_i, i = 1, \ldots, n\}$. Lesser weight is assigned to redundant data. The sum of the weights is one: $\sum_{i=1}^{n} w_i = 1.0$. The cumulative distribution function $F(z)$ is inferred directly using the weights:

$$F(z) \approx \text{sum of weights to data with } z_i \leq z \quad (2.2)$$

A nonparametric cumulative distribution function $F(z)$ inferred with [see Eq. (2.2)] is a series of step functions (see solid line in Figure 2.9). The subscript $z_{(i)}$ indicates the order in the data set, 1 being the smallest and n being the largest. Five equal-weighted data are illustrated in Figure 2.9. Some form of interpolation may be used to provide a more continuous distribution $F(z)$ that extends to arbitrary minimum $z_{\min}$ and maximum $z_{\max}$ values. Linear interpolation is often used. More complex interpolation models (see Chapter 5 in GSLIB (Deutsch and Journel, 1998)) could be considered for highly skewed data distributions with limited data.

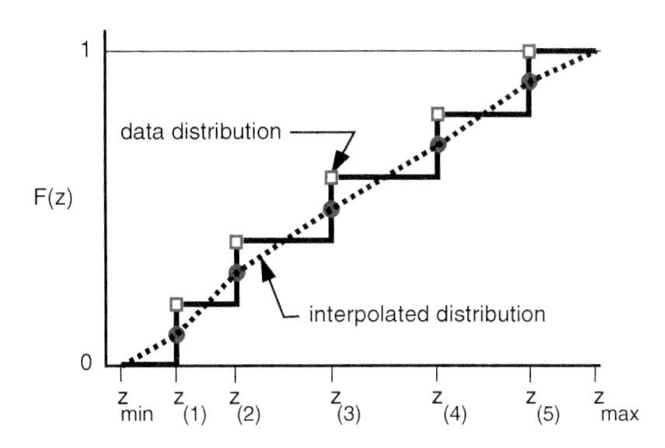

FIGURE 2.9: A Nonparametric Distribution Inferred from Sample Data. The solid line is from the ordered data $z_{(1)}, \ldots, z_{(n)}$ and the definition of a probability [see Eq. (2.1)]. The dashed line is a linear interpolation of the cdf to provide a continuous relation between $F(z)$ and z.

Quantiles and Probability Intervals

A *quantile* is a z value that corresponds to a fixed cumulative frequency. For example, the 0.5 quantile, also called the median and denoted $q(0.5)$, is the z-value that separates the data into two equal halves. $q(0.25)$ is the lower quartile. $q(0.75)$ is the upper quartile. The quantile function $q(p)$ is the inverse of the cumulative distribution function:

$$q(p) = F^{-1}(p) \quad \text{such that} \quad F(q(p)) = p \ \in [0, 1]$$

The quantile function $q(p)$ contains the same information as the cdf $F(z)$; it is just another way of looking at a probability distribution.

A symmetric *p-probability interval* is defined by the following lower and upper quantiles:

$$q\left(\frac{1-p}{2}\right) \quad \text{and} \quad q\left(\frac{1+p}{2}\right)$$

For example, the 90% symmetric probability interval is bounded by $q(0.05)$ and $q(0.95)$. Any probability interval can be determined once the cdf $F(z)$ or the quantile function $q(p)$ have been established, parametrically or nonparametrically.

Distributions of Discrete Variables

The probability distribution of a discrete or categorical variable is defined by the probability or proportion of each category, that is, $p_k, k = 1, \ldots, K$, where there are K categories. The probabilities must be positive $p_k \geq 0, \forall k = 1, \ldots, K$ and sum to 1.0: $\sum_{k=1}^{K} p_k = 1.0$. A table of the p_k values summarizes the data distribution adequately. At times, however, we can consider a histogram and cumulative histogram format (see Figure 2.10).

The cumulative histogram in Figure 2.10 is a series of step functions for an arbitrary ordering of the discrete categories. Such a cumulative histogram is not useful for descriptive purposes, but is needed for many calculations such as Monte Carlo simulation and data transformation (see below). In general, the ordering is not important. We will discuss cases when the ordering affects the results.

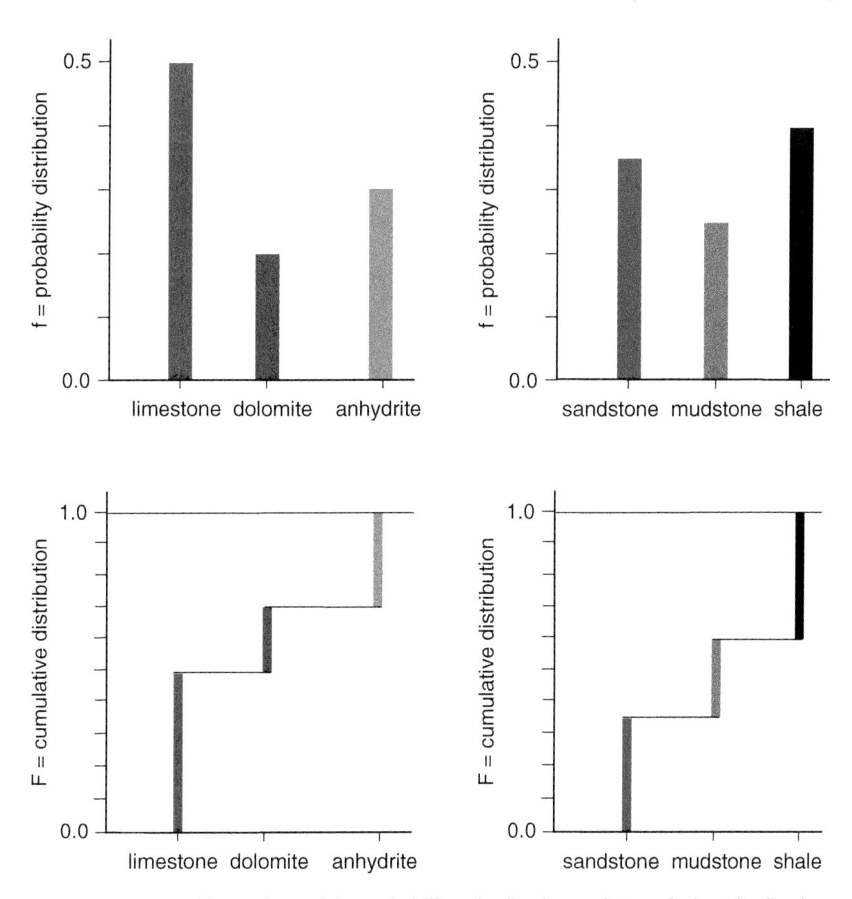

FIGURE 2.10: Two Illustrations of the Probability Distribution and Cumulative Distribution of a Categorical Variable.

Continuous Variable Statistics

Continuous data are often summarized by a *central* value such as the average or *mean*:

$$\hat{m} = \frac{\sum_{j=1}^{N} w(\mathbf{u}_j)z(\mathbf{u}_j)}{\sum_{j=1}^{N} w(\mathbf{u}_j)} \tag{2.3}$$

where there are N data values $z(\mathbf{u}_j), j = 1, \ldots, N$, possibly weighted differently by $w(\mathbf{u}_j), j = 1, \ldots, N$. The notation $\hat{m}$ refers to the average or sample mean of some limited number of observations. The notation m is used for the true mean of a statistical model.

The spread of the data is often measured by the average squared difference from the mean:

$$\hat{\text{Var}}\{Z\} = \hat{\sigma}^2 = \sum_{j=1}^{N} w(\mathbf{u}_j)\left(z(\mathbf{u}_j) - \hat{m}\right)^2 \tag{2.4}$$

The notation $\hat{\sigma}^2$ refers to an observed variance, and the notation σ^2 is used for the true variance of a statistical model. The square root of the variance is the standard deviation (σ).

The coefficient of variation (CV) is the ratio of the standard deviation and the mean, that is, $CV = \sigma/m$. The CV is only defined for variables that are strictly positive. The CV is a useful measure of spread since it is unit-free. The magnitude of the CV, however, does not tell us how difficult it will be to model a variable; the spatial variability must be considered. Nevertheless, it is generally understood that a CV of more than 2.0 indicates that a variable is highly *heterogeneous*. The CV could be high due to mixing of data from different facies with "homogeneous" values in the facies. This may cause us to reconsider our decision of stationarity, since we should probably not mix homogeneous facies together.

There are other less commonly used measures of central tendency such as the median (0.5 quantile), mode (most common observation), and geometric mean ($[\prod_{j=1}^{N} z_j]^{\frac{1}{N}}$). Less commonly used measures of spread include the range (difference between the largest and smallest observation), mean absolute deviation $\sum_{j=1}^{N} w(\mathbf{u}_j)|z(\mathbf{u}_j) - \hat{m}|$, and interquartile range (0.75 quantile minus 0.25 quantile).

Other high-order summary statistics intended to measure features such as shape, skewness, or modality are rarely, if ever, used in geostatistics, they are also very sensitive to outlier and/or limited data.

Categorical Variable Statistics

Consider K mutually exclusive categories $s_k, k = 1, \ldots, K$. This list is also exhaustive; that is, any location $\mathbf{u}$ belongs to one and only one of these K categories. Let $i(\mathbf{u}; s_k)$ be the indicator variable corresponding to category s_k, set to 1 if location $\mathbf{u} \in s_k$, zero otherwise, that is,

$$i(\mathbf{u}_j; s_k) = \begin{cases} 1, & \text{if location } \mathbf{u}_j \text{ in category } s_k \\ 0, & \text{otherwise} \end{cases}$$

Mutual exclusion and exhaustivity entail the following relations:

$$i(\mathbf{u}; s_k) \cdot i(\mathbf{u}; s_{k'}) = 0, \qquad \forall\, k \neq k'$$

$$\sum_{k=1}^{K} i(\mathbf{u}; s_k) = 1$$

The mean indicator for each category $s_k, k = 1, \ldots, K$, is interpreted as the proportion of data in that category—for example, estimated with declustering weights such as

$$\hat{p}_k = \sum_{j=1}^{N} w_j i(\mathbf{u}_j; s_k)$$

The variance of the indicator for each category $s_k, k = 1, \ldots, K$, is a simple function of the mean indicator:

$$\widehat{\text{Var}}\left\{i(\mathbf{u}; s_k)\right\} = \sum_{j=1}^{N} w_j \left[i(\mathbf{u}_j; s_k) - \hat{p}_k\right]^2$$

$$= \hat{p}_k\,(1.0 - \hat{p}_k)$$

2.2.3 Bivariate Distributions

Until now the discussion of preliminary statistical concepts has focused on one variable. Now we introduce two variables. Figure 2.11 shows a cross plot of permeability versus porosity. The vertical lines divide the data into six classes, each with an equal number of porosity values. The horizontal lines within each porosity class divide the data into six subclasses with an equal number of permeability values. The result is 36 classes with an equal number of data.

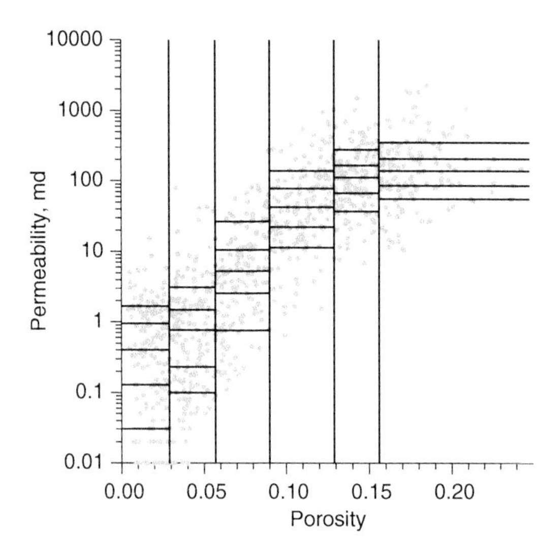

FIGURE 2.11: Cross Plot of Permeability versus Porosity. The vertical lines divide the data into six classes with an equal number of data points. The horizontal lines within each porosity class divide the data into six subclasses with an equal number of data. The result is 36 classes with equal number of data.

The correlation coefficient, ρ, measures the degree of correlation between paired values $z(\mathbf{u}_j), y(\mathbf{u}_j)$, $j = 1, \ldots, N$. It is calculated from the variance of each variable σ_z^2 and σ_y^2 [see Eq. (2.4)] and the covariance between the two variables. The sample covariance is defined as

$$\hat{c}_{z,y} = \sum_{j=1}^{N} w(\mathbf{u}_j)\left(z(\mathbf{u}_j) - \hat{m}_z\right) \cdot \left(y(\mathbf{u}_j) - \hat{m}_y\right)$$

where $\hat{m}_z$ and $\hat{m}_y$ are the sample means of the z and y data, from Eq. (2.3). The sample linear correlation coefficient is then defined as

$$\hat{\rho} = \frac{\hat{c}_{z,y}}{\sqrt{\sigma_z^2 \cdot \sigma_y^2}} \tag{2.5}$$

ρ is valued between -1 and 1, where -1 entails perfect negative correlation, 0 entails no correlation, and 1 entails perfect positive correlation.

Note that the covariance $c_{z,y}$ and correlation coefficient ρ measure "linear" correlation. The scatter on nonlinear or nonmonotonic relationships is not adequately reflected by the correlation coefficients in Figure 2.12.

The GSLIB program `scatplt` creates a cross plot of paired values with either arithmetic or logarithmic scaling of both axes. The means, variances,

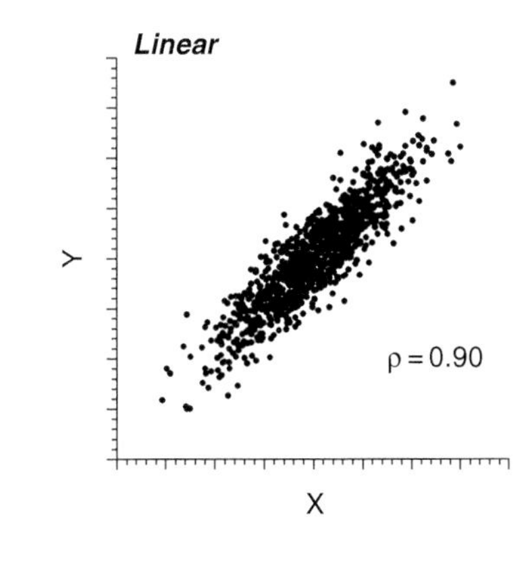

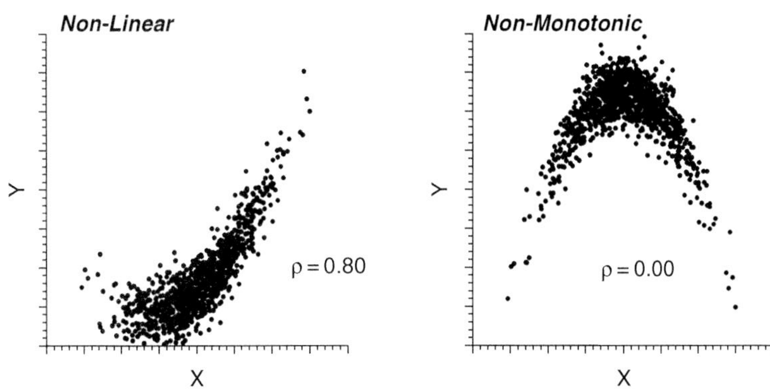

FIGURE 2.12: Linear, Nonlinear, and Nonmonotonic Relationships with Essentially the Same "Scatter." The correlation coefficient is significantly reduced for nonlinear and nonmonotonic relationships.

and correlation coefficient are reported. A "rank correlation" is also presented. This rank correlation is calculated the same as above [refer back to Eq. (2.5)] except that the cdf values $F_Z(z(\mathbf{u}_j))$ and $F_Y(y(\mathbf{u}_j))$ are used in place of the raw data values. This alleviates the impact of large outlier z- or y-data values; indeed, the histogram of both sets of values $F_Z(z(\mathbf{u}_j))$ and $F_Y(y(\mathbf{u}_j))$ are uniform between 0 and 1, hence they do not show any specific tail shape.

2.2.4 Q–Q Plots and Data Transformation

We turn our attention to comparing two different data distributions and data transformation.

Q–Q and P–P Plots

A *Q–Q plot* is a refined graphical tool for comparing two or more distributions. Of course, we could simply look at the two histograms side-by-side and compare the summary statistics. A Q–Q plot has advantages. A Q–Q plot is a cross plot of matching p-quantile values $q_1(p)$ versus $q_2(p)$ from the two different distributions.

For example, Figure 2.13 shows a histogram of core porosity and a histogram of well-log-derived porosity. The core porosity has a different histogram than well-log-derived porosity. Figure 2.14 shows the corresponding two cumulative distributions and the Q–Q plot. As highlighted in the figure, the 0.6 quantile of the core porosity is 0.309 and the 0.6 quantile of the log porosity is 0.279. The pair (0.309, 0.279) is one point on the Q–Q plot. The other points are associated with the other 406 quantiles determined from the data distributions.

When all points on a Q–Q plot fall on the 45° line, the two distributions are exactly the same. Departure from the 45° line indicates a difference in the two histograms, in particular,

1. A systematic departure above or below the 45° line indicates that the *center* or mean of the distributions is different. A shift above the 45° line implies that the Y-distribution is higher valued than the X-distribution. A shift below the 45° line entails that the X-values are higher.

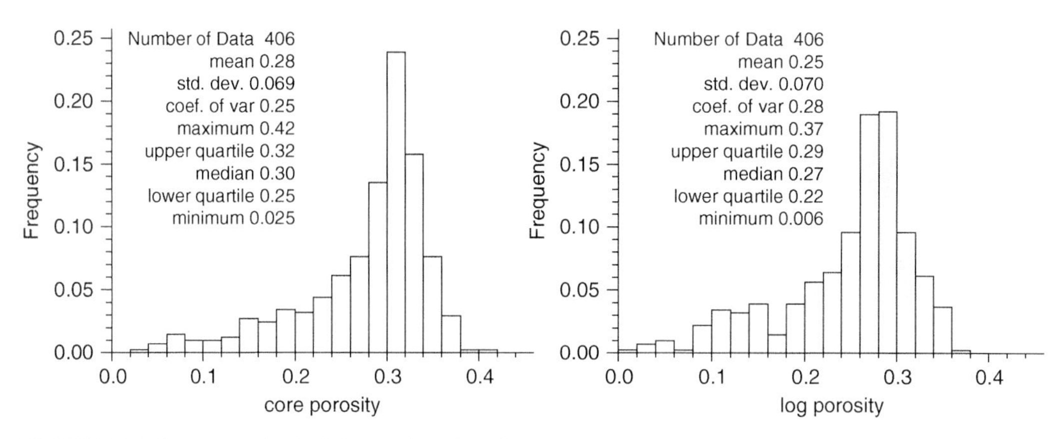

FIGURE 2.13: Histograms of Paired Core and Log Porosity Data.

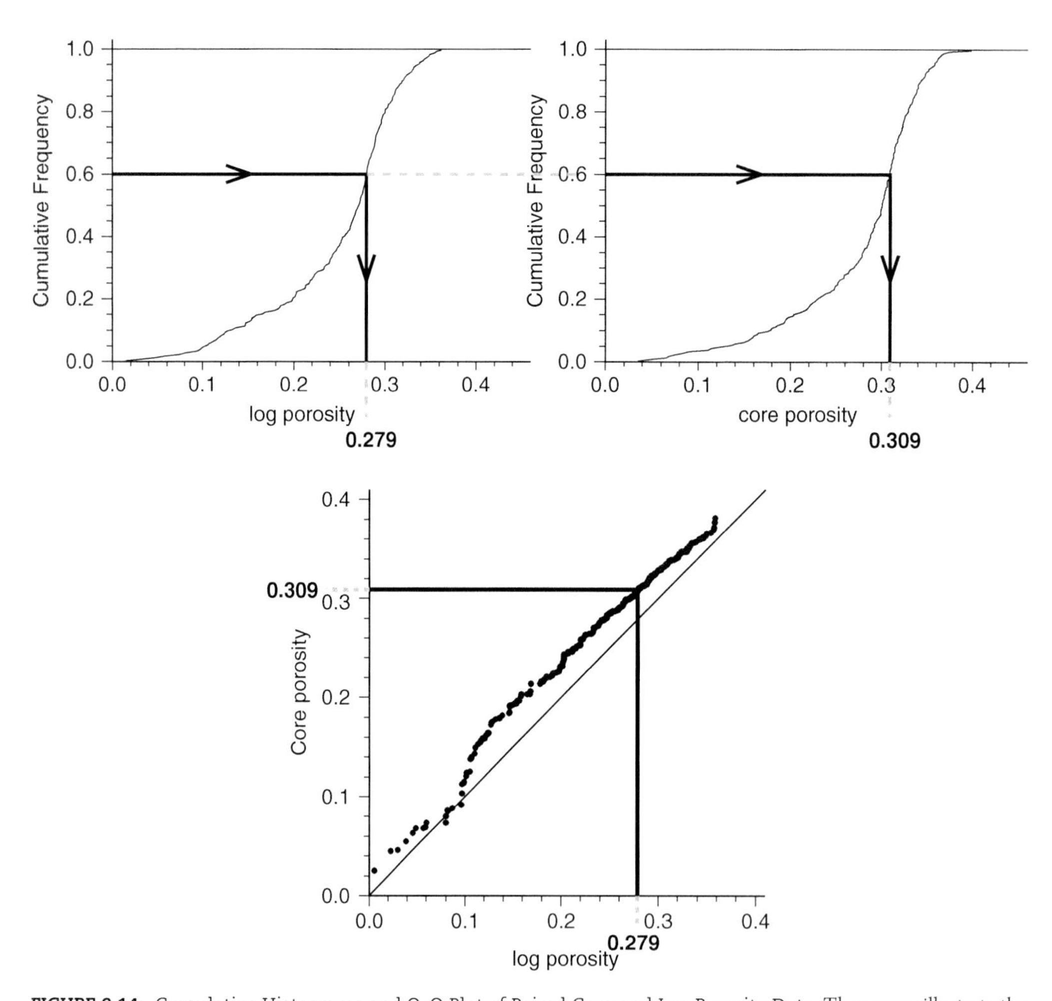

FIGURE 2.14: Cumulative Histograms and Q–Q Plot of Paired Core and Log Porosity Data. The arrows illustrate the steps to get the 0.60 cumulative frequency point on the Q–Q plot: (1) The 0.6 quantile of the core porosity distribution is determined to be 0.309, (2) the 0.6 quantile of the log porosity distribution is 0.279, and (3) the values 0.309 and 0.279 are cross plotted on the Q–Q plot.

2. A slope different from 45° indicates that the *spread* or variance of the two distributions is different. A slope greater than 1 (or 45°) indicates that the Y-variance is higher than X. A slope less than 1 indicates that the X-variance is greater.
3. Curvature on a Q–Q plot indicates that the two distributions have different shape.

In practice, differences in all three features (center, spread, and shape) are observed. Looking at Figure 2.14 we note (1) the difference in average porosity from core to log (the vertical shift) and (2) the different shape of the histograms for porosity values less than 0.15.

2.2.5 Data Transformation

It is often necessary to transform data from one histogram to another histogram, for example, we may want to transform (1) well-log-derived porosity to the histogram of core porosity to correct for the approximative nature of log interpretation and the deemed more reliable histogram of core data, (2) a simulated realization of porosity values to the approximate sample data distribution, or (3) the available data to the congenial normal/Gaussian distribution for subsequent geostatistical analysis. In data transformation, the ordering of the data should be preserved; that is, high values before transformation should remain high after transformation. The transformation method described below preserves such rank ordering.

The traditional method of standardizing data is to calculate the standard residual, $Y = (Z - m)/\sigma$; where m and σ^2 are the mean and variance of the original variable Z. The mean and variance of the new Y data are 0 and 1; however, the shape of the distribution remains unchanged. A straightforward transformation procedure that also corrects the entire shape of the histogram is the rank-preserving quantile transform.

The idea of quantile transformation is to match the p-quantile of the data distribution to the p-quantile of the target or reference distribution. Consider the data variable z with cdf $F_Z(z)$. This will be transformed to a y-value with cdf $F_Y(y)$ as follows:

$$y = F_Y^{-1}\left(F_Z(z)\right)$$

A graphical representation of this procedure is easier to understand.

The example shown in Figure 2.14 can be used as an illustration. We want to transform the well-log-derived porosity values to have the same histogram as the core porosity values to (1) put the two data types on the same basis for further geostatistical analysis and (2) correct the histogram for the scale (support volume) difference between the two data types, thus removing any bias introduced by the well log interpretation software. As stated above, the transformation procedure amounts to *match quantiles*. That is, a well-log-derived porosity of 0.279 is transformed to 0.309 (the corresponding 0.6 quantile on the core porosity distribution). Any z value can be transformed, provided that some within-class interpolation scheme is considered to fully specify the original distribution $F_Z(z)$.

For some applications, there are far more z data to transform than there are "hard" y data for defining the target cdf $F_Y(y)$. The Q–Q plot is built at the resolution of the smallest data set and, then, interpolation options are used in the lower tail, between quantile pairs, and in the upper tail. Often, linear interpolation is used because of its simplicity. More elaborate alternatives are discussed in GSLIB (Deutsch and Journel, 1998). The `trans` program in GSLIB implements the quantile transformation procedure described above.

The normal scores transformation has the standard normal distribution as a target distribution. A normal distribution with mean m and standard deviation σ has the following probability density:

$$f(z) = \frac{1}{\sigma\sqrt{2\pi}} e^{-\frac{1}{2}\left(\frac{z-m}{\sigma}\right)^2}$$

Figure 2.15 illustrates the quantile transformation procedure to the standard normal distribution. The methodology is the same as the quantile transformation procedure described above; the target distribution is now the standard normal distribution. Note that although the cumulative distribution for the standard normal distribution has no closed-form analytical solution, there are excellent numerical approximations (Kennedy Jr. and Gentle, 1980). The `nscore` program in GSLIB implements normal scores transformation.

Despiking

The quantile transformation or normal scores transformation procedures described above are reversible, provided that there are no "spikes" in either

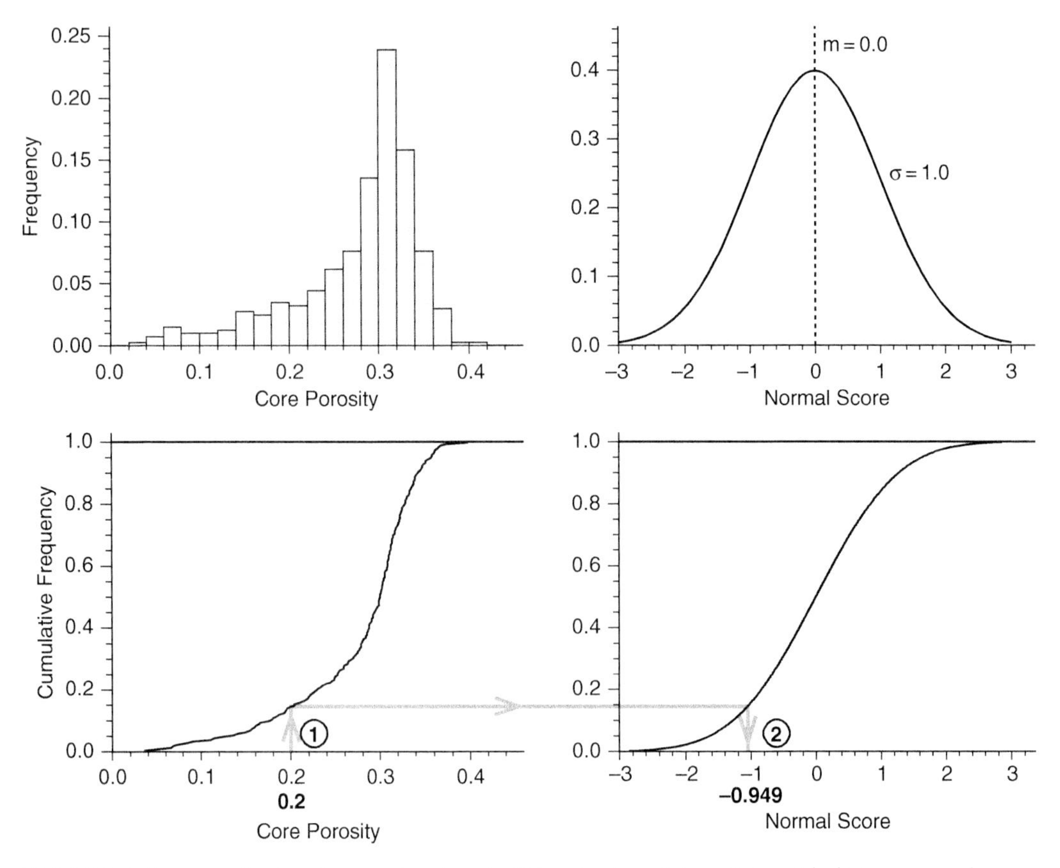

FIGURE 2.15: Procedure for Transforming Core Porosity Values to Normal Score Values. The histograms are shown at the top of the figure. The cumulative distributions, shown at the bottom, are used for transformation. To transform any core porosity (say 0.2): (1) Read the cumulative frequency corresponding to the porosity and (2) go to the same cumulative frequency on the normal distribution and read the normal score value (−0.949). Any porosity value can be transformed to a normal score value in this way, and vice versa.

the starting histogram or the target histogram. For example, there would be no unique transformation to any target distribution if there are 10% original zero values. There is a need to break the ties prior to transformation.

The simplest approach is to break the ties randomly. Often, the ordering of ties is left to the sorting algorithm or the order the values are encountered in the data file. More sophisticated random despiking would consist of adding a small random number to each tie and then ordering from smallest to largest. In either case, random despiking could have an effect on subsequent modeling. In particular, the variogram calculated in the transformed data space (often, after normal score transformation) could show a too-high nugget effect[2] and unrealistic short-scale variations. Notwithstanding the limitations of random despiking, it is acceptable when there is a small proportion of constant values.

It should be noted that there is no need for despiking with indicator-based methods, that is, where the range of variability is naturally divided by a series of thresholds. Each set of constant values or "spike" could constitute a separate geological population. For example, a spike of zero porosity values could correspond to completely cemented rock or shales. The cemented population might be considered a distinct geological population and modeled as a separate facies; the properties within certain geological populations can be held constant.

Another approach must be considered when random despiking is considered unacceptable and there is no geological basis to isolate the spike as a separate population. A straightforward and effective approach for despiking was proposed by Verly (1984). The

[2] Details of variogram terminology and inference are presented in Section 2.3, Quantifying Spatial Correlation.

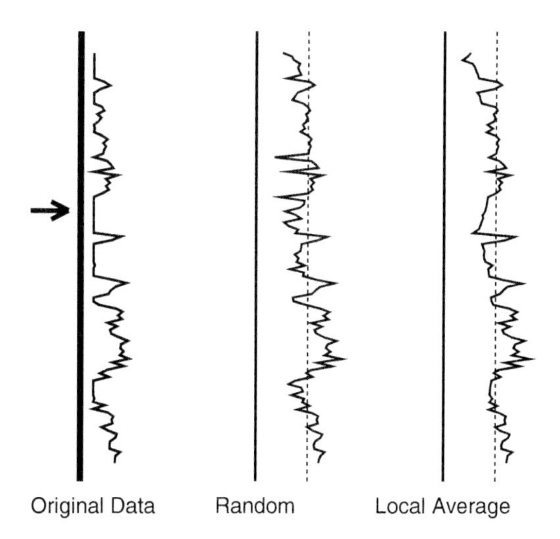

FIGURE 2.16: 1-D Example of Despiking: the Original Porosity Values Are on the Far Left, Normal Score Transform of the Porosity Values with Random Despiking, and, on the Right, Normal Score Transform with Local Average Despiking. Note that there is less randomness in the regions of constant 0.0 porosity, for example, at the location of the arrow.

idea is to compute local averages within local neighborhoods centered at each tied data value. The data are then ordered or despiked according to the local averages; high local averages rank higher. The `despike` program will perform the local average despiking of 1-, 2-, or 3-D data. The user must choose the size of the window. Clearly, if the window is too large or too small, then no despiking will be done. There is little benefit to study the window size in great detail; the despiking results are relatively robust with respect to the window size. Figure 2.16 shows an example of despiking. The normal score transformed values with the local average despiking show less small-scale fluctuations (in the constant valued zones) than the results with random despiking.

Consideration of the "despiking" problem marks a shift in emphasis between *sample* statistics and *population* statistics. Although we only have limited data, we are interested in the statistical parameters that represent the entire population or reservoir volume. A concern much more consequential than "spikes" is clustered/nonrepresentative data.

2.2.6 Declustering and Debiasing

Data are rarely collected with the goal of statistical representativity. Wells are often drilled in areas with a greater probability of good reservoir quality. Core measurements are taken preferentially from good-quality reservoir rock. These "data collection" practices should not be changed; they lead to the best economics and the greatest number of data in portions of the reservoir that contribute the greatest flow. There is a need, however, to adjust the histograms and summary statistics to be representative of the entire volume of interest.

The volume of interest is most often a particular facies of a particular stratigraphic sequence. The fact that no core porosity and permeability data have been taken from shale is accounted for in the construction of a prior sand/shale model. A concern is that, even in a particular facies, there may be clustering of data in the low- or high-valued areas.

Most contouring or mapping algorithms automatically correct this preferential clustering. Closely spaced data inform fewer grid nodes and, hence, receive lesser weight. Widely spaced data inform more grid nodes and, hence, receive greater weight. A map constructed by ordinary kriging is effectively declustered. Even though modern stochastic simulation algorithms are built on the mapping algorithm of kriging, they do *not* correct for the impact of clustered data on the target histogram; these algorithms require a distribution model (histogram) that is representative of the entire volume being modeled. Simulation in an area with sparse data relies on the global distribution which must be representative of all areas being simulated.

Polygonal and Cell Declustering

Declustering techniques assign each datum a weight, $w_i, i = 1, \ldots, n$, based on its closeness to surrounding data. Then the histogram and summary statistics are calculated with these declustering weights. The mean and standard deviation are calculated as

$$\bar{z} = \sum_{i=1}^{n} w_i z_i \quad \text{and} \quad s = \sqrt{\sum_{i=1}^{n} w_i (z_i - \bar{z})^2}$$

where the weights $w_i, i = 1, \ldots, n$, are between 0 and 1 and add up to 1.0. The most intuitive approach to declustering is to base the weights on the volume of influence of each sample.

As a first example, consider the 122 well locations shown in Figure 2.17. The underlying gray scale values represent the "true" porosity values, which were simulated for this example. The mean and standard deviation of the true distribution are 18.72% and

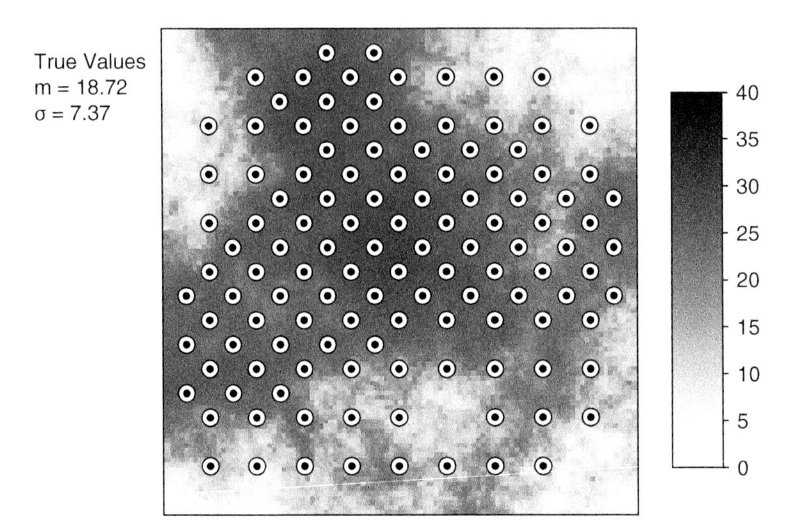

FIGURE 2.17: Location Map of 122 Wells. The gray-scale code shows the underlying "true" distribution of porosity. In practice, we may know only the central trend of high porosity.

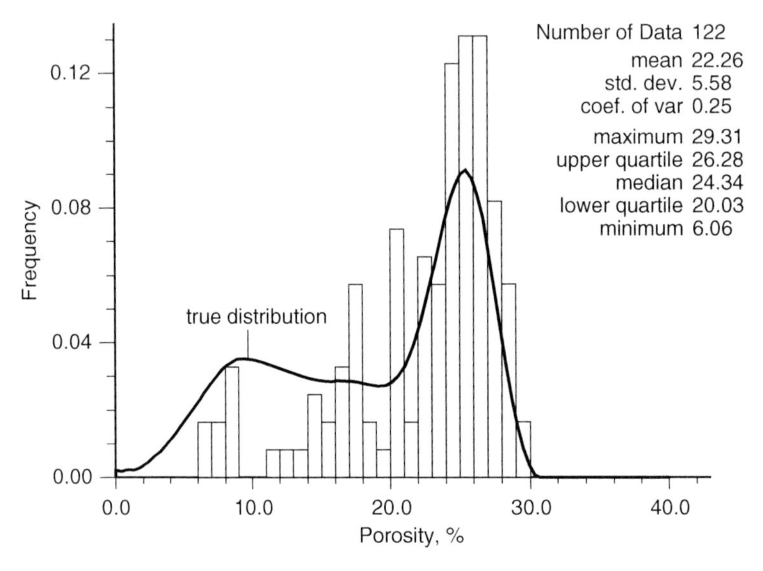

FIGURE 2.18: Equal-Weighted Histogram of 122 Well Data with the True Reference Histogram Shown as the Black Line. Note the greater proportion of data between 25% and 30% and the sparsity of data in the 0% to 20% porosity range.

7.37%. Although hand contouring and a knowledge of the regional geology would reveal the central trend of high porosity, the details of the true porosity distribution would be inaccessible in practice. Figure 2.18 shows an equal-weighted histogram of the 122 well data. The sample equal-weighted mean (22.26%) significantly overestimates the true mean, and the standard deviation (5.58%) underestimates the true standard deviation.

The polygonal areas of influence are shown in Figure 2.19. There are many algorithms to establish the polygons around each datum (Rock, 1988) and calculate the area of influence $A_i, i = 1, \ldots, n$. These algorithms become CPU-intensive and somewhat unstable in 3-D. Declustering weights are taken proportional to the areas:

$$w_i^{(p)} = \frac{A_i}{\sum_{j=1}^{n} A_j}$$

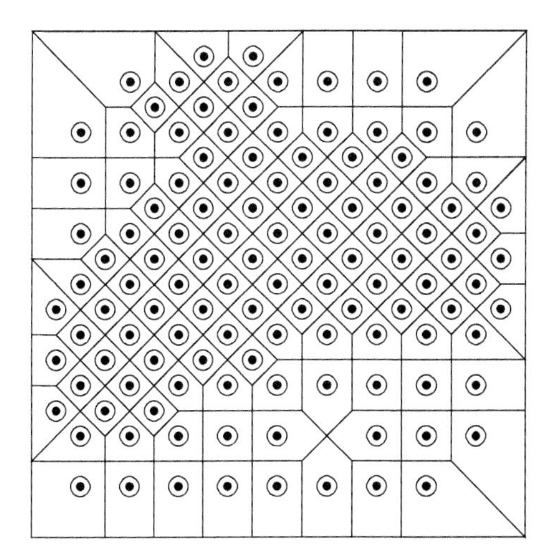

FIGURE 2.19: Location Map of 122 Wells with Polygonal Areas of Influence.

where $w_i^{(p)}, i = 1, \ldots, n$, are the polygonal declustering weights. The histogram and summary statistics using these weights are shown on Figure 2.20. The proportion of data between 25% and 30% has been corrected downward, and the weight given to data in the 0% to 20% porosity range has been increased. The summary statistics are now much closer to the true values (see Table 2.5). The technique of *cell declustering* is another commonly used declustering technique (Deutsch, 1989b; Journel, 1983). Cell declustering works as follows:

1. Divide the volume of interest into a grid of cells $l = 1, \ldots, L$.
2. Count the occupied cells L_o, $L_o \leq L$, and the number of data in each occupied cell n_{l_o}, $l_o = 1, \ldots, L_o$, where $\sum_{l_o=1}^{L_o} n_{l_o} = n =$ the number of data.
3. Weight each data according to the number of data falling in the same cell, for example, for datum i falling in cell l, $l \in [0, L_o]$, the cell declustering weight is

$$w_i^{(c)} = \frac{1}{n_l \cdot L_o}$$

The weights $w_i^{(c)}, i = 1, \ldots, n$, are between $(0, 1]$ and sum to one, $\sum_{i=1}^{n} w_i^{(c)} = 1.0$. These weights are inversely proportional to the number of data in each cell: Data in cells with one datum receive a weight of $1/L_o$, data in cells with two data receive a weight of $1/(2 \cdot L_o)$, data in cells with three data receive a weight of $1/(3 \cdot L_o)$, and so on.

Figure 2.21 illustrates the cell declustering procedure. The area of interest is divided into a grid of $L = 36$ cells, there are $L_o = 33$ occupied cells, the number of data in each occupied cell is established by arbitrarily moving data on the grid boundaries to the right and down, and the weights for data falling in the column of cells at the far right are shown in Figure 2.21. The weights depend on the cell size and

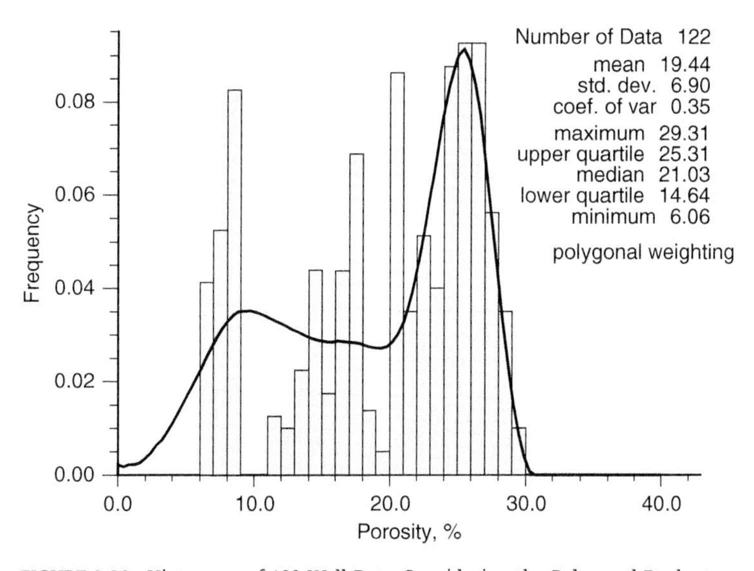

FIGURE 2.20: Histogram of 122 Well Data Considering the Polygonal Declustering Weights with the True Reference Histogram Shown as the Black Line. The proportion of data between 25% and 30% has been corrected, and the weight given to data in the 0% to 20% porosity range has been increased.

TABLE 2.5. MEAN AND VARIANCE FOR EXAMPLE ILLUSTRATED IN FIGURE 2.17

	Mean		Standard Deviation	
Reference	18.72		7.37	
Equal weighted	22.26	+18.9%	5.58	−24.3%
Polygonal declustering	19.44	+3.8%	6.90	−6.4%
Cell declustering	20.02	+6.9%	6.63	−10.0%

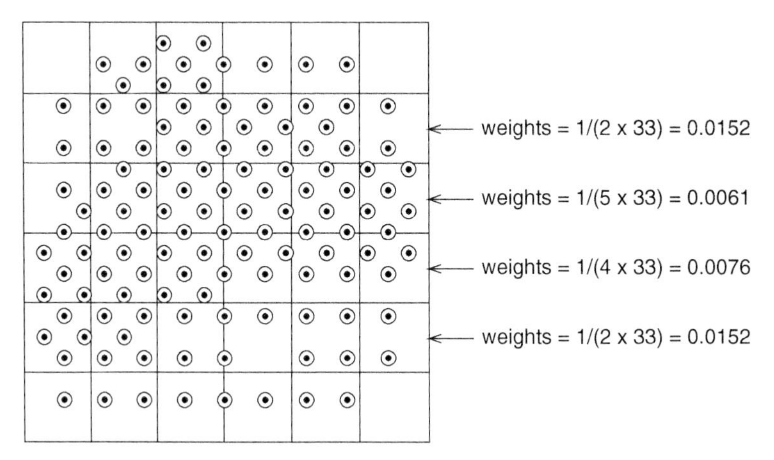

FIGURE 2.21: Illustration of Cell Declustering. The weights for the 13 data in the last column are shown. Weights for all other data are established in the same way.

the origin of the grid network. It is important to note that the cell size for declustering is *not* the cell size for geologic or flow modeling; it simply defines an intermediate grid that allows assigning declustering weight. When the cell size is very small, each datum is in its own cell ($L_o = n$) and receives an equal weight. When the cell size is very large, all data fall into one cell ($L_o = 1$) and are equally weighted. Choosing the *optimal* grid origin, cell shape, and size requires some sensitivity studies; consider the following guidelines:

- Perform an areal 2-D declustering when the wells are vertical (or nearly vertical). Consider 3-D declustering when there are horizontal or highly deviated wells that may preferentially sample certain stratigraphic intervals.
- The shape of the cells depends on the geometric configuration of the data; at this point we are not seeking to quantify the spatial continuity. Thus, adjust the shape of the cells to conform to major directions of preferential sampling. For example, if the wells are more closely spaced in the X direction than in the Y direction, the cell size in the X direction should be reduced.
- Choose the cell size so that there is approximately one datum per cell in the sparsely sampled areas. Check the sensitivity of the results to small changes in the cell size; large changes in the result most likely indicate that the declustering weight is changing for one or two anomalously high or low wells.
- In mature fields where there are many data and it is known that the high- or low-valued areas have been oversampled, then the cell size can be selected such that the weights give the minimum (or maximum) declustered mean of the data. Plot the declustered mean versus the cell size for a range of cell sizes and choose the size with the lowest declustered mean (data clustered in high-valued areas) or the size with the highest declustered mean (data clustered in low-valued areas). Cell size based on data spacing in sparsely sampled areas is preferred because this method is known to typically overcorrect the distribution.

- The origin of the cell declustering grid and the number of cells L must be chosen such that all data are included within the grid network. Fixing the cell size and changing the origin often leads to different declustering weights. To avoid this artifact, a number of different origin locations N_{origin} should be considered for the same cell size (Deutsch, 1989b; Deutsch and Journel, 1998). The declustering weights are then averaged for each origin offset.

Returning to the 122 well example, an optimal cell size of 280 grid units was determined by trying 95 different cell sizes from 75 through 1025 grid units in increments of 10 with 20 different origins per cell size (GSLIB program `declus` does this automatically and very fast). The plot of declustered mean versus cell size is shown in Figure 2.22. A cell size of 280 units with declustered mean 20.02% was chosen. The histogram of the 122 well data considering cell declustering weights is shown in Figure 2.23.

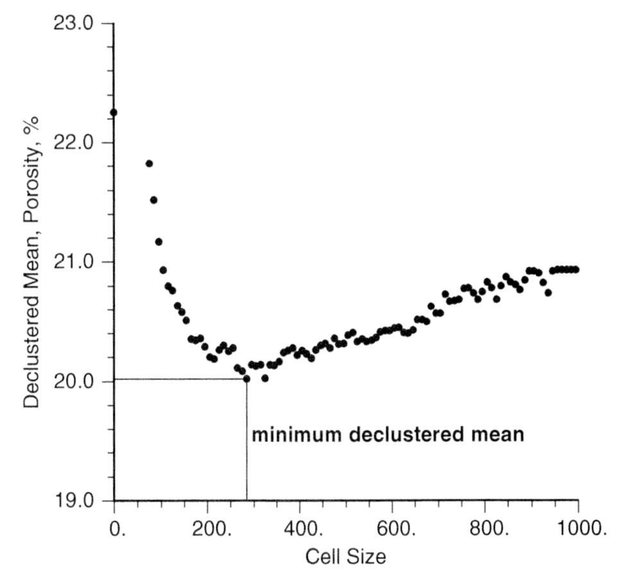

FIGURE 2.22: The Declustered Mean versus the Cell Size. Note the equal-weighted mean of 22.26%.

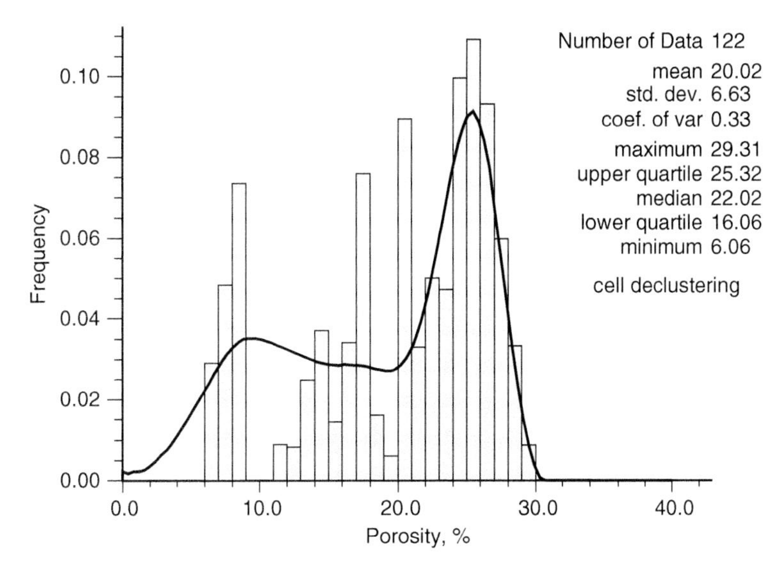

FIGURE 2.23: Histogram of 122 Well Data Considering Cell Declustering Weights. The true reference histogram is shown as the black line. The proportion of data between 25 to 30% has largely been corrected and the weight given to data in the 0 to 20% porosity range has been increased.

The proportion of data between 25 to 30% has largely been corrected and the weight given to data in the 0 to 20% porosity range has been increased. The summary statistics are shown in Table 2.5.

Polygonal declustering led to statistics closer to the reference histogram in this 122 well example. This is not a general conclusion. It has been observed that polygonal declustering works well when the limits (boundaries) of the volume of interest are well-defined and the polygons do not vary in size by more than a factor of, say, 10 (largest area/smallest area). Due to the difficulty of determining Voronoi volumes in influence in 3-D, cell declustering is also preferred in complex 3-D situations. In spite of the results shown in Table 2.5, polygonal declustering is known not to perform well beyond the mean. The important point is that both declustering procedures lead to a histogram closer to the truth.

Another Example

The 63 wells shown in Figure 2.24 are from a West Texas carbonate reservoir. The wells have been shown with a gray-scale coding that represents the vertical average of porosity. The cluster of wells in the northeast corner appears to be in a high-porosity area. The reservoir layer of interest is an average of 52 feet thick. An equal-weighted histogram of well-log-derived porosity sampled at a 1-ft spacing is shown in Figure 2.25. Note the average porosity of 8.33% and the standard deviation of 3.37%.

Applying cell declustering with an areal cell size of 1500 ft and a vertical cell size of 10 ft yields the histogram shown on Figure 2.26. The declustered mean is 7.99% and the standard deviation is 3.23%. The average porosity has decreased by 0.34 porosity units or 4.1%, and the standard deviation has decreased by 0.14 porosity units or 4.2%. A Q–Q plot of the equal-weighted porosity distribution and the cell declustered porosity distribution is shown in Figure 2.27. The declustered distribution gives less weight to the porosity values between 8% and 16%.

For some applications a decrease of 4% in the average may be inconsequential. There are times, however, when 4% represents an important volume of hydrocarbon.

Declustering with Multiple Variables

Declustering weights are largely determined on the basis of the geometric configuration of the data; therefore, only one set of declustering weights is calculated in the presence of multiple variables that have been equally sampled. Different declustering weights will need to be calculated when there is unequal sampling. For example, there are often fewer permeability data than porosity data, which would require two sets of declustering weights.

Declustering weights are primarily used to determine a representative histogram for each variable; however, we also require the correlation between multiple variables. The same set of declustering

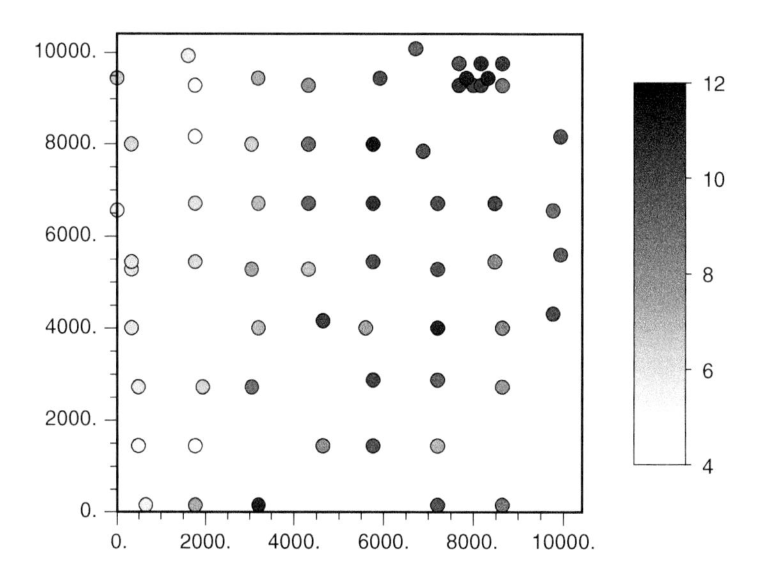

FIGURE 2.24: Location Map of 63 Wells. The well locations are shown with a gray-scale circle that represents the vertical average of porosity in that well. Note the higher porosity and the clustering of wells in the northeast corner.

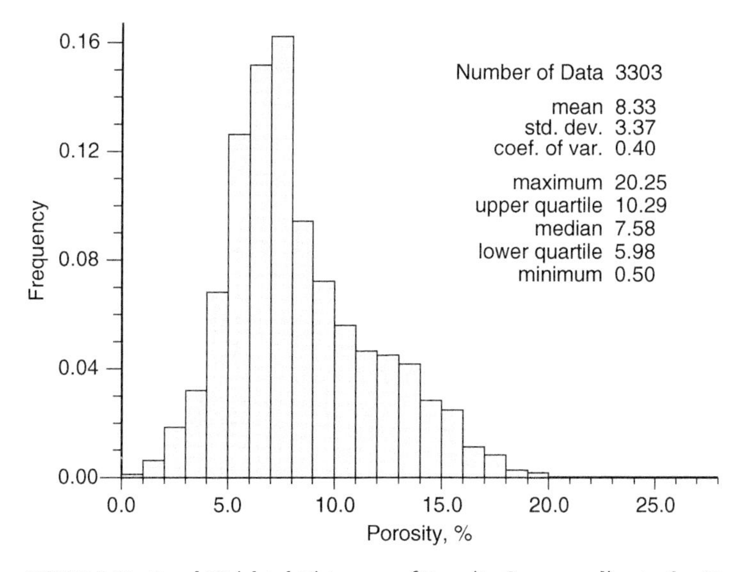

FIGURE 2.25: Equal-Weighted Histogram of Porosity Corresponding to the 63 Wells Shown in Figure 2.24.

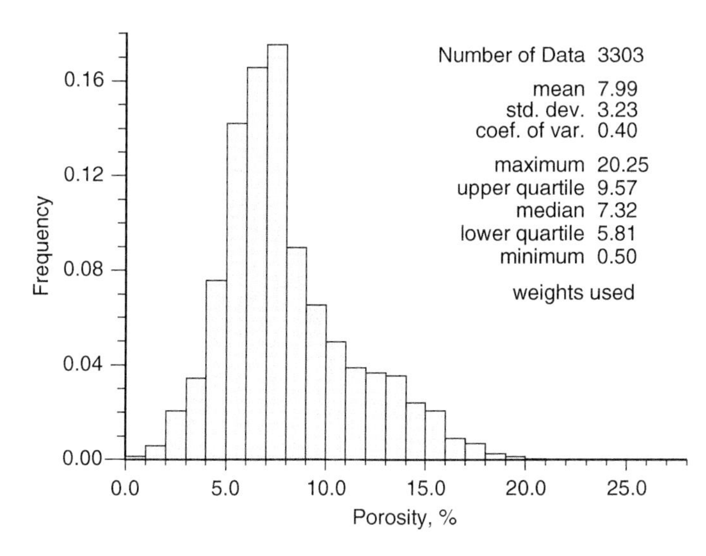

FIGURE 2.26: Weighted Histogram of Porosity Where the Cell Declustering Weights Were Derived from with an Areal Cell Size of 1500 ft and Vertical Cell Size of 10 ft.

weights can weight each pair contributing to the correlation coefficient (the `scatplt` program in GSLIB permits such weighting).

Correcting Nonrepresentative Data

Declustering only works when there are enough data to assign greater and lesser weights. The conventional polygonal and cell declustering methods are inadequate to determine a representative distribution unless there is adequate data coverage in both *good* and *poor* areas of the reservoir. There may be sufficient seismic or geological data to know a priori where the high-pay *good* areas are; the first wells would be located in the best areas. Then, later in reservoir development, a geostatistical model may be constructed for further development planning. At that time, unbiased statistics are required. The central idea in this section is to use the secondary seismic or geological data to determine representative statistics and probability distributions—in other words, correct for the nonrepresentative data (Pyrcz et al., 2006).

The first requirement is a spatial distribution of the secondary variable. The secondary variable could be a seismic attribute, a hand-contoured reservoir quality (RQ) variable, or simply structural depth if reservoir quality is linked to depth. An early application of the method presented below was to a Danish

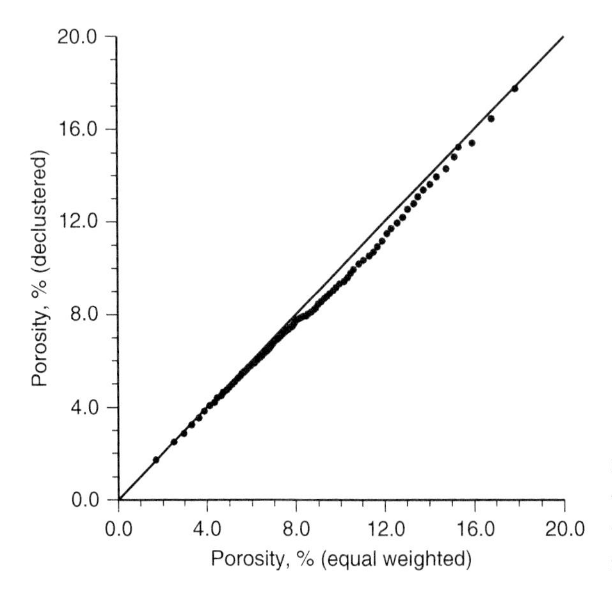

FIGURE 2.27: Q–Q Plot of Porosity Values with and without Declustering. Declustering has decreased the probability to be in the 8% to 16% porosity range.

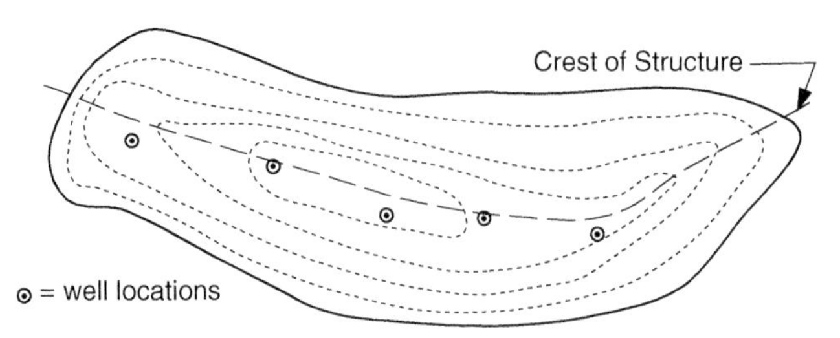

FIGURE 2.28: An Illustration of Wells Drilled along the Crest of the Reservoir Structure. At times, there is a systematic degradation in reservoir quality with depth. The dashed lines represent contour lines of decreasing depth away from the center. In this case, the distributions of porosity and permeability must be corrected.

chalk reservoir, where the porosity exhibits a clear decrease with depth.

Figure 2.28 shows a schematic areal view of a reservoir where the wells are located near the crest of the structure. This is done to maximize production from a limited number of wells early in development. The dashed lines on this figure represent depth contours, which decrease away from the center. A priori knowledge of systematic trends is common; such knowledge is often used for the placement of early production wells.

In addition to an exhaustive map of the secondary variable, a "calibration" relationship between the secondary variable (say, depth) and the reservoir property under consideration (say, porosity) is required. A schematic calibration relationship is shown in Figure 2.29. It would be preferable to have more actual data to support the trend shown in Figure 2.29. In this case, with five wells at the crest, the data would have to come from some external source of information—for example, experience with similar reservoirs in similar depositional settings.

The information illustrated in Figures 2.28 and 2.29 could be merged to provide a representative distribution of porosity. The two required pieces of information are:

1. a secondary variable—for example, depth (Y) at all locations—and
2. a bivariate relationship between the secondary variable Y and the Z variable of interest.

The region of interest, denoted A, is discretized by a number of locations $\mathbf{u}$, $\mathbf{u} \in A$. The notation $\mathbf{u}$ represents the 3-D coordinates of a geological modeling cell. The geological modeling cells are all about the same volume. The secondary variable is available at all locations, $y(\mathbf{u}), \mathbf{u} \in A$. For each secondary variable $y(\mathbf{u})$-value there is a corresponding probability distribution of z, denoted $\hat{f}_{Z|Y}(z|Y = y(\mathbf{u}))$. These conditional probability distributions come from the calibration relationship. In the schematic example of Figures 2.28 and 2.29, the conditional distributions would show high z-values for shallow (low) depth y values and low z-values for greater depth.

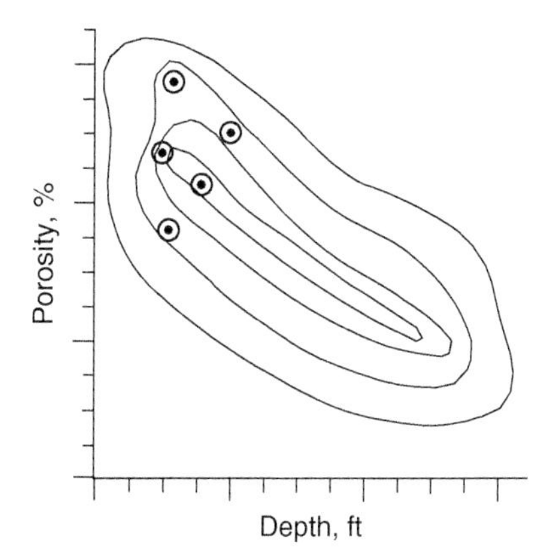

FIGURE 2.29: A Calibration Cross Plot between the Depth (See Figure 2.28) and the Porosity. The five well data are at relatively shallow depths whereas most of the reservoir is deeper. The contour lines are equal-probability lines.

The distribution of all y-values (all $\mathbf{u} \in A$) is representative of the entire area of interest; therefore, a representative distribution of the primary Z value may be constructed by accumulating all of the conditional distributions:

$$f_Z^*(z) = \sum_{\text{all } \mathbf{u} \in A} \frac{1}{C} f_{Z|Y}\left(z|Y = y(\mathbf{u})\right) \tag{2.6}$$

where C is a normalizing constant to ensure $f_Z^*(z)$ sums to one. Figure 2.30 gives an illustration of this procedure: The data distributions (probability distributions with solid lines) are biased due to the location of the wells, the secondary data are known everywhere and a representative distribution may be established (dashed line below y axis), and Eq. (2.6) may be used to calculate the representative distribution shown by a dashed line to the side of the vertical z axis. This simple summation procedure is encoded in a number of software. We see that the final representative or declustered distribution $f_Z^*(z)$ depends entirely on the secondary variable and the calibration. The actual z data only contribute to establishing the calibration relationship. Of course, these z data are also used as local data in subsequent geostatistical modeling.

The secondary data and calibration could come from seismic data. Figure 2.31 illustrates the calculation of a representative histogram using an exhaustive grid of seismic data (top left), the corresponding histogram of the seismic attribute (top right), and the bivariate relationship between the seismic attribute and porosity (middle) to arrive at a distribution of porosity (bottom). The "vertical lines" on the central calibration cross plot arise from the vertical wells; there are many porosity values down each well that correspond to a single seismic attribute value.

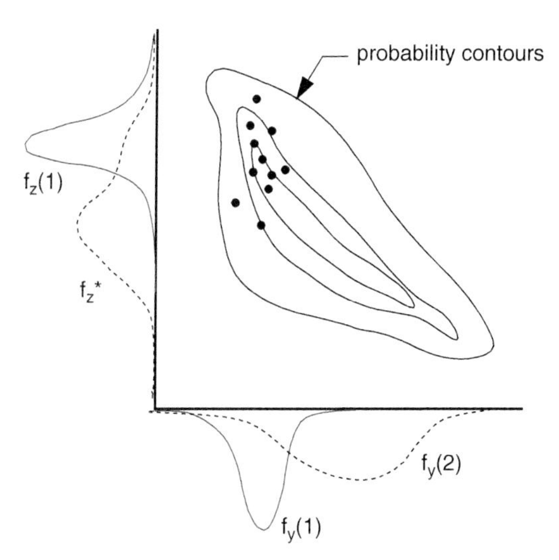

$f_z(1)$, $f_y(1)$ – distributions of primary (z) and secondary (y) variables at well locations

$f_y(2)$ – distribution of secondary (y) variable representative of entire area

f_z^* – derived distribution of z variable representative of entire area

FIGURE 2.30: Illustration of the Calibration Procedure of Deriving Representative Probability Distribution. Distribution $f_Z^*(z)$ is calculated with Eq. (2.6).

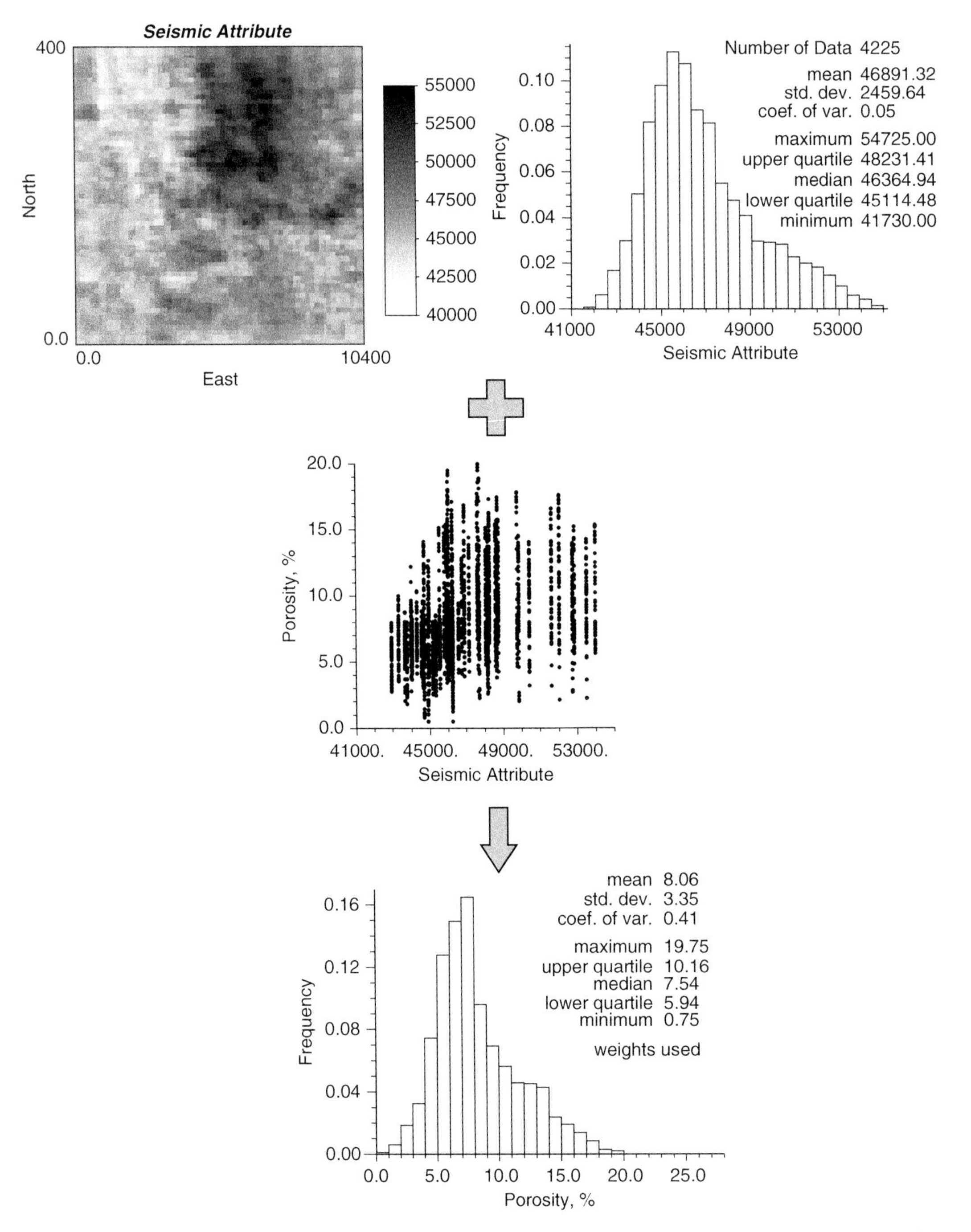

FIGURE 2.31: Establishing a Representative Histogram by Using an Exhaustive Secondary Variable and a Calibration Cross Plot.

Another method to create a representative histogram or, at least, mean value is to construct a geologic trend model. The average of the trend model provides a reasonable estimate of the overall average. The data-derived histograms or proportions could be corrected to reproduce the average from the trend model.

Skipping declustering or some other form of correction for nonrepresentative data can lead to biased reservoir models. In addition to correcting bias through declustering, it is sometimes necessary to *smooth* the histograms and cross plots.

2.2.7 Histogram and Cross-Plot Smoothing

Saw-tooth-like spikes appear in histograms when there are few sample data. If more data were available, these spikes would not likely appear; they are an artifact of data paucity and should be smoothed out. The problems of lack of resolution and spikes in the univariate histogram become far worse in the case of a bivariate histogram (cross plot) due to the number of bivariate class that must be informed. For example, if 100 classes are required for the porosity and permeability histograms, then 10,000 classes would be required for the bivariate histogram. This is known as the "curse of dimensionality." There is rarely enough data to reliably inform the bivariate histogram without modeling or *smoothing*. Typically, histogram smoothing is applied to improve the visual presentation of distributions and has no significant impact on model results; therefore, significant effort and time should not be invested.

A first conventional approach to smooth univariate and bivariate distributions is to fit a parametric probability distribution model such as a normal, lognormal, or power-law distribution to the sample data (Johnson and Kotz, 1970; Scott, 1992). These parametric models overcome all problems related to resolution and spikes in the sample histogram. The problem, however, is that real earth-science data can rarely be fitted with parametric distributions with few parameters.

A second approach is to replace each datum with a kernel function (Scott, 1992; Silverman, 1986)—that is, a parametric probability distribution with a mean equal to the datum value and a small variance. The *smooth* distribution is obtained by adding up these kernel functions. The resulting distribution does not, in general, simultaneously honor the mean, variance, and important quantiles of the sample data and may show negative values although the variable is positive.

A third approach is to pose the histogram smoothing problem as an optimization problem (Deutsch, 1996a; Journel and Xu, 1994; Xu and Journel, 1995). Critical summary statistics, that are deemed reliably informed by the sample data are reproduced. For example, data limits, the declustered mean and variance, certain quantiles (such as the median), linear correlation coefficients, bivariate quantiles (for nonlinear behavior in the scatterplot), and measures of smoothness can be imposed. The methodological details of all methods are available in the references.

Histogram Smoothing

Four examples of histogram smoothing are shown in Figure 2.32. The `histsmth` program from GSLIB (Deutsch and Journel, 1998) was used for all four unrelated examples. In general, there are no fixed guidelines about the best parameters for smoothing. An attempt at optimizing the smoothing parameters on the basis of cross-validation scores may be possible (Scott, 1992) but probably not worth the effort. Significant details may be lost if the distribution is smoothed too much. On the other hand, artifacts of scarce data are preserved if the smoothing is too "small."

Scatterplot Smoothing

Figure 2.33 shows a smoothed scatterplot for 243 porosity/permeability data pairs. The smoothed porosity and permeability distributions, shown below the porosity axis and to the left of the permeability axis, were calculated first. The bivariate distribution was then smoothed to honor these two marginal distributions. The gray scale is white for low probability and black for high probability. The final smoothed distribution shown has been constrained to the marginal distributions, some selected bivariate quantiles, and a measure of smoothness.

The `histsmth` and `scatsmth` programs from GSLIB (Deutsch and Journel, 1998) were used for this example. Once again, there is no objective estimate of the exact amount of smoothing for scatterplots. A visual verification must suffice.

2.2.8 Monte Carlo Simulation

The application of Monte Carlo or stochastic methods is ubiquitous in modern science. In the present

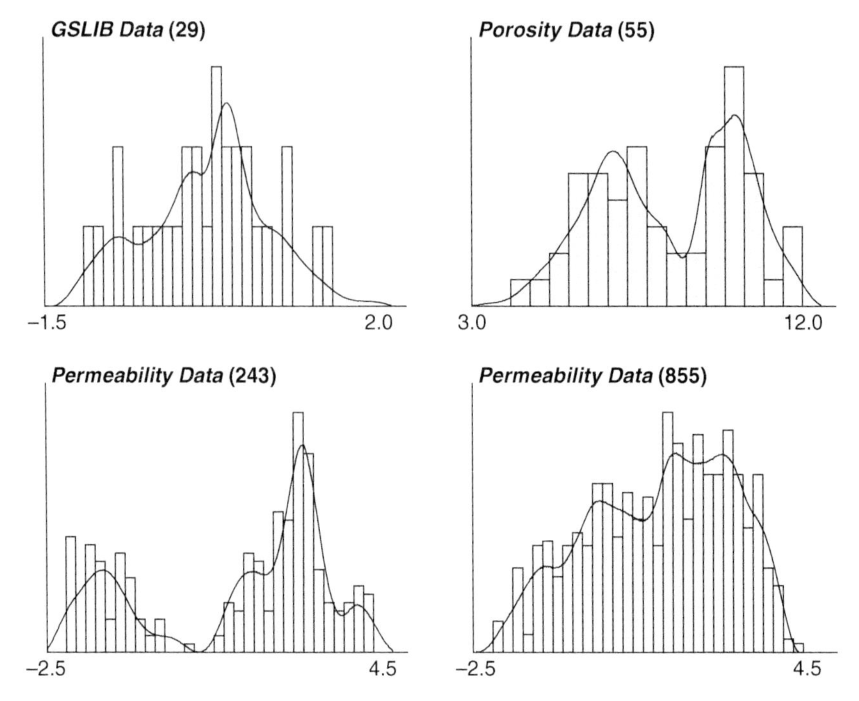

FIGURE 2.32: Four Examples of Histogram Smoothing, Clockwise from the Upper Left: 29 data taken from GSLIB (Deutsch and Journel, 1998), 55 porosity data, 243 permeability data, and 855 permeability data. The four data sets are unrelated.

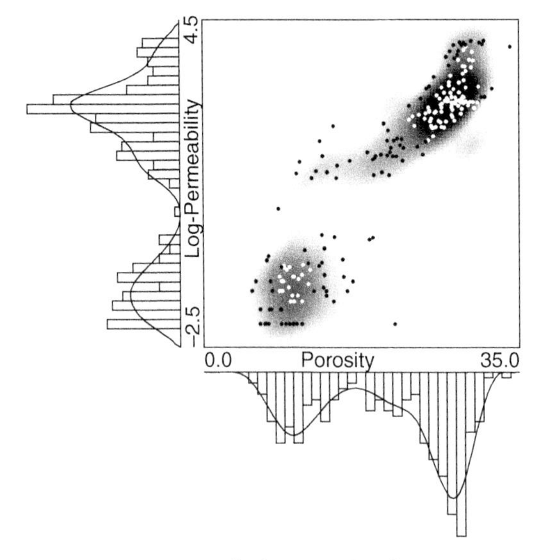

FIGURE 2.33: A Smoothed Scatterplot for 243 Porosity/Permeability Data.

section we limit ourselves to introducing the notion of Monte Carlo simulation and discussing a particular implementation of it, the bootstrap. Later sections will fully develop the concept of geostatistical simulation—that is, an extended form of Monte Carlo simulation that accounts for spatial constraints as imposed by a variogram model.

Monte Carlo simulation is the terminology used for the process of drawing realizations with specified probability. Drawing colored stones from an urn or picking pieces of paper (with different outcomes written on them) from a hat are classic illustrations of Monte Carlo simulation. The *Monte Carlo* prefix is dropped from "simulation" for brevity. Mechanical and electronic devices for generation of realizations were quickly replaced by numerical algorithms more than 50 years ago. These algorithms were devised to generate pseudo-random numbers that have the mathematical properties of random numbers—that is, uniformly distributed between zero and one with no correlation. The *pseudo* prefix is dropped from "random numbers" for brevity. There is a rich and varied history and literature related to random number generation. One of the latest random number generators is the ACORN generator (Wikramaratna, 1989) that is commonly used in GSLIB. Marsaglia's random number generator (Marsaglia, 1972) is also used. The book of Ripley (1981) presents some additional background and testing criteria for randomness. No major problems have been yet documented with the random number generators currently used in geostatistics.

The quantile method of simulating of realizations from arbitrary probability distributions is accomplished by (1) generating a random number, p, uniformly distributed between zero and one, and (2) calculating the inverse of the cumulative probability distribution (cdf) function:

$$y = F^{-1}(p)$$

The cdf $F(y)$ and its inverse $F^{-1}(p)$ are defined for both categorical and continuous variables. A large number of simulated realizations is typically required; that is, $y^{(l)} = F^{-1}(p^{(l)}), l = 1, \ldots, L$, where L is a large number and $p^{(l)}, l = 1, \ldots, L$, is a set of random numbers generated with an algorithm and a particular "seed" number. Regardless of the number L, the set of random numbers can always be regenerated with the algorithm and seed number.

This simple yet powerful concept of Monte Carlo simulation is at the heart of most modern geostatistical techniques. Accounting for spatial correlation and integrating data of different types in simulation of reservoir models are explained in detail in later chapters. The simple bootstrap technique and other methods, described next, provide a powerful tool for assessing uncertainty in input statistics.

2.2.9 Parameter Uncertainty

Geostatistical work flows rely on multiple equiprobable realizations to characterize uncertainty. In the past, many held the input statistics constant and utilized ergodic fluctuations between realizations alone to characterize uncertainty. Yet, this approach typically underestimates uncertainty (Wang and Wall, 2003). Uncertainty is reduced because local fluctuations above and below the average cancel out between simulated realizations when constant input statistics are applied and the realizations generally imply a very small uncertainty (Babak and Deutsch, 2007a).

It is now recognized that the uncertainty in the statistical inputs should also be quantified and integrated. Often, uncertainty in the input statistics is a significant source of uncertainty with considerable impact on uncertainty in reservoir prediction. For example, uncertainty of ±3% for the porosity distribution mean may result in a ±3% change in reservoir volume (assuming no complicating factors).

Yet, the assessment of parameter uncertainty remains as a challenging topic in geostatistics for which there is no panacea solution. To address this important topic, a variety of methods are presented; each has their own limitations, assumptions, and components of uncertainty that are addressed. For example, while bootstrap is a rigorous tool for assessing parameter uncertainty, it does not account for spatial correlation between the data. While spatial bootstrap accounts for spatial correlation of data, it only accounts for data redundancy and not data correlation with the domain of interest. We present two additional methods that account for the domain, but at a loss of simplicity and flexibility. The right method will depend on careful specification of the problem. For rigorous uncertainty modeling, we need to check for all sources of uncertainty.

It may be tempting to assess the uncertainty in the uncertainty model. Given the limitations and challenges in the inference of first-order uncertainty, further investigation into second-order uncertainty (i.e., uncertainty in the uncertainty) and beyond will not be feasible nor practical. The results of such an effort will not be reliable nor interpretable.

Bootstrap

A popular application of the Monte Carlo simulation technique is the *bootstrap* method developed by Efron (1982) and Efron and Tibshirani (1993). The bootstrap is a statistical resampling technique that permits the quantification of uncertainty in statistics by resampling from the original data—in other words, "pulling yourself up by your bootstraps."

Consider the 17 permeability data shown at the top of Figure 2.34. There is significant uncertainty in the mean permeability since we only have 17 data. The bootstrap can be used to quantify uncertainty in the mean by the following procedure.

1. Draw (simulate) 17 values from the distribution of 17 permeability values. This can be seen as drawing with replacement; that is, some values may be chosen more than once and other values may never be chosen.
2. Calculate the average of the permeability values $\overline{K}$ and save it as one possible average.
3. Go back to step 1 many times to assess the uncertainty in the mean.

The histogram at the bottom of Figure 2.34 shows the uncertainty in the average permeability if the procedure is repeated 1000 times.

The same Monte Carlo procedure can be used to assess uncertainty in any calculated statistic, including the correlation coefficient, for example, (1) draw N pairs from the N paired observations, (2) calculate ρ, and (3) repeat many times. Figure 2.35 shows

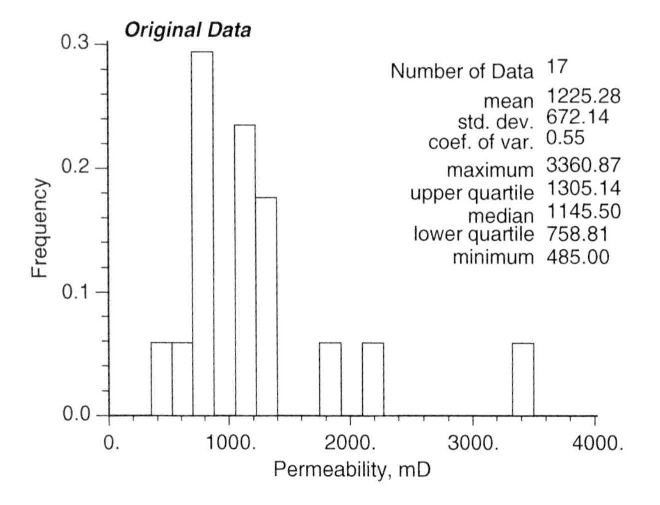

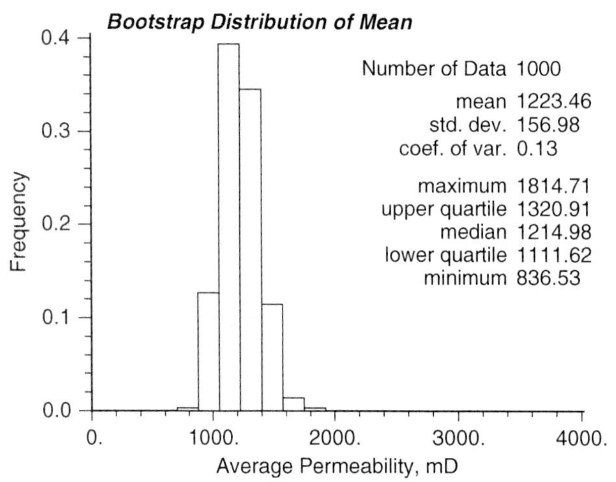

FIGURE 2.34: A Histogram of 17 Permeability Data Is Shown at the Top. The Bootstrap Was Applied to Arrive at a Distribution of Uncertainty in the Mean Permeability, Shown at the Bottom. Note that the sample mean permeability 1225 remains unchanged, the bootstrap simply quantifies the uncertainty in the average permeability.

a calibration cross plot between a seismic attribute and porosity with a linear correlation coefficient of 0.54. Applying the bootstrap procedure to assess the uncertainty in the correlation coefficient leads to the distribution below.

It is also possible to use the bootstrap to assess uncertainty in more complex statistics such as recoverable hydrocarbon. The procedure in this case is as follows:

1. Quantify the uncertainty in gross rock volume by modeling the surfaces and the fluid contacts stochastically (techniques for this are developed in later chapters).
2. Quantify the uncertainty in (a) net-to-gross ratio, (b) porosity in the net reservoir rock, and (c) oil saturation by application of the bootstrap, as presented above.
3. Then, perform a bootstrap simulation where a gross rock volume, net-to-gross ratio, net porosity, and oil saturation are drawn and then multiplied together to get an oil volume. The simulation is repeated to lead to a full distribution of uncertainty in oil-in-place.

Such uncertainty assessment may be valuable early in reservoir appraisal where no flow simulation is being considered. More detailed geostatistical modeling, as presented in this text, is required for reservoir performance predictions.

The bootstrap assumes that the data are *independent* one from another and representative of the underlying population. The independence assumption may be acceptable early in reservoir appraisal with widely spaced wells; however, highly correlated data (such as nearby measurements along a well) and close wells do not meet this assumption.

The bootstrap provides uncertainty for any parameter of interest at the scale of the samples, given the data are independent. As demonstrated, this

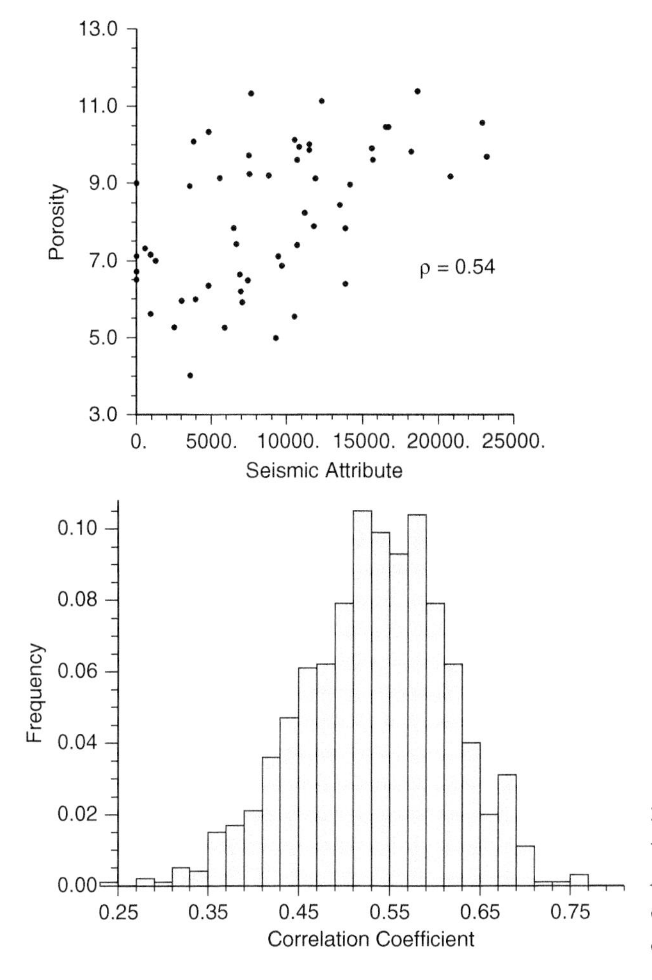

FIGURE 2.35: Calibration Cross Plot on the Top with a Linear Correlation Coefficient of 0.54. Applying the bootstrap procedure to assess the uncertainty in the correlation coefficient leads to the distribution of uncertainty shown below.

approach is very flexible and can be applied to any parameter, yet sample independence is not commonly a reasonable assumption in reservoir modeling problems. Even if the data are widely spaced, trends commonly impart some form of dependence or in some cases negative dependence may exist (i.e., low values related to high and vice versa). In spatial problems, a "spatial" bootstrap or resampling from stochastic simulations (Journel, 1994) could then be considered to account for these dependencies between samples.

Spatial Bootstrap and Alternatives

Correlation often exists spatially between our data samples for a single variable and between our variables. If we apply bootstrap and ignore this correlation, then bias may result in the assessment of uncertainty. This bias is often a significant underestimation of uncertainty. *Spatial or correlated bootstrap* is a method of bootstrap in which the correlation is honored between subsequent draws with replacement of the sample distribution (Feyen and Caers, 2006; Journel and Bitanov, 2004; Journel and Kyriakidis, 2004; Kedzierski et al., 2008; Maharaja, 2007).

There are a variety of practical methods to accomplish this, including simulating realizations at the data locations or scanning a well template through a set of realizations with the right spatial continuity and histogram. An important feature of spatial bootstrap is that its results are independent of the model domain; therefore, uncertainty increases as the range of correlation increases. Bras and Rodrıguez-Iturbe (1982) explain the latter feature as the increase in spatial continuity decreases the number of effective data.

The spatial bootstrap is the method of choice for uncertainty at the sample scale without consideration of the volume of interest; that is the locations of the samples within the volume or the size of the volume.

There are some alternatives that have been recently proposed to assess parameter uncertainty in the presence of spatial continuity between data. It has been suggested by Babak and Deutsch (2007a) that the size of the domain and the position of the data within this domain should also be considered in the resulting uncertainty model. One could imagine that the histogram uncertainty model may be different if the entire domain is sampled as opposed to if only a portion of the domain is sampled and that as the spatial continuity increases the information content from each data increase with respect to modeling domain. In other words, better data coverage and increased correlation between the data and the domain result in the data being more informative of the domain and should reduce uncertainty. This would not be captured with spatial bootstrap.

Babak and Deutsch (2007a) proposed an alternative to spatial bootstrap called the *conditional finite domain* method. The method is based on recursive simulation that accounts for the domain relative to the data locations and results in decreasing parameter uncertainty with increasing spatial continuity. In the first step a set of conditional realizations are simulated. The statistic from each realization is applied to simulate each a new set of realizations. This process is repeated until the uncertainty model stabilizes due to conditioning constraints and limited model size. The final set of realization distributions represent uncertainty in the distribution.

Deutsch and Deutsch (2010) proposed a kriging based method to calculate uncertainty in the mean from the kriging estimation variance, known as *global kriging*. The method is attractive because it provides a model of uncertainty in the mean while explicitly accounting for spatial correlation between the data and the effects of the domain. The method is simple and noniterative and relies on the kriging system presented in the next section. Yet, the method is inflexible because it only provides uncertainty in the distribution mean, unlike the previously presented methods that provide uncertainty in the full distribution or any other statistic.

Global kriging is useful for estimating the uncertainty in the distribution mean at the entire model volume support size, given the sample locations, spatial continuity, and volume of interest.

Villalba and Deutsch (2010) proposed the *stochastic trend* method for assessing uncertainty in the trend. With this methodology, multiple realizations of the trend model are fit to the data by randomizing the trend coefficients. This method accounts for data locations and the volume of interest and provides the uncertainty in the mean at the volume support size.

Integrating Parameter Uncertainty

The previously described methods, such as spatial bootstrap, may be applied to calculate multiple realizations of input statistics such as the histogram, semivariogram, and correlation coefficient. These statistics may be matched to specific realizations to generate an uncertainty model that accounts for uncertainty in the statistical inputs. The parameter uncertainty model should be checked to ensure that it does not violate known logical and physical constraints on the data distributions. For example, the porosity distribution should have an appropriate shape and non-negative values that are not larger than physical constraints on porosity determined from facies, compaction, and cementation.

At times, a scenario approach may be needed. For example, instead of a set of histogram realizations to characterize the histogram uncertainty, there may be low and high histogram cases. When scenarios are applied, a decision is required concerning the probability of each scenario and thus the proportion of realizations that utilize each scenario. For more detailed discussion see Section 5.3, Uncertainty Management.

2.2.10 Bayesian Statistics

Bayesian statistics are a major branch of statistics with a fundamentally different frame of reference than the *frequentist* perspective. Bayesian statistics utilizes belief and probability logic to build useful probability relationships (Sivia, 1996). This is a powerful tool for considering probability problems for which we do not have replicates to build a distribution, when we need to combine multiple sources of information or invert a conditional probability from an accessible $P(A|B)$ to an inaccessible $P(B|A)$. The basic building block of probability logic in the Bayesian framework is the product rule

$$P(A \text{ and } B) = P(A|B)P(B)$$

We can rearrange the product rule to express Bayesian updating

$$P(A|B) = \frac{P(A \text{ and } B)}{P(B)}$$

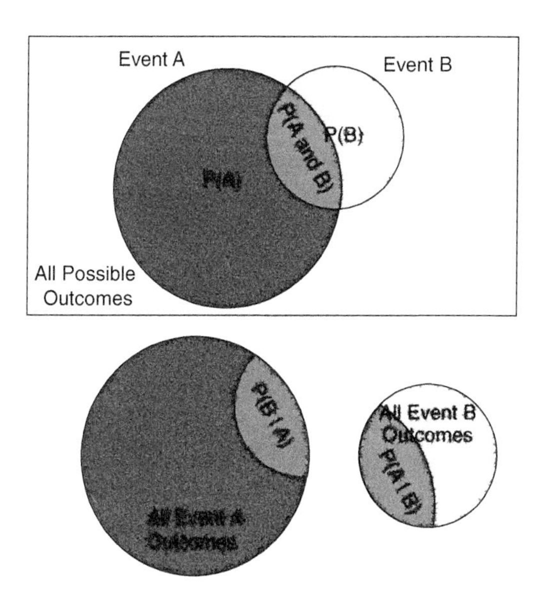

FIGURE 2.36: Venn Diagram Illustrating Bayesian Updating and the Associated Component Probabilities. The probabilities of events, A and B, P(A) and P(B), respectively, along with the joint probability P(A and B), are represented above in the space of all possible outcomes. The conditional probabilities, P(A|B) and P(B|A), are represented by only considering the space where events B and A occur, respectively.

It may be helpful to visualize this relationship in a Venn diagram. Event A and event B are represented as areas within circles, and the overlap is the joint occurrence of events A and B. The probabilities of each of these are their respective areas divided by the total area in the rectangle representing all possible events. Bayesian updating solves for the probability of A given B, $P(A|B)$, by standardizing the joint outcome of A and B over the area of event B (the orange area divided by the yellow and orange area in the event B circle in Figure 2.36).

Bayesian inversion is another useful interpretation of Bayesian updating. Given $P(A \text{ and } B) = P(B \text{ and } A)$, we can substitute another product rule for $P(A \text{ and } B)$ as follows to formulate the relationship:

$$P(A|B) = \frac{P(B|A) \cdot P(A)}{P(B)}$$

This may be restated as the posterior probability is equal to the likelihood probability multiplied by the prior probability divided by the evidence probability:

$$\text{Posterior} = \frac{\text{Likelihood} \cdot \text{Prior}}{\text{Evidence}}$$

This result provides a logically consistent methodology to integrate information. For example, consider that event A is a porosity value greater than 20% at a location within our reservoir and $P(A)$ is the associated probability of this occurrence. Now consider that we have some form of secondary data available such as seismic that may aid in the inference of porosity greater than 20%. For example, event B is a specific seismic derivative result for an indicator of porosity, such as acoustic impedance, at the same location in our reservoir and $P(B)$ is the associated probability of this event. It stands to reason that if events A and B are not independent (or conditionally independent), then integration of information related to B will improve our estimation of A. The Bayesian inversion equation above provides a method to accomplish this. $P(B)$ is the probability of the positive seismic indicator at the location, often the global probability of a positive seismic indicator is applied, $P(A)$ is the probability of porosity greater than 20% without considering seismic, and $P(B|A)$ is the probability of a specific acoustic impedance response given porosity greater than 20%. Note that in the case where seismic adds no information, $P(B|A) = P(B)$ and the updated probability or posterior probability is the same as the prior.

We can extend Bayesian updating to the case where we jointly consider multiple information sources related to event A. For example, we may have multiple indicators of porosity at a location $B_1, \ldots, B_n$ In this multivariate setting, this updating is more difficult because there is often no access to the multivariate distributions required:

$$P(A|B_1 \ldots, B_n) = \frac{P(A) \prod_{i=1}^{n} P(B_i|A, B_1, \ldots, B_{i-1})}{P(B_1, \ldots, B_n)}$$

This exact expression requires knowledge of the various joint probabilities. These joint distributions are often unknown, and some simplification is required. Several models of dependency exist that allow integration of multiple sources of information accounting for their redundancy. In Section 2.4, Preliminary Mapping Concepts, these methods are introduced as probability combination schemes.

Returning to the case of Bayesian updating with a single source of information, Neufeld and Deutsch (2004) provide a simplification of the Sivia (1996) analytical solution for Bayesian updating for the case of Gaussian distributions. In this case, we consider the entire prior and likelihood distributions to calculate the posterior distribution, parameterized by mean and variance. The mean of the updated distribution is

$$\bar{x}_{\text{updated}} = \frac{\bar{x}_{\text{likelihood}}(\mathbf{u}) \cdot \sigma^2_{\text{prior}}(\mathbf{u}) + \bar{x}_{\text{prior}}(\mathbf{u}) \cdot \sigma^2_{\text{likelihood}}(\mathbf{u})}{[1 - \sigma^2_{\text{likelihood}}(\mathbf{u})][\sigma^2_{\text{prior}}(\mathbf{u}) - 1] + 1}$$

and the variance for the updated distribution is

$$\sigma^2_{\text{updated}}(\mathbf{u}) = \frac{\sigma^2{}_{\text{prior}}(\mathbf{u})\, \sigma^2{}_{\text{likelihood}}(\mathbf{u})}{[1 - \sigma^2{}_{\text{likelihood}}(\mathbf{u})][\sigma^2{}_{\text{prior}}(\mathbf{u}) - 1] + 1}$$

where $\bar{x}$ and σ^2 are the mean and variance for each distribution.

It is instructive to apply these equations to various examples to compare prior, likelihood, and posterior distributions (see Figure 2.37). Firstly, note that the updated variance is only impacted by the variance and not the mean of either the prior or likelihood distributions. This is expected as the Gaussian distribution is homoscedastic. Secondly, note the decrease in the posterior distribution variance relative to the prior distribution variance. Finally, the posterior mean may be between the means of the prior and likelihood distributions. However, when the prior is high or low and the likelihood is also high or low, respectively, the updated mean will be even higher or lower. This is expected because both data sources corroborate one another.

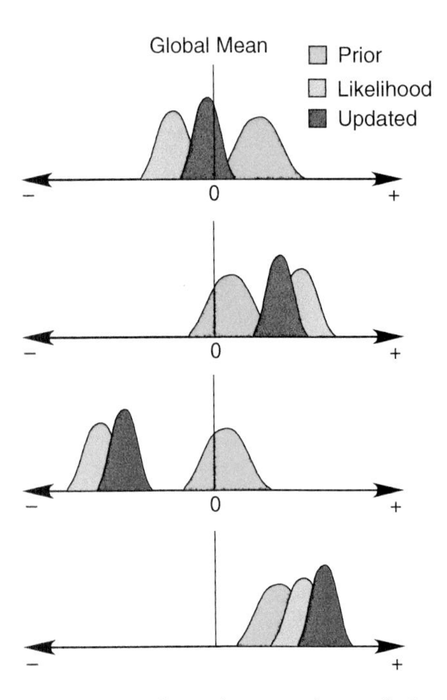

FIGURE 2.37: Illustration Bayesian Updating of Gaussian Distributions (Modified from Ren (2007)). Note that the variance of the updated distribution is less than the prior and likelihood distributions' variances and that the variance is not impacted by the mean. Also, the updated mean is between the prior and likelihood means, unless the prior mean departs from the mean significantly and the likelihood departs even further.

2.2.11 Work Flow

Figure 2.38 illustrates a work flow to separate different rock types or facies. An iterative scheme is considered where different rock-type distinctions are used, statistics and statistical displays are created, and then geologically or statistically significant distinctions are kept. This may include the need to work hierarchically with multiple facies in each depositional facies or rock type, depending on specific nomenclature scheme and level of detail required in the model.

Figure 2.39 illustrates the work flow to establish representative global facies proportions for subsequent facies modeling. As we discussed earlier in this chapter, to naïvely consider equal weighting is often unsatisfactory. Declustering may be required. Geological trends may need to be considered. Seismic data may also be used.

Figure 2.40 illustrates the work flow to determining representative histograms for porosity and permeability variables. Preferential sampling affects the histograms and statistics of continuous variables just like facies proportions. The required statistics for subsequent modeling are different, but the declustering steps are similar to those needed for facies.

Figure 2.41 illustrates the work flow for data transformation. There are a number of situations where data transformation is required: (1) correcting bias in well-log-derived measurements, (2) fixing the distribution of values from a geostatistical simulation, (3) transformation to the normal or Gaussian distribution prior to spatial statistics and geostatistical modeling, and (4) conversion to a standard distribution so that purely spatial differences can be discerned. The transformation is straightforward, but checking must follow.

Figure 2.42 illustrates the work flow to establish a bivariate calibration relationship for subsequent cosimulation or modeling. The use of seismic data,

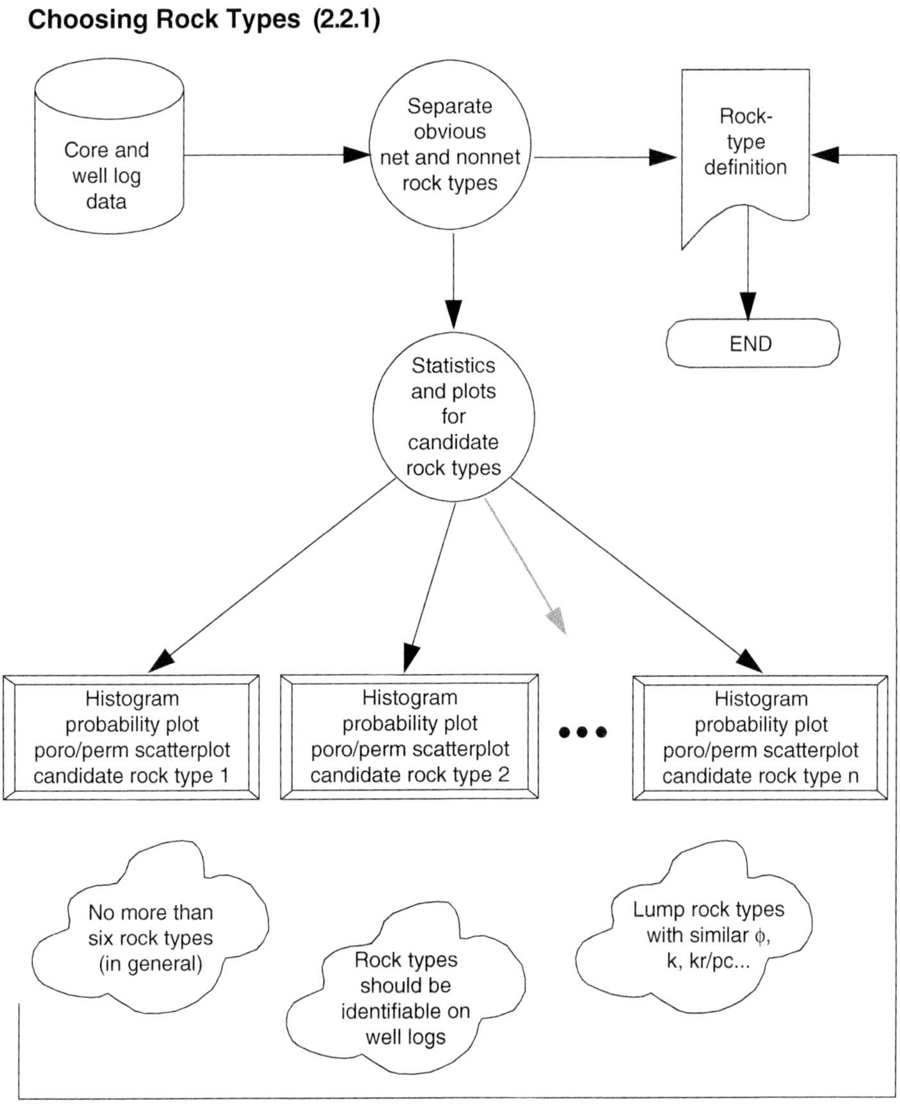

Choosing Rock Types (2.2.1)

Core and well log data

Separate obvious net and nonnet rock types

Rock-type definition

END

Statistics and plots for candidate rock types

Histogram probability plot poro/perm scatterplot candidate rock type 1

Histogram probability plot poro/perm scatterplot candidate rock type 2

...

Histogram probability plot poro/perm scatterplot candidate rock type n

No more than six rock types (in general)

Rock types should be identifiable on well logs

Lump rock types with similar ϕ, k, kr/pc...

FIGURE 2.38: Work Flow for Choosing Facies or Rock Types for Geostatistical Modeling.

production data, or a geological trend variable is common in geostatistical reservoir modeling. In such cases, the calibration between the variable of interest and the secondary variable is critical. The main steps in this work flow are to get the data pairs and smooth the corresponding cross plot.

Figure 2.43 illustrates the work flow to check histogram and statistics reproduction of a geostatistical model. We must compare statistical displays and interpret Q–Q plots.

Figure 2.44 illustrates the work flow to calculate uncertainty in global statistics. One can apply bootstrap to calculate the uncertainty in a global statistic. The global statistics one would usually consider include (1) the mean and variance, (2) the entire histogram, and (3) the correlation between two variables of interest. The bootstrap should not be applied to data with significant spatial correlation—hence the check for spatial correlation. In the presence of spatial correlation, spatial bootstrap or other alternatives may be applied.

2.2.12 Section Summary

We require a decision of stationarity to pool data from multiple locations for the inference of representative statistics and to apply these as input statistics over a volume of the reservoir that is deemed to be sufficiently similar. These statistics include a variety

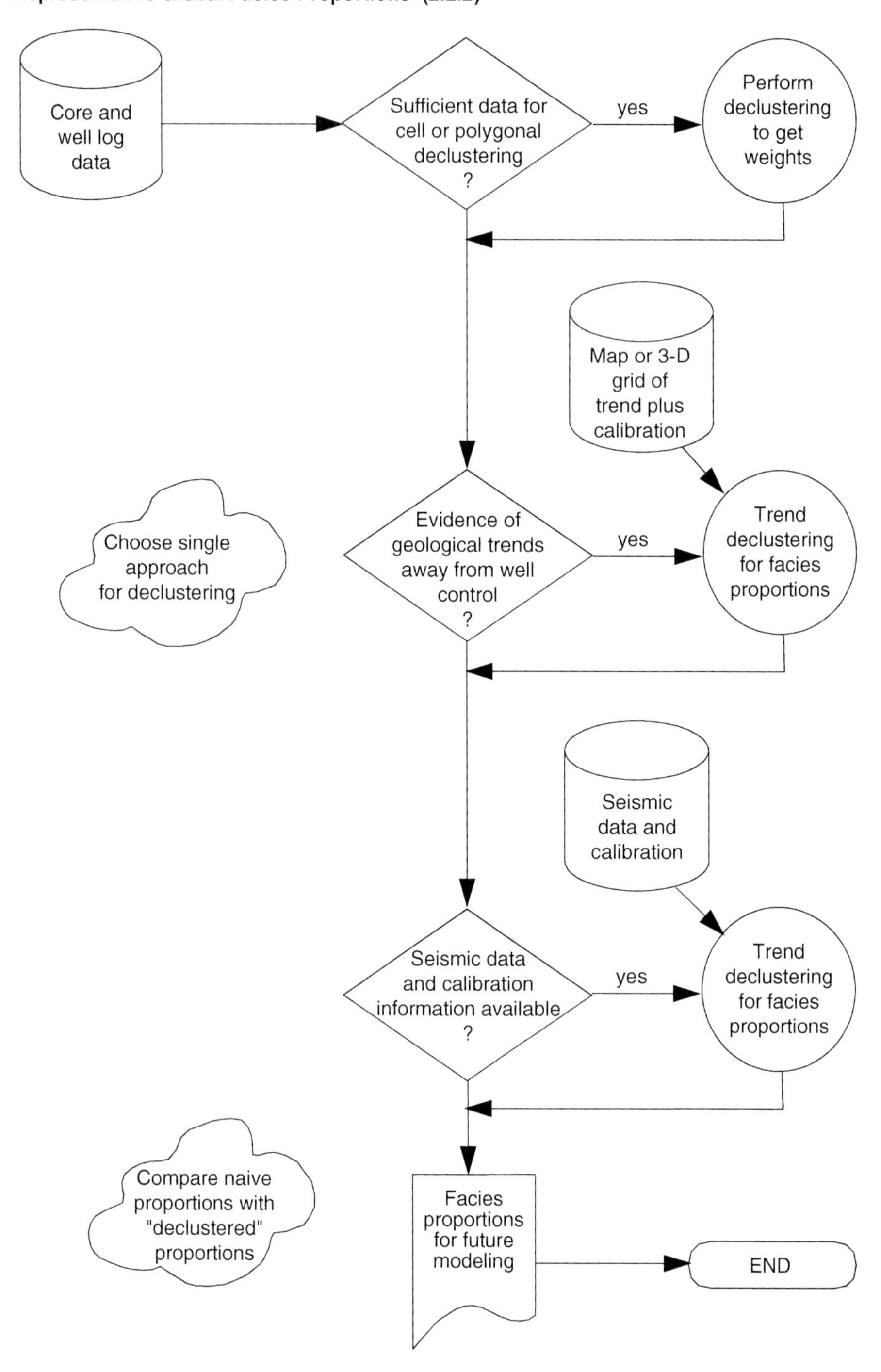

FIGURE 2.39: Work Flow for Determining Representative Global Facies Proportions for Subsequent Facies Modeling.

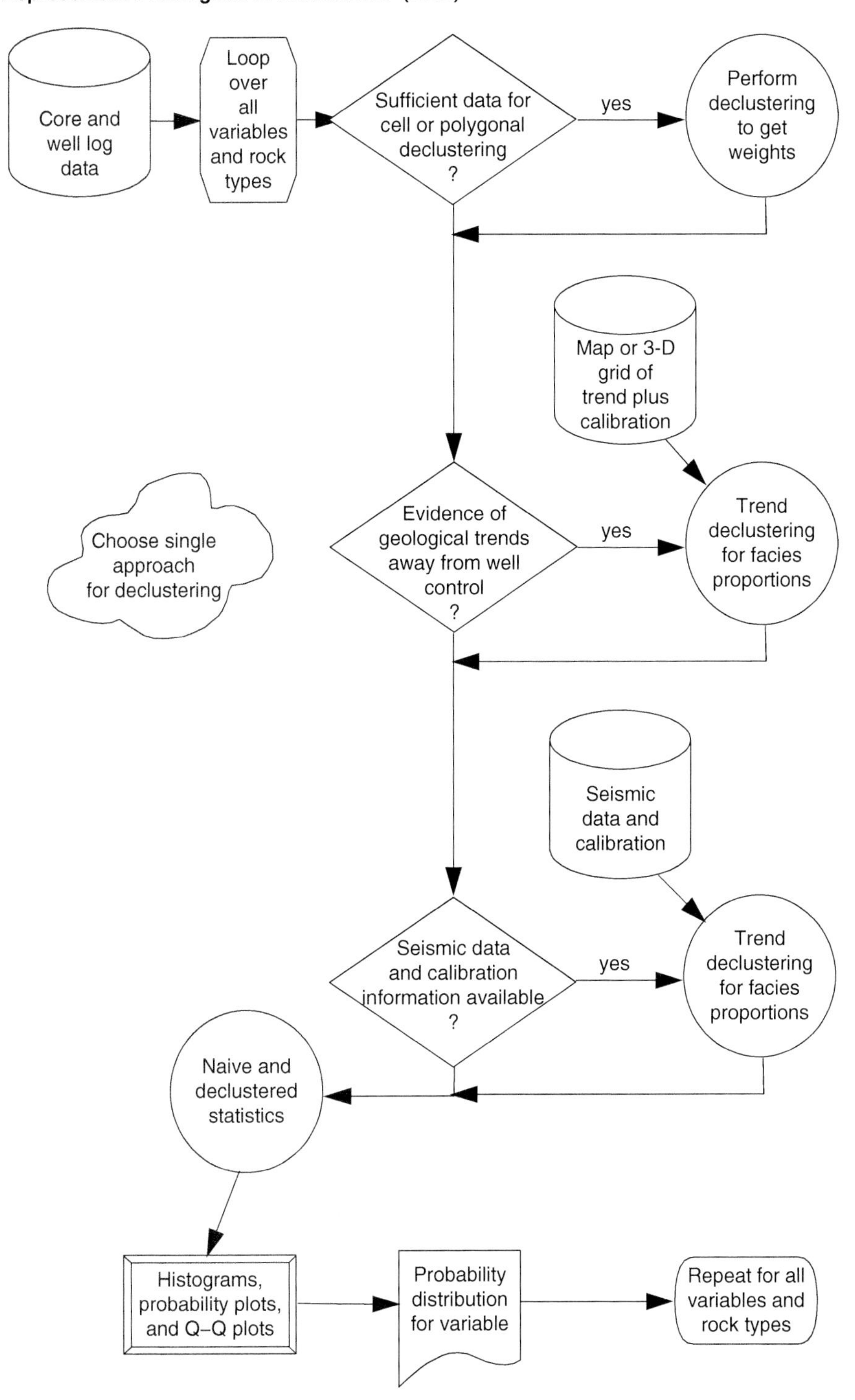

FIGURE 2.40: Work Flow for Determining Representative Histograms for Porosity and Permeability Variables for Subsequent Facies Modeling.

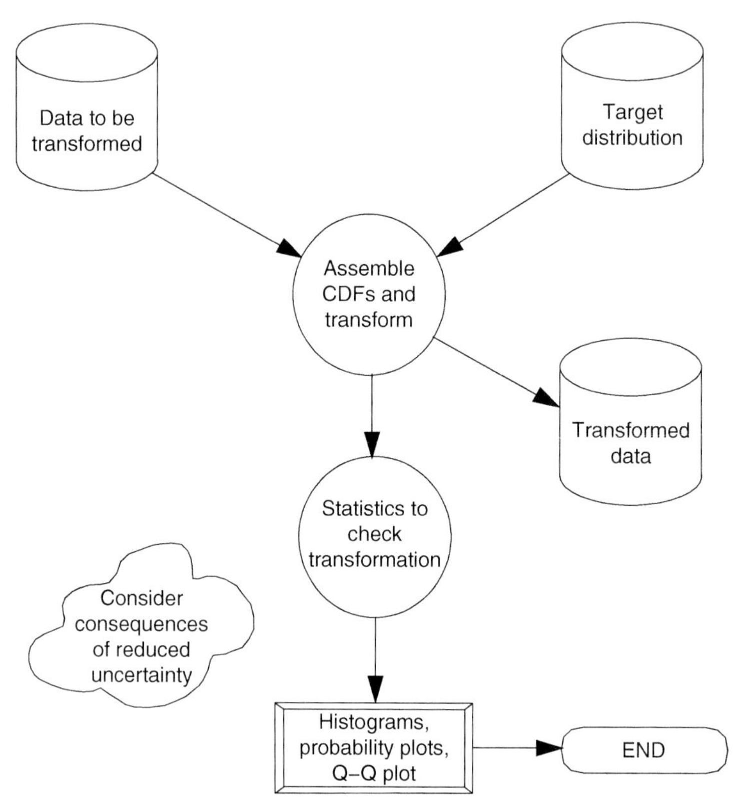

FIGURE 2.41: Work Flow for Data Transformation.

of univariate, bivariate, and spatial statistics (covered in the next section). Geostatistical algorithms reproduce this statistics with some limitations.

Data transformations provide us with methods to correct our distributions (e.g., log to more reliable core porosity) or to transform our distributions to those required by the modeling methods (e.g., Gaussian for Gaussian methods discussed later). Declustering and debiasing provide tools that improve the representativity of our input statistics in the face of biased data collection practices.

Uncertainty modeling requires assessment of uncertainty in the input statistics. Tools such as bootstrap, spatial bootstrap, conditional finite domain, and global kriging quantify uncertainty in input statistics that may be carried through the model uncertainty by varying these input parameters over multiple reservoir realizations. Careful consideration and determination of the specific uncertainty problem is required to select the appropriate method.

Bayesian statistics provides a useful frame of reference for interpreting probability and probability logic to combine or update with new sources of information. The multivariate extension requires a challenging inference of joint probabilities; yet, in the next section, probability combination schemes are presented to approximate these probabilities under specific assumptions.

Spatial statistics were conspicuously absent from this section. These are central to construction of spatial models and are discussed in detail in the next section.

2.3 QUANTIFYING SPATIAL CORRELATION

An essential aspect of geostatistical modeling is to establish quantitative measures of spatial correlation to be used for subsequent estimation and simulation. Spatial variability is different for each variable (facies indicators, porosity, and permeability) within each reservoir layer.

The spatial correlation for object-based modeling (introduced in the next section) is quantified through object shapes, sizes, and relationships.

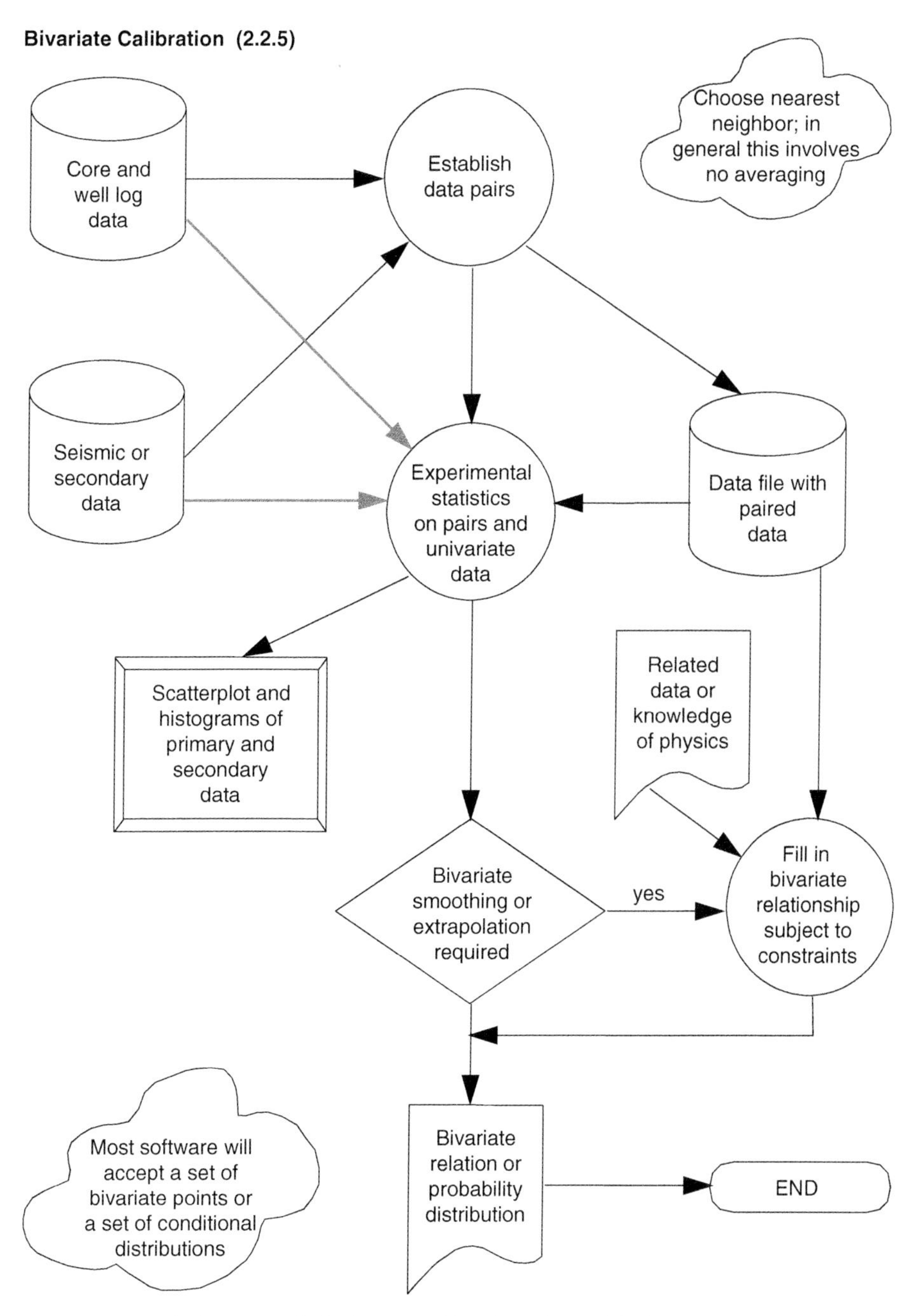

FIGURE 2.42: Work Flow to Establish Bivariate Calibration Relationship for Subsequent Cosimulation or Modeling.

These spatial measures are inseparable from the object-based modeling approach (see Section 4.4, Object-Based Facies Modeling, for detailed discussion on object-based modeling).

The variogram is the most commonly used measure of spatial correlation for porosity and permeability modeling and is commonly applied for cell-based facies modeling. The random function (RF) concept and theoretical background for variogram-based measures of spatial correlation are presented in the *Random Function* subsection. The *Calculating Experimental Variograms* subsection describes the steps involved in variogram calculation. Careful selection of variogram calculation parameters is critical for obtaining a clean, interpretable, sample variogram.

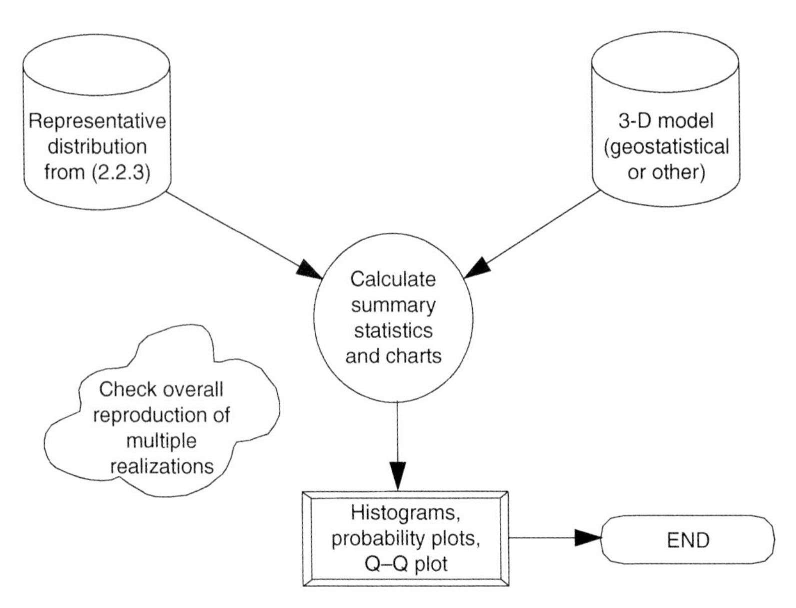

FIGURE 2.43: Work Flow to Check Histogram and Statistics Reproduction of Model.

Principles of variogram interpretation are described in the *Interpreting Experimental Variograms* subsection. The effect of stratigraphic cyclicity, vertical and areal trends, and stratification are described in detail. Often, there are inadequate data to reliably calculate and interpret a reservoir-specific variogram. The *Horizontal Variograms* subsection presents analog data and procedures to supplement limited well data by more abundant "soft" geological data. Modeling experimental variograms, consistent with soft geological data, is a necessary step prior to kriging-based techniques; the *Variogram Modeling* subsection describes variogram modeling in detail.

The *Cross Variograms* subsection presents the calculation, interpretation, and modeling of cross variograms between multiple variables. Depending on software and data availability, inference and modeling of cross variograms can be tedious. Approximations as provided by the various Markov-type models and the model implicit to collocated cokriging are discussed.

Recently, multiple-point statistics-based simulation has become widely applied for categorical variables. The *Multiple-Point Statistics* subsection extends the spatial continuity model provided by variograms to multiple-point statistics. It is not practical to infer these statistics from available data; therefore, current practice is to extract multiple-point statistics from an exhaustive analog model known as a training image. Considerations with regard to construction of training images are discussed. Section 2.4, Preliminary Mapping Concepts, introduces a practical method to integrate multiple-point statistics into a reservoir facies model, while Section 4.3, Multiple-point Facies Modeling, covers implementation details and provides examples.

Models of spatial continuity are useful for purposes other than simply as the statistical inputs for geostatistical reservoir modeling. As mentioned in the previous section, they are inputs for various methods to assess uncertainty in statistics, such as spatial bootstrap, conditional finite domain and kriging-based uncertainty. In addition, spatial continuity models are applied for the purpose of scaling distributions and variograms to the appropriate model scale. The *Volume Variance Relations* subsection discusses the relationship between the size support of data or model cells and variability between these samples. This is critical for understanding how statistical parameters such as the histogram change between support sizes such as from core sample size to reservoir model cell size. Without explicitly accounting for this change in size support, the reservoir property variance may be too high in the reservoir model. The prerequisite concepts of volume-averaged variogram values (gamma bar values) and

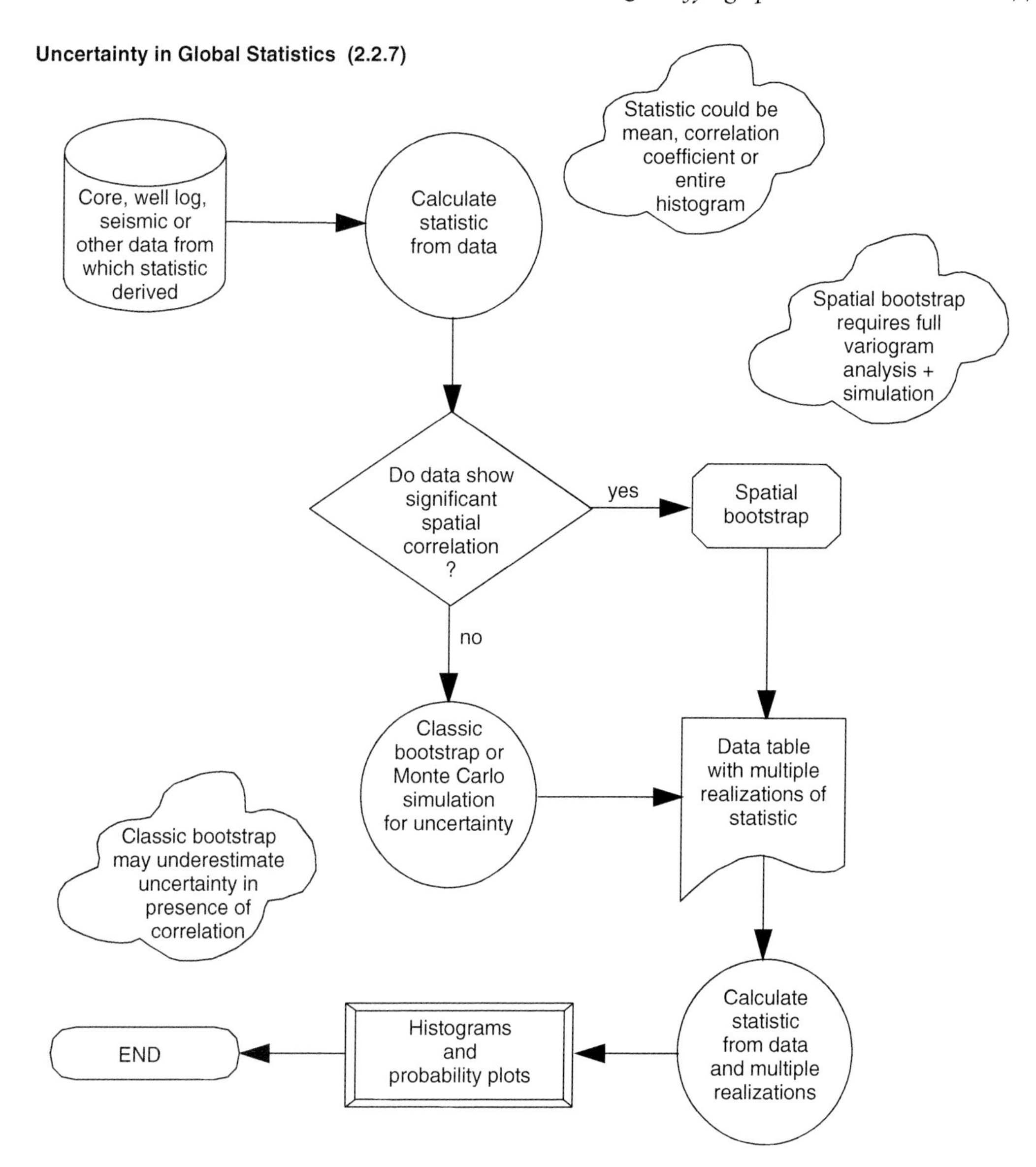

FIGURE 2.44: Work Flow to Calculate Uncertainty in Global Statistic.

dispersion variances are presented. Practical methods for dealing with data and model scale are covered in Section 5.2, Model Post-processing.

The novice modeler will benefit most from discussion of spatial continuity modeling, including methods to calculate, interpret, and model variograms and the basics of multiple-point statistics. The novice may skip the subsections on random functions and volume variance relations. Intermediate and expert modelers may benefit from the rigor available in the random function and model scale subsections.

2.3.1 The Random Function Concept

The material in this section is more completely covered in other geostatistical texts such as *Mining Geostatistics* (Journel and Huijbregts, 1978), *An Introduction to Applied Geostatistics* (Isaaks and Srivastava, 1989), *Geostatistics for Natural Resources Evaluation* (Goovaerts, 1997), *Geostatistical Simulation: Models and Algorithms* (Lantuéjoul, 2001), Statistics for Spatial Data (Cressie, 1991), Spatial Variation (Matern, 1980), or *Geostatistics: Modeling Spatial Uncertainty* (Chilès and Delfiner, 2012). Nevertheless, this section is included for completeness. It may be skipped at first reading. These details will become more relevant to answer questions such as "Why are conventional geostatistical methods limited to two-point variogram statistics?"

The uncertainty about an unsampled value z is modeled through the probability distribution of a random variable (RV) Z. The probability distribution of Z after data conditioning is usually location-dependent; hence, the notation $Z(\mathbf{u})$, with $\mathbf{u}$ being the coordinate location vector. A random function (RF) is a set of RVs defined over some field of interest—for example, $\{Z(\mathbf{u}), \mathbf{u} \in \text{ study area } A\}$ also denoted simply as $Z(\mathbf{u})$. Usually the RF definition is restricted to RVs related to the same attribute, say z; hence, another RF would be defined to model the spatial variability of a second attribute, say $\{Y(\mathbf{u}), \mathbf{u} \in \text{ study area}\}$.

Just as a RV $Z(\mathbf{u})$ is characterized by its cdf, an RF $Z(\mathbf{u})$ is characterized by the set of all its N-variate cdfs for any number N and any choice of the N locations $\mathbf{u}_i, i = 1, \ldots, N$, within the study area A:

$$F(\mathbf{u}_1, \ldots, \mathbf{u}_N; z_1, \ldots, z_N) = \text{Prob}\{Z(\mathbf{u}_1) \leq z_1, \ldots, Z(\mathbf{u}_N) \leq z_N\} \quad (2.7)$$

Just as the univariate cdf of the RV $Z(\mathbf{u})$ is used to characterize uncertainty about the value $z(\mathbf{u})$, the multivariate cdf [see Eq. (2.7)] is used to characterize joint uncertainty about the N values $z(\mathbf{u}_1), \ldots, z(\mathbf{u}_N)$.

The bivariate ($N = 2$) cdf of any two RVs $Z(\mathbf{u}_1), Z(\mathbf{u}_2)$, or more generally $Z(\mathbf{u}_1), Y(\mathbf{u}_2)$, is particularly important since conventional geostatistical procedures are restricted to univariate ($F(\mathbf{u}; z)$) and bivariate distributions:

$$F(\mathbf{u}_1, \mathbf{u}_2; z_1, z_2) = \text{Prob}\{Z(\mathbf{u}_1) \leq z_1, Z(\mathbf{u}_2) \leq z_2\} \quad (2.8)$$

One important summary of the bivariate cdf $F(\mathbf{u}_1, \mathbf{u}_2; z_1, z_2)$ is the covariance function defined as

$$C(\mathbf{u}_1, \mathbf{u}_2) = E\{Z(\mathbf{u}_1)\, Z(\mathbf{u}_2)\} - E\{Z(\mathbf{u}_1)\}\, E\{Z(\mathbf{u}_2)\} \quad (2.9)$$

However, when a more complete summary is needed, the bivariate cdf $F(\mathbf{u}_1, \mathbf{u}_2; z_1, z_2)$ can be described by considering binary *indicator* transforms of $Z(\mathbf{u})$ defined as

$$I(\mathbf{u}; z) = \begin{cases} 1, & \text{if } Z(\mathbf{u}) \leq z \\ 0, & \text{otherwise} \end{cases}$$

Then, the previous bivariate cdf shown in Eq. (2.8) at various thresholds z_1 and z_2 appears as the non-centered covariance of the indicator variables:

$$F(\mathbf{u}_1, \mathbf{u}_2; z_1, z_2) = E\{I(\mathbf{u}_1; z_1) I(\mathbf{u}_2; z_2)\} \quad (2.10)$$

Relation (2.10) is the key to the indicator geostatistics formalism (Journel, 1986a): It shows that inference of bivariate cdfs can be done through sample indicator covariances.

The probability density function (pdf) representation is more relevant for categorical variables. For example,

$$f(\mathbf{u}_1, \mathbf{u}_2; k_1, k_2) = \text{Prob}\{Z(\mathbf{u}_1) \in k_1, Z(\mathbf{u}_2) \in k_2\}, \quad k_1, k_2 = 1, \ldots, K$$

is the bivariate or two-point distribution of $Z(\mathbf{u}_1)$ and $Z(\mathbf{u}_2)$. This two-point distribution, when established from experimental proportions, is also referred to as a *two-point histogram* (Farmer, 1992).

Recall that a categorical variable $Z(\mathbf{u})$, which takes K outcome values $k = 1, \ldots, K$, may arise from a naturally occurring categorical variable or from a continuous variable discretized into K classes.

The purpose for conceptualizing an RF as $\{Z(\mathbf{u}), \mathbf{u} \in \text{ study area } A\}$ is never to study the case where the variable Z is completely known. If all the $z(\mathbf{u})$'s were known for all $\mathbf{u} \in$ study area A, there would be neither any problem left nor any need for the concept of a random function. The ultimate goal of an RF model is to make some predictive statement about locations $\mathbf{u}$ where the outcome $z(\mathbf{u})$ is unknown.

Inference of any statistic requires some repetitive sampling. For example, repetitive sampling of the variable $z(\mathbf{u})$ is needed to evaluate the cdf

$$F(\mathbf{u}; z) = \text{Prob}\{Z(\mathbf{u}) \leq z\}$$

from experimental proportions. However, in many applications, at most one sample is available at any single location $\mathbf{u}$ in which case $z(\mathbf{u})$ is known (ignoring sampling errors), and the need to consider the RV model $Z(\mathbf{u})$ vanishes. The paradigm underlying statistical inference processes is to trade the unavailable replication at location $\mathbf{u}$ for another replication available somewhere else in space and/or time. For example, the cdf $F(\mathbf{u}; z)$ may be inferred from the sampling distribution of z-samples collected at other locations, $\mathbf{u}_\alpha \neq \mathbf{u}$, within the same field.

As discussed in Section 2.2, Preliminary Statistical Concepts, this trade of replication corresponds to the decision of stationarity. Stationarity is a property of the RF model, not of the underlying physical spatial distribution. Thus, it cannot be checked from data. The decision to pool data into statistics across rock types is not refutable a priori from data; however, it can be shown to be inappropriate a posteriori if differentiation per rock type is critical to the undergoing study. For a more extensive discussion see Isaaks and Srivastava (1989) and Journel (1986b).

The RF $\{Z(\mathbf{u}), \mathbf{u} \in A\}$ is said to be stationary within the field A if its multivariate cdf [see Eq. (2.7)] is invariant under any translation of the N coordinate vectors $\mathbf{u}_k$, that is,

$$F(\mathbf{u}_1, \ldots, \mathbf{u}_N; z_1, \ldots, z_N) = F(\mathbf{u}_1 + \mathbf{l}, \ldots, \mathbf{u}_n + \mathbf{l}; z_1, \ldots, z_n), \quad \forall \text{ translation vector } \mathbf{l}$$

Invariance of the multivariate cdf entails invariance of any lower-order cdf, including the univariate and bivariate cdfs, and invariance of all their moments, including all covariances of type [see Eq. (2.9) or (2.10)]. The decision of stationarity allows inference. For example, the unique stationary cdf

$$F(z) = F(\mathbf{u}; z), \qquad \forall\, \mathbf{u} \in A$$

can be inferred from the cumulative sample histogram of the z-data values available at various locations within A. The stationary mean and variance can then be calculated from that stationary cdf $F(z)$:

$$E\{Z(\mathbf{u})\} = \int z\, dF(z) = m, \qquad \forall\, \mathbf{u}$$

$$E\left\{[Z(\mathbf{u}) - m]^2\right\} = \int [z - m]^2\, dF(z) = \sigma^2, \qquad \forall\, \mathbf{u}$$

The decision of stationarity also allows inference of the stationary covariance

$$C(\mathbf{h}) = E\left\{Z(\mathbf{u} + \mathbf{h})Z(\mathbf{u})\right\} - \left[E\left\{Z(\mathbf{u})\right\}\right]^2 \quad \forall \mathbf{u}, \mathbf{u} + \mathbf{h} \in A$$

from the sample covariance of all pairs of z-data values approximately separated by vector $\mathbf{h}$. At $\mathbf{h} = 0$ the stationary covariance $C(0)$ equals the stationary variance σ^2, that is,

$$\begin{aligned} C(0) &= E\left\{Z(\mathbf{u} + 0)Z(\mathbf{u})\right\} - \left[E\left\{Z(\mathbf{u})\right\}\right]^2 \\ &= E\left\{Z(\mathbf{u})^2\right\} - \left[E\left\{Z(\mathbf{u})\right\}\right]^2 \\ &= Var\left\{Z(\mathbf{u})\right\} = \sigma^2 \end{aligned}$$

The notation $C(0)$ will often be used for the variance.

In certain situations the standardized stationary correlogram is preferred:

$$\rho(\mathbf{h}) = \frac{C(\mathbf{h})}{C(0)}$$

In other cases another second-order (two-point) moment called the variogram is considered:

$$2\gamma(\mathbf{h}) = E\left\{\left[Z(\mathbf{u} + \mathbf{h}) - Z(\mathbf{u})\right]^2\right\} \quad \forall \mathbf{u}, \mathbf{u} + \mathbf{h} \in A \tag{2.11}$$

Under the decision of stationarity the covariance, correlogram, and variogram are equivalent tools for characterizing two-point correlation:

$$C(\mathbf{h}) = C(0) \cdot \rho(\mathbf{h}) = C(0) - \gamma(\mathbf{h})$$

This relation depends on the stationarity decision implying that the mean and variance are constant and independent of location. These relations are the foundation of variogram interpretation. That is, (1) the sill plateau value of the stationary variogram is the variance, which is the variogram value that corresponds to zero correlation, (2) the correlation between $Z(\mathbf{u})$ and $Z(\mathbf{u} + \mathbf{h})$ is positive when the variogram value is less than the sill, and (3) the correlation between $Z(\mathbf{u})$ and $Z(\mathbf{u} + \mathbf{h})$ is negative when the variogram exceeds the sill.

These three principles depend on knowledge of the "model" variance σ^2, which must be finite. The variance of petrophysical properties in petroleum reservoirs is indeed finite. Moreover, the variance is known more precisely than any experimental variogram value; therefore, the experimental variance may be associated to the model variance. The bootstrap procedure of Section 2.2 could be considered to assess uncertainty in the variance.

h-scatterplots may be used to illustrate these principles. An **h**-scatterplot is a cross plot of the "head" value $Z(\mathbf{u})$ with the "tail" value $Z(\mathbf{u} + \mathbf{h})$. The use of the words "head" and "tail" relate to the vector **h**. Figure 2.45 shows three **h**-scatterplots corresponding to three lag vectors on a typical semivariogram. The horizontal line at 1.0 on the semivariogram is at the model variance of 1.0.

The decision of stationarity is critical for the appropriateness and reliability of geostatistical simulation methods. Pooling data across geological facies may mask important geological differences; on the other hand, splitting data into too many

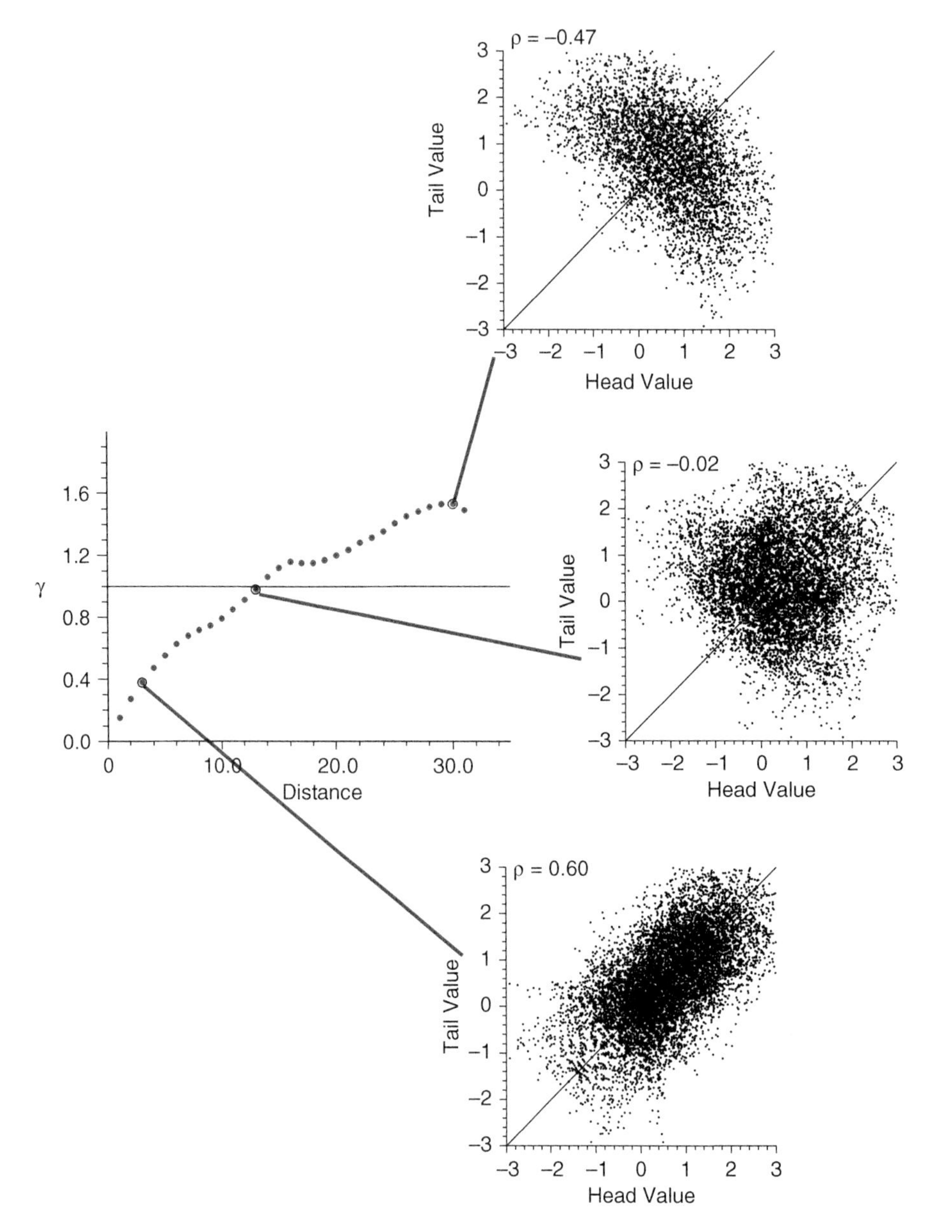

FIGURE 2.45: Semivariogram of the Normal Score of Real Porosity Data with the **H**-Scatterplots Corresponding to Three Different Lag Distances. Note that the correlation on the **h**-scatterplot is positive when the semivariogram value is below the sill, zero when the semivariogram is at the sill, and negative when the semivariogram is above the sill.

subcategories may lead to unreliable statistics based on too few data per category. The rule in statistical inference is to pool the largest amount of *relevant* information to formulate predictive statements.

Stationarity is a property of the RF model; thus, the decision of stationarity may change if the scale of the study changes or if more data become available. If the goal of the study is global, then local details may be unimportant; conversely, the more data available, the more statistically significant differentiation becomes possible.

2.3.2 Calculating Experimental Variograms

In probabilistic notation, the variogram is defined as the expected value (refer back to Eq. (2.11):

$$2\gamma(\mathbf{h}) = E\left\{[Z(\mathbf{u}) - Z(\mathbf{u}+\mathbf{h})]^2\right\}$$

The variogram is $2\gamma(\mathbf{h})$. The *semi*variogram is one-half of the variogram, that is, $\gamma(\mathbf{h})$. The word "variogram" is often used interchangeably for the semivariogram. Experimentally, the semivariogram for lag distance **h** is defined as the average squared difference of values separated approximately by **h**:

$$\hat{\gamma}(\mathbf{h}) = \frac{1}{2N(\mathbf{h})} \sum_{N(\mathbf{h})} \left[z(\mathbf{u}) - z(\mathbf{u} + \mathbf{h}) \right]^2 \qquad (2.12)$$

where $N(\mathbf{h})$ is the number of pairs for lag **h**. Some questions that must be addressed before calculating experimental variograms are as follows:

- Does the data variable require transformation or the removal of an obvious trend? See Section 2.4, Preliminary Mapping Concepts, which follows.
- Do we have the correct geological or stratigraphic coordinate system for locations **u** and distance vectors **h**? See below.
- What lag vectors **h** and associated tolerances should be considered? See below.

There are many different types of variogram-type measures of spatial variability that could be considered (Armstrong, 1984; Cressie and Hawkins, 1980; Deutsch and Journel, 1998; Omre, 1985; Srivastava, 1987b; Srivastava and Parker, 1989). The practice proposed here is use the traditional semivariogram with the correct data variable in the appropriate stratigraphic coordinate system. The theory behind kriging and simulation requires the use of either the covariance or the variogram. Some alternative variogram-type measures among the most robust to outlier values are only acceptable to help identify the range of correlation and anisotropy.

Establish the Correct Variable

Variogram calculation is preceded by selection of the Z variable to use in variogram calculation. The choice of the variable is evident in conventional kriging applications; however, data transformation is common in modern geostatistics. The use of Gaussian techniques requires a prior normal score transform of the data (see Section 2.4) and the variogram of those transformed data. Indicator techniques require an indicator coding of the data prior to variogram calculation (see Section 2.4).

An important consideration that precedes transformation is the volume scale of the data. It is common to block or scale-up the data to a scale more relevant for geological modeling. For example, the vertical resolution of many geological models is 0.5 m; thus the original data (often at a 0.1-m resolution) are scaled up to this larger scale. The most common approaches for different variables are: Take the most common facies, simply average porosity, and geometrically average permeability. Often the data are averaged over the entire 0.5-m thickness regardless of the small-scale facies. Care must be taken to avoid the inadvertent introduction of a bias. The variogram must be of the variable at the scale of modeling.

The correct variable also depends on how trends are going to be handled in subsequent model building. Often, clear areal or vertical trends are removed prior to geostatistical modeling and then added to geostatistical models of the residual (original value minus trend). If this two-step modeling procedure is being considered, then the variogram of the residual data is required. There is a risk, however, of introducing artifact structures in the definition of the trend and residual data.

Another aspect of choosing the correct variable is outlier detection and removal. Extreme high and low data values have a large influence on the variogram since each pair is squared in variogram calculation. Erroneous data should be removed. Of greater concern are legitimate high values that may mask the spatial structure of the majority of the data. Logarithmic or normal score transformation mitigates the effect of outliers, but are only suitable if an appropriate back transform is being considered in later geostatistical calculations.

Coordinate Transformation

The coordinate transformations presented in Section 3.2, Conceptual Model, are necessary before variogram calculation. In the presence of vertical wells, the vertical variogram does not depend on the stratigraphic coordinate transform, as long as calculations are limited to data within the appropriate stratigraphic layer and facies type. The horizontal variogram, however, is very sensitive to the stratigraphic z_{rel} coordinate transform. Attempting variogram calculation prior to such stratigraphic coordinate transformation can lead the modeler to the erroneous conclusion that the data have no horizontal correlation. Data sparsity may also lead to the same erroneous conclusion.

A characteristic feature of geological phenomena is spatial correlation. Coordinate transform errors, sparse data, errors in calculation parameters, and many other factors could lead to the wrong conclusion that there is no spatial correlation. A pure-nugget model should not be adopted.

When the geological formation has been extensively folded, a more elaborate coordinate transform that allows following the curvilinear structure may be required (Dagbert et al., 1984). The gOcad group has developed such elaborate structural unfolding schemes (Mallet, 1999, 2002).

Data and coordinate transformation are necessary prerequisites to variogram calculation and interpretation. Once the data are prepared for variogram calculation, it is necessary to choose the distance lags, **h** values, to consider.

Choosing Variogram Directions and Lag Distances

Anisotropy

Variograms are rarely isotropic; geologic continuity and variogram continuity are direction-dependent. In sedimentary structures, continuity in the vertical direction is typically less than that in the horizontal direction. Moreover, horizontal continuity depends on the direction of deposition and subsequent diagenetic alteration. Directions of continuity are most often known from geological interpretation or preliminary contouring of the data. Rarely is it a good idea to "search" for the principal directions of continuity by calculating the variogram in many directions.

A critical first step is to identify the "vertical" direction. This direction is perpendicular to the time-stratigraphic correlation and often has the least continuity. This should normally be the Z_{rel} direction calculated with the considerations outlined in Section 3.2, Conceptual Model. One complication is when there are clinoform structures within the stratigraphic layer (see Figure 2.46). In this case the "vertical" directions will be at some azimuth and dip from the vertical direction.

Anisotropy or directional continuity in geostatistical calculations is typically *geometric*. Three angles define orthogonal x, y, and z coordinates, and then the components of the distance vectors are scaled by three range parameters to determine the scalar distance, that is,

$$h = \sqrt{\left(\frac{h_x}{a_x}\right)^2 + \left(\frac{h_y}{a_y}\right)^2 + \left(\frac{h_z}{a_z}\right)^2}$$

where h_x, h_y, and h_z are the components of a vector **h** in 3-D coordinate space and a_x, a_y, and a_z are the scaling parameters in the principal directions. Contour lines of equal "distance" must follow ellipsoids. This concept of geometric anisotropy is covered in greater detail in the Variogram Modeling subsection.

The anisotropy in petroleum reservoirs is defined by a single angle that identifies the "major" and "minor" horizontal directions of continuity. The vertical direction is then assumed to be perpendicular to the horizontal direction.

Most often, sound geological principles should be used to build a depositional and diagenetic understanding of the formation; then, the major and minor directions of continuity are evident. In the case of ambiguity, the variogram may be calculated in a number of directions to observe directions of greater or lesser continuity, but there are rarely enough data to do so.

The *variogram map* takes the idea of calculating the variogram in a number of directions to its logical extreme. The variogram is calculated for a large number of directions and distances; then, the variogram values are posted on a map where the center of the map is the lag distance of zero (see Figure 2.47). Selection of the lag spacing or the size of the pixels on a variogram map is dictated by the same considerations as discussed below. The variogram

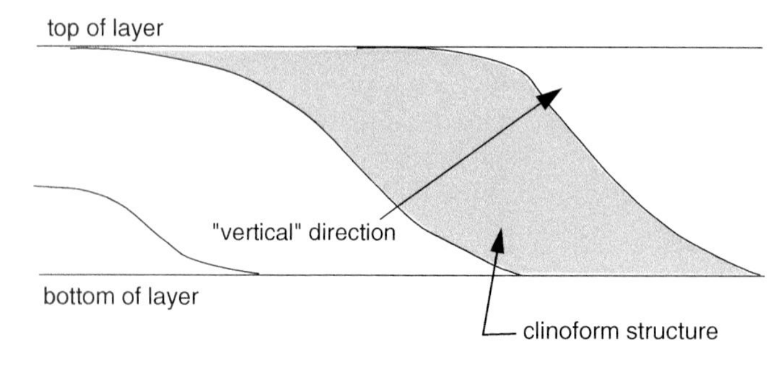

FIGURE 2.46: Flattened Stratigraphic Layer with Clinoform Structures. The geologic continuity follows the curvilinear directions within each clinoform.

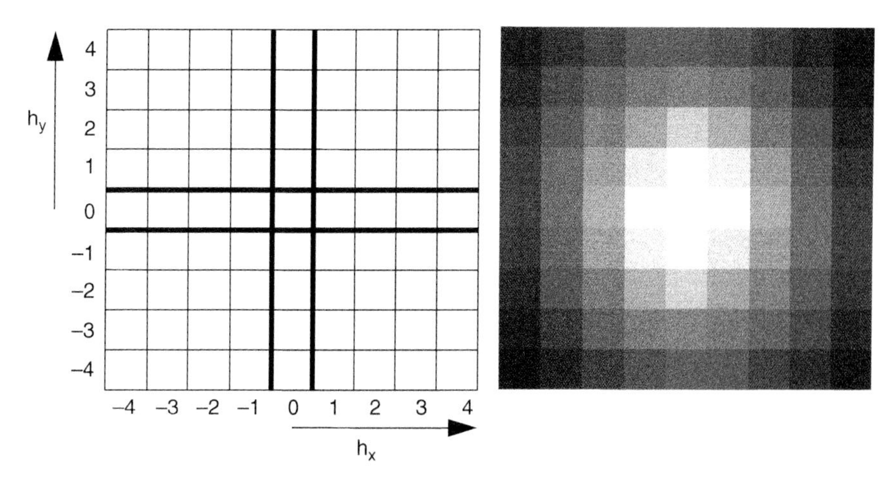

FIGURE 2.47: Illustration of a Variogram Map. A square lag template is shown to the left. The variogram is calculated for all $nx \cdot ny$ pairs and then plotted as a gray-scale map on the right (white is low and black is high values).

map is discussed extensively by Isaaks and Srivastava (1989). The `varmap` program in GSLIB (Deutsch and Journel, 1998) will calculate a variogram map with either gridded or scattered data. The main use of the variogram map is to detect the major and minor directions of continuity. Note that the variogram map will be very noisy and of little use in the presence of sparse data, which is precisely the case when the directions of anisotropy are poorly understood. Once the "vertical" and two "horizontal" directions are chosen, the next decision is the variogram lag distances to consider.

Lag Distance

Once the direction for variogram calculation has been chosen, it is necessary to choose the distance lags. This is not a problem with regularly gridded data; the spacing between the grid nodes in the direction of interest is the distance lag. In the presence of irregularly located data, the specification becomes more complex.

For data not on a regular grid, distance and direction tolerances must be permitted to have enough data; that is, $N(\mathbf{h})$ in Eq. (2.12) should be large enough to permit reliable variogram calculation. The vertical direction is considered separately since its distance scale is significantly different from the horizontal. At the end, the vertical and horizontal directions will be modeled together; however, calculation proceeds separately.

There is an important trade-off in the selection of tolerance parameters for both the vertical and horizontal directions. We want to make the number of data in each lag, $N(\mathbf{h})$, as large as possible so that the variogram is as reliable as possible. At the same time, we would like to restrict the tolerance parameters as much as possible so that the distance resolution and directional anisotropy are resolved in as much detail as possible. Inevitably, achieving the correct trade-off in any particular situation requires iterative refinement of the chosen tolerance parameters.

Vertical Lag Distance

Figure 2.48 shows the tolerance parameters for a vertical lag: a distance, h, a distance tolerance, h_{tol}, an angle tolerance, a_{tol}, and a bandwidth, b. Some guidance on the selection of these parameters:

- The lag separation distance h is usually chosen to coincide with the data spacing. Well log porosity values separated by 0.5 ft would naturally lead to a unit lag distance of $h = 0.5$ and multiples of 0.5. As mentioned above, this should be the blocked upscale of the data. It is not a good idea to resort to the original data because there are more values; the *right* variogram is that of the data being used in modeling at the correct scale.
- The distance tolerance, h_{tol}, is often chosen at one-half of the unit lag distance h. This could be reduced to, say, one-quarter of the unit lag distance when there are many data on a nearly regular grid. The tolerance, h_{tol}, could also be

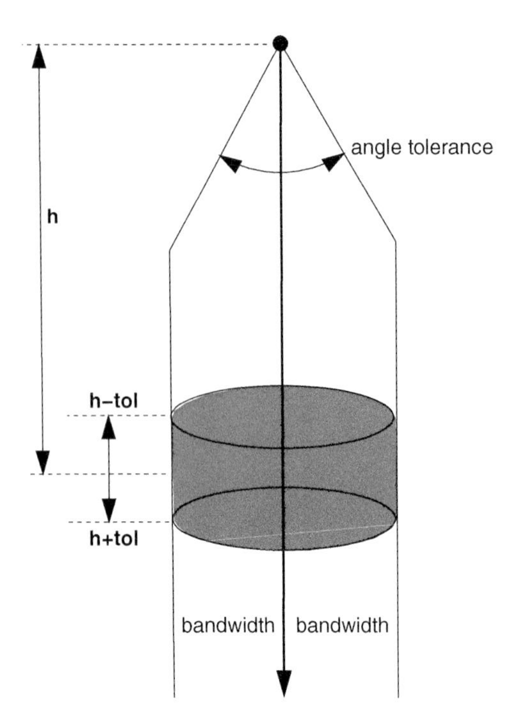

FIGURE 2.48: Illustration of a Vertical Lag Distance Specified by (1) a distance, h, (2) a distance tolerance, (3) an angle tolerance, and (4) a bandwidth.

increased to, say, three-quarters the unit lag distance in presence of few data. Increasing the tolerance beyond one-half the unit lag distance causes some data pairs to contribute to multiple lags; this is recommended when there are less than, say, 50 data pairs reporting to any given lag.

- The angle tolerance, a_{tol}, is needed when the wells are not truly vertical. A tolerance of 10 to 20 degrees is often used. Highly deviated and horizontal wells should not contribute to the calculation of the vertical variogram.
- A bandwidth parameter, b, is sometimes used to limit the maximum deviation from the vertical direction. The bandwidth could be considered with deviated wells. It is much more common to need a vertical bandwidth parameter in the calculation of horizontal variograms.

The number of distance lags must also be chosen. The number of lags, $n_{\mathbf{h}}$, is chosen so that the total distance, $n_{\mathbf{h}} \cdot h$ is about one-half of the reservoir size in the direction being considered. For example, 70 lags would be chosen with a lag separation distance of 0.5 ft in a reservoir 70 ft thick with data every 0.5 ft. Distance lags greater than one-half of the reservoir size do not permit data in the center of the reservoir to be used. The variogram becomes erratic and not representative of the entire reservoir layer. Moreover, the variogram for such long distances is typically not needed for later geostatistical modeling.

Horizontal Lag Distance

Figure 2.49 shows the tolerance parameters for a horizontal lag: a distance, h, a distance tolerance, h_{tol}, a horizontal angle tolerance, a^h_{tol}, a horizontal bandwidth, b_{hor}, a vertical angle tolerance, a^v_{tol}, and a vertical bandwidth, b_{ver}. The guidance given above for the distance, distance tolerance, and number of lags applies here. Additional considerations are as follows:

- In the presence of no horizontal anisotropy the parameter of the horizontal angle tolerance, a^h_{tol}, may be set to 90° or greater, which amounts to pool all horizontal directions. This yields an *omnidirectional* horizontal variogram.
- In the presence of horizontal anisotropy, the tolerance a^h_{tol} must be restricted as much as possible. Too small and there are too few pairs for reliable variogram calculation; too large and the result is a *blurred* picture of the anisotropy. Starting at 22.5° and performing a sensitivity study is recommended.
- The bandwidth parameter, b_{hor}, is sometimes used to limit the maximum deviation from the horizontal direction. It is set to a large value for omnidirectional variogram calculation or when there are few data. For directional variograms in presence of sufficient data, it can be set small, say, to 1 to 3 times the unit lag distance.
- The vertical angle tolerance, a^v_{tol}, should be set quite small due to the much larger variability in the vertical direction. Normally, a combination of a small-angle tolerance, say $a^v_{tol} = 5°$, and a vertical bandwidth, b_{ver}, set to a small value, say $b_{ver} = 2$ ft, is effective at limiting the calculation to data at approximately the same stratigraphic position.

As with the vertical variogram, the number of distance lags is chosen so that the total distance, $n_{\mathbf{h}} \cdot h$, is about one-half of the dimension of the area represented by the variogram. For example, 5 lags would

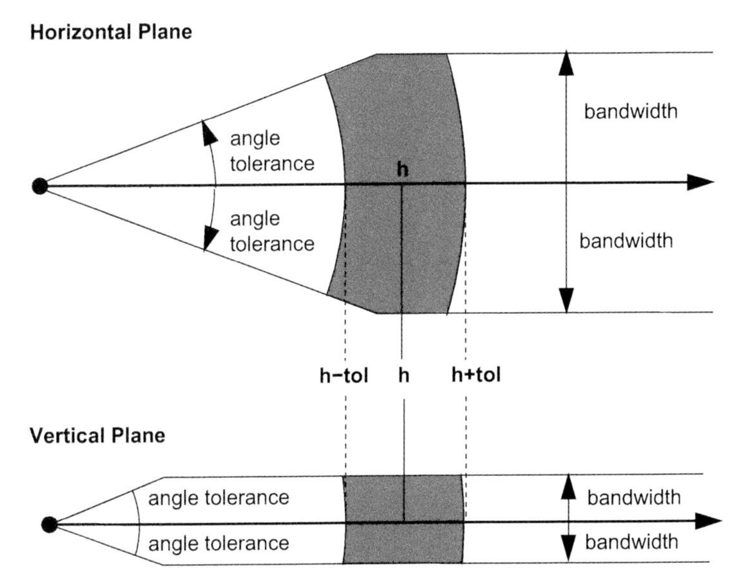

FIGURE 2.49: Illustration of a Horizontal Lag Distance Specified by (1) a Distance, **h**, (2) a Horizontal Distance Tolerance, (3) a Horizontal Angle Tolerance, (4) a Horizontal Bandwidth, (5) a Vertical Angle Tolerance, and (6) a Vertical Bandwidth.

be chosen with a lag separation distance of 1000 ft in a reservoir 10,000 ft in size with wells about every 1000 ft.

The data spacing in the horizontal direction is often more irregular than the vertical direction, making it more difficult to establish the unit lag separation distance. A "lag histogram" may be a useful tool to help choose the lag distances in the horizontal direction.

A lag histogram shows the number of pairs falling into different distance classes. Figure 2.50 shows example lag histograms for two horizontal directions. In this case, natural lag spacing for variogram calculation are indicated by the period of peaks in these histograms—for example, in the Y direction at 2000, 3000, 3800, 4500, 5600, and 7500 ft. Choosing a lag spacing of 1000 ft with a tolerance of 500 ft would result in a maximum of pairs in each lag. An arbitrary choice, usually too small, could lead to a noisy variogram.

2.3.3 Interpreting Experimental Variograms

Variogram interpretation is important. The calculated variogram points are not directly usable since (1) noisy results should be discounted, (2) geological interpretation should be used in the construction of the final variogram model, and (3) we need a licit variogram measure for *all* distances and directions. For these reasons, the experimental directional variogram must be understood and then modeled appropriately.

An apparently simple and widely misunderstood concept in variogram interpretation is the *sill*. In an ideal theoretical world, the sill is either the stationary infinite variance of the random function or the dispersion variance of data support volumes within the volume of the study area. In a naïve practical world, the sill is the where the variogram appears to flatten off. Reservoirs are not infinite and variograms do not flatten off where they are supposed to. In practical geostatistical reservoir modeling, the sill is the equal-weighted variance of the data entering variogram calculation (1.0 if the data are normal scores with *no* declustering weights).

There is no universal agreement among geostatisticians regarding this definition of the sill. Some would maintain that the sill is where the variogram plateaus; however, large-scale "nonstationarities" such as trends and zonal anisotropy cause the variogram points at large distances to be very erratic. Although there are more calculated variogram values at large distance than with the equal-weighted variance, the equal-weighted variance is always more reliable than trying to fit a plateau from experimental points. There is more discussion on this in Armstrong (1984), Cressie and Hawkins (1980), and Gringarten and Deutsch (1999).

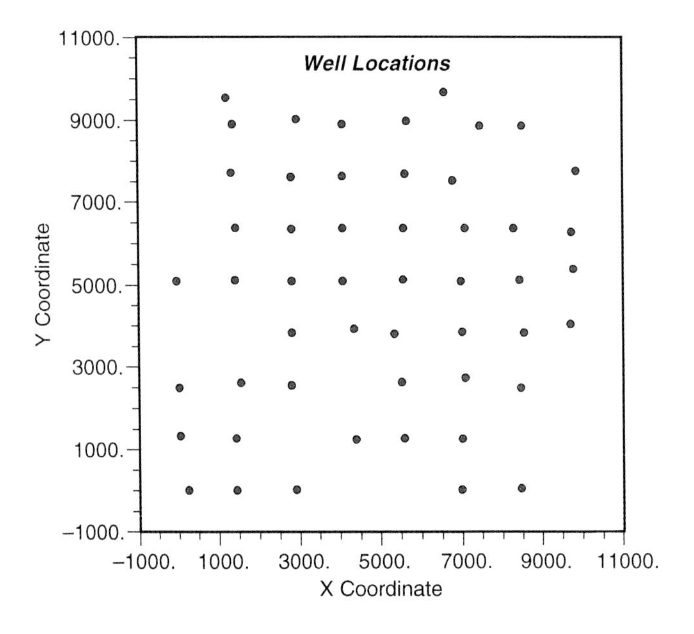

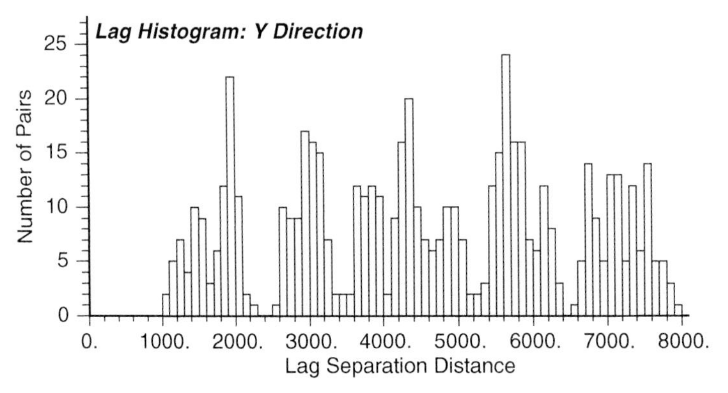

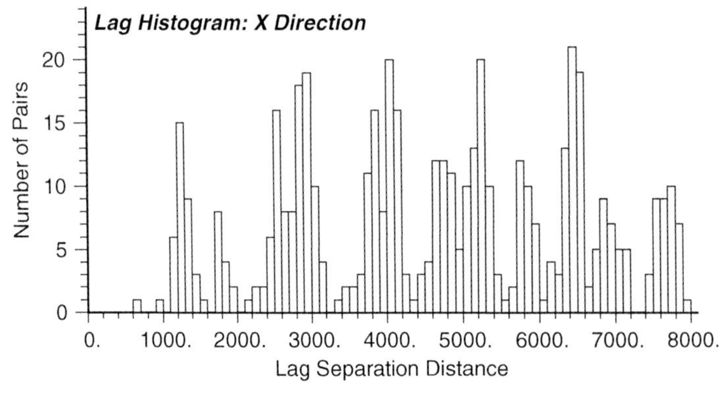

FIGURE 2.50: Location Map of 51 Wells and a Histogram of the Distance between Well Pairs in the X and Y Directions. A 45° angle tolerance was considered in both cases. These histograms can be used to select the lag distances and tolerances for variogram calculation. This case is quite straightforward.

The *range* is the distance at which the variogram reaches the sill. The range typically depends on direction, with horizontal directions showing greater continuity (larger ranges). At times, the continuity is so great that there is no apparent range. The range is merely one parameter of the variogram; the entire shape of the variogram is important.

Geological Variability

Variogram interpretation consists of explaining variability over different distance scales. The variogram is a chart of *variance versus distance* or *geological variability versus direction and Euclidean distance*. Different features are observed at different distance scales. The "nugget effect" is the behavior at distances less than the smallest experimental lag.

Nugget Effect

The nugget effect is the apparent discontinuity at the origin of the variogram. This discontinuity is the sum of measurement error and geological variability at scales smaller than the smallest experimental lag. This short-scale variogram structure is important for scale-up and flow studies.

Any error in measurement values or the location assigned to the data translates to a higher nugget effect. Sparse data may also lead to an apparently high nugget effect. A "real" geological nugget effect of more than 30% is unusual. In fact, with most petroleum data there is no nugget effect. This is because most variables in a sedimentary environment are locally continuous. Moreover, it is common that well log data measure a larger volume than the spacing that the measurements are reported; this leads to some smoothing in the data and an absence of a nugget.

Geometric Anisotropy

Most depositional processes impart spatial correlation to facies, porosity, and other petrophysical properties. The magnitude of spatial correlation decreases with separation distance until a distance at which no spatial correlation exists, the range of correlation. The length-scale or range of correlation depends on direction; typically, the vertical range of correlation is often much less than the horizontal range due to the larger lateral distance of deposition. Although the correlation range varies with direction, the nature of the decrease in correlation is often the same in different directions. The reasons for this similarity are the same reasons that underlie Walther's Law, that is, geological variations in the vertical direction are similar to those in the horizontal direction. This type of variogram behavior is called geometric anisotropy.

Cyclicity

Geological phenomena often occur repetitively over geologic time, leading to repetitive or cyclic variations in the facies and petrophysical properties. This imparts a cyclic behavior to the variogram; that is, the variogram will show positive correlation, followed by negative correlation, at the length-scale of the geologic cycles. These cyclic variations often dampen out over large distances because the size or length-scale of the geologic cycles is not perfectly constant. Such cyclicity is sometimes referred to as a hole effect, from the field of spectral analysis of the shape of the covariance.

Figure 2.51 shows a vertical semivariogram of permeability through a succession of fluvial channels. The high semivariogram value at a distance

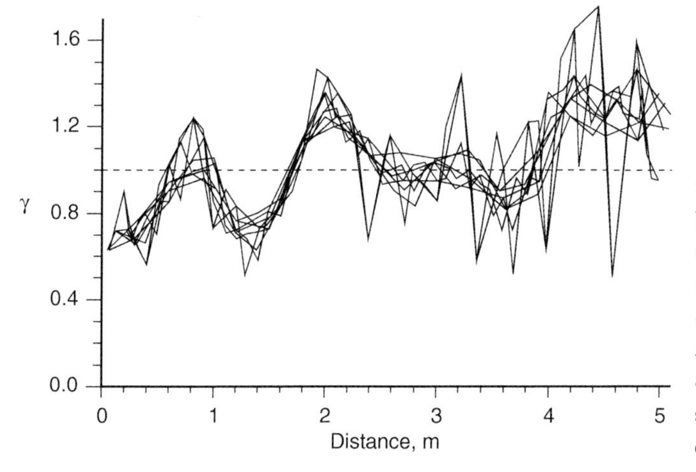

FIGURE 2.51: Semivariogram of Permeability Through an Alternating Succession of Fluvial Channels. The high semivariogram value at a distance of 0.8 m indicates a channel thickness of about 0.8 m. At a distance of 1.2 to 1.6 m the semivariogram, on average, is comparing values at similar stratigraphic positions in different channels.

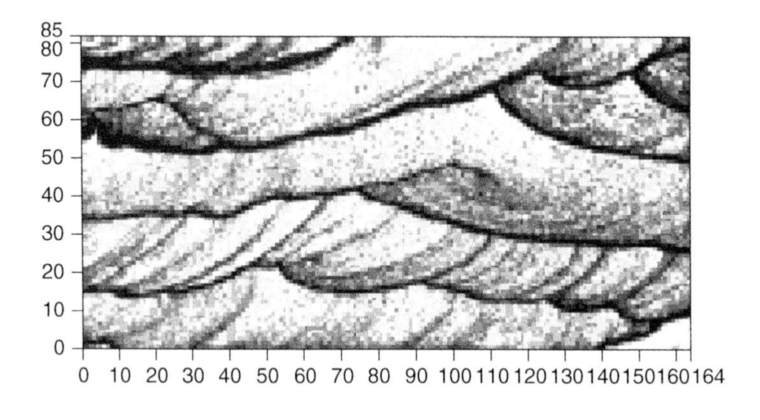

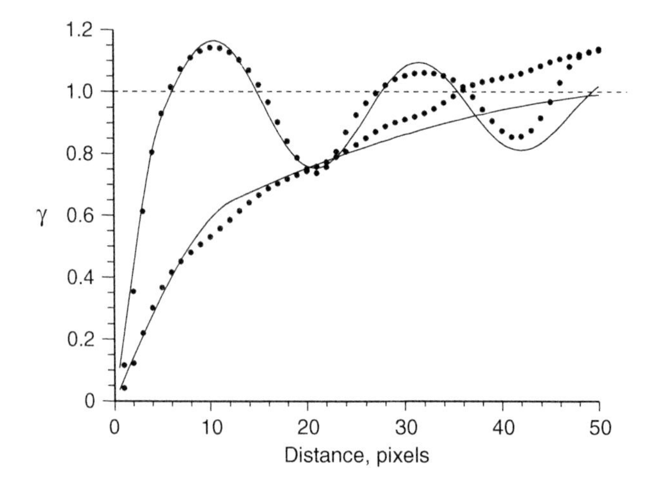

FIGURE 2.52: Gray-Scale Image of an Aeolian Sandstone and the Corresponding Vertical and Horizontal Semivariograms. The semivariogram was calculated on the normal score transform of the gray-scale level (finer-grained low-permeability sandstone appears darker). Note the cyclic behavior in the vertical direction.

of 0.8 m indicates an average channel thickness of about 0.8 m. At a distance of 1.2 to 1.6 m, the semivariogram, on average, is comparing values at similar stratigraphic positions in different channels.

Figure 2.52 shows a gray-scale image of an aeolian sandstone and the corresponding vertical and horizontal semivariograms. The semivariogram was calculated on the normal score transform of the gray-scale level (finer-grained low-permeability sandstone appears darker). Note the cyclic behavior in the vertical direction.

Large-Scale Trends

Virtually all geological processes impart a trend in the petrophysical property distribution—for example, fining or coarsening upward vertical trends or the systematic decrease in reservoir quality from proximal to distal portions of the depositional system. Such trends cause the variogram to show a negative correlation at large distances. In a fining upward sedimentary package, the high porosity at the base of the unit is negatively correlated with low porosity at the top. The large-scale negative correlation indicative of a geologic trend shows up as a variogram that increases beyond the sill variance $C(0) = \sigma^2$. As we will see later, it may be appropriate to remove systematic trends prior to geostatistical modeling. The trend is of no consequence if at scales smaller than the modeling cells or if there are many data. The trend would be reproduced in the final models by the conditioning data. Most practical reservoir modeling is done in the presence of sparse well data; therefore,

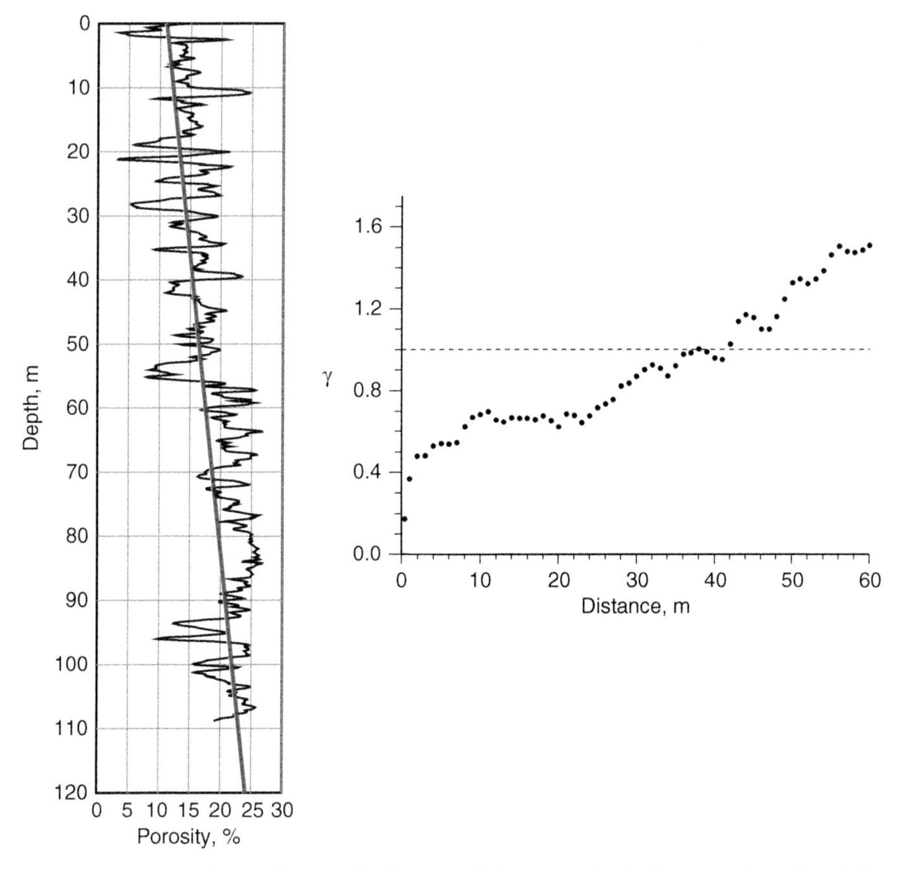

FIGURE 2.53: A Porosity Log (Note Scale) from a Deltaic Succession is Shown on the Left and the Corresponding Normal Scores Variogram Is Shown on the Right. The fining upward trend in porosity is revealed by the variogram points climbing above the expected sill of 1.0.

trend modeling is important. Then, we must infer the residual variogram which will typically show less structure than the variable with a trend.

Figure 2.53 shows an example. A porosity well log (note scale) from a deltaic succession is shown on the left, and the corresponding normal scores variogram is shown on the right. The fining upward trend in porosity is revealed by the variogram points climbing above the expected sill of 1.0. Semivariogram points above the sill imply a negative correlation at that distance, which is correct; a high porosity value at the bottom entails a low porosity at the top.

In some cases, a trend could be accounted for by subdividing the data differently—that is, reconsidering the decision of stationarity. There is a hint of this in Figure 2.53 where a subdivision about midway though the interval would separate the lower more variable values at the top from the higher more continuous values at the bottom.

Zonal Anisotropy

Zonal anisotropy is a limit case of geometric anisotropy where the range of correlation in one direction exceeds the field size, which leads to a directional variogram that appears not to reach the sill or variance. In practice, a zonal anisotropy is modeled as an extreme case of geometric anisotropy; however, there are two important special cases, which follow.

Areal Trends

Areal (horizontal) trends have an influence on the vertical semivariogram, in that the vertical semivariogram will not encounter the full variability of the petrophysical property. There will be positive correlation (semivariogram $\gamma(\mathbf{h})$ below the sill variance $C(0) = \sigma^2$) for large distances in the vertical direction. A schematic illustration of this is given in Figure 2.54. A vertical variogram for such a case is given in Figure 2.55.

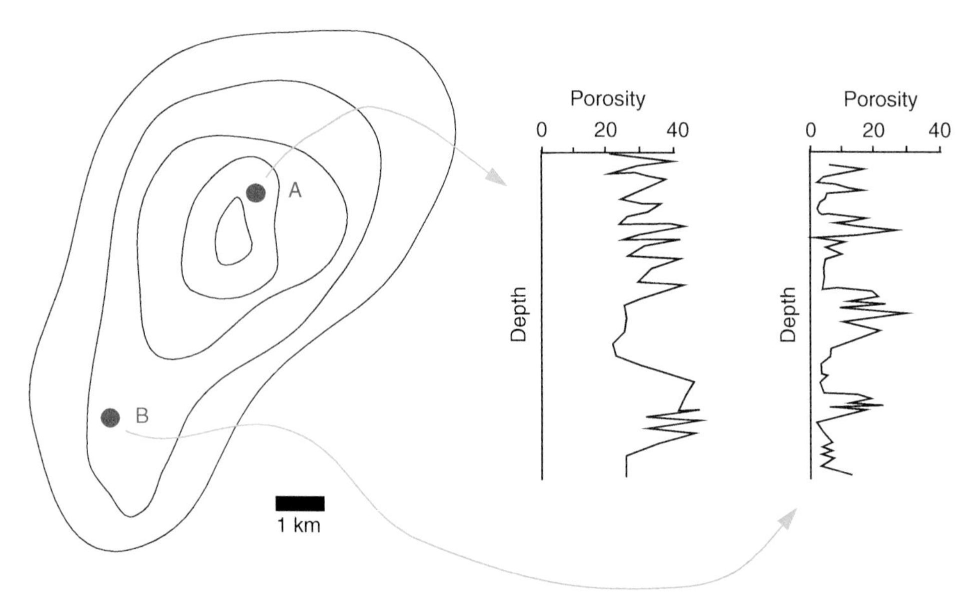

FIGURE 2.54: In the Presence of Areal Trends (Illustrated at the Left), Each Well Will Not "See" the Full Range of Variability; That Is, Wells In the Higher-Valued Areas (for example, Well A) Encounter Mostly High Values, Whereas Wells in the Lower-Valued Areas (For Example, Well B) Encounter Mostly Low Values. The vertical variogram in this case does not reach the total variability; that is, it shows a zonal anisotropy.

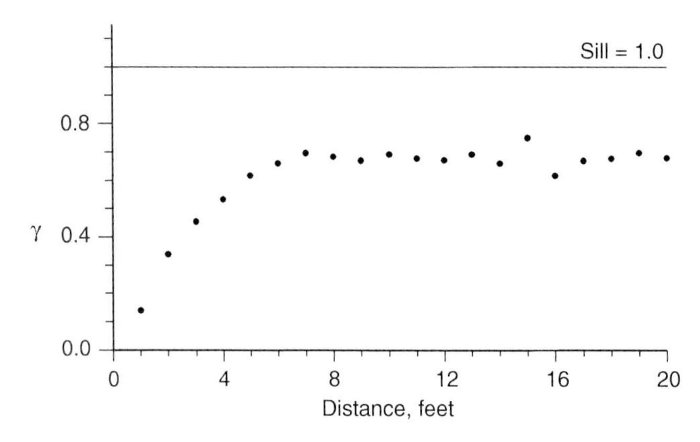

FIGURE 2.55: The Vertical Semivariogram of the Normal Scores of Porosity Showing a Zonal Anisotropy, That Is, the Semivariogram Points Flatten Off at a Variance Different than the Expected Variance. This corresponds to the schematic interpretation given on Figure 2.54.

Layering

There are often stratigraphic layer-like features or vertical trends that persist over the entire areal extent of the reservoir—even within the subdivided layers that we are considering. This leads to positive correlation (semivariogram $\gamma(\mathbf{h})$ below the sill variance $C(0) = \sigma^2$) for large horizontal distances. Although large-scale geological layers are handled explicitly in the modeling, there can exist layering and features at a smaller scale that cannot be handled conveniently by deterministic interpretation. This type of variogram behavior is also called zonal anisotropy because it is manifested when some directional variograms do not reach the expected sill variance.

Common Problems

The single biggest problem in variogram interpretation is a lack of data to calculate a reliable sample or experimental variogram; there is too little to interpret. Too few data for reliable variogram interpretation does not mean that there is no variogram; it is just hidden or masked by data paucity. The use of analog data and familiarity with other reservoirs of similar depositional setting is critical to make up for too few data.

Clustered data are often present in anomalous areas—for example, regions of high reservoir quality. Such data can cause the first variogram lags to be too high or too low, which could lead to a misinterpretation of the variogram structure. The most common misinterpretations would be (1) a too-high nugget effect, and (2) a short-scale cyclic characteristic. Solutions to this problem include (1) removing the clustered data, (2) discrediting unusually high variogram points for the short distances, or (3) using more robust two-point statistics such as the sample correlogram rather than the variogram. The sample correlogram is defined as

$$\rho(\mathbf{h}) = \frac{C(\mathbf{h})}{\sqrt{\sigma \cdot \sigma_{\mathbf{h}}}}$$

where

$$C(\mathbf{h}) = \frac{1}{N(\mathbf{h})} \sum_{N(\mathbf{h})} \left[z(\mathbf{u}) \cdot z(\mathbf{u}+\mathbf{h}) \right] - m \cdot m_{\mathbf{h}}$$

$$m = \frac{1}{N(\mathbf{h})} \sum_{N(\mathbf{h})} z(\mathbf{u}), \qquad m_{\mathbf{h}} = \frac{1}{N(\mathbf{h})} \sum_{N(\mathbf{h})} z(\mathbf{u}+\mathbf{h})$$

$$\sigma = \frac{1}{N(\mathbf{h})} \sum_{N(\mathbf{h})} \left[z(\mathbf{u}) - m \right]^2$$

$$\sigma_{\mathbf{h}} = \frac{1}{N(\mathbf{h})} \sum_{N(\mathbf{h})} \left[z(\mathbf{u}+\mathbf{h}) - m_{\mathbf{h}} \right]^2$$

This measure is robust because of the use of lag-specific mean and variance values.

Outlier values may cause the variogram to be noisy and difficult to interpret. A lag scatterplot—that is, a cross plot of the $(Z(\mathbf{u}), Z(\mathbf{u}+\mathbf{h}))$ pairs—sometimes reveals problem data that can be removed from the data set for variogram calculation. Figure 2.45 shows three **h**-scatterplots corresponding to three different lag values.

Removing Trends

As mentioned above, the first important step in all geostatistical modeling exercises is to establish the correct variable to model and to make sure (inasmuch as possible) that this property can be modeled as stationary over the domain of the study. Indeed, if the data show a systematic trend, this trend must be modeled and removed before variogram modeling and geostatistical simulation. Variogram analysis and all subsequent estimations or simulations are then performed on the residuals. The trend is added back to the estimated or simulated values at the end of the study.

There are problems associated with defining a reasonable trend model and removing the deterministic portion of the trend; however, it is essential to consider deterministic features such as large-scale trends deterministically. The presence of a significant trend makes the variable nonstationary; in particular, it is unreasonable to expect the mean value to be independent of location. Residuals even from some simple trend model are easier to consider stationary.

Trends in the data can be identified from the experimental variogram, which keeps increasing above the theoretical sill (see earlier discussion). In simple terms, this means that as distances between data pairs increase, the differences between data values also systematically increase.

To illustrate the above, consider the porosity data shown in Figure 2.56, which exhibit a clear trend in the porosity profile along a well. Porosity increases with depth due to a fining-upwards of the sand sequence. The (normal-score) variogram corresponding to this porosity data is shown at the upper right of Figure 2.56. It shows a systematic increase significantly above the theoretical sill of 1[3]. A linear trend was fitted to the porosity profile and then removed from the data. The resulting residuals constitute the new property of interest, and their profile is shown at the lower left of Figure 2.56. The (normal score) variogram of the residuals, shown on the right, now exhibits a clearer structure reaching the theoretical sill of 1 at about 7 distance units.

Examples

Experimental variograms almost always reflect a combination of these different behaviors. Considering the three images and their associated variograms presented in Figure 2.57, we see evidence of all the behaviors: nugget effect, geometric anisotropy, and zonal anisotropy, a vertical trend in the middle example: and cyclicity most pronounced on the bottom image.

Figure 2.58 shows eight vertical normal score variograms of well log porosity from eight different layers in a reservoir. The variogram for layer 4

[3] One could fit a "power" or "fractal" variogram model to the experimental variogram at the upper right of Figure 2.56; however, since these models do not have a sill value, they cannot be used in simulation algorithms such as sequential Gaussian simulation.

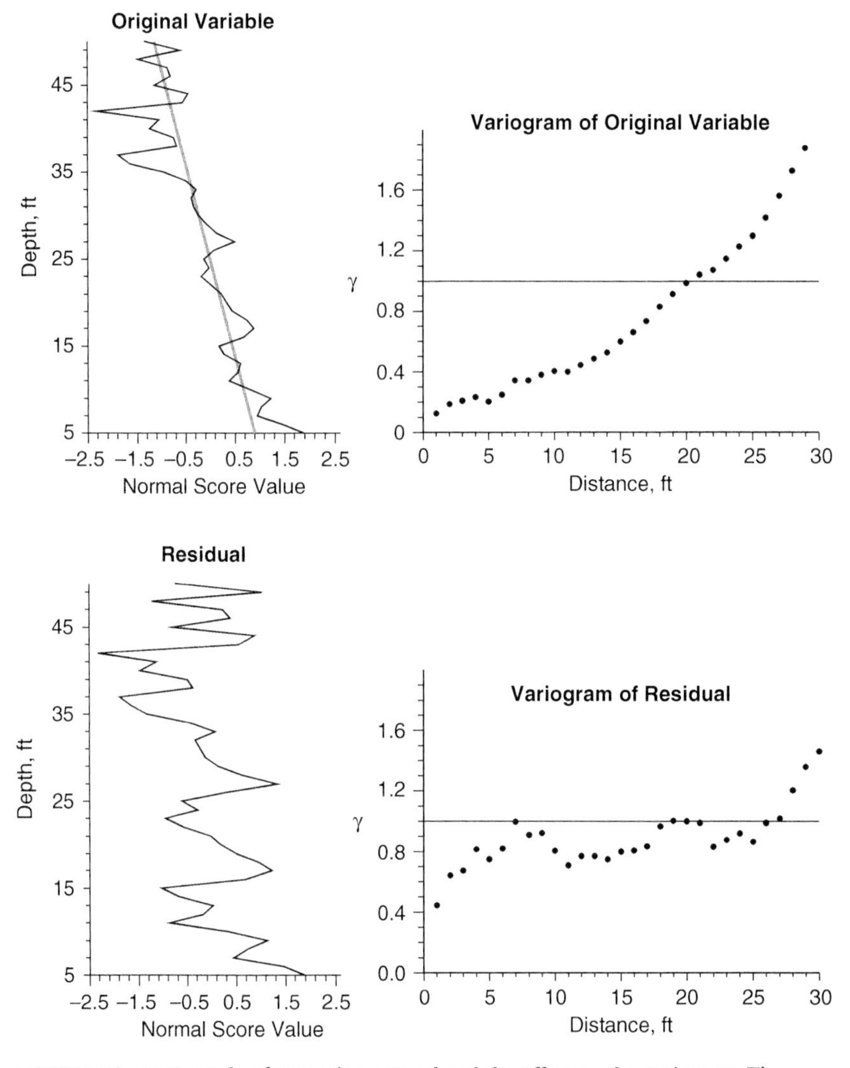

FIGURE 2.56: An Example of Removing a Trend and the Effect on the Variogram. The upper left shows the normal score transform of the original porosity values; the upper right figure shows the variogram of these data, which clearly shows the trend; the lower left shows the normal score transform of the residual values (from the linear trend shown on the upper left): and the figure on the lower right shows the variogram of the residual values.

appears *classical* in the sense that it shows a low nugget effect and rises to the theoretical sill of 1.0. All other cases show combinations of cyclicity (layers 1 and 8), vertical trend (layers 2 and 7), and areal trends (layers 3, 5, and 6). Figure 2.59 shows well log data from 3 layers illustrating cyclicity, vertical trend, and an areal trend. More examples are shown in the subsection on variogram modeling.

2.3.4 Horizontal Variograms

Geostatistical reservoir modeling faces a unique problem. Most wells (particularly exploration wells) are vertical. This makes it straightforward to infer the vertical variogram, but difficult to infer a reliable horizontal variogram. Given the overwhelming noise content in sample horizontal variograms, one evident error is to adopt a pure nugget model, which appears to closely fit the experimental variogram values. This is a convenient, but unrealistic, alternative. Our goal is to infer the best parameters for the underlying phenomenon; it is not to obtain a best fit to unreliable experimental statistics. We must consider secondary information in the form of horizontal wells, seismic data, conceptual geological models,

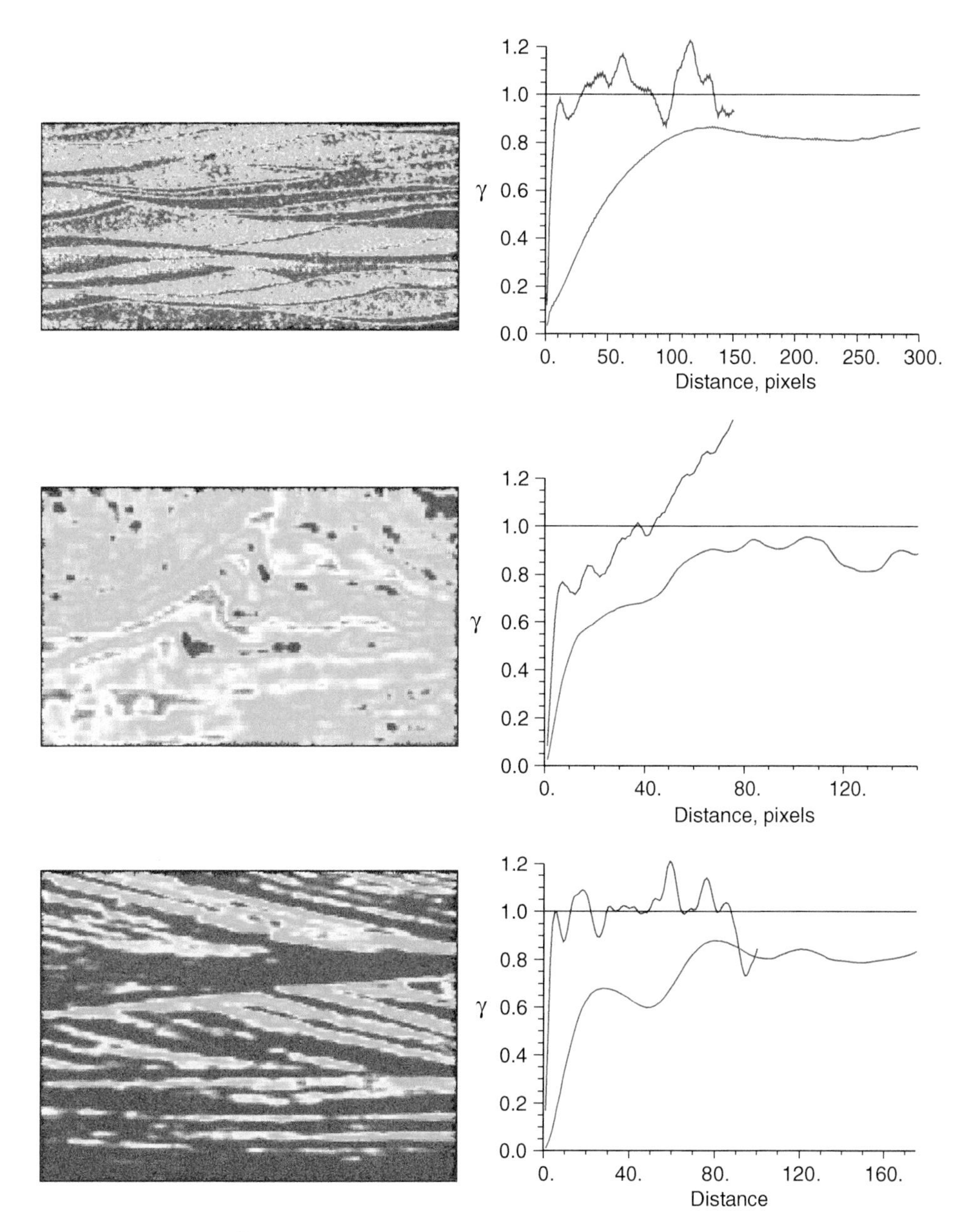

FIGURE 2.57: Three Different Geological Images with the Corresponding Directional Variograms. Note the cyclicity, trends, geometric anisotropy, and zonal anisotropy. The top image is an example of migrating ripples in a man-made aeolian sandstone (from the U.S. Wind Tunnel Laboratory), the central image is an example of convoluted and deformed laminations from a fluvial environment. The original core photograph was taken from page 131 of Sandstone Depositional Environments (Scholle and Spearing, 1982). The bottom image is a real example of large-scale cross laminations from a deltaic environment. The original photograph was copied from page 162 of Sandstone Depositional Environments (Scholle and Spearing, 1982).

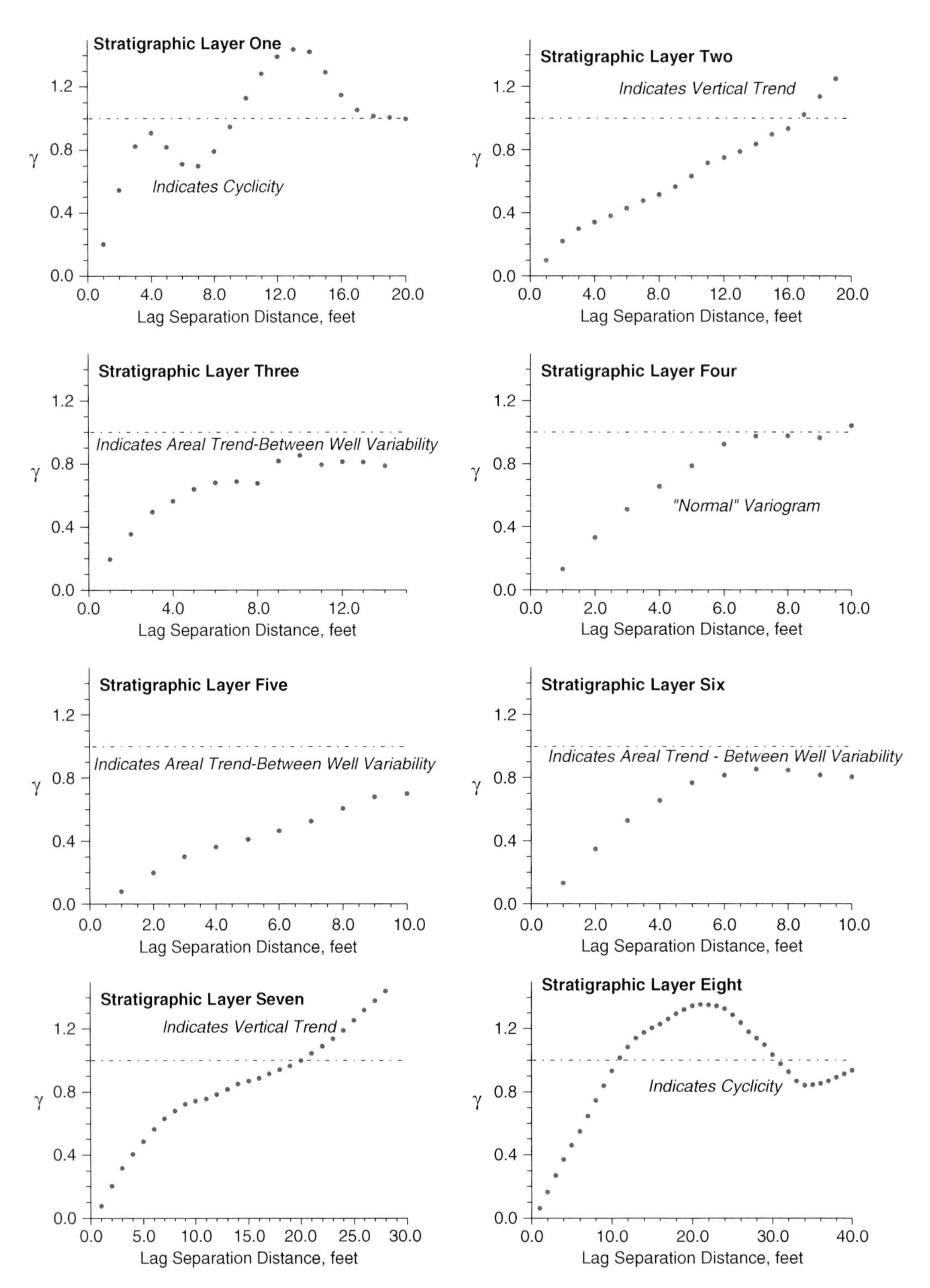

FIGURE 2.58: Eight Variograms, of the Normal Score Transform of Porosity, for Eight Different Stratigraphic Layers within the Same Reservoir.

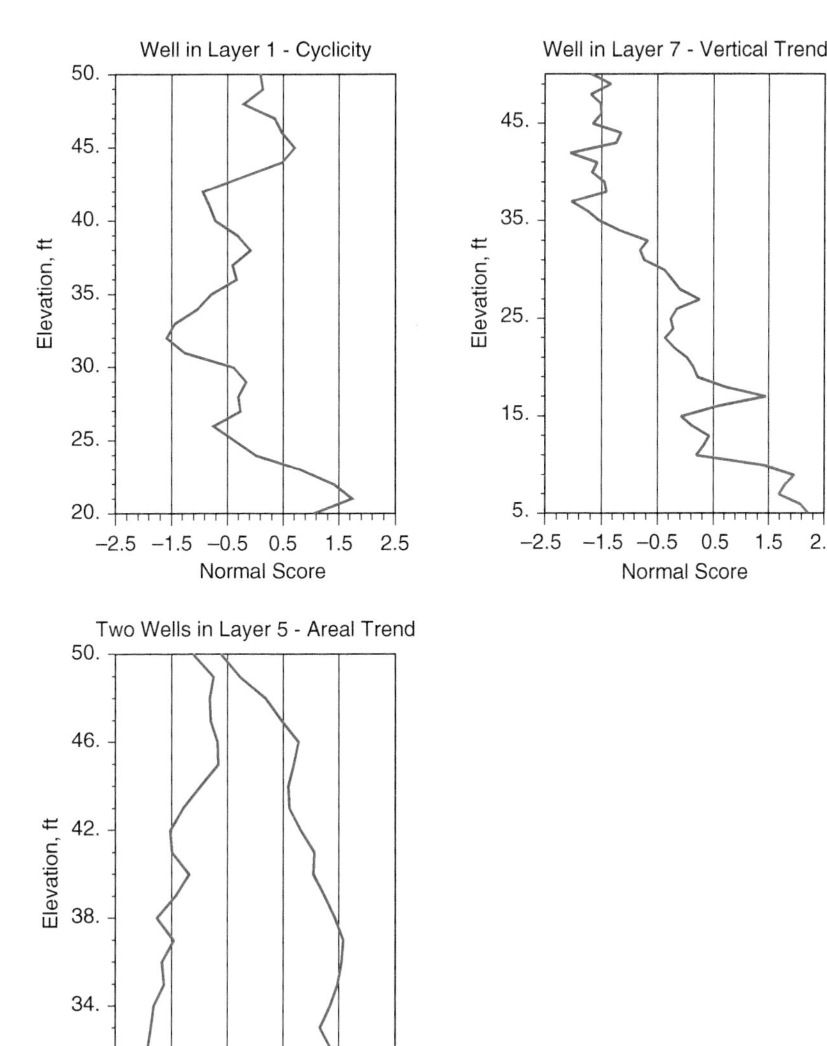

FIGURE 2.59: Well Log Data from Three Layers Illustrating Cyclicity, Vertical Trend, and an Areal Trend.

and analog data. In all cases, however, expert judgment is needed to integrate global information from analog data with sparse local data.

In the presence of sparse horizontal data, there are two critical steps to synthesize a horizontal variogram:

1. Determine what fraction of the variance may be explained by zonal anisotropy, that is, stratification that leads to persistent positive correlation in the horizontal direction.
2. Establish the horizontal-to-vertical anisotropy ratio based on secondary data; the synthetic horizontal variogram consists of the zonal anisotropy (step one) and the scaled vertical variogram.

Figure 2.60 shows a schematic illustration of these two parameters. Both the zonal anisotropy contribution and the horizontal-to-vertical anisotropy ratio will have considerable uncertainty. A sensitivity study should be performed in subsequent geostatistical simulation.

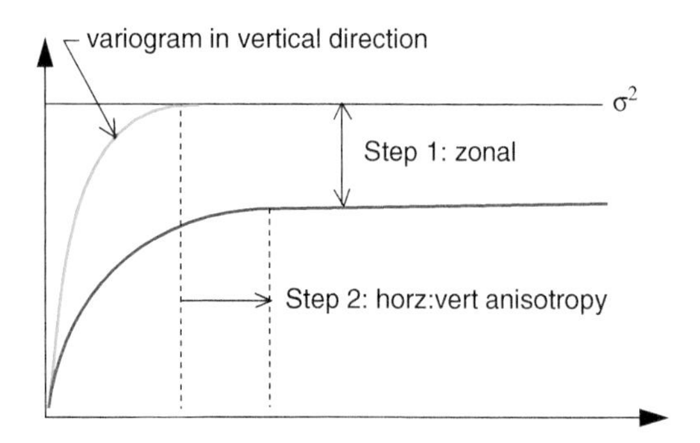

FIGURE 2.60: Schematic Illustration of the Two Decisions Required for Horizontal Variogram Inference in the Presence of Sparse Data: (1) the Amount of Zonal Anisotropy and (2) the Horizontal-to-Vertical Anisotropy Ratio.

Zonal Anisotropy

There may be persistent stratification that would make the horizontal variogram to plateau different than the sill variance—that is, show zonal anisotropy. In the presence of sufficient data, this could be observed on directional variograms. In the presence of sparse data, that zonal anisotropy must be inferred either from the available well data or from a conceptual geological model.

A *conceptual geological model* is essentially an expert opinion or judgment regarding the spatial variability. No reservoir stands in complete isolation. There may be reservoirs in the same sedimentary basin, reservoirs of similar type in different parts of the world, or modern analogs that give some indication of patterns of spatial variation. The "zonal anisotropy" considered here is a specification of the fraction of the variability explained by reservoir-wide stratification. Normally, this is between 0% and 30% of the total variability.

The *available data* may be too few and too widely spaced to calculate a reliable horizontal variogram. Nevertheless, zonal anisotropy consists of reservoir-wide patterns of variation and may be observed with as few as two wells. A horizontal **h**-scatterplot of values paired at the same stratigraphic vertical coordinate can be constructed; the correlation coefficient is an estimate of the relative magnitude of the zonal anisotropy; a large correlation coefficient indicates a large zonal anisotropy. A correlation coefficient of zero would indicate no zonal anisotropy. This value cannot be used blindly. Outliers or insufficient data would be reason to discredit this approach. Figure 2.61 shows an example with two vertical wells.

Horizontal-to-Vertical Anisotropy Ratio

A horizontal variogram model could consist of the vertical variogram points scaled (1) to show the correct plateau, determined previously and (2) to show the correct distance scale. The distance scale is determined by establishing a horizontal-to-vertical anisotropy ratio. Secondary information such as horizontal wells, seismic data, conceptual geological models, and analog data are considered to establish the required anisotropy ratio.

The increasing popularity of horizontal wells has not significantly helped with horizontal variogram inference. Horizontal wells rarely track the stratigraphic "time lines" or stratigraphic coordinates described in Section 3.2, Conceptual Model. When wells are close to "stratigraphic" horizontal, experimental horizontal variograms should be calculated. Note that even small departures from horizontal can lead to drastically reduced correlation—that is, an underestimation of the horizontal continuity.

Seismic data do not directly measure the facies, porosity, or permeability variables being considered in variogram analysis; in particular, the vertical scale of observation is much larger. Nevertheless, the acoustic responses of seismic are related to these petrophysical properties. One could consider the seismic data as a nonlinear average of the underlying petrophysical variables with additional error. The volume-scale difference leads to predictable changes to the variogram behavior if the averaging is linear (see volume variance discussion latter in this section); the range of correlation is increased by the scale of averaging. This is very important for the vertical variogram since the vertical scale of averaging is often many times larger than the vertical range. The large-scale nature of seismic data is not as critical in the horizontal direction since the horizontal range of correlation is many times larger than the horizontal scale of averaging. The idea, then, is to use the horizontal range of correlation from seismic data.

The extensive lateral coverage of seismic data permits calculation of a clear and easy-to-interpret horizontal variogram and horizontal range of correlation. The horizontal-to-vertical anisotropy ratio is calculated from the horizontal seismic variogram and the vertical variogram for the variable being studied.

Conceptual geological models consist of an expert opinion regarding the depositional system and the consequent spatial correlation (see Section 3.2, Conceptual Model, for more complete discussion).

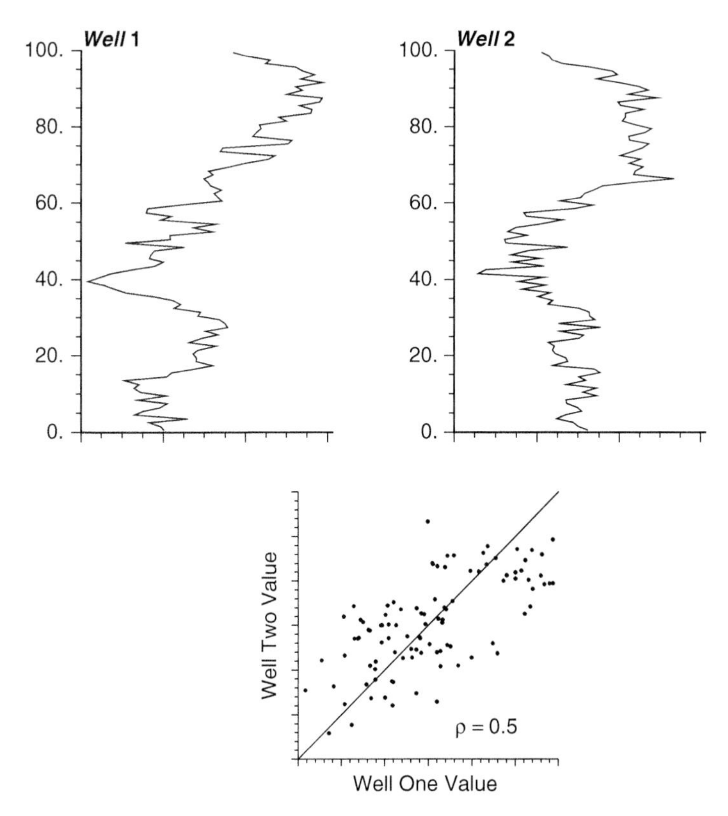

FIGURE 2.61: Porosity Values at Two Vertical Wells and a Cross Plot of the Matching (Same Stratigraphic Position) Porosity Values from One Well to the Other.

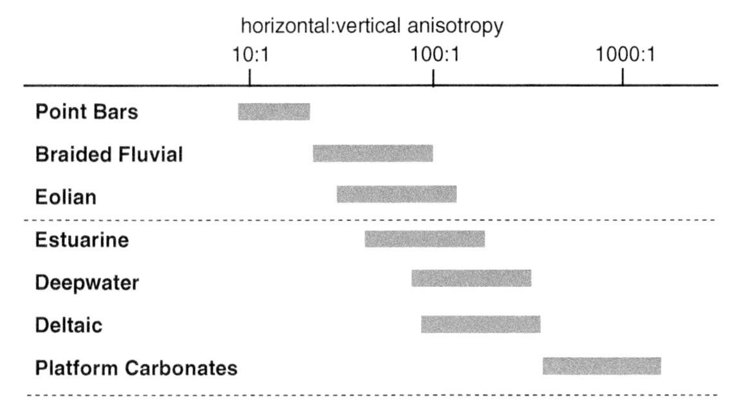

FIGURE 2.62: Some Typical Horizontal-to-Vertical Anisotropy Ratio *Conceptualized* from Available Literature and Experience. Such generalizations can be used to verify actual calculations and supplement very sparse data.

The following are some considerations for the application of conceptual models to aid in variogram inference. Such conceptual models are usually based on analog data, which may come from (1) other, more extensively sampled, reservoirs, (2) geological process simulation, or (3) outcrop measurements. In all cases, expert judgment is needed to integrate global information from analogs with sparse local data. There are published sources of such data—for example, the compilation of Kupfersberger and Deutsch (1999). Some typical anisotropy ranges by depositional setting are shown in Figure 2.62.

Geological-based process models simulate physical processes such as sediment transport, deposition, and erosion to characterize property variations over the time scale in which a reservoir evolved. Such process models are not yet viable reservoir modeling tools; however, we expect them to show the general character of reservoir heterogeneity (Davis et al., 1992; Webb, 1992). Essential features such as the existence of preferential flowpaths or sedimentological organization may be revealed (Anderson, 1991). Geological process models are constrained by the geological variables that create the stratigraphic record such as amount and type of sediment supply. Even if the observed location and geometry of facies units cannot be reproduced exactly, the general character may be appropriate for purposes of horizontal variogram inference (Ritzi and Dominic, 1993). They could be used to construct a "training image" from which horizontal variograms are extracted. This would require significant effort (Michael et al., 2010).

Experimental variograms are calculated. Analog data may be considered to supplement inadequate data. These variograms must be consistent with the geological model of the reservoir. Lastly, the variograms must be modeled.

2.3.5 Variogram Modeling

The experimental variogram points are not used directly in subsequent geostatistical steps such as kriging and simulation; a parametric variogram model is fitted to the experimental points. There are three reasons why experimental variograms must be modeled:

1. The variogram function is required for all distance and direction vectors **h** within the search neighborhood of subsequent geostatistical calculations; however, we only calculate the variogram for specific distance lags and directions (often, only along the principal directions of continuity). There is a need to interpolate the variogram function for **h** values where too few or no experimental data pairs are available. In particular, the variogram is often calculated in the horizontal and vertical directions, but geostatistical simulation programs require the variogram in off-diagonal directions where the distance vector simultaneously contains contributions from the horizontal and vertical directions.
2. There is a need to integrate additional analog geological knowledge that is not evident in the calculated experimental variograms (see the discussion above).
3. The variogram measure, $\gamma(\mathbf{h})$, must have the mathematical property of "positive definiteness" for the corresponding covariance model; that is, we must be able to use the variogram and its covariance counterpart in kriging and stochastic simulation. A positive definite model ensures that the kriging equations can be solved and that the kriging variance is positive.

For these reasons, geostatisticians have fit sample variograms with specific known positive definite functions like the spherical, exponential, Gaussian, and hole effect variogram models. It should be mentioned that any positive definite variogram function can be used, including tabulated variogram or covariance values (Yao and Journel, 1998) or even based on geometric offsets of a volume (Pyrcz and Deutsch, 2006b). The use of an arbitrary function or nonparametric table of variogram values would require a preliminary check to ensure positive definiteness (Myers, 1991). In general, the result will not be positive definite, and some iterative procedure is required to adjust the values until the requirement for positive definiteness is met.

Traditional parametric models permit all geological behaviors discussed above to be fit and allows straightforward transfer to existing geostatistical simulation codes.

A variogram model can be constructed as a positive sum of known positive definite licit variogram functions:

$$\gamma(\mathbf{h}) = \sum_{i=1}^{nst} C_i \Gamma_i(\mathbf{h}) \tag{2.13}$$

where $\gamma(\mathbf{h})$ is the variogram model for any distance and direction vector **h**, nst is the number of variogram functions or "nested structures," $C_i, i = 1, \ldots, Nst$, are the variance contributions for each nested structure, and $\Gamma_i, i = 1, \ldots, nst$, are the elementary licit variogram functions. The variance or "sill" of each elementary structure $\Gamma_i, i = 1, \ldots, nst$, is one; the sum of the variance contributions $\sum_{i=1}^{nst} C_i$ is then the variance implicit to the variogram model.

For the sake of simplicity and because there are often few well data, some would prefer to simply use one structure. This is rarely good practice because the presence of short-range and long-range structures may be observed from the vertical variogram alone. Also, more than one structure is required to account for zonal anisotropy that may also be identified with few well data.

In general, variograms are systematically fit to the theoretical sill. The commonly used licit variogram models for each nested structure will be presented, followed by some details about how to combine them together into a final variogram model.

Licit Variogram Models

Figure 2.63 shows six variogram models commonly used in practice. The nugget effect, spherical, exponential, and Gaussian models have a distinct plateau or sill. Each of these structures will be multiplied by a variance C_j value as shown in Eq. (2.13).

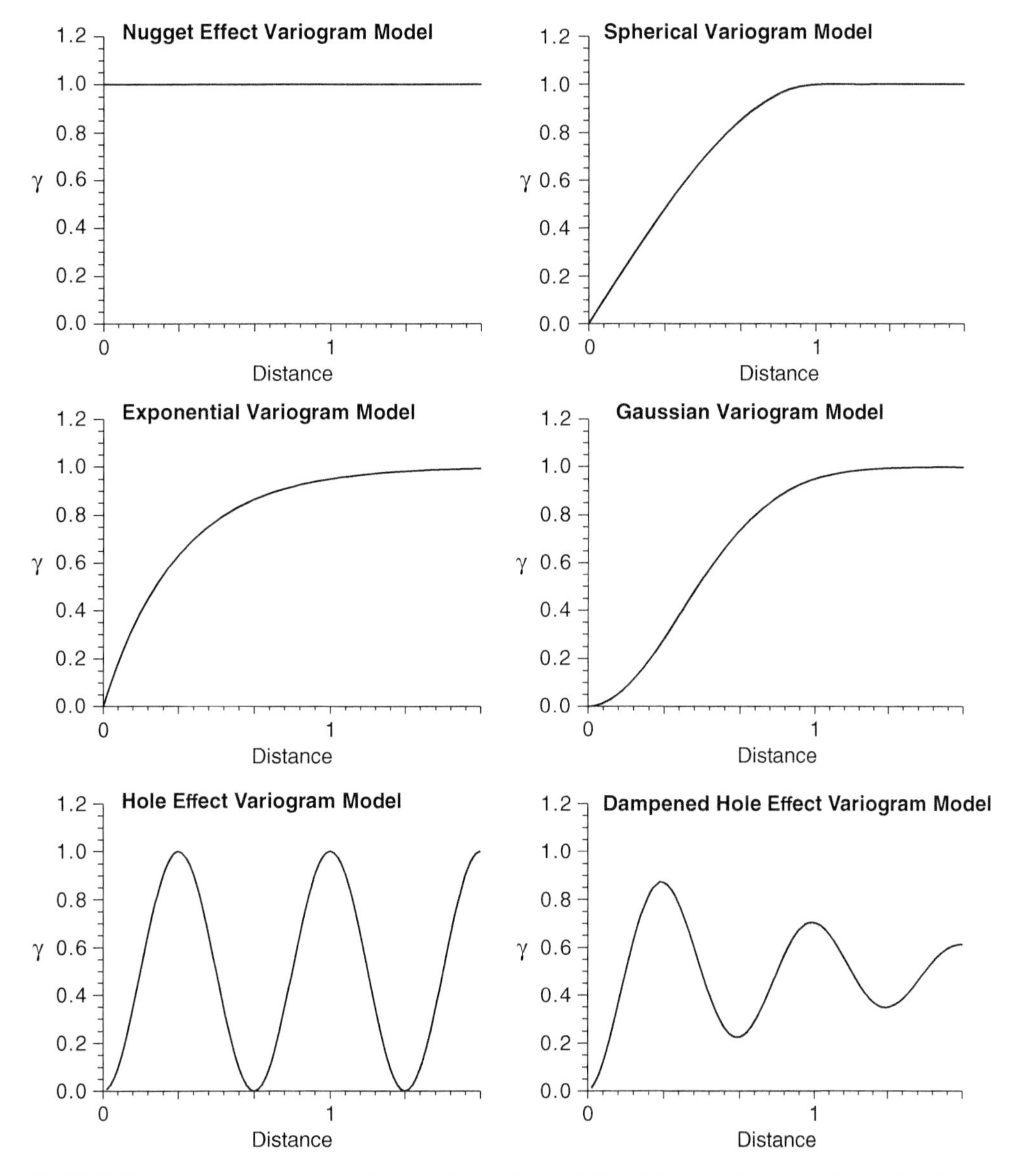

FIGURE 2.63: Six Commonly Used Variogram Models: Nugget Effect, Spherical, Exponential, Gaussian Model, Hole Effect Model, and Dampened Hole Effect Model.

Each variogram model has the distance h as a parameter. This scalar distance is calculated by decomposing the distance vector **h** into its three principal components, h_{vert}, $h_{h-major}$, and $h_{h-minor}$. By convention, *vert* is the vertical direction or that direction aligned with the stratigraphic Z_{rel} coordinate, *h-major* is the horizontal direction of greatest continuity aligned perpendicular to the Z_{rel} coordinate, and *h-minor* is the horizontal direction of least continuity aligned perpendicular to the Z_{rel} coordinate. Then, the distance is calculated as:

$$h = \sqrt{\left(\frac{h_{vert}}{a_{vert}}\right)^2 + \left(\frac{h_{h-major}}{a_{h-major}}\right)^2 + \left(\frac{h_{h-minor}}{a_{h-minor}}\right)^2} \quad (2.14)$$

In all generality, three angles are needed to define the three directions in Eq. (2.14); however, we usually only need one azimuth angle to define the *major* and *minor* directions since the vertical direction is defined stratigraphically. The distance *range* parameters, a_{vert}, $a_{h-major}$, and $a_{h-minor}$, may be different for each nested structure. They are calculated or iteratively adjusted so that all directional sample variograms are fit correctly. Considering all range distance parameters in the calculation of h [see Eq. (2.14)] amounts to scale h to be dimensionless with a range of 1. The variance contributions multiply the nested structures [see Eq. (2.13)]; therefore, the following nested structures also have a dimensionless sill value of 1:

Nugget effect: a constant of 1 except at h equal to zero, where the variance is zero:

$$\Gamma(h) = \begin{cases} 0 & \text{if } h = 0 \\ 1 & \text{if } h > 0 \end{cases}$$

The nugget effect is due to measurement error and geological small-scale structures—that is, features occurring at a scale smaller than the smallest data separation distance. Recall that the nugget should explain a small fraction of the total variability.

Spherical: a commonly encountered variogram shape, which increases in a linear fashion and then curves to sill of 1 at a distance of 1:

$$\Gamma(h) = Sph(h) = \begin{cases} \left[1.5h - 0.5h^3\right], & \text{if } h \leq 1 \\ 1 & \text{if } h \geq 1 \end{cases}$$

This variogram equation is related to the volume of intersection of two spheres separated by some distance. Any shape could be considered. The spherical variogram is commonly used because it rises linearly, which is common in practice; however, the slope is not as steep as the exponential variogram described below.

Exponential: a variogram shape similar to the spherical shape—that is, linear near the origin and curving toward a sill of 1. The main differences are that it rises more steeply than the spherical shape and reaches the sill value asymptotically:

$$\Gamma(h) = Exp(h) = 1 - e^{-3h}$$

Gaussian model: the squared term in the exponential causes this variogram structure to have a parabolic shape at short distances:

$$\Gamma(h) = Gau(h) = 1 - e^{-3h^2}$$

The implicit continuity at short distances is typical of continuous phenomena such as structural surfaces and thickness.

Hole effect model: a periodic shape:

$$H_a(h) = 1.0 - \cos(h \cdot \pi)$$

A hole effect variogram can act in one direction only, that is, two of the three range parameters are set to (essentially) infinity ∞. It is uncommon to use this variogram even in cases where the data show cyclicity.

Dampened hole effect model: the product of the exponential covariance and the hole effect model:

$$DH_{d,a}(h) = 1.0 - \exp\left(\frac{-3h \cdot a}{d}\right) \cdot \cos(h \cdot a)$$

Because geological periodicity is rarely at the same distance scale, the dampened hole effect model is more commonly used than the hole effect model. A dampened hole effect variogram can also act in one direction only.

There are many other variogram models. The *powerlaw* variogram model, $\Gamma(h) = h^{\omega}$ with $0 < \omega \leq 2$, is one example that is characteristic of trends or fractal-type behavior.

Summary of Variogram Modeling

Variogram modeling is facilitated by interactive software. Whether or not such software is available, the principles and procedures for variogram modeling are the same if the results are to be transferred to conventional geostatistical modeling software.

1. Determine the number of nested structures—that is, the nst of Eq. (2.13). This number should be as small as possible while retaining the flexibility to capture significant behaviors in different variogram directions.
2. For each nested structure, $i = 1, \ldots, \text{nst}$, choose (1) the variogram model type—that is, the Γ_i elementary model; (2) the variance contribution, C_i for this nested structure; and (3) the anisotropy parameters to define distance [Eq. (2.14)]—that is, horizontal azimuth rotation angle and a_{vert}, $a_{h\text{-major}}$, and $a_{h\text{-minor}}$.
3. Display the fitted model with the calculated points in all directions. Iteratively refine parameters as necessary to achieve a "good" fit. The short-scale behavior and significant anisotropy must be fit well for a good fit. Slight periodicity, minor zonal anisotropy, and long-range trends are often of little consequence.

The presentation of final variogram model fits in the following examples does not reveal the iterative nature of variogram fitting.

Examples

Figure 2.64 shows horizontal and vertical variograms from a Canadian reservoir. These variograms are indicator variograms standardized to a sill of 1.0. The

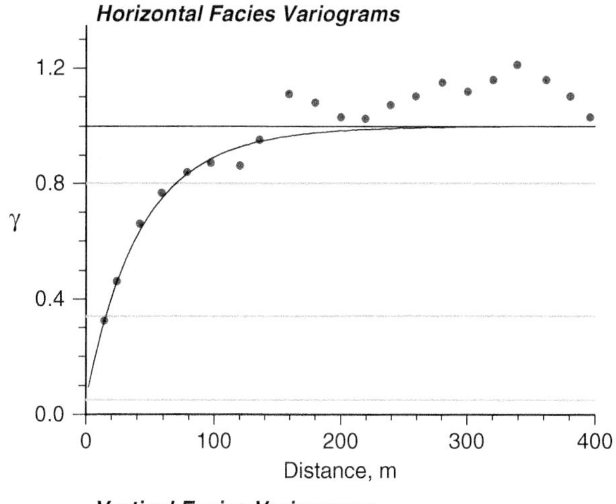

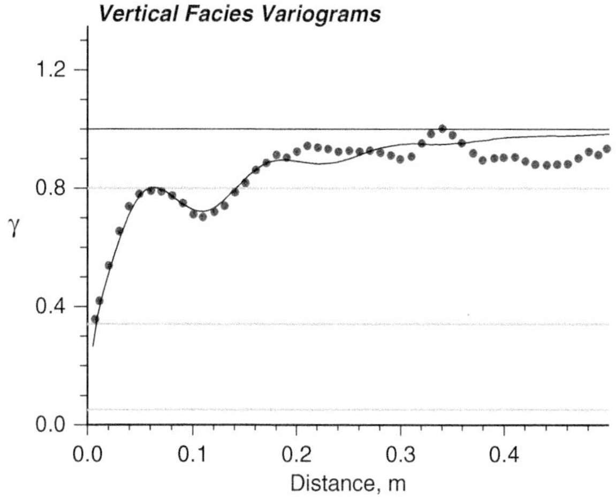

FIGURE 2.64: Horizontal and Vertical Indicator Variogram for Presence/Absence of Sand in a Canadian Reservoir. Note the excellent horizontal variogram that could be calculated with eight horizontal wells.

TABLE 2.6. VARIOGRAM PARAMETERS CORRESPONDING TO VARIOGRAM SHOWN IN FIGURE 2.64

Variance Contribution	Type of Variogram	Horizontal Range, m	Vertical Range, m
0.05	Nugget		
0.29	Exponential	100.0	0.015
0.46	Exponential	175.0	0.450
0.20	Exponential	100.0	0.500
0.20	Dampened hole effect	∞	0.060

horizontal distance scale is in meters; the vertical distance scale is relative to the thickness. Five variance regions can be defined for the horizontal and vertical variograms (see Table 2.6 and Figure 2.64). The first small component (5% of variance) is an isotropic nugget effect. The next three are exponential variogram structures with different range parameters. Three exponential structures are required to capture (a) the inflection point at a variance value of about 0.3 on the vertical variogram and (b) the long-range structure in the vertical variogram in the variance region 0.8 to 1.0. The fifth dampened hole effect variogram structure only applies in the vertical direction and adds no net contribution to the variance. The dampening component is five times the vertical range of 0.06. The dampened hole effect is largely aesthetic, since it would have little affect on subsequent geostatistical calculations.

Figure 2.65 shows a horizontal and vertical variogram of the normal score transform of porosity for a West Texas reservoir. This data are the "Amoco" data provided to Stanford University for testing and developing geostatistical methods. The horizontal variogram has a range of 5000 ft. The vertical variogram shows a clear additional variance component (zonal anisotropy), which is due to the systematic areal variations in average porosity. Three variance regions could be defined for the horizontal and vertical variogram of Figure 2.65 (see Table 2.7). The first two components are spherical variogram structures with geometric anisotropy. The last spherical structure captures the zonal anisotropy in the vertical direction.

Figures 2.66 and 2.67 show facies (presence of limestone coded as 1, dolomite and anhydrite coded as zero) and porosity variograms calculated from a major Saudi Arabian reservoir, see Benkendorfer et al. (1995) for the original variograms. Note the interpretable horizontal variograms and the consistent vertical and horizontal variograms. We also note the presence of a zonal anisotropy in the case of porosity, but not for the facies indicator variogram.

Two variance regions were identified for the facies variogram in Figure 2.66 (see Table 2.8). Note that the sill in this case is 0.24 (related to the relative proportion of limestone and dolomite). Both are anisotropic exponential structures. Three variance regions were identified for the porosity variogram in Figure 2.67 (see Table 2.9). The third region defines the zonal anisotropy in the vertical direction.

2.3.6 Cross Variograms

In virtually all reservoir modeling cases, there is a need to model the joint distribution of multiple variables—for example, seismic impedance, facies, porosity, permeability, and residual water saturation (Behrens, 1998). Conventional practice is to model them in a sequential fashion considering pairwise correlation:

1. Establish the *best* seismic attribute, possibly as a complex combination of the different seismic attributes,
2. Create a geostatistical model of facies considering the seismic data as a secondary variable,
3. Model porosity within each facies considering the seismic data as a secondary variable,
4. Model permeability within each facies considering porosity as a secondary variable.
5. Continue with other variables such as residual water saturation using the "best" secondary variable—that is, the one most correlated with the variable being modeled.

Geostatistical models have been devised for any number of variables; however, they are difficult to

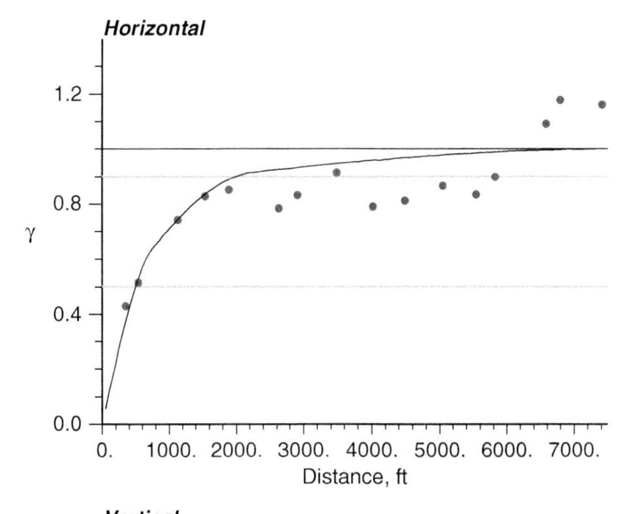

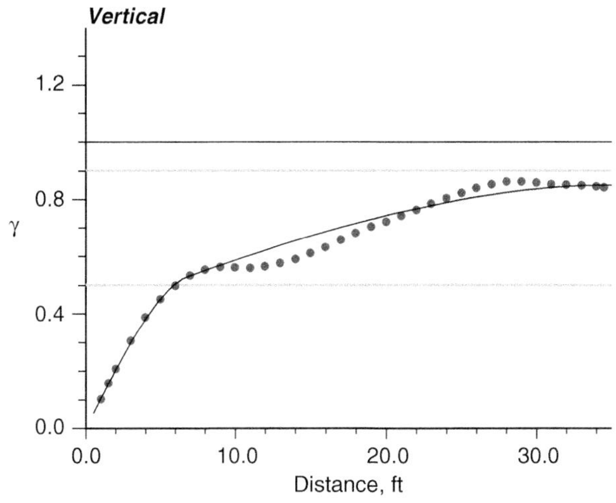

FIGURE 2.65: Horizontal and Vertical Variogram of Normal Score of Porosity for West Texas Reservoir. These data are the "Amoco" data provided to Stanford University for testing and developing geostatistical methods.

TABLE 2.7. VARIOGRAM PARAMETERS CORRESPONDING TO VARIOGRAM SHOWN IN FIGURE 2.65

Variance Contribution	Type of Variogram	Horizontal Range, m	Vertical Range, m
0.50	Spherical	750.0	6.0
0.40	Spherical	2000.0	50.0
0.10	Spherical	7000.0	∞

apply in practice. We only consider the spatial correlation between two variables Z and Y. Practical methods to integrate large numbers of variables is discussed in Section 4.1, Large-Scale Modeling.

The cross plot of collocated Z and Y values provides the first assessment of the correlation between the two variables. In general, the relationship between collocated or paired values $z(\mathbf{u}_i), y(\mathbf{u}_i), i = 1, \ldots, n$, does not suffice to capture the full spatial relationship between two variables. The cross-spatial relationship between pairs of value separated by some lag distance $z(\mathbf{u}_i), y(\mathbf{u}_i+\mathbf{h}), i = 1, \ldots, n(\mathbf{h})$, must also be considered. The cross variogram, defined below, measures this cross spatial dependence between two variables.

It is possible to avoid cross variograms. A model of cross-spatial relationship can be built from the correlation of collocated values. These collocated

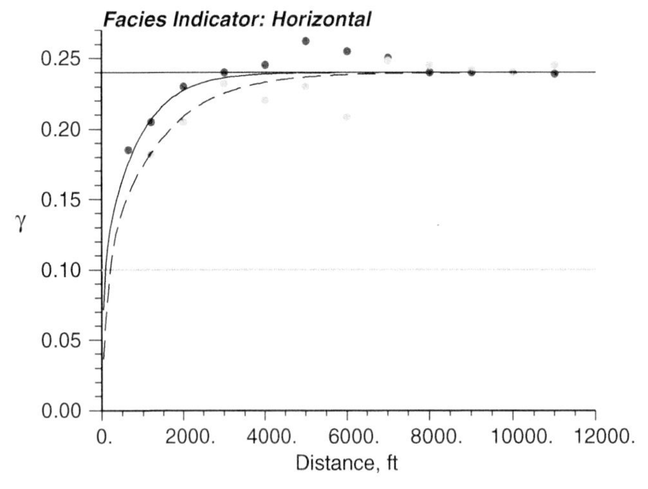

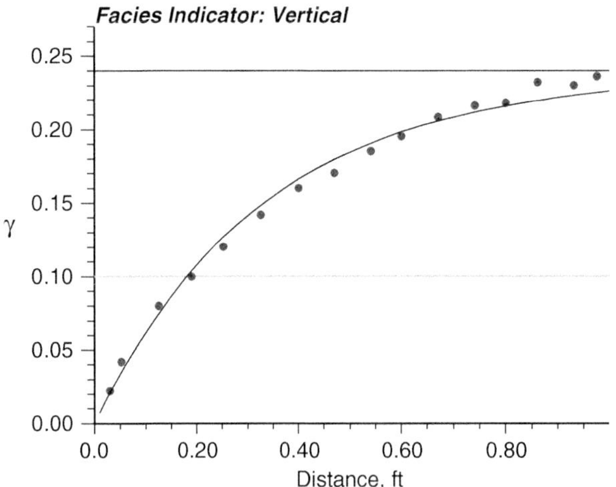

FIGURE 2.66: Horizontal and Vertical Facies Variograms from a "Major Arabian Reservoir." The top figure shows two directional horizontal variograms (see Table 2.8) and the bottom figure is the vertical variogram.

or Markov models (Zhu, 1991) have gained wide popularity because of their simplicity. The implicit assumptions underlying such models are reasonable in the presence of an exhaustively sampled secondary variable from either measurement (seismic) or prior geostatistical simulation. These models will be developed after the more general cross variogram is defined, interpretation principles are discussed, and modeling with the linear model of coregionalization is presented.

Recall the classical definition of the semivariogram for one variable (Z):

$$\gamma_{Z,Z}(\mathbf{h}) = \frac{1}{2}E\left\{\left[Z(\mathbf{u}) - Z(\mathbf{u}+\mathbf{h})\right]^2\right\}$$
$$= \frac{1}{2}E\left\{(Z(\mathbf{u}) - Z(\mathbf{u}+\mathbf{h}))\,(Z(\mathbf{u}) - Z(\mathbf{u}+\mathbf{h}))\right\}$$

This is extended to the cross semivariogram for two variables (Z and Y):

$$\gamma_{Z,Y}(\mathbf{h}) = \frac{1}{2}E\left\{(Z(\mathbf{u}) - Z(\mathbf{u}+\mathbf{h}))\,(Y(\mathbf{u}) - Y(\mathbf{u}+\mathbf{h}))\right\}$$

The variogram is the average product of the difference $z(\mathbf{u}) - z(\mathbf{u}+\mathbf{h})$ multiplied by itself. The cross variogram is the average product of the z difference, $z(\mathbf{u}) - z(\mathbf{u}+\mathbf{h})$, and the y-difference for the same location and lag distance, $y(\mathbf{u}) - y(\mathbf{u}+\mathbf{h})$. The cross variogram is a straightforward extension of the variogram.

To better understand the cross variogram, consider standardized variables (mean and variance are 0 and 1, respectively) and expand the squared term:

$$\gamma_{Z,Y}(\mathbf{h}) = \frac{1}{2}E\left\{Z(\mathbf{u})\cdot Y(\mathbf{u}) - Z(\mathbf{u}+\mathbf{h})\cdot Y(\mathbf{u}) - Z(\mathbf{u})\cdot Y(\mathbf{u}+\mathbf{h}) + Z(\mathbf{u}+\mathbf{h})\cdot Y(\mathbf{u}+\mathbf{h})\right\}$$

Recall the definition of the covariance and correlation coefficient for normalized variables from Section 2.2:

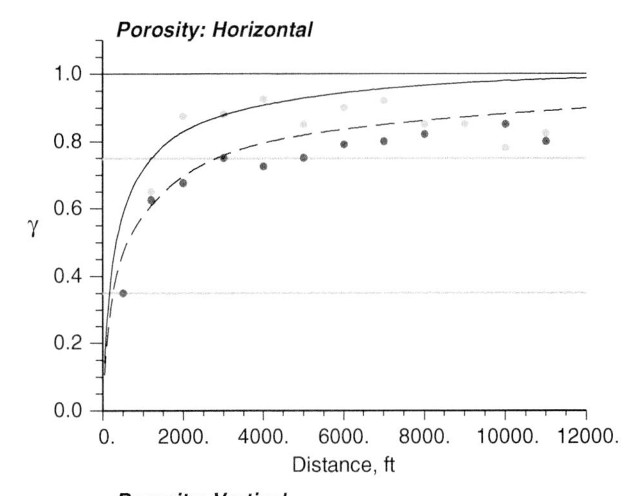

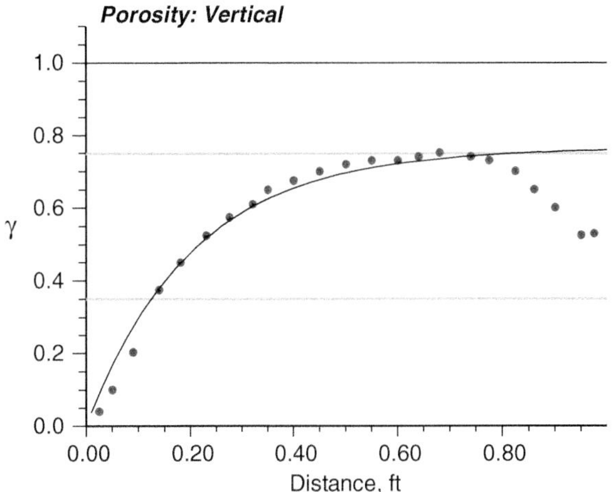

FIGURE 2.67: Horizontal and Vertical Porosity (Normal Score) Variograms from the Limestone Facies in a "Major Arabian Reservoir." The top figure shows two directional horizontal variograms (see Table 2.9) and the bottom figure is the vertical variogram. Note the zonal anisotropy in the vertical direction, which indicates the presence of areal trends.

TABLE 2.8. VARIOGRAM PARAMETERS CORRESPONDING TO VARIOGRAM SHOWN ON FIGURE 2.66

Variance Contribution	Type of Variogram	Horizontal Range NW-SE, ft	Horizontal Range NE-SW, ft	Vertical Range Relative
0.10	Exponential	150.0	400.0	0.8
0.14	Exponential	2500.0	4000.0	1.2

$$\begin{aligned}\gamma_{Z,Y}(\mathbf{h}) &= \frac{1}{2}\left[C_{ZY}(0) + C_{ZY}(0) - C_{ZY}(\mathbf{h}) - C_{YZ}(\mathbf{h})\right] \\ &= C_{ZY}(0) - \frac{1}{2}\left[C_{ZY}(\mathbf{h}) + C_{YZ}(\mathbf{h})\right] \\ &= \rho_{ZY}(0) - \frac{1}{2}\left[\rho_{ZY}(\mathbf{h}) + \rho_{YZ}(\mathbf{h})\right] \\ &= \rho_{ZY}(0) - \rho_{ZY}(\mathbf{h}) \qquad (2.15)\end{aligned}$$

The replacement of covariances with correlation coefficients—for example, $C_{ZY}(\mathbf{h})$ with $\rho_{ZY}(\mathbf{h})$—is possible only with standardized variables where $\sigma_Z^2 = \sigma_Y^2 = 1$. Setting $\rho_{ZY}(\mathbf{h}) = \rho_{YZ}(\mathbf{h})$ amounts to assume that the Z and Y covariances are "symmetric."[4]

[4] Nonsymmetric cross covariances are encountered when correlated Z and Y variables are somehow "offset" from one another. This sometimes happens with well logs. Fitting and using nonsymmetric cross covariances is not conveniently handled by any software; therefore, the data should be depth-shifted and modeled with conventional symmetric cross covariances or cross variograms.

TABLE 2.9. VARIOGRAM PARAMETERS CORRESPONDING TO VARIOGRAM SHOWN ON FIGURE 2.67

Variance Contribution	Type of Variogram	Horizontal Range NW-SE, ft	Horizontal Range NE-SW, ft	Vertical Range Relative
0.35	Exponential	400.0	500.0	0.6
0.40	Exponential	2000.0	4000.0	0.6
0.25	Exponential	12000.0	40000.0	∞

The cross semivariogram at $\mathbf{h} = 0$ is equal to 0, just like the semivariogram, because $\rho_{ZY}(\mathbf{h} = 0) = \rho_{ZY}(0)$ [see Eq. (2.15)]. At large lag distances the Z and Y variables often show no correlation; that is, $\rho_{ZY}(\mathbf{h})$ approaches zero. Therefore, the sill of the cross semivariogram is the correlation coefficient of collocated Z and Y values, $\rho_{ZY}(0)$ [see Eq. (2.15)]. This assumes that the Z and Y variables have been standardized. The sill of a semivariogram is the variance; the sill of a cross semivariogram is the covariance (correlation coefficient for standardized variables) at lag 0.

The covariance or correlation at lag 0 can be negative and so can the cross variogram. The relation in Eq. (2.15) can be used to interpret the cross-variogram. Like the variogram, values above the sill imply negative correlation and values below the sill imply positive correlation. In fact, all of the interpretation principles discussed above including geometric anisotropy, trends, cyclicity, and zonal anisotropy are applicable to cross variograms.

Example

As an example of correlated variables, consider 1396 measurements of % bitumen and % fines from an oil sands operation in Northern Alberta. These data are distributed in the 2-D horizontal plane over an area about 10 km square. % bitumen and % fines are the two critical factors affecting recovery of hydrocarbon in oil sands operations. Evaluating their spatial variability is important for process control in the extraction plant. Consider the normal score transforms of each variable. The cross plot of the collocated measurements is shown in Figure 2.68. Note that the correlation $\rho_{ZY}(0)$ is -0.73, which implies that the sill of the cross-variogram model should be about negative 0.73. This negative correlation is geologically reasonable; that is, the more fines there are, the less space there is for bitumen.

The variogram of the normal score transform of % bitumen is shown in Figure 2.69, the variogram of the normal score transform of % fines is shown

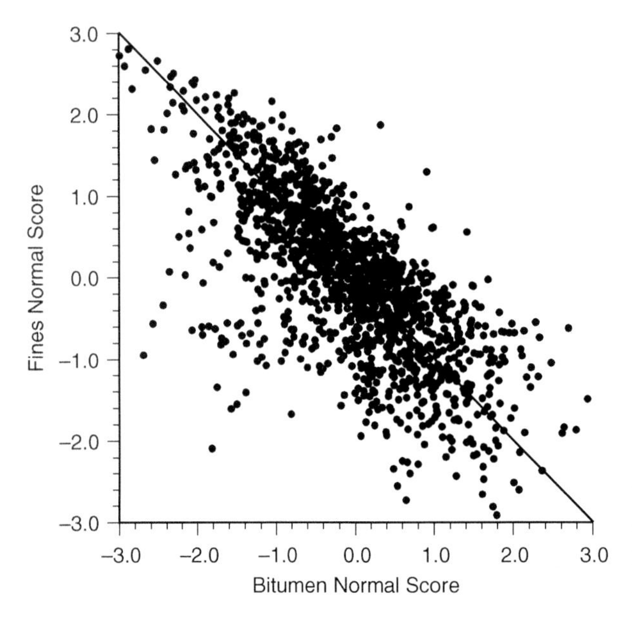

FIGURE 2.68: Cross Plot of % Fines (Normal Score Transform) Versus % Bitumen (Normal Score Transform). The correlation $\rho_{ZY}(0)$ is about −0.73.

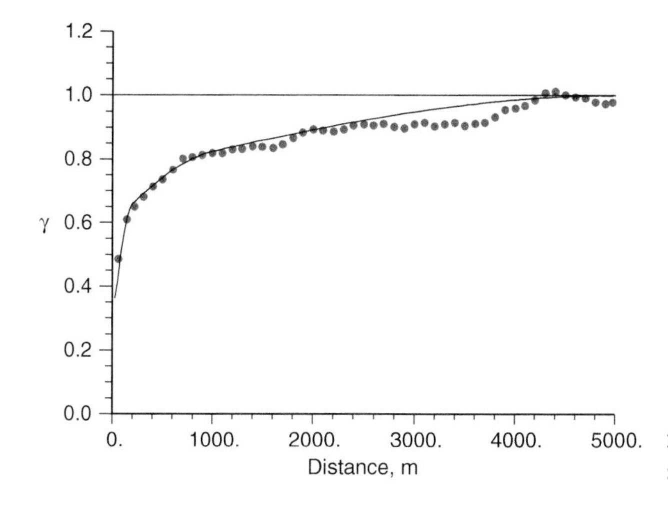

FIGURE 2.69: Variogram of % Bitumen (Normal Score Transform).

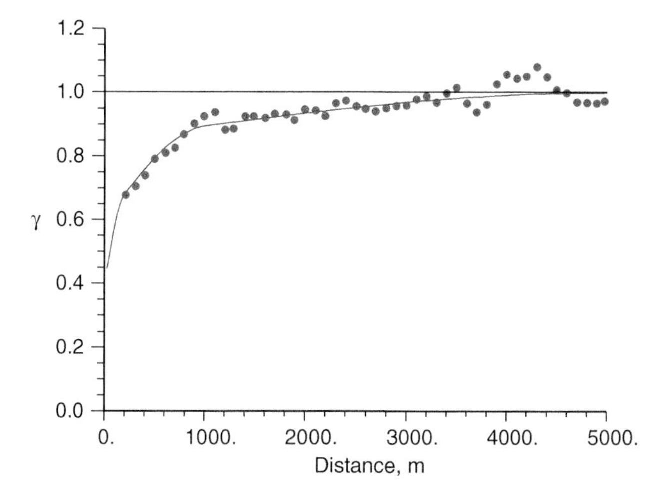

FIGURE 2.70: Variogram of % Fines (Normal Score Transform).

in Figure 2.70, and the cross variogram between the two is shown in Figure 2.71.

One cannot fit variograms independently since the corresponding (cross) covariances must be jointly positive definite. The Linear Model of Coregionalization (LMC) is used almost exclusively for modeling variograms of two or more variables. Other models of coregionalization related to the collocated cokriging and other such models are limit cases of this linear model of coregionalization. The linear model of coregionalization (LMC) takes the following form:

$$\gamma_{Z,Z}(\mathbf{h}) = b^0_{Z,Z} + b^1_{Z,Z} \cdot \Gamma^1(\mathbf{h}) + b^2_{Z,Z} \cdot \Gamma^2(\mathbf{h}) \ldots$$
$$\gamma_{Z,Y}(\mathbf{h}) = b^0_{Z,Y} + b^1_{Z,Y} \cdot \Gamma^1(\mathbf{h}) + b^2_{Z,Y} \cdot \Gamma^2(\mathbf{h}) \ldots$$
$$\gamma_{Y,Y}(\mathbf{h}) = b^0_{Y,Y} + b^1_{Y,Y} \cdot \Gamma^1(\mathbf{h}) + b^2_{Y,Y} \cdot \Gamma^2(\mathbf{h}) \ldots$$

where the $\Gamma^i, i = 1, \ldots, \text{nst}$ are nested structures made up of common variogram models, as presented earlier. So, the LMC amounts to model each direct and cross variogram with the same variogram nested structures. The sill parameters (the b-values) are allowed to change within the following constraints:

$$\left.\begin{array}{c} b^i_{Z,Z} > 0 \\ b^i_{Y,Y} > 0 \\ b^i_{Z,Z} \cdot b^i_{Y,Y} \geq b^i_{Z,Y} \cdot b^i_{Z,Y} \end{array}\right\} \quad \forall\, i$$

The first step in fitting a LMC is to choose the pool of nested structures: $\Gamma^i, i = 1, \ldots, \text{nst}$. Some considerations are as follows:

- Each nested structure is defined by a variogram type (shape) and anisotropic ranges (angles defining directions of anisotropy and respective ranges in each direction).
- The nested structures should be chosen so that all of the *deemed* significant features on the experimental variograms can be modeled.

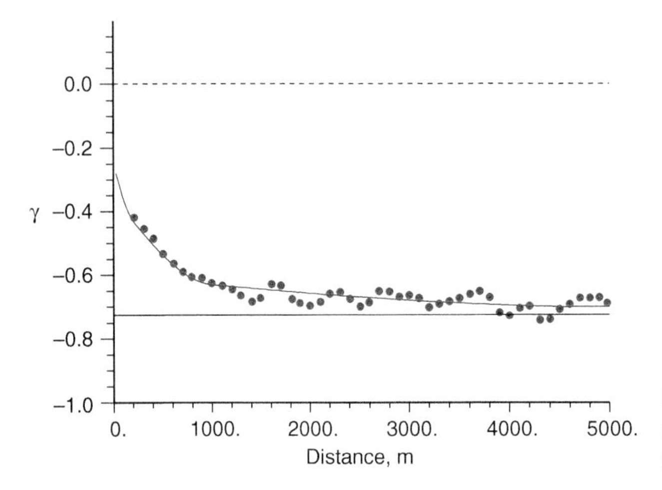

FIGURE 2.71: Cross Variogram of % Fines (Normal Score Transform) and % Bitumen (Normal Score Transform).

An automatic fitting algorithm could be considered, there are many softwares for this (Goovaerts, 1997).

The fitted variogram curves in Figures 2.69, 2.70, and 2.71 are from the following model:

$$\begin{aligned}\gamma_Z(\mathbf{h}) &= 0.3 + 0.3 \cdot \Gamma^1(\mathbf{h}) + 0.25 \cdot \Gamma^2(\mathbf{h}) \\ &\quad + 0.15 \cdot \Gamma^3(\mathbf{h}) \\ \gamma_{Z,Y}(\mathbf{h}) &= -0.25 - 0.1 \cdot \Gamma^1(\mathbf{h}) - 0.25 \cdot \Gamma^2(\mathbf{h}) \\ &\quad -0.1 \cdot \Gamma^3(\mathbf{h}) \\ \gamma_Y(\mathbf{h}) &= 0.4 + 0.2 \cdot \Gamma^1(\mathbf{h}) + 0.25 \cdot \Gamma^2(\mathbf{h}) \\ &\quad + 0.15 \cdot \Gamma^3(\mathbf{h})\end{aligned}$$

where $\Gamma^1(\mathbf{h})$ is spherical with range 200 m, $\Gamma^2(\mathbf{h})$ is spherical with range 1000 m, and $\Gamma^3(\mathbf{h})$ is spherical with range 5000 m. This is a licit model of coregionalization since $0.3 \cdot 0.4 \geq (-0.25)^2, 0.3 \cdot 0.2 \geq (-0.1)^2$, $0.25 \cdot 0.25 \geq (-0.25)^2$, and $0.15 \cdot 0.15 \geq (-0.1)^2$.

The linear model of coregionalization could be extended to three or more variables; however, that is rarely done in practice. As mentioned at the start of this section, variables are typically considered pairwise.

Markov Models of Coregionalization

Many consider the linear model of coregionalization cumbersome to apply. A simpler alternative consists of modeling the cross variogram with a single correlation coefficient parameter. This simplification is reasonable when the cross variogram cannot be fit reliably from the data. The simplification results from a Markov-type assumption whereby one data type prevails completely over collocated hard and soft data. The classical Markov model, sometimes called Markov Model I, assumes that hard Z data prevail over soft Y data. The cross variogram is given by

$$C_{ZY}(\mathbf{h}) = B \cdot C_Z(\mathbf{h}), \qquad \forall \mathbf{h}$$

where $B = \sqrt{\sigma_Z^2/\sigma_Y^2} \cdot \rho_{ZY}(0), \sigma_Z^2, \sigma_Y^2$ are the variances of Z and Y, and $\rho_{ZY}(0)$ is the correlation coefficient of collocated $z - y$ data. Application of this model of coregionalization requires the covariance of the Z variable $(\sigma_Z^2 - \gamma_Z(\mathbf{h}))$ and the correlation coefficient of collocated data.

A second Markov model, sometimes called Markov Model II (Journel, 1999; Shmaryan and Journel, 1999), assumes that soft Y data prevail over hard Z data. The cross variogram is given by

$$C_{ZY}(\mathbf{h}) = B \cdot C_Y(\mathbf{h}), \qquad \forall \mathbf{h}$$

where B is the same as above. Application of this model of coregionalization requires the covariance of the Y variable $(\sigma_Y^2 - \gamma_Y(\mathbf{h}))$ and the correlation coefficient of collocated data.

The full LMC gives greater flexibility and better results than either Markov model, but at the expense of greater effort. The choice of Markov model depends on whether the cross variogram looks more like the Z variogram or Y variogram. When the Y data are at a larger volume support—for example, coming from seismic—then the second Markov Model II is likely most appropriate.

There are implementation details to be addressed in the Markov approach including the *other* variogram; that is, the Y variogram if Markov Model I is used or the Z variogram if Markov Model II is used. There is no problem if a single collocated datum is retained for estimation. These and other application details are addressed in subsequent chapters.

2.3.7 Multiple-Point Statistics

The following provides motivation for extending our spatial statistics from the previously discussed two-point variogram, defines a multiple-point statistic, and provides methods to calculate them. Detailed discussion on sequential methods to build models with multiple-point statistics is presented in the next section, and implementation details and examples for modeling with multiple-point statistics are covered in Section 4.3, Multiple-Point Facies Modeling.

Motivation for Multiple-Point Statistics

Two-point statistics are unable to characterize curvilinear features and ordering relationships (with the exception of Truncated Gaussian Simulation discussed in Section 4.2, Cell-Based Facies Modeling). Curvilinear features are common in naturally occurring architectures. For example, consider sinuous fluvial and deepwater channels, dispersive patterns in distributary lobes, and the mounds associated with patch reefs. One may argue that linear features are likely the exception while more complicated curvilinear architectures are the norm. Two-point statistics are parameterized by range and direction and consider only two points at a time in each lag vector **h**; therefore, they are unable to characterize curvilinear features. The practical result of this limitation is that variogram-defined random functions have a general form that may be characterized as the product of isolated or amalgamated ellipsoids, resulting from linear features potentially with anisotropy.

Ordering relationships describe the transition probabilities from a specific category to another or from specific continuous value range to another. For example, consider natural transitions in channel axis to off-axis to margin and from upper shore face to lower shore face and to marine facies. Specific transitions in categorical properties are common.

While the variogram does not characterize curvilinear features and ordering relations, these features may be still imposed into the resulting models through conditioning and nonstationary statistics models. In mature reservoir settings, dense data conditioning may impose more complicated structure beyond those in the variogram. In dense data settings, this may be sufficient, yet the practitioner must be careful because this may result in unanticipated artifacts—for example, changes in heterogeneity away from data. In addition, any of the input statistics may be considered nonstationary and modeled over all locations. Methods to characterize nonstationary locally variable anisotropy models may be a sufficient work around (see Discussion in Section 2.4, Preliminary Mapping Concepts).

Definition of Multiple-Point Statistics

The following presents a definition of multiple-point statistics and associated illustration and nomenclature. Strebelle (2002) provides an introduction to the terminology and nomenclature for multiple point statistics, along with a thorough coverage of the development of the technology and previous work.

Consider an attribute Z with K possible categories, or continuous with $K-1$ thresholds z_K, $k=1,\dots,K$. A multiple point data event, d_n, may be characterized by an unknown value $z(u)$, and n known values separated from the unknown value location as characterized by lag vectors $\mathbf{h}_\alpha, \alpha=1,\dots,n$, with data values represented by $z(\mathbf{u}+\mathbf{h}_\alpha), \alpha=1,\dots,n$. The configuration of this data event is known as the data event template. While the schematic in Figure 2.72 indicates 4 known locations (a 5-point multiple-point statistic including the unknown location at **u**), in practice many points, possibly on the order of 10–40 points, are required in 3-D to characterize practical heterogeneities. With the assumption of stationarity previously described in this section, for the purpose of spatial inference, we need the conditional probability distribution function for the unknown value, given the data event $P(Z(\mathbf{u})=z_k|d_n)$.

Inference of Multiple-Point Statistics

At first glance this is a very difficult inference problem as described by Strebelle (2002). Firstly, consider the number of conditional probabilities that must be calculated. For a case with K categories

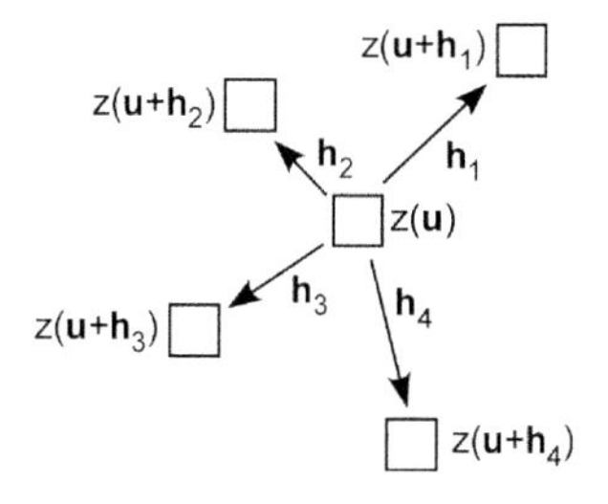

FIGURE 2.72: Illustration of a Multiple-Point Event and Associated Nomenclature.

and n data event locations in the template, there are K^{n+1} conditional probabilities. As the number of categories exceeds a couple and/or the number of data event locations exceeds 20, the number of possible conditional probabilities becomes impractically large (e.g., the minimum number of categories, $k = 2$, and a very small template in 3-D, $n + 1 = 20$, results in over 1 million possible conditional probabilities already). Secondly, consider the difficulty in inference of just the two-point statistic for the variogram with the typically sparse sampling density of the available data. There is often a paucity of experimental pairs for specific directions and distances due to limited available data spacing and orientations. For multiple-point statistics, real data will not likely inform a significant proportion of all possible multiple-point configurations. This former issue is addressed by (a) limiting the number of categories and data events in the template, (b) search trees for efficient storage and retrieval of conditional probabilities (Strebelle, 2002), and (c) multiple grid and hierarchical methods with a sequence of binary categories the latter is addressed by borrowing statistics from a training image (Guardiano and Srivastava, 1993) (discussed in Section 4.3).

In current practice the multiple-point model of spatial continuity is the training image, an exhaustive 3-D representation of the inferred heterogeneity at the scale of the model. The training image should be limited to the categories considered in the multiple-point statistic and should have features that are judged to have a significant impact on the transfer function and that have a good likelihood to be reproduced with the selected data event template. Given that the training image is large enough, sufficient replicates should be available to calculate the required multiple-point conditional probabilities.

In the multiple-point methods, the spatial continuity inference step becomes a modeling step. This results in some fundamental differences with the previously discussed spatial continuity inference central to the variogram-based methods with regard to repeatability, implicit assumptions, mathematical expression, and reproduction. These should be considered along with the previously discussed unique abilities of multiple-point statistics to characterize and impose curvilinear and ordering features that are not possible with the variogram.

Some consider the utilization of the variogram as more repeatable, since the experimental variogram is calculated from available data and the positive definite variogram model is fit largely aided by the experimental variogram. In most cases, the model of spatial continuity comes from the available data. There is no practical work flow currently to build the training image from the data. Boisvert et al. (2007b) suggest methods to check training image compatibility with available data, but still the training image construction is largely subjective.

A natural response is that the variogram-based methods rely strongly on implicit random function models. In the case of Gaussian simulation the implicit random function model with its maximum entropy features (discussed in the cost of Gaussianity discussion previously in this section). Some would argue that we consider an explicit image-based random function model, albeit subjective, than to rely on a potentially poorly understood implicit random function model.

A variogram model is a compact analytical model of spatial continuity, mathematically valid for all directions and distances. This model may be efficiently utilized to infer covariance between any two locations under the assumption of stationarity. These covariances are guaranteed to be positive definite; therefore, they may be utilized in the kriging equations to minimize the estimation variance. In addition, as discussed in the Volume Variance Relations subsection, the variogram may be utilized to model the change in variance with change in size support, and the variogram model itself may be scaled to account for the change in support size. Whereas variogram-based methods may be generalized to account for variable model cell support size, as is the case with unstructured grids, there is no method with the current implementation of multiple-point simulation to account for variable cell size.

The multiple-point statistic random function continuity model is a discrete list of conditional probabilities. There are currently no methods available to interpolate new consistent conditional probabilities from this list or to change the support size. That is, if a data event was not extracted from the training image, it is not available and there is no way to assess a conditional probability consistent with the rest of the list. Even changes in lower-order statistics, such as the global PDF, are not straightforward to integrate into the multiple-point statistic random function continuity model.

A variogram model may be applied to categorical and continuous variables without limitation. Current practical multiple-point methods are limited to categorical variables (see Section 4.3 for comments on continuous MPS). Continuous reservoir properties within the categorical facies must be modeled with other methods, generally semivariogram-based methods. This may not be a major concern because in many settings, facies is the major constraint on heterogeneity; therefore, modeling of continuous reservoir properties within facies may be considered a secondary concern. As discussed in Section 4.3, there remain challenges in placing continuous properties in multiple-point statistic facies models.

Typically, with variogram-based methods there is a strong confidence that the positive definite variogram model will be reproduced in the final model. In fact, in the minimum acceptance criteria for model checking, Leuangthong et al. (2004) suggest a check of variogram reproduction, because poor reproduction suggests an implementation issue that should be addressed (discussed in Section 5.1, Model Checking). Yet, any features may be put into the training image without guarantee of reproduction in the final model. The reproduction is conditional on the data event template design and specific implementation issues discussed in Section 4.3, Multiple-Point Facies Modeling. Training images should be designed with consideration of these limitations, or the result may be wasted effort and/or miscommunication. Consider the time required to place high resolution, contiguous and long range features that will not likely be reproduced, but communicate the expectation of their inclusion in the final model to the project team.

2.3.8 Volume Variance Relations

Reconciling data from different scales is a longstanding problem in reservoir characterization. Data from core plugs, well logs of different types, and seismic data must all be accounted for in the construction of a geostatistical reservoir model. These data are at vastly different scales, and it is wrong to ignore the scale difference when constructing a geostatistical model (Tran, 1996).

Geostatistical scaling laws were devised in the 1960s and 1970s primarily in the mining industry where the concern was mineral grades of selective mining unit (SMU) blocks of different sizes. These techniques can be extended to address problems of core, log, and seismic data; however, there are limitations in the context of petroleum reservoir modeling: (1) the ill-defined volume of measurement of well log and seismic measurements, (2) uncertainty in the small-scale variogram structure, and (3) nonlinear averaging of many responses, including acoustic properties and permeability. The essential results of classical scaling laws are recalled here, but additional research is needed to overcome these limitations.

The first important notion in volume variance relations is the spatial or dispersion variance. The dispersion variance $D^2(a,b)$ (Journel and Huijbregts, 1978) is the variance of values of volume a in a larger volume b—that is, in the classical definition of the variance:

$$D^2(a,b) = \frac{1}{n}\sum_{i=1}^{n}\left(\underbrace{z_i}_{\text{Support } a} - \underbrace{m_i}_{\text{Support } b}\right)^2$$

In a geostatistical context, almost all variances are dispersion variances. Often, the volume a is the quasipoint hard data $(\cdot)$ and the volume b is the entire area of interest (A); thus σ^2 could be reported as $D^2(\cdot,A)$.

An important relationship between dispersion variances is the "additivity of variance" relation that is sometimes referred to as Krige's relation:

$$D^2(a,c) = D^2(a,b) + D^2(b,c)$$

The additivity of variance notion is explained in detail in any statistics book dealing with analysis of variance (ANOVA). In the context of scaling from core scale to geological modeling scale, consider $a = \cdot =$ the volume of the hard core data, $b = v =$ the size of the geological modeling cells, and $c = A =$ the area of interest (reservoir). Then

$$D^2(\cdot,A) = D^2(\cdot,v) + D^2(v,A)$$

In words, the variance of core values in the reservoir is the variance of core values in the geological modeling cells plus the variance of the geological modeling cells in the reservoir. It is important to note that this depends on linear averaging, which is correct for porosity and facies proportions but incorrect for permeability.

Before going further with the development of dispersion variances, it is necessary to discuss the meaning and calculation of "gamma bar" values. The "gamma bar" value represents the mean value of $\gamma(\mathbf{h})$ when one extremity of the vector $\mathbf{h}$ describes the domain $v(\mathbf{u})$ and the other extremity independently describes domain $V(\mathbf{u}')$. The size of volumes

$v(\mathbf{u})$ and $V(\mathbf{u}')$ need not be the same. The locations $\mathbf{u}$ and $\mathbf{u}'$ need not be collocated. Of course, the volumes and locations could be the same $v(\mathbf{u}) \equiv V(\mathbf{u}')$. In mathematical notation the gamma bar value is expressed as

$$\bar{\gamma}\left(v(\mathbf{u}), V(\mathbf{u}')\right) = \frac{1}{v \cdot V} \int_{v(\mathbf{u})} \int_{V(\mathbf{u}')} \gamma(y - y')\, dy dy'$$

where v and V are the volumes of $v(\mathbf{u})$ and $V(\mathbf{u}')$, respectively. In three dimensions this integral expression is a sextuple integral, which was infamous among early practitioners of geostatistics.

Although there exist certain analytical solutions to $\bar{\gamma}\,(v(\mathbf{u}), V(\mathbf{u}'))$, we systematically estimate $\bar{\gamma}\,(v(\mathbf{u}), V(\mathbf{u}'))$ by discretizing the volumes $v(\mathbf{u})$ and $V(\mathbf{u}')$ into a number of points and simply averaging the variogram values:

$$\bar{\gamma}\left(v(\mathbf{u}), V(\mathbf{u}')\right) \approx \frac{1}{n \cdot n'} \sum_{i=1}^{n} \sum_{j=1}^{n'} \gamma(\mathbf{u}_i - \mathbf{u}'_j)$$

where the n points $\mathbf{u}_i, i = 1, \ldots n$, discretize the volume $v(\mathbf{u})$ and the n' points $\mathbf{u}_j, j = 1, \ldots n'$, discretize the volume $V(\mathbf{u}')$. The discretization is a regular spacing of points where each point represents the same fractional volume. The number of points n or n' is chosen large enough to provide a stable numerical approximation of $\bar{\gamma}\,(v(\mathbf{u}), V(\mathbf{u}'))$. In practice, $n = n' = 100$ to 1000 is more than adequate. 100 points in 1-D, $10 \cdot 10$ in 2-D, and $10 \cdot 10 \cdot 10$ in 3-D is sufficient and provides stable estimates of the gamma bar value.

There are two critical relationships that can be derived from the definition of dispersion variance and gamma bar values:

$$D^2(v, V) = \bar{\gamma}(V, V) - \bar{\gamma}(v, v)$$

and the special case where $v = \cdot$ and $V = v$:

$$D^2(\cdot, v) = \bar{\gamma}(v, v)$$

These tell us that we can calculate any particular dispersion variance (or volume-related variance) once we have a variogram model for the spatial continuity. The derivation is given in Mining Geostatistics (Journel and Huijbregts, 1978).

In the presence of variables that average linearly, the change of volume from quasi-point scale to a geological modeling scale v entails the following:

- The mean remains unchanged regardless of scale.
- Variance reduces to predictable value $\sigma^2 - \bar{\gamma}(v, v)$.
- The shape of the histogram of v-size volumes *should* become more symmetric.

The remaining problem is to establish a "change of shape" model. There are a number of analytical models (Isaaks and Srivastava, 1989) that make specific assumptions about how the distribution changes shape. Experience has shown that these analytical models work when the reduction in variance is small, say, less than 30%. In general, however, a numerical approach must be used to establish the shape of the histogram of geological modeling cell values; that is, we simulate a distribution of quasi-point scale values, average to the desired block scale v, and directly observe the resulting histogram shape.

Relations can also be derived for the change in variogram range and sill parameters [see Deutsch and Kupfersberger (1993); Frykman and Deutsch (1999); Journel and Huijbregts (1978)]. As mentioned at the start of this subsection, there are limitations to these existing procedures that must be addressed by additional research before systematic application in the practice of petroleum reservoir modeling. An important limitation is that these procedures apply to stationary random function models of variables that average linearly and that are characterized by a variogram alone.

One critical limitation is the assumption that the variogram characterizes spatial variability. Geological features are often much more complex than can be captured by a variogram alone. A striking feature of sedimentary deposits is a geometry defined by stratigraphic bounding surfaces; a promising new area of development is surface-based modeling.

These principles are critical prerequisites for model scaling discussion in Section 5.2, Model Postprocessing.

2.3.9 Work Flow

The central idea of this section is to quantify spatial correlation to be used in subsequent geostatistical modeling. A number of work flows are related to this quantification.

Figure 2.73 illustrates the work flow to prepare data for variogram calculation. A categorical data variable must be coded as an indicator. A continuous variable must have deterministic trends removed, normal score transformed for subsequent Gaussian simulation, and indicator transformed

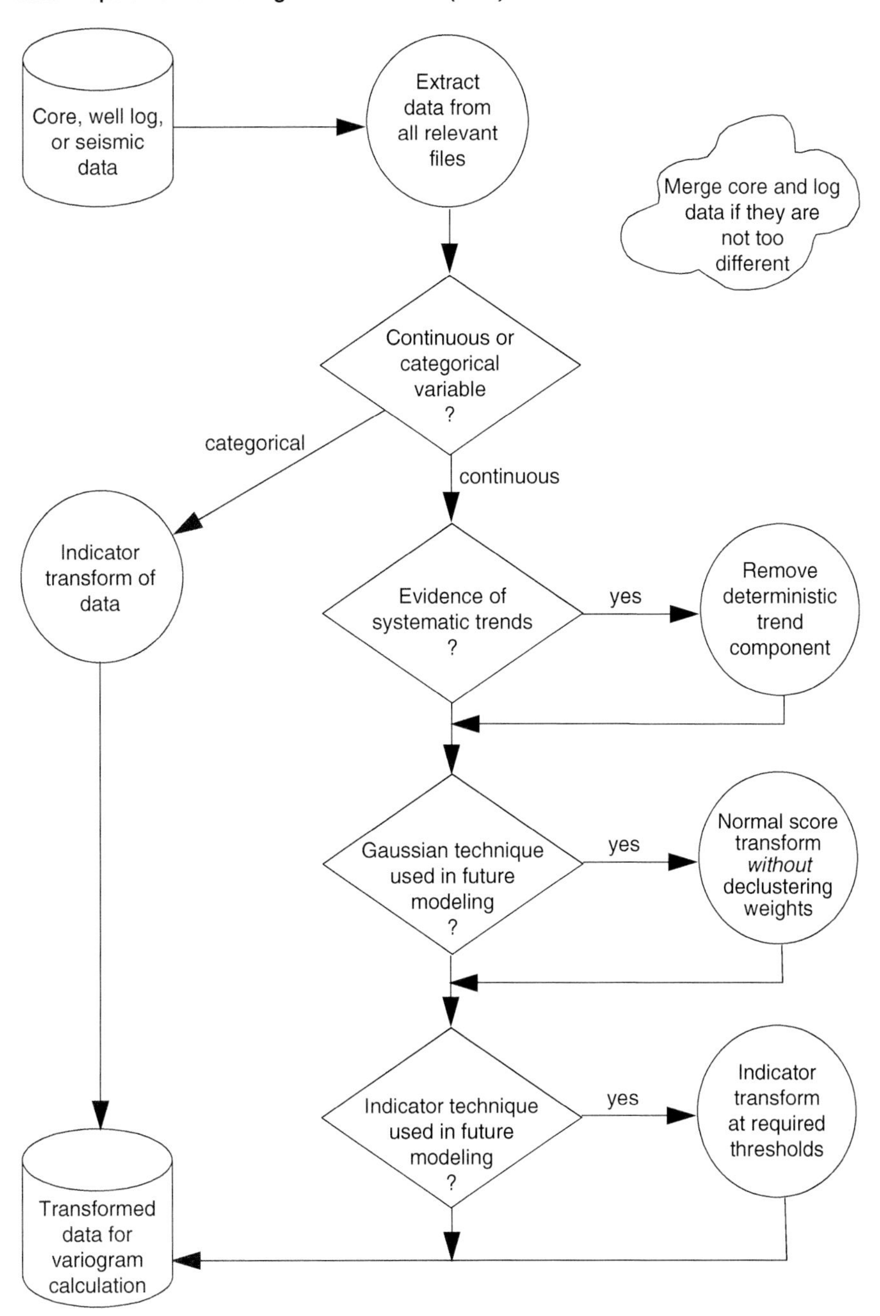

FIGURE 2.73: Work Flow for Data Extraction and Transformation for Variogram Calculation.

for subsequent indicator simulation of continuous variables.

Figure 2.74 illustrates the work flow to calculate the variogram of scattered data. The common case of vertical wells permits the calculation of a vertical variogram. Horizontal variograms must be inferred from analog data when there are few wells. In the presence of sufficient data, we can calculate the omnidirectional horizontal variogram and then directional variograms.

Figure 2.75 illustrates the work flow to create a 3-D variogram model. A 3-D model is assembled once the three principal directional variograms have been determined. The range parameters in the three directions can be changed to achieve good fits in all directions.

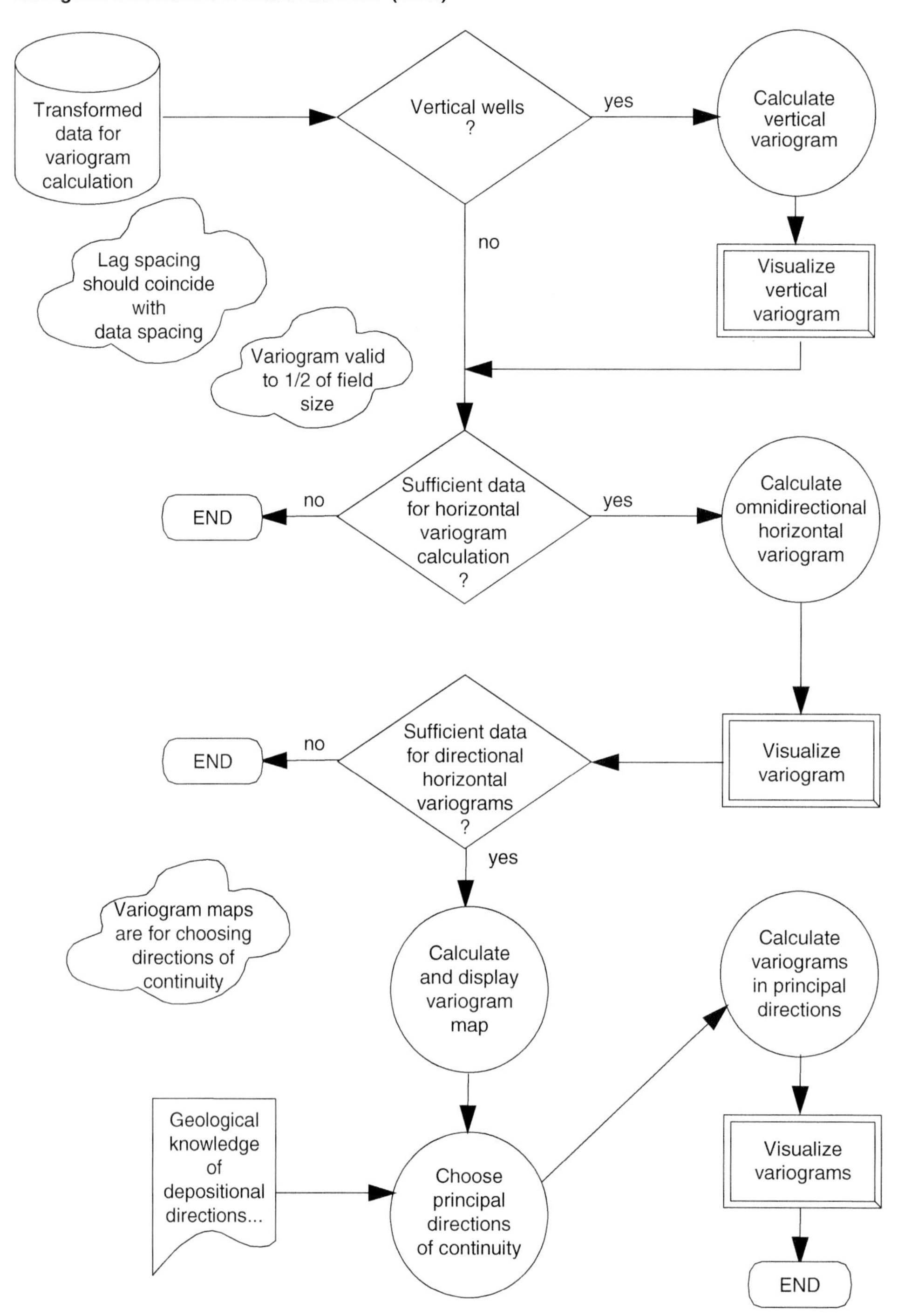

FIGURE 2.74: Work Flow for Calculation of Variograms of Scattered Data.

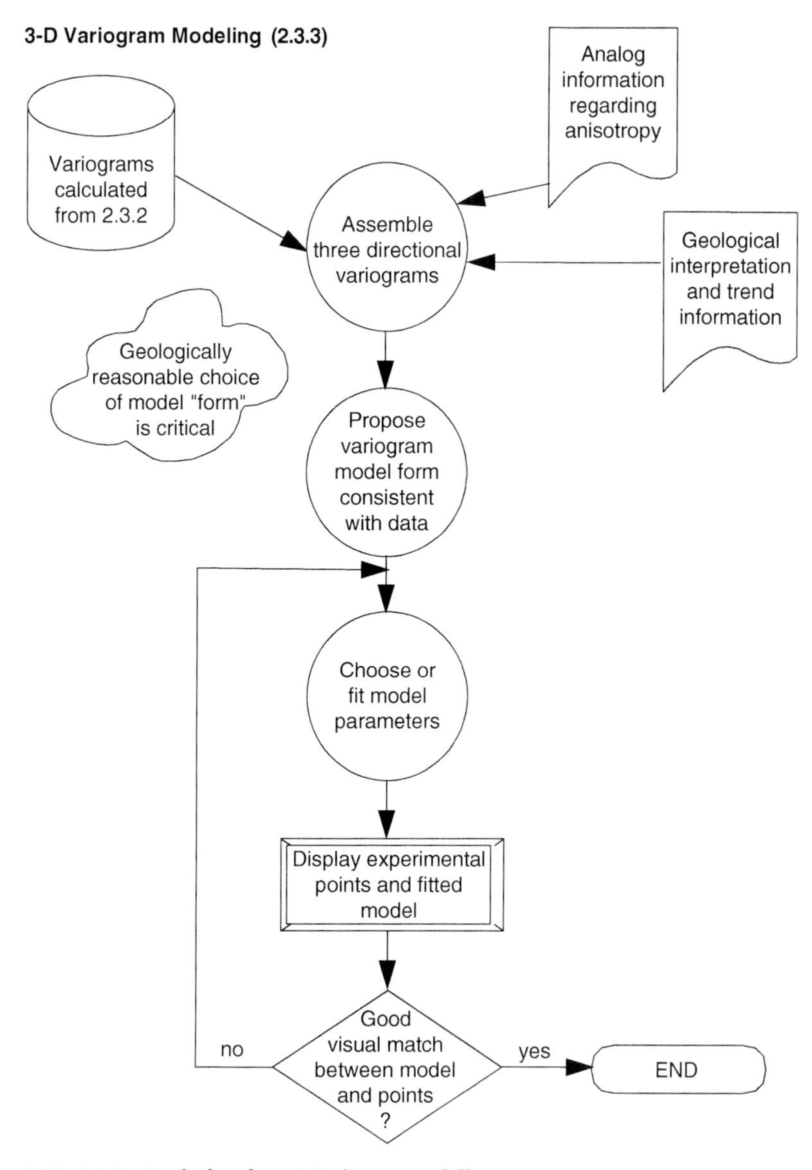

FIGURE 2.75: Work Flow for 3-D Variogram Modeling.

Figure 2.76 illustrates the work flow to fit a linear model of coregionalization. Collocated cosimulation does not require a linear model of coregionalization, but full cokriging (or cosimulation) requires such a model. The same pool of nested structures is used with coefficients that must meet the criteria for positive definiteness.

2.3.10 Section Summary

This section has covered the fundamental concepts and methods to characterize spatial continuity. Along with the various univariate and multivariate descriptions from Section 2.2, these measures are the fundamental input statistics that constrain the estimation and simulation methods discussed in the next section.

Variogram calculation, interpretation, and modeling have been discussed in detail. In many sparse data settings, these tasks are difficult, but nevertheless essential for geostatistical modeling. Multiple-point statistics have been introduced as a method to capture curvilinear and ordering patterns for categorical variables. This requires a greater level of inference and, therefore, reliance on an exhaustive training image. Implementation details are discussed in Section 4.3, Multiple-Point Facies Modeling.

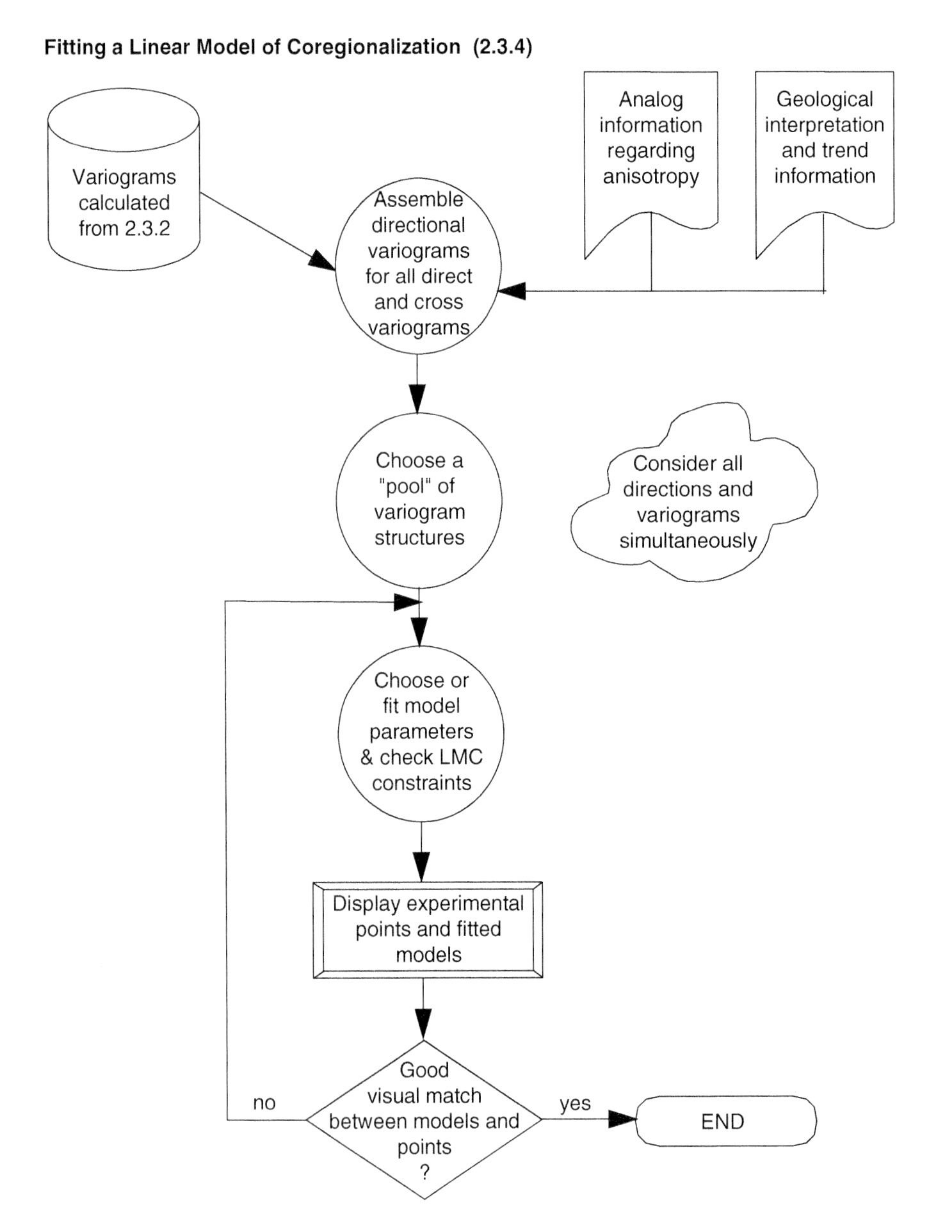

FIGURE 2.76: Work Flow for Fitting a Linear Model of Coregionalization.

Volume variance relations are introduced here because they are based on the variogram model and are prerequisites to model scaling discussed in Section 5.2.

With the prerequisite geological concepts from Section 2.1, along with statistical concepts from Section 2.2 and this section, we are now ready to cover the application of these concepts for mapping in the next section.

2.4 PRELIMINARY MAPPING CONCEPTS

Geostatistics is concerned with a wide variety of techniques for spatial data analysis, estimation, simulation, and decision making. The field of geostatistics is extensive; the goal here is a pragmatic presentation of the concepts used in reservoir modeling. Kriging, cokriging, and simulation concepts are condensed into a single reservoir-oriented section. Short discussions are provided to link non-kriging-based simulations methods such as multiple-point simulation and object-based simulation. This section also provides context for the discussion on the Conceptual Model in Section 3.2 and Problem Formulation in Section 3.3. More complete implementation details and examples are provided in Chapter 4.

Kriging is the workhorse of traditional mapping applications and an essential component of variogram-based geostatistical simulation methods. For this reason, the *Kriging and Cokriging* subsection gives a detailed description of the estimation variance, the simple kriging equations, and cokriging. Although this section could be skipped at first reading, an understanding of this material is required for an understanding of the commonly used variogram-based simulation methods presented in Chapter 4.

The *Sequential Gaussian Simulation* subsection presents this important popular algorithm for continuous property modeling. The popularity of this algorithm is due to its relative simplicity and effectiveness at creating numerical models with correct spatial statistics. Discussion includes the motivation, implications and limitations of the Gaussian assumption. The multivariate Gaussian distribution permits the calculation of local distributions of uncertainty from simple kriging estimates and variances. Global distributions are appropriately reproduced by these conditional distributions. This convenience of the multivariate Gaussian distribution comes at a cost of a maximum entropy model that may have an impact on the transfer function.

The *Indicator Formulism and Simulation* subsection extends the simulation paradigm to the indicator transform of continuous and categorical variables. This results in improved control on the spatial continuity of each category and across the continuous distribution and a very natural framework for the integration of soft data.

Probability or *p*-field simulation methods have some popularity because of their flexibility in accounting for secondary information such as seismic data; *Probability Field Simulation* presents the *p*-field methods.

Multiple-point statistics could be used to calculate the required local conditional probability distributions in the sequential simulation paradigm. The *Multiple-Point Simulation* subsection develops this methodology. This results in categorical simulation models that integrate curvilinear and ordering features. Implementation details and alternative methods for multiple-point methods are deferred to Section 4.3, Multiple Point Facies Modeling.

The *Object-Based Simulation* subsection introduces a simulation method that stochastically places parametric geometries in the model. While full details are provided in Section 4.4, a short introduction is provided here for completeness. The *Optimization Algorithms for Modeling* subsection introduces the construction of a reservoir model as an optimization problem. The reservoir model under statistical and conditioning constraints is treated as a classic combinatorial minimization problem. A simulated annealing framework and demonstration is presented.

The *Alternatives for Secondary Data Integration* subsection collects a variety of methods to integrate secondary data. This includes an example of Bayesian updating, the commonly applied collocated cokriging, stepwise conditional transform, cloud transform, probability combination schemes, and the use of combined "super variables." Fairly accounting for all secondary data is a challenging component of geostatistical modeling with diverse circumstances and different solutions are applicable in different situations.

Mapping geological variables becomes complex due to many special considerations such as discontinuities, trends, and changing directions of continuity. The *Accounting for Trends* subsection presents a number of these special considerations. In many respects, this section is one of the most important of the book. The considerations discussed here are relevant even *before* statistical analysis and spatial statistics. The need, however, to discuss trends or locally varying means comes up after a presentation of key kriging and simulation concepts. Details on methods to practically construct trend models is postponed to Section 4.1, Large-Scale Modeling.

The mathematics presented in this chapter are not difficult, but they may appear odd to the newcomer to geostatistics or statistical inference. Little of this is required from a practical perspective; however, some of the implementation details and limitations of geostatistical reservoir models can only be appreciated after working through these basics. The novice modeler may review the basic concepts of each modeling method and return for more in-depth understanding. The intermediate level modeler may immediately appreciate the method steps and mathematical descriptions. Finally, the expert modeler may find some new ideas in the Alternative for Secondary Data Integration subsection.

2.4.1 Kriging and Cokriging

A common problem in earth sciences is creating a map of a regionalized variable from limited sample data. This common problem was initially addressed by hand-contouring, which provides insight into trends and the uncertainty in the data. Hand-contouring important variables such as the thickness of net pay should always be performed as a means to investigate trends, anisotropy, and other essential features of the data. Early machine contouring algorithms evolved from principles of hand-contouring. The goal of such algorithms was to create smooth maps that would reveal the same geological trends as hand-contouring. Such algorithms remain popular.

There were others, primarily mathematicians and engineers, who wanted to create maps fit for a particular purpose. That is, they considered that the mapped values should be optimal in some objective sense. One measure of optimality is *unbiasedness*; that is, taken all together, the map should have the correct average. It turns out that global unbiasedness is straightforward to achieve; however, with simplistic estimation methods, global unbiasedness is achieved at the expense of overestimating low values and underestimating high values. The estimates are said to have *conditional bias*. Daniel Krige, a South African mining engineer, was interested in correcting this bias because it was unacceptable to have mine stopes with predicted high grades systematically have lower true grades. The interpolation method of kriging developed by Georges Matheron (1961/62), was named for Daniel Krige's pioneering work in the early 1950s (Krige, 1951).

Least squares optimization has been used for more than 200 years. The idea proposed by early workers in geostatistics (Goldberger, 1962; Matheron, 1962) was to construct an estimator $Z^*(\mathbf{u})$ that is optimum in a minimum expected squared error sense, that is, minimize the average squared difference between the true value $z(\mathbf{u})$ and the estimator:

$$SE = \left[z(\mathbf{u}) - z^*(\mathbf{u})\right]^2$$

Of course, the true values are only known at the data locations and *not* at the locations being estimated. Therefore, as is classical in statistics, the squared error SE is minimized in expected value over all locations $\mathbf{u}$ with the same data configuration within a study area A, deemed stationary.

Most estimators in geostatistics are linear. The original data may have been transformed in a nonlinear fashion into polynomial coefficients, indicators, or power-law transforms; however, the estimates of the transformed variables remain linear, that is,

$$z^*(\mathbf{u}) - m(\mathbf{u}) = \sum_{\alpha=1}^{n} \lambda_\alpha \cdot \left[z(\mathbf{u}_\alpha) - m(\mathbf{u}_\alpha)\right]$$

where $z^*(\mathbf{u})$ is the estimate at unsampled location $\mathbf{u}$, $m(\mathbf{u})$ is the prior mean value at unsampled location $\mathbf{u}$, $\lambda_\alpha, \alpha = 1, \ldots, n$, are weights applied to the n data, $z(\mathbf{u}_\alpha), \alpha = 1, \ldots, n$, are the n data values, and $m(\mathbf{u}_\alpha), \alpha = 1, \ldots, n$, are the n prior mean values at the data locations. All prior mean values could be set to a constant mean $m(\mathbf{u}) = m(\mathbf{u}_\alpha) = m$ if no prior information on trends is available.

From a classical regression estimator perspective, we could imagine adding terms involving jointly two or more data such as the product $\left[z(\mathbf{u}_\alpha) - m(\mathbf{u}_\alpha)\right]\left[z(\mathbf{u}_\beta) - m(\mathbf{u}_\beta)\right]$; however, this is not done in practice. The most important reason to limit ourselves to linear estimates is for simplicity. Incorporation of many data simultaneously would require a measure of correlation between these multiple data and the property being estimated, which is a multiple-point measure of correlation and not just the two-point variogram or covariance we have considered thus far.

The following classical regression approach to arrive at the weights $\lambda_\alpha, \alpha = 1, \ldots, n$, is known as kriging for historical reasons.

A principle of numerical modeling is that known geological and geometrical constraints should be accounted for deterministically. For that reason, we systematically consider residuals from known trends, that is, we work with residual data values:

$$Y(\mathbf{u}_\alpha) = Z(\mathbf{u}_\alpha) - m(\mathbf{u}_\alpha), \qquad \alpha = 1, \ldots, n$$

The subject of trend modeling—that is, determining the trend or prior mean $m(\mathbf{u})$ for all locations $\mathbf{u}$ in the reservoir—comes at the end of this section. When there are no large-scale predictable trends, the mean could be considered known and fixed at the global mean: $m(\mathbf{u}) = m$ for all locations in the area or reservoir $\mathbf{u} \in A$.

Regardless of the decision for the mean m, the residual variable Y is deemed stationary; that is,

$E\{Y\mathbf{u})\} = 0, \ \forall \ \mathbf{u} \in A$, with stationary covariance $C_Y(\mathbf{h})$ and variogram $2\gamma_Y(\mathbf{h})$. The adoption of Gaussian techniques would involve a normal score transform of the Y residual data variable. The decision of stationarity would apply to the normal score variable, and the variogram would be calculated after normal score transformation. As mentioned before, stationarity assumes that the statistics apply over the study area—for example, the porosity within a particular facies in a particular reservoir layer.

Although it is somewhat repetitive, there is value in reviewing the needed statistics. The stationary mean or expected value of the Y variable is 0; that is, $E\{Y\} = 0$. The stationary variance is $\mathrm{Var}\{Y\} = E\{Y^2\} = C(0) = \sigma^2$. The variogram of the stationary residuals is defined as

$$2\gamma(\mathbf{h}) = E\left\{\left[Y(\mathbf{u}) - Y(\mathbf{u}+\mathbf{h})\right]^2\right\}$$

and the covariance is defined as

$$C(\mathbf{h}) = E\left\{Y(\mathbf{u}) \cdot Y(\mathbf{u}+\mathbf{h})\right\}$$

The link between the variogram and the covariance is

$$C(\mathbf{h}) = C(0) - \gamma(\mathbf{h})$$

This result is important since it permits us the flexibility of calculating and modeling experimental variograms and then using the covariance in subsequent mathematical calculations such as kriging.

Now, consider a linear estimator at a location where no data are available:

$$Y^*(\mathbf{u}) = \sum_{\alpha=1}^{n} \lambda_\alpha \cdot Y(\mathbf{u}_\alpha) \qquad (2.16)$$

where $Y^*(\mathbf{u})$ is the estimated value and $\lambda_\alpha, \alpha = 1, \ldots, n$, are weights applied to the n data values $Y(\mathbf{u}_\alpha), \alpha = 1, \ldots, n$. The error variance for this estimator may be defined as

$$E\left\{\left[Y^*(\mathbf{u}) - Y(\mathbf{u})\right]^2\right\} = \sigma_E^2(\mathbf{u})$$

Although we do not know the true value $y(\mathbf{u})$, it is possible to expand this error variance term using expected values[5], and the error variance is expanded as

$$
\begin{aligned}
&= E\left\{\left[Y^*(\mathbf{u})\right]^2\right\} - 2 \cdot E\left\{Y^*(\mathbf{u}) \cdot Y(\mathbf{u})\right\} + E\left\{\left[Y(\mathbf{u})\right]^2\right\} \\
&= \sum_{\alpha=1}^{n}\sum_{\beta=1}^{n} \lambda_\alpha \lambda_\beta E\left\{Y(\mathbf{u}_\beta) \cdot Y(\mathbf{u}_\alpha)\right\} - \\
&\quad 2 \cdot \sum_{\alpha=1}^{n} \lambda_\alpha E\left\{Y(\mathbf{u}) \cdot Y(\mathbf{u}_\alpha)\right\} + C(0) \\
&= \underbrace{\sum_{\alpha=1}^{n}\sum_{\beta=1}^{n} \lambda_\alpha \lambda_\beta C(\mathbf{u}_\beta - \mathbf{u}_\alpha)}_{\text{redundancy}} - \\
&\quad \underbrace{2 \cdot \sum_{\alpha=1}^{n} \lambda_\alpha C(\mathbf{u} - \mathbf{u}_\alpha)}_{\text{closeness}} + \underbrace{C(0)}_{\text{variance}} \qquad (2.17)
\end{aligned}
$$

This final equation for the error variance is very interesting; it is a mathematical expression for the error variance for any set of weights $\lambda_\alpha, \alpha = 1, \ldots, n$. Clearly, the estimation variance depends on the model of covariance or variogram. There are three terms in the equation for the estimation variance:

1. *Redundancy:* The more redundant the data (the covariances between them, $C(\mathbf{u}_\alpha, \mathbf{u}_\beta)$, will be higher) the larger the estimation variance.
2. *Closeness:* The closer the data to the location being estimated (the covariances $C(\mathbf{u} - \mathbf{u}_\alpha)$ will be higher), the smaller the estimation variance.
3. *Variance:* The estimation variance is equal to the variance or $C(0)$ when all data are too distant to receive any weight; that is, $\lambda_\alpha = 0, \alpha = 1, \ldots, n$, and the estimate is the local mean $Y^*(\mathbf{u}) = m(\mathbf{u})$ assumed known.

Equation (2.17) allows us to calculate the estimation variance for any set of weights $\lambda_\alpha, \alpha = 1, \ldots, n$; however, our goal is to calculate the weights that minimize this estimation variance. A classical minimization procedure is followed.

Simple Kriging

The partial derivatives of Expression (2.17) with respect to each of the weights $\lambda_\alpha, \alpha = 1, \ldots, n$, are calculated and set to zero. This leads to equations that can be solved for the weights that minimize the estimation variance. The second derivative could be taken to show that we have a minimum; however, the maximum estimation variance is infinity, so any solution to kriging must be a minimum. This

[5] This expression for the estimation variance is derived using (1) the linearity of the expected value operator, that is, $E\{A + B\} = E\{A\} + E\{B\}$, and (2) the implicit assumption of stationarity—for example, $E\{Y(\mathbf{u}_\beta) \cdot Y(\mathbf{u}_\alpha)\} = C(\mathbf{h} = \mathbf{u}_\beta - \mathbf{u}_\alpha)$.

procedure is classical in mathematics. The partial derivatives of Expression (2.17) with respect to the weights $\lambda_\alpha, \alpha = 1, \ldots, n$, are calculated as

$$\frac{\partial \left[\sigma_E^2(\mathbf{u})\right]}{\partial \lambda_\alpha} = 2 \cdot \sum_{\beta=1}^{n} \lambda_\beta C(\mathbf{u}_\beta - \mathbf{u}_\alpha) - 2 \cdot C(\mathbf{u} - \mathbf{u}_\alpha), \qquad \alpha = 1, \ldots, n$$

These may be set to zero to calculate the weights that minimize the estimation variance:

$$\sum_{\beta=1}^{n} \lambda_\beta C(\mathbf{u}_\beta - \mathbf{u}_\alpha) = C(\mathbf{u} - \mathbf{u}_\alpha), \qquad \alpha = 1, \ldots, n \tag{2.18}$$

This system of n equations with n unknown weights is known as the simple kriging (SK) system. These equations are also called the normal equations in optimization theory (Luenberger, 1969).

As an example, consider a case with three data values:

$$C(\mathbf{u}_1 - \mathbf{u}_1) \cdot \lambda_1 + C(\mathbf{u}_1 - \mathbf{u}_2) \cdot \lambda_2 + C(\mathbf{u}_1 - \mathbf{u}_3) \cdot \lambda_3 = C(\mathbf{u} - \mathbf{u}_1)$$

$$C(\mathbf{u}_2 - \mathbf{u}_1) \cdot \lambda_1 + C(\mathbf{u}_2 - \mathbf{u}_2) \cdot \lambda_2 + C(\mathbf{u}_2 - \mathbf{u}_3) \cdot \lambda_3 = C(\mathbf{u} - \mathbf{u}_2)$$

$$C(\mathbf{u}_3 - \mathbf{u}_1) \cdot \lambda_1 + C(\mathbf{u}_3 - \mathbf{u}_2) \cdot \lambda_2 + C(\mathbf{u}_3 - \mathbf{u}_3) \cdot \lambda_3 = C(\mathbf{u} - \mathbf{u}_3)$$

which may be written in matrix notation as

$$\begin{bmatrix} C(\mathbf{u}_1 - \mathbf{u}_1) & C(\mathbf{u}_1 - \mathbf{u}_2) & C(\mathbf{u}_1 - \mathbf{u}_3) \\ C(\mathbf{u}_2 - \mathbf{u}_1) & C(\mathbf{u}_2 - \mathbf{u}_2) & C(\mathbf{u}_2 - \mathbf{u}_3) \\ C(\mathbf{u}_3 - \mathbf{u}_1) & C(\mathbf{u}_3 - \mathbf{u}_2) & C(\mathbf{u}_3 - \mathbf{u}_3) \end{bmatrix} \begin{bmatrix} \lambda_1 \\ \lambda_2 \\ \lambda_3 \end{bmatrix} = \begin{bmatrix} C(\mathbf{u} - \mathbf{u}_1) \\ C(\mathbf{u} - \mathbf{u}_2) \\ C(\mathbf{u} - \mathbf{u}_3) \end{bmatrix}$$

$$\mathbf{C}\boldsymbol{\lambda} = \mathbf{c}$$

The weights are obtained by inverting the left-hand-side $\mathbf{C}$ matrix and multiplying it to the right-hand-side $\mathbf{c}$ vector. Note that the left-hand-side $\mathbf{C}$ matrix contains all of the information related to redundancy in the data and the right-hand-side $\mathbf{c}$ vector contains all of the information related to closeness of the data to the location being estimated.

Kriging solves this system of linear equations. The matrix $\mathbf{C}$ is symmetric. The matrix is positive semidefinite, provided the covariance (or the corresponding variogram) is fit with a positive definite function and no two data of the same support are co-located.

Figure 2.77 shows a map of kriged estimates using 11 data. The variogram in this case is a Gaussian variogram with no nugget effect and no anisotropy.

The Kriging Variance

The equation for the estimation variance may be simplified if simple kriging is used to calculate the weights; that is,

$$\sigma_E^2(\mathbf{u}) = \sum_{\alpha=1}^{n} \lambda_\alpha \underbrace{\sum_{\beta=1}^{n} \lambda_\beta C(\mathbf{u}_\beta - \mathbf{u}_\alpha)}_{\text{to be replaced}} - 2 \cdot \sum_{\alpha=1}^{n} \lambda_\alpha C(\mathbf{u} - \mathbf{u}_\alpha) + C(0)$$

which simplifies to

$$\sigma_E^2(\mathbf{u}) = C(0) - \sum_{\alpha=1}^{n} \lambda_\alpha C(\mathbf{u} - \mathbf{u}_\alpha)$$

when the "to be replaced" portion is replaced with the right-hand side of the kriging equations, Eq. (2.18).

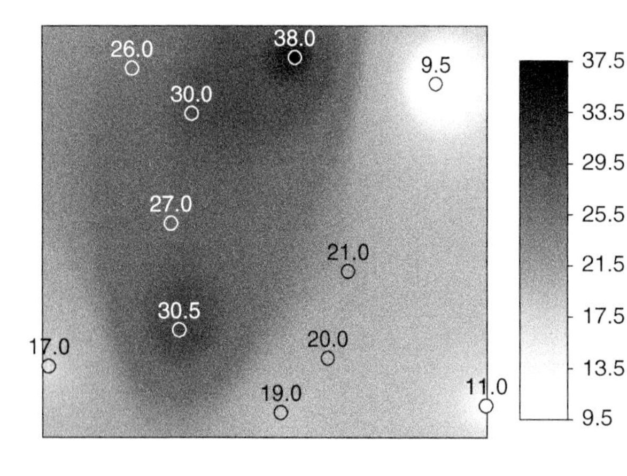

FIGURE 2.77: A Kriged Map of Thickness Using a Gaussian Variogram. Note the smooth interpolation between the data points.

Kriging may be considered a spatial regression, thus it creates an estimate that is often smooth. The combination of n data by a weighted linear sum tends away from low and high estimates; there are more near the mean. Kriging accounts for redundancy between the data and a measure of distance specific to the data considered, that is, the variogram model. Another significant advantage of kriging is that it provides a quantification of how smooth the estimates are, that is, the variance of the kriging estimate may be calculated as

$$\begin{aligned}\mathrm{Var}\{Y^*(\mathbf{u})\} &= E\left\{\left[Y^*(\mathbf{u})\right]^2\right\} - E\left\{y^*(\mathbf{u})\right\}^2\\ &= \sum_{\alpha=1}^{n}\sum_{\beta=1}^{n}\lambda_\alpha\lambda_\beta E\left\{Y(\mathbf{u}_\beta)\cdot Y(\mathbf{u}_\alpha)\right\} - 0^2\\ &= \sum_{\alpha=1}^{n}\lambda_\alpha C(\mathbf{u}-\mathbf{u}_\alpha) \qquad (2.19)\end{aligned}$$

Once again, the last substitution in Eq. (2.19) is the right-hand side of the kriging equation, (2.18). Equation (2.19) tells us the variance of the kriging estimator $Y^*(\mathbf{u})$ at any location $\mathbf{u}$. This variance is the stationary variance $C(0)$ when kriging at a data location since $\lambda_\alpha = 1$ for the data location $\mathbf{u} = \mathbf{u}_\alpha$, all other weights are zero, and $C(\mathbf{u}-\mathbf{u}_\alpha) = C(0)$; the stationary variance $C(0) = \sigma^2$. The variance far from local data is zero since no data are given any weight; that is, all weights are zero.

A map of kriged estimates is smoother than the real variable. This smoothing is particularly noticeable away from the data locations. The smoothing of kriging is directly proportional to the kriging variance, there is no smoothing at data locations where the kriging variance is zero, the estimate is completely smooth far away from data where all estimates are equal to the mean, and the kriging variance is the stationary variance $C(0)$. We can write the "missing" variance or the smoothing effect as

$$\text{Missing variance} = C(0) - \sum_{\alpha=1}^{n}\lambda_\alpha C(\mathbf{u},\mathbf{u}_\alpha)$$

This rather unusual concept can help our understanding when we describe the theoretical background behind simulation.

Other Forms of Kriging

There are other versions of kriging that make different assumptions regarding the mean [see Deutsch and Journel (1998), Goovaerts (1997), and Journel and Huijbregts (1978) for more information]. Simple kriging of residuals from a geologically interpreted trend is the preferred approach because the theory of simulation requires simple kriging. Other types of kriging may be used for estimation or in special cases.

Ordinary kriging (OK) estimates a constant mean value m from the data used in the kriging neighborhood; then, simple kriging is done with residuals from this implicitly estimated mean. Ordinary kriging has seen wide application in map making because it is robust with respect to trends. There is a need, however, to have sufficient data to reliably estimate the mean at each location. This is not the case early in a reservoir lifecycle.

Universal kriging (UK) or kriging with a trend model (KT) assumes that the mean follows a particular polynomial shape, for example, linear in the N30E direction. That mean "surface" is fitted as a scaling of the polynomial trend. Then, simple kriging is performed with residuals from the implicitly calculated mean values. Universal kriging is unstable in extrapolation because the coefficients of the fitted mean surface become unstable. As with ordinary kriging, universal kriging requires sufficient data to reliably estimate the mean.

Kriging with an external drift considers a nonparametric trend shape that could come from a secondary variable such as seismic. The mean "surface" is fitted as a linear scaling of the external drift variable. Then, simple kriging is performed with residuals from the implicitly calculated mean values.

Cokriging

The term *kriging* is traditionally reserved for linear regression using data with the same attribute as that being estimated. For example, an unsampled porosity value $z(\mathbf{u})$ is estimated from neighboring porosity sample values defined on the same volume support.

The term *cokriging* is reserved for linear regression that also uses data defined on different attributes. For example, the porosity value $z(\mathbf{u})$ may be estimated from a combination of porosity samples and related acoustic impedance values.

In the case of a single secondary variable (Y_2), the simple cokriging estimator of $Y(\mathbf{u})$ is written as

$$Y^*_{\mathrm{COK}}(\mathbf{u}) = \sum_{\alpha_1=1}^{n_1}\lambda_{\alpha_1}Y(\mathbf{u}_{\alpha_1}) + \sum_{\alpha_2=1}^{n_2}\lambda'_{\alpha_2}Y_2(\mathbf{u}'_{\alpha_2}) \qquad (2.20)$$

where the λ_{α_1} are the weights applied to the n_1 Y samples and the λ'_{α_2}s are the weights applied to the n_2 Y_2 samples. Expression (2.20) is written as simple cokriging for standardized variables; that is, the means of Y and Y_2 are assumed known and zero. Kriging requires a model for the Y covariance. Cokriging requires a *joint* model for the matrix of covariance functions including the Y covariance $C_Y(\mathbf{h})$, the Y_2 covariance $C_{Y_2}(\mathbf{h})$, the cross Y–Y_2 covariance $C_{YY_2}(\mathbf{h})$, and the cross Y_2–Y covariance $C_{Y_2Y}(\mathbf{h})$.

When K different variables are considered, the covariance matrix requires in all generality K^2 covariance functions. Such inference becomes extremely demanding in terms of data and the subsequent joint modeling is particularly tedious (Goovaerts, 1997). This is the main reason why cokriging has not been extensively used in practice. Algorithms such as collocated cokriging and Markov models (see hereafter) have been developed to shortcut the tedious inference and modeling process required by cokriging.

Other than tedious variogram or covariance inference, cokriging is the same as kriging. The cokriging equivalents of OK and UK exist where the mean values are implicitly estimated from the neighborhood data. The reader is referred to Carr and Myers (1985); Chilès and Delfiner (2012); Doyen (1988); Goovaerts (1997); Journel and Huijbregts (1978); Myers (1982, 1984); Wackernagel (1988) for details.

2.4.2 Sequential Gaussian Simulation

There are many algorithms that can be devised to create stochastic simulations: (1) matrix approaches (LU decomposition), which are not extensively used because of size restrictions (an $N \times N$ matrix must be solved where N, the number of locations, could be in the millions for reservoir applications), (2) turning bands methods (Emery and Lantuéjoul, 2006) where the variable is simulated on 1-D lines and then combined into a 3-D model; not commonly used because of artifacts, (3) spectral methods using FFTs can be CPU fast, but honoring conditioning data requires an expensive kriging step, (4) fractals, which are not used extensively because of the restrictive assumption of self-similarity, and (5) moving average methods, which are infrequently used due to CPU requirements.

The common approach adopted in recent times for reservoir modeling applications is the sequential Gaussian simulation (SGS) approach (Hu and Ravalec-Dupin, 2005; Isaaks, 1990). This method is simple, flexible, and reasonably efficient. Let's review the theory underlying SGS. Recall the simple kriging estimator:

$$Y^*(\mathbf{u}) = \sum_{\beta=1}^{n} \lambda_\beta \cdot Y(\mathbf{u}_\beta)$$

and the corresponding simple kriging system:

$$\sum_{\beta=1}^{n} \lambda_\beta C(\mathbf{u}_\alpha - \mathbf{u}_\beta) = C(\mathbf{u} - \mathbf{u}_\alpha), \qquad \mathbf{u}_\alpha = 1, \ldots, n$$

The covariance between the kriged estimate and one of the data values can be written as

$$\begin{aligned} \mathrm{Cov}\{Y^*(\mathbf{u}), Y(\mathbf{u}_\alpha)\} &= E\{Y^*(\mathbf{u}), Y(\mathbf{u}_\alpha)\} \\ &= E\left\{\left[\sum_{\beta=1}^{n} \lambda_\beta \cdot Y(\mathbf{u}_\beta)\right] \cdot Y(\mathbf{u}_\alpha)\right\} \\ &= \sum_{\beta=1}^{n} \lambda_\beta \cdot E\{Y(\mathbf{u}_\beta) \cdot Y(\mathbf{u}_\alpha)\} \\ &= \sum_{\beta=1}^{n} \lambda_\beta C(\mathbf{u}_\alpha - \mathbf{u}_\beta) \\ &= C(\mathbf{u} - \mathbf{u}_\alpha) \end{aligned}$$

The covariance is correct! Note that the last substitution comes from the kriging equations shown above. The kriging equations force the covariance between the data values and the kriging estimate to be correct; however, the variance is too small, and the covariance between the kriged estimates themselves is incorrect. We can imagine fixing the covariance between the kriged estimates by proceeding sequentially—that is, to use previously predicted values in subsequent predictions.

Although the covariance between the kriged estimates and the data is correct, the variance is too small. The variance of the stationary random function should be $\sigma^2 = C(0)$. This stationary model variance should be constant everywhere:

$$\sigma^2(\mathbf{u}) = \sigma^2, \qquad \forall\, \mathbf{u} \in A$$

The smoothing effect of kriging amounts to reduce this variance, particularly at locations far away from data values. Another interesting property of kriging is that the variance of the kriged estimate is known:

$$\mathrm{Var}\{Y^*(\mathbf{u})\} = C(0) - \sigma^2_{SK}(\mathbf{u})$$

This tells us how much variance is *missing*: the kriging variance $\sigma^2_{SK}(\mathbf{u})$. This missing variance must be added back in without changing the covariance reproduction properties of kriging.

An independent component with zero mean and the correct variance can be added to the kriged estimate:

$$Y_s(\mathbf{u}) = Y^*(\mathbf{u}) + R(\mathbf{u})$$

The covariance between the simulated value $Y_s(\mathbf{u})$ and one of the data values used in the estimation may be calculated:

$$\begin{aligned}\text{Cov}\{Y_s(\mathbf{u}), Y(\mathbf{u}_\alpha)\} &= E\{Y_s(\mathbf{u}), Y(\mathbf{u}_\alpha)\}\\ &= E\left\{\left[\sum_{\beta=1}^{n} \lambda_\beta \cdot Y(\mathbf{u}_\beta) + R(\mathbf{u})\right] \cdot Y(\mathbf{u}_\alpha)\right\}\\ &= \sum_{\beta=1}^{n} \lambda_\beta \cdot E\{Y(\mathbf{u}_\beta) \cdot Y(\mathbf{u}_\alpha)\} + E\{R(\mathbf{u}) \cdot Y(\mathbf{u}_\alpha)\}\end{aligned}$$

Note that $E\{R(\mathbf{u}) \cdot Y(\mathbf{u}_\alpha)\} = E\{R(\mathbf{u})\} \cdot E\{Y(\mathbf{u}_\alpha)\}$, since $R(\mathbf{u})$ is independent of any data value. The expected value of the residual, $E\{R(\mathbf{u})\}$, is 0; therefore, $E\{R(\mathbf{u}) \cdot Y(\mathbf{u}_\alpha)\} = 0$.

Thus the covariance between the simulated value and all data values is correct; that is, $\text{Cov}\{Y_s(\mathbf{u}), Y(\mathbf{u}_\alpha)\} = \text{Cov}\{Y^*(\mathbf{u}), Y(\mathbf{u}_\alpha)\} = C(\mathbf{u}, \mathbf{u}_\alpha)$.

In light of these two features, (1) the covariance reproduction of kriging and (2) unchanged covariance by adding an independent residual, the SGS algorithm is as follows:

- Transform the original Z data to a standard normal distribution (all work will be done in "normal" space). This is step (a) in Figure 2.78. We will see later why this is commonly performed.
- Place conditioning data into the model.[6] This is step (b) in Figure 2.78.
- Go to a random location $\mathbf{u}$ and search for all neighbouring data and previously simulated nodes. This is step (c) in Figure 2.78.
- Perform kriging with these data and previously simulated nodes to obtain kriged estimate and the corresponding kriging variance (this is step (d) in Figure 2.78):

$$Y^*(\mathbf{u}) = \sum_{\beta=1}^{n} \lambda_\beta \cdot Y(\mathbf{u}_\beta)$$

$$\sigma^2_{SK}(\mathbf{u}) = C(0) - \sum_{\alpha=1}^{n} \lambda_\alpha C(\mathbf{u}, \mathbf{u}_\alpha)$$

- Draw a random residual $R(\mathbf{u})$ that follows a normal distribution with mean of 0.0 and variance of $\sigma^2_{SK}(\mathbf{u})$.
- Add the kriged estimate and residual to get the simulated value:

$$Y_s(\mathbf{u}) = Y^*(\mathbf{u}) + R(\mathbf{u})$$

  Note that $Y_s(\mathbf{u})$ could be equivalently obtained by drawing from a normal distribution with mean $Y^*(\mathbf{u})$ and variance $\sigma^2_{SK}(\mathbf{u})$. This is step (e) in Figure 2.78. The independent residual $R(\mathbf{u})$ is drawn with classical Monte Carlo simulation. Any reputable pseudo-random number generator may be used. It is good practice to retain the "seed" to the random number generator to permit reconstruction of the model if required.
- Add $Y_s(\mathbf{u})$ to the set of data to ensure that the covariance with this value and all future predictions is correct. This is step (f) in Figure 2.78. As stated above, this is the key idea of sequential simulation, that is, to consider previously simulated values as data so that we reproduce the covariance between *all* of the simulated values.
- Visit all locations in a random order. This is step (g) in Figure 2.78. There is no theoretical requirement for a random order or path; however, practice has shown that a regular path can induce artifacts (Isaaks, 1990).
- Back transform all data values and simulated values when model is populated. This is step (h) in Figure 2.78.
- Create any number of realizations by repeating with different random number

[6] Previously, this step required some type of scaling relationship to move from well support to model grid support. Recent work (Deutsch, 2005c) has suggested the work flow of modeling at well support at model nodes with upscaling prior to transfer functions that require model cell averaged reservoir properties, such as flow simulation and some complicated volumetric calculations (e.g., involving thresholds).

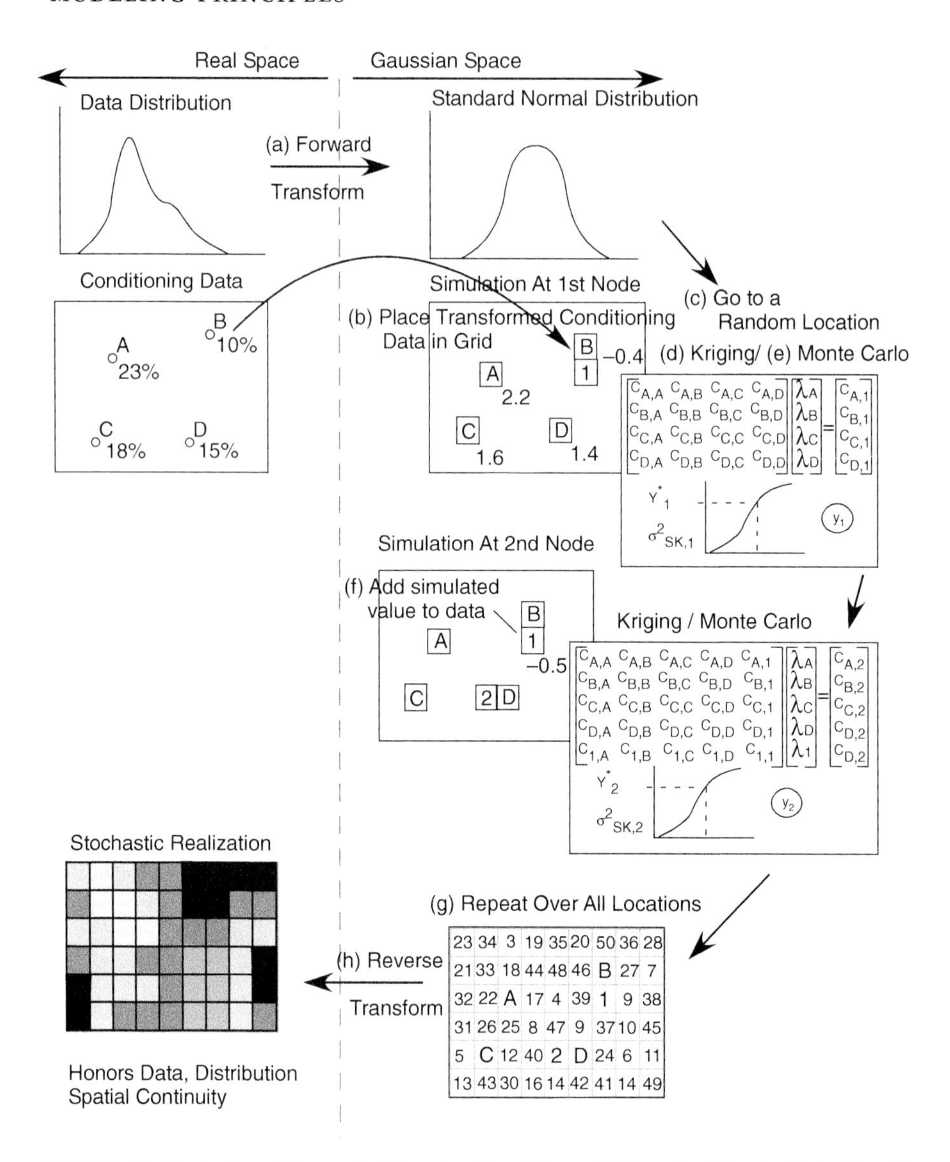

FIGURE 2.78: Illustration of Sequential Gaussian Simulation with Four Conditioning Data (A,...,D) Expanded for the First 2 Simulated Nodes (1 and 2). Note that the left side represents original data space (untransformed) and the right side represents Gaussian space (after transform and assuming all distributions are Gaussian). The steps (a) through (g) are itemized in the associated text.

seeds. A different seed leads to a different sequence of random numbers and, as a consequence, a different random path and different residuals for each simulated node. Each realization is "equally likely to be drawn" and often called equiprobable.

Gaussian Simulation Theory

The following discussion dives deeper into the theory that supports Gaussian simulation, including motivation for the use of the Gaussian distribution, simulation from the high-dimensional distributions that describe our reservoirs, the cost of Gaussianity, and the scale of simulation.

Why Gaussian?

The key mathematical properties that make sequential Gaussian simulation work are not limited to the Gaussian distribution. The covariance reproduction of kriging holds regardless of the data distribution; the correction of variance by adding a random residual works regardless of shape of the residual distribution (the mean must be zero and the variance equal to the kriging variance).

The use of any distribution other than the Gaussian distribution will lead to an incorrect global distribution of simulated values. The mean may be correct, the variance is correct, the variogram of the values taken all together is correct, but the "shape" will not be. This is not a problem in "Gaussian space" since all distributions are Gaussian (Anderson, 1958), including that of the final simulated values.

There is another reason why the Gaussian distribution is used: The central limit theorem tells us that the sequential addition of random residuals to obtain simulated values leads to a Gaussian distribution.

Monte Carlo Simulation

Simulation can be described as joint Monte Carlo simulation from a multivariate distribution where N is the number of locations simulated. Monte Carlo simulation from a univariate distribution is straightforward (see Section 2.2 for discussion), but Monte Carlo simulation from very high-dimensional distributions is more complicated. Consider that the spatial relations and local data conditioning must all be honored. This requires a multivariate distribution that describes the relationships between all locations in the model that may be sampled. Considering the size of the typical reservoir model (i.e., millions of grid nodes), the Gaussian model is the only practical multivariate model that is available. As discussed before, we apply a univariate transform of all variables to Gaussian distributions and assume all multivariate statistics are Gaussian.

Sequential simulation can be thought of as a recursive decomposition of a multivariate Gaussian distribution through Bayes' Law (see Section 2.2 for descriptions and definitions). Recall the product rule:

$$P(A,B) = P(B|A) \cdot P(A)$$

where the joint probability, $P(A,B)$, is calculated from the conditional probability, $P(B|A)$, and the marginal probability, $P(A)$. Through this relation, we sample from the joint probability with the univariate probability of A and the conditional probability of B given A. We may continue to expand this relation to a trivariate distribution.

$$P(A,B,C) = P(C|B,A) \cdot P(B|A) \cdot P(A)$$

The recursive application of Bayes' Law may be posed as follows:

$$\begin{aligned} P(A_1,\ldots,A_N) &= P(A_N|A_1,\ldots,A_{N-1}) \cdot P(A_1,\ldots,A_{N-1}) \\ &= P(A_N|A_1,\ldots,A_{N-1}) \cdot P(A_{N-1}|A_1,\ldots,A_{N-2}) \\ &\quad \cdot P(A_1,\ldots,A_{N-2}) \\ &= P(A)\prod_{i=1}^{n} P(B_i|A,B_1,\ldots,B_{i-1}) \end{aligned}$$

This may be described as the simulation of N of A_j joint events proceeding sequentially as follows:

- Draw A_1 from the marginal distribution $P(A_1)$.
- Draw A_2 from the conditional distribution $P(A_2|A_1 = a_1)$.
- Draw A_3 from the conditional distribution $P(A_3|A_1 = a_1, A_2 = a_2)$.
- Draw A_N from the conditional distribution $P(A_N|A_1 = a_1, A_2 = a_2, \ldots, A_{N-1} = a_{N-1})$.

Importantly, this recursive application of Bayes' Law to sample from the high-order joint probability distribution is theoretically valid and does not require approximations or assumptions. Now we can recall sequential Gaussian simulation in the context of recursive applications of Bayes' Law. Sequential simulation requires knowledge about the $N-1$ conditional distributions where N is the number of grid nodes in the model. The first location along the random path is simulated from the marginal distribution. If data are available, then the first value is conditional to the data and the marginal distribution. As we proceed in simulation sequence along the random path over N grid nodes, there is an increase in conditioning due to the sequential inclusion of previously simulated nodes. We are drawing from the conditional distributions, given all previously simulated nodes and data. It is common to limit the search distance to remove previously simulated nodes and data that are not significantly correlated with the location being simulated. This greatly speeds up the simulation, while having no significant influence on the results.

The Cost of Gaussianity

The cost of the mathematical simplicity of Gaussian techniques is the character of maximum entropy (Journel and Deutsch, 1993), which implies maximum spatial disorder beyond the imposed variogram correlation. One consequence is that there is reduced probability of simulating connected paths of extreme values. This will have an affect on the results of flow simulation, but it is impossible to know how significant this will be without performing flow simulation.

The consequence of maximum entropy is that an SGS realization will likely have less spatial structure than the real reservoir. This may be important for variables like permeability since the continuity of high- and low-permeability values has a significant impact on fluid flow predictions. Other simulation methods such as indicator geostatistics or simulated annealing could be considered when additional knowledge of the spatial structure can be quantified.

Notwithstanding the consequences of Gaussianity, the sequential Gaussian simulation algorithm is widely used. Figure 2.79 illustrates two SGS realizations of reservoir thickness. The spatial correlation matches the input variogram; high and low values are grouped together.

Scale of Simulation

Now we should consider the scale of the simulation. While all estimation and simulation methods assume scale, simulation is expected to reproduce the input statistics and these statistics have an assumed support size. For example, as discussed in Section 2.3 with the concept of volume variance relations, the histogram and variogram of well support, grid cell support and reservoir support are usually very different. Traditionally, simulation models are assumed to be at grid cell support size. That is, facies, porosity, permeability, and other reservoir properties are the effective properties over the entire grid cell for each in the model. This work flow requires that the data and associated statistics are scaled up to the grid cell support size and then simulation proceeds at this support size. When models are viewed, the cells are labeled or "painted" with a single set of reservoir properties representative of the entire cell. For example, at each location $\mathbf{u}_{\alpha=1,\ldots,N}$ in the model a single model cell may be represented by a simulated result of facies, $f(\mathbf{u}_{\alpha=1})$ = sand, with porosity, $\phi(\mathbf{u}_{\alpha=1}) = 15\%$,

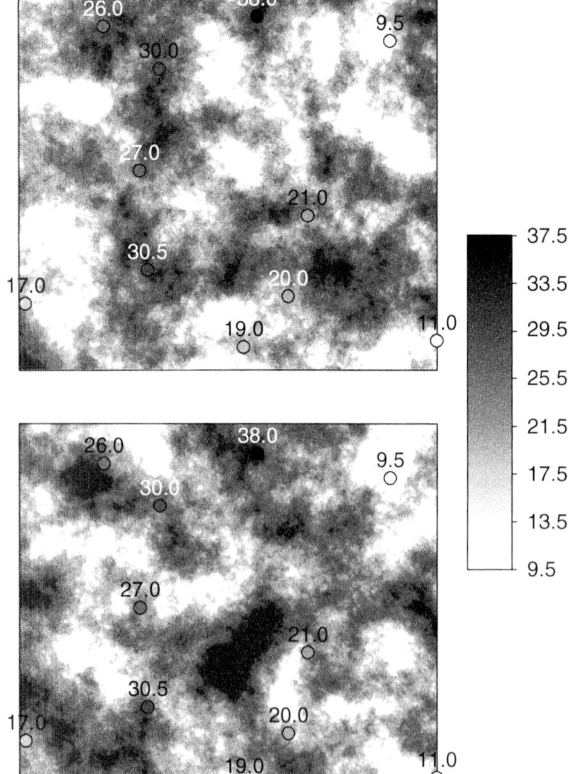

FIGURE 2.79: Two SGS Realizations Created for Illustration Purposes. Note how the data are reproduced in both cases. There is significant uncertainty in the realizations since the range of correlation is less than the average data spacing.

and horizontal permeability, $k_h(\mathbf{u}_{\alpha=1}) = 550\,\text{mD}$ and vertical permeability, $k_v(\mathbf{u}_{\alpha=1}) = 40\,\text{mD}$. In practice though, the scale up from data support to grid support is challenging (see scaling discussion in Section 5.2). In fact, often scale up is not rigorously dealt with or even ignored. This issue has been labeled the *missing scale*. In addition, seismic integration requires non-unique downscaling for consistent scale, which is often also ignored (Kalla et al., 2008).

An alternative work flow is to simulate at data support scale (Deutsch, 2005c). For example, wells data are composited to a single value in the grid cell, but are considered to represent well support along the typically vertical or subvertical trajectory across the grid cells and seismic data is retained at seismic resolvability without down scaling. Statistics, including correlations, are calculated at the data support size. The simulated realizations are considered to be at point support. Such a model could be accurately visualized as a mesh of simulated values in the model space, in that simulated model does not inform the locations between the simulated nodes. Yet, it is more convenient to visualize simulated realizations with cells painted; therefore, it is important to explicitly state the assumed data scale support to avoid misunderstanding. This work flow has avoided the previously mentioned scale issues. In fact, for many volumetric transfer functions, numerical integration over the mesh is sufficient and there is no need to apply scaling. For example, oil in place may be calculated by integrating over the porosity and saturation meshes, given the mesh is fine enough to describe the reservoir. Yet, flow simulation does require effective properties at the scale of the flow grid cells; therefore, scale-up is required from the simulated mesh at data support size to the flow grid support size. Even the traditional work flow above often required a second scale-up from reservoir model scale to flow scale size. With the new work flow the scale-up from simulated data support to flow grid support size is readily dealt with as the mesh should provide good constraints for flow based upscaling.

While the second method is convenient, in that it reduces work flow complexity by postponing or avoiding scaling, the first method is more commonly applied; therefore, both are discussed with respect to model scale. In Section 2.3, statistical methods for scaling statistical inputs is described; in Section 5.2, model scaling is introduced as a form of model post-processing.

2.4.3 Indicator Formalism

The key idea behind the indicator formalism is to code all of the data in a common format—that is, as *probability* values (Alabert, 1987; Gómez-Hernández and Srivastava, 1990; Journel, 1983; Journel and Alabert, 1988, 1990; Journel and Gómez-Hernández, 1993; Sullivan, 1985). The two main advantages of this approach are (1) simplified data integration because of the common probability coding, and (2) greater flexibility to account for different continuity of extreme values. The comparative performance of indicator methods has been studied extensively by Goovaerts (1994a–c).

The indicator approach for continuous data variables requires significant additional effort versus Gaussian techniques. The indicator formalism applied to categorical data has seen wide application in facies modeling, particularly for carbonate reservoirs and high net-to-gross siliciclastic reservoirs. Regardless of the variable type, the indicator approach leads directly to a distribution (histogram) that is a model of uncertainty at each unsampled location estimated. Nevertheless, the implementation details for continuous data are quite different from categorical variables; the two approaches are discussed separately.

Continuous Data

Indicator methods for continuous variables are somewhat out of favor due to know artifacts discussed in Section 4.6. For completeness and due to their wide application, the method is presented here. The aim of the indicator formalism for continuous variables is to directly estimate the distribution of uncertainty $F_Z(\mathbf{u})$ at unsampled location $\mathbf{u}$. The cumulative distribution function is estimated at a series of threshold values: $z_k, k = 1, \ldots, K$. As an example, Figure 2.80 shows $K = 5$ probability values at five threshold values that provide a distribution of uncertainty. The probability values are evaluated by first coding the data as indicator or probability values. The indicator coding at location $\mathbf{u}_\alpha$ is written as

$$i(\mathbf{u}_\alpha; z_k) = \text{Prob}\{Z(\mathbf{u}_\alpha) \le z_k\} = \begin{cases} 1, & \text{if } z(\mathbf{u}_\alpha) \le z_k \\ 0, & \text{otherwise} \end{cases} \tag{2.21}$$

The expected value of this indicator variable is the stationary prior probability to be less than the

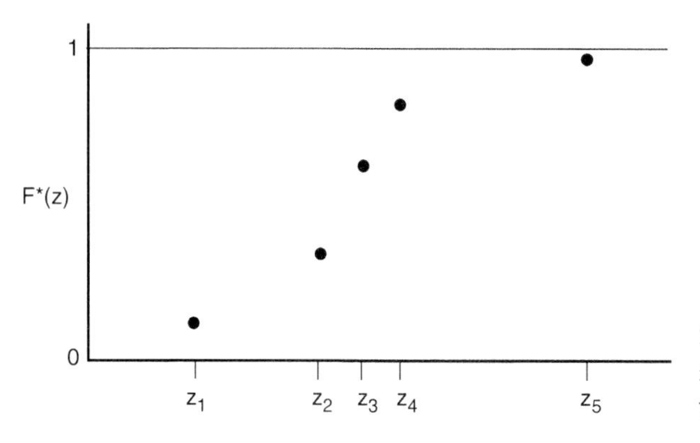

FIGURE 2.80: Schematic Illustration of Probability Distribution F(z) at a Series of Five Threshold Values, z_k, $k = 1, \ldots, 5$.

threshold, that is, $F_Z(z_k)$. As in Eq. (2.21), above, we consider residual data:

$$Y(\mathbf{u}_\alpha; z_k) = i(\mathbf{u}_\alpha; z_k) - F(z_k),$$
$$\alpha = 1, \ldots, n, \; k = 1, \ldots, K$$

As in Eq. (2.16), a linear estimate of this residual is considered:

$$Y^*(\mathbf{u}; z_k) = \sum_{\alpha=1}^{n} \lambda_\alpha(z_k) \cdot Y(\mathbf{u}_\alpha; z_k) \qquad (2.22)$$

where $Y^*(\mathbf{u}; z_k)$ is the estimated residual from the prior probability at threshold z_k; $F(\mathbf{u}; z_k)$, $\lambda_\alpha(z_k)$, $\alpha = 1, \ldots, n$, are the weights; and $Y(\mathbf{u}_\alpha; z_k), \alpha = 1, \ldots, n$, are the residual data values.

The indicator kriging derived cumulative distribution function at an unsampled location at threshold z_k is calculated as

$$F_{\text{IK}}(\mathbf{u}; z_k) = \sum_{\alpha=1}^{n} \lambda_\alpha(z_k) \left[i(\mathbf{u}_\alpha; z_k) - F(z_k) \right] + F(z_k)$$

In the presence of no data ($n = 0$) the indicator kriging (IK) estimate of the distribution of uncertainty is simply the prior mean at that threshold, that is, $F(z_k)$.

The IK process is repeated for all K threshold values $z_k, k = 1, \ldots, K$, which discretize the interval of variability of the continuous attribute Z. The distribution of uncertainty, built from assembling the K indicator kriging estimates, represents a probabilistic model for the uncertainty about the unsampled value $z(\mathbf{u})$. This indicator kriging procedure requires a variogram measure of correlation corresponding to each threshold $z_k, k = 1, \ldots, K$ so that the weights $\lambda_\alpha(z_k), \alpha = 1, \ldots, n, k = 1, \ldots, K$, can be determined.

The correct selection of the threshold values z_k for indicator kriging is essential: Too many threshold values and the inference and computation becomes needlessly tedious and expensive; too few, and the details of the distribution are lost. Between 5 and 11 thresholds are normally chosen. The thresholds are rarely above the 0.9 quantile or below the 0.1 quantile since it becomes difficult to reliably infer the corresponding indicator variograms. The thresholds are often chosen to be equally spaced quantiles, for example, the nine deciles are often chosen. Implementation details will be covered in more detail in Section 4.6, Porosity and Permeability Modeling.

The indicator residual data in the IK expression [see Eq. (2.22)] originate from hard data $z(\mathbf{u}_\alpha)$ that are deemed perfectly known; thus, the indicator data $i(\mathbf{u}_\alpha; z)$ are hard in the sense that they are valued either 0 or 1 and are available at any cutoff value z. There are many applications where some of the z-information is due to soft secondary data such as seismic data. This soft data may enter indicator kriging in a variety of ways.

The first approach for soft data is to use them for locally variable prior probability values $F(\mathbf{u}; z_k)$. The soft data would be calibrated to hard data by considering collocated hard and soft data. The hard data are cross plotted against the soft data, and "conditional" distributions of the original data variable for classes of the soft data variable are extracted. The $F(\mathbf{u}; z_k)$ values are taken from the conditional distributions considering the soft data variable at each location $\mathbf{u}$. This is illustrated by looking ahead to the bottom of Figure 2.81. A second approach is to consider the $F(\mathbf{u}; z_k)$ values as a covariate in a cokriging context rather than as a prior mean. Any licit model of coregionalization can be used including the linear model of coregionalization or a collocated model. The Markov–Bayes model (Zhu, 1991) is a model of coregionalization specially formulated for soft and hard indicator data.

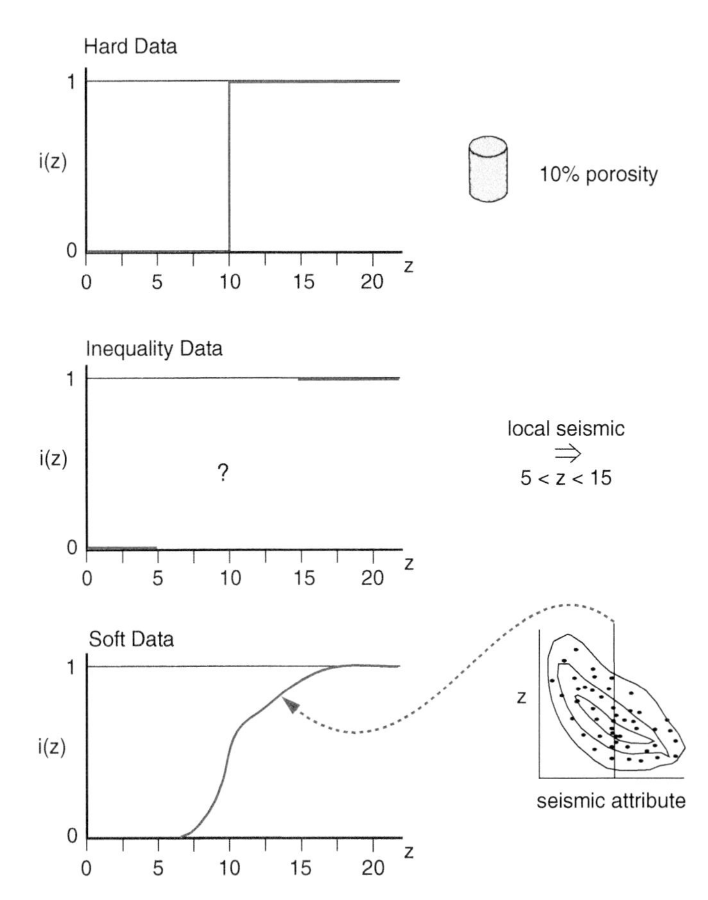

FIGURE 2.81: Schematic Illustration of (top) Hard Data Where the Indicator Transform i Is Zero or One, (Middle) Inequality Data Where the Indicator Data Are Zero Below the Lower Limit and One Above the Upper Limit, and (Bottom) a Soft Data Derived Distribution.

A third approach for soft data is to consider them as inequality constraints or soft indicator data. At location $\mathbf{u}_\alpha$, inequality data tell us that z-value is between a lower limit $z_{\text{low}}(\mathbf{u}_\alpha)$ and an upper limit $z_{\text{upper}}(\mathbf{u}_\alpha)$, that is,

$$i_{\text{inequality}}(\mathbf{u}_\alpha; z_k) = \begin{cases} 0, & \text{if } z_k < z_{\text{low}}(\mathbf{u}_\alpha) \\ \text{undefined}, & \text{if } z_{\text{low}}(\mathbf{u}_\alpha) \geq z_k \leq z_{\text{upper}}(\mathbf{u}_\alpha) \\ 1, & \text{if } z_k > z_{\text{upper}}(\mathbf{u}_\alpha) \end{cases}$$

The inequality constraint data can be used for thresholds that are defined; otherwise, there are no data for thresholds that are "missing or undefined." Figure 2.81 illustrates the indicator representation of hard data, inequality constraints, and soft data.

Indicator Variograms

One of the primary advantages of the indicator formalism is the ability to specify different spatial continuity at each threshold. This amounts to specifying problem-specific continuity for the low, median, and high values. Figure 2.82 presents indicator variograms at seven thresholds. These variograms were calculated from a 2-D exhaustive image of laminated sandstone. Note the changing zonal anisotropy and nugget effect for the different thresholds. The calculated variogram values and the models should be standardized to a unit variance by dividing by $p(1-p)$, where $p = F(z_k)$ is the global mean of the indicator variable or proportion of 1s. The parameters (relative nugget effect, ranges, and so on) of the indicator variograms are related since they arise from a common continuous-variable origin.

Order Relations

The series of indicator-derived probability values $F_{\text{IK}}(\mathbf{u}; z_k), k = 1, \ldots, K$, must meet the requirements of a cumulative distribution function—that is, a nondecreasing function between zero and one. This is not guaranteed by indicator kriging; therefore, an a posteriori correction is applied. The corrections required are usually small. Large deviations imply a problem with the data or inconsistency with the indicator variograms. The correction method

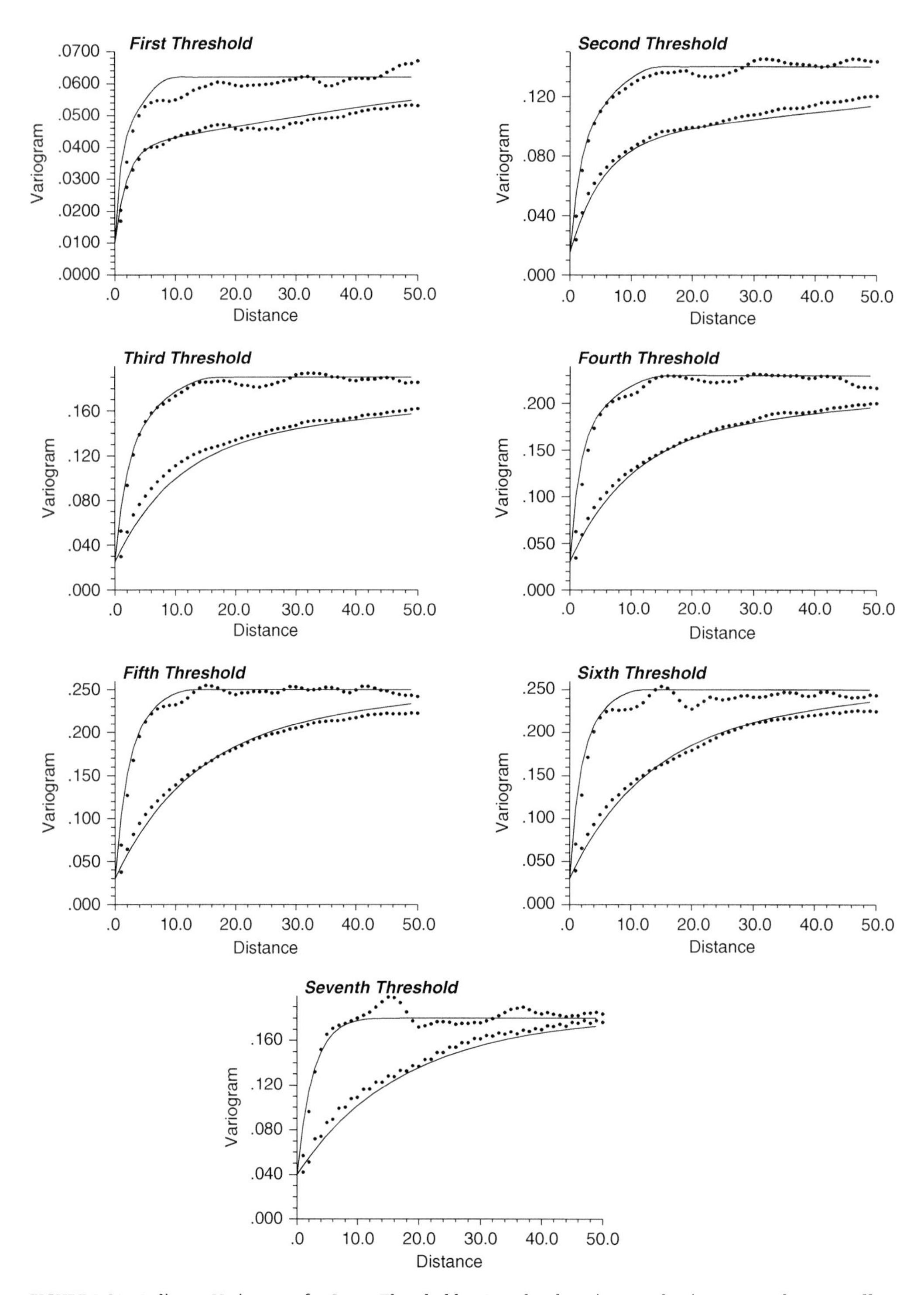

FIGURE 2.82: Indicator Variograms for Seven Thresholds. Note the changing zonal anisotropy and nugget effect for the different thresholds. Also note how the variogram is different at low and high thresholds.

documented in GSLIB (Deutsch and Journel, 1998) is simple and effective.

Categorical Data

The aim of the indicator formalism for categorical variables is to directly estimate the distribution of uncertainty in the categorical facies variable. The probability distribution consists of estimated probabilities for each category: $p^*(k), k = 1, \ldots, K$. As an example, Figure 2.83 shows $K = 5$ probability values for five facies. The probability values are estimated by first coding the data as indicator or probability values:

$$i(\mathbf{u}_\alpha; z_k) = \text{Prob}\{\text{facies } k \text{ being present}\}$$
$$= \begin{cases} 1, & \text{if facies } k \text{ is present at } \mathbf{u}_\alpha \\ 0, & \text{otherwise} \end{cases}$$

The expected value of this indicator variable is the stationary prior probability of facies k, that is, $p(k)$. As above, we consider residual data:

$$Y(\mathbf{u}_\alpha; z_k) = i(\mathbf{u}_\alpha; k) - p(k),$$
$$\alpha = 1, \ldots, n, \quad k = 1, \ldots, K$$

Kriging of these residual data is used to derive the probability of each facies $k = 1, \ldots, K$ at an unsampled location. Once again, a variogram measure of correlation is required for each facies $k = 1, \ldots, K$. The result of indicator kriging is a model of uncertainty at location $\mathbf{u}$:

$$p_{\text{IK}}(\mathbf{u}; k) = \sum_{\alpha=1}^{n} \lambda_\alpha(k) \cdot \left[i(\mathbf{u}_\alpha; k) - p(k) \right] + p(k),$$
$$k = 1, \ldots, K$$

Seismic and other secondary data sources may be coded as soft probabilities for cokriging; details of this are covered later in Section 4.6, Porosity and Permeability Modeling.

Order Relations

The estimated probabilities $p_{IK}(k), k = 1, \ldots, K$, must meet the requirements of a probability distribution—that is, being non-negative and sum to one. As with continuous variables, these order relation requirements are not guaranteed by indicator kriging; therefore, an a posteriori correction is applied. The probabilities are set to zero if they are negative and then reset according to

$$p^*(k) = \frac{p_{\text{IK}}(k)}{\sum_{k=1}^{K} p_{\text{IK}}(k)}, \qquad k = 1, \ldots, K$$

which ensures they sum to one. Large deviations from licit probabilities imply a problem with the data or the variograms.

Sequential Indicator Simulation

The concept of sequential simulation, described in context of Gaussian (SGS), can be extended to the indicator-based model of uncertainty. The grid nodes are visited sequentially in a random path. At each grid node:

1. Search for nearby data and previously simulated values.
2. Perform indicator kriging to build a distribution of uncertainty (correcting order relation problems if necessary).
3. Draw a simulated value from the distribution of uncertainty.

Multiple realizations are then generated by repeating the entire procedure with a different random number seed. Implementation details are included in Section 4.2, Variogram-Based Facies Modeling. This indicator method provides a flexible variogram-based simulation method for categorical variables, along with the ability to constrain spatial continuity for each category. Yet, the indicator approach does not account for ordering relationships between categories; the indicator variogram is only a measure of probability of transition from the current category to any other category. To reproduce category ordering relationships, truncated Gaussian simulation or multiple-point methods are required.

Truncated Gaussian Simulation

The truncated Gaussian simulation method is another variogram-based methodology to generate categorical variable realizations (Armstrong et al., 2003). This methodology may be viewed as a post-processing (i.e., truncation) of continuous Gaussian simulation (or multivariate Gaussian simulations in the case of Plurigaussian simulation). The algorithm steps include:

1. Truncate a continuous Gaussian distribution to honor category proportions and ordering relations (categories in contact with each other should be in adjacent bins).
2. Infer a single continuous variogram to represent the spatial continuity of all the facies.

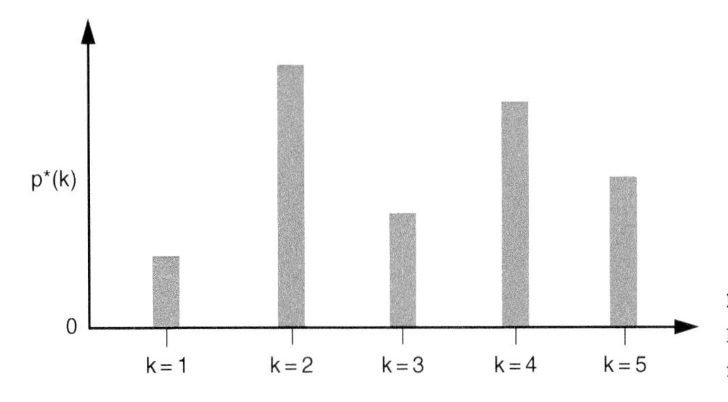

FIGURE 2.83: Schematic Illustration of Probability Distribution $p^*(k), k = 1, \ldots, 5$, for Five Facies Types, $k, k = 1, \ldots, 5$.

3. Calculate a continuous simulation conditioned to the categorical data transformed into the continuous variable (typically with sequential Gaussian simulation). Common practice is to assign the centroid of each bin in the continuous variable to the associated category. For example, if "facies 1" includes the lower tail up to −0.25 (in Gaussian space) or the lower 40% of the continuous distribution, then facies 1 at the well locations may be coded as −0.84 or the P20 of the standard normal distribution for the continuous property simulation.
4. Truncate the continuous simulation into a categorical simulation. The thresholds may vary locally to reproduce locally variable category proportions.

Multiple realizations are then generated by repeating the entire procedure with a different random number seed. The critical feature is that unlike indicator simulation, truncated Gaussian simulation reproduces ordering relationships, but loses explicit control over the spatial continuity of each category. While this simple description of basic implementation is a sufficient introduction for the context required for discussion on algorithm selection in Section 3.2, Conceptual Modeling, there are more details and more advance implementations that are discussed in Section 4.2, Variogram-Based Facies Modeling.

2.4.4 P-Field Methods

Sequential simulation proceeds by (1) constructing a conditional distribution model that accounts for all original data and previously simulated grid nodes and (2) drawing with Monte Carlo simulation from the conditional distributions. Multiple realizations are generated by repeating the procedure. The CPU cost of such methods is mostly due to the cost of constructing each local conditional distribution over again for each new realization.

In sequential simulation, correlation between simulated values is achieved in the construction of the conditional distributions. R. M. Srivastava proposed a different approach (Froidevaux, 1993; Srivastava, 1992). The two key aspects of his proposal were to (1) construct the conditional distributions using *only* the original data so that it can be done only once instead of repeatedly for each realization and (2) draw from the conditional distributions using correlated probability values instead of the random numbers used in traditional Monte Carlo simulation.

- Spatial correlation between the probability values imparts spatial correlation in the resulting simulated values.
- The original data are honored since the conditional distributions at data locations have zero variance and the original data values are retrieved regardless of the simulated probability value.

There is a significant CPU advantage to p-field simulation when multiple realizations must be constructed.

Figures 2.84 and 2.85 present a 1-D schematic illustration of the procedure. The construction of the conditional distributions of uncertainty at each location, using the original data alone, is the first step, (see Figure 2.84). Three correlated probability fields and the corresponding simulated values are shown in Figure 2.85.

The *p*-field simulation methodology requires unconditional realizations of probability values. There

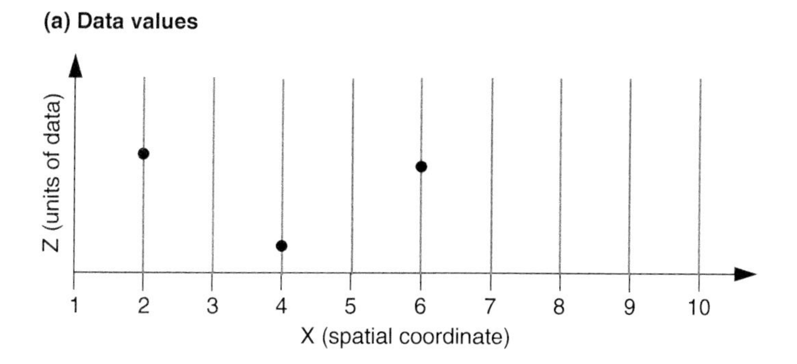

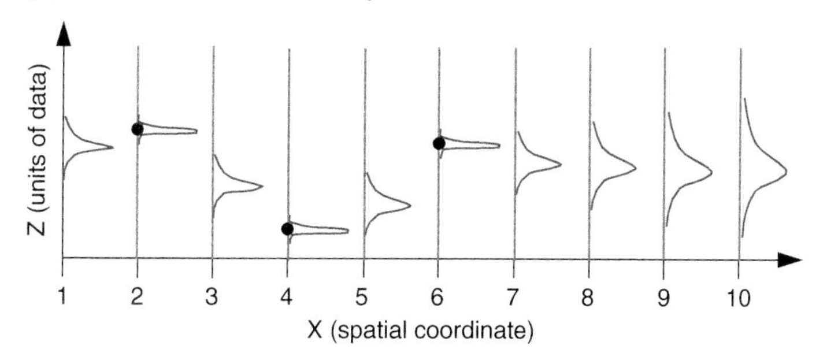

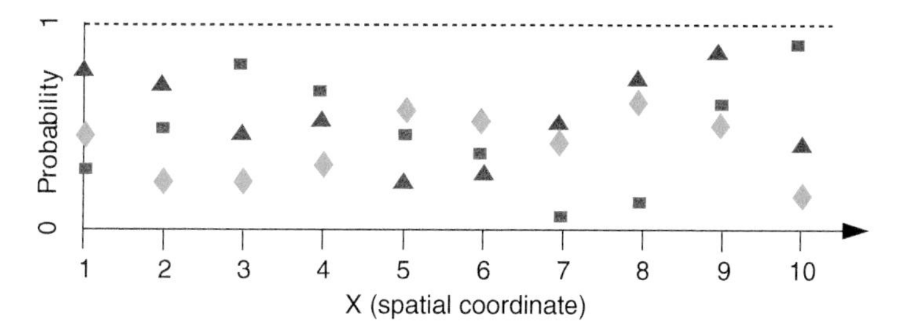

FIGURE 2.84: The First Step in p-Field Simulation, That Is, the Construction of the Conditional Distribution of Uncertainty at Each Location Using the Original Data: (a) 1-D example conditioning data, (b) distributions of uncertainty, and (c) probability values from unconditional simulation. Note how the variance of the conditional distributions in (b) is lower near the data values.

are a number of numerical methods that are suited to fast simulation of uncorrelated spatial fields such as FFT, fractal, moving average, and turning bands methods (Chu and Journel, 1994; Gutjahr, 1989; Hewett, 1986, 1993, 1995). These will not be recalled here; however, we note that there are more efficient techniques than the common SGS method.

In spite of the apparent advantages of the p-field simulation methodology, it should be used only for special applications. Froidevaux (1993), Pyrcz and Deutsch (2002), and Srivastava and Froidevaux (2005) pointed out two artifacts. The artifacts are as follows: (1) The conditioning data appear as local minima and maxima, and (2) the spatial continuity is not reproduced next to the conditioning data; the values are too smooth or too well correlated near the conditioning data.

The first artifact is caused by the separation of conditioning and the Monte Carlo simulation. The local distributions collapse to a step function at the local hard data locations. This causes the simulated values to "pinch" toward the hard data values in the vicinity of the hard data. The artifact becomes more significant as the probabilities depart from 0.5 in the

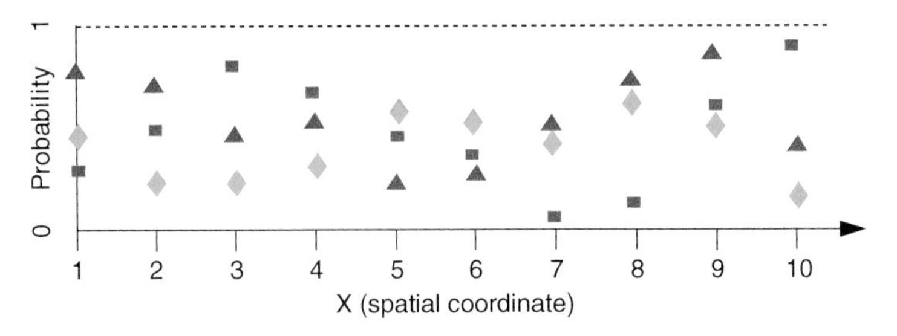

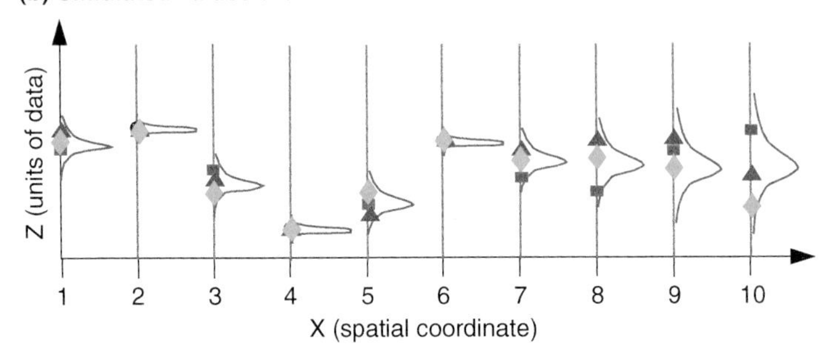

FIGURE 2.85: The Second Step in *p*-Field Simulation Is the Generation of Correlated Probability Fields and the Simultaneous Drawing from the Conditional Distributions of Uncertainty: (a) three sets of correlated probability values, and (b) distributions of simulated values drawn from the distributions of uncertainty shown in Figure 2.84.

vicinity of the data location. Probability field values greater than 0.5 in the vicinity of the data cause a local minimum. Probability values less than 0.5 cause a local maximum. The only instance in which this artifact will not occur is when the nonconditional correlated probability value is exactly 0.5 at the data location, which is unlikely.

The second artifact is that there is increased continuity next to conditioning data. Although expected from theory, this bias causes practical problems. It should be mentioned that these artifacts do not appear significant when dealing with categorical variables, where the method is equivalent to the truncated Gaussian method (see Section 4.2, Variogram-Based Facies Modeling).

A popular variant of p-field is the cloud transform simulation. In this method the local distributions of uncertainty are unconditional (derived by fitting a bivariate distribution between primary and secondary data) and the p-field is conditioning to draw the data values at the data locations; therefore, this method does not suffer from the previously mentioned artifacts. This method is discussed in detail below in Alternatives for Secondary Data Integration.

2.4.5 Multiple-Point Simulation

The previously described sequential simulation method may be applied with the substitution of multiple-point statistics introduced in Section 2.3. The same steps—including assigning data to the model at data locations, simulating along a random path at all other model locations, Monte Carlo simulation from local distributions of uncertainty, and sequential placement of the simulated values into the model—are applied. The only difference is that the local conditional probability distributions are found by sampling the training image with a multiple-point template.

After the construction of a training image (analogous to inferring a variogram for Gaussian simulation), the basic multiple point simulation method proceeds in a fashion similar to that of Gaussian simulation, although there is no forward transform to Gaussian space and no back transform to original units. The steps are as follows:

- Place conditioning data into the model. This step requires some blocking of the well data over the grid cell to represent the cell scale effective property within the cells.
- Go to a random location **u** and search for all neighbouring data and previously simulated nodes.
- Scan the relevant training image for replicates with the data configuration observed in the model. Pool these to calculate the conditional probability density function at location **u** [in the Strebelle (2002) method, this is completed up front and stored in a convenient search tree]. The conditional distribution amounts to a probability for each facies given the available data template, d_n: $P(Z(\mathbf{u}) = z_k|d_n), k = 1, \ldots, K$.
- Perform Monte Carlo simulation from this conditional probability density function. Generate a random number and choose the facies where the cumulative conditional probability is less than or equal to that facies.
- Visit all locations in a random order. There is no theoretical requirement for a random order or path; however, practice has shown (Isaaks, 1990) that a regular path can induce artifacts.
- Create any number of realizations by repeating with different random number seeds. A different seed leads to a different sequence of random numbers and, as a consequence, a different random path and different facies at each location.

Modern methods diverge from these basic steps to improve run time, pattern reproduction, and memory storage efficiency. These are considered part of the various implementation details related to computational efficiency and pattern reproduction, such as template design and search trees along with alternative methods to integrate multiple-point statistics that are covered in Section 4.3, Multiple-Point Facies Modeling.

2.4.6 Object-Based Simulation

Object-based simulation is another framework for building simulated realizations of facies. The previous methods may all be classified as "cell-based"; that is, these methods utilize statistical constraints such as the variogram and multiple-point statistics from a training image to constrain the spatial continuity between grid cells; the resulting models do not have direct control on large-scale geometry and, as a result, do not reproduce clean geometries.

Object-based simulation produces attractive (visually clean) models that honor the idealized geometries often interpreted from outcrop and high-resolution seismic data. This is accomplished by sequentially placing geometries into the model. When specific reservoir geometries are readily identified and parameterized and have significance to the transfer function, these geometries may be integrated directly with object-based methods.

A short introduction to object-based models is provided here while background, implementation details and examples are presented in Section 4.4. Two major considerations for object-based modeling are geometric parameterization and geometry placement.

Geometrical Parameterization

A prerequisite for object-based modeling is that the reservoir heterogeneity can be characterized by parametric geometries. For example, a distributary lobe complex is composed of multiple lobate features, a fluvial channel complex may be composed of multiple channel fills, and related gull wing levees and lobate crevasse splays or a massive sand with shale lenses. In each of these cases, the architecture is converted to parametric geometries.

A second prerequisite is that information is available on the appropriate parameters. In general, these parameters are considered uncertain and/or related to other parameters. For example, lobe geometries may be characterized by uniform distribution of length with min and max of 500 m and 2 km, respectively, while the depth and width of the lobe are related to the length by ratios of 1/100 and 1/2, respectively. The possible set of geometries are limitless when one considers common shapes, their 3-D equivalents, and the combination/merging of shapes and modifications that may be applied to these shapes, such as increase in undulation or sinuosity, thinning/thickening in any set of dimensions. Combinations may include the association of multiple linked geometries as distinct facies, such as channels with attached levees, crevasse splays, and distributary lobes (see examples in Figure 2.86).

Geometry Placement

This is another major distinction with the cell-based methods. Whereas cell-based methods assign data to the model cells and proceed along a random path

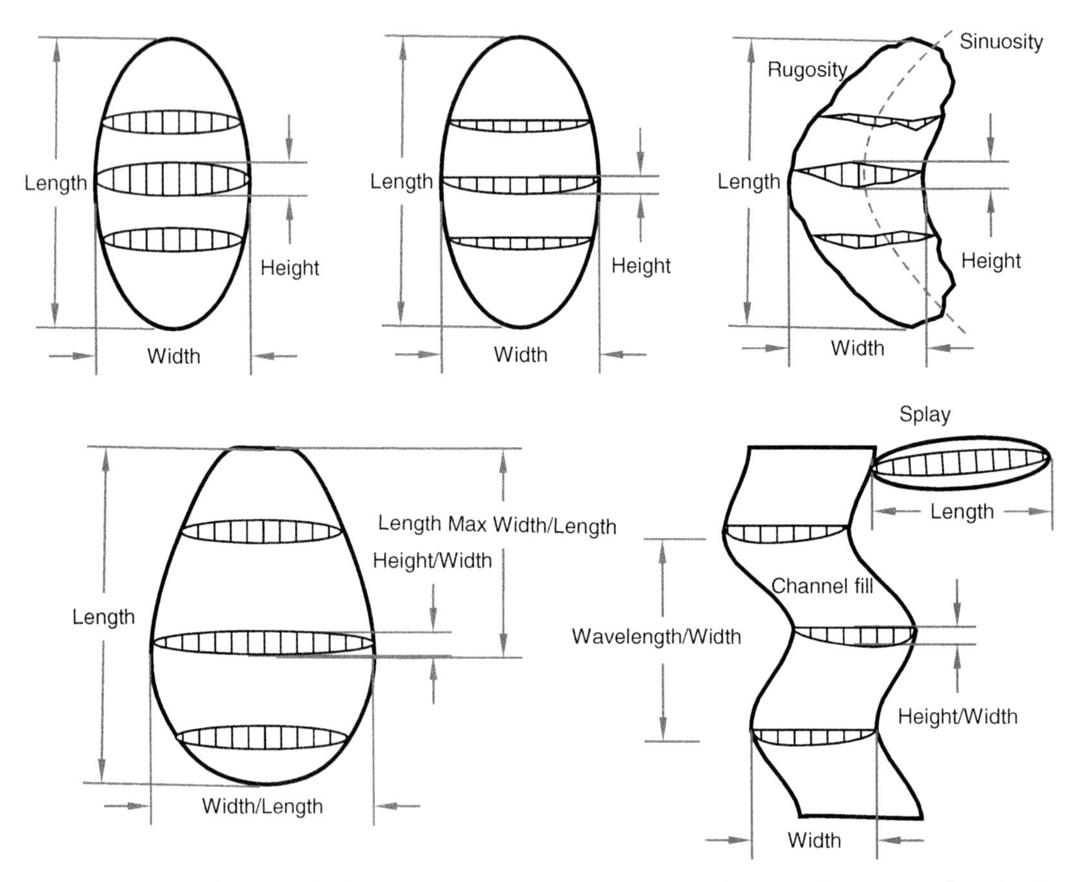

FIGURE 2.86: Example Geometric Objects and Associated Parameterization. Object complexity can vary (consider the simple ellipsoid) with length, maximum width, and height. Removal of the top half results in a concave upward fill, and further additions of rugosity and sinuosity result in a complicated geometry. In addition, various relationships between parameters may be known and utilized; for example, consider either the width-to-length and height-to-width ratios of a lobe or the height-to-width and wavelength-to-width ratios of a channel fill. In addition, multiple objects may be linked, such as splay deposits attached to channel fills.

simulating at all other cells, object-based modeling initializes the model space with a background facies and places geometries randomly until criteria such as a global proportion of objects are met. The first consequence of this method is that conditioning to well data is not automatic and guaranteed as is the case with cell-based methods. In fact, object-based conditioning remains as a major challenge as is discussed in Section 4.4. The second consequence is the need for statistical constraints on the placement of the parametric geometries.

In the simplest case the objects are placed by a Poisson point process; that is, the object centroids are located randomly within the model. In more complicated cases constraints are included between objects, for example, objects must have a minimum spacing or objects cluster or the density of specific geometries vary over the model. One could consider a wide variety of methods to place the geometries (see Figure 2.87). More complicated parameterization and placement methods are available in the related Process-Mimicking Modeling Subsection in Section 4.5. Also, surface-based modeling is closely related to object-based modeling with the key difference being a focus on surfaces the bound the geometries.

There are many practical situations that are well-suited for object-based simulation. See Section 4.4 for implementation details and examples. Conditioning challenges limit the applicability, although some success has resulted by utilizing optimizations algorithms to match data. Actually, any

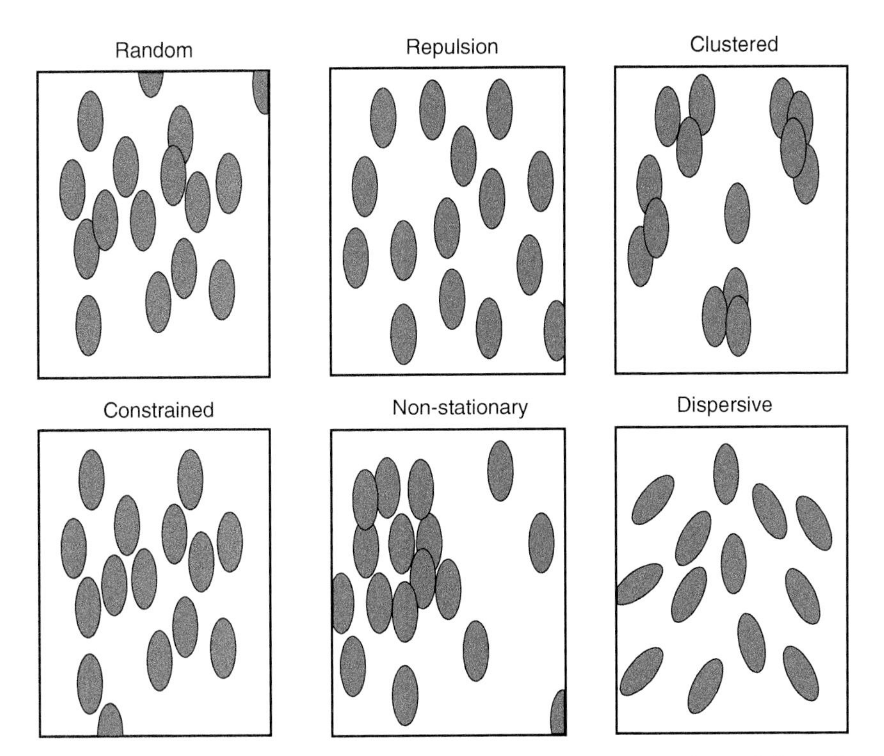

FIGURE 2.87: Illustration of Various Methods for Placement of Objects. These include simple random placement, repulsion, and clustering and more complicated constraints (no overlap in this case), nonstationary object density, and orientation.

modeling method and associated input statistics and conditioning data may be posed as an optimization problem.

2.4.7 Optimization Algorithms for Modeling

Data integration is of fundamental importance in reservoir modeling. The optimization technique of simulated annealing has found application in the integration of diverse data types. The key concept is iterative correction of an initial model to impart desired spatial features and to honor data conditioning. There are a number of issues related to (1) the initial image, (2) the perturbation mechanism, and (3) the objective function used in simulated annealing.

Simulated annealing is a solution method in the field of combinatorial optimization that has been adapted to geostatistical applications; it is not a stochastic simulation algorithm. This requires careful implementation to avoid artifacts such as (1) edge effects, (2) discontinuities at conditioning data, and (3) a restricted space of uncertainty. In spite of the problems of simulated annealing, it has found a place in reservoir characterization. More details are presented in Section 4.7, Optimization for Model Construction.

2.4.8 Accounting for Trends

Any input statistical constraint may be varied over all locations ($\mathbf{u}_\alpha$, where $\alpha \in V$) within the volume of interest. Most commonly the locally varying directions of continuity and the locally variable mean or categorical proportions are modeled and integrated into the reservoir model. Some details on this follow, and specifics for each modeling method are covered subsequently in Sections 4.2 to 4.7.

Locally Varying Directions of Continuity

Reservoir facies and petrophysical properties may exhibit changing directions of continuity; that is, the principal direction of continuity may depend on location. There are two issues associated with locally variable direction of continuity, including: (1) calculation of the locally variable direction field over the reservoir and (2) integration of the locally variable field in estimation and simulation. Boisvert (2010) has dealt with these in detail in his PhD thesis.

Calculation of Locally Variable Directions

One of the problems (as usual) is the data to evaluate the local variations. In many sparse data settings, inference is not possible given the data paucity, although geological concepts and analogs may provide some gross constraints on local directions. As with all other model parameters, if the inputs are uncertain, then this uncertainty should be passed through the modeling with multiple parameter scenarios or realizations. In dense data settings, calculation of locally variable directions is possible directly from the data. This may be accomplished by (1) manual mapping, (2) moment of inertia of local covariance maps, (3) direct estimation, and (4) automatic feature interpolation. If expert knowledge is available, then the manual mapping is preferable because it provides the ability to freely integrate this information, although manual methods may be challenging in complicated 3-D settings or to efficiently generate multiple realizations of locally variable direction fields. If exhaustive data are available, then the local moment of inertia method works well. If direct measures of direction are available, then direct estimation may be applied. The only issue is to perform estimation such that the unique nature of azimuths and angles is honored (e.g., 0 azimuth = 360 azimuth). This is usually accomplished by summing vectors with orientation from data and length from data weights. Automatic feature detection links similar values with polylines and interpolates orientation between without magnitude.

Figure 2.88 shows an example data set where the underlying detailed pixel plot shows locally varying directions of anisotropy. These data are the Walker Lake data from (Isaaks and Srivastava, 1989). Direction could be modeled manually or with moment of inertia methods.

A number of authors have presented approaches to handle locally varying directions of continuity. The central idea is to modify the kriging equations locally. Deutsch and Lewis (1992) setup the local kriging equations using the local direction of continuity. Xu (1995) implemented a similar procedure in the context of mimicking the facies distribution in a sinuous fluvial channel setting. Horta et al. (2010) implemented a procedure where only the right-hand side to the kriging equations were altered; the left-hand side, which accounts for data redundancy, used a global measure of anisotropy.

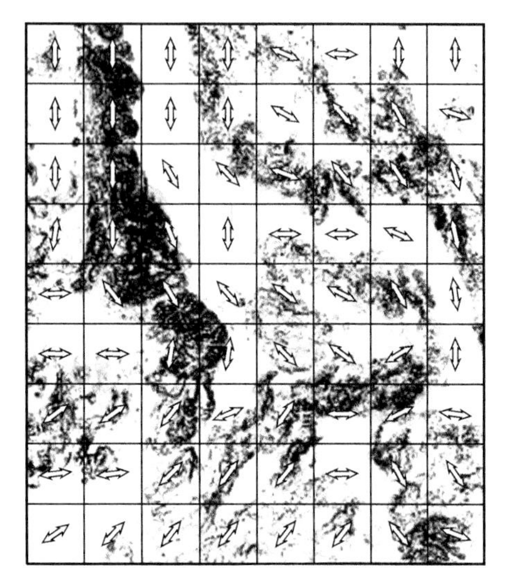

FIGURE 2.88: Specification of Direction Vectors within a Large-Scale Grid Network Superimposed on the Fine-Scale Distribution of Properties. The fine-scale distribution is the Walker Lake data from Isaaks and Srivastava (1989).

The typical approach to implement kriging and simulation with variable anisotropy for Gaussian and indicator sequential simulation are straightforward. The idea is to locally adjust to a constant direction of anisotropy within each search neighborhood. The problem of calculating distances in the presence of locally varying anisotropy and ensuring positive definiteness of the resulting kriging matrices is much simplified by retaining a single direction of anisotropy for each kriging. That direction of anisotropy is then changed from location to location.

Figure 2.89 shows an example of sequential indicator simulation with two local directions of anisotropy; the top half is at an azimuth angle of 60° and the bottom half is at an angle of 120°. Any complex grid of local directions could be used; however, the reproduction of the input direction grid depends on the kriging search radius and the conditioning data. Complex small-scale features will not be well reproduced because of the key assumption that the anisotropy does not change within a kriging neighborhood.

Boisvert (2010); Boisvert and Deutsch (2011) suggest a more rigorous methodology for accounting for locally variable directions beyond the previously mentioned local rotations. In this methodology, the

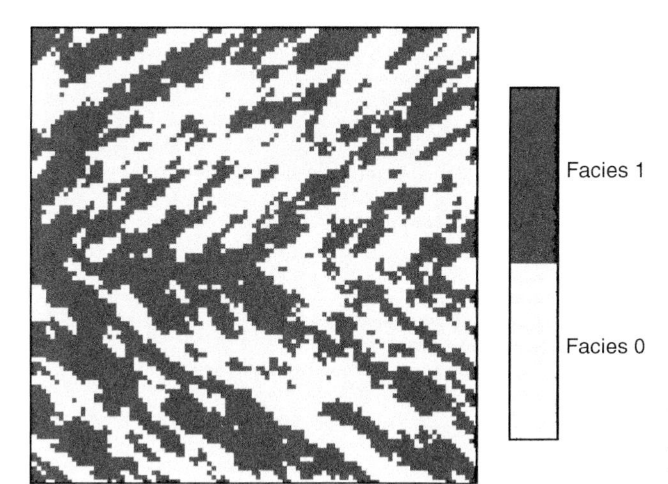

FIGURE 2.89: Example of Sequential Indicator Simulation with Two Local Directions of Anisotropy. The top half is 60° different from the bottom half.

locally variable direction model is accounted for completely. Nearest-neighbor data and previously simulated nodes and their respective distances are calculated accounting for shortest anisotropic distance to find the associate covariances to populate the kriging matrices. This is calculated by exploring the various possible paths integrating anisotropic distance from cell to cell between two cells in the model. This difficult problem is equivalent to the problem of routing aircraft and ships in atmospheric and oceanic currents. By contrast, the straight line path is assumed with the previous methods that assume constant anisotropy in the local neighborhood. Boisvert (2010) applied Dijkstra's algorithm to solve the shortest anisotropic distance.

These shortest anisotropic distances are directly applied to kriging and simulation; but while they are a valid distance measures, they are not guaranteed to result in a positive definite covariance matrix. Boisvert (2010) applied multidimensional scaling to embed the geostatistical grid into a high-dimensional Euclidean space where known positive definite covariance functions can be applied to guarantee positive definiteness. While this results in positive definite covariance matrices for kriging and simulation, the drawback is that the shortest anisotropic distances are only honored approximatively. Also, this more rigorous approach adds considerable computational effort to estimation and simulation.

The problem of calculating a reliable variogram in the presence of locally varying directions of continuity is significant. An isotropic variogram must be extracted from the local data and locally variable direction field. Conventional software does not permit using such information. The common alternative is to calculate the variogram in directions or subareas where the anisotropy follows a nearly constant direction. The actual anisotropy will be more than observed because of mixing the directions.

Locally Varying Mean

A *trend* is defined as a general tendency or pattern that is a deterministic or predictable aspect of the spatial distribution of reservoir facies or properties. Different authors have different definitions.

Virtually all natural phenomena exhibit trends. For example, the vertical profile of porosity and permeability may "fine upward" within each successive strata simply because coarse particles settle first within a depositional event. Moreover, there may be other large-scale trends related to long-term changes in the sediment source, space for deposition, or climate changes during deposition. The geological details of why there are systematic large-scale trends are less important than the fact that there are such trends and that they often affect reservoir performance predictions.

Any discussion on trends is subjective. All geostatisticians struggle with the separation of deterministic features, like trends, and stochastic correlated features, which are left to simulation. This separation is truly a decision of the modeler since reservoirs are deterministic; a "trend" and "residual" are meaningless if the reservoir is known exactly. Trends are necessarily dependent on the available data. This is illustrated in Figure 2.90, where the trend appears (1) different at different scales and (2) different with different data. Most of the variation appears

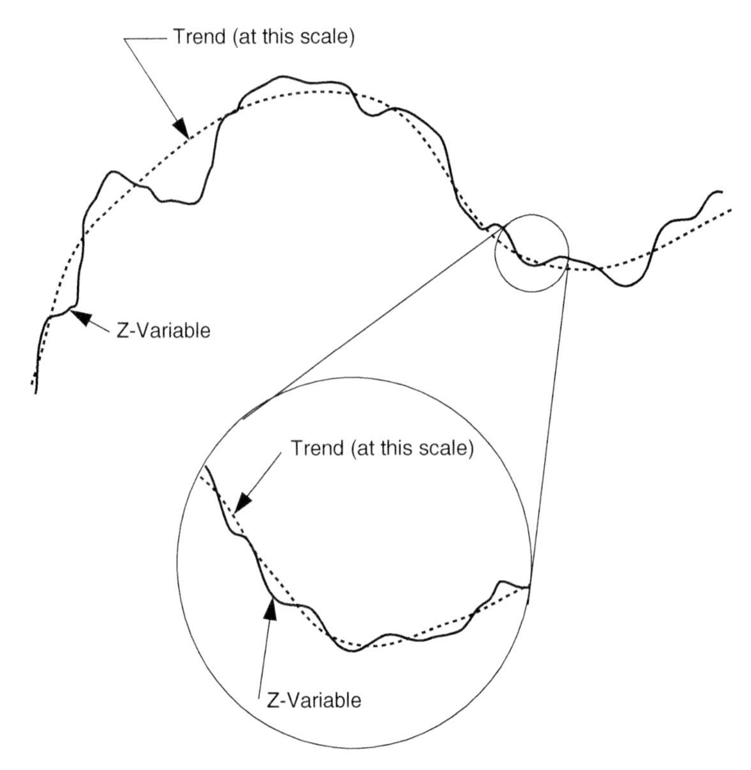

FIGURE 2.90: Schematic Illustration of How Trends Appear Different at Different Scales. A decision must be made about how much of the variability to put in the trend and how much to consider stochastic variations.

stochastic in presence of sparse data; as more data becomes available, the more refined the trend model and less of the intrinsic variation is left to geostatistical modeling.

Certain trends may be inferred from general geological knowledge. For example, the porosity within individual sedimentary beds in a deepwater depositional system is known to fine upward. This is due to well-understood physical principles and does not depend on an abundance of data. In some situations the presence of a trend is observed through the data. There is an apparent contradiction with many well data. Abundant data permit the trend to be identified, but it is less important since the data impose their trend features as local conditioning data on the resulting model; therefore, explicitly modeling a trend may be considered less important. In all realistic reservoir scenarios (versus some mining or environmental studies) there are rarely enough well data to enforce a trend throughout the study area. Trends are only reproduced in the immediate vicinity of the wells. Some variant of the following procedures must be considered in the realistic case of widely spaced well data.

An implicit assumption of all geostatistical algorithms such as kriging is that the variable being considered is "stationary." This assumption entails that spatial statistics do not depend on location, for example, the mean and variogram are constant and relevant to the entire study area. Special efforts must be made to account for trends or so-called nonstationarities when they are known to exist from our geological understanding of the data.

Constraints have been added to kriging in an attempt to capture trends. Kriging with a trend model (KT) permits a trend of polynomial shape to be fitted within each local kriging neighborhood. Kriging with an external drift permits an independent variable to define the shape of the local trend surface. Experience has shown that these forms of kriging work when there are adequate data within each local neighborhood to define the trend. For practical reservoir modeling, where data are often sparse, these methods do not satisfactorily account for the trend.

Practical guidance on trend model construction and implementation is provided in Section 4.1, Large-Scale Modeling.

2.4.9 Alternatives for Secondary Data Integration

Secondary data integration is a challenging component of modeling with diverse circumstances and no panacea solution. There are various methods, each with their own assumptions, strengths, and limitations. In the following discussion we cover collocated cokriging, Bayesian updating, stepwise conditional transform, cloud transform, and probability combination schemes.

Collocated Cokriging

A reduced form of cokriging consists of retaining only the collocated secondary data $y_2(\mathbf{u})$ or relocated data $y_2(\mathbf{u}')$ to the nearest node $\mathbf{u}$ being estimated (Doyen, 1988; Xu et al., 1992). This is not a problem if the distance $|\mathbf{u} - \mathbf{u}'|$ is small with respect to the volume of influence of $y_2(\mathbf{u})$. The cokriging estimator is written as

$$Y^*_{\mathrm{COK}}(\mathbf{u}) = \sum_{\alpha_1}^{n_1} \lambda_{\alpha_1} Y(\mathbf{u}_{\alpha_1}) + \lambda' Y_2(\mathbf{u})$$

The corresponding cokriging system requires knowledge of the Y covariance $C_Y(\mathbf{h})$ and the Y–Y_2 cross covariance $C_{YY_2}(\mathbf{h})$. The latter can be approximated through the following model:

$$C_{YY_2}(\mathbf{h}) = B \cdot C_Z(\mathbf{h}), \qquad \forall \mathbf{h}$$

where $B = \sqrt{C_Y(0)/C_{Y_2}(0)} \cdot \rho_{YY_2}(0)$, $C_Y(0)$, $C_{Y_2}(0)$ are the variances of Y and Y_2, and $\rho_{YY_2}(0)$ is the linear coefficient of correlation of collocated $y - y_2$ data.

The approximation consisting of retaining only the collocated secondary datum does not affect the estimate (close-by secondary data are typically very similar in values), but it may affect the resulting cokriging estimation variance: that variance is overestimated, sometimes significantly. In a kriging (estimation) context this is not a problem, because kriging variances are of little use. In a simulation context (see Section 4.6, where the kriging variance defines the spread of the conditional distribution from which simulated values are drawn), this may be a problem. The collocated cokriging variance should then be reduced by a factor (assumed constant for all locations) to be determined by trial and error. Babak and Deutsch (2007b) have shown that by adopting the intrinsic correlation model (all direct and cross-covariance functions are all proportional to the same underlying spatial correlation function and all secondary data at primary locations are retained), variance inflation is no longer an issue.

The Bayesian Updating Approach

Bayesian updating, as presented in Section 2.2, provides a straightforward framework for integrating various forms of secondary data. In Section 4.1 a Bayesian framework is provided for large-scale mapping. For example, one can further simplify collocated cokriging, into the *Bayesian Updating Approach* (Doyen et al., 1994, 1996, 1997). This method is gaining in popularity because of its simplicity and ease with which seismic data is accounted for.

At each location along the random path, indicator kriging is used to estimate the conditional probability of each facies, $i^*(\mathbf{u}; k), k = 1, \ldots, K$, from hard i data alone, and then Bayesian updating modifies or updates the probabilities as follows:

$$i^{**}(\mathbf{u}; k) = i^*(\mathbf{u}; k) \cdot \frac{p(k|\mathrm{ai}(\mathbf{u}))}{p_k} \cdot C, \qquad k = 1, \ldots, K$$

where $i^{**}(\mathbf{u}; k), k = 1, \ldots, K$, are the updated probabilities for simulation, $p(k|\mathrm{ai}(\mathbf{u}))$ is the seismic-derived probability of facies k at location $\mathbf{u}$ being considered, p_k is the overall proportion of facies k, and C is a normalization constant to ensure that the sum of the final probabilities is 1.0. The factor $p(k|\mathrm{ai}(\mathbf{u}))/p_k$ operates to increase or decrease the probability, depending on the difference of the calibrated facies proportion from the global proportion. No change is considered if $p(k|\mathrm{ai}(\mathbf{u})) = p_k$, which is the case where the seismic value $\mathrm{ai}(\mathbf{u})$ contains no new information beyond the global proportion. An additional step is required to ensure exactitude—that is, exact reproduction of facies observations at well locations.

The simplicity and utility of this approach is appealing. There are two implicit assumptions behind Bayesian updating that may be important: (1) the collocated seismic data screens nearby seismic data and (2) the scale of the seismic is implicitly assumed to be the same as the geological cell size. A relatively simple check on the simulated realizations will judge whether these assumptions are causing artifacts. The calibration table (just like Table 4.1) can be checked using the seismic data and the SIS realization; a close match to the probabilities calculated from the original data indicates that the approach is working well. Problems would be revealed by high probabilities (large $p(k|\mathrm{ai})$) becoming exaggerated to even higher values in the final model; low values would be even lower.

Cloud Transform

Previous discussion on probability field simulation introduced the concept of separating the construction of local conditional probability distributions and imposing spatial continuity. This can be a valuable tool for integrating secondary data in an algorithm known as cloud transform. Given an exhaustive model of a secondary reservoir property (such as porosity) and the bivariate relationship between secondary and a primary reservoir property (such as permeability), one can simulate the primary reservoir property, namely, permeability conditional to porosity. The advantage of this approach is that the bivariate relationship, regardless of form or complexity, is generally well reproduced. Given the importance of the complex porosity and permeability relationship, this is often a method of choice for simulating permeability. As a disadvantage, care must be taken to ensure that the primary variable distribution and spatial continuity are reasonable.

The permeability value at a location **u** can be drawn by Monte Carlo simulation from the conditional distribution of permeability, given the porosity at that location $f(k|\phi(\mathbf{u}))$. As a prerequisite, a series of conditional distributions are constructed (see the three distributions in Figure 2.91). In general, 10 or more conditional distributions are used. The porosity "windows" used to construct the conditional distributions can overlap (recall that the details of Monte Carlo simulation were covered in Section 2.2).

The spatial continuity of permeability is imposed through the variogram of the p-field that is applied to draw from the conditional distributions. Often it is difficult to infer a permeability variogram due to limited data, measurement accuracy, and scale issues. A common practice is to apply the porosity variogram with the range reduced by $\frac{1}{3}$ to $\frac{1}{2}$ because permeability often has a shorter range than porosity. This method assumes that the spatial heterogeneity of porosity and permeability are related and would not be appropriate if this is not the case—for example, in strongly diagenetically altered rock or dual perm systems with matrix and fracture permeability. Regardless, variogram reproduction is not precise with this method due to features imposed by the exhaustive porosity and associated conditional distributions.

The permeability conditioning is imposed by conditioning the p-field such that the right permeability values are drawn from the conditional distributions at the data locations. This is required, contrary to p-field simulation, with cloud transform the local distributions from the conditional distributions of permeability, given that porosity does not express reduced variance near data locations; conditioning is not built into the local distributions. As a result the two previously mentioned artifacts of *p*-field simulation are not present.

Stepwise Conditional Transform

Stepwise conditional transform presents an interesting avenue for joint simulation of reservoir properties with potentially complicated multivariate relationships, including complicated constraints, curvilinear and heteroscedastic features (Leuangthong, 2003; Leuangthong and Deutsch, 2003). With this technique, each variable is transformed sequentially and conditionally to previous variables into an independent multivariate Gaussian distribution. As each transformed variable is independent, they may be simulated separately and then back transformed to re-impose their multivariate relationships.

The stepwise-conditional technique is identical to the normal score transform in the univariate case. For bivariate problems, the normal transformation of the second variable is conditional to the probability class of the first variable. Correspondingly, for k-variate problems, the kth variable is conditionally transformed based on the $(k-1)$th variable; that is,

$$Y_1 = G^{-1}[\mathrm{Prob}\{Z_1 \le z_1\}]$$

$$Y_{2|1} = G^{-1}[\mathrm{Prob}\{Z_2 \le z_2 | Y_1 = y_1\}]$$

$$Y_{3|2,1} = G^{-1}[\mathrm{Prob}\{Z_3 \le z_3 | Y_2 = y_2, Y_1 = y_1\}]$$

For example, Z_1 can be determined from Y_1 with the correct conditional distribution; Z_2 can be calculated from Z_1 and the simulated value of Y_2. Conditional transformation of the data results in transformed secondary variables that are now artificial variables with little physical interpretation. It is a combination of both the primary and the secondary variable. Also, the multivariate spatial relationship of the original model variable is not transformed for $\mathbf{h} > 0$, that is, there is no modification of bivariate spatial distributions $Y(\mathbf{u})$ and $Y(\mathbf{u}+\mathbf{h})$, trivariate distributions $Y(\mathbf{u})$, $Y(\mathbf{u}+\mathbf{h}_1)$ and $Y(\mathbf{u}+\mathbf{h}_2)$, and so on.

The result of this transformation is independence of the transformed variables at $\mathbf{h} = 0$. Since each class of Y_2 data is independently transformed

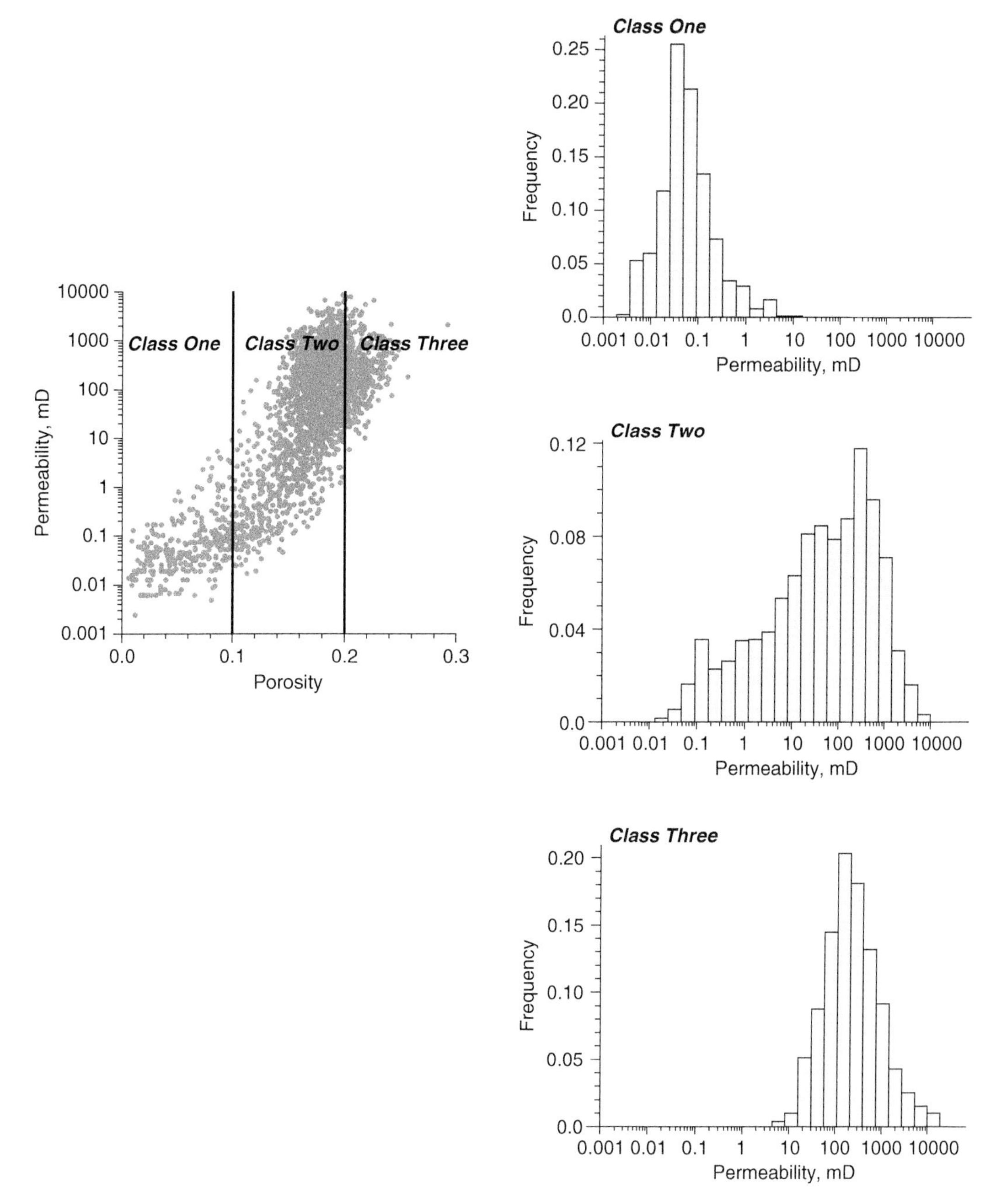

FIGURE 2.91: Porosity/Permeability Cross Plot with Three Conditional Distributions for Monte Carlo Simulation. More conditional distributions would be used in practice; three are shown for illustration.

to a normal distribution, correlation between $Y_{2|1}$ and Y_1 is removed at $\mathbf{h} = 0$. Consequently, the simulation of multivariate problems does not require cosimulation due to the independence of the transformed variables. This is the primary motivation for transforming multiple variables in a stepwise conditional fashion.

Some limitations with stepwise conditional transform include: The method requires sufficient data to characterize the multivariate relationship and the simulation of the secondary variables is conducted in a nonphysical transformed space; as a result, there are difficulties in inferring the appropriate spatial continuity model.

Probability Combination Schemes

Probability combination schemes (PCS) have been developed independently in many research areas in order to find the integrated probability, given several potentially redundant sources of derived probabilities (Benediktsson and Swain, 1992; Journel, 2002; Lee et al., 1987; McConway, 1981; Rasheva and Bratvold, 2011; Winkler, 1981). Detailed overview of this topic is provided by Hong and Deutsch (2009a). With PCS we can approximate the probability of a primary outcome, given multiple secondary data, $p(P|S_1, S_2, \ldots, S_n)$, as a function of the probability computed using each secondary data separately, $p(P|S_1)$, $p(P|S_2)$ through $p(P|S_n)$, where primary and secondary data are denoted as P and S_i for $i = 1, \ldots, n$ and there are n types of secondary data. The work flow is illustrated in Figure 2.92.

$$\frac{p\{P|S_1, S_2, \ldots, S_n\}}{p\{P\}} \sim \Phi\left[\frac{p\{P|S_1\}}{p\{P\}}, \ldots, \frac{p\{P|S_n\}}{p\{P\}}, C\right]$$

Probability terms independent of primary variable P are absorbed in a normalizing term C. $p(P)$ is a global probability of the primary outcome. If the primary variable is categorical, then this probability is the proportion of a specific category, k. Function $\Phi[\ldots]$ is a generic notation of probability combination model. In the following, specific forms including the permanence of ratios, conditional independence, and weighted combinations are discussed.

Permanence of Ratios (PR-Model)

Journel (2002) developed a permanence of ratios model that approximates the probability, assuming that ratios of probability increments from different sources are constant. This method is known as the naïve Bayes model in machine learning.

The estimated probability by the PR model meets closure condition and positiveness regardless of the number of data S_i, although the PR model is limited to only the binary case for the primary

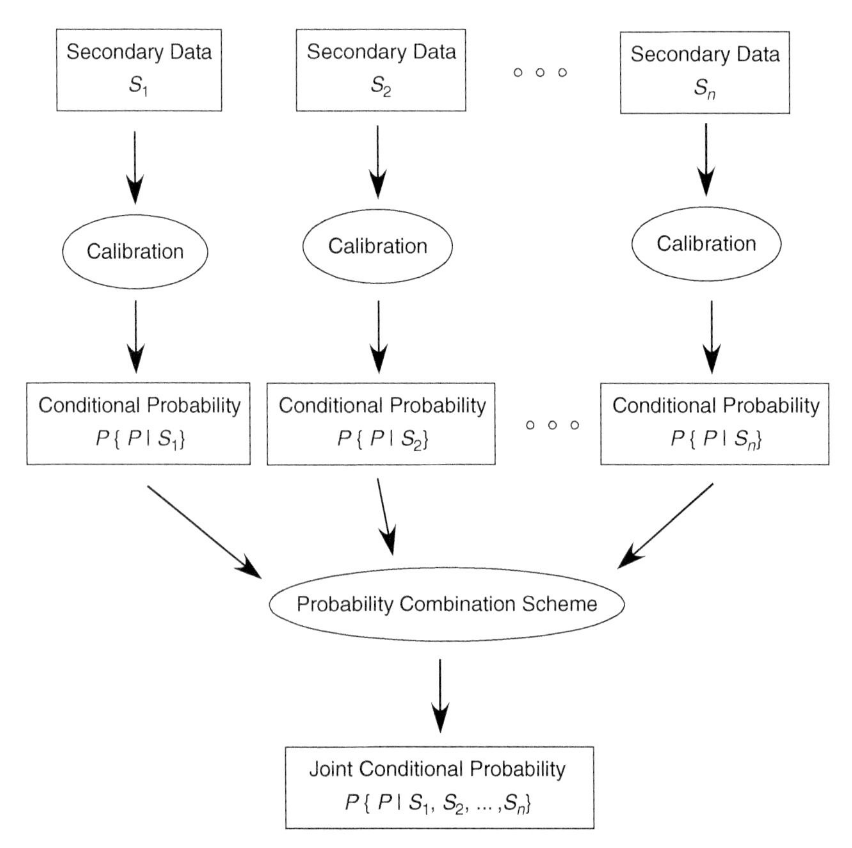

FIGURE 2.92: Work Flow for Probability Combination. First apply calibration to calculate the probability distributions for the primary variable, given each secondary information source individually. Utilize a probability combination scheme to combine these for the probability distribution of the primary variable, given all secondary information.

Modified from Hong and Deutsch (2009a).

variable. For instances, the sum of $P_{PR}\{k|S_i, \ldots, S_n\}$ over $k = 1, \ldots, K$ categories where K is greater than 2 does not enforce closure (conditional probabilities do not always sum to 1). Therefore, we represent the PR model estimates with respect to categories as the probability:

$$p_{PR}\{P|S_1, \ldots, S_n\} = \frac{\left(\frac{1-p\{k\}}{p\{k\}}\right)^{n-1}}{\left(\frac{1-p\{k\}}{p\{k\}}\right)^{n-1} + \prod_{i=1}^{n}\left(\frac{1-p\{k|S_i\}}{p\{k|S_i\}}\right)} \in [0.1]$$

Conditional Independence

Hong and Deutsch (2009a) demonstrated the equivalence of PR and conditional independence for binary variables. Also, they showed that conditional independence enforces the closure condition and recommend its use over PR. Under the conditional independence assumption of $(S_1, \ldots, S_n)$, the generic equation for combining conditional probabilities becomes

$$p\{k|S_1, \ldots, S_n\} = p\{k\} \cdot \prod_{i=1}^{n}\left(\frac{p\{k|S_i\}}{p\{k\}}\right) \cdot C \in [0, 1]$$

where C is a normalizing term to meet the closure condition. It is independent of the primary variable, P.

Weighted Combination

Permanence of ratios and the conditional independence model assume independence among secondary data. By adopting the independence assumption, combining probabilities is simplified into the product of individual probabilities, $p\{P|S_i\}$. In some cases, the simplified model could result in a serious bias because multiplication of each probability will result in a very high combined probabilities potentially very different from the global probabilities or prior model $p\{P\}$ and all the individual conditional probabilities for each secondary data individually, $p\{P|S_i\}, i = 1, \ldots, n$. Specifically, the resulting probability may approach 1 or 0 if many secondary data are considered for the integration and they are highly redundant.

Weighted combination approaches are advanced to adjust the influence of elementary probabilities. The probability is approximated by combining individual probabilities with data-specific weights:

$$\frac{p\{P|S_1, S_2, \ldots, S_n\}}{p\{P\}} = \Phi\left[\left(\frac{p\{P|S_1\}}{p\{P\}}\right)^{w_1}, \ldots, \left(\frac{p\{P|S_n\}}{p\{P\}}\right)^{w_n}, C\right]$$

One of the most common weighted models is the Tau model for combining data-specific probabilities. It has been widely applied since its first development by Journel (2002) (Caers et al., 2006; Castro et al., 2006; Chugunova and Hu, 2008; Krishnana, 2004). The Tau model approximates $p(P|S_i), i = 1, \ldots, n$, by imposing the power τ_i and is expressed for facies as follows:

$$p_\tau\{P|S_1, \ldots, S_n\} = \frac{\left(\frac{p\{k\}}{1-p\{k\}}\right) \cdot \left(\frac{1-p\{k\}}{p\{k\}}\right)^{\sum_{i=1}^{n}\tau_i}}{\left(\frac{p\{k\}}{1-p\{k\}}\right) \cdot \left(\frac{1-p\{k\}}{p\{k\}}\right)^{n-1} + \prod_{i=1}^{n}\left(\frac{1-p\{k|S_i\}}{p\{k|S_i\}}\right)^{\tau_i}} \in [0.1]$$

The Tau model has the same form as the permanence of ratios model except power weights τ_i are introduced that control the contribution of elementary probabilities $p\{k|D_i\}$. Similar to the PR model, the Tau model only works for the binary case for the primary variable, P. There is no guarantee that sum of $P_\tau\{P_k|S_i, i = 1, \ldots, n\}$ over $k = 1, \ldots, K$ is 1.

The lambda model was developed by Hong and Deutsch (2007). It can be interpreted as an expanded model of conditional independence with introducing data-specific weights, λ_i. If $\lambda_i = \tau_i$ for $i = 1, \ldots, n$ and binary category is only considered, then P_{PR} and P_λ are exactly same. A λ model is not limited to the binary case. Thus, the λ model is a more generalized weighted combination model.

In weighted combination models, choosing appropriate weights is a critical issue (Krishnana, 2004). Optimal weights are found if data redundancy among secondary data are fully characterized and can be expressed just by power. However, this is almost impossible. As an alternative to direct quantifying redundancy, a calibration method may be considered. For instance, weights are obtained in order to minimize the errors between the true value at well locations and the estimated probability P_λ of true value at that location (Hong and Deutsch, 2007).

Recall, all of these probability combination schemes are attempting to calculate conditional probabilities $p\{P|S_1, \ldots, S_n\}$ from assessable conditional probabilities for each secondary data individually. But, if the full multivariate distribution $p\{P, S_1, S_2, \ldots, S_n\}$ was available, then all the required joint and conditional probabilities would be available to solve for these directly by sampling the multivariate distribution. This is known as the *multidimensional estimation* (MDE) (Hong, 2009). This method is challenging, given the typically limited data and the curse of dimensionality (i.e., if 100 data are required to model a univariate distribution, 10,000 are required to model a bivariate distribution and 1,000,000 are required to model a trivariate distribution).

2.4.10 Work Flow

Figure 2.93 illustrates a work flow for kriging to build a map. The kriging type chosen determines whether the mean must be explicitly mapped. The results are appropriate for visualizing trends in data and, perhaps, for smoothly varying surfaces.

Figure 2.94 illustrates the work flow for sequential Gaussian simulation (SGS) of a stationary random function. A declustered representative distribution and variogram model are the key inputs. There are a number of other details such as search, but those are essentially the same as required for kriging.

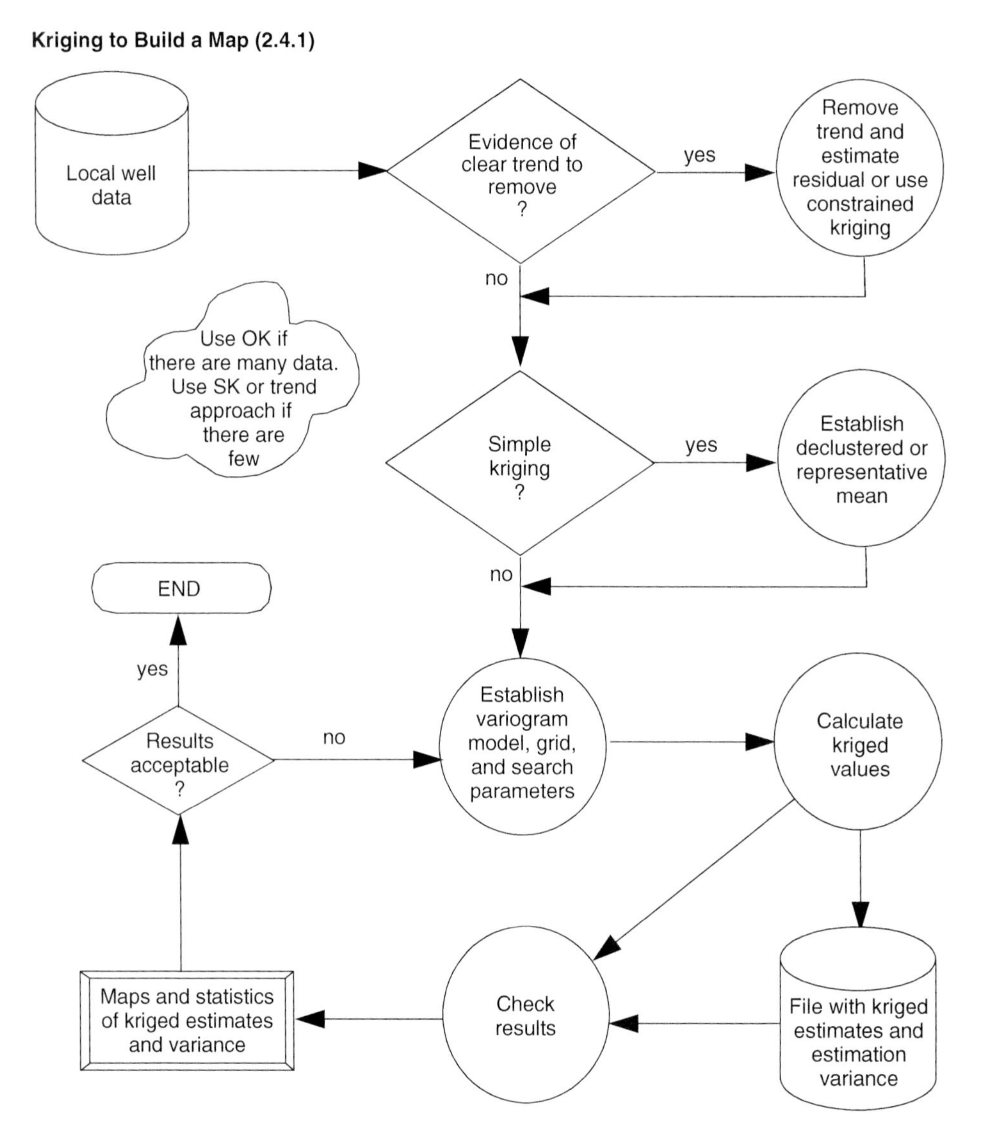

FIGURE 2.93: Work Flow for Kriging to Build a Map.

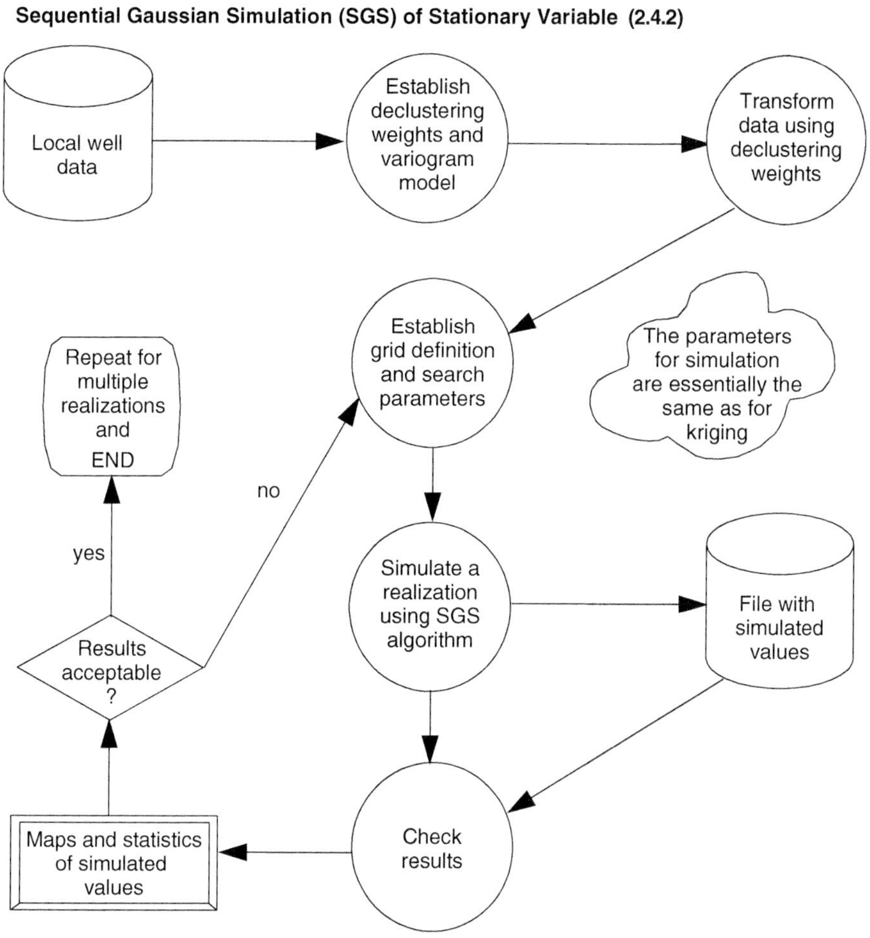

FIGURE 2.94: Work Flow for Sequential Gaussian Simulation (SGS) of a Stationary Random Function.

Figure 2.95 illustrates the work flow for indicator kriging for mapping local uncertainty. A very similar procedure could be used for indicator simulation of categorical or continuous variables.

Figure 2.96 illustrates the work flow to build a trend model for porosity, permeability, or facies modeling. For convenience, a 1-D vertical trend and 2-D areal trend are constructed first and then combined into a 3-D trend model.

2.4.11 Section Summary

Building from the statistical prerequisites in Section 2.2 and 2.3, we now have a toolbox of various methods to address geostatistical reservoir modeling. Kriging, based on the variogram, introduced in Section 2.3, provides a linear unbiased estimator away from data locations. The sequential Gaussian simulation method corrects for the smoothness of kriging, reproduces the global statistics while sacrificing local accuracy, but provides an uncertainty model through multiple equiprobable realizations and parameter uncertainty.

Indicator methods provide a useful framework to characterize and simulate categorical variables or to improve control over spatial continuity of low, median and high values. In addition, the indicator coding allows for direct input of soft data, coded as probabilities of being each category.

P-field simulation is novel simulation method that divorces the calculation of local distributions of uncertainty and simulation with the correct spatial continuity and global distribution. This provides a significant speed up because these local distributions need only be calculated once for each location to produce multiple realizations, but also results in increased flexibility and ease to integrate information.

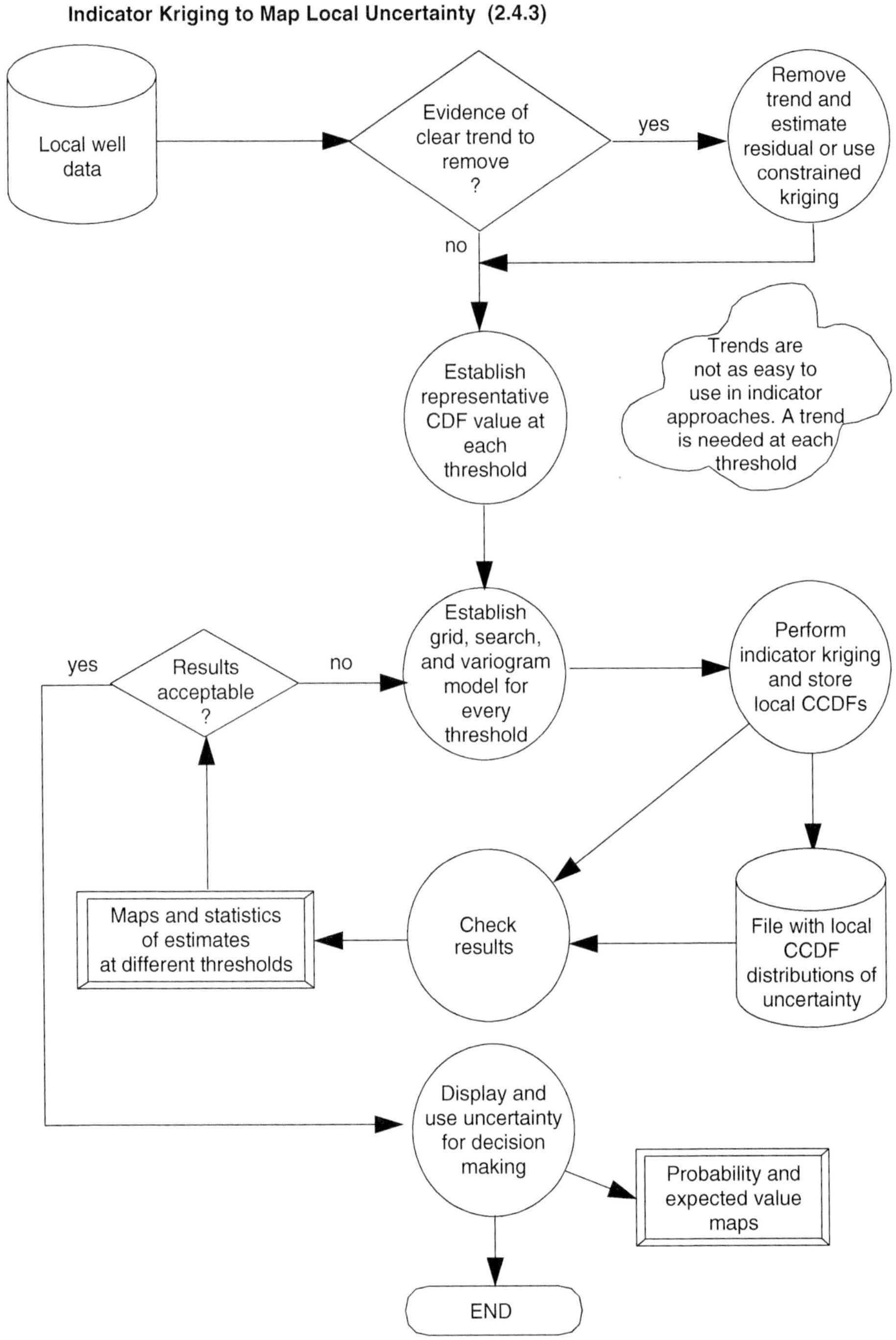

Indicator Kriging to Map Local Uncertainty (2.4.3)

Local well data

Evidence of clear trend to remove ?

yes

Remove trend and estimate residual or use constrained kriging

no

Establish representative CDF value at each threshold

Trends are not as easy to use in indicator approaches. A trend is needed at each threshold

yes

Results acceptable ?

no

Establish grid, search, and variogram model for every threshold

Perform indicator kriging and store local CCDFs

Maps and statistics of estimates at different thresholds

Check results

File with local CCDF distributions of uncertainty

Display and use uncertainty for decision making

Probability and expected value maps

END

FIGURE 2.95: Work Flow for Indicator Kriging for Mapping Local Uncertainty.

Multiple-point simulation attempts to increase the complexity of the spatial continuity model by using more than the two points considered in the variogram-based methods. This method requires a training image to infer these statistics and is currently limited to categorical variables.

Object-based simulation allows for the reproduction of crisp reservoir element geometries. There is a great deal of flexibility in the parameterization of these geometries and constraints on their placements within the model. Yet, unlike all the previous methods (that are cell-based), object-based

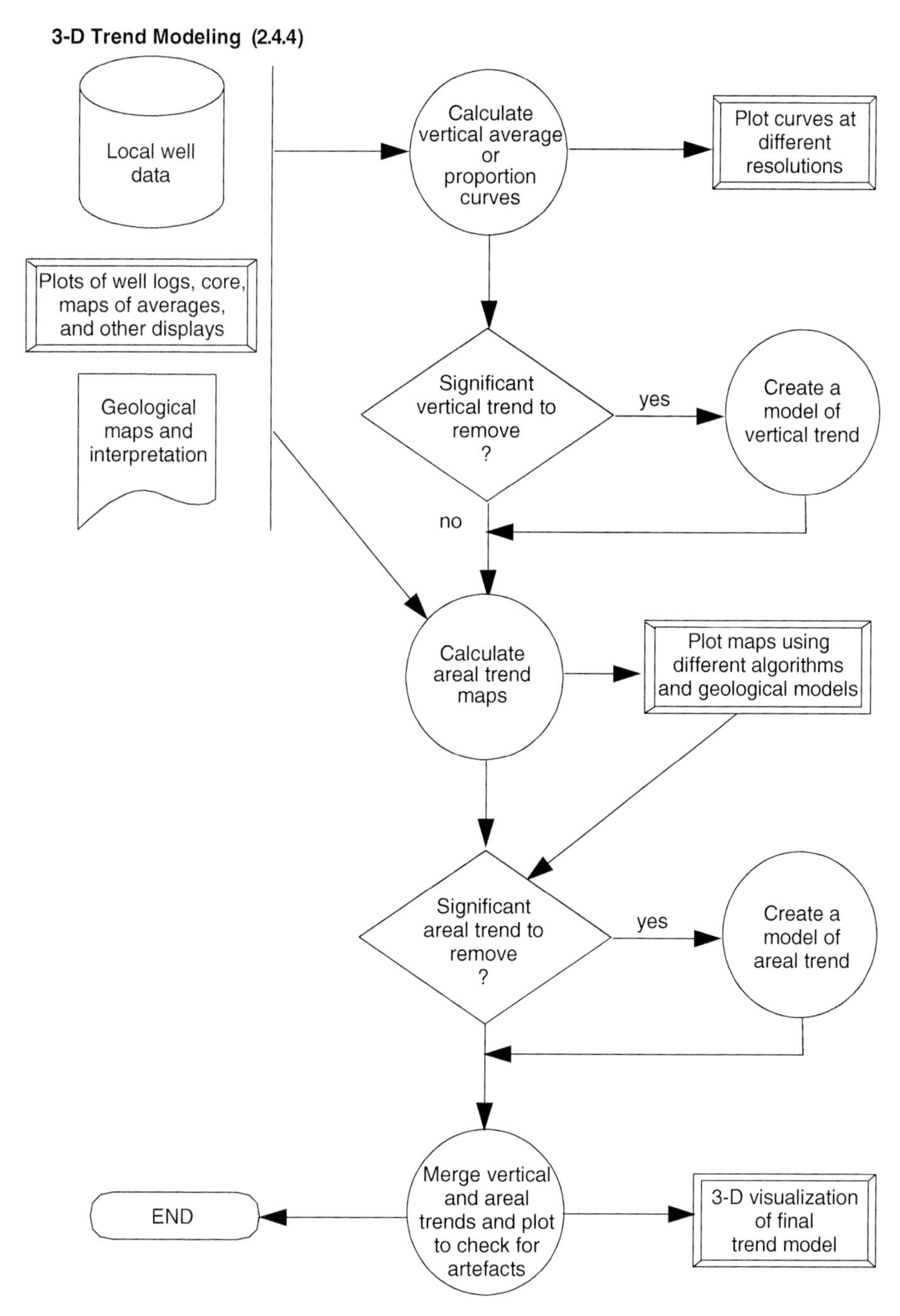

FIGURE 2.96: Work Flow to Build a Trend Model for Porosity, Permeability, or Facies Modeling.

conditioning is not guaranteed and is often challenging in dense data settings and with the presence of detailed trends and secondary data. A variant of object-based modeling, known as process-mimicking or event-based modeling, is presented in Section 4.5. These models work with objects, but include concepts of temporal sequences and process-mimicking rules for object geometries and placement.

Optimization algorithms provide flexible methods to impose almost any constraints on a model. As opposed to the sequential methods presented in this section, optimization relies on iteration and can perturb a model to match any desired criteria.

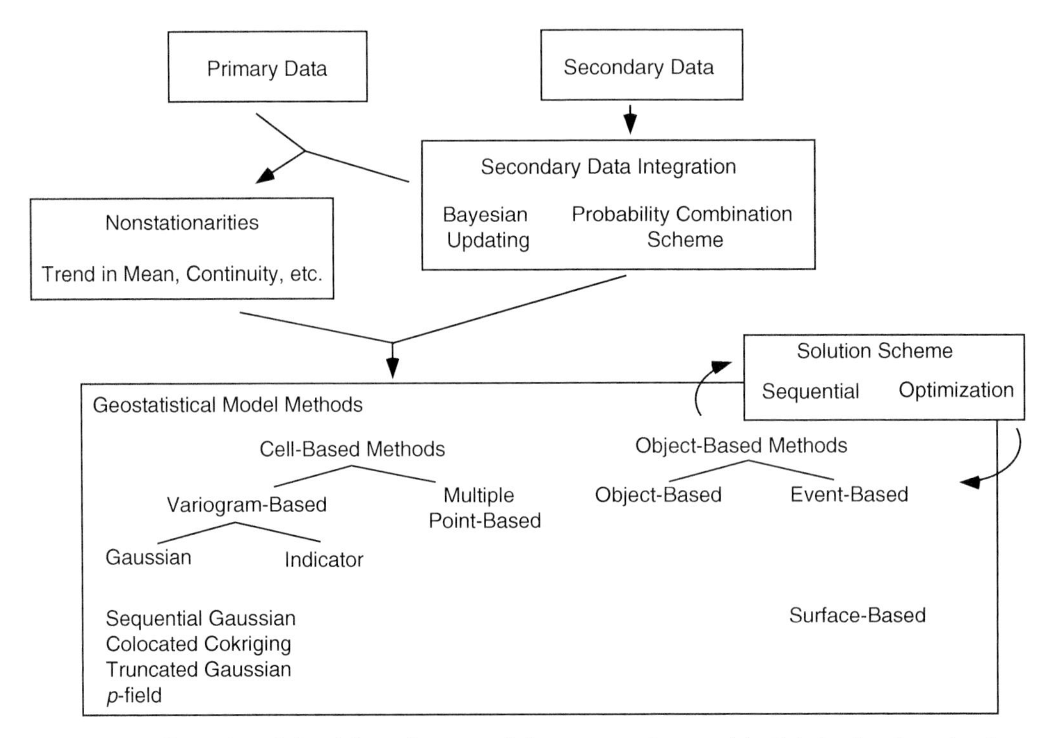

FIGURE 2.97: Illustration of the Linkages between of the Concepts Presented in This Section. Secondary integration methods provide methods to prepare secondary data for modeling, while nonstationarity modeling methods provide methods that integrate all available information, including primary and secondary data. There are a variety of modeling methods that may be broken up by the statistics that they reproduce. While sequential solution methods are typically employed for their efficiency and simplicity, optimization methods provide greater flexibility.

For example, optimization could be applied to place objects to match conditioning in an object-based model, to correct higher-order statistics in a variogram-based model, or to match flow performance. From this perspective, optimization is not a type of geostatistical modeling method, categorized by the statistics that are honored, but is instead a method to match any of these statistics that could be applied within any of the geostatistical modeling methods. While greater flexibility is possible with optimization, the requirement of iteration is typically more CPU-intensive and convergence is not guaranteed.

Within any of these methods, nonstationarities may be considered. These may include locally variable means, proportions, continuity or object density, geometry, or orientation. Since stationarity is a decision and not a hypothesis, this choice cannot be tested. Nevertheless, this is an important decision and opportunity to integrate geologic information into the model. If we skip this step, then the global statistics are assumed to be constant over the entire model with the exception of feature imposed by conditioning data and secondary data.

Secondary data integration is a broad topic with various methods available. Collocated cokriging, Bayesian updating, stepwise conditional transform, cloud transform, and probability combination schemes have their various unique capabilities and limitations and provide tools to handle secondary data in any reservoir modeling setting.

To link the various methods discussed in this section together, we provide this summary figure that puts each of their methods in their place within geostatistical mapping toolkit (see Figure 2.97). Importantly, we have now provided the context central to subsequent discussions on data inventory, conceptual model, and problem formulation in the next chapter. More details, implementation issues, and examples for these methods are postponed to discussion in Chapter 4.

3

Modeling Prerequisites

Chapter 2 covered the fundamental model principles related to geological and statistical concepts, methods to quantify spatial correlation, and methods to construct estimation maps and simulation models. This chapter takes those fundamental concepts and brings them together in the context of geostatistical reservoir modeling. Firstly, we discuss the available data in the *Data Inventory* section, then we provide guidance on converting geological and statistical descriptions into reservoir models in the *Conceptual Model* section, and finally we discuss the general modeling strategy in *Problem Formulation.* Chapter 4 then proceeds with details and example on the various modeling methods.

3.1 DATA INVENTORY

Geostatistical reservoir models aim to reproduce all local data accounting for specific data scale and accuracy. There are various sources and types of local information, including direct and indirect local measurements and analog conceptual information. Local information is a measurement of the reservoir properties of interest at a location in the reservoir and with a specific scale and accuracy of measurement. Utilization of this information away from the sampled locations requires a decision of stationarity and a pattern of spatial correlation. Direct measurements include core data where a part of the subsurface is directly analyzed and interpreted, providing, for example, porosity and permeability data. Indirect measurements utilize a measurement of rock response, without extracting the sample from the subsurface that then must be modeled to infer site-specific measurement, such as well logs and seismic attributes. Analog conceptual information is acquired from sources outside the reservoir, such as mature reservoirs, outcrops, flume experiments, and physical-based models. The integration of all of this information is one of the greatest challenges and strengths of geostatistical reservoir modeling.

The *Data Events* subsection introduces the concept of a data event. This extends the typical definition of data as a value at a location with documentation of vintage, quality control and checking processes, limitations and issues, and scale and geological population. Through this increased rigor, improvements may be realized with a reduction in data related errors and time for subsequent data checking.

The *Well Data* subsection describes the direct sampling of the reservoir in the form of cores and a detailed suite of indirect measurements in the form of well logs with high-resolution indication of facies and fluids. Modern image logs are providing additional resolution with information on sedimentary structures and faults.

Seismic Data provide valuable information on large-scale reservoir geometry and, when resolution is sufficient, information on reservoir architecture and perhaps even an indication of reservoir properties.

Dynamic Data are from well tests, historical production, and 4-D seismic data. These data sources are emerging and providing valuable information about the flow response of the reservoir. They introduce (a) new challenges with complicated inversion problems and (b) large-scale support size and complicated relationships with reservoir properties.

Various types of *Analog Data* are often available. These data are external to the reservoir, but provide information, concepts, and statistics that may be applied to the reservoir of interest with the appropriate decision of stationarity. These types of data include mature fields, outcrops, geomorphology, high-resolution seismic, experimental stratigraphy, and numerical process models.

The discussion in this section covers general concepts of the various types of data, where they are used and general data considerations. Discussion on how these data are applied within different modeling methods in covered in Chapter 4.

The novice modeler will benefit from this basic coverage of the data available for reservoir models. The intermediate and expert modelers may benefit from the introduction of the concept of data events and consideration of analog sources such as geomorphology, experimental stratigraphy and numerical process models that are not commonly applied to assist with reservoir modeling.

3.1.1 Data Events

The concept of *data events* introduced by Alahaidib and Deutsch (2010) is a useful formalization of data in all possible forms. In standard geostatistics data are denoted as, $z(\mathbf{u}_\alpha), \alpha = 1, \ldots, n$, where $\mathbf{u}$ indicates the location of the measurements at each of n locations. This simple concept is insufficient because modern geostatistics is almost always related to multiple correlated variables each with their own scale and with different error content, and site and non-site specific. In this context, data are described more completely as data events, $\alpha = 1, \ldots, n$. Each data event, α, is characterized by a set of attributes shown in Figure 3.1. The list of attributes in Figure 3.1 may be augmented by additional properties. A common set of data attributes includes:

1. There may be a single measurement or possibly multiple related measurements. The measurements relate to the same location, scale, and geological population (see below). If any of these attributes vary, then the measurement should be considered a unique data event. Every measurement includes: the measurement(s) with appropriate units, a measurement-type description, and a measure of the uncertainty inherent to the measure. The measurement type should include specifics on the measurement method and/or tool used. Information relevant to data quality such as vintage, steps taken to check the measurement, limitations or issues with the measurement, and any corrections to the measurement should be listed to ensure appropriate use of the measurement. Any subsequent processes such as transformations, cleaning, and filters should be also included. Finally, if declustering techniques are applied, declustering weights may be associated with each measurement for the calculation of more representative statistics.
2. For site-specific information a location or anchor location for the data event is required

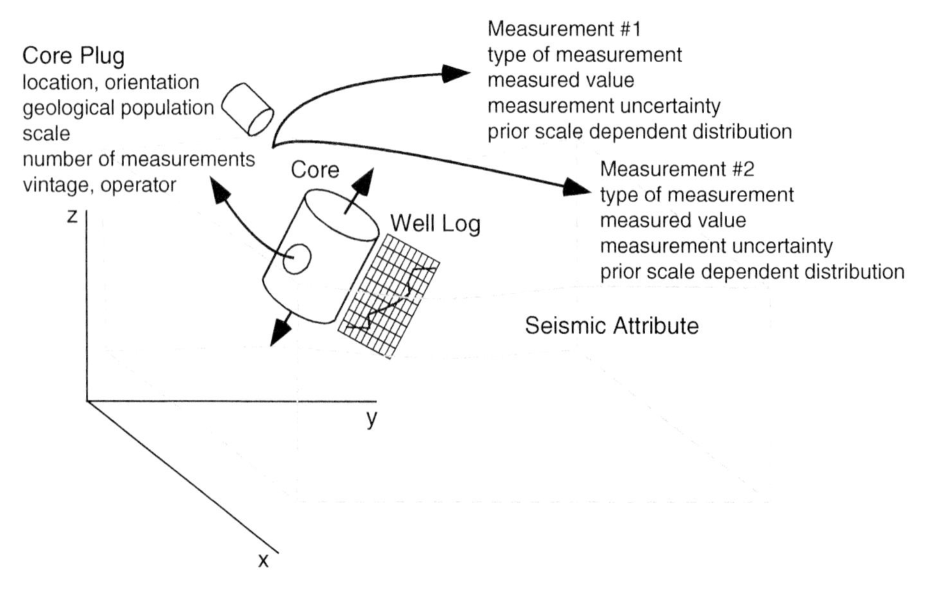

FIGURE 3.1: Illustration of Data Events. At a location within a reservoir, various data types exist including core, well log, and seismic. Inclusion of a core plug from the core results in a further set of details including location, a geological population, scale, and so on, and from this core plug two separate measurements with their associated details including type of measurement, measured value and associated uncertainty etc. This could be exhaustively expanded for all shown data types (not drawn to scale).

to place the data event within the reservoir. This may require the location in multiple coordinate systems such as original and stratigraphic coordinates. For non-site-specific analog data a region of applicability may be included to describe subsets of the reservoir that are considered analogous to the information.

3. The scale of the data will be identified by a volume and a shape. The shape is typically described by an anisotropy ratio with orientation to indicate the sample window size. For core data, scale is the size of the core, while for well logs the scale is determined by the resolution and depth of penetration of the tool.
4. The geological population associated with the measurement, such as a zone, facies, architectural hierarchy, geological unit, or estimation domain, should be identified. This is important because it provides information on how to pool specific measures from different data events for modeling away from data events.

The raw data will contain only some of this information. Yet, the missing information will be generally available among the experts of the project team. While compilation requires an increased effort, a complete data event description will ensure best use of the data, reduce the occurrences of data-related blunders, and provide for improved communication and faster cycle time for data integration. Data are sometimes referred to as "hard" or "soft." Hard data suggest a local measure of the reservoir property of interest without significant uncertainty in the measure. Geostatistical models honor these data exactly at the data locations; this property is referred to as *exactitude*. Soft data suggest a more imprecise local measure of the reservoir property of interest with significant uncertainty. It is typical to utilize soft data to calibrate constraints or probabilities with respect to hard data or to inform hard data trends. Some modeling methods, such as indicators (see Section 2.2, Preliminary Statistical Concepts), may directly account for data uncertainty.

The following discussion provides a brief coverage of the various data types available for geostatistical reservoir modeling. Of course, there is a great body of knowledge available on each of these, but the focus here is on a basic description of what may be integrated into the model, scale and coverage. In addition, some forms of data were omitted as their interpretation and application require highly specialized knowledge not typically available to reservoir geologists and engineers, including geochemistry. Although, it is acknowledged that these are important sources of information for developing a conceptual model and stratigraphic correlation.

3.1.2 Well Data

In this discussion, well data are limited to measurements made in or near the well bore. This includes various site-specific measures, including direct measurements made on extracted cores of the well bore and indirect measurements made by well logging and image tools. Production data from well tests are considered separately, and well cuttings are not discussed.

Well data are typically the only source of hard data within a reservoir. All other data are considered soft. Yet, the subsequent discussion should demonstrate the challenges associated with well data. Cores have their sampling issues and are typically very limited in coverage; sometimes core is not taken due to cost and technical difficulties of extraction. Well logs are highly dependent on calibration and formation assumptions. As a result, well data have unavoidable uncertainty and sample representativity issues that should be considered.

To illustrate the importance of well data, we rely on a study of the Cook Formation, Norwegian North Sea by Folkestad et al. (2012). This study is of interest because (1) in the absence of sufficient seismic information, well data are the principle source of information concerning architecture along with reservoir properties, (2) sparse core information is utilized to calibrate well logs, and (3) modern image logs are utilized to infer sedimentary structures and greatly aid in the interpretation. We thank Elsevier for permission to reprint figures and A. Folkestad for review and high-quality images.

Core Data

For geologists, core data are valued as the only opportunity to directly sample and visually inspect the reservoir. Core is generally applied to calibrate all other measures and reservoir conceptual models. Core and associated sampled core plugs may be subjected to physical experiments to determine and porosity, permeability that provide information on reservoir quality and mineralogical analyses to determine provenance and age. In addition, textures and sedimentary structures (Boggs, 2001) and ichnofacies (Walker and James, 1992) to identify depositional and post-depositional process and microfossils aid in determining time sequence. Yet, core collection greatly increases the cost of drilling, so they are generally only present for a fraction of well trajectories, compounding the scarcity of this important form of information.

With the scarcity of core samples, issues related to sample representativity are acute. Firstly, as mentioned in Section 2.2, Preliminary Statistical Concepts, discussion on declustering, wells are generally drilled preferentially in the best rock. Then core may be selectively sampled from the best-quality part of the well. Core from poor-quality rock may not even be recoverable. Finally, core plugs are selected by visual inspection of the core to collect reservoir property measurements. Given the expense of these laboratory tests, poor-quality parts of the core are not generally sent for tests. The tests may not be feasible on very fine-grained or cemented rock. Each of these steps potentially compounds the sampling bias associated with core measurements of reservoir properties.

Despite the potential for bias and the necessity to manage the data carefully, core data are very valuable. An example set of photographs from sectioned core is shown in Figure 3.2 from the Cook Formation in the North Sea. The interpretation by Folkestad et al. (2012) indicate a variety of structures that aid in the determination of the depositional environment.

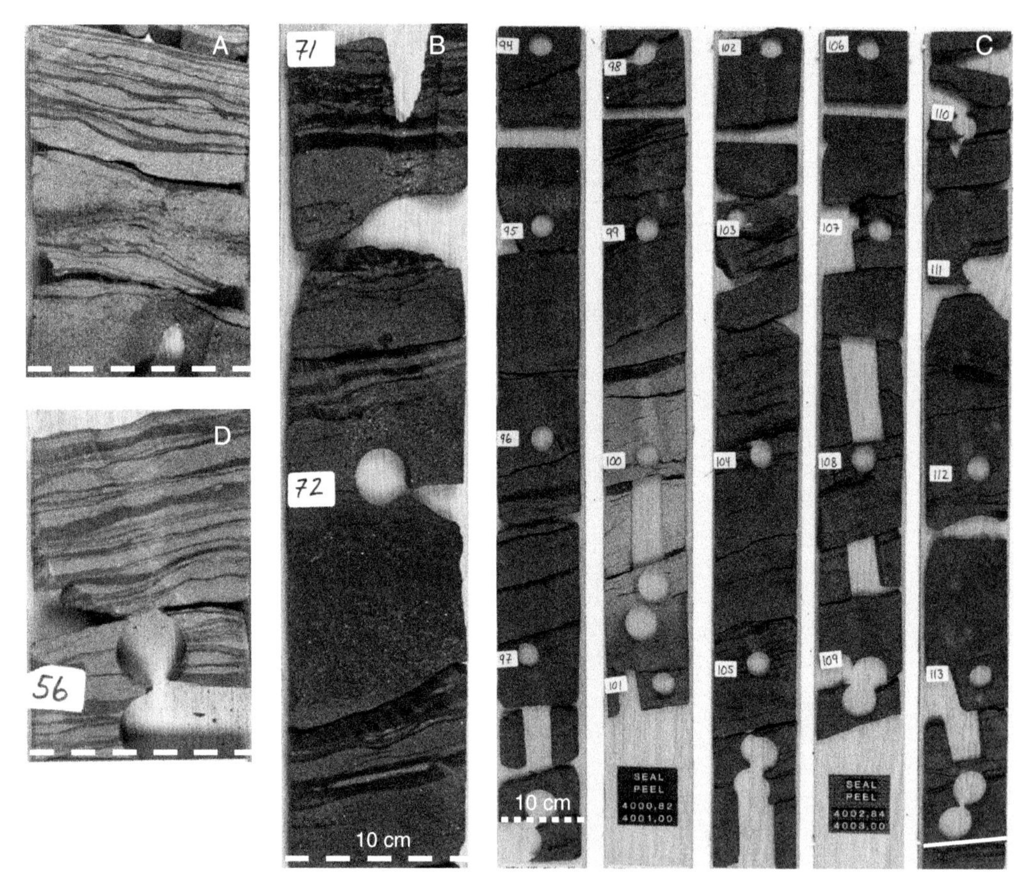

FIGURE 3.2: Sectioned Core Photographs of the Cook Formation, a Shallow Marine Sandstone Reservoir from the North Sea. The core data have been interpreted as a fluvial/deltaic depositional setting with general progradation upward. A summary of the core interpretation by Folkestad et al. (2012) follows. A: Medium-grained sandstone with asymmetrical ripples, flaser bedding, coal, and mud clasts. B: Sandstone units with fining upward trends capped by mudstones. C: Cross-stratified sandstone beds and D: Alternating mudstones and very fine grained sandstones with asymmetrical ripples and mud drapes.

For example, the trace fossils indicate a shallow marine setting, flaser bedding suggests a tidal flat, and cross-stratified beds indicate high energy. It is apparent that these observations are critical to piecing together the geological story of the reservoir.

Well Logs

After drilling there are a variety of methods available to collect information concerning the subsurface in the vicinity of the well. These include wireline well logs such as traditional caliper, resistivity, electrical potential, neutron density, and gamma ray. A suite of these tools, along with expert interpretation, comparison between logs, calibration to core, assumptions on rock properties, and mathematical models provide a measure of porosity, perhaps soft measurements of permeability and lithology at decimeter vertical scale and variable depth of penetration away from the well bore. For more complete details on well log acquisition and interpretation the reader is referred to Krygowski et al. (2004) and Eillis and Singer (2010).

For example, the caliper log provides a measure of well bore diameter and can indicate well bore issues that may interfere with the other well logs (collapse and wash out that result in the tools losing contact with the rock). The passive gamma ray tool provides a direct estimation of shale or clay content. The gamma ray, neutron density, and sonic tools provide a measure of formation bulk density and can be calibrated to infer porosity. This analysis may be complicated by the interactions of formation lithology and fluids (including gas). Interpretation from well logs are calibrated with core and known lithology and fluids and supported by multiple logs and experience concerning log response in analogous fields. An example suite of well logs is provided in Figure 3.3 from the same shallow marine North Sea reservoir along with the associated core description from collocated core. The well logs include caliper (CALI), gamma ray (GR), resistivity (RT), bulk density (RHOB), and neutron porosity (NPHI) along with descriptions from the collected core and interpretation.

In addition, traditional well logs provide measurements that may aid in interpretation of stratigraphic surfaces, faults, and reservoir fluids (e.g., the oil–water contact). Also, a variety of well bore imaging tools have recently entered common practice. These tools such as oil-based microimager (OBMI) and fullbore formation microimager (FMI), utilize many microresistivity sensors oriented along the circumference of the tool to image of the entire well bore (360 degrees) at a millimeter resolution. These tools provide additional information such as bedding structures and dips that may further assist in inference of the geological conceptual model and assigning facies at the well locations. This is a major advantage because this information was only previously available from the typically sparsely sampled cores. These image logs often have much better coverage because core extraction is very expensive and slows down drilling. Returning to the shallow marine North Sea reservoir studied by Folkestad et al. (2012), image logs play a critical role, providing detailed sedimentary structures at up to 5-mm scale. An FMI, which is used in the identification of various structures such as laminated, low-angle laminated, cross-bedded, deformed, and mottled along with ichnofacies such as planolites, was applied. An example FMI log from this work is shown in Figure 3.4.

With good coverage of available wells, well logs calibrated to core data often provide the overall reservoir bounding surfaces and stratal correlation style (see Section 4.1, Large-Scale Modeling) and may provide valuable information on heterogeneity and trends in reservoir properties.

3.1.3 Seismic Data

Seismic data cover all the types of information gathered by inducing sound waves through the reservoir and recording and processing the reflections and is a focus of geophysics. Geophysics is a broad scientific discipline; the following discussion is quite cursory, and the reader is referred to reference texts on seismic acquisition, processing, and interpretation by Gadallah and Fishe (2010) and seismic interpretation by Posamentier et al. (2007).

The *sweep* is the set of sound frequencies directed into the subsurface. Seismic reflections indicate potential rock transitions, and the properties of the reflections (amplitude and polarity) provide information on the relative nature of the transition. Inversion of these recorded reflections provides an indirect measure of acoustic impedance or transitions in the rock density and sound velocity (Bosch et al., 2010; Buland et al., 2003; Dubrule, 2003).

With appropriate time to depth conversion and geometric correction for the spherical sound fronts, *migration*, these reflections are applied to map out large-scale structures, including strata correlation styles and bounding surfaces. The uncertainty in the

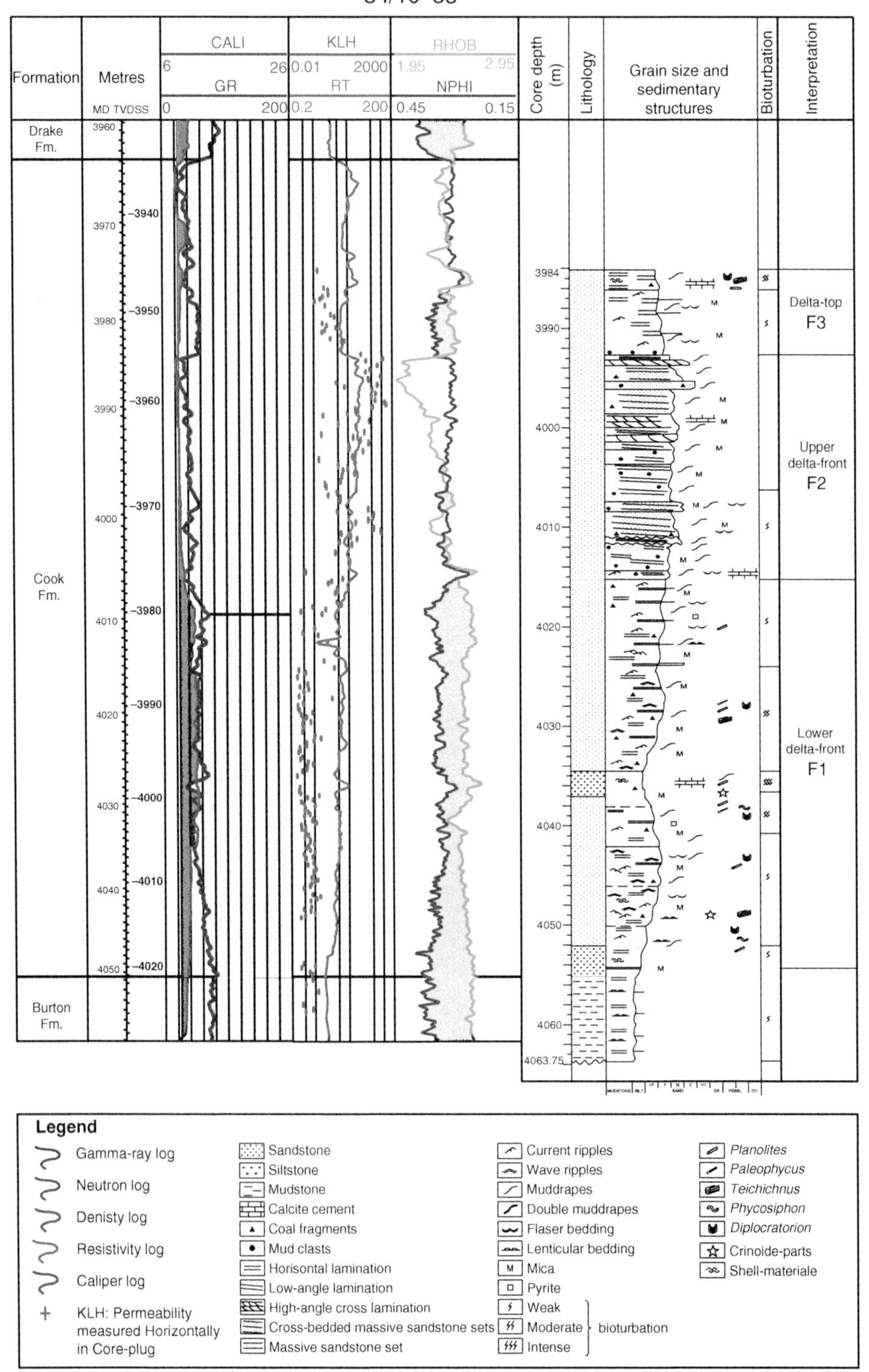

FIGURE 3.3: Suite of Well Logs with Interpreted Structures from the Core Data and Stratigraphic Units Form the Cook Formation, a Shallow Marine Sandstone Reservoir from the North Sea. The core has been interpreted as a fluvial/deltaic depositional setting with general progradation upward (note the general coarsening upward) and used to calibrate the log response.

Reprinted from Folkestad et al. (2012) with permission from Elsevier.

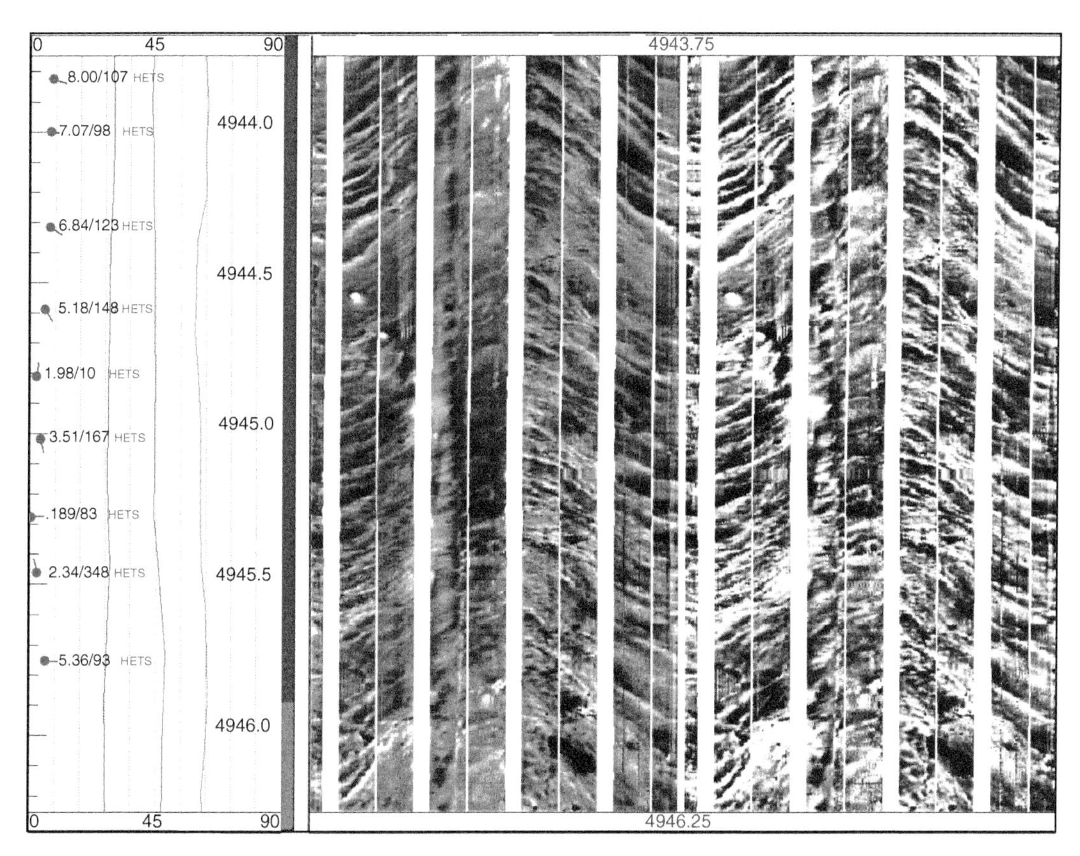

FIGURE 3.4: FMI Image Log Included from a Shallow Marine Sandstone Reservoir from the North Sea. Polarity is normal with resistive rock shown as bright and conductive rock shown as dark. Note: There are two sets of four tracks; the left set has an absolute color scale, while the right has a relative color scale to amplify local features. The four tracks provide a 360-degree view of resistivity, and the sinusoidal pattern is indicative of bed dip. Folkestad et al. (2012) interpreted thin laminae (brown bar) and small-scale ripple lamination.

Reprinted from Folkestad et al. (2012) with permission from Elsevier.

resulting framework is a function of seismic resolvability linked to the interpretation and error in the rock velocity model.

The resolution of seismic data is variable because the resolvable limit of seismic data is a function of noise, the quality of data, and the interpreter's skill (Bertram and Milton, 1996; Sharma, 1986). The rule of thumb is that the resolvable limit is $\frac{1}{4}\lambda$, where λ is the wavelength of the p-wave. Resolution decreases with depth due to increased sonic velocity with depth and the preferential filtering of the high frequencies of the sweep. In addition, artifacts may occur due to multiples, reflected waves reflecting again, and errors in migration and in the velocity field.

$$\lambda = \frac{v}{f}$$

where λ is the wavelength, v is the sonic velocity, and f is dominant reflected frequency. The increase in velocity and decrease in frequency with depth increase wavelength and diminish resolution. Example vertical resolvable limits are shown below.

- Shallow loosely consolidated sandstone

$$v \approx 1800\ \text{m/s},\ 60\ \text{Hz},\ \frac{1}{4}\lambda = 7.5\ \text{m}$$

- Deeper Paleozoic carbonate

$$v \approx 4500\ \text{m/s},\ 15\ \text{Hz},\ \frac{1}{4}\lambda = 75\ \text{m}$$

There are four modes for considering seismic data in reservoir modeling that may be roughly considered along a continuum of scale. The first is seismic-derived structural interpretation—that is, surface grids and fault locations that define large-scale reservoir geometry (large-scale reservoir modeling is discussed in Section 4.1). The second is the interpretation of reservoir depositional setting and architecture from seismic (this information is considered specifically in Section 3.2, Conceptual Model). The third is seismic-derived reservoir properties in the form of variations in facies proportions and porosity that may be applied over individual model cells. Often seismic resolution is lower than that of individual cells; for example, vertical averaging is assumed (Sections 4.2–4.6 consider the input of this information into facies and continuous reservoir property models). The fourth is microseismic monitoring to resolve induced fractures and fluid flow (Maxwell and Urbancic, 2001). Integration of this emerging technology is not covered in this text.

In general, the associated difficulty in seismic data integration and uncertainty in the results increases as this scale decreases from large-scale structure to reservoir depositional setting to seismic-derived reservoir properties. For example, in many reservoir settings, seismic data provides good constraint on reservoir geometry, some indication on reservoir architecture, and potentially weak constraint on reservoir properties.

Seismic-derived reservoir properties are more difficult to infer than structure and architecture. This mode requires strong assumptions on rock acoustic properties and expert calibration. The final result is an indication of large-scale variations, trends, in facies proportions and porosity. Post-processing will likely be required to deal with noise in the calibrated results due to noise and artifacts in the seismic data.

An example of a seismic line is shown in Figure 3.5 from the lower Cretaceous of the north coast of Alaska. This line includes a set of prograding clinoforms associated with shallow deltaic and fluvial depositional setting. Such seismic information provides valuable information on potential reservoir locations, geometry, connectivity, and quality of the reservoir.

3.1.4 Dynamic Data

A critical aspect of data integration is consideration of well test, historical production, and 4-D seismic data. These data are time-varying responses that are nonlinearly related to the static facies, porosity, and permeability. Consideration of dynamic data integration in reservoir characterization is an active area of research. Following is a brief presentation of where research and development related to production data appear to be going.

Early approaches to numerical history matching by mathematical inversion have been hampered by: (1) excessive CPU execution time for even relatively small reservoir models; (2) limiting, simplifying assumptions—for example, a linear relationship between pressure and reservoir properties; (3) consideration of one data type at the expense of information coming from other data sources such as seismic and geology; and (4) significant difficulty with multiphase flow data. Notwithstanding some research advances in this area, the most common history matching procedure is manual iteration.

Production data measurements are essentially at a single point in space that yield effective properties over a large volume. Attempts to build a detailed subsurface model from production data alone are not likely to be successful. A promising avenue is to use the production data at wells to establish coarse-scale maps of permeability. These maps are then used as constraints along with those from other available data in geostatistical reservoir modeling.

Production data could be used to improve the estimation of global reservoir parameters, such as the mean and variogram of reservoir properties.

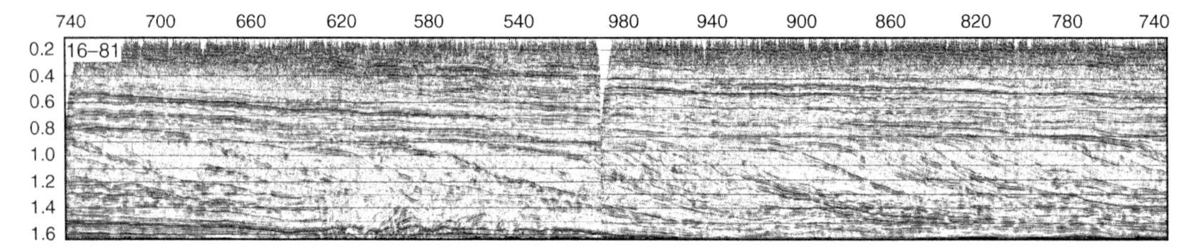

FIGURE 3.5: A Seismic Line for the Torok Formation Clinoforms of the Lower Cretaceous in the National Petroleum Reserve of Alaska, USA Just South of the Harrison Bay on the Coast of the Beaufort Sea. Seismic lines are shot by the USGS and are available in public domain.

Figure provided in high resolution by Professor Chris Kendall, available at the Society for Sedimentological Research Stratigraphy Web and reprinted with permission from the Society of Sedimentological Research.

Once estimated, such statistical parameters are used as inputs in geostatistical techniques to construct reservoir models. The contribution of production data lies in the improvement in the estimation of statistical parameters describing the reservoir heterogeneity.

In general, from a provenance perspective, production data may be summarized into three main groups: single-well test data, multiple-well test data, and multiple-well multiple-phase historical production data.

Mathematical inversion methods for single well test pressure data and interpretive tools are largely in place (Earlougher, 1977; Horne, 1995; Oliver, 1990a,b; Raghavan, 1993; Sabet, 1991). This is the subject of well test analyses research. Some of the typical single well test data are RFT data, drawdown/buildup test data, variable rate test data, production logs, and permanent pressure gauges. The main result of these mathematical inversion methods are large-scale effective porosity and permeability values. To a large extent, these large-scale properties can be accounted for in geostatistical modeling (Alabert, 1989; Deutsch, 1992a; Deutsch and Journel, 1995).

Compared to single-well test data, multiple-well test data are more extensive in terms of areal coverage and provide specific connectivity information between wells. An important issue is whether sufficient data are available to establish unambiguous connections between multiple locations.

According to production mechanism, historical production data arise from different sources. The classification can be with respect to reservoir depletion with or without water drive, with gas-cap drive, water injection, or gas injection. Each of these has unique implementation and interpretation issues.

A thorough review of the subject of parameter identification in reservoir simulations is not possible. Some reasonable literature reviews include Jacquard and Jain (1965), Gavalas et al. (1976), Watson et al. (1980), Feitosa et al. (1993a,b), and Oliver (1994); Yeh (1986), and Carrera and Neuman (1986) have prepared similar reviews in groundwater hydrology.

3.1.5 Analog Data

There is a broad variety of analog data available. By definition, these data are external to the reservoir and are not site-specific, although they may be deemed applicable to specific regions of the reservoir. The determination of appropriate analog data is based on the interpretation of the reservoir story and associated geological concepts as described in Section 2.1, including large-scale basin formation and filling and reservoir architectural concepts. Once the story is determined, available mature fields, outcrops, and high-resolution seismic data with similar reservoir stories may be judiciously classified as acceptable analogs. In some cases, analogs may be extended to numerical process models and experimental stratigraphy. Regardless of the analog and the level of inferred similarity to the reservoir of interest, the analogous nature and any associated caveats concerning possible differences should be preserved in the analog data events (see previous discussion on data events in this section).

Mature Fields

Mature fields offer a good analog option with the advantage of a high sampling rate for reliable reservoir property distributions and production experience. The data sampling may not be dense enough (and given sparse sampling of the reservoir of interest) to completely eliminate uncertainty with regard to the reservoir story; therefore, there may be uncertainty with regard to the appropriateness of the mature field as an analog. This uncertainty may be accounted for with multiple analog scenarios (see Section 5.3, Uncertainty Management). In addition, data may not be sufficient to determine high-resolution element and subelement architecture.

Yet, mature fields provide a valuable opportunity to validate our judgment in reservoir inference and exploitation decisions in the presence of sparse data. At one time, the mature field was immature with a similar level of data and uncertainty as the reservoir of interest. *Look backs* in analogous fields may provide an indication of potential mistakes and systematic biases.

Outcrops

Outcrops provide a unique opportunity to observe rock, with relatively straightforward observation and sampling with no limit on the fine-scale resolution and with the large-scale features limited by the extent of the outcrop exposure. Various world class outcrops are famous for exposure extents in the range 10–100's kilometers. For example, consider the Karoo Sub-basin of South Africa (Prélat et al., 2010; Pringle et al., 2010) and the Book Cliffs of Utah, USA (Howell and Flint, 2003). It is of note that in rare cases the reservoir of interest outcrops at some distance from the active reservoir. Williams Fork Formation in Wyoming, USA, is an example of this, and the outcrop offers a potentially strong analog data set (Pranter and Sommer, 2001).

The typical method for surveying outcrops includes photomosaics, measured sections, and coring. In the case of photomosaics, multiple photographs are shot and merged together with care to limit distortion due to perspective and outcrop face irregularity and to maximize clarity with proper lighting. For measured sections, survey lines are walked and interpreted at high resolution with information such as grain size and type, sedimentary structures, faults, and so on. With a sufficient density of measured sections and aided by photomosaics, full interpretation of the outcrop is determined over a full range of hierarchies (see Figure 3.6). At times, attempts are made to directly measure reservoir properties from an outcrop. This typically involves utilization of coring tools to extract core for similar analysis as those performed on well core to assess grain size, mineralogy, porosity, and permeability.

Modern developments in outcrop studies have resulted in two potentially significant changes in outcrop survey. These include LIDAR (light detection and ranging) survey and advance quantification. LIDAR is a laser-based survey tool that is able to rapidly map outcrops at centimeter scale in 3-D space. In addition, the LIDAR provides information about the reflection that can be calibrated to provide information on lithology and grain size. It is common to anchor photomosaics on the LIDAR data set to construct a high-resolution virtual outcrop that may be efficiently interpreted. Consider the ease to observe

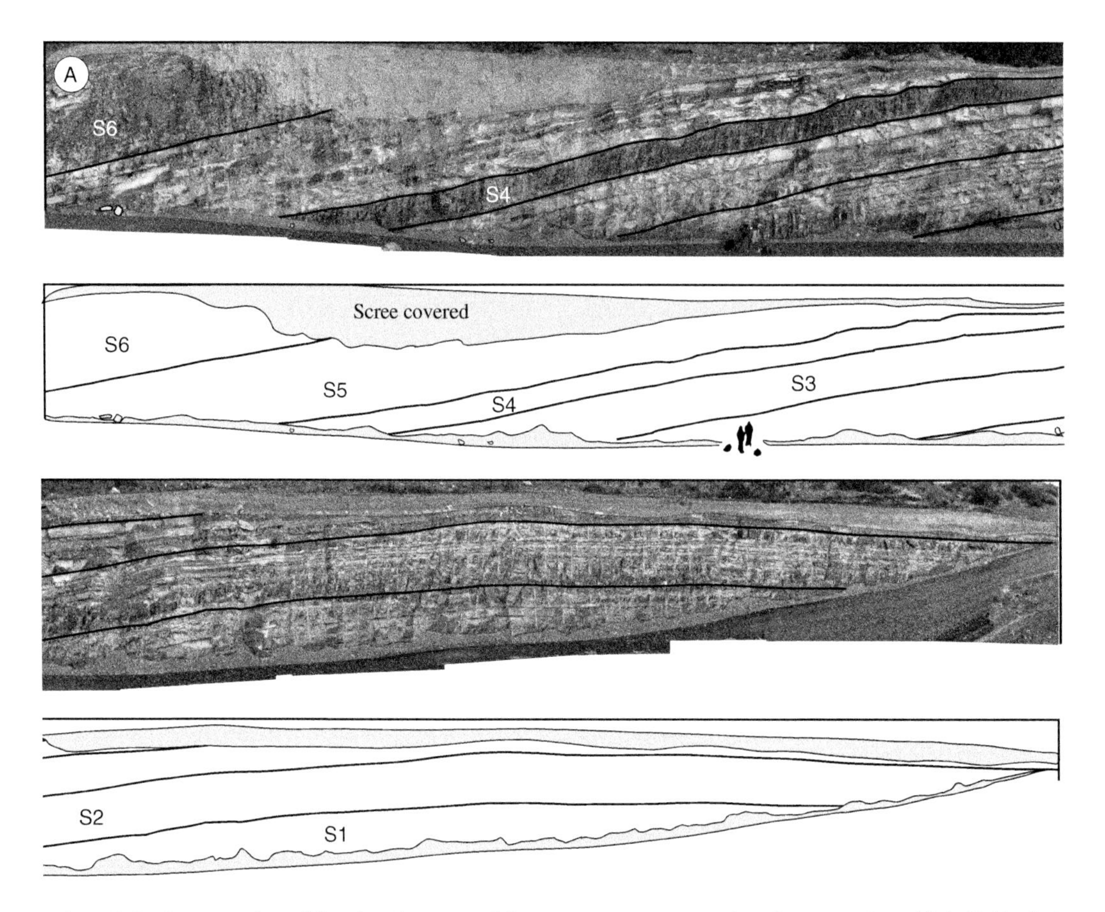

FIGURE 3.6: Photomosaic and line drawing trace of the Punta Barrosa Formation sheet complex with individual elements denoted as sheet elements S1–S6. This detailed field study identified the overall laterally extensive, continuous sheets with limited erosion and various facies resulting from the transport and depositional processes. Figure reprinted with permission from the Society for Sedimentary Research, from Figure 2.1 in Fildani et al. (2009) and high-resolution figure provided by Dr. Andrea Fildani.

from any perspective, over distance, and out of line of sight in addition to the ability to digitize features and measure distance and angles (Pyles and Jennette, 2009). Admittedly, due to limits in resolution and information, it is not anticipated that virtual outcrops will remove the need to visit the real rocks.

Also, there has been a trend for outcrop studies to apply improved quantification. For example, it is common for outcrop studies to include various distributions describing the geometrical parameters of various reservoir elements and their associated proportions and trends in these parameters linked to allogenic forcing. This has been a determined effort to make outcrop studies more directly applicable to reservoir modeling. Various authors have taken this a step further with the application of spatial statistics to outcrops to measure reservoir element stacking with compensation index (Wang et al., 2011) and Ripley's K functions (Hajek et al., 2010).

There are many examples that may illustrate the value of outcrop studies to aid in reservoir inference. For example, consider the Karoo Formation of South Africa that provides exceptional exposures of deepwater channels and sheets. Detailed models concerning the stacking and internal heterogeneity of channels and sheets have been developed to improve understanding of the channels and sheets present in the deepwater Gulf of Mexico (Prélat et al., 2010; Pringle et al., 2010).

There are some significant limitations to outcrop information (Lantuéjoul et al., 2005). Firstly, the data is typically 2-D. Plan view irregularity may provide some information on the missing dimension, and some efforts have been made to drill behind outcrops or to use ground penetrating radar (Li and White, 2003). In addition, outcrops present an interesting observational bias. Consolidated rock is preserved in outcrop while weaker rocks (e.g., shales) are eroded and, if dominant, prevent the formation of outcrops in the first place. Also, weathering and unloading of the rocks may change the rocks and obscure observation of features relevant in the in situ state.

Shallow, High-Resolution Seismic Analogs

High-resolution seismic analogs commonly result from the imaging of shallow substrate while surveying deeper objectives. While these shallow data sets do not likely image actual reservoirs, they may be useful analogs to the deeper objective or other reservoirs deemed to be analogous. This use of high-resolution seismic analogs to characterize plan view architecture has been termed *seismic geomorphology*, in contrast to traditional seismic stratigraphy that is focused on the interpretation of vertical seismic sections. A good overview of seismic geomorphology is provided by Posamentier et al. (2007).

The advantages of high-resolution seismic analogs were previously discussed. In comparison to previously discussed outcrops, the ability to observe full 3-D and evolution through time (represented by vertical dimension in aggrading systems) over large areas is quite attractive and has been fueling the growth of seismic geomorphology.

The Joshua channel system basinward of the Mississippi Delta in the Desoto Canyon of the Gulf of Mexico described by Posamentier (2003) is a good example of the use of high-resolution seismic data (see Figure 3.7). Shallow seismic studies may be useful for understanding and predicting the reservoir heterogeneities of channelized Gulf of Mexico reservoirs. We thank Elsevier for permission to reprint the figure, and we thank Dr. Posamentier for high-resolution figure.

Shallow seismic data are useful for improved understanding of architecture, but limits in seismic resolvability prevent the detailed study possible with outcrops. In addition, these seismic analogs will not likely have core to calibrate architectural concepts, and shallow seismic data are subject to the "high stand bias." Geologically speaking, we are currently in a high-stand systems tract, with the associated aggradation of fluvial valleys, storage on continental shelves and quiescence in deepwater basins: therefore, all shallow and recent deposits are part of the global high stand. It is likely that falling stage and low-stand geomorphology would be quite different.

Geomorphology

Geomorphology, as mentioned in Section 2.1, is the study of the evolution of landforms. This includes the processes that erode, transport, and deposit rock. Fundamental to geomorphology is the characterization of sediment transport laws that are parameterized from field measurements and tested in physical models (Dietrich et al., 2003). Geomorphology plays an important role in understanding the storey behind the formation of the reservoir. While this is a valuable field of study, care must be taken to transform geomorphology into sedimentology, the preservation of these features in the rock record.

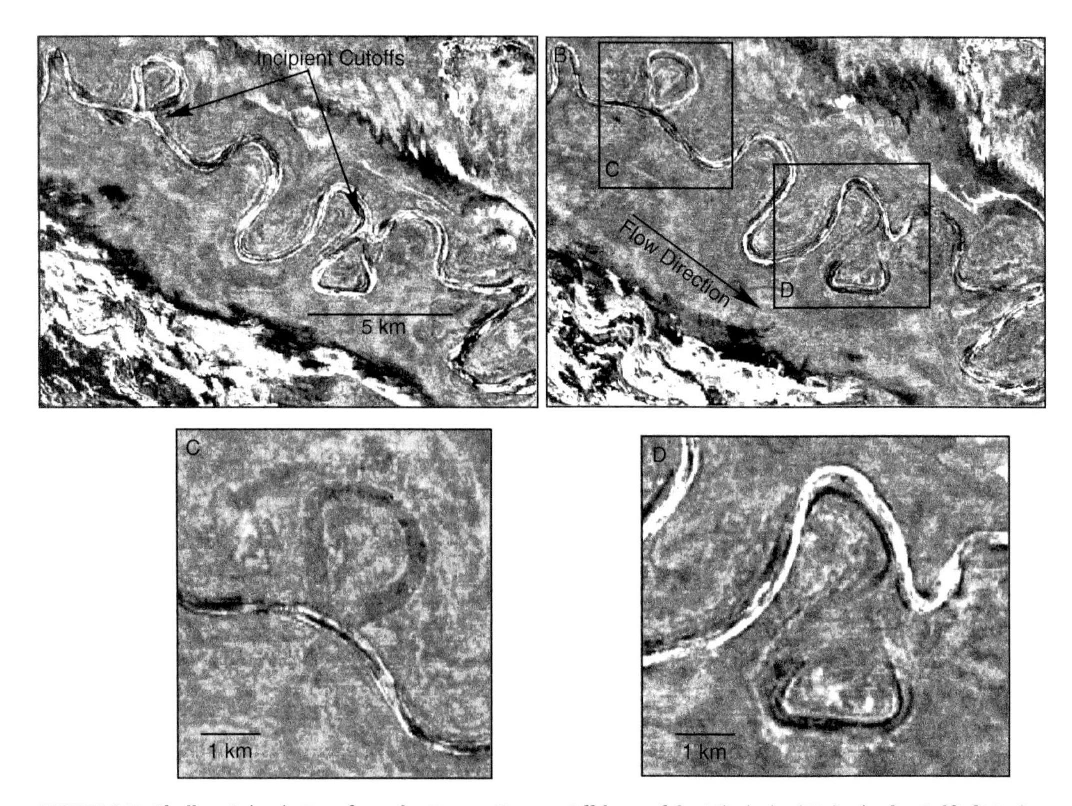

FIGURE 3.7: Shallow Seismic Data from the Desoto Canyon Offshore of the Mississippi Delta in the Gulf of Mexico. By taking a sequence of time horizons from the seismic data, one can observe the evolution of the Joshua channel. This includes downdip translation of the channel and neck cutoffs. In (A), two neck cutoffs are imminent; in (B) the cutoffs are complete; and in (C) and (D) the abandonment of the cutoffs is complete.

High-resolution figures provided by Dr. Henry Posamentier; reprinted from Posamentier (2003) with permission from Elsevier.

This can be challenging, considering that a meandering river channel may actually be preserved as a large uniform sand unit. It is natural that such a transformation of concepts is required, given that reservoir geostatistical modeling is focused on the quantification reservoir heterogeneity of the current reservoir heterogeneity, while geomorphology is focused on the evolution of landforms over significant spatial and temporal scales. With expert application, geomorphometric understanding is useful for building a complete reservoir story. These concepts may be explored through experimental stratigraphy and numerical process models.

Experimental Stratigraphy

Experimental stratigraphy is an attempt to recreate unscaled models of geologic processes. For example, Paola (2000) and Paola et al. (2001) provide an overview of the methodology, capabilities, and limitations. In general, these models are considered weak analogs or even "metaphors," because of intractable issues with respect to representing spatial scale, time scale, process, and material behavior. However, experimental models can be applied to answer specific questions on the behaviors of a system such as the general relationship between allogenic controls and autogenic response or the depositional trends.

Experimental EarthScape (Jurassic Tank) at Saint Anthony Fall Laboratory is a well-known example of experimental stratigraphy. Jurassic tank includes the ability to precisely control the subsidence of the basin, sediment, and water inputs and sea level and results in preservation of realistic sediment structures (see Figure 3.8). These features include clinoforms, channel scours, various depositional bedform types, avulsion, and valley erosion.

Experimental stratigraphy is yielding new understanding on potential reservoir heterogeneity. For example, the work of Paola et al. (2009) has added insight into the deposition grain size and thickness distributions associated with deepwater channel bends. This work demonstrates significant

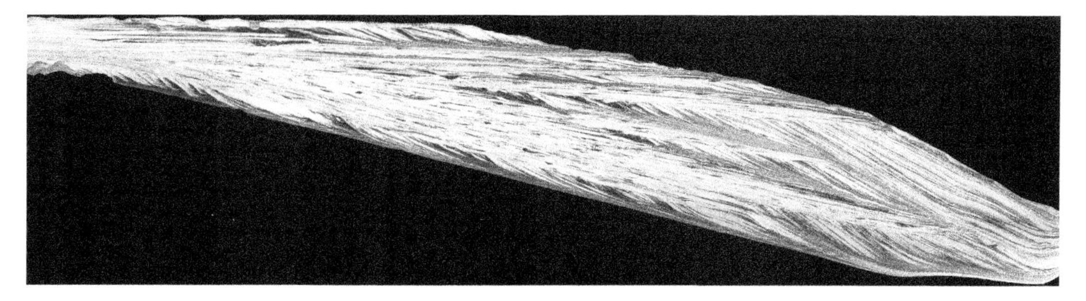

FIGURE 3.8: A Dip Section from the Experimental EarthScape (Jurassic Tank) Experiment from Saint Anthony Falls Laboratory. The details of this experiment are available in Paola et al. (2001).

Figure provided by and reprinted with permission from Professor Chris Paola. For details about the facility see Paola et al. (2001).

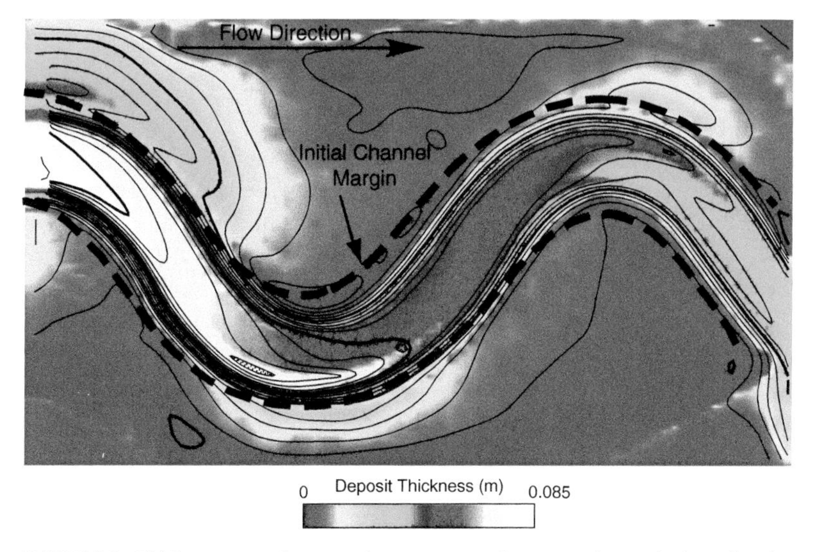

FIGURE 3.9: Thickness Map from a Laboratory Experiment on the Evolution of a Sinuous Submarine Channel Due to Sedimentation from 24 Turbidity Currents. Figure provided by and reprinted with permission from Professor Kyle Straub, based on experimental work documented in Straub et al. (2008).

asymmetries in deposition with coarser and thicker deposition on the outer bank, in fact, the outer levees had grain sizes comparable to the center of the channel. This work provides support for within channel grain size trends and evidence of reservoir quality sand outside of channels in deepwater settings (see depositional thickness plot provided by Prof. Kyle Straub in Figure 3.9). Such concepts may be readily adapted into statistical and non-stationarity descriptions for geostatistical reservoir models.

There are various limitations with current experimental stratigraphy models. They are generally related to scaling and cohesion. There is no current method that allows for rigorous scaling of these systems. Under more rigorous scaling constraints the experiments would not run over practical time scales (the discharge rates of water and sediments to match the properly scaled flow conditions would be very low). In addition, scaled [down] fine-grained particles become so small that they do not readily settle out. On a related note, it is very difficult to recreate the cohesion observed in natural systems that results in bank stability and resistance to erosion. As a result, most experiments lack confined channel flow and appear more braided with large wetted fractions of the model. Hoyal and Sheets (2009) have described unique material mix that result in cohesive behavior. General discussion on the applications and limitations of experimental stratigraphy are provided by Paola et al. (2009). Nevertheless, with recognition of the limitations, these experiments are yielding valuable insights.

Numerical Process Models

Numerical models of earth processes may be characterized as three types: landscape evolution models, sediment transport models, and morphodynamic models. Landscape evolution models attempt to capture the large-scale changes in topography, such as erosion of valleys and progradation of deltas. Sediment transport models focus on the movement of sediments in rivers, debris flows, or turbidity currents. Morphodynamic models attempt to model the complicated coupling of flow and substrate (Slingerland and Kump, 2011; Syvitski, 2012).

Common to all of these models are limitations in the process understanding, computational facilities, and inaccessibility of initial and boundary conditions. For example, while the process of channelized flow is well understood, entrainment of grains into the flow can only be described by empirical models as there is no first principles derived erosion equations (Parker, 2012). Due to computational cost, modeling is conducted generally at large scales (space and time) and often do not generate high-resolution heterogeneity over reservoir scale models. Finally, these models are generally highly sensitive to initial and boundary conditions that are very difficult or impossible to infer for a reservoir. This would require paleo-restoration of the basin to the time of reservoir formation and the inference of the influx rates and type from a source that likely no longer exists. As a result, these models are not likely to provide reservoir prediction with local accuracy.

Nevertheless, these models are valuable tools for learning the complex interplay of various allogenic and autogenic processes and the resulting reservoir geometry, trends, and perhaps even flow heterogeneity. These process models are discussed in Section 4.5 as the source of process rules for process-mimicking modeling.

Applications of Analog Data to Reservoir Modeling

There are limitations and uncertainty inherent to the integration of analog data. Some examples of the integration of information into geostatistical models are provided in Section 3.2, Conceptual Model. In general, there are two typical modes for the application of analog data to reservoir modeling: filling in consistent details in the conceptual model and constraining statistical inputs.

In the first case, the analog data provide consistent details that would not be available from the data and from other information from the reservoir of interest. For example, limited well data may suggest various reservoir channels, and shallow seismic analogs may identify confined channel complexes spatially confined by paleovalleys. In this case, the analog may be employed to constrain the areal extent of the channels.

In the second case, input statistics are utilized from the reservoir analog. This may include various forms, including validation, constraint, and adoption. In the case of validation, data distributions and concepts from the reservoir of interest are validated by comparison to the analog. For example, the porosity distribution inferred from sparse well data is supported by similarity to an analogous mature field. The constraint form utilizes the analog statistics to improve the target reservoir statistics. This may take the form of defendable modification, such as scaling or Bayesian updating with the analog supplying a prior distribution. In the final case, the statistics are exported from the analog to the reservoir of interest. This requires a strong choice of stationarity between the reservoir of interest and the analog. In the presence of little or no local information as in the case of some exploration settings, this type of reliance on analog information is unavoidable.

3.1.6 Data Considerations

There are various general considerations associated with data integration. Firstly, all data types have different scales, accuracies, and coverage. Secondly, data should be checked, cleaned, and prepared for statistical analysis and application to modeling. Finally, data may require calibration including conversion from secondary information to a conditional probability associated with the reservoir property of interest.

Scale, Accuracy, Coverage, and Application

All of the previously listed data sources have their associated resolution or scale of measurement, accuracy, coverage, and application. In general, there is a scale and coverage trade-off in reservoir data sources. For example, core data provide very high resolution information, but with very low coverage, while seismic data may provide very good coverage of the reservoir, but generally with very low resolution.

In addition, there is also a trade-off between accuracy and coverage. In this case, we consider accuracy with respect to reservoir properties such as

TABLE 3.1. A SUMMARY OF THE AVAILABLE DATA AND INFORMATION FOR THE CONSTRUCTION OF RESERVOIR MODELS[a]

Type	Resolution	Coverage	Application
Core	$\simeq\infty$	V. Low	Sample/Calibration
Well log	10 cm	Low	Reservoir properties/trends, surfaces
Image log	$\simeq$ 5 mm	Low	Sedimentary structures/faults
Seismic	10 m	$\leq$ Complete	Trends, secondary, surfaces
Production data	10–100 m	Medium	Reservoir quality
Analog data			
Mature fields	10–100 m	$\leq$ Complete	Validation, prior all
Outcrop	$\simeq\infty$	None	Concepts, input statistics
Geomorphology	$\simeq\infty$	None	Concepts
Shallow seismic	$\geq$ Element	None	Concepts, input statistics
Experimental stratigraphy	$\simeq\infty$	None	Concepts
Numerical process	$\geq$ Complex	None	Concepts

[a]These values represent approximate measures. Greater detail is provided in the text.

porosity. Well cores provide the highest accuracy for reservoir properties, but at the lowest coverage, while well logs sacrifice some accuracy for improved coverage and once again seismic data may have very good coverage, but the accuracy in measuring reservoir properties is much lower than well derived data. This is summarized in Table 3.1.

Data Checking

An observation of experienced geomodelers is that the prerequisite data cleaning, formatting, and checking is about 80% of the effort in a reservoir modeling project. This is due to the complexity and massive size of modern reservoir data sets. This often includes data of various diverse types and sources that are collected. To put this data maintenance effort in perspective, consider the cost of collecting the data in the first place and the cost of mistakes due to mishandling the data that would likely be expressed as lost opportunity cost in exploitation of the reservoir.

The first type of data issue is data blunder. This includes issues in the measurement, transcription, or storage that results in some or all values being incorrect. For example, a mix-up in units may result in exaggeration of reservoir thickness or a well location being shifted. Care must be taken with unusual unit conventions such as feet for vertical dimension and meters for area dimensions. Geostatistical tools do not identify these issues, although the summarization and visualization associated with geostatistical analysis may indicate an issue.

There are various methods that may be applied to check for data blunders. Simple methods such as searching the database for invalid values (values with non-numeric values or outside a reasonable range), missing values, and sequences of repeating values may be fruitful in efficiently finding obvious blunders. Plotting data histograms and scatter plots is useful for identifying outliers that may require attention. Posting data values to maps may aid in checking spatial coherency and identifying "out of place" data values.

A second type of data issue is bias. More subtle issues such as bias may require careful investigation and scrutiny of the database. For example, one may review statistics for each sample campaign, operator, or vintage to identify bias. Also, there may be a need to return to the raw data and check the associated interpretations to ensure accuracy. These issues may be mitigated with the use of more complete data documentation through the application of data events described previously in this section.

A third type of data issue is consistency between data types. For example, core permeability data may imply an average permeability of, say, 50 mD and well test data may imply an average of 300 mD. Perhaps there are fractures or other high-permeability features causing the large-scale effective permeability to be higher than the measured core data; it may be necessary to correct or complete the core data before modeling. Alternatively, an inappropriate interpretation model may have been used in the well test analysis program; reinterpretation may be necessary before using the data. Inconsistency may also include correlation between data types that should not be correlated. Geostatistical tools, at best, identify inconsistent data. It does not help to resolve any discrepancies or inconsistencies; the reconciliation of the different data types must be done by the various professionals involved in the project team. Most often, inconsistencies are reduced by interpreting each source of raw data in light of all other data types.

It is relatively easy to construct a numerical reservoir model that reproduces only some of the available data; for example: (1) A hand-contoured map of average porosity would reproduce averaged well data and geologic trends, but would not represent the details of the well data nor flow properties from well test and historical production data, (2) a statistical model would reproduce all local well data and certain statistical measures of heterogeneity, but may not represent important geologic features that could impact flow predictions, and (3) a large-scale "inversion" from production or seismic data would reproduce the production or seismic data, but would not represent small-scale heterogeneities and other possibly important geologic constraints.

The challenge addressed by geostatistical reservoir modeling, and this book, is to simultaneously account for all relevant geological, geophysical, and engineering data of varying degrees of resolution, quality, and certainty. The idea of a creating a "shared earth model" that represents all relevant data is not new (see Gawith et al. (1995)).

Data Calibration

Data calibration is a valuable tool with many applications in geostatistical modeling. Central to the method is the modeling of a relationship between a measured property and a property of interest and then utilizing this relationship to map each measured soft data value into a soft indication of the inferred property. For example, calibration may be applied to map a seismic attribute into a probability of a particular facies. In another example, calibration may be applied to fill in missing facies assignments with available reservoir properties (see Section 4.6) (and Figure 3.10). In Section 2.2, calibration (in the form of debiasing) was applied to correct a bias distribution with a secondary data and modeled relationship between the secondary and primary data.

The calibration method is general and may be applied with any combination of continuous and categorical, any number of secondary variables (one variable is illustrated here is simplicity), measured and inferred properties, and scales as long as consistency is maintained for each property. The general steps are to bin the measured data (not required if categorical) to allow for the pooling of multiple samples in each bin. Bin size is determined to provide sufficient samples in each for interpretable results and not too large as to mask important information. The result is a conditional distribution (continuous probability density function or categorical proportions) in each bin. There may be important constraint information or expert judgment not captured in the calibration results, or noise due to too few data. It may be necessary to smooth or modify the results in a defendable manner.

Mapping the measured variable through the calibrated relationship will result in a soft indication of the inferred variable. For example, if fraction of shale (V_{sh}) is the measured variable, then a V_{sh} may

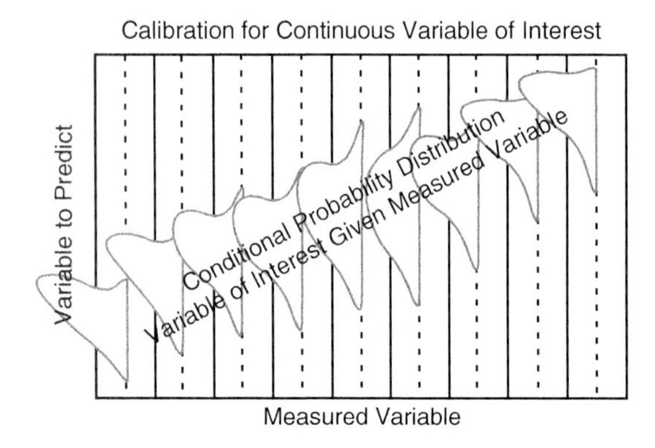

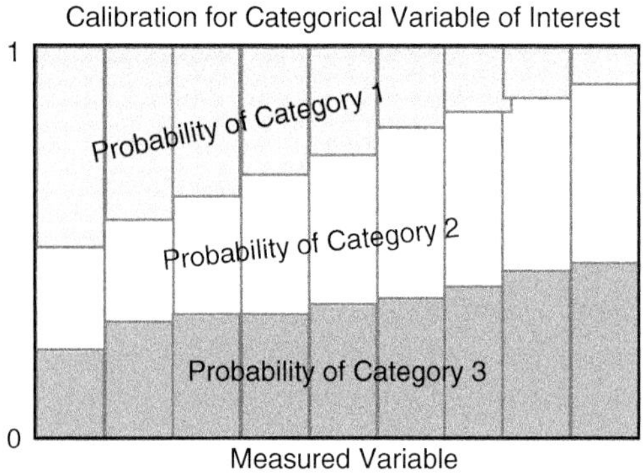

FIGURE 3.10: Illustration of Calibration for a Continuous or Categorical Variable of Interest. In the continuous case for each bin or category of the measured variable, a probability density function is inferred; and in the categorical case, a proportion illustrated by the relative length of the bars is inferred.

indicate a high probability of sandy facies, a low probability of inter-bedded sand and shale, and a very low probability of shale. This information could be integrated directly as soft data in the indicator framework (see Section 2.2 and a demonstration in Section 4.6).

3.1.7 Section Summary

The great challenge of geostatistics is the integration of various data sources into a consistent numerical geological model. These data sources include direct measurements, indirect measures, and analog information.

The data event methodology requires greater rigor in handling data, but will improve the management of the large amounts of data inherent to many geostatistical studies. In addition, care must be taken to adequately check data and information that is integrated into the project. Mistakes during the data preparation and cleaning stage will often dramatically and negatively impact model and resulting decision quality. In the next section, the geological conceptual model is introduced as a framework for the integration of these data. Then Section 3.3 will formulate important reservoir modeling problems and associated work flows.

3.2 CONCEPTUAL MODEL

This section covers general concepts related to the formulation of the conceptual model and considerations for integration of these concepts into a geostatistical model. The primary focus is on large-scale framework modeling and then modeling reservoir properties within this framework with the considerations for selection of suitable geostatistical modeling algorithms and assessment of required statistical inputs.

This section takes concepts covered in Section 2.1, Preliminary Geological Modeling Concepts, and discusses how they are encoded into statistical constructs described in Sections 2.2, Preliminary Statistical Concepts, and 2.3, Quantifying Spatial Correlation, and input into heterogeneity modeling methods described in Section 2.4, Preliminary Mapping Concepts. This provides a foundation for the application of large scale modeling described in Section 4.1, Large-Scale Modeling and reservoir scale modeling described in Sections 4.2–4.7 that cover the algorithms for geostatistical simulation of reservoir properties. This section also prepares us for the more detailed discussion on work flow formulation in Section 3.3, Problem Formulation. In bridging these topics, this section provides general guidance and comparison of potential methods with a focus on what spatial variables we need to model to meet the project goals, in the context of what we are trying to model, while Chapter 4 focuses on how specific models are constructed.

The *Conceptual Geological Modeling* subsection covers the general concepts related to integration of information from the conceptual geologic model. These concepts include fit-for-purpose modeling, feasibility, what information to integrate, the limits in integrating this information and the cost of ignoring information.

The *Modeling Framework* subsection discusses the basics in reservoir framework construction. These are practical concepts related to stratigraphic grid construction, such as accounting for strata correlation style and faulting. This is a prerequisite for algorithms related to large scale framework modeling in Section 4.1, Large-Scale Modeling.

The *Modeling Method Choice* subsection discusses the trade-offs in selecting between the available geostatistical reservoir modeling algorithms to fill reservoir properties into a framework model. Considerations such as objectives, software, professional expertise, data density, and geological complexity are discussed and specific examples are provided with suggested algorithms. More detailed discussion on these considerations with regard to work flow formulation is provided in Section 3.3, Problem Formulation.

The *Statistical Inputs and Geological Rules* subsection discusses the general form of inputs for geostatistical algorithms. Specifics inputs for each geostatistical reservoir modeling algorithm are deferred to their associated sections (Sections 4.2–4.6); nevertheless, this discussion focuses on the importance of the conceptual geologic model in constraining these algorithm specific parameters.

This section will likely be most useful for novice modelers, specifically for the connections that are made between conceptual model and the choice of modeling method and inference of input statistics. Once this is understood, data collection and interpretation needs can be planned and communicated concisely with other members of the project team. For the intermediate level modeler, this is likely understood, but this section may be useful for communication and educating other members of the project team. Expert modelers may choose to skip this section.

3.2.1 Conceptual Geological Model

It is of primary importance to understand the phenomenon that we are modeling. Section 2.1 deals with a very short description of broad scientific discipline and terminology related to the genesis of a petroleum reservoir. The conceptual geologic model is the integration of the discussed physical processes, local information, and available analogs into a single or multiple (as part of a comprehensive uncertainty model) consistent qualitative models of the reservoir.

As discussed in Section 2.1, Preliminary Geological Concepts, it is essential to reconstruct the story behind the genesis of the reservoir. Ideally, it should be our scientific interest to seek a deeper understanding concerning the phenomenon that we are modeling. Practically, this is essential due to the modern transition to discipline integration. Earth science and engineering project team members are expected to cross disciplinary boundaries within integrated project teams. With this integration, the work is no longer "passed over the fence" between geophysists, geologists, geostatisticians, reservoir, and drilling engineers. Efficiency and improved decision quality are often realized by blurring these boundaries. A schematic of the interaction of reservoir geostatistics with adjacent scientific disciplines is shown in Figure 3.11.

One could pose the conceptual geological model as a movie of the original depositional processes and subsequent processes such as the structural and diagenetic alteration (see Figure 3.12). Imagine playing this geological movie in reverse from modern day, stripping away millions of years of sedimentation

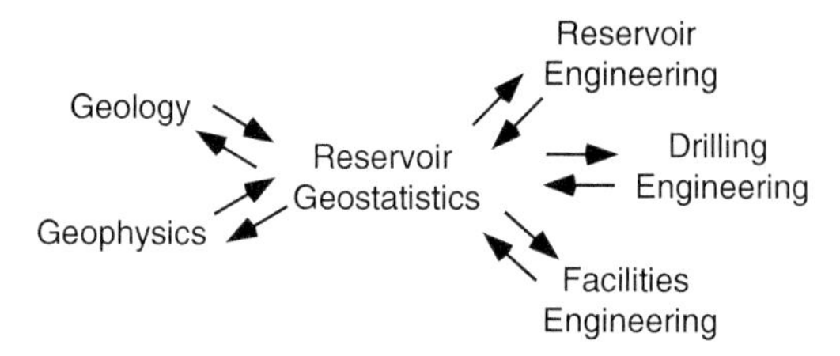

FIGURE 3.11: Linkages between Reservoir Geostatistics and the Adjacent Earth Science and Engineering Disciplines on a Project Team.

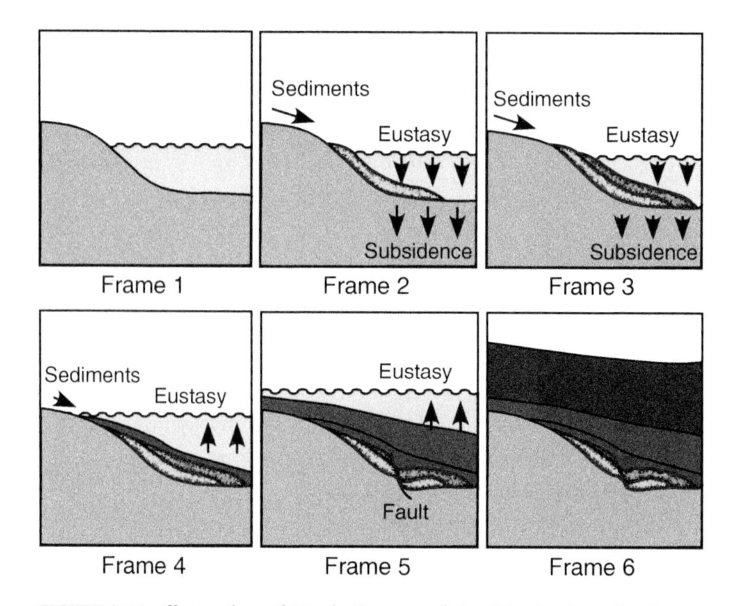

FIGURE 3.12: Illustration of Movie Frames of the Dip Section of a Reservoir Formation Movie. In frame 1 we start with available accommodation. In frames 2–4 terrigenous sediments the basin subsides and sea level cycles and the deposited reservoir sands (yellow and orange) are eventually buried by marine shales (dark gray). In frames 5–6 the reservoir is faulted and further buried by sealing shales.

and restoring eroded sediments, faults and compaction. This ability is central to geological sciences and represents an attempt to more completely understand the geological phenomenon. Although the geomodeler may not possess the skill sets necessary to unravel this movie from the limited data sources, the best possible appreciation of this conceptual model is important.

While the reservoir is dynamic over geological time, our goal is to pause this movie and model the time snapshot of the subsurface with our geostatistical tools that have the ability to integrate spatial data and statistics, and quantify uncertainty, but do not directly consider these physical processes. Yet, knowing the movie, including the frames before our moment in paused geological time, will likely improve the quality of the model and associated model uncertainty.

By understanding the movie outside our paused frame, we move from a static, superficial appreciation to a more complete understanding of earth dynamics. We appreciate the origin of the reservoir architecture and the potential post-depositional alterations such as faulting and diagenesis. This is analogous to the improvement from lithostratigraphy to sequence stratigraphy. In extending to a temporal understanding, our models face more rigorous accuracy and consistency checks. Our burden is to not only interpolate architectures and reservoir quality between the wells, but to ensure that these features are plausible over a sequence of events, thereby, meeting a higher standard of defendability.

This section is concerned with the translation of these qualitative concepts, our paused movie, into a reservoir model. This includes a geological framework and the heterogeneities placed in this framework with specific geostatistical facies and property simulation algorithms. Our challenge is to distill the salient features from this geological conceptual model and to integrate them into the geostatistical reservoir model.

General guidance and comparison is provided in this section in the context of what we are attempting to integrate. Specific implementation details on how we integrate for framework modeling are provided in Section 4.1; on various facies simulation algorithms in Sections 4.2–4.5; on various property simulation algorithms in Section 4.6; and with optimization in Section 4.7.

In the design of a modeling work flow and the selection of specific steps discussed in Section 3.3, Problem Formulation, the following concepts are considered:

1. What information to consider?
   (a) What is the purpose of the model?
   (b) Is the information supported by data?
   (c) Are the concepts consistent?
   (d) What information can be integrated?
2. What happens if the information is not explicitly integrated into the model?
3. Is there uncertainty in the conceptual model?
4. Is the work flow feasible?

A more specific description of each follows.

What information to consider? A decision will be required on the scale and type of conceptual geological information to integrate into the model. This will be a function of purpose and feasibility mentioned above and data availability, quality, and coverage as discussed in Section 3.1 and the capabilities and limitations of the geostatistical algorithms. In making this choice, we have the following questions to guide us: What is the purpose of the model? Is the information supported by data? Is the result consistent? What information can be integrated? This decision may be complicated by the expectations of other members of the project team that may not appreciate the capabilities and limitations of geostatistics.

What is the purpose of the model? As discussed in Section 1.2, the geostatistical model is typically only an intermediate product and the final product is the result of a transfer function applied to the geostatistical model. For example, the transfer function may be flow simulation and the ultimate result is a recovery factor. In Section 3.3, Problem Formulation, we will explain that geostatistical models should be fit-for-purpose. In the current context this requires the information under consideration to have impact on the transfer function. The input information should potentially influence the transfer function to justify the effort to integrate it in the final model.

The following examples may illuminate this point. If levee elements do not add significantly to reservoir volume and do not change reservoir connectivity, then it may not be beneficial to differentiate between levee and overbank in the reservoir model. In some tight reservoirs, depositional architecture may not have a significant impact as porosity is dominated by diagenetic alteration and permeability is constrained by fractures. In this case, detailed architectural modeling may not be helpful.

Is the information supported by data? Due to data scarcity, often the local data are not sufficient to observe all aspects of the conceptual model. That is, the conceptual model moves beyond the data, by integrating all other information. For example, large-scale trends are anticipated from process and modern analogs, but may not be resolvable in available seismic. High-resolution facies and porosity trends are observable in well exposed outcrops, but often not resolvable in well data due to amalgamation, erosion, and interpretation imprecision and the 1-D nature of well data. In this case a decision may be made to omit this information, to impose this information even without local data support or to include this information in specific realizations to evaluate the risk through the impact of this conceptual feature on the transfer function.

Are the concepts consistent? The conceptual model should be compared and tested against the local data. It would not be defendable to integrate a conceptual model that contradicts this local data. This check may not always be possible or may be quite uncertain, given the limitations in data scale and coverage. In addition, in a complicated geological setting the conceptual model may include components that are inferred from broad set of disciplines including sequence stratigraphy, geochemistry, geophysics, and so on. All components must be formulated into a consistent set of concepts for integration into the model.

What information can be integrated? The conceptual information must be translated into a quantitative description of the current reservoir architecture, and then this quantitative description must be encoded as a framework or input statistics for the best-suited geostatistical algorithm. This is important because geostatistical models are limited to a set of input statistics, and they slavishly honor these input statistics, within gridding and trend constraints and modeling algorithm limitations. For example, variogram-based methods are limited to spatial concepts that can be described by a variogram. The model of heterogeneity is described by the inputs and implicit assumptions; therefore, conceptual information related to the final products (our paused movie) of the geological processes must be integrated explicitly. Also, concepts related to the sequence of geological processes responsible for the reservoir must be converted into a conceptual model of final preserved products; otherwise they will not be reproduced in the model. Process-mimicking geostatistics offers some greater flexibility in the information that can be integrated, but at a cost, as discussed in Section 4.5.

What happens if the information is not explicitly integrated into the model? As mentioned, features that are not imposed through input parameters or conditioning will not be reproduced. In the absence of constraint, the output heterogeneity will be determined by stationarity and the implicit assumption of the modeling method. For the former, due to the assumption of stationarity the statistics will be the same over the model if unconstrained. For the latter, the heterogeneity beyond that constrained by input statistics will tend toward a maximum entropy model. This model may be characterized by maximum disorder with practical consequence on the connectivity. This is discussed further in the discussion on the Gaussian assumption in Section 4.6.

Is there uncertainty in the conceptual model? The conceptual model may be considered uncertain. As mentioned previously, data are not likely to be sufficient to establish all of the components of the conceptual model. If there is significant uncertainty, and given the previous constraints concerning fit-for-purpose and feasibility, then this uncertainty may be transferred through a geostatistical model by allowing for multiple scenarios of the conceptual model. This is discussed further in Section 3.3, Problem Formulation.

Is the work flow feasible? Considerations include the importance of this information, cost of its omission, and cost and risk associated with this increase in complexity. This choice may be based on expert judgment and experience because it may be difficult to quantify value and increase in complexity associated with a specific piece of information.

All these questions are related to the capabilities of the modeling professionals and the time and resources available to the modeling project. The inclusion of a specific type of conceptual information may result in a significant increase in work flow complexity. With greater complexity there is greater chance for mistakes and recycle, especially if the capability of the available modeling professionals is exceeded. In general, more complexity results in more time to complete the project. Realistic accounting and expectations need to be applied to allow for success.

Finally, there are trade-offs in the selection of framework and geostatistical methods. Specific work flows and methods have information that they can readily integrate, information that they can approximately integrate, and information that they cannot integrate. For example, object-based models can easily integrate geometric information, but can only reproduce trend models in an approximate manner and in general cannot reproduce dense well or detailed trend information.

Nevertheless, there are various ways to constrain geostatistical heterogeneity with geological information. The following discussion covers the use of a conceptual model to constrain a large-scale model framework, selection of geostatistical algorithms, and formulation of statistical inputs to constrain these algorithms. These steps are central to converting the conceptual model, based on data and analogs, to a geostatistical reservoir model for the application of a transfer function for optimal decision making in the face of uncertainty (see Figure 3.13).

3.2.2 Model Framework

A first step in reservoir modeling is to establish the major reservoir architecture—that is, the geometry of the hydrocarbon-bearing formation and important adjacent geological formations. The reservoir "volume" must be filled with a grid network that makes it straightforward to account for the variety of available data, in particular the geological information concerning layering and stratigraphic correlation.

This subsection is concerned with establishing stratigraphic, layer-specific, Cartesian grid systems—that is, the appropriate grid system within which facies, porosity, and permeability will be modeled. Techniques for facies, porosity, and permeability modeling are presented in Chapter 4. This is only a basic summary on gridding considerations; for more detailed discussion see Mallet (2002) and Caumon et al. (2005b). Often, the stratigraphic grid system for geostatistical modeling is not the grid system chosen for flow simulation; for the latter, well placement, fluid flow processes, numerical accuracy, and CPU cost must also be considered. Scale averaging for flow simulation is discussed in Section 5.2, Model Post-processing.

Framework modeling has various considerations, including gridding for geological modeling, stratigraphic correlation and coordinates, accounting for faults and uncertainty in reservoir geometry. These are each discussed below.

Gridding for Geological Modeling

The basic topology used in most reservoir modeling and throughout this book is *Cartesian*, defined by corner-point areal grids and isochore vertical grids (see Figure 3.14). An elevation is associated to each grid node or cell corner. Elevations associated to entire grid cells would constitute a "cell-centered" specification (see Figure 3.15). The areal spacing of the grid nodes is typically regular, that is, with constant X and Y intervals; in fact, most modern geomodels are built with square areal cells.

The vertical coordinate is measured perpendicular to the horizontal plane—that is, *isochore* thickness, which is measured vertically even when the

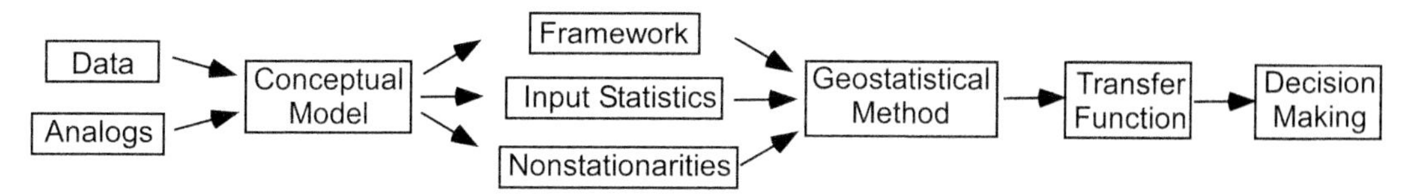

FIGURE 3.13: General Work Flow Including Formulation of a Conceptual Model from Data and Analogs Conversion of This Conceptual Model into a Framework, Input Statistics, and Nonstationarities to Constrain the Geostatistical Method to Build Models for a Transfer Function for Optimal Decision Making.

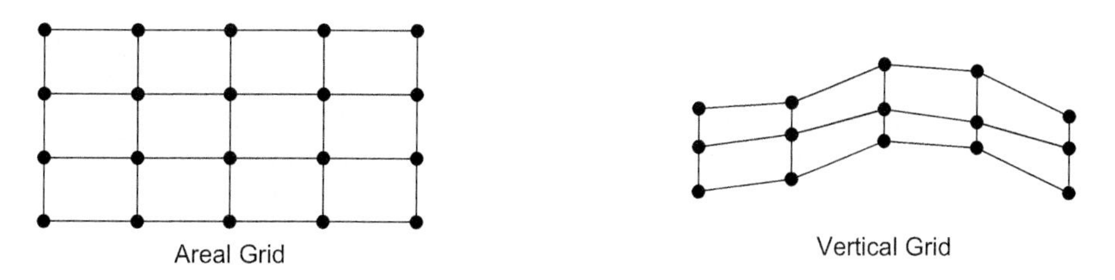

FIGURE 3.14: Basic Topology Used for Geostatistical Model Grid: Corner-Point Surface and Isochore Grids.

FIGURE 3.15: Cell-Centered Grids Are Not Connected According to the Z-Coordinate Elevation (Left-Hand Side), but Rather According to Cell Index Number (See Dashed Lines on Right-Hand Side).

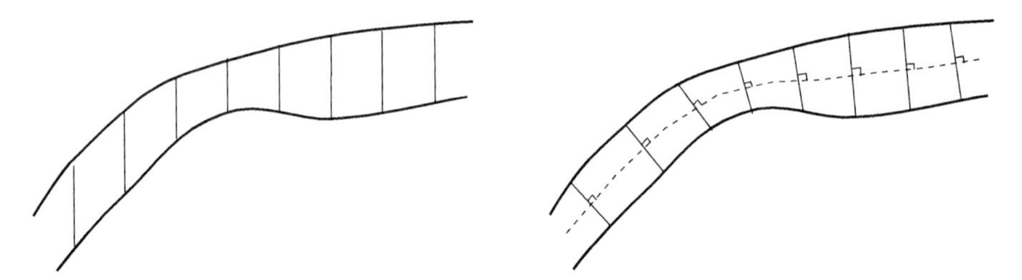

FIGURE 3.16: Two Approaches to Establish Areal Gridding: on the Left the Cells/Nodes Are Stacked Vertically, and on the Right the Cells/Nodes Are Aligned on Lines Perpendicular to the Bounding Surfaces. The first approach, with a universal vertical coordinate, is usually adopted because of computational simplicity.

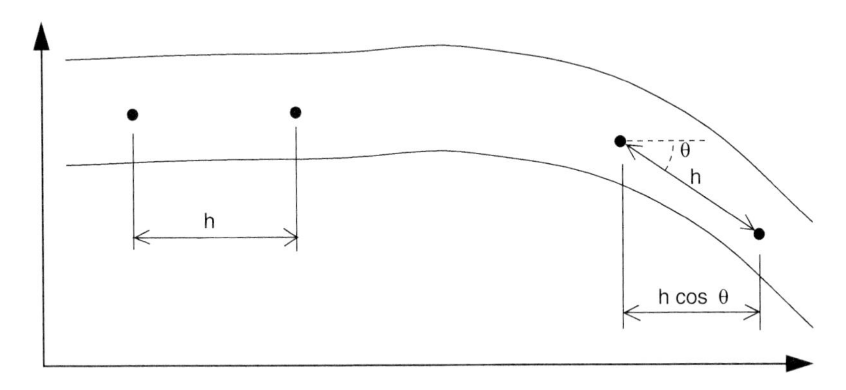

FIGURE 3.17: Both Pairs of Data Points Are at the Same Stratigraphic Z Coordinate and Separated by the Same Distance **h**; However, the Pair of Points to the Right Appear Closer.

stratigraphic formation dips. At times, the vertical coordinate is measured perpendicular to stratigraphic dip; that is, *isopach* thickness is considered. Very steep dips that may be encountered in areas of significant structural deformation must be dealt with specially.

Standard practice is to have the cells aligned along the vertical Z axis. Figure 3.16 shows this approach (on the left) and an alternative, where the cells are aligned on lines perpendicular to the bounding surfaces. This latter alternative would allow more correct horizontal distances at an added computational cost.

Figure 3.17 illustrates the distortion in calculated horizontal distances that results from having all cell boundaries line up vertically. To alleviate the consequences of this decision, and to make the grid more efficient, the primary X, Y, and Z coordinates could be rotated to conform to the overall strike and dip of the reservoir. A new set of orthogonal coordinates can be established if the dip is significant. Although a single-step transformation could be defined, a two-step procedure is clearer: (1) Rotate the horizontal X and Y coordinates to be aligned along the strike and dip direction of the stratigraphic unit (see Figure 3.18), and (2) rotate the new X direction to

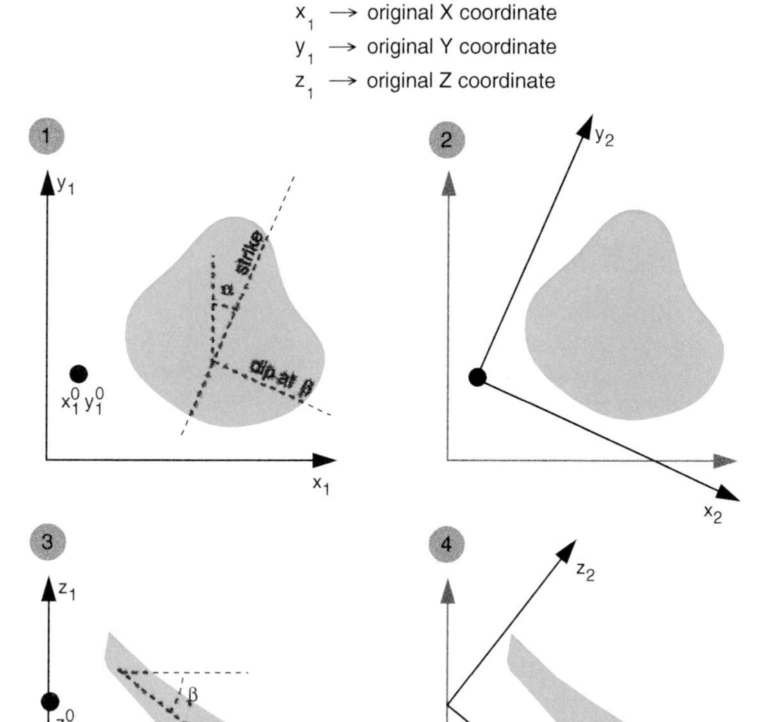

FIGURE 3.18: Rotated Z Coordinate System for a Dipping Stratigraphic Unit.

be aligned with the dip direction. The first translation (to an origin at x_1^0, y_1^0) and clockwise rotation of the original X (x_1) and Y (y_1) by α is written as

$$\begin{bmatrix} x_2 \\ y_2 \\ z_2 \end{bmatrix} = \begin{bmatrix} \cos\alpha & -\sin\alpha & 0 \\ \sin\alpha & \cos\alpha & 0 \\ 0 & 0 & 1 \end{bmatrix} \begin{bmatrix} x_1 - x_1^0 \\ y_1 - y_1^0 \\ z_1 - z_1^0 \end{bmatrix}$$

$$\text{or} \quad [X_2] = \left[R_{Z_1}\right][X_1]$$

The new x_2 and y_2 are aligned in the dip and strike directions respectively (see Figure 3.18). The second translation (to an origin at z_1^0) and rotation by dip β is written as

$$\begin{bmatrix} x_3 \\ y_3 \\ z_3 \end{bmatrix} = \begin{bmatrix} \cos\beta & 0 & -\sin\beta \\ 0 & 1 & 0 \\ \sin\beta & 0 & \cos\beta \end{bmatrix} \begin{bmatrix} x_2 \\ y_2 \\ z_2 \end{bmatrix}$$

$$\text{or} \quad [X_3] = \left[R_{Y_2}\right][X_2]$$

So, $[X_3] = \left[R_{Y_2}\right]\left[R_{Z_1}\right][X_1]$, or multiplying the two rotation matrices:

$$\begin{bmatrix} x_3 \\ y_3 \\ z_3 \end{bmatrix} = \begin{bmatrix} \cos\alpha\cos\beta & -\cos\beta\sin\alpha & -\sin\beta \\ \sin\alpha & \cos\alpha & 0 \\ \sin\beta\cos\alpha & -\sin\beta\sin\alpha & \cos\beta \end{bmatrix} \begin{bmatrix} x_1 - x_1^0 \\ y_1 - y_1^0 \\ z_1 - z_1^0 \end{bmatrix}$$

The 3-D rotation matrix may be inverted for back transformation.

Cell Size

There are a number of considerations in choosing the grid size. An overriding consideration is the ultimate size of the model and available computer resources. At present, a 100 million cell model is near the upper limit of practicality; larger models are expensive to store and slow to visualize and manipulate. A simplistic approach to choosing the cell size is as follows:

1. Establish the gross reservoir volume of interest. As an example, consider a total envelope of 10,000 ft by 10,000 ft by 200 ft.
2. Choose the vertical "stratigraphic" cell size ds needed to represent the heterogeneities observed in the wells. Consider, say $ds = 2$ ft.
3. Pick a target model size, say 10 million cells.
4. Calculate the consequent areal grid size, da, in this case:

   $$\frac{10,000}{da} \cdot \frac{10,000}{da} \cdot \frac{200}{ds} = 10,000,000$$

   yielding $da = 31.6$. Rounding da to 50 ft would yield a 200 by 200 by 100 cell model with a total of 4 million cells.
5. Iterate as necessary to achieve a balance between the total number of cells, the stratigraphic cell size, and the areal cell size. Of course, there is no need to use square areal cells. Pronounced areal anisotropy could be accommodated by rectangular cells, but this is uncommon.

Choosing the cell size according to a preconceived maximum model size is not completely satisfactory. It is worth considering more *reservoir-specific* considerations. Perhaps the cell size could be much coarser and yet meet all objectives of the reservoir modeling study? Alternatively, if the cell size *should* be a lot smaller, a high-resolution model over a small area (a segment model) could be considered to complement the full reservoir model.

The first consideration is that the model should be suitable for the specific project goals. Relatively coarse models are required to assess reservoir volumetrics. Detailed models are required to study water and gas coning near production wells (perhaps a small-area segment model could be considered for this goal). Often, the project goals tend to change and the models are used for more than originally intended. For this reason, it is appropriate to err on the side of building in too much detail even when not explicitly called for by the initial modeling objectives.

A second consideration, aside from model size, is that the areal and "stratigraphic" grid must be chosen so that important features (reservoir boundaries, faults, significant internal lithofacies changes, and petrophysical variations) can be resolved with the final model. For example, a 1-ft vertical cell size would be required to represent relatively thin shales (between 1 and 3 ft). There is a necessary element of judgment required as to what constitutes a "significant" feature. The only definitive approach to determine whether a specific feature is important is to build it into the model and then consider the best alternative. The feature is important if the flow response of the simplistic model is far from the more "correct" model.

A third consideration is the resolution required to ensure meaningful scale-up from the geological model to the flow simulation model. In general, there should be at least three geological cells, in each coordinate direction, within the scaled-up flow simulation model. Otherwise, the results will be too erratic. This consideration is based on discretization errors in flow-simulation-based scale-up programs. Some scale-up algorithms, such as power averaging, are not as sensitive to the number of geological cells but are less accurate.

Triangulated Grids

Numerically representing geologic surfaces with triangular facets is very flexible for complex folded and faulted surfaces. Classical Cartesian grids are not appropriate for surfaces with significant folding, for example, with multiple Z values for a given X–Y location. The gOcad software (Mallet, 2002) among others implement triangulated grids and can easily handle such complexity.

A tetrahedra-based gridding scheme is used for geological cells bounded by triangulated surface grids. This tetrahedra-based topology is not yet used in geostatistical calculations for a number of reasons:

- The tetrahedra do not all have the same volume. This makes it more difficult to apply geostatistical tools and integrate hard and soft data.
- That topology is less commonly accepted by finite-difference scale-up and flow simulation programs.

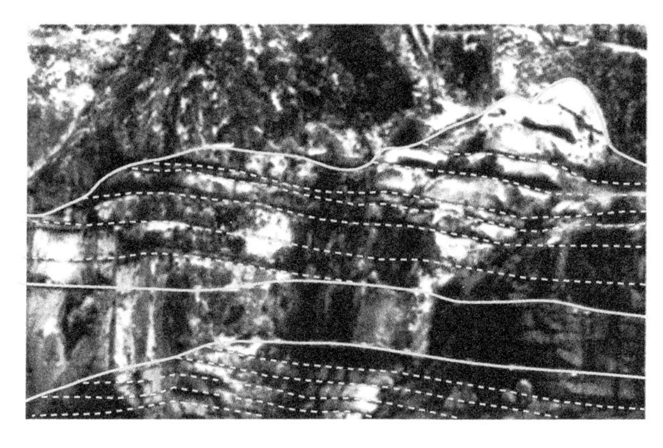

FIGURE 3.19: Photograph of Channel Features at an Outcrop of Deep Sea Sediments (Turbidite). The area photographed is about 69 m long by 43 m high. The lower photo shows geological surfaces that have been identified at the outcrop.

Presently in the oil industry, Cartesian grids are used almost exclusively because of their direct relation to finite-difference flow scale-up and simulation programs. The complex geometry defined by the tetrahedra grids are often transformed to a Cartesian grid for geostatistical calculations. The resulting models are then back transformed. Some techniques have been developed for direct population of tetrahedra-based and other non-Cartesian grids, yet many technical and software issues have prevented their adoption into practice (see Chapter 6).

Stratigraphic Correlation/Coordinates

Reservoirs consist of a series of genetically related strata that may be correlated over the areal extent of the reservoir (Wagoner et al., 1990). The surface grids that define this geologic correlation correspond to a chrono-stratigraphic or sequence stratigraphic framework, that is, the bounding surfaces between the layers correspond to a specific geologic time that separates two different periods of deposition or a period of erosion followed by deposition. As discussed in Section 2.1, Preliminary Geological Modeling Concepts, it is essential to have a sound geological framework prior to geostatistical modeling. Some references on the determination and importance of stratigraphic frameworks include (Bashore and Araktingi, 1995; Catuneanu, 2006; Caumon et al., 2005b; Weber, 1982; Weber and Van Geuns, 1990).

Figure 3.19 shows a photograph of channel features at an outcrop of deep sea sediments (turbidite). The area photographed is about 69 m long by 43 m high. The lower photo shows geological surfaces that have been identified at the outcrop; four (partial) geological layers are defined. Figure 3.20 shows two possible Cartesian gridding schemes. The top scheme is appropriate for modeling the geological details. The simpler bottom scheme may be enough for larger-scale 3-D modeling. The procedure to establish the "stratigraphic" coordinates on the top of Figure 3.20 is presented below.

Reservoirs are made up of a number of reservoir layers. Sequence stratigraphic analysis reveal that each layer corresponds to a particular time period in the creation of the reservoir. The surfaces that separate these layers relate to a significant geological change. The definition of a "layer" is ambiguous. In the present context, a geological layer is a reservoir unit between 5 and 100 ft thick that can be correlated between multiple wells and mapped over

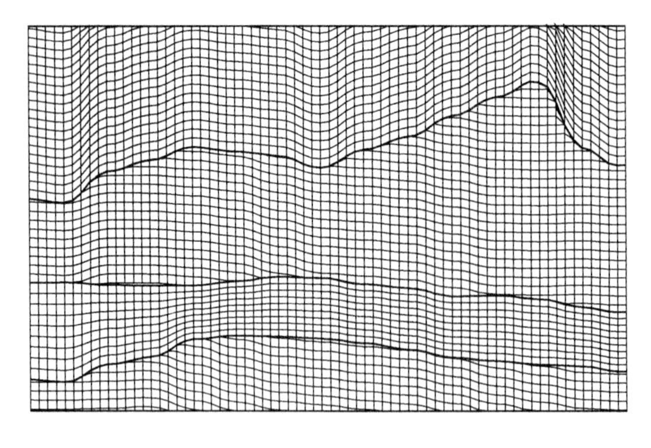

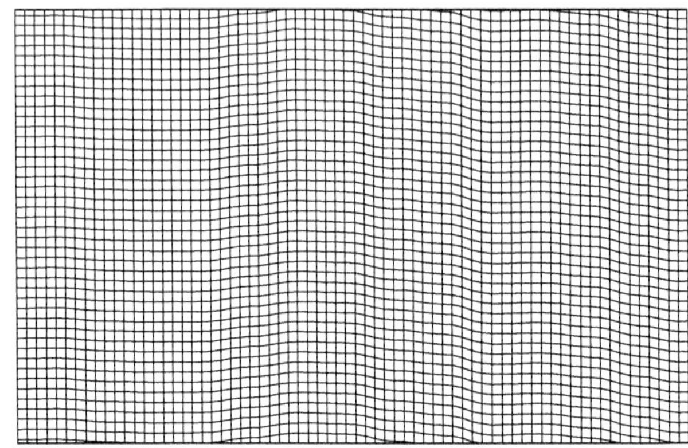

FIGURE 3.20: Two Possible Cartesian Gridding Schemes for the Geological Features in Figure 3.19. The top scheme is appropriate for modeling the geological details. The simpler bottom scheme may be enough for larger-scale 3-D modeling.

a substantial areal extent. The layers are defined to provide a large-scale subdivision of the reservoir into geologically homogeneous units. Most reservoirs are divided into 5 to 10 layers. These layers are sometimes called "sequences" after sequence stratigraphy; however, the layers considered in geostatistical modeling need not correspond directly with the formal definition of geological sequences.

Each layer is defined by existing top and base surface grids: $z_{et}(x,y)$ and $z_{eb}(x,y)$, where x and y are areal coordinates; et refers to existing or present top and eb refers to existing (present) base. These surface grids are not flat because of differential compaction and subsequent structural deformation. Also, the layer may have been eroded by later depositional events or the sediments may have "filled" existing topography. The continuity of the facies and reservoir properties within a layer do not necessarily follow grids based on the existing (present) bounding surfaces. Additional "correlation grids" may be required to define the stratigraphic continuity or "correlation style" within each layer. A common approach to capturing stratigraphic continuity is to model the continuity of the facies and reservoir properties as proportional between top and base *correlation surface grids*. These grids may not correspond to any existing or past surface; however, they are useful for subsequent geostatistical modeling. The upper correlation grid accounts for erosion at the top of the layer. The lower correlation grid accounts for onlap or "filling" geometry at the base of the layer. The correlation top and correlation base grids have the same format as other surface grids, but they come from geological interpretation more than direct interpretation from seismic and well data.

Figure 3.21 shows four common correlation styles. The solid lines represent the existing top and base surface grids (same in all four cases) and the dashed lines represent correlation grids. The continuity of the sediments (strata) is proportional between the upper and lower correlation grids:

Proportional: The strata conform to the existing top and base. The strata may vary in thickness because of differential compaction or sedimentation rate and may be structurally deformed and faulted; however, the correlation grids coincide with the existing grids.

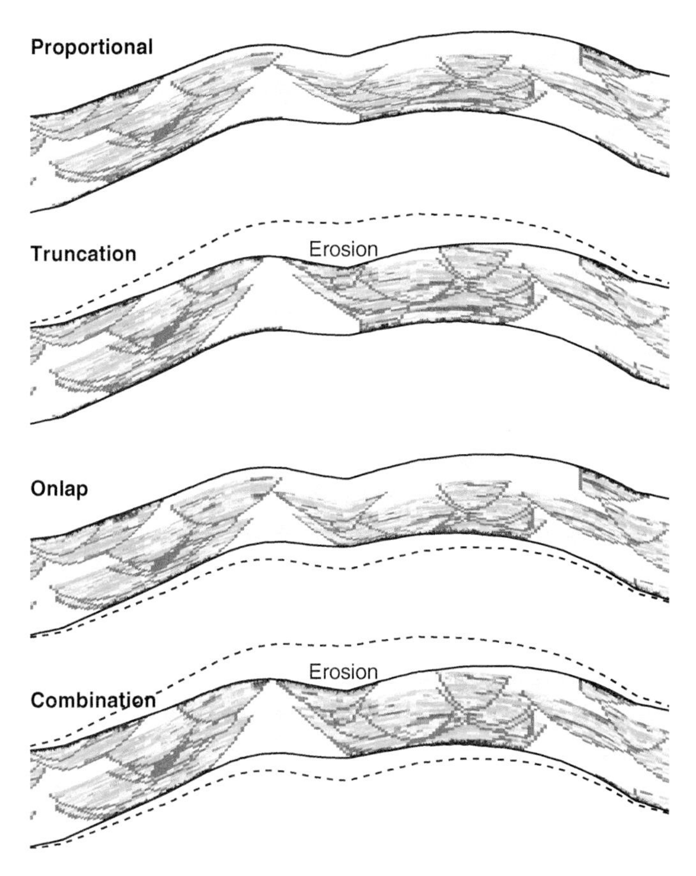

FIGURE 3.21: Example of Different Stratigraphic Correlation Styles. Internal channels illustrate proportional deformation, erosion, and onlap. Note that the solid lines, representing the existing reservoir layer, are the same in all four cases. The dashed lines in the cases of truncation, onlap, and combination represent geological correlation grids.

Truncation: The strata conform to the existing base but have been eroded at the top. The lower correlation grid coincides with the existing base. The upper correlation grid defines the areally varying amount of erosion.

Onlap: The strata conform to the existing top (no erosion) but have "filled" existing topography so that the base correlation grid does not coincide with the existing base.

Combination: The strata neither conform to the existing top nor the existing base. Two additional correlation grids are required.

A vertical coordinate will be defined as the relative distance between a correlation top and correlation base grid. This will make it possible to infer "natural" measures of horizontal correlation and to preserve the geologic structure in the final numerical model.

Each stratigraphic layer is modeled independently with a relative stratigraphic coordinate z_{rel} derived from four surface grids: (1) the existing top z_{et}, (2) the existing base z_{eb}, (3) the correlation top z_{ct}, and (4) the correlation base z_{cb}:

$$z_{rel} = \frac{z - z_{cb}}{z_{ct} - z_{cb}} \cdot T$$

The coordinate z_{rel} is 0 at the stratigraphic base and T at the stratigraphic top. T is a thickness constant chosen to yield reasonable distance units for the z_{rel} coordinate; most commonly, T is the average thickness of the layer using the z_{et} and z_{eb} surface grids. This transform may be reversed by

$$z = z_{cb} + \frac{z_{rel}}{T} \cdot (z_{ct} - z_{cb})$$

Converting all depth measurements to z_{rel} permits modeling of each reservoir layer in regular Cartesian x, y, z_{rel} coordinates. The locations of the facies and other reservoir properties are converted back to real z coordinates before visualization, volumetric calculations, or input to flow simulation. There will be no back-transformed z-values outside the existing interval $z \notin (z_{eb}, z_{et})$ because the locations above the existing top and below the existing base are known ahead of time and excluded from modeling. Figure 3.22 illustrates heterogeneities between existing and restored grids. The white regions between

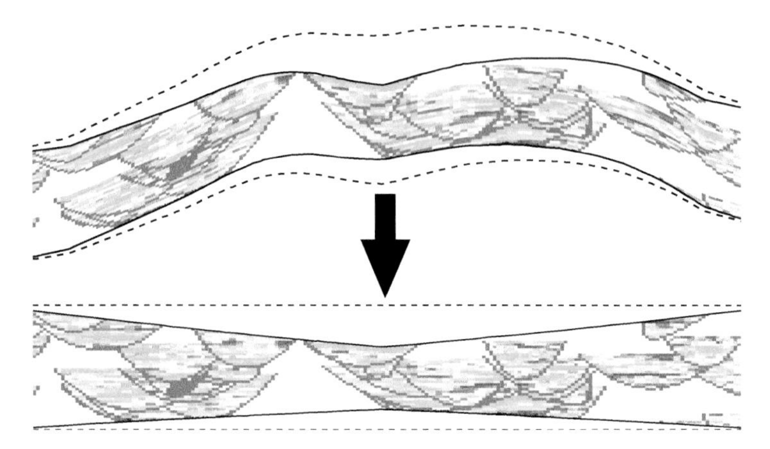

FIGURE 3.22: Representation of Heterogeneities between Existing Surface Grids and Converted to z_{rel} Coordinate Space. White regions between dashed and solid grids are present for geological correlation but not in the reservoir.

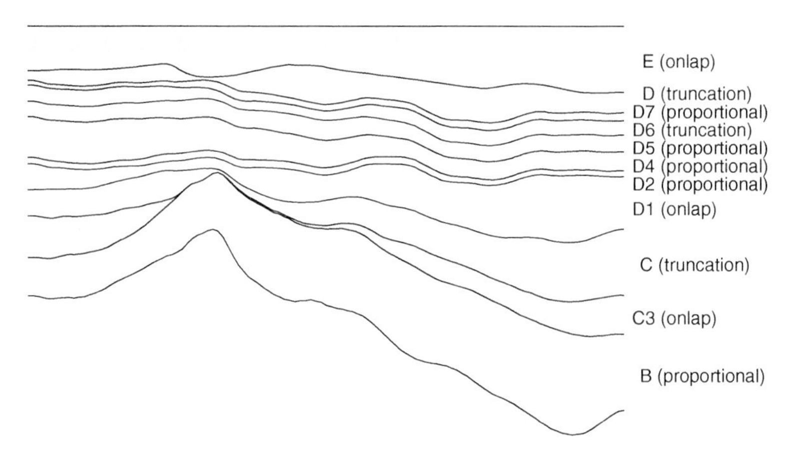

FIGURE 3.23: Cross Section through a Reservoir with 11 Different Stratigraphic Layers of Differing Thickness and Correlation Style.

dashed and solid grids are present for geological correlation but are not in the reservoir; they would be excluded from modeling; that is, facies and petrophysical properties would not be assigned in those regions.

Figure 3.23 shows 11 stratigraphic layers of differing thickness. The view in Figure 3.23 has been flattened according to the top of the reservoir. Each stratigraphic layer has a different stratigraphic correlation style.

Straightening Functions

As stated above, two motivations for coordinate rotation and stratigraphic coordinate transformation are to (1) allow the calculation of *natural* "horizontal" measures of correlation, and (2) reflect the geological structure in the final numerical model. Other transformations could be considered in cases where there is large-scale undulation or curvature that is difficult to honor with geostatistical modeling techniques that assume a single direction of continuity. As an example, Figure 3.24 illustrates the *straightening* of a large channel or valley-type structure. The Y coordinate along the primary direction of continuity is left unchanged. The X coordinate is corrected to be the distance from the centerline, that is, for any point x, y the new X coordinate is defined as

$$x_1 = x - f^c(y)$$

where $f^c(y)$ is the deviation of the undulating centerline from a straight constant-X reference line (see Figure 3.24 for a picture). The main advantage of this correction is that it can be easily reversed. The

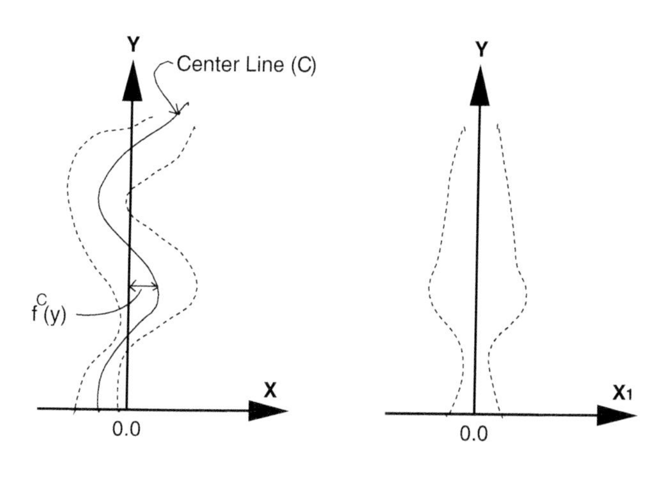

FIGURE 3.24: Transformation from the Channel Direction to a Channel Coordinate System Following the Sinuous Channel Centerline.

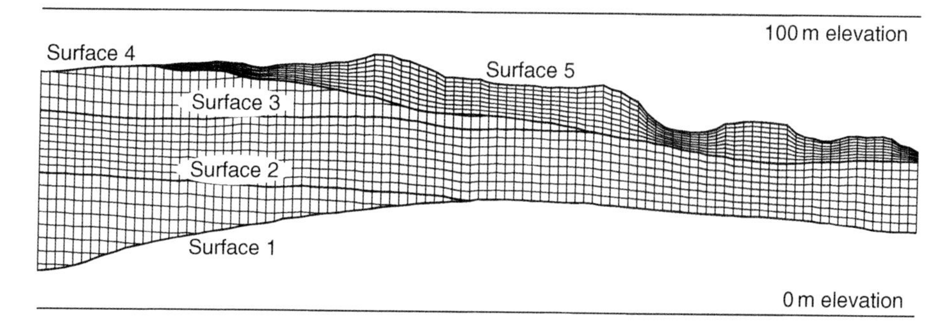

FIGURE 3.25: Cross Section through an Outcrop with Four Different Stratigraphic Layers of Differing Thickness and Correlation Style. The correlation styles cannot be seen clearly since no within-layer properties are shown.

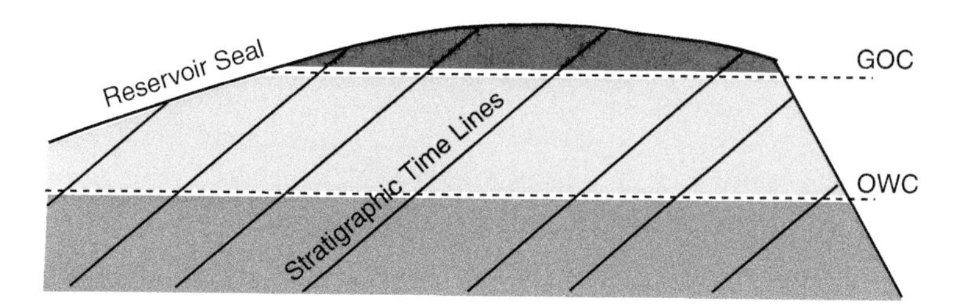

FIGURE 3.26: Illustration of *Steeply* Dipping (a Steep Dip May Be Only 1 to 5 Degrees; Shown Here with Significant Vertical Exaggeration) Strata with Gas Oil (GOC) and Water Oil (WOC) Contact.

center line for the large-scale undulation *must* be known beforehand for all y values over the extent of the reservoir. Also, the curvature cannot be too great; the main direction cannot meander too much or reverse itself otherwise the function $f^c(y)$ will not be uniquely defined.

Figure 3.25 shows four stratigraphic layers of differing thickness. The view in Figure 3.25 is of an actual outcrop.

Special Considerations

The z_{rel} vertical coordinate requires top and base correlation grids over the areal extent of the reservoir layer. These are poorly defined when the stratigraphic dip is "steep." Figure 3.26 shows a schematic example with fluid contacts defining the layer of interest. It may be more straightforward to simply use the original Z vertical coordinate and consider a dip angle in subsequent geostatistical modeling steps.

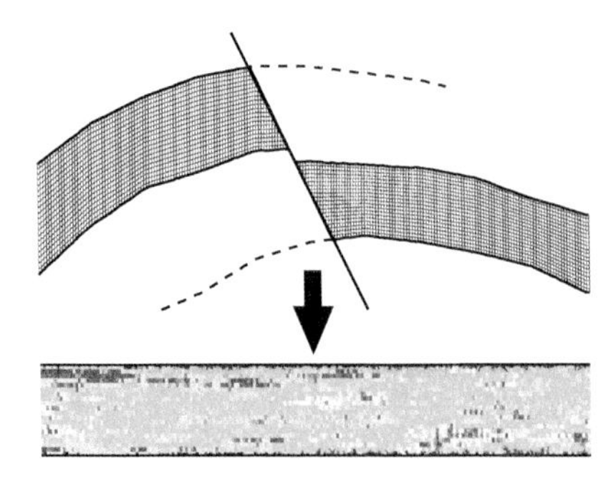

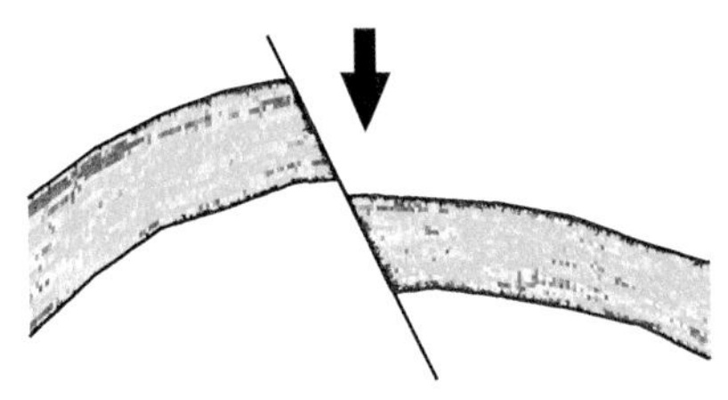

FIGURE 3.27: Modeling a Faulted Stratigraphic Layer: (i) Top figure shows the fault location and the existing surfaces extended as restored surfaces, (ii) the middle figure shows the restored grid with assigned petrophysical properties, and (iii) the bottom figure shows the properties within the existing faulted grids.

Top and base correlation surface grids for Figure 3.26 would be separated by a large distance. The top correlation grid would start close to the existing top on the left-hand side; however, it would be at a very high elevation on the right-hand side. The base correlation grid would have the same "tilt" as the correlation top, but it would be at a very low elevation on the left-hand side. The separation between the two grids could be very large. Constant elevation grids aligned with the OWC or GOC could be used for the top and base correlation grids. These grids would only be separated by the thickness of the reservoir. As indicated above, the direction of greatest continuity would not be "horizontal" in the x, y, z_{ref} coordinate space; strike and dip angles would need to be considered in subsequent modeling.

Faults

The problems presented by faults are diverse, challenging, and somewhat outside the scope of this book. Straightforward normal faults, such as illustrated by the cross section in Figure 3.27, can be handled by correlation grids and stratigraphic coordinate transformation as described above.

Reverse faults cause problems because there are multiple surface values at some x, y locations (see Figure 3.28). One solution to this problem is to assume that the faults are vertical through each geological layer. Then, we are back to the case of simply requiring four surface grids for each layer. Such fault "verticalization" appears crude but is a reasonable approximation if the faults are relatively steep and the offset is not too great. As an example, Figure 3.29 shows a schematic illustration of a geological layer 50 ft thick and 2000 ft across with a normal fault at 70° with an offset of 150 ft. The horizontal separation is $\cos(70) \cdot 150 = 51.3$ ft, which is small relative to the total areal extent of the geological layer. Assuming a vertical fault in the center would cause a mismatch of less than 26 ft. The areal size of geological modeling cells often exceeds this size.

There are cases, in thick formations with significant offset, where fault verticalization does not adequately capture the geometry of the reservoir layers.

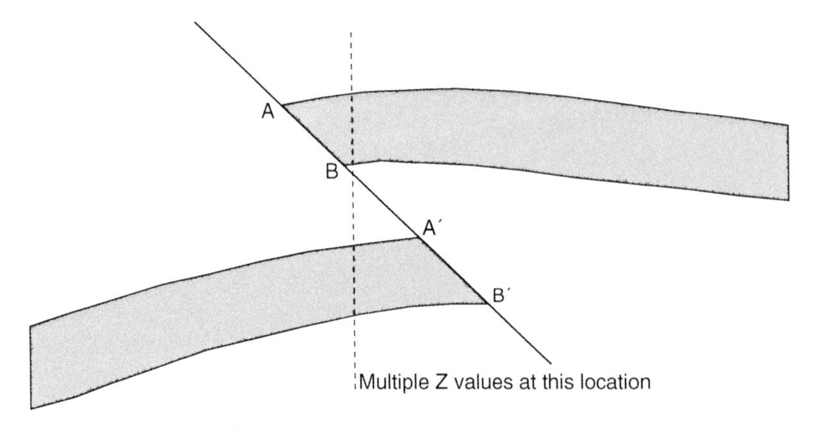

FIGURE 3.28: Example of a Reverse Fault Illustrating the Need for New Horizontal Coordinates. Multiple z-coordinate values are needed for some locations.

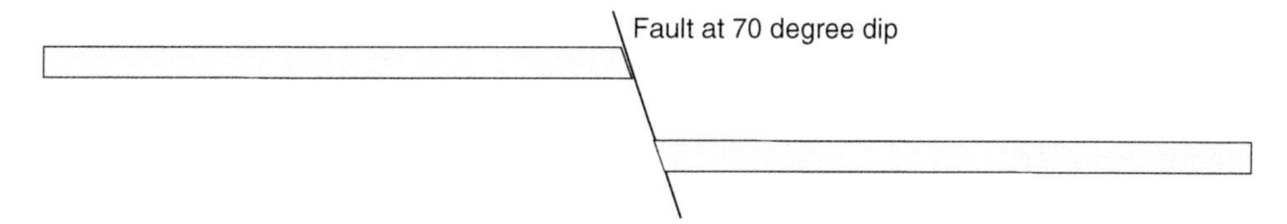

FIGURE 3.29: Faulted Geological Layer Shown with No Vertical Exaggeration. It is assumed that the vertical fault would have little effect on the final geological model.

There are a number of solutions: (1) Each fault block can be modeled separately, or (2) the horizontal coordinate can be expanded to remove overlaps. The first alternative is simpler and does not distort horizontal distances. The correlation of petrophysical properties across fault blocks can be enforced by sequential modeling where results from previously simulated fault blocks are considered.

The procedure in more detail is as follows:

- Identify fault locations and model the fault surfaces.
- Flag all geological modeling cells and well data by fault block.
- Construct Z_{rel} coordinate system in each fault block.
- Construct a geostatistical model of facies, porosity, and permeability (as described in later chapters) for each fault block. Data from previously modeled fault blocks could be considered if correlated to the fault block currently being considered.

The two alternatives presented above will be used throughout this text; that is, (1) calculate Z_{rel} with modified restored grids or vertical faults, or (2) model each fault block separately. It is important to note that geostatistical methods are *not* limited to Cartesian geometry and gridding. In fact, at the time of writing this book, research into geostatistical modeling of facies and properties on unstructured grids is very active. The future of geostatistical reservoir modeling will be to model the complex geometry directly rather than with the coordinate "fixes" that have been introduced here (Mallet, 2002).

Uncertainty in Reservoir Geometry

Uncertainty in the large-scale reservoir geometry is often the most consequential when assessing uncertainty in reservoir volumetrics. Each geostatistical reservoir model should consider stochastic surface grids consistent the well data and seismic data (Hamilton and Jones, 1992; Jones et al., 1986). The procedure for stochastic modeling of surface grids will be developed in Section 4.1, Large-Scale Modeling.

3.2.3 Modeling Method Choice

Geostatistical reservoir modeling is reliant on a toolbox of modeling algorithms that all have their unique features. In Section 2.4, Preliminary Mapping Concepts, the categories of methods were introduced, including variogram-based, multiple point-based, object-based, and so on. These methods vary in the types of input statistics, their ability to honor these inputs, and ease of implementation. The following discussion focuses on method selection, with general considerations such as amount and type of data available, type of reservoir fluid, geological heterogeneity, and input statistics. Discussion on individual algorithms, examples, and implementation details are later provided in specific sections on each method (Sections 4.2 to 4.7).

The following provides a linkage between the geological concepts covered in Section 2.1 and these algorithms. In addition, a high-level comparison between these algorithms and their associated tradeoffs is provided. This guidance is somewhat qualitative. This is necessary due to the complexities in the methods and complexities in the interaction of overall work flow construction and modeling purpose (covered in Section 3.3, Problem Formulation). These choices are best supported by the tradecraft that is earned through experience and with consideration of modeling objectives.

The most important consideration in algorithm choice is feasibility. The software and professional expertise must be present. Given these criteria are first met, the following are considerations for method selection.

Fluid Considerations

Central to the reservoir modeling method choice is a decision of the reservoir heterogeneity features

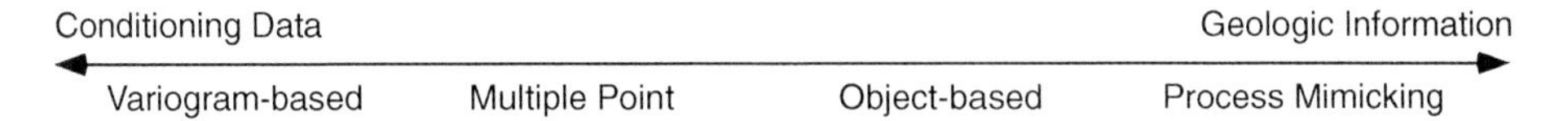

FIGURE 3.30: Continuum of Geostatistical Algorithm Types on a Continuum Indicating the Inherent Ability to Reproduce Conditioning Data (Wells and Trends) and Geologic Information.

that impact the transfer function. When the transfer function is based on fluid flow, fluid considerations are important. For example, a gas reservoir may be relatively insensitive to flow baffles, and significant barriers are necessary to impede flow, while a heavy oil reservoir may be sensitive to baffles, and even small changes in reservoir properties and heterogeneity.

Geological Complexity

Geological complexity is defined as reservoir heterogeneity with short-scale features and/or long-range features that depart from linear and/or random and result in highly nonuniform flow behavior. An example of geological complexity would be contiguous shale drapes that act as significant barriers or baffles to fluid flow or long-range contiguous channel fills that result in piping of reservoir and injected fluid flow.

This is a functional definition that recognizes geologic complexity as features that may be difficult or impossible to characterize with limited statistics and random function models such as those introduced in Section 2.4, Preliminary Mapping Concepts for the cell-based methods such as any variogram-based and MPS. These complexities are often more amendable to geometric and rule-based parameterization that is utilized as inputs for object-based and process-mimicking methods. Yet, it is recognized that data conditioning immediately becomes an issue as discussed below.

Data Considerations

A major constraint in modeling method selection is the amount of conditioning data and the level of detail or resolution of other data and concept derived constraints such as trends. There is a wide spectrum in the availability of hard well data and seismic constraint for a reservoir model. In some exploration or early development settings only a couple of wells may be available, while in mature fields thousands of wells may be available at very dense spacing along with high resolution seismic with reliable calibration of reservoir properties. The objectives of reproducing data and imposing a model of geologic complexity are often competing objectives.

The methods described in Section 2.4, Preliminary Mapping Concepts could be placed on a continuum, at one end is the ability to honor very dense data and at the other end is the ability to honor geologic complexity as defined above (see Figure 3.30). The variogram and multiple-point algorithms are cell-based; therefore, have limitless ability to honor well data and trends. While multiple-point algorithms can reproduce more geological complexity than the variogram-based methods as more statistics are integrated, but they cannot match the geological details including crisp geometries, long-range structures, interrelationships and internal trends that may be honored with object-based and process-mimicking algorithms. This capability is accepted with the sacrifice in flexibility to honor dense well data and very specific categorical trend information. In fact, in some applications process-mimicking algorithms become like process-based models and lose their ability to directly condition to well data and categorical trend information.

Consider that dense data generally precludes the application of object-based and process-mimicking methods. In such settings, typical object-based methods will often fail to condition to the data and / or produce models with conditioning artifacts. If a reservoir model fails to honor local conditioning data, it firstly will lose credibility and secondly will likely be unfit to meet the study objectives (i.e., result in significant error in the transfer function). Yet, dense data have benefits as the data themselves may be able to impose important long-range, complicated features, allowing the application of methods that may not adequately impose these features, such as variogram-based and multiple-point methods, to reproduce these features through conditioning instead of input statistics.

Scale Considerations

As discussed in Section 3.3, Problem Formulation, the scale and resolution of the model may vary dramatically from micro-borehole models at lamina resolution to basin scale models with very low resolution. The heterogeneity expressed at lamina scale is dramatically different from the heterogeneity expressed at bed scale, element scale, complex scale, complex set, and so forth. In all natural settings, heterogeneity is related to scale. Self-similar scale invariant (sometimes called fractal) heterogeneity is limited. While there may be some interesting examples of ripples, bars, and dunes superimposed across scales, lobes within lobes in a distributary lobe complex set, or channels within channel complexes, the flow significant heterogeneity is likely very different at each scale. For example, the shales associated with fine-scale lamina are much less likely to be preserved than thick shale units that separate architectural complexes. Scale must be specified when discussing heterogeneity, for the subsequent discussion typical reservoir scale is assumed.

While modeling method selection is influenced by available data and scale, the conceptual model is also important. The general work flow is to map relevant conceptual model components to input statistics (specific to the chosen modeling method) and then to use these input statistics to constrain the chosen modeling method. Both the conceptual model and type of input statistics may be applied as criteria for selection and are non-unique as various types of inputs statistics may result from a conceptual model and various conceptual models may exhibit similar statistics; therefore, in the following discussion we consider first the modeling method choice from the perspective of the conceptual model, including mapping to input statistics, and then we consider model selection by the available input statistics with some comments on conceptual model.

Modeling Method Selection by Conceptual Models

In the following discussion we consider a variety of common conceptual models by depositional setting and then map them to input statistics and then to the geostatistical methods that utilize these inputs. This summary is very cursory and illustrative; detailed discussion of reservoir architecture is available in Reading (1996) and in Walker and James (1992), along with a concise description of implications for reservoir connectivity in Galloway and Hobday (1996). Fluvial, deepwater, and carbonates depositional settings are discussed to provide a reasonable range of geological complexity, and the provided concepts are readily extendable to other depositional settings.

Fluvial

Fluvial depositional settings result in a wide variety of architectures (Galloway and Hobday, 1996; Miall, 1996; Walker and James, 1992). The following describes alluvial fans, as well as braided and meandering channels.

Alluvial fans form through a combination of mass and water transport (see Figure 3.31). When both mechanisms are present, there is a proximal to distal trend with the proximal area of the fan dominated by debris flows (products of mass transport) and the distal by water-laid deposits with debris flow inter-fingering.

Texture is the major difference between debris flows and water-laid deposits. Debris flows result in poorly sorted unstratified textures (within a single event). These deposits are comprised of cobbles and boulders supported by a matrix of fines. The water-laid deposits result in fluvial bedload with moderate- to well-sorted and well-bedded gravels with tabular cross bedding and slurry imbricated clast supported with poor size separation (Wasson, 1977).

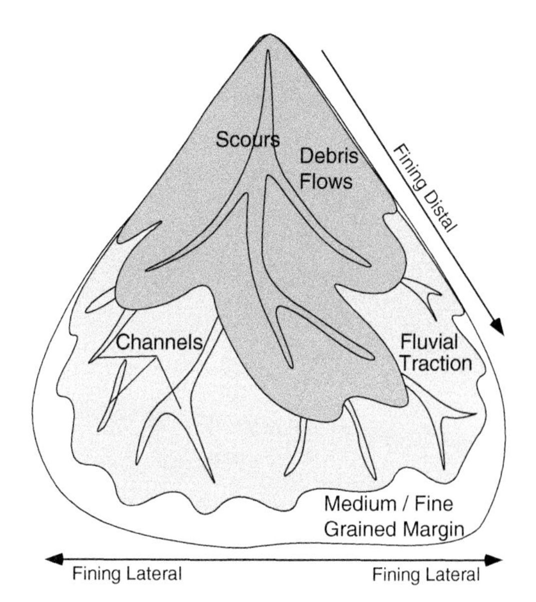

FIGURE 3.31: Schematic of an Alluvial Fan with Transition from Poorly Sorted Debris Flow to Well Sorted Fluvial Traction Deposits. There is a general grain size fining lateral and distal within the fan. Individual beds are imbricated, reducing the scale of the preserved beds and elements.

Due to frequent avulsions of the flows, the preserved heterogeneity features are imbricated short-scale beds, with large-scale trends of fining distally with decrease in gradient. These concepts may be mapped directly into a detailed nonstationary facies proportion model, and spatial continuity may be represented well with cell-based, variogram-based continuity models. If short-range facies ordering is significant, then truncated Gaussian or MPS may be applied to impose this in addition to the long-range proportion model in the trend. Within facies, reservoir properties may be simulated with sequential Gaussian simulation with proximal distal trends to honor a general grain size fining and decreasing porosity and permeability basinward.

In braided channels there is some preservation of long-range channel features, yet channel imbrication and channel migration and avulsion may break up these features. Complicated baffles may form due to local paleosol formation, and barriers may be broken up by local scours. If channel fill and paleosol contiguity is significant, they may be best represented by geometric parameters within an object-based model method. If avulsions and imbrication are significant, these features may only occur over short scales and may be captured with multiple-point statistics or semivariogram. Reservoir properties within channels may be simulated with sequential Gaussian simulation; but if channels are highly contiguous, a locally variable azimuth model may be required to impose property continuity along the channel.

Meandering channels may preserve extensive point bar-derived inclined heterolithic strata with complicated internal heterogeneities that act as flow conduits, baffles, and barriers (see Figure 3.32). The abandoned shale filled channels may be preserved as large contiguous barriers that interrupt the inclined heterolithic strata. Object-based modeling is required to reproduce the long range abandoned channel features and general point bar geometries. More complicated process-mimicking is likely required to capture realistic inclined heterolithic strata geometries and the internal heterogeneities (Hassanpour et al., 2013) (see Figure 3.33).

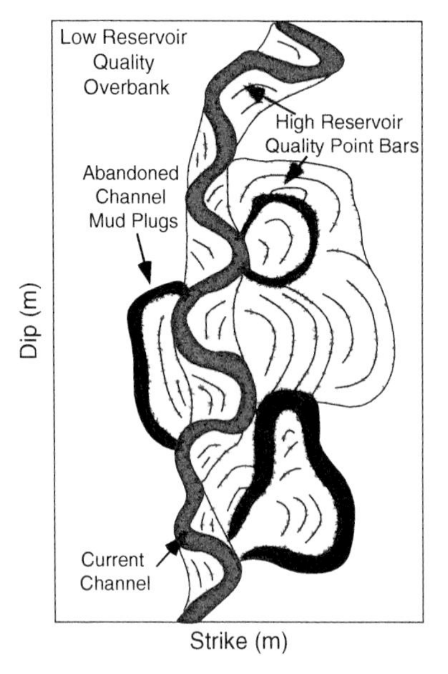

FIGURE 3.32: Schematic of a Meandering Fluvial Architecture. Note the imbrication of multiple point bars results in a preserved extensive sheet of sand punctuated by abandoned mud-filled channels. While the individual channels had long range structure, the preserved architecture is spatially broken up.

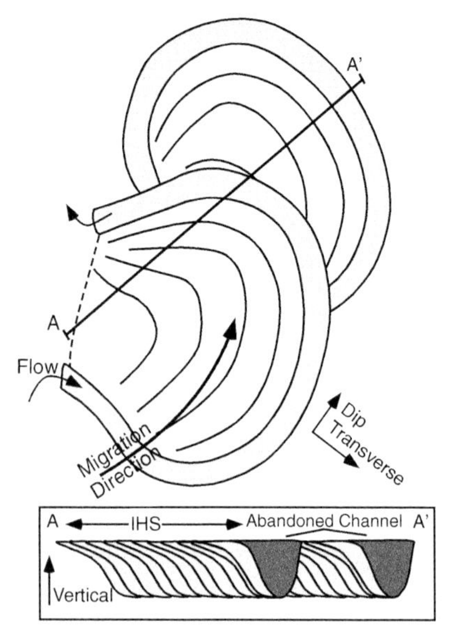

FIGURE 3.33: Schematic of a Set of Imbricated Fluvial Inclined Heterolithic Strata (From Point Bars). The abandoned channels are preserved as shale-filled non-reservoir barriers or baffles and the IHS have various property trends identified by Thomas et al. (1987).

Deepwater

Deepwater depositional settings generally result in channelized and sheet architectures. Deepwater channels are generally highly contiguous low-sinuosity sand-filled features surrounded by overbank shales. They typically exhibit either organized or disorganized stacking patterns. For the former, a subsequent channel is the result of translation of the previous channel; for the latter, all channels are unrelated to each other. Disorganized or simple organized patterns may be reproduced with object-based methods. Strongly disorganized and highly amalgamated channels may reduce geological complexity with long-range features broken up, allowing for modeling with MPS. Strongly organized channels may be reproduced with process-mimicking methods coded with a channel stacking rule. Either object-based or process-mimicking methods may also account for the within-channel facies association transition from axis, off-axis to margin commonly observed in deepwater channels (see Figure 3.34).

Deepwater sheets are formed through the repeated deposition of compensational distributary lobes (see Figure 3.35). Channels that feed sediments to the lobes may be preserved in the proximal lobes. The lobes have variable reservoir quality generally related as axis to margin and proximal to distal fining of grain size. The resulting architecture is stacked lobes with proximal channels and variable reservoir quality and potentially contiguous shale drape baffles or barriers separating the lobes. One approach is to model the shale drapes within the lobe sand matrix as opposed to modeling the lobes. If the shales are highly contiguous or geometrically complicated, then they may be reproduced by object-based simulation of the shale drapes within a sand background (see Figure 3.36). If the shale drapes are less contiguous, then sequential indicator simulation or MPS may be applied to simulate the shales. Then truncated Gaussian simulation or MPS may be applied to simulate the facies transitions from inner lobe to outer lobe. For both approaches, reservoir properties may be simulated within facies using sequential Gaussian simulation.

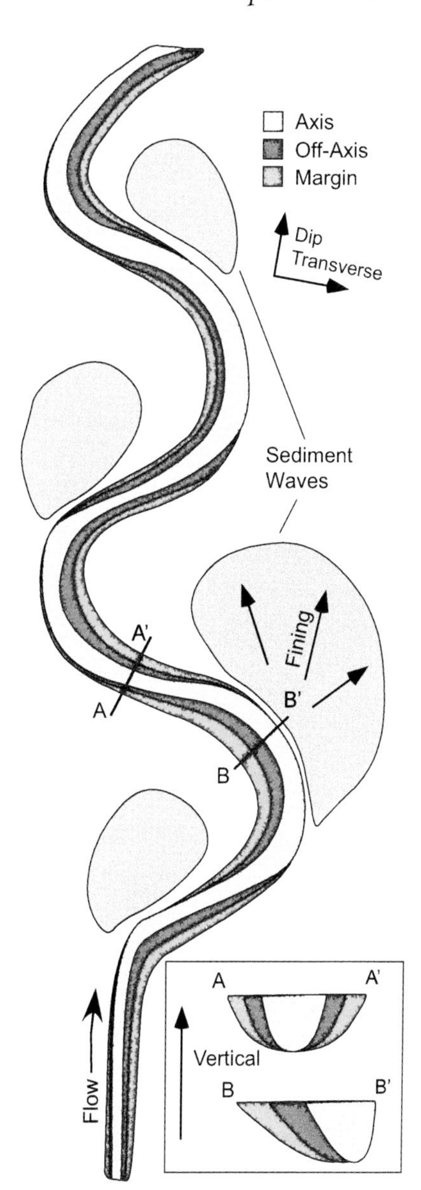

FIGURE 3.34: Schematic of a Single Deepwater Turbidite Channel. Channels are typically composed of axis, off-axis, and margin facies associations, ordered with decreasing reservoir quality. The asymmetry of internal fills and external geometry are typically directly related to the channel planform morphology. In addition, flow stripping around bends results in the deposition of coarser overbank facies (sediment waves) that may be important for flow.

Carbonate Platforms

Carbonate platforms include multiple cycles of growth and common facies transition from back to fore-reef, with back-reef preserving finer sediments due to lower energy. Yet, these facies trends are often significantly modified by digenesis that overprints the expected long-range reservoir property trends. In settings where the facies-based reservoir properties are still expressed, the cycles and facies transitions may be mapped and modeled as locally variable facies proportions with attention to trajectory of the

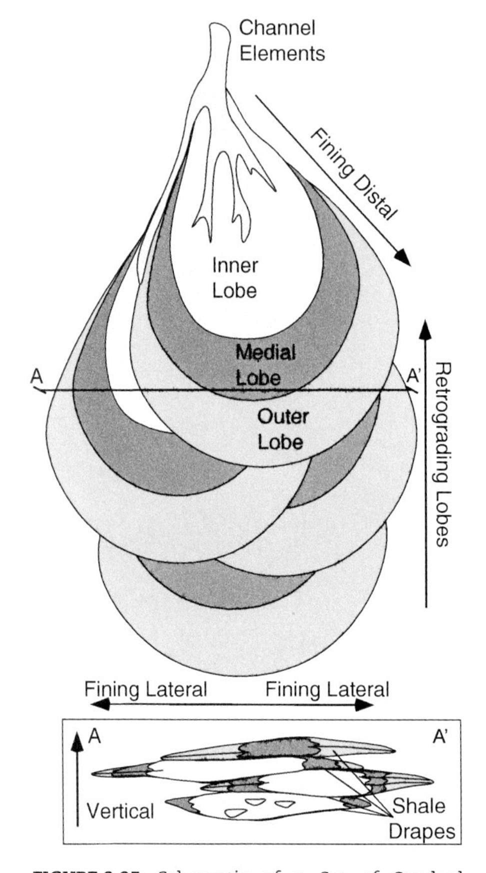

FIGURE 3.35: Schematic of a Set of Stacked Deepwater Lobes Channel Embedded in Overbank Shales. The lobes are typically composed of axis, medial, and margin facies associations, ordered with decreasing reservoir quality. While the pattern of decrease in reservoir quality distal in each lobe is straightforward, the stacking of these lobes results in potential preservation of mud drapes between lobes and juxtaposition of facies associations.

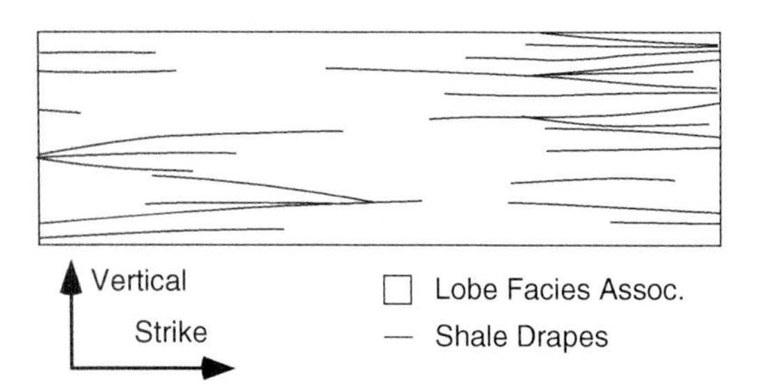

FIGURE 3.36: Schematic of a Cross Section of a Set of Stacked Deepwater Lobes with a Lobe Facies Association Set as Background and Shale Drapes Modeled as Objects. The shales have greater abundance near lobe margins.

growth due to aggradation, retrogradation, or progradation. When these are uncertain, then multiple realizations may be utilized. Then facies may be simulated with MPS or truncated Gaussian to preserve the ordering relationships. MPS is preferred when complicated curvilinear features are present in the facies. In settings where diagenesis has disrupted facies control on continuous reservoir properties, continuous reservoir properties may be simulated directly with their inferred trends (if any). When diagenetic alteration is minor, process-mimicking methods may be applied to produced models with complicated facies geometries, transitions, and cycles inspired by process models such as DIONISOS (Granjeon, 2009) and local constraints such as sediment supply, eustasy, and accommodation discussed in Section 2.1.

Modeling Method Selection by Input Statistics

The following discussion itemizes the types of statistical information that may be integrated into a reservoir model, the implications for method selection, and with reference to conceptual model (i.e., features commonly observed in specific depositional settings).

Complexity of Facies Features

The overall complexity of the heterogeneity features is an important consideration. A cursory inspection of almost any outcrop, high-resolution seismic or conceptual model (such as those illustrated in Walker and James (1992)) confirms that geologic heterogeneity is inherently complicated. The fundamental question is, How much of this complexity is required in the reservoir model to meet the project goals? The answer to this question is a major consideration in reservoir modeling method selection. Variogram-based methods are typically limited to linear features, and various implementations with nonstationary statistics may be applied to relax this limitation. Multiple-point methods may also reproduce a variety of repeating small-scale geometries; but as the scale, complexity, and nonstationarity in these heterogeneities increases, the reproduction becomes more approximate and reproduced features may become an ambiguous mixture without explicit accounting for nonstationarity. Object-based methods offer flexibility to impose crisp geometric complexity, while process-mimicking provides almost limitless capability to impose complexity.

Amalgamation and Isolation

Complexity is often influenced by the degree of architectural amalgamation or isolation. In this context, amalgamation and isolation refers to the degree to which unique architectural elements are preserved in the final reservoir model. In highly amalgamated reservoirs, long-range features are not typically preserved due to a high degree of erosion, reworking, and imbrications. For example, a channel complex with highly amalgamated channels preserves a large sand body with the potential for some remnants of non-sand-associated with overbank or abandonment fills. Due to many channel fill on channel fill erosional contacts, individual channels are not apparent. While the intermediate processes and forms were complicated (coeval braiding or an evolving meandering active channel) the preserved features may be simple (refer back to Figure 3.32). In these settings, application of variogram-based or multiple-point methods may be sufficient to capture these preserved features instead of attempting to explicitly model the channels with object-based methods. Given the combined effect of multiple channel erosions the intrachannel may not have readily parameterized geometries and may be readily modeled with variogram-based and multiple-point statistics. Once again, in high net to gross settings, it may be advantageous to focus on modeling the non-net features.

Facies Proportion Trend Models

Often information is available with respect to local changes in facies proportions; facies proportion trends exist in almost all reservoirs. Seismic data-derived trends may provide control on facies, or geological concepts may be informative. Vertical trends are commonly related to allogenic and autogenic controls on the reservoir. For example, cycles in reservoir quality may be related to cycles of eustasy or major avulsions or sedimentological processes as in the case of fining upward profiles of turbidity beds. Areal trends are often related to sediment source and transport and depositional processes, as was the case with the deepwater distributary lobe shown in Figure 3.35 and more specifically with the areal trend shown in Figure 3.37.

Areal and vertical trends may be combined together. For example, in the areal view, fairway may be inferred in a deepwater slope valley as a region with a high proportion of good reservoir-quality channel facies. At the margin of the fairway the proportion of poor reservoir facies may increase due to the loss in flow energy. Vertically, channels are usually abundant from margin to margin near the base and focused along the axis near the top (see Figure 3.37). A structure map in a carbonate ramp reservoir may be utilized to infer full 3-D effects of paleo-water depth and wave energy to constrain local carbonate

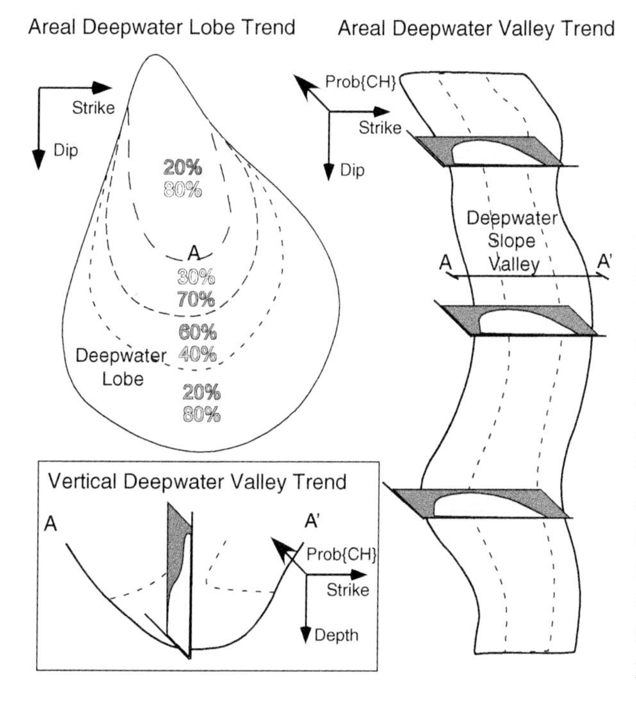

FIGURE 3.37: Schematic of Deepwater Trend Models. The areal deepwater lobe facies association trends are defined by regions with variable proportions of channel (gray), axis (yellow), medial (orange), and margin (green) shown in plan view. The trend includes transitions from channels to lobe axis to lobe medial to lobe margin distally. The deepwater slope valley trend model is represented by combined areal and vertical continuous model of probability of channel (CH) facies. Channels are more likely at the base and along the axis of the slope valley in plan view. Dashed line indicates the mapped channel fairway, the region with high proportion of channel facies.

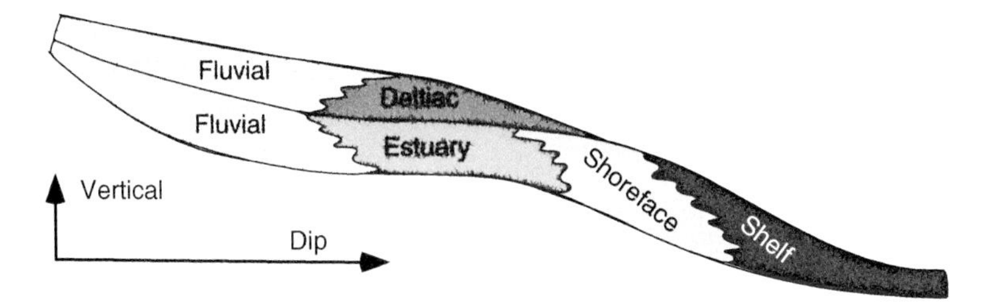

FIGURE 3.38: Cross Section of Retrograding and Then Prograding Coast with Two Systems Tracts and the Associated Juxtaposition of Depositional Settings.

production potential and the resulting locally variable proportions of course-gained facies. When both vertical and areal trend information is available, they should be combined into a single consistent 3-D trend model (see Section 4.1).

When trend information is at high resolution (at or below the resolution of individual reservoir elements), cell-based modeling is preferred because object-based methods are not be able to reproduce high-resolution trends. Conversely, when trend information is less specific, object-based and process-mimicking methods are again practical.

Facies-Ordering Relationships

Some geological settings result in consistent facies-ordering relationships. For example, a retrograding coast results in a natural upward transition from fluvial to shoreface to marine (MacDonald and Aasen, 1994) (see Figure 3.38). When these features are important at reservoir scale, the modeler may choose to apply truncated Gaussian simulation or multiple-point methods for their ability to reproduce ordering relationships. In another case of facies ordering, a fluvial reservoir may be composed of isolated sinuous channels such as within-channel transitional facies including lags, axial sands, marginal sands, and abandonment and overbank facies transition related to levees and crevasse splays. Due to the high level of geometric complexity, these transitions may be best modeled with nested object-based methods (Deutsch and Wang, 1996).

Curvilinear Facies Features

Various scales of curvilinear features may be expected in a reservoir. Bathymetric constraint such as antecendent salt-induced relief and valley confinement may interact with large-scale deposition, resulting in large-scale curvilinearity. Smaller-scale curvilinearity is inherent at the scale of depositional elements due to the interactions of depositional processes. For example, helical flow in channels and variability in the erodibility in substrate result in sinuous channel forms, and dissipation of energy after loss of confinement results in lobate shapes. Large-scale curvilinearity may be constrained by a trend model or locally variable azimuth fields for any cell- or object-based methods. When these features are at a smaller scale, multiple-point algorithms are convenient to reproduce the general form of these small-scale curvilinear features.

Long-Range Facies Continuity

Some reservoir settings are characterized by very-long-range continuity structures. For example, sand-filled slope valley channels may form isolated, contiguous pipes for flow in a muddy overbank (see Figure 3.39). In these extreme cases, only

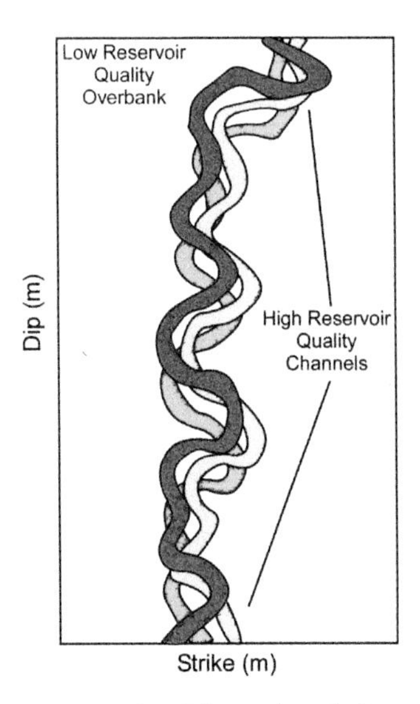

FIGURE 3.39: Schematic of Isolated Channels with an Organized Stacking Pattern as Often Observed in Deepwater Channels.

object-based or process-mimicking methods will reproduce these features. Variogram-based and multiple-point methods are not amenable to the reproduction of these long-range features, with reproduction including artificial breaks. This may have a significant impact on flow behavior of the model as long-range contiguous features tend to limit sweep while broken-up features tend to improve sweep.

Porosity and Permeability Relationship

It may be essential for flow modeling to capture the specific form of the porosity and permeability bivariate relationship (see Figure 3.40). In this situation, a specific algorithm for secondary data integration methods, cloud transform, is preferred for the ability to honor the precise bivariate relationship. Cosimulation based on variogram-based methods will only at best approximately honor the bivariate relationship. Yet, the cloud transform may not honor the spatial continuity and histogram as well as the variogram-based cosimulation methods.

In addition to modeling method choice, specific algorithmic parameters and statistical inputs for the selected modeling method are also based on the geological conceptual model.

3.2.4 Statistical Inputs and Geological Rules

Statistical inputs and geological rules are the constraints on the specific modeling method. Statistical inputs are discussed first because they are common to all geostatistical reservoir modeling methods. Geological rules, a more general form of input specific to some object-based and all process-mimicking methods, are discussed afterwards.

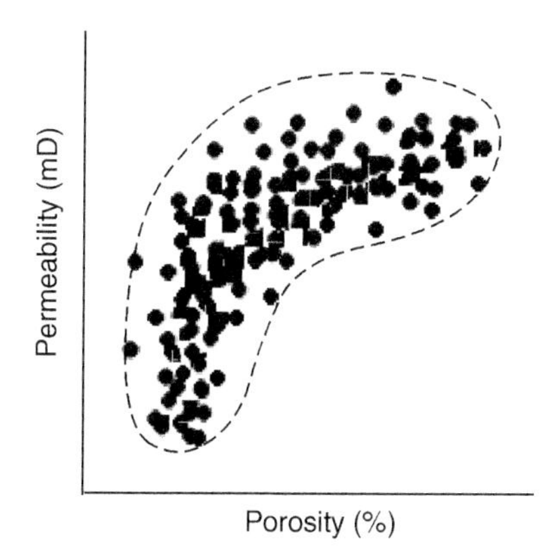

FIGURE 3.40: Schematic of a Permeability and Porosity Relationship with Nonlinear Features.

Statistical Inputs

Statistical inputs directly constrain the form of the model within the limits of the specific modeling algorithm. These inputs rely on the quantification of the conceptual model (supported by local data). Types of statistical inputs include a reservoir property distribution in the form of a continuous cumulative density function or categorical probability density function (including the specification of the number of categories), local trend information in the form of a locally variable mean or proportion model, continuity described by a variogram for variogram-based methods, a training image for multiple-point simulation, or geometric parameters for object-based simulation. In addition, there are various specific parameters related to algorithm implementation; these are discussed in the sections associated with each algorithm.

The approach for quantification of the conceptual model to constrain statistical inputs depends on the availability of local data. In sparse data settings, the conceptual model may be heavily relied on, while in dense data settings the conceptual model may be applied simply as a consistency check. For example, in a sparsely sampled reservoir, a conceptual model may be applied to select densely sampled mature reservoir analogs for inference of statistics not available for the actual reservoir. A conceptual model could indicate carbonate ramp, and the required statistics such as a global histogram could be exported from more completely sampled, similar carbonate ramps. When there is more local data available, the conceptual model may be applied to fill in or check input statistics. For example, the conceptual model may be applied to select analog outcrops and high-resolution seismic data to constrain the geometries of channel elements in an object-based model. In yet another case, geological concepts related to sediment source and paleo-flow directions are applied to constrain the continuity directions of the reservoir model. In all of these examples the decisions related to the conceptual model will potentially have a significant impact on the input statistics and eventually on the transfer function.

In some cases these parameters may include geological rules beyond the statistical inputs mentioned above. Note that geological rules may be defined as a generalization of all possible constraints that may be applied to reservoir model; yet in subsequent discussion, for clarity, rules exclude the statistical inputs discussed above and refer to constraints beyond these statistical inputs. Examples of geological rules include constraints such as: Channel objects repulse each other and may not cross, channels are attracted to previous channels and cluster, and injectites form attached to thick sands in isolated channels. Geological rules allow for further integration of geological information related to the sequence of events by which a reservoir is constructed. Note that these types of rules typically relate to the interactions of reservoir elements and complexes and, therefore, are only applicable to object-based and process-mimicking methods. When geological rules rely on depositional sequence, they are only applicable to process-mimicking methods. More detailed discussion on rules is discussed in Section 4.5.

Coding the Conceptual Model into Statistical Inputs

There are various methods to code conceptual information into quantitative statistical inputs. While local data provide the opportunity for quantitative inputs, in general, the conceptual geologic model will be qualitative and globally applicable, not locally precise.

A common approach is to utilize well data to infer statistical inputs such as distributions, trends, and spatial continuity as variograms and geometries and then to apply conceptual model to improve the inference. For example, parameters such as the minimum and maximum of a distribution, horizontal variogram, or full 3-D trend may be difficult to infer directly from available data. A conceptual model may provide information to improve the reliability of these statistical inputs.

A Bayesian approach (see Section 2.4, Preliminary Mapping Concepts, for the formulation) may be applied to utilize the conceptual model (and associated analogs) to formulate prior models of the required statistical inputs such as distributions and trend models (posed as local distributions). Then, the local data could be used to formulate the likelihood function. The application of Bayesian updating may be applied to calculate the posterior distribution that integrates the conceptual model and local data. If the prior distribution was heavily influenced by the local data, then the model may not be fair because Bayesian updating assumes updating with new information.

Other Considerations

In selecting these inputs, the following concepts should be considered, such as statistical representativity, scale, input statistic uncertainty, and the use of analog information.

As discussed in Section 2.2, Preliminary Statistical Concepts, reservoir data are rarely collected to be statistically representative. In fact, often the available sample data are collected in a biased manner. Tools such as declustering and debiasing and other methods for correcting for nonrepresentative data should be applied to construct representative inputs for simulation, because geostatistical simulation simply reproduces these inputs and does not correct the representativity of these statistics within the algorithms.

Furthermore, we must be aware that it is very unlikely that the limited data have sampled the extremes of any distribution. Extrapolation of appropriate input statistic minimum and maximum values (e.g., distribution tails) remains challenging. Typically, physical constraints such as zero porosity and maximum possible porosity given lithology and compaction may be useful to provide ultimate distribution bounds.

Any input statistic has an implicit scale, and all modeling methods assume a model scale. The distributions of a reservoir property at core, well log, and seismic scales are not the same. The reservoir property measurements must be compiled and adjusted to the scale assumed in the model. Otherwise, the distribution variability, spatial continuity, correlations, and so on, observed at the measurement scale (e.g., cores or logs) will be incorrectly assumed at the reservoir model scale (typically many times larger). The tools for change in volume support are presented in Section 2.3, Quantifying Spatial Correlation, and model scaling is discussed in Section 5.2, Model Post-processing.

Section 2.2, Preliminary Statistical Concepts, introduces concepts and tools for estimation of input statistic uncertainty, such as bootstrap, conditional finite domain, and kriging methods. These tools may be applied to assess the uncertainty in input statistics due to limited sample data, spatial continuity resulting in redundancy between samples, data locations within the model, and the model size. Once this uncertainty is quantified, a decision can be made whether to include this input statistic uncertainty in the modeling work flow. More details on uncertainty management are provided in Section 5.3, Uncertainty Management.

The use of analog reservoirs based on a conceptual model has been discussed. It would be tempting to excessively rely on a decision of parameter invariance between reservoirs with similar depositional settings. For example, all Devonian carbonate platform reservoirs have the same or very similar input statistics. While judicious pooling of analogs is useful, reservoirs tend to be somewhat unique due to a multitude of local factors that are often not apparent. Variations in provenance, tectonics, basin form, and post-depositional history may result in dramatic differences even in similar depositional systems of similar geological age. All of this underscores the need to integrate all available local information for the reservoir of interest and learn the unique story. In utilizing analogs, the critical decision is the choice of those reservoirs that are analogous. This is a strong decision of stationarity. As discussed in Section 2.2, Preliminary Statistical Concepts, it is not a hypothesis and therefore cannot be tested.

3.2.5 Work Flow

The central idea of this section is to determine the geological story of the reservoir. Figure 3.41 illustrates a work flow to determine the geological story of the reservoir of interest. The critical first step is the integration of all available information, including well data, seismic data, and conceptual information and experience into a consistent sequence of events responsible for constructing the reservoir. From this sequence of events, the framework of stratigraphic layering and regions are established. Then facies and continuous reservoir properties are modeled within this large-scale framework.

3.2.6 Section Summary

This section has covered general concepts related to the formulation of the conceptual model and considerations for integration of these concepts into a geostatistical model. This step is required to encode the geological concepts into a modeling framework, input statistics, and method choices.

The decision of what geological concepts to integrate into the model must be made. Considerations such as fit-for-purpose modeling, feasibility, the algorithm limits in integrating this information, and the cost of ignoring information are presented.

The modeling framework relates to practical concepts to stratigraphic grid construction, such as accounting for strata correlation style and faulting. It is essential that grid design has the appropriate resolution to capture essential heterogeneity efficiently, along with stratal correlations and transformations to best represent spatial continuity in the reservoir.

The choice between various geostatistical tools is important. Each has their trade-offs, and an appropriate choice will result in the best models to meet the project goals. Then, the statistical inputs and quantified geological rules must be inferred specific to the modeling method. There are various considerations such as statistical representativity, scale, input statistic uncertainty, and the use of analog information.

Now that we have covered the prerequisite modeling knowledge, compiled our data inventory, and mapped our data and geological concepts into modeling choices, we can next put this all together into a modeling work flow. The next section (Section 3.3, Problem Formulation) deals with modeling goals, modeling constraints, and work flow design.

3.3 PROBLEM FORMULATION

This section covers general concepts related to problem formulation. Given careful treatment of the data inventory (discussed in Section 3.1, Data Inventory) and the development of a conceptual model (discussed in Section 3.2, Conceptual Model), problem formulation is the development of work flow plan for the geostatistical reservoir modeling project. This is the *fit-for-purpose* design of a work flow to meet the established goals and purposes under the specific modeling constraints. Rigorous reporting and documentation are central to success of the project and any future work. This section is strongly based on the discussion in Deutsch (2010a, 2011).

The *Goal and Purpose* subsection covers the reason for conducting the geostatistical reservoir modeling project and determination of the final product from the effort. Aside from the typical goal of quantifying uncertainty in the resource, there is a wide range of modeling goals such as determining if more data are required, planning future developments, or even building a model for the opportunity to integrate data and concepts and for application as a visualization and communication tool. The goals and purpose must be clearly established, and the work flow must be designed to meet these goals under the modeling constraints (Koltermann, 1996).

The *Modeling Work Constraints* subsection discusses the constraints that limit the available resources to meet the project goals. Professional time,

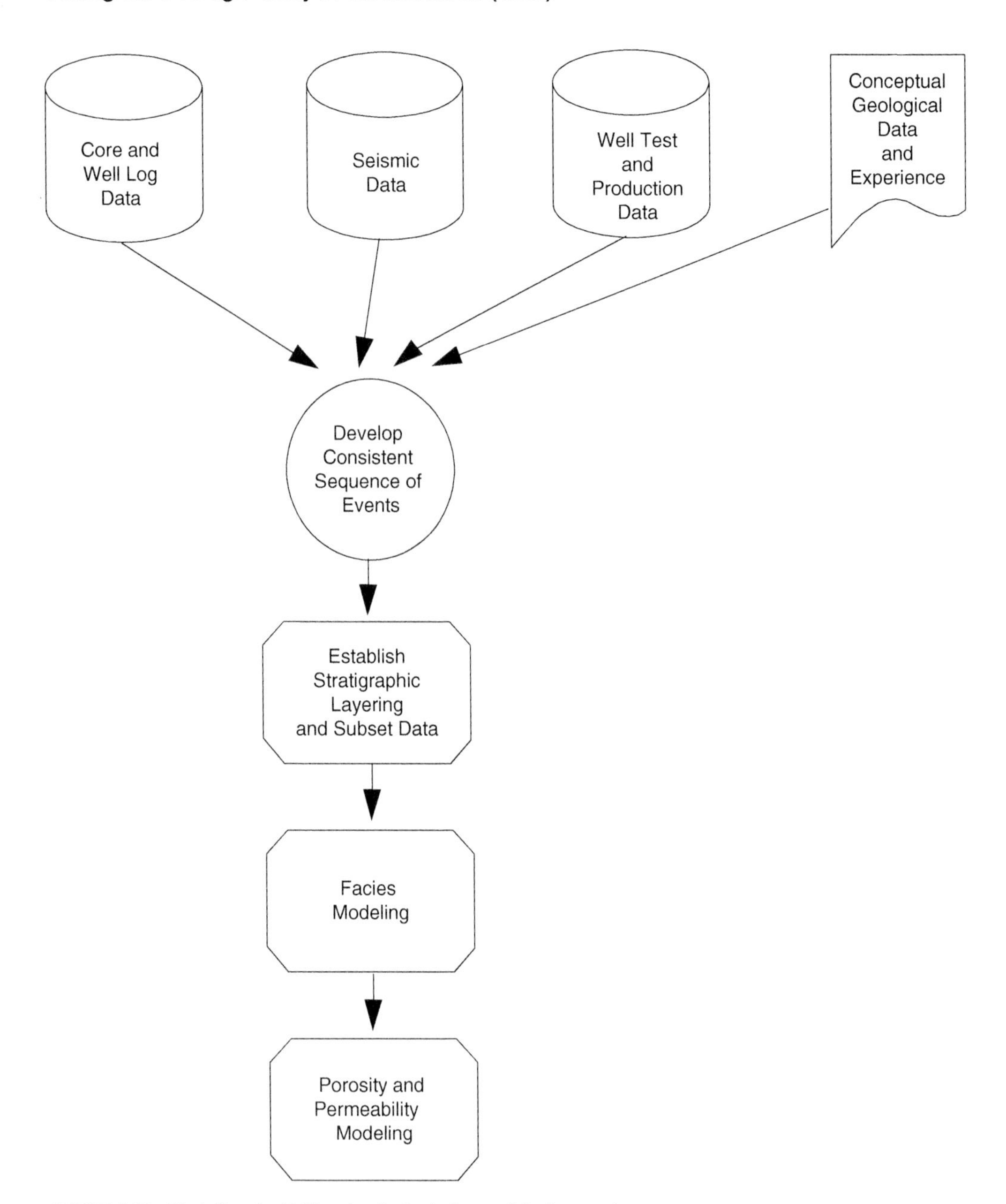

FIGURE 3.41: Work flow for Telling the Geologic Story of the Reservoir.

computational facilities, and capital are all valuable and their availability will be limited and shared over multiple competing projects. In addition, the expertise required for a project may not be available. The work flow must be designed under these constraints to ensure feasibility.

The *Synthetic Paleo-basin* subsection introduces a synthetic basin and associated reservoir data sets that are utilized throughout the remainder of the book to demonstrate reservoir work flows and specific modeling methods. These synthetic reservoir data sets provide reasonably realistic examples along with flexibility to cover various scales, data densities, and depositional settings.

The *Work Flow* subsection discusses work flow design that is strongly influenced by the model scale, the required variables, and their interrelationships. A common work flow is presented with typical modeling steps, and specific work flows are illustrated. Yet, it is important to recognize that the work flow must be carefully designed fit-for-purpose and under the modeling constraints. We provide examples to be illustrative, but not to promote a *cookbook* approach to modeling.

The *Reporting and Documentation* subsection discusses the importance of and considerations for rigorous reporting and documentation. With most projects the amount of information and data dramatically increases during the analysis and execution of the work flow. The combinatorial of multiple iterations of multiple operations on multiple variables in multiple regions results in a significant data management challenge. Without rigorous documentation, mistakes will occur and the project memory will be lost as project team members cycle to new projects.

The project goals and purpose, the work flow constraints, and the planned work flow and documentation represent the *project scope*. It is essential to define this scope at the start. Some consideration will be given to reducing the scope of the study; perhaps the objectives of the study could be achieved with a smaller study. The limitations must be understood; perhaps the objectives could only be achieved by a study much larger in scope. Perhaps a study is needed to even understand what could be done. In addition, new information, constraints, and project goals during the project may require reevaluation of project scope.

> This section will likely be most useful for the intermediate level modeler because they will have context and experience with the referenced steps and challenges. New modelers will read this without context, but will benefit from application of the common work flow and established structure, terminology, and warnings. Expert modelers will likely have already internalized most of these principles and may only find this useful for communication and educating other members of the project team.

3.3.1 Goal and Purpose Definition

Geostatistical models are always built for a purpose. In academia, small models may be built to teach the theory, implementation, and limitations of a particular technique. Models may be built to assist in the development of a new technique. The purpose may not be to maximize economic gain or make the best development decisions, but should be clear from the start. In practice, there are many varied objectives and, often, models are built for multiple purposes. Best practice requires an understanding of the study objectives. The study objectives will influence the study volume or area of interest, the variables to be modeled, the scale of modeling, the data preparation, and model post-processing. The purpose of most geostatistical studies is often a variant or combination of the following:

1. *Build a numerical model* of a site that accounts for all available data. A numerical model that integrates all available data provides a unified understanding of the site and provides some numerical support for further investigations. The model will be used for visualization of geological trends, understanding where the data are reliably, informing the spatial distribution, and understanding where there are too few data to reveal much about the true distribution. This *common earth model* acts as a valuable communication tool between the various project team members and between the team and management and other stakeholders.
2. *Assess the resources* of a site; that is, compute the gross volume of interest, the spatial arrangement of the interesting fraction within that gross volume, and attributes of the interesting fraction. Some consideration is given to an ultimate extraction technology and economics. There are spatial scales of interest to the recovery process and thresholds on rock properties of value and rock properties that are deleterious.
3. *Quantify uncertainty in the resource* at a specified scale. The measures of uncertainty could be used to support technical decisions or for classification and public disclosure. The assessment of uncertainty at relatively small scale (specific locations or grid blocks) is relatively straightforward, but the quantifying the uncertainty in global resources at a large scale will require uncertainty in parameters to be quantified and transferred through high-resolution modeling.
4. *Investigate geological-risk* associated with a range of geological scenarios that may

significantly impact connectivity and compartmentalization. The models integrate a range of specific geological factors such as sealing and flowing faults, contiguous shale drapes, and significant architectural trends that may exist in the reservoir. Although these features may not be deemed likely enough to include in the regular uncertainty model, this investigation provides quantification of the impact of these heterogeneities and justification for mitigation plans, such as further sampling or exploitation and facilities contingencies.

5. *Export statistics* for description, characterization, classification, or comparison of heterogeneities or for consideration in another model. The model of reservoir heterogeneity with or without local information may be constructed to build statistical descriptions of heterogeneity. These may be applied to explore constraints between heterogeneity components or to classify heterogeneity types. For example, it may be useful to statistically quantify and compare architectural geometry and heterogeneity with connectivity. In addition, models may provide statistics to support other models as is the case of training images for MPS or even numerical analogs to calculate relevant variograms in sparse data settings. For example, one may build channelized continuous property models to export variogram parameters for Gaussian simulation.
6. *Decide if additional data are required* or determine the value of additional data. Collecting closely spaced data are expensive, but making the wrong development decision or being unable to supply the required product for downstream hydrocarbon processing is also very expensive. The valuation of additional data, from a statistical perspective, could be assessed by a carefully designed geostatistical study. This type of simulation study will not help with uncertainty associated with the conceptual model.
7. *Assess the reserves*; that is, calculate the resources that would be extracted after applying economic thresholds and technical limits of the extraction methodology. This valuation could be based on multivariate evaluation and the application of economic thresholds in particular areas of the deposit. The scale of extraction and the possibility of future data becoming available would have to be considered.
8. *Evaluate different recovery processes* that may be suited for the specific geological heterogeneity. The efficacy and economic performance of different recovery schemes could be evaluated to help choose candidates for further consideration. The focus would be on realistic models of geological variability and the uncertainty.
9. *Make plans for future development* and provide the engineering and economic input for project valuation. In many cases, only one model will be subjected to detailed evaluation to see that other aspects of the project such as facilities, processing, waste management, and so on, all work together. The geostatistical model supporting future decisions would consider the possibility of additional information becoming available before final decisions.
10. *Make final decisions* about infill drilling locations or the positioning of facilities such as wells and facilities. Final decisions require the best possible local estimates with all of the available data at the present date; no consideration is given to additional data, because it would be too late.

The details of the geostatistical study undertaken will largely be determined by the specific study objectives and the designed work flow; that is, the study must be *fit-for-purpose*. There are other important modeling work constraints that influence the design of the modeling work flow.

3.3.2 Modeling Work Constraints

Clearly, the goals and purpose of the study should have an influence on the required resources and expertise, but it is not always that simple. Resources and technical expertise are limited, and their limits must be considered in planning a work flow.

Due to the need to maintain capital stewardship and maximize profitability, the resources allocated to a study will be limited. These resources include *professional time* (i.e., number of professionals and time each may contribute), *project timeline, computational facilities*, and *total budget*.

Professional time is valuable. Many organizations face difficulties in maintaining operational capabilities through hiring individuals with critical skill sets, developing these individuals and retention. In addition, even when there are professionals available, there are various projects competing for professional time. These all combine to limit the number of professionals and their time allocated to each project.

In addition, there are often absolute time deadlines related to maximizing project value, overall project planning, and agreements with partners and internal and external service providers (e.g., seismic acquisition, drilling, facilities, lease terms etc.). Project delays are expensive with delay in future positive cash flows and potential penalties; time is limited.

With the accelerating increase in computational power, processing and storage are generally not the limiting factors; instead the issues are often related to data management and visualization. Large, very-high-resolution models of 10s to 100s million cells and with many realizations of several variables of interest over various regions and facies, over multiple iterations and updates with new information, become increasingly difficult to manage and interrogate.

The needs of the project will have to be prioritized. Ultimately a total budget will generally provide a limit that will require a balancing to time, data collection, and utilization of facilities.

In addition, the expertise level of the available professionals to complete the study is important. There is a wide variation in the expertise of geomodeling professionals. A large proportion of these professionals are competent geologists and engineers with years of experience in the oil and gas industry and broad experience in exploration, development, data acquisition, seismic interpretation, well logging, production optimization and so on. These professionals are excellent generalists who may be competent to utilize tools available in well-documented and well-tested conventional earth reservoir modeling commercial software. They may or may not understand the numerical details of the tools they use. We recommend at least a basic understanding of "what goes on under the hood," because this is necessary to avoid many potential practical pitfalls and to select methods and infer input parameters.

Conversely, the professional may be an expert. The reservoir modeling professional may be current in the peer review literature, contributing to the science with their own work, writing their own modeling algorithms, and internally and externally recognized. These individuals are able to put together custom work flows to fit the unique needs of the study. At the other extreme case, the reservoir modeling professional may have little experience and may be only comfortable to implement established and documented work flows.

The available professionals will likely have skills somewhere along the continuum defined by the previous two end members. Care and audit procedures would be established to avoid blunders or theoretical mistakes leading to incorrect results. Due to available expertise, the project will often be limited to well-established methodologies and work flows.

Finally, regardless of available resources and expertise, there are basic, competing constraints on projects. The Project Management Institute refers to a "project triangle" that has been simplified to the conservation of quality, time, and cost (various versions include more corners). In general, it is not possible to meet all three competing components: One must choose only two. A project can be good and fast, but it will not be cheap. A project can be good and cheap, but it will not be fast. A project can be fast and cheap, but unfortunately it will not likely be good. These competing constraints must be considered in the design of a project work flow for geostatistics.

3.3.3 Synthetic Paleo-basin

A number of example data sets are utilized to demonstrate work flows and individual algorithms. We utilize a combination of real and synthetic data sets. Four synthetic reservoir examples have been developed. The data sets are synthetic for the following reasons: (1) The truth models of actual reservoir properties are known for each data set, removing well log and seismic interpretation issues that are not the subject of this book, (2) the data sets can be designed to demonstrate specific work flows at various scales and over various depositional settings, and (3) the data sets may be applied and shared freely without confidentiality issues.

We have designed a set of data sets from fluvial to deltaic to carbonate platform to deepwater turbidites denoted as the *synthetic paleo-basin*. A location map of the synthetic basin is shown in Figure 3.42. This example is intentionally simple, with

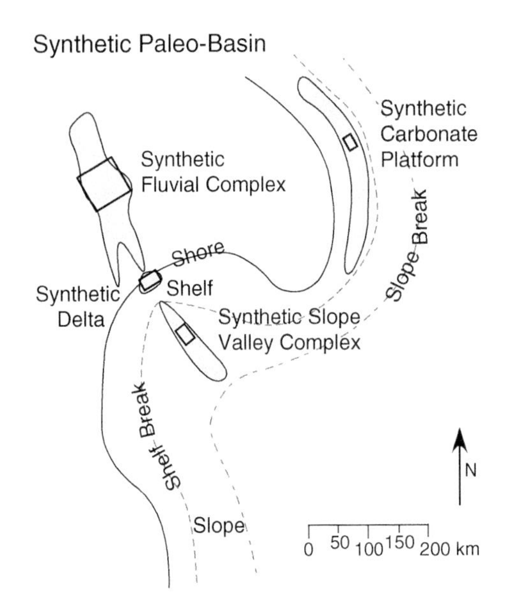

FIGURE 3.42: Illustrative Location Map of Synthetic Paleo-Basin with the General Topographic Features and the Locations of the Synthetic Reservoir Data Sets.

the depositional systems all positioned within a single systems tract with perfect preservation and without faulting.

Interpretation of reservoir properties is assumed error-free unless otherwise noted—for example, as in the case to demonstrate the integration of data softness. For each data set well log and in some cases seismically derived reservoir properties porosity, permeability and saturations are available. While the data sets provide dense coverage, when required they are filtered to provide sparse or incomplete data examples. The following is a brief introduction to each synthetic reservoir example.

Fluvial Reservoir

The *fluvial reservoir* is a typical meandering fluvial depositional system with two channel complexes with avulsions and imbricated lateral accretion, including: channel lag, lateral accretion, abandoned mud plug, and overbank facies. Reservoir properties are constrained strongly by facies with good reservoir properties in the channel lag and lateral accretion and mud plugs potentially acting as baffles or barriers to flow. Reservoir quality in the overbank is poor (see Figure 3.43).

Deepwater Reservoir

The *deepwater reservoir* is a typical deepwater slope valley channelized depositional system with shoe strings of reservoir sand in overbank mud. Due to the segregation of grain sizes associated with depositional turbidity currents, distinct facies have predictable patterns within the channels, with decreasing reservoir quality from channel axis, channel off-axis, to channel margin. The channels within a channel

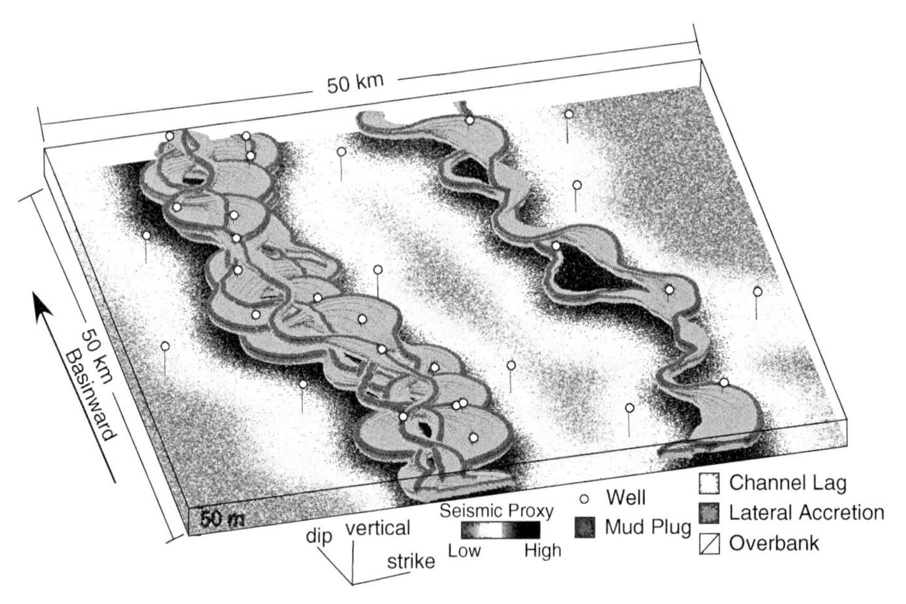

FIGURE 3.43: Oblique View of the Fluvial Reservoir Truth Model with Facies Painted on the Reservoir net Region, Well Locations, and a Seismic Attribute Painted on a 2-D Plane.

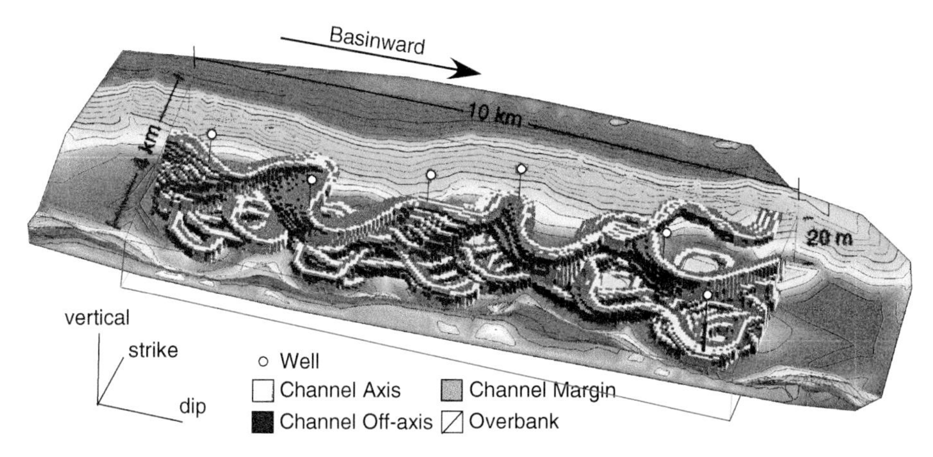

FIGURE 3.44: Oblique View of the Deepwater Reservoir Truth Model with; Facies Painted on the Reservoir Region, Well Locations and the Seismic Derived Slope Valley Boundary Surface.

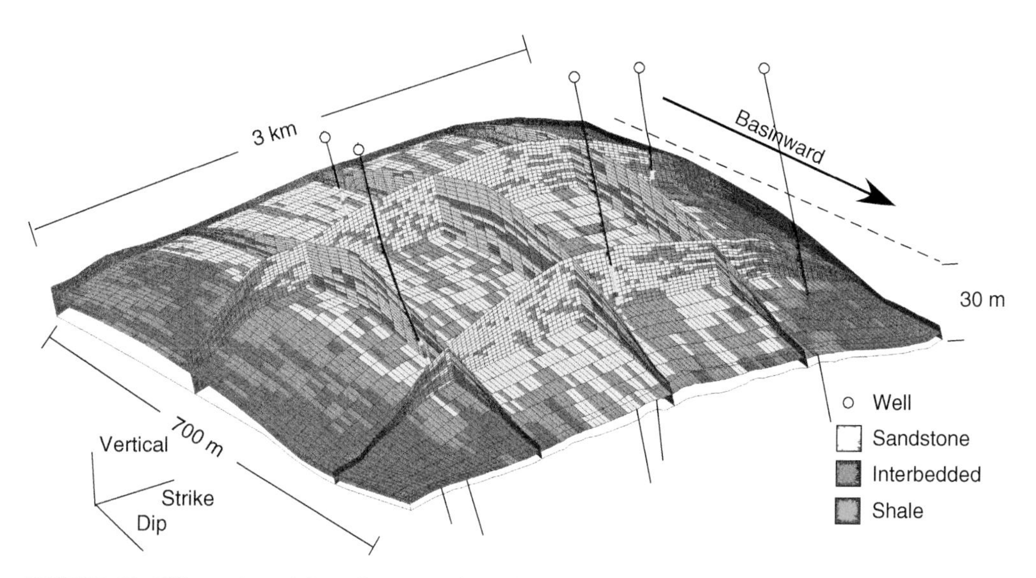

FIGURE 3.45: Oblique View of the Delta Reservoir Truth Model With; Facies Painted on Dip, Strike and a Plan Section and Well Locations.

complex commonly stack with disorganized and organized patterns. The deepwater reservoir includes one disorganized channel complex at the base of the slope valley with an organized channel complex juxtaposition (see Figure 3.44).

Delta Reservoir

The *delta reservoir* includes multiple delta lobe elements with facies transitioning from sandstone to interbedded sandstone and shale. Multiple compensationally stacked lobe elements and allogenic cycles result in complicated preserved heterogeneity, but overall good connectivity. Strong trends exist with transition from sand dominate to interbedded to shale toward the delta margins (lateral and basinward) and vertical cycles tied to eustatic cycles. Reservoir properties are strongly controlled by the facies (see Figure 3.45).

Carbonate Reservoir

The carbonate reservoir is a shelf-edge aggradational platform. Shelf-edge carbonate production and high energy results in a boundstone core bounded by slope failure sourced breccias and finer-grained grainstones and a basinward transition to marine mudstones. There is a strong correlation between

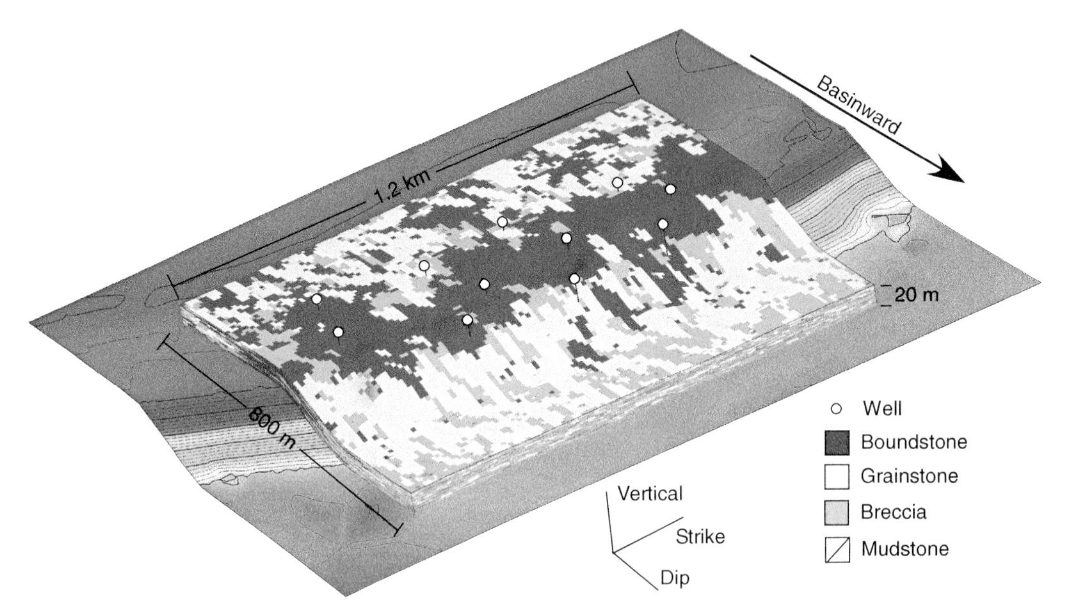

FIGURE 3.46: Oblique View of the Carbonate Reservoir Truth Model with; Facies Painted on a Reservoir Region with Mudstones Hidden, the Base Surface Mapped from Seismic and Well Locations.

facies and reservoir properties with the best reservoir quality in the boundstones, then grainstones, and finally breccias (see Figure 3.46).

3.3.4 Modeling Work Flows

A work flow is a sequence of modeling steps that are designed to meet the modeling objectives under the modeling constraints. This work flow may represent the steps to realize the concepts discussed up to this point in Chapter 3, Modeling Prerequisites, proceeding from data and expert knowledge to conceptual geological model to numerical representations, transfer function to access model performance and ultimately optimal decision making.

As mentioned previously, this work flow should be fit-for-purpose. Implicit to fit-for-purpose is the concept of appropriate scale, perhaps with the need to work across multiple scales and various variables, and the customized combination of steps that match model objectives. Each of these is subsequently covered in detail.

Model Variables

The objectives of the study will dictate the variables to be modeled. These should be enumerated early with an assessment of the available data (see Section 3.1, Data Inventory). The dependencies between the data should be understood, because the need to reproduce these relationships will constrain the modeling work flow. Some variables are modeled hierarchically one after another in order to impose relationships while some are modeled simultaneously if no significant relationships exist. In addition, all data sources have their own resolution and coverage. As a result of all these considerations, project complexity greatly increases with highly multivariate problems—that is, when there are many variables to be modeled. Simplifications such as utilizing empirical relationships between variables or deciding to omit variables may improve project feasibility with the potential to degrade the model results.

In general, the addition of variables adds significantly to the work flow complexity. In addition, the inference of multivariate statistics generally requires an exponential increase in the number of data to infer the multivariate relations; often, some multivariate distributional assumptions must be made.

Model Scale

The objectives of the study will also dictate the scale of modeling. Issues of scale will be faced throughout a geostatistical study, and some early choices have a large effect on downstream calculations. Reservoirs are often very large (10^8 m^3 or more) and it is not possible to create a numerical model at the scale of the data measurements. Also, considering the purposes described in the Goal and Purpose Definition subsection, a model may be required of a well bore, a subset of the reservoir, the reservoir, or even larger scales. These scales vary from millimeters to kilometers. Small-scale models can be scaled up; however, the size of numerical models in terms of the number of grid cells must be manageable—there would be about 10^{10} dm^3 in a reasonably thick reservoir interval over a section ($3\,km^2$). The computer memory required to manage 100 realizations of three reservoir properties for a model this size would exceed 10,000 gigabytes. Even with vast computer resources, it would not be practical to work with models this large. Moreover, a dm^3 is quite large for the small-scale features seen in core, and a section is small for the large areas under evaluation.

It is not simply a matter of computer resources. The data are not available to support high-resolution models for the entire volume of interest. There is no supporting information for very-high-resolution models in the vast interwell region. Perhaps very-high-resolution models make sense near well bores due to the availability of mm^3 scale information from core photos and image logs.

For these reasons, multiple models at different scales may be required. Three important model types are advocated:

1. Large-scale multivariate regional mapping
2. Conventional 3-D reservoir modeling
3. High-resolution mini modeling near wells

Regardless of model scale, four scales within the model should be understood and documented: (1) the area/volume of interest, (2) the target resolution to represent the rock properties or variables within the volume of interest, (3) the discretization required within the target resolution to provide meaningful upscaling to the target resolution, and (4) the data scale that will also be used to assign values at the discretization scale.

Some details of the heterogeneity will inevitably be lost. Variability seen between core samples will not be observed between large blocks in a block model; however, the unbiased transfer of the influence of the small-scale variability to larger scales is critical. This relates to the choice of the four scales mentioned above and the suitability of those scales for the study objectives.

These multiple-scale models are fit for different purposes in reservoir management. Certain geostatistical tools are used regardless of scale and purpose; the generic geostatistical work flow is explained first. Then, the specific details of regional mapping, reservoir modeling, and mini modeling are presented with example work flows.

As discussed in Section 2.4, Preliminary Mapping Concepts, an alternative methodology to handling model scale is to view the model as a mesh of estimated or simulated values at the original data scale (Deutsch, 2005c). Conversely, in conventional work flows, the reservoir properties are considered at and representative of the model cells. In this method, the models are constructed at data scale, but only informed at the centroids of the model cells. In this framework, scale-up is procrastinated to immediately before the application of the transfer function. For example, flow simulation assumes that the cell reservoir properties are representative and homogeneous over the flow simulation cell. Care must be taken to upscale with sufficient number of model cells within each flow grid block and to scale each property to represent effective reservoir properties over the flow blocks.

The Common Work Flow

The common work flow was introduced and illustrated with a simple example in Chapter 1.0, Introduction. Given all of the modeling prerequisites provided in Chapter 2 we are now prepared to further develop this common work flow and then provide some examples. The work flow is developed with a focus on study objectives together with previously discussed issues related to the resources and expertise available to the study. It may not be necessary to utilize all steps, and the level of effort in each step may vary significantly for fit-for-purpose modeling. Regardless of the project goals and scale, the fundamental approach and many of the tools are the same (Chilès and Delfiner, 2012; Yarus and Chambers, 2006). A generic work flow for geostatistics could be summarized by six steps.

1. *Specify* goals and data.
2. *Develop* a conceptual model.
3. *Infer* statistical parameters.
4. *Estimate* each variable.
5. *Simulate* multiple realizations.
6. *Post-process* models.

A summary of these steps follows. (1) Specify the goals of the study and take inventory of the available measurements. (2) Develop a conceptual model for the area/volume of interest including the story, division into subsets that are relevant for the specific situation, and choosing how the mean of each variable depends on location within each chosen subset. (3) Infer all required statistical parameters for creating spatial models of each variable within each subset and integrating any trend models. (4) Estimate the value of each variable at each unsampled location. (5) Simulate multiple realizations to assess joint uncertainty at different scales. (6) Finally, post-process the statistics, estimated models, and simulated realizations to provide decision support information. The detailed implementation of these steps will depend on the purpose of the study. Details on each step are provided below.

First: The goals of the study must be specified to determine the scope; that is, the work effort required for the study, the variables to be predicted, the scale relevant for evaluation and the specific estimation, simulation, and post-processing steps (as discussed previously in this section). As described in Section 3.1, Data Inventory, a data inventory must be taken to review all available measured data from wells, seismic data, and production data. The numerical models should reproduce all of these measured data within the scale and accuracy of the data. These two steps are integrated with goals potentially identifying the need for more data and data influencing the project goals.

This step is important and requires significant effort. Often, more time is spent on getting ready to do geostatistics than on actually applying specific geostatistical tools. It takes significant time to understand the data, the study objectives, and ensure that the modeling work flow is designed to meet those objectives. Cleaning the data takes a lot of time. Often, the data are not dirty or incorrect, but the format is different and inconsistent, there are missing data, there are different vintages of data, there are different companies involved, and so on. Preparing the site-specific data takes significant time. Yet, there must be a balance between satisfying prerequisites and getting on with the geostatistics to meet the study objectives.

Second: A conceptual model must also be developed including a geological understanding of the spatial distribution and analog data, distinct subsets, and transitions within the model. Understanding the geological context of the data is essential to supplement sparse data and to make good choices of model setup and modeling work flow, as described in Section 3.2, Conceptual Model.

The entire volume being modeled is rarely modeled in one step. There are logical subsets based on geological zones, rock types and facies. If possible, rock that genetically belongs together is kept together. The subdivisions must be large enough to contain sufficient data for reliable statistics, yet small enough to isolate geological features for local accuracy. A hierarchical system (for example, see Section 2.1, Preliminary Geological Modeling Concepts) is chosen where the site is divided into different zones and, perhaps, into areas where the deposition was controlled by different processes. The surfaces and geometric limits that separate these zones may be modeled deterministically or by geostatistical tools. Then, the facies within each zone are modeled at the chosen grid resolution potentially with local changes in proportions. Finally, continuous rock properties are assigned within each facies. The choice of the hierarchical subdivision for modeling has a significant influence on the final numerical geological model.

The average value of each continuous variable may depend on location within the chosen subset. There are often significant trends in the distribution of rock type and facies proportions within a geological zone. These trends are understood even with few data. Categorical variables almost always have a locally varying mean model. Continuous variables within categories are more likely to have a constant mean model. The result of this step is a subset of the volume for geostatistical analysis and modeling the location dependence of the mean. The choice of subsets and trend model amounts to a decision of stationarity. The logical test for the final model is its ability to demonstrate consistency with this conceptual model across scales and variables.

Third: Infer all required statistical parameters. The required statistical parameters will depend on the chosen technique that, in turn, depends on the conceptual model chosen for each stationary subset of the domain, as discussed in Section 3.2, Conceptual Model. Almost always, there will be a need to infer the univariate proportions of each category and the histograms of each continuous variable within each category. These univariate distributions are computed from the data and calculated to be representative of the entire subset. Some measures of spatial variability must also be inferred. In traditional Matheronian geostatistics (Matheron, 1971), variograms are the measures that quantify the spatial variability of each category and rock property.

There are other techniques that require size distributions (object-based modeling) or training images (multiple-point statistics). These statistics and methods have been discussed in Chapter 2. In the presence of sparse data, these statistical parameters are considered uncertain and a number of scenarios are documented. Depending on the purpose of the study, the uncertainty in the parameters is quantified and accounted for in subsequent geostatistical modeling.

Fourth: Calculate an estimate of each variable at each unsampled location. These estimates are based on the data and do not involve a random number generator. The estimation is commonly a form of kriging considering indicators, data transformation, cokriging, and/or a locally varying means as required. Whenever possible, the uncertainty is estimated directly with indicators for categorical variables and normal scores in a multivariate Gaussian context for continuous variables. This provides a single best estimate at each unsampled location together with a measure of uncertainty. This is based entirely on the data and decisions taken in the first three steps. The results are useful for resource assessment, and checking as simulations (discussed next) may mask inference and implementation issues. Some geostatistical techniques such as object-based modeling and multiple-point statistics techniques are only used in simulation mode. In these cases, the practitioner would run multiple realizations and summarize the results as facies proportions.

Fifth: Multiple realizations of all surfaces, volumes, facies, and reservoir property variables are simulated to quantify joint uncertainty and to provide a model of heterogeneity suitable for flow simulation. The simulation techniques are often closely linked to the estimation techniques. The estimation results are used for checking the realizations and for a first estimate of the resource/reserve. Uncertainty over a large volume depends on the simultaneous uncertainty at many locations; simulating multiple realizations is the only practical approach to quantify large-scale uncertainty. Also, the details of the geological heterogeneity are likely to have a large influence on recovery, reserve, and production calculations.

Sixth: Post-process the model results. Models of different variables must be combined. Models are an intermediate product in most studies and are usually applied to a transfer function to support decision making. At a minimum they are visualized and summarized to understand the subsurface.

In some cases, the ultimate value of the geostatistical model is as a common earth model shared by all the project team members. In the mode, the model is aimed at data integration, communication, consensus building, and/or training tool. The numerical reality of the model forces detailed quantification of geological concepts and integration of various data sources enabling improved communication and scientific debate in the face of limited data and inevitable uncertainty.

Sometimes, the statistical parameters are useful in themselves; variogram ranges may be used to understand data spacing and expected length scales of geological features. These models may be applied to support global development decisions without the need for a transfer function. In other cases, the estimated model from the estimation step provides expected results at unsampled locations and measures of local uncertainty that are useful for data collection and uncertainty management.

Most models are subjected to a transfer function that provides a model of uncertainty for a response of interest that is then utilized for decision making. These transfer functions may measure the following: static response, as in the case of volumetrics and connectivity analysis; dynamic response, as in the case of recovery factor, flow rates, and flow components; mechanical response of the reservoir, such as compaction and artificial fractures; response to thermal or pressure stimulation; and so on. The simulated models provide large-scale uncertainty and input to subsequent engineering design. Often, multiple realizations must be ranked to permit a few to be selected for more computationally expensive transfer functions (see Section 5.2, Model Post-processing).

Example Work Flows

The following are illustrative work flows. For each work flow the applicable steps of the common work flows are described, including: specify goals and data, develop conceptual model, infer statistical parameters, estimate variables, and simulate multiple realizations and post-process model.

2-D Mapping for Volumetrics

The goal of this work flow is to generate an estimate of the oil (or gas or other resource) in place. The typical data includes stratigraphic thickness, average porosity, and average oil saturation. The thickness data are derived from wells, and the average porosity and average oil saturation are averaged over the wells from the well logs calibrated to core information.

The conceptual model is a simple, unfaulted reservoir where the stratigraphic units extend over large distances. The thickness, porosity, and saturations are expected to vary smoothly over the reservoir area. Statistical inference is not a major issue for this problem. The estimation method that is applied to interpolate between the data for mapping will intrinsically decluster, resulting in reproduction of a representative mean. The spatial continuity model will be assigned with high continuity such that the resulting maps are smooth.

Only an estimate model is required for this problem. Kriging is applied to construct a 2-D map of each reservoir property; thickness, average porosity, and saturation. If there are significant relationships between these three variables, then cokriging may be applied to impose these relationships. Also, if trends are identified in the conceptual model and are not likely to be imposed by the data (data density is not sufficient), then the trend may be modeled for each property and imposed by kriging with a trend.

Model post-processing is simply a volumetric calculation based on the summation of the product of the three maps at each location.

$$\text{OIP} = \sum_{\alpha=1}^{n} t(\mathbf{u}_\alpha) \cdot \phi(\mathbf{u}_\alpha) \cdot s_o(\mathbf{u}_\alpha)$$

where there are $\alpha = 1, \ldots, n$ locations within the map.

Regional Mapping

The goal of regional mapping is to understand the resource, defined by a presence of reservoir index based on a definition such as net continuous bitumen or permeability height (kh), with uncertainty over a reasonably large area. This will guide a number of reservoir management decisions including the valuation of different lease areas, acquisition of new data, sequencing of development, and layout of surface facilities. Detailed examples of this work flow are provided by Deutsch (2010a) and Ren et al. (2006).

The areal scale of these models is many sections and perhaps townships (1000s of square kilometers). The vertical scale is the entire net thickness that could be produced by the anticipated recovery mechanism. The gross volume being modeled could be 10s of km by 10s of km by 10s of m thick—that is, billions of cubic meters of rock. The detailed vertical distribution of facies and other properties are important, but that information is collapsed to relevant 1-D measures that include structural markers, gross intervals, net intervals, net properties, and complicating factors such as thief zones and bottom water. These 1-D measures computed from the wells are modeled at a square areal grid resolution of about 50 to 400 m; thus there are 100s of thousands of locations and 10s of variables to predict in the geostatistical model. This is very reasonable and practical.

The goal is to provide uncertainty maps for the presence of reservoir index based on a definition such as net continuous bitumen or permeability height (kh). In addition, a goal is to integrate a wide variety of regionalized data to provide the best uncertainty model, but also to understand the multivariate relationships and the value of each data source. At this large scale, there is no specific interest in reservoir heterogeneity; therefore, a 2-D model with large cell size is sufficient.

The data available include a reasonably dense set of well data with good measures of the primary variable and reservoir index, allowing for calibration between the many secondary data and the primary variable. The secondary data include various geological variables mapped from the conceptual model and well data, including depth of top, depth of base, thickness, distance from depositional axis and source, and various seismic attributes.

The conceptual model for this reservoir suggests a structurally simple, amenable to 2-D mapping, laterally extensive reservoir with locally focused good reservoir quality that may exist between the nominal well spacing.

The inferred statistics include distributions and spatial continuity for the primary variable based on the wells. The distributions of the exhaustive secondary data are inferred for data checking, and all data are transformed to Gaussian for the multivariate work flow described in Section 4.1, Large-Scale Modeling. The result of this step is a model of the multivariate relationships of the primary and secondary data and local distributions of uncertainty given each data source and combined uncertainty.

Simulation proceeds from the local distributions honoring the multivariate and spatial relationships with the methods described in Section 4.1. Multiple 2-D models of the primary variable are calculated. Post-processing includes summarization of the uncertainty in the reservoir index with maps of the local P10, P50, and P90 and conditional variance (variance of all the realizations at a location). These uncertainty models may be applied to determine locations of potential reservoir sweet spots. In addition, these maps may be calculated iteratively with specific data omitted and compared to determine the value of specific information sources.

Micro Modeling

The goal of micro modeling is to quantify small-scale geological heterogeneities and transfer their influence to reservoir modeling (Deutsch, 2010a). This is closely related to the calculation of representative vertical permeability values and supplements a limited number of core plug measurements. The core plugs are often preferentially located to avoid shale/mudstone where no flow would be measured and avoid small-scale heterogeneities where the measured values would be unstable; thus high-resolution geological heterogeneity may not be completely captured in core data. Core plugs are usually sparsely sampled and likely located in the high-quality reservoir. It is very difficult to representatively sample the small-scale mudstone features.

The results of micro modeling are used to improve the statistical inputs to reservoir scale modeling. The details of vertical permeability can be important for flow simulation. It would be reasonable to undertake micro modeling before reservoir modeling. In practice, these multiple-scale models are constructed iteratively and in a repeated fashion as additional data becomes available.

Micro modeling is defined as modeling approximately a cubic decimeter scale with a cubic millimeter resolution. This is aimed at upscaling from core photos and image logs to the elementary data scale. The goal of micro modeling is to correct for nonrepresentative core data. Often, well test or production data are used to calibrate the geological model to correct for the concerns of representivity and scale; however, the viscosity of many heavy oils makes well testing impossible, and the models must be constructed before enough production data are available for correction.

Image logs provide data on a cylinder, and core photos provide data on a plane. Due to data density, trends are not commonly modeled and the statistical parameters are assumed to be stationary. The statistical parameters are relatively easy to infer given the dense data distribution. The range of correlation is often large with respect to the size of the models being constructed. This can lead to significant fluctuations in the model results.

Estimation is only undertaken for checking. Simulation is used to create models that reproduce the heterogeneity observed in the well data. Micro models are simulated conditional to photos/image logs. Simulation of micro models proceeds in three steps: (1) assigning a sand/mud indicator at the mm^3 scale with sequential indicator simulation or multiple point simulation, (2) assigning porosity on a by-facies basis using estimates of properties, and (3) assigning permeability on a by-facies basis correlated to porosity using estimates of properties. The estimates are based on clean sand and pure mud. The permeability at the very small mm^3 scale should be approximately isotropic.

The post-processing consists of upscaling the results to effective properties at the larger scale. The effective porosity is computed as an arithmetic average of the small-scale values. The directional effective permeability is computed by direct solution of the pressure equation given arbitrary boundary conditions such as a constant pressure gradient in the direction of flow and no-flow in the other directions. The difference between vertical and horizontal permeability will increase with scale because of the anisotropy of the small-scale geological features. Wen (2005) documents methods and software for modeling high-resolution sedimentary structures that could be useful in micro modeling, and Massart et al. (2011) provide an example for tide-dominated heterolithic sandstone with elements of micro and mini modeling.

Mini Modeling

Mini modeling also quantifies small-scale geological heterogeneities and transfers their influence to reservoir modeling. Yet, mini modeling is concerned with upscaling from the data scale to the nominal resolution of the reservoir model. For this goal, a slightly larger scale is utilized, defined as modeling the cubic meter scale at a cubic decimeter resolution.

Like micro modeling, data density is high, allowing for relatively straightforward statistical inference. However, it may still be challenging to characterize heterogeneity horizontal away from the well. Also, estimation is applied for model checking and may not be applicable if object-based or multiple-point simulation is applied.

Mini models are simulated with the statistical parameters from the micro modeling and from well logs. Mini models are often not conditional to specific well data; instead, the unconditional realizations are used to understand the scale-dependent relationship between porosity, vertical permeability, and horizontal permeability.

Like micro models, the post-processing consists of upscaling the results to effective properties at the larger scale. This provides a numerical basis for the scaling relations of the reservoir properties and identifies small-scale features that may be impactful on reservoir-scale reservoir properties.

Reservoir Modeling

The goal of reservoir modeling is to provide geological input for connectivity calculations and flow simulation. Flow simulation is often used to understand and optimize different reservoir development options. Estimates of production and injection rates are required for economic analysis. The timing and costs of development will also be estimated from the results of flow simulation.

The areal scale of these models is limited to a relatively small number of sections (10s of square kilometers). The vertical scale is the formation thickness that will be considered in the flow simulator. The vertical resolution in the numerical model is often between 0.25 and 0.5 m because this permits the model to resolve important barriers that would affect flow simulation. The areal resolution of the model is often between 10-m and 50-m square grid cells. A gross volume of one billion cubic meters of rock would therefore require 10 million grid cells. Relatively few variables are modeled: facies, water saturation, porosity, vertical permeability, and horizontal permeability. This model size and number of variables is reasonable and practical with available computer resources.

The data include facies, porosity, saturation, and directional permeability derived primarily from well data. The vertical scale is between 0.25 m and 1.0 m. Some vertical upscaling or reblocking is necessary because the scale of the raw data is often 0.1 m. Facies are upscaled based on the most common, saturation and porosity are arithmetically averaged, and permeability is normally computed as a transform based on porosity. If there are high-frequency variations in facies at a smaller scale, then a combined heterolithic facies should be defined.

The reservoir is almost always subdivided into multiple stratigraphic layers based the conceptual model. The stratigraphic layers are 10s of meters thick and continuous across most or all of the reservoir area. Then, facies are modeled within the stratigraphic layers before the saturation, porosity, and permeability variables are modeled within each facies. The number of facies is often between three and eight; too few facies and important features are not represented, and too many facies leads to excessive uncertainty in the assignment of facies.

A trend is almost always modeled for facies. A vertical proportion curve is calculated from the well data, and areal proportions may be computed from the well data or seismic data. These are combined into a 3-D proportion cube for facies modeling. A trend may also be modeled for water saturation since it often has a strong dependence on elevation. A trend is not usually considered for porosity or permeability because subsetting the reservoir by facies captures the main trends (see Section 4.1, Large-Scale Modeling).

The statistical parameterization for reservoir modeling consists of (1) determining representative proportions of each facies within each stratigraphic zone—this is often done by trend modeling, (2) determining a histogram for each continuous reservoir property within each facies and stratigraphic zone—declustering is considered if trend modeling is not being used, (3) determining indicator variograms for each facies and variogram for the normal score transform of each primary variable, and (4) calculating the correlation between each continuous reservoir property (in normal score units). Proportions, histograms, and variograms are all discussed in Chapter 2. There are other techniques that could be used for facies modeling, such as object-based modeling and multiple-point statistics. The input parameters for these techniques would be size/shape distributions and training images, respectively.

Estimation would take place for facies proportions and the continuous variables in the same multivariate Gaussian context as for regional mapping. Such estimation is not common because a central feature of these conventional reservoir models is that they are to represent heterogeneity and quantify the influence of that heterogeneity on the flow response for different processes. The primary reason to estimate the variables is to check the simulation results (next step). Commonly, practitioners simply check that the simulated realizations reproduce the specified statistical parameters and appear reasonable; estimation is not an important step in reservoir modeling.

Multiple realizations (typically 100) of the facies and reservoir properties are simulated in sequence to reproduce the local uncertainty and the correlation between the variables. The maximum number of facies modeled is normally limited to seven. The distribution of facies is used to constrain the distribution of saturation and porosity. The porosity values are simulated within each facies, and permeability values are simulated at each grid cell based on the collocated facies and porosity values. The realizations are heterogeneous and provide a measure of uncertainty. These multivariate realizations are considered as possible realizations of the true distribution and are post-processed accordingly.

The resources can be assessed on each realization. One or a few realizations are selected for flow simulation since the computational and professional time required to process flow simulation results is significant. The process of ranking and selecting the realizations is important when the resulting uncertainty will be used for reservoir management. A proxy model or response surface model could be constructed between summary statistics of the realizations and the flow properties. The aim of reservoir modeling is to understand the influence of heterogeneity on recovery predictions. The spatial distribution of vertical and horizontal permeability has a large influence on the flow predictions.

Sector Modeling

The goal of sector modeling is to further resolve and understand flow behavior of reservoir architecture. A sector model is an architectural high-resolution model of a subset of the reservoir model (see above). The cell size within the sector model are usually reduced in size by some factor such as 1/4 to 1/10, resulting in vertical cell extents of 10 cm to even a couple centimeters and areal extents of 10 m to a meter. Because the sector subset is typically a small fraction of the reservoir, the cell count is manageable. The mini models mentioned above are constructed within homogeneous facies to establish properties for larger-scale modeling. A sector model differs in that this model may span multiple facies and is intended for high-resolution flow modeling, often near a vertical or horizontal well.

In a sector model, the architecture, including specific flow conduits, baffles and barriers that were omitted from the reservoir model due to scale and algorithms limits are careful imposed and tested.

The data are similar to reservoir modeling with perhaps an effort to resolve the data to a higher resolution with facies, porosity, saturation, and directional permeability derived primarily from well data.

The conceptual model is utilized for the general stratigraphic layering and trends, but also for various high-resolution architectural scenarios. For example, the conceptual model may suggest the possibility of lateral contiguous shale drapes barriers or sand lag conduits. These unique features may be added, even if deemed unlikely to provide a measure of risk and to support mitigation.

The statistical parameterization includes those for the reservoir model at the appropriate higher resolutions (smaller scale) and the addition of parameters that describe the high-resolution architectural features. These may be difficult to infer from data and may require strong reliance on analog data sets.

Estimation and simulation considerations are similar to reservoir modeling. Post-processing generally includes (a) connectivity analysis and (b) flow simulation to assess the impact of detailed reservoir architecture. The results may provide support for the reservoir model scale or demonstrate that greater resolution is required, or may identify the presence or absence of significant architectural risks.

Additional Work Flow Considerations

In putting together and executing a geostatistical reservoir modeling work flow, there are various additional considerations.

Sufficient time must be allocated to sort out the study objectives, site-specific data, analog data, and a conceptual understanding of the site. There is a balance between how much time to front end load the project versus getting down to work. Of course, time must be left to perform the geostatistical study and meet the study objectives. Often, some data must be left out, some risk of error in the database must be accepted, some geological scenarios must be ignored, and an incomplete understanding of the geological context must also be accepted. Compromise is inevitable.

Most geostatistical studies are repeated as more data become available or the objectives change. It is rare that a particular geostatistical study is the first analysis of brand new data for a reservoir that has never been modeled before. Best practice is to assemble and review all relevant prior work such as reports, maps, models, and data files. Contact those that have studied the reservoir in the past to avoid making preventable mistakes and to address improvements that previous studies never had the time, data, or resources to address.

Most organizations have established processes related to the execution of a reservoir modeling work flow. These include identified milestones, periodic reviews, and decision gates. With consistent project milestones, project progress and project phase can be readily tracked. In addition, this provides a communication tool for describing and comparing progress between projects. There is typically an associated formal process for project review conducted within the project team and periodically by stakeholders external to the project team for critical, arm's length, peer review. Technical input during these reviews provides opportunities for improvement and may result in adjustments to the work flow or even major changes and recycle of the work flow. Established decision gates formalize the process of evaluating the project and the decision to continue or abandon the effort. While all of this is challenging and a significant burden on the project team, this oversight and, even at times, the choice to recycle work or kill a project is essential to optimize project value.

3.3.5 Reporting and Documentation

While we provide practical advice on internal project reporting and documentation, no effort is made in this book to address formal reserves and resource reporting regulations (e.g., as set by the United States Securities and Exchange Commission). For information on this topic see the current published standards and regulations in the appropriate jurisdiction.

Reservoir modeling project reporting and documentation is labor-intensive and does not seem to advance the project objectives. Also, considering all the combinations and iterations of data, data transformations and calibration, geological concepts developed, and statistical parameters inferred during initial modeling, it is difficult to keep documentation current and organized. Without appropriate documentation, there is a high risk of blunders or needless repeating of work. This often occurs, even during the same modeling session and almost certainly in a new round of modeling. When documenting, always imagine trying to recreate the project years after the original project team has moved on. The assumptions and limitations must be stated, the decisions must be defended, and the steps must be repeatable by a reasonably skilled professional.

Data Documentation

Many geostatistical reservoir modeling projects integrate large databases and require various data cleaning, formatting, calibration, conversion, and scaling operations. Best practice is to carefully document the data inventory and the limitations that exist in the database and conceptual understanding. Utilization of the concept of data events, described in Section 3.1, Data Inventory, will help here. Each data set and even datum within a data set have metadata that detail their origin, vintage, operators applied to them such as transforms, and calibration, along with potential data issues that may be addressed in the future. This will greatly improve model quality and revision.

This is essential because it is common for difficulties to exist in locating data and/or to mix up data versions and the modifications applied to the data. In many projects, especially during initial analysis, data versions multiply as various combinations of operations are applied in an effort to distill data relationships and trends over each variable or between variables. For example, various filters may be applied to a map of seismic response in an effort to reveal relationships with reservoir properties, or various trends may be fit to the data in an effort to separate heterogeneity into rational deterministic and stochastic components that match the conceptual model. Mixing up these data and data derivative versions may lead to significant error. In addition, efforts to apply quality control to the database should be documented to provide quality assurance and prevent readdressing these issues latter.

For documentation, a complete written report is favored, although it is recognized that presentation-based documentation is becoming increasingly common in many organizations. We suggest caution, because presentation-based formats do not typically promote adequate detail to understand and reproduce the previous project work flows.

Geological Concepts

Geological concepts tend to be qualitative and/or semiquantitative; therefore, they are not amenable to storage in typical reservoir modeling applications that typically store data sets, project files, scripts, and parameter sets. They are best presented in drawn maps, illustrations, and diagrams. For this reason, additional effort must be made to provide a detailed report retained with the project with links to the actual data and data derivatives. This will allow for communication of the data and analogs supporting the geological conceptual model, associated uncertainty model, and alternatives considered. This document should refer to the database and associated data events with the same nomenclature utilized in the modeling project.

Geostatistical Algorithms

Geostatistics is an evolving science with a toolbox with many versions of each tool. These tools or algorithms often change over time with new developments. These changes may be obvious and expressed as a new algorithm or new parameter or may be more subtle with hidden algorithm changes internal to the program. For example, many new versions of MPS have been developed over the last few years with various implementation details and capabilities. Contrast this with subtle algorithm changes such as "hidden" methods to automatically assign optimum search parameters in kriging and simulation or the automatic application of a distribution transformation for forced histogram reproduction for each realization after simulation.

To ensure repeatability, it is important to indicate the version of the geostatistical algorithm and the specific implementation details (in addition to the input statistics, random number seed, and implementation parameters). When the programs are open source, they may be included with the project or with commercial software the software version with attention to software update information from the vendor.

Uncertainty Model

In geostatistics, the uncertainty model is represented as an ensemble of realizations that result from the utilization of parameter, input statistic, and conceptual uncertainty (more will be discussed in Section 5.3, Uncertainty Management). It is best practice to include details of this uncertainty model in the documentation, including assessed uncertainties for each input and decision, and the method for projecting these uncertainties into the multiple realizations. This may include details on a design of experiments approach such as full factorial or Plackett–Burman partial factorial and efforts to rank realizations.

Although it may be challenging to retain the actual realizations, that would be the preference. Theoretically, retaining the inputs should allow for regeneration of the realizations on demand at a later date. This decision should be made with the following considerations. Geostatistical realizations result in very good compression ratios when archived, and it is much faster and convenient to compress and expand than to re-simulate the realizations. Also, the previously discussed issues with algorithm version may impede re-simulation.

Ranking presents unique challenges. Percentile rank has extreme "stickiness" when applied to a model realization. It is essential that the precise criteria for ranking be included with the ranked models. Otherwise, misuse of the ranked models will likely occur. For example, a model realization that ranked 90th in global recovery factor may be labeled as *P90* in the project. Subsequently, this realization may be incorrectly applied as a P90 result for connectivity, volumetrics, local flow between two wells, and so on, when clearly it is a P90 result only for the specific criteria applied in the original ranking.

Lessons Learned and Future Work

Documentation should consider the future audience composed of a new project team that is picking up the project again. Lessons learned are often gained through significant time investment and may greatly improve future work quality and efficiency. These nuggets should be clearly documented to aid any future efforts by the current or next project team. Also, consider sharing lessons learned between project teams because they may benefit similar projects. Examples of lessons learned may include: a specific series of steps to interpret or improve the interpretability of data, the ability to remove a variable and simplify the project work flow, or even the level of effort required to clean the data. All of these may be valuable considerations for any additional work.

Invariably, the project team will compile a list of opportunities for future work that were not within current project scope or possible with the available resources. This future work may potentially address limitations identified by the project team that may increase project value, and could be considered as proposed additions to the current scope or for the next project cycle. This future work could include: additional data collection to test or prove a geological concept or modeling choice, efforts to improve data quality with analysis and reprocessing, investigation of additional volumes of interest, application of new modeling methods to capture specific geologic features, changes to the transfer function to account for different reservoir conditions, and so on. Many projects recycle over long periods of time and have benefited from mining the future work sections from previous project reports.

Work Flow Automation

A decision may be made to automate the modeling work flow. In many geostatistical modeling software packages, this capability is available. Even in the application of standalone public domain modeling applications, scripting and batching are available to automate the work flow. In many cases, this adds some time to the work flow construction, but dramatically decreases the effort required to modify and rerun the work flow. As a rule, it is rare for a work flow to be executed only once. Considering this, it is advisable (when possible) to automate, recognizing the future cost savings. Also, the automation, usually in the form of a script, provides a clear audit trail for the work flow, documentation of the work flow steps, algorithms applied, parameters, and input statistics. It is a good idea to include comments in the script or the work flow automation routine to improve readability and to also explain major choices, unorthodox methods, and even alternative ideas or suggestions.

3.3.6 Work Flow

The central idea of this section is to construct fit-for-purpose modeling work flows to meet specific project goals with the available project resources. Figure 3.47 illustrates a common work flow that may be applied to address a broad variety of reservoir modeling problems. This includes the following steps: specify the goals and data, develop a conceptual model, infer statistical parameters, estimate each variable, simulate multiple realizations and post-process models. Depending on project needs, steps may be omitted or added into this common work flow for any required customization.

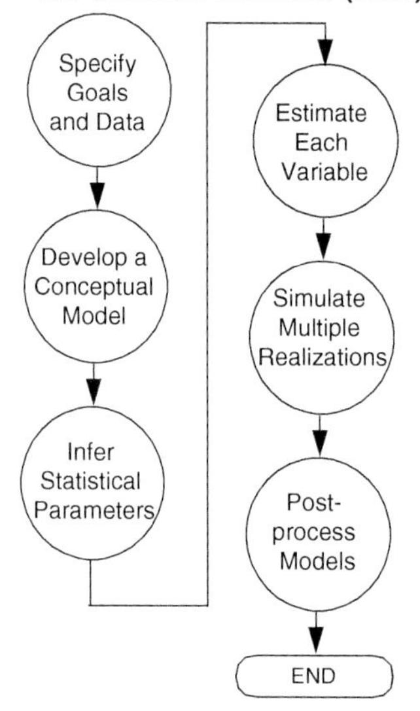

FIGURE 3.47: Work Flow That May Be Applied to Address a Broad Variety of Reservoir Modeling Problems.

3.3.7 Section Summary

This section has covered the concepts necessary to scope and plan a reservoir modeling project. This is accomplished on a fit-for-purpose basis with consideration of goals and purpose and modeling constraints. These modeling goals may include: construction of a reservoir model for data integration and communication as a common earth model or for extracting statistics, assessment of resources, quantification of reservoir uncertainty, evaluation of different recovery processes, decision on additional data collection, and development of future plans. This must be conducted within the modeling constraints, including: project deadlines, professional time, computational facilities, total budget, and available expertise.

While various goals and purposes have been identified along with various scales of investigation, it is still possible to provide a common work flow for guidance with the following steps: Specify goals and data, develop a conceptual model, infer statistical parameters, estimate each variable, simulate multiple realizations, and post-process the models. Example work flows have been provided to illustrate the application of these steps.

The synthetic paleo-basin is introduced as a set of reservoir data sets from various depositional settings within a systems tract to provide context for subsequent examples in this book.

Finally, reporting and documentation is essential to avoid blunders and to enable understanding of the modeling results and future work. While this effort does not immediately move a project forward, the absence of a significant effort in this will surely impede project progress.

In Chapter 2, Modeling Principles, we developed the fundamental principles and introduced many of the common modeling tools. In this chapter we described the considerations and methods for building a consistent work flow with the common modeling tools. In Chapter 4, Modeling Methods, we cover the various modeling methods with implementation details and examples.

4

Modeling Methods

This chapter covers the variety of methodologies and work flows for building reservoir models. The *Large-Scale Modeling* section refers to the modeling and mapping at scales that are normally above that of geostatistical reservoir simulation. This includes bounding surface, domains, regions, trends, and methods for large-scale integration of diverse data sources.

The *Variogram-Based Facies Modeling* section comments on facies and facies modeling and covers the traditional variogram-based methods for facies simulation. These methods are widely applied and very robust.

The *Multiple-Point Facies Modeling* section covers MPS that has recently become popular for facies modeling. This method offers improved flexibility to honor geological concepts while preserving the cell-based ability to honor conditioning.

The *Object-Based Facies Modeling* section discusses a more geometrical method for building facies models. Object-based models are visually attractive because the resulting facies realizations mimic idealized geometries observed in outcrops, high-resolution seismic analogs, and modern analogs.

The *Process-Mimicking Facies Modeling* section covers methods that attempt to reproduce even more geological complexity by coding the geological process, through rules, into the simulation algorithms.

The *Porosity and Permeability Modeling* section discusses the various methods for simulating continuous reservoir properties such as porosity and permeability, usually within a facies framework.

The *Optimization for Model Construction* section presents optimization methods as an alternative for the integration of complicated constraints and conditioning.

4.1 LARGE-SCALE MODELING

Large-scale modeling refers to the modeling and mapping at the scales above that normally associated with geostatistical reservoir simulation, that are discussed in Sections 4.2 to 4.6. This may include the modeling of areal limits and bounding surfaces that limit the reservoir volume, the structural framework and stratal correlation, the region identification for separation of distinct reservoir subsets, trend construction to model and constrain within-region changes, and mapping techniques that integrate various types of data to improve understanding of large-scale behavior. These all represent important initial decisions in reservoir modeling that may even be relevant to large-scale exploration modeling.

The *Structure and Bounding Surfaces* subsection covers the establishment of the domain of interest for modeling. This is a critical gridding framework and is part of the stationarity decision concerning the volume of the subsurface over which statistics may be pooled and applied. While practical considerations are presented, clearly this part of modeling is inseparably tied to good geological mapping and judgment.

The domain of interest is commonly subdivided into regions with distinct stationary reservoir properties. In many cases this is a decision to model separate facies or even larger-scale subsets that exist within the same gridded domain framework. The *Identification of Regions* subsection discusses the decisions associated with regions, such as methods to subset data and boundary choices.

Section 2.4, Preliminary Mapping Concepts, introduced the application of trend models and methods to impose trends within geostatistical algorithms. The *Trend Model Construction* subsection covers the practical methods and implementation details for trend construction. This includes methods such as hand mapping, inverse distance and moving window methods, and trend checking and other considerations.

The *Multivariate Modeling* subsection introduces a work flow for large-scale integration of various secondary data sources to provide information on primary variables of interest and methods to construct uncertainty models that honor the multivariable relationships between the primary variables. These tools provide a powerful integration tool to aid in formulating large-scale models, such as the domains, regions, and trends covered in this section.

This section is useful to modelers of various experience levels. The structure and bounding surfaces and identification of regions subsections should be useful introductions to novice modelers while the definitions may be useful to intermediate modelers. Ideas for trend modeling, checks, and considerations will likely be most useful for intermediate modelers. All of this is likely apparent to expert modelers, but may be useful for communication and educating other members of the project team. The multivariate modeling methodology has recently been developed and demonstrated in various settings and may be a useful addition to the expert modeler's toolbox.

4.1.1 Structure and Bounding Surfaces

Bounding surfaces define reservoir structure and are generally the outer constraints on the entire reservoir model or a stratigraphically unique unit within the reservoir (Caumon et al., 2009). This amounts to the decision of the domain or volume to subsequently apply geostatistical simulation. This is perhaps the most important aspect of the stationarity decision and has a first-order impact on reserves, production planning, and economic forecasting. The volume within the bounding surfaces (and associated regions when applied) is a stationary domain represented by model grid framework (discussed in Section 3.2, Conceptual Model) and stationary input statistics and nonstationary trends when required.

This approach subsets the reservoir model into distinct groups with their own gridding system. This may be applied on various scales such as systems, complex sets, and complexes. Typically, elements and subelement architectures are more readily modeled with regions using facies cell or object-based simulation methods as discussed in Sections 4.2 to 4.6 (covered below). The utilization of bounding surface modeling at these smaller scales would be time-consuming, inefficient, and likely not possible with available information.

Consider a bounding surface model for the Delta Reservoir. The boundary represents the transition from one domain to another. For example, the boundary may define the extents of an entire deltaic reservoir with appropriate grid framework, input statistics, and nonstationary trend models defined inside, or multiple boundaries may be defined to separate out various components of the delta, each with appropriate grid framework, input statistics, and nonstationary trend models. A dip section from the Delta Reservoir is shown in Figure 4.1 with a single domain and domains representing the delta topsets and forsets.

The decision of structure and bounding surface is an essential component of the decision of

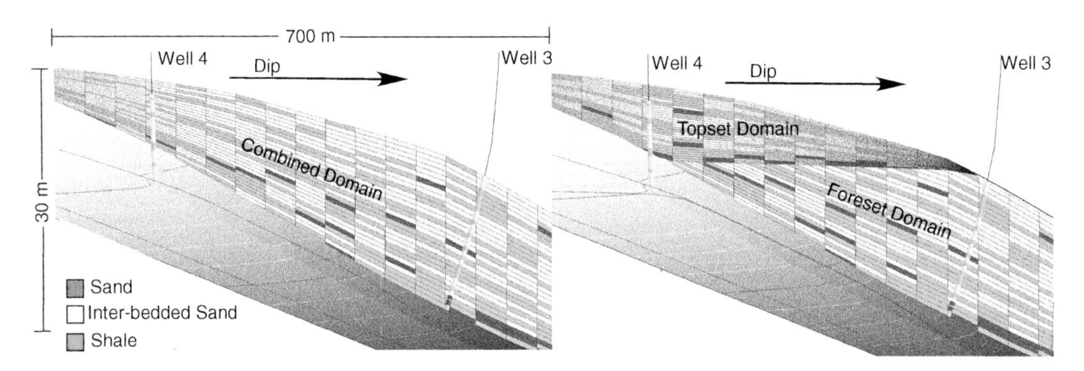

FIGURE 4.1: Dip Section of Facies for Two Models from the delta reservoir. On the Left, a Single Domain is Applied. On the right, two domains are applied representing the topsets and foresets of the delta. By using two domains, the stratal correlations and erosional discontinuity are represented, but at a cost of more complexity and modeling time.

stationarity. An important aspect of stationarity is that the statistical properties within the domains are established separately, and the spatial extent of the limits must be easier to model than the variables of interest directly. If the spatial extent of the boundaries is completely unclear and there is no reliable conceptual geological knowledge about the nature of the limits, then perhaps the domains should not have been divided. Information in the form of data and concepts must be available to assist.

Reservoir modeling presents various challenges with regard to bounding surfaces and domains. Reservoirs may represent various structural and stratigraphic traps with various fluid contacts. The associated bounding surfaces may represent the entire reservoir or specific reservoir units related to facies, elements, complexes, complex sets, or distinct reservoir fluids as in the case of oil–water and gas–oil contacts. Bounding surfaces may represent top and base stratigraphic unconformities, sealing faults, or faulted transitions to juxtaposed nonreservoir rock. There are also potential implications of bounding surface models on reservoir volumetrics and exploitation.

Implications on Volumetrics

Bounding surfaces represent the ultimate extents and, therefore, the volumetrics of each reservoir domain (or the entire reservoir if modeled as a single domain). In many reservoirs a vertical shift of a domain bounding surfaces by several meters or laterally by tens to hundreds of meters may make the difference between an economic and an uneconomic reservoir accumulation. Geometrical complexity related to stratigraphic and structural pinching and fluid contacts may further amplify the importance of bounding surface locations on volumetrics.

Implications on Exploitation

Bounding surfaces provide models of reservoir structure and extent. This is important for the design of reservoir development plans. For example, for designing a water flood, the position of the injectors may be optimized with placement close to reservoir extents, or producer placement may be optimized close to a dipping margin. Improvements in directional drilling have increased the importance of good bounding surface and bounding surface uncertainty models. These models are essential to optimize and mitigate risks associated with advanced exploitation methods.

Methodology

The general methodology for modeling bounding surface is as follows:

1. Pool all relevant hard data.
2. Apply the conceptual model and secondary data to map away from hard data.
3. Design an appropriate uncertainty model.

Pool Hard Data

The first step is the decision of what data to pool together for meaningful analysis, with consideration of the conceptual model. Although the following discussion is focused on the choice of domain, these concepts extend to the next scale for the choice of regions within the domain. Common to various stationarity decisions in geological modeling this decision is the typical lumping versus splitting dilemma. The following are some considerations that may help with this decision on how much to pool—or in other words, how far to extend stationarity. In some circumstances, it may be appropriate to model uncertainty in data pooling with multiple scenarios or realizations of domains or regions along wells (see uncertain bounding surface well intercepts on the right-hand side of Figure 4.2).

The uncertainty in the domain bounding surfaces—that is, the domain position and volume—must be modeled. There is a trade-off in this decision analogous to the balance of deterministic trend, and stochastic residual, discussed later in this section.

There is a partitioning of uncertainty and variability within domains and between domains. There is a decrease in model uncertainty when the domain model is made more specific. On the other hand, larger domains result in a more significant stationarity assumption; domain statistics are considered location-independent over larger volumes. In general, with larger domains, less local information is imposed and the resulting models demonstrate less organization and more mixing of features. A central premise is that all valuable and reliable information should be imposed on the model to produce the most informative model possible.

Wilde (2011) proposed the utilization of Q–Q plots to visually aid in the decision of stationarity. In this approach the distribution of various subsets are compared to the global distribution. With this visualization, one can readily observe the similarity and difference in continuous variable distributions between the subsets and between each subset and the global.

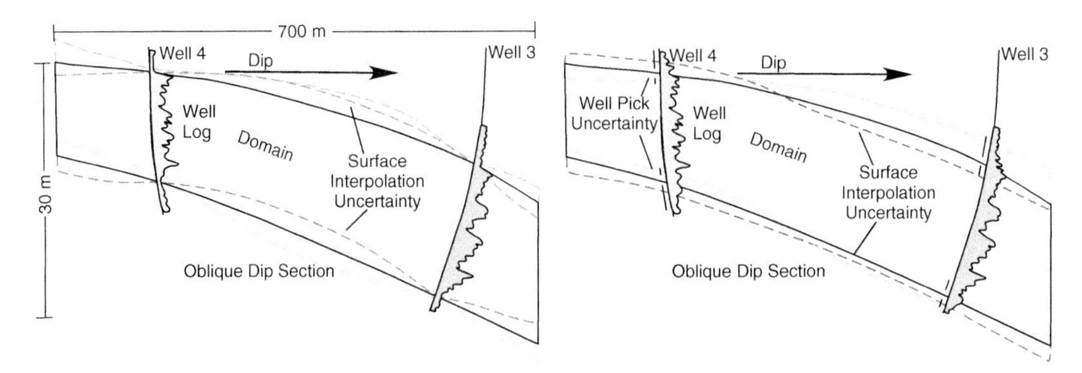

FIGURE 4.2: Schematic of Nearly Dip Section from the Delta Reservoir with Two Wells and Associate Well Log Indicating Reservoir Quality. On the left, the decision of how to pool the data is straightforward due to the sharp contacts at the top and base of the reservoir unit. On the right, the choice of how to pool the data is more challenged due to gradational transitions at the top and base of the reservoir. The contact for the domain may be considered uncertain. In both cases uncertainty in the surface interpolation may be added.

Subsets that vary significantly from the 45° line may be set as their own domains, and geologically reasonable subsets that are juxtapositioned to each other may be pooled together. It is important to note that univariate distributions are not the only consideration for separation of domains: Spatial heterogeneity, map-ability, and gridding are also important considerations.

The choice of domains along the wells is related to the choice of how far to extend the reservoir away from the wells. In many situations the reservoir extents are represented by clear unconformities with contacts between reservoir quality facies and conspicuous poor reservoir rock. However, these facies may exhibit heterogeneity without a clear indication of interwell connectivity, leading to difficult choices on whether or not to include them in the reservoir domain. In other cases, the facies transitions may be transitory, resulting in a gradual decrease in reservoir quality perhaps represented by overall system transgression or retrogradation vertically. In these cases, there is a difficult choice of where to place the reservoir domain boundary. Clearly, expanding the domain at well locations vertically and lumping more transitional rock in with the reservoir domain will result in more volume, but should simultaneously decrease the average reservoir quality over the reservoir domain.

Map Domains Away from Hard Data

It is often challenging to establish the extent of the stationary domains away from the data locations. The boundaries/limits that separate the chosen domains away from the data locations may be unambiguous and modeled deterministically or they may be quite uncertain and modeled with stochastic techniques. As mentioned above, an important aspect of stationarity is that the statistical properties within the domains are distinct and the spatial extent of the limits must be easier to model than the variable directly. If the spatial extent of the boundaries is completely unclear and there is no good conceptual geological knowledge about the nature of the limits, then perhaps the domains should not be subset.

The most common deterministic approaches to model boundaries/limits are to digitize them on seismic sections, construct surfaces interpolated between the sections, or utilize guided auto-picking routines that identify surfaces from 3-D seismic to directly model the bounding surfaces.

The stationary domain is required to map the overall extents of the reservoir. Clearly, this is an important decision that impacts reservoir volumetrics and development decisions. This decision is best addressed by good geological mapping based on the integration of all relevant information. The following are some considerations that may assist.

Similar to domain assignment at well locations, the trade-off between average reservoir quality and volume should be recognized. For example, consider the areal trend map of reservoir quality from the Delta Reservoir (see Figure 4.3). It may be difficult to place the reservoir boundaries between the conspicuous reservoir and nonreservoir boundaries. Clearly, extending the reservoir areally will increase reservoir volume with a decrease in average reservoir quality.

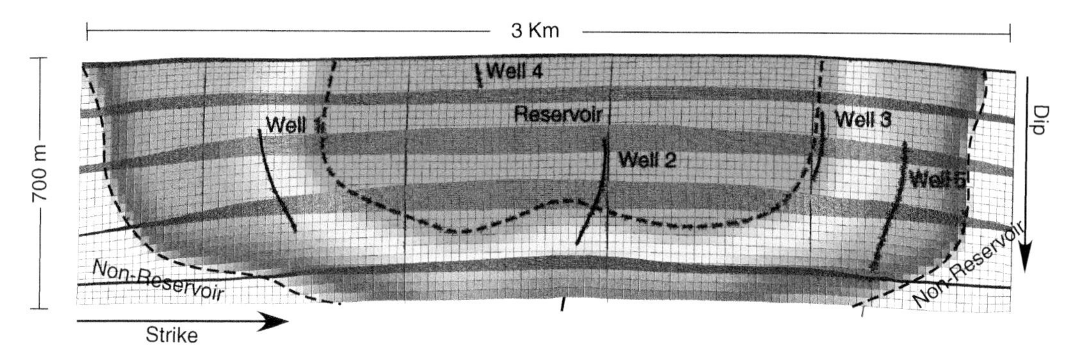

FIGURE 4.3: Areal Trend Map of Reservoir Quality from the Delta Reservoir in the Paleo-Basin Synthetic Models. Cold colors represent good reservoir quality, and hot colors represent poor reservoir quality.

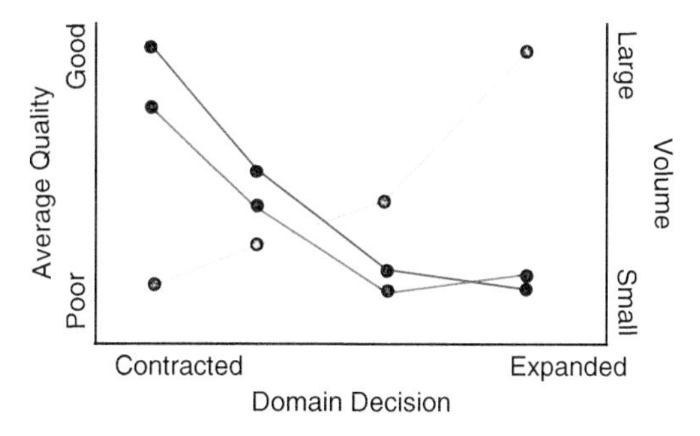

FIGURE 4.4: Areal Trend Map of Reservoir Quality from the Delta Reservoir in the Paleo-Basin Synthetic Models. Blue and red lines are two average reservoir quality properties while the green line is reservoir volume.

The trade-off between the volume of rock and the quality of rock is recognized and expressed relative to cutoff grade as the "grade tonnage curve" in mining (see schematic in Figure 4.4). While not commonly applied in reservoir geostatistics, some concepts are transferable. As the reservoir is extended further into transitory facies, reservoir volume increases, but average reservoir quality decreases. It would not be appropriate to extend the statistical distributions from primary reservoir rock over the transitory rock; efforts must be made to decluster/debias the statistics. When making decisions with regard to reservoir domain, this effect may be observed by comparing volumetric and reservoir property statistics for various scenarios of data pooling.

Bounding Surface Uncertainty Models

Constructing deterministic limits between stationary domains is best practice when they can be modeled with reasonable confidence. If uncertainty quantification is part of the study objectives, then uncertainty in the limits must often be considered. This may be assessed by considering the significance of bounding surface uncertainty on project goals. An illustration of surface uncertainty is shown in Figure 4.2.

Given the first-order control on reservoir volumetrics and potential scarcity and limited resolution of data control on boundaries, this uncertainty is often an important consideration. Also, given the various assumptions and potential for error involved in bounding surface interpretation, it may not be reasonable to assume perfect knowledge, or no uncertainty even with good data control. For the seismic-based surface mapping methods, there is the potential for systematic and local error in interpretation of the velocity model that may result in significant bias.

Stochastic deviations in surfaces could be considered by simulating a difference value over surfaces (Deutsch, 2011; McLennan and Deutsch, 2007). A typical work flow would be to model the top structure (or the surface best informed by seismic data) and work with isochore thickness up/down through the stratigraphic column. Care should be taken to ensure data conditioning and reasonable deviations from the base case surfaces at each step. Uncertainty in the average deviation of the surface should be used; a bootstrap resampling procedure (see Section 2.2, Preliminary Statistical Concepts) could be applied to calculate uncertainty in average deviation. Care should be taken to account for all error sources. For example, limits of seismic resolvability may result in a small-magnitude, short-range model of error while uncertainty in the velocity field may be modeled as a large-magnitude, long correlation/consistent shift in the bounding surfaces. The following summarizes one reasonable procedure to account for this consideration.

The base case surface is assumed to be unbiased, and it is assumed that deviations from the base case follow a Gaussian distribution. This is likely reasonable. If the distribution is known better, then a normal score transformation and back transformation would replace the simple (non)standardizing approach taken below. The base case value (structure or thickness) is required: $(z_b(\mathbf{u}), \mathbf{u} \in A)$, a 2-D grid of values coming from the seismic. In general these values are fitted to the well data: a global estimate of the uncertainty in the base case surface, σ_δ, a single number established from time interpretation uncertainty, and time to depth uncertainty (sum of these errors if they are considered to be independent). This would be based on a review of the seismic data and, perhaps, differences between different interpretations. The uncertainty in the mean is calculated from standard error and the number of independent data (widely spaced wells); or if wells are correlated, then the spatial bootstrap is required. A map of locally varying uncertainty could be considered. The continuity of surface fluctuations may be calculated by performing semivariogram analysis on the well picks and associated isopach thickness.

These parameters must be established from the available data. The simulation proceeds by establishing a target mean that could be different from zero, simulating the deviations and adding them to the base case surface. Details of this simulation are found in the thesis of Alshehri (2009). If the target mean is held constant at zero, then the resulting bounding surface realizations locally fluctuate above and below the base case, but in expectation there is no net change in reservoir volume. If these fluctuations have short scales relative to reservoir architecture and trends, then it is possible that this will result in very little or no impact on total reservoir uncertainty, that is, reductions in reservoir thickness in one location are balanced by increase in reservoir thickness in another location. Therefore, it is recommended to consider the potential for global shifts with a target mean deviation of zero. Also, significant short-scale fluctuations may result in isolated islands in the domain that may not be realistic.

Instead of the previous simulation method, bounding surfaces may be represented by expert-designed surface scenarios. This method may be the best when there is a high degree of surface geometrical and constraint complexity and may maximize the integration of geological expert knowledge. In this case, various sets of surfaces are designed to span the uncertainty in surface interpretation. As with scenario-based uncertainty discussed in Section 5.3, Uncertainty Management, a sufficient number of scenarios are required and the probability of each scenario must be assigned.

In some cases, boundary modeling may be automated with distance-based methods proposed by Wilde and Deutsch (2011) and McLennan (2008). This method has the advantage of being able to deal with complicated surfaces typical of mining geobodies. In these methods, boundaries are interpolated between data with a smooth assumption and a distance function is established based on distance from the most likely boundary inside domain (with negative coordinates decreasing from zero at most likely boundary) and outside domain (with positive coordinates increasing away from most likely boundary). Boundary realizations are generated by systematically applying a distance function threshold to globally dilate or erode the domain or by generating a field of distance values thresholds that locally dilate or erode the domain. If automated methods are applied, we highly recommend expert guidance be applied and careful checking of the results.

4.1.2 Identification of Regions

Within a model domain defined by bounding surfaces, there is often a need to further subset the model into stationary regions. The term *region* could represent any subset within the domain grid framework that is modeled separately. This usually includes mappable reservoir units with significant changes in univariate, multivariate, or spatial statistics that are not readily accounted for with a nonstationarity trend model, but would be tedious due to discontinuity, scale, and irregularity to model as separate domains. Also, regions are conformable to a single grid framework; therefore, it can be included in a single domain.

A commonly used region is facies (see Sections 4.2–4.6 for information on facies modeling). Facies categories have distinct statistical behavior, yet they are often heterogeneous and discontinuous; therefore, they should not be modeled as their own domains. Yet, they should be separated; otherwise reservoir properties will mix between all the facies, resulting in a loss of important information.

Regions may exist at various scales—for example, representing complexes and complex sets. The decision between subsetting as a domain or region is one of convenience and gridding. Once again, if the subset represents a contiguous, large-scale change in stratal correlation style or required grid resolution to represent the heterogeneity, then a new domain should be assigned; otherwise, regions are an efficient means to model separate stationary subsets of the model.

All of the considerations discussed in bounding surfaces for domains are relevant for regions. Volumetrics of the reservoir are directly impacted by the proportion of good reservoir quality facies, and the extents are impacted by the locally variable proportion models for facies (or regions). Region decisions impact reservoir continuity; therefore, they may impact reservoir development plans. Finally, regions represent a significant part of the stationarity model. Once again, a choice is made on how much variability to include within versus between regions. For example, a highly specific facies scheme may have very little reservoir quality variability within each facies, and the majority of variability is between facies. With such a model, facies modeling has the most important constraint on reservoir property heterogeneity, and subsequent continuous reservoir property simulation has a diminished significance. Conversely, a less specific facies model may pool a wide range of reservoir quality rock, resulting in likely fewer facies categories and significant within-facies variability and less between facies variability. In this setting, continuous reservoir property simulation has a greater significance.

There may be a temptation to apply purely statistical tools for region assignment. Certainly, facies assignment that does not consider reservoir properties (that is, the primary drivers for reservoir volumetrics and heterogeneity) is inappropriate. For example, a facies scheme based on source, mineral composition, and texture, but with weak linkage to porosity and permeability, should be avoided. Yet, relying on statistical clustering and binning of well-derived reservoir properties may miss important information related to spatial continuity and geologically informed mapping. For example, a facies grouping that may not be the most efficient statistical clusters may result in more geologically realistic and mapable architectures.

Region Boundaries

Region boundaries may be categorized as hard or soft. Hard boundaries are commonly assumed with region-based modeling. With this assumption there is no continuity across boundaries, and data conditioning is only imposed within the data event's identified region. For example, in a sand, interbedded sand, and shale model with hard boundaries there is no continuity of porosity from sand to interbedded sand, sand to shale or interbedded sand, to shale. A high-porosity zone in sand may be next to a low-porosity zone in the interbedded sand. Also, the porosity samples in sand will have no impact on the neighboring porosity simulation in interbedded sand or shale. This is a strong assumption that may have significant effects on the resulting reservoir heterogeneity model.

Soft boundaries relax this assumption in various ways. One method is to keep the global statistics as defined within each region, but to allow neighboring data from other regions to condition across the region boundaries. More complicated methods have been proposed by Larrondo (2004) and Larrondo and Deutsch (2004), which involve mixing of region statistics over a "boundary region."

The region-boundary-type assumption is important. Contact profiles are proposed by Larrondo (2004) and Wilde (2011) as a means to test boundary types directly from well data (see Figure 4.5). Once the regions are defined, the reservoir property of interest is plotted relative to the distance from the boundary. An expectation line or curve may be fit to illustrate the nature of the boundary. Hard boundaries should be represented by a discrete change in the property across the boundary, while soft boundaries should indicate a transitional zone near the boundary.

Hard and soft boundaries may be applied across the bounding surfaces of the domains discussed above. This is less common because domain bounding surfaces commonly represent stratigraphic discontinuities; therefore, hard boundaries are expected. However, soft boundaries may provide methods to improve the model by accounting for information outside the model domain when appropriate. Properties within domains and regions may be further constrained by trend models.

4.1.3 Trend Model Construction

Section 2.4, Preliminary Mapping Concepts, discusses (a) general methods for accounting for trends in input statistics such as directions of continuity, continuous means, and categorical proportions and (b) methods for including trends by decomposition of deterministic trends and stochastic residual. These methods provide the means to impose trends

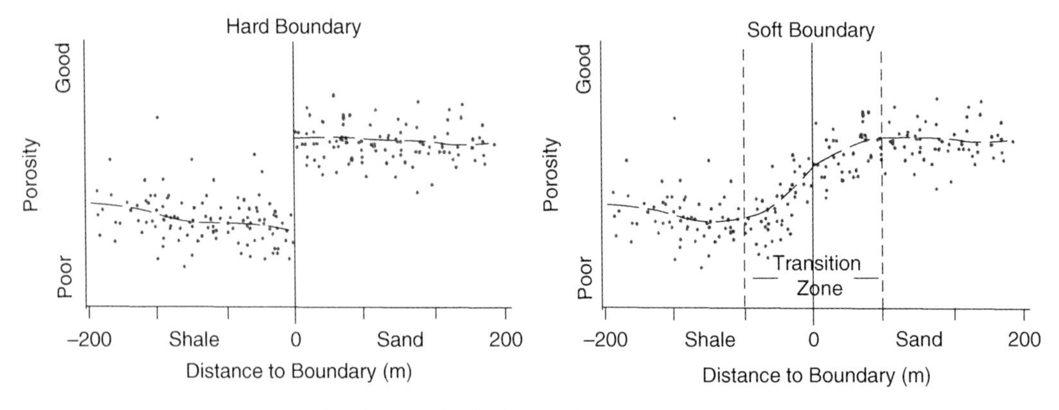

FIGURE 4.5: Example Contact Profiles for Porosity in Shale and Sand Facies. A hard boundary is shown on the left with discrete change in expected porosity at the boundary. A soft boundary is shown on the right with a gradual transition in porosity between shale and sand facies.

within modeling region(s) and within the bounded domain(s), discussed above. In this subsection the mechanics of trend construction are discussed.

General guidance on trend modeling is provided in other geostatistics books, such as Caers (2005), Chilès and Delfiner (2012), Deutsch and Journel (1998), Goovaerts (1997), and Isaaks and Srivastava (1989). Sound geological judgment and the careful consideration of all available information is important. There is no purely objective method for trend modeling; as with all other matters of stationarity, trend modeling is a decision that can be defended, but not tested nor proven.

The separation of deterministic and stochastic variations must be done with care. Overfitting the deterministic component leads to reduced uncertainty and overfit property models with too little variation in heterogeneity. Not fitting deterministic aspects of reservoir property variation leads to property models that do not reproduce real geological trends, ignore important information, and thus may lead to poor reservoir forecasts.

Our understanding of the trend will change with the amount of data we have and can be made more precise with more coverage and higher-resolution data. The scale of categorical proportions or continuous mean variations should generally be significantly larger than the data spacing and be backed up by the geological conceptual model. Any higher-resolution trends, not supported by data, should be applied with care and with strong justification from concepts and analogs.

Philosophically, there is a compromise in trend modeling. It would not be practical to view our reservoir as too complicated and variable for any stationarity assumption, nor would it be reasonable to group complex sets with widely different depositional and diagenetic controls together and assume predictive models will emerge.

The following focuses on methods to build categorical trends, known as locally variable facies proportion models. We focus on the categorical case because it is more commonly applied than the continuous non-stationary mean application, but all the concepts are general and may be extended.

Methods to Build Trends

There are a variety of methods available to construct trend models (Gonzalez et al., 2007; Kedzierski et al., 2008). In general, trends are typically calculated for areal, 2-D, and vertical, 1-D, and then combined into a full 3-D trend model, although at times data density and specific heterogeneity may support the direct calculation of 3-D trends. More information on formulating 3-D trends is provided after this list. The advantage of working with 2-D and 1-D trends is that overfitting is less of a concern.

Hand Mapping

Traditional hand mapping is a valuable method for integrating geological knowledge. These methods are extremely flexible and allow for the integration of any smooth trend features derived from the geologic conceptual model. However, these methods are generally time-consuming and are not generally practical for the direct construction of a full 3-D trend model. Modern methods allow for guided machine contouring that optimizes integration of conceptual information and efficiency. This is the recommended method in sparse data because any of the subsequently described machine fit methods do not perform well with sparse data.

Moving Window

Moving window averaging provides a flexible tool for calculating a trend model. A window size is specified and then scanned over the domain or region with the average of the data within the window assigned to the centroid of the window. This requires some assumption over locations without data (expand the window or assume global statistics) and may result in domain edge artifacts because fewer data are available to average at the edges. Typically, the only parameter available is the window size. Increasing the window size, will result in a smoother trend, while decreasing the window size will result in more local variability.

Manchuk and Deutsch (2011) have suggested a more flexible trend modeling approach including window anisotropy and nonuniform weights. Window anisotropy may be imposed based on a spatial continuity and possible anisotropy of the property of interest and data density in each direction (See Figure 4.6). For example, in typical well-derived data sets the vertical (or near vertical) sampling rate is much higher than the areal sampling rate. A uniform moving window would average out potentially important vertical information.

Gaussian weighting may be applied as an alternative to the typical uniform weighting. Uniform weighting results in an arithmetic average that has the advantage that it does not introduce information

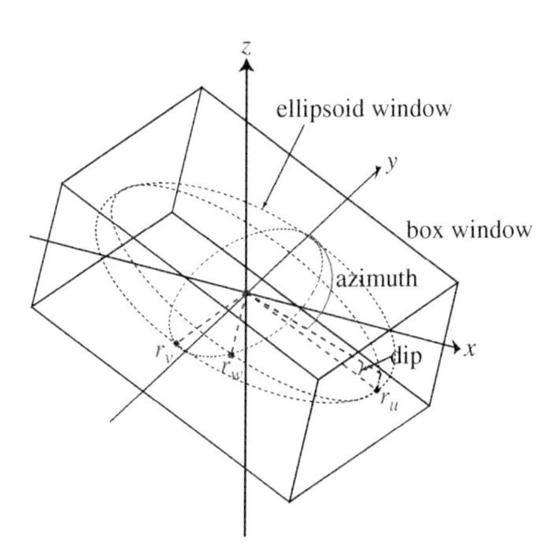

FIGURE 4.6: Box and Ellipsoid Averaging Windows and Associated Parameters (Manchuk and Deutsch, 2011).

to the trend that is not necessarily presented in the data. With no additional information, the assumption that all data be treated equal is unbiased. However, equal weighting has the disadvantage that it can lead to discontinuous trends, which appear as artifacts. Alternatively, a Gaussian weighting function calculates weights using a function similar to the Gaussian distribution function:

$$w_i = \exp\left(\frac{-9R_i^2}{r^2}\right)$$

where R_i geometric distance (accounting for anisotropy in the manner discussed in Section 2.3, Quantifying Spatial Correlation) is the distance from $\mathbf{u}_\alpha$ to the ith datum in the window. Using Gaussian weights assumes that data that are closer to $\mathbf{u}_\alpha$ are more important than those further away, which is often the case. If the window is defined explicitly, the r is an ultimate anisotropic range. If windows are defined implicitly, r is defined as the maximum distance observed among the n points found near $\mathbf{u}_\alpha$. With either implicit or explicit windows, the choice of r ensures a continuous trend even when a small shift in the window leads to an exchange of data. Examples of both uniform and Gaussian weights are shown in Figures 4.7 and 4.8, respectively.

Inverse Distance

Inverse distance calculates a local trend value based on the inverse distance weighted average of the data. A power is applied to the distance to account for the penalty of distance away from the estimate location. By increasing the power, the trend becomes more specific (data further away receive less weight); and by decreasing the power, the trend becomes more general. Anisotropy may be imposed through a geometric transform of the distance (same as that explained in Section 2.3, Quantifying Spatial Correlation). A constant should be added to the distance prior to inverse distance weighting to avoid overfitting and to enhance the smoothing of the trend.

Kriging

Kriging as discussed in Section 2.4, Preliminary Mapping Concepts, provides an efficient method to calculate estimates accounting spatial continuity. The user may vary the input variogram model to directly impact the generality of the trend model. A short continuity range results in a trend model that approaches the global mean more quickly away from

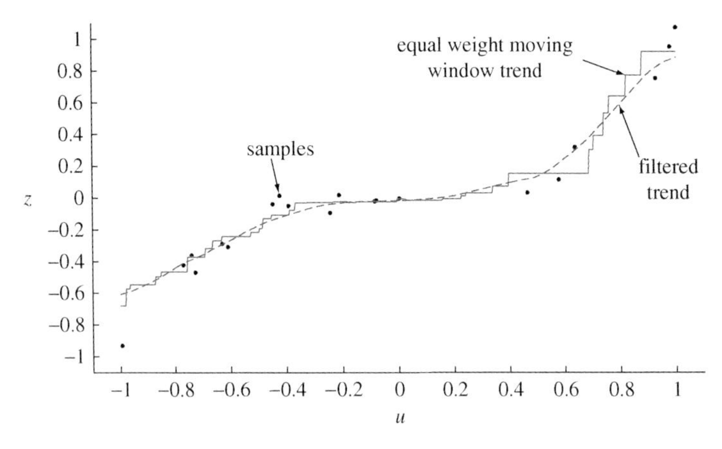

FIGURE 4.7: Example of a Discontinuous Trend Resulting from Equal Weighted Moving Window Averaging. A window radius of 0.24 units was used. Discontinuities can be filtered out (Manchuk and Deutsch, 2011).

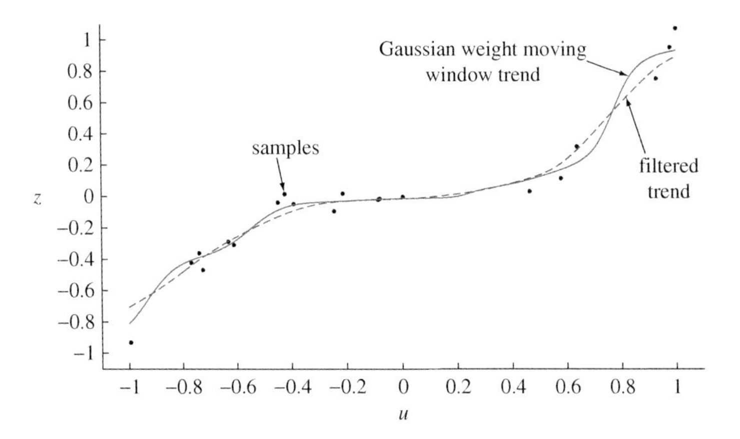

FIGURE 4.8: Example of a Continuous Trend Resulting from Gaussian-Weighted Moving Window Averaging. A window radius of 0.5 units was used. Filtering can be used to smooth the result (Manchuk and Deutsch, 2011).

data. In general, block kriging should be considered with a variogram that has a reasonably high nugget effect (30–40% relative nugget effect). A large search should also be considered to avoid artifacts.

Building a Deterministic Trend Model

For any of the previously mentioned methods it is essential to avoid the tendency to overestimate the spatial variability and to ensure that the trend is not overfit. These concepts are related to each other; both result in too much variance described by the trend and too little remaining to be modeled stochastically as a residual. Trends should generally be smooth features that can be reasonably informed with local information and geological concepts. Also, the trend should not exactly reproduce the data. This would result in no residual for residual statistical analysis.

To prevent trend exactitude for trend applications with inverse distance, the data should be averaged first (e.g., vertical averaging for an areal trend or areal averaging for a vertical trend) or a constant should be added to the distances; with kriging-based trends the data may be similarly averaged, or blocked ordinary kriging with grid cell discretized may be applied. To ensure smoothness, the moving window should be reasonably large, inverse distance weighting is implemented with lower than normal inverse distance powers, and block kriging may be applied with a nugget effect.

In practice, we consider the vertical trend first and then map areal trends. The first step in vertical trend analysis is visual inspection of the well data. Most trends will be known geologically before any geostatistical modeling is considered. A vertical trend plot for continuous variables or vertical proportion curve for facies should be constructed to look for and quantify vertical trends. Figure 4.9 shows an example of each.

There are a number of comments to be made regarding the continuous trend curve on the left side of Figure 4.9. A vertical trend does not need to be considered in modeling when all points fall close to the same average or there is considerable overlap in the simplified "box plots." A trend should be considered when there are distinct features on this plot. These points can be used directly for the trend; that is, the trend or mean is considered constant within each stratigraphic interval equal to the average of the data in the interval. It takes more time to fit a polynomial model and check that the fit is reasonable; however, it may be worthwhile to avoid discontinuity artifacts at the boundaries of the chosen intervals.

Similar comments can be made about the facies proportion curve on the right side of Figure 4.9. Different techniques are used to account for trends in continuous and categorical variables (see below).

Consider the sketch in Figure 4.10. If the wells are relatively far apart—that is, distances *d*1 and *d*2 are large with respect to the horizontal range of correlation–then locations *A*, *B*, and *C* will have the same average over many realizations. This would be wrong; the average at *A* should be more than *C* even when there are no nearby well data. Fitting the vertical trend and working with residuals enforces this trend.

Areal trends are usually modeled separately from vertical trends. It is hard to visualize and fit a 3-D trend model all at once, and there is a danger of overfitting; it is just easier to fit the vertical trends

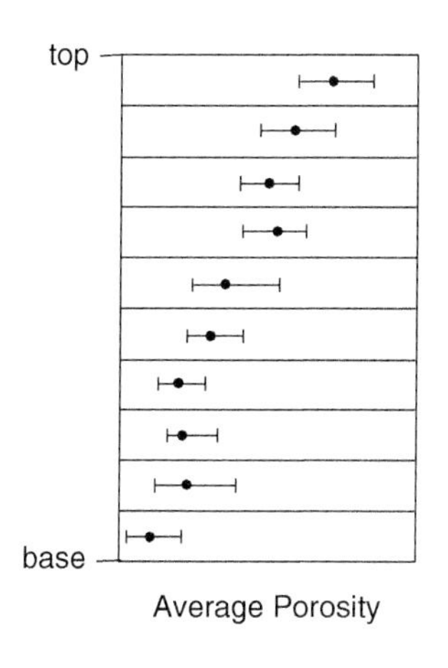

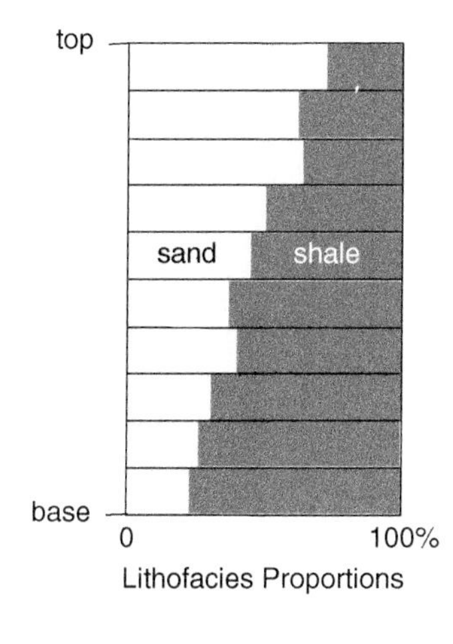

FIGURE 4.9: Example of Vertical Trends within a Stratigraphic Layer. The layer is partitioned into equal stratigraphic intervals (10 in this case); the average of the continuous property is calculated, as is the proportion of different facies. The black dots are the average of all porosity measurements falling in that stratigraphic interval. The attached lines are the 80% (or some other arbitrary) probability interval of the values falling in the interval.

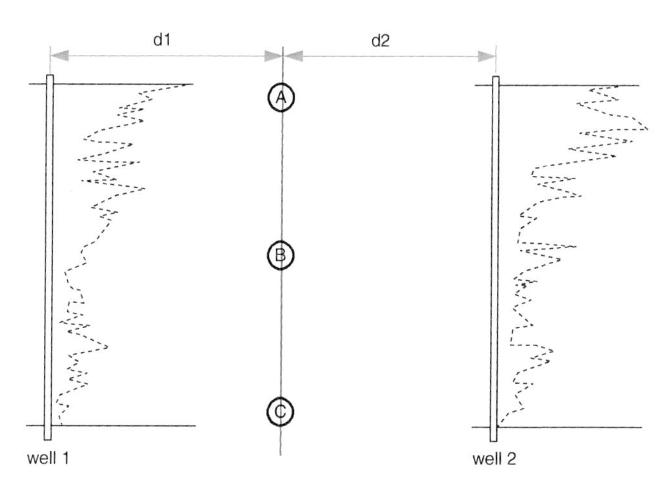

FIGURE 4.10: Illustration of Two Wells Displaying an Obvious Vertical Trend and a Location between Them. The expected average or mean values at the three locations A, B, and C should be different; however, they will be the same if d1 and d2 are beyond the range of correlation from the wells and the trend is not modeled explicitly.

and then areal trends. Of course, the areal and vertical trends must be reconciled into a consistent 3-D trend model prior to any geostatistical modeling (see below).

For facies indicator variables, the proportion of each facies is calculated over the vertical extent of the stratigraphic layer. For continuous variables, the average of the variable (within each facies) is calculated over the vertical extent of the stratigraphic layer. These values are displayed on a 2-D map and contoured by hand or with the computer. Figure 4.11 gives an illustration of two maps. The trend on the left should be used. The trend on the right should not. We look for trends or variations at scales larger than the data spacing. In general, when there is doubt as to the need for a trend model, we should not model the trend. Models constructed without a trend could be reviewed to confirm that the large-scale features are reproduced acceptably.

The areal trend is modeled by a gridded map at the same grid spacing as the geostatistical facies and property models to be built later. The areal trend map $m(x, y)$ and vertical trend curve $m(z)$ must now be merged to a 3-D trend model.

There is no unique way to merge a trend map with a trend curve. Ideally, additional geological knowledge should be brought to bear to determine the best approach. A simple and practical approach is to scale the vertical curve to the correct areal average:

$$m(x, y, z) = m(z) \cdot \frac{m(x, y)}{\overline{m}} \tag{4.1}$$

where $m(x, y, z)$ is the final 3-D trend values, $m(z)$ is the vertical trend for the z location, $m(x, y)$ is the areal trend value for this x, y location, and $\overline{m}$ is the global average of the variable being considered. This approach amounts to scale the vertical trend curve to the correct areal average. "The resulting 3D trend model, m(x,y,z), should be checked for nonphysical results." One possible problem is that high values are too high and low values are too low. Upper and lower limits to $m(x, y, z)$ can be imposed.

An alternative to Eq. (4.1) is to average the trend values—that is, $m(x, y, z) = (m(z) + m(x, y))/2$—but this often smoothes the trends so much that neither can be reproduced in subsequent modeling. Hong and Deutsch (2009b) suggest a combination method for categorical proportions models based on weighted probability combination schemes.

$$p_k(x, y, z) = \left(p_k \frac{p_k(x, y)}{p_k}\right)^{w_1} \left(\frac{p_k(\mathbf{z})}{p_k}\right)^{w_2} \in [0, 1]$$

where w_1 and w_2 are weights imposed on each proportion ratio to constrain the relative importance of areal and vertical components. The weighted model reverts to the conditional independence model with letting $w_1 = w_2 = 1$. The proportion ratio has a minimum bound of zero, but theoretically it has no maximum bound. Closure is enforced on the resulting trend model by normalizing the proportions to sum to 1.0.

Figure 4.12 illustrates a trend model of the logarithm of permeability (from a real reservoir study). The vertical trend is very pronounced. Figure 4.12 shows two cross sections through a 3-D trend model for the entire reservoir layer.

These procedures lead to a field-wide deterministic trend model for all x, y, z locations. The procedures for locally variable proportion models are the same. We now consider how to use the trend models. There are three approaches for continuous variables: (1) decomposition into a mean and residual, (2) treating the trend as a secondary variable, and (3) modeling the variable of interest conditional to the trend. These approaches are discussed first and then categorical variables are discussed.

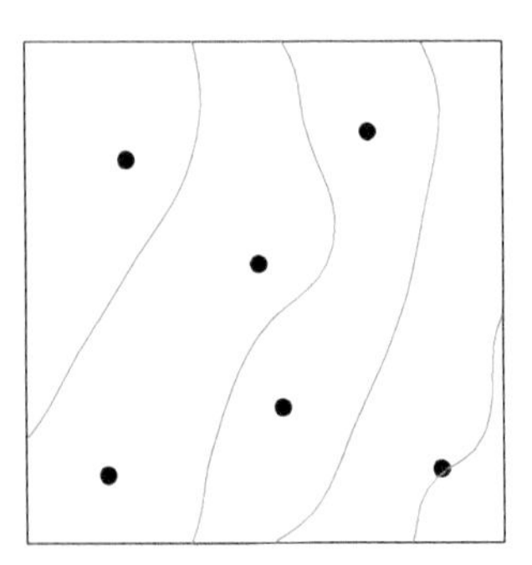

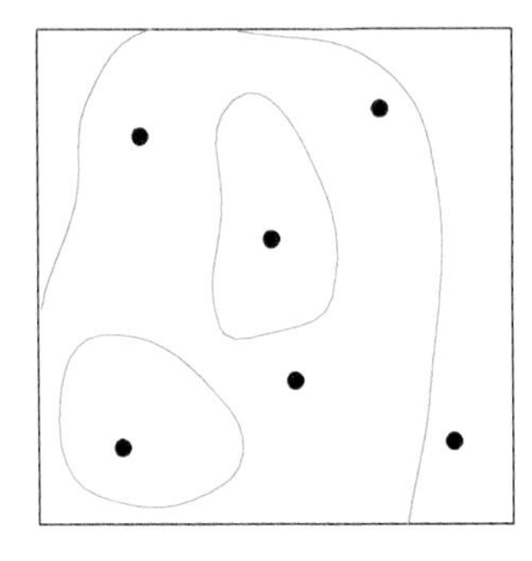

FIGURE 4.11: Illustration of Two Maps for Areal Trends. The example on the left shows a trend at a scale much larger than the data spacing. The example on the right is not so clear. We may choose not to model areal trends in this case (right side).

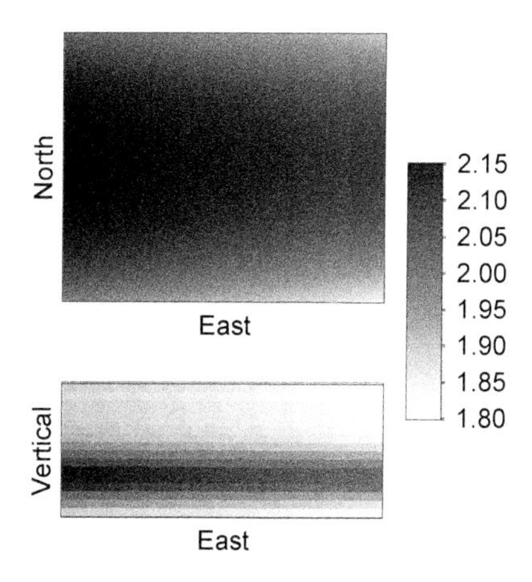

FIGURE 4.12: Horizontal and Vertical Cross Section through a 3-D Trend Cube Constructed for the Log Permeability within a Reservoir Layer (Stratigraphic Vertical Coordinate). Note the high-permeability layer and the decrease in permeability to the east of the reservoir.

Decomposition into Mean and Residual

The idea is to decompose the original continuous Z variable (porosity or permeability) into a locally varying mean and a residual:

$$Z(\mathbf{u}) = m(\mathbf{u}) + r(\mathbf{u})$$

The locally varying mean $m(\mathbf{u})$ or $m(x, y, z)$ is defined at all locations over the entire reservoir. The residual $r(\mathbf{u})$ is only defined at data locations. The residual may be estimated at all locations and added to the mean for a numerical model; however, it is more appropriate to simulate multiple realizations of the residual and add them to the mean to create multiple realizations of the underlying continuous variable.

The arbitrary separation of the data into trend and residual can lead to artifact nonphysical Z values, that is, negative porosities resulting from large negative residuals added to low mean values or too-high values resulting from large positive residuals and high mean values. Most artifacts are a consequence of modeling the residuals independently from the prior mean.

It is good practice to calculate the variance of the z values, the m values, and the r values. By definition of the variance and covariance (see Chapter 2) we know that

$$\sigma_Z^2 = \sigma_M^2 + \sigma_R^2 + 2C_{\text{R-M}}(0)$$

where σ_Z^2, σ_M^2, and σ_R^2 are the variances of the z, m, and r values and $C_{\text{R-M}}(0)$ is the covariance between m and r. $C_{\text{R-M}}(0)$ can be negative or positive. If it is relatively close to zero, then there should be few artifacts of considering them independently. The importance of the trend can be determined by the ratio σ_M^2/σ_Z^2, which is the fraction of variability explained by the trend or local mean values.

If the relation between R and M is significant—that is, $C_{\text{R-M}}(0)/\sigma_Z^2 > 0.15$ (in standardized units)—then artifacts are likely to be significant and the remnant correlation between the mean and residual must be accounted for.

The simple kriging formalism presented in Section 2.4 implicitly assumes that a locally variable mean $m(\mathbf{u})$ is available at all locations.

Secondary Variable

The central idea here is to treat the locally varying mean $m(\mathbf{u})$ as a secondary data variable $Z_2(\mathbf{u})$ to be used in cokriging or cosimulation of the residual $r(\mathbf{u})$ or the original data variable $z(\mathbf{u})$. Considering the Z variable as the primary variable has the advantage of direct enforcement of the Z histogram and variogram.

The local mean or secondary variable $Z_2(\mathbf{u})$ is available at all locations by mapping (see Section 2.4). A conventional cokriging formalism would require a linear model of coregionalization (LMC) between the primary variable and the $Z_2(\mathbf{u})$ variable. The variogram of $Z_2(\mathbf{u})$ is likely to be very smooth due to the origin of this variable. The variogram of Z and R will not be as smooth. The LMC will require a mixture of nested variogram structures: some with great continuity (Gaussian) and some with less (spherical).

Since the secondary variable (the local mean) is available at all locations, a collocated alternative could be considered. As described in Section 2.4, this requires only the correlation coefficient between the primary and secondary variables. The smoothness of the secondary variable, however, may cause trouble in collocated cokriging.

Conditional to Trend

The problems associated with trend modeling result from modeling of a separation of trend and residual and failure to account for the relationship between the trend and residual, most importantly heteroscedasticity. Deutsch (2011) suggested the method of working directly with the property of interest (no residual) and modeling the property conditional to the trend model. In this framework we construct a model.

$$F(Z|t_i < T < t_{i+1})$$

for $i = 1, \ldots, n-1$, where n are truncations of the trend model. This modeling and method is equivalent to conditional cloud transform simulation described in Section 2.4, Preliminary Mapping Concepts. While this method reproduces the relationships between trend and variable of interest and because the variable of interest is modeled directly, this method prevents unrealistic values, but there may be issues related to histogram and variogram reproduction common to conditional cloud transform simulation.

Decomposition into mean and residual is, perhaps, the simplest approach unless significant problems arise in reproducing the histogram of the original Z values with simulation.

The way that the trend models enter subsequent modeling depends on the type of facies modeling adopted. Sections 4.2 and 4.3 discuss cell-based modeling. Section 4.4 discusses object-based modeling and Section 4.5 discusses process mimicking methods. The specific approach to use these proportion values will be discussed in these sections.

Categorical Variables

Locally variable facies proportions can be used with the indicator transformation of categorical variables presented in Section 2.3. Trend models in average facies proportions are geologically reasonable and expected. There are both vertical and areal trends. These trend models are constructed in the same way as for continuous variables.

Only one trend model is needed when there are two facies types; the proportion or probability of the second facies is one minus the proportion of the first. When there are three or more $(K > 2)$ facies, the sum of the proportions should equal one. A restandardization step may be required:

$$p_k^* = \frac{p_k}{\sum_{k'=1}^{K} p_{k'}}$$

The way that the trend models enter subsequent modeling depends on the type of facies modeling adopted (i.e. cell-based modeling, object-based, or process mimicking). The specific approach to use these proportion values will be discussed in their respective sections.

Building an Uncertainty Model for Trend

If there is significant uncertainty in the trend model, a choice may be made to not trend model or, when possible, to model this uncertainty so that it may be carried through the modeling work flow. This may be accomplished be retaining multiple trend scenarios, based on multiple interpretations. If the trend model has been calculated with a machine method, then the trend parameters such as window size or variogram model may be changed to vary the trend fit. It may be more meaningful to locally or globally shift the trend to account for uncertainty. This would require direct editing of the trend model which is available in most reservoir modeling software packages and geological and data justification for these edits. Given the large degree of subjectivity in trend modeling, uncertainty in trend models is best assessed with expert judgment.

Trend Correction

There may be a need to correct a trend model. The following discussion considers constraints and global proportions. Care should be taken to correct trends while preserving the trend model's spatial features.

There may be constraints that must be imposed on a trend model. A common example is the requirement for locally variable proportions to sum to 1.0 at each location, also known as *closure*. Only one trend model is needed when there are two facies types; the proportion or probability of the second facies is one minus the proportion of the first. Thus, closure is only an issue for three or more categories. Some methods such as moving window trend with uniform weights should automatically honor this constraint, yet for other trend calculation methods this is not guaranteed. Typical practice is to locally standardize categorical trend models to ensure closure.

$$p^{\text{stand}}(\mathbf{u}_\alpha; k) = \frac{p^{\text{stand}}(\mathbf{u}_\alpha; k)}{\sum_{k=1}^{k} p^{\text{stand}}(\mathbf{u}_\alpha; k)}$$
$$\forall \mathbf{u} \in A, \quad k = 1, \ldots, K$$

In addition, the trend values may exceed known constraints—for example, a negative porosity. Firstly, careful trend calculation, not overfitting the trend model and avoiding excessive extrapolation, should minimize these issues. If issues remain, as a last resort, truncation of the trend model with local smoothing may be required.

A trend model should honor the global categorical proportions or continuous mean. If the trend model deviates significantly, then the resulting model may be biased by the trend model. A simple global shift of the trend model may be applied for small changes.

$$p^{\text{cor}}(\mathbf{u}_\alpha; k) = p(\mathbf{u}_\alpha; k) + (p^{\text{target}}(k) - p^{\text{current}}(k)), \quad \forall \mathbf{u} \in A, k = 1, \ldots, K$$

where $p^{\text{cor}}(\mathbf{u}_\alpha; k)$ is the corrected local proportion, $p(\mathbf{u}_\alpha; k)$ is the initial biased trend model, $p^{\text{current}}(k)$ is the initial global proportion, and $p^{\text{target}}(k))$ is the target global proportion for category k. Of course, this correction is somewhat dangerous because this systematic shift may result in implausible local values and will disrupt closure and require a closure correction. Care should be taken to ensure a close match of global proportions during trend calculation, and the results should be checked after any correction.

Trend Goodness

No objective criterion has been developed to judge the appropriateness of a trend model. Yet, there are some metrics to check for unreasonable trend features. These criteria include the following: The trend should match the global mean or proportions (mentioned above); and when a residual is applied, the residual should have a mean of zero and the trend and residual should not be correlated. If the trend varies from the global input statistics or the mean of the residual is not zero, then the trend is likely to introduce bias during simulation. If correlation between trend and residual exists, this will likely introduce artifacts as the decomposition of trend, and the residual assumes independence.

Given that these metrics have been met, the trend may be checked with respect to performance of the resulting simulated realizations. Leuangthong et al. (2004) have proposed a set of minimum acceptance criteria checks to be applied to simulated realizations, including reproduction of global distributions and variograms (see Section 5.1, Model Checking). This is important, because the decomposition of trend and residual can result in extreme values that disrupt these statistics. Hong and Deutsch (2009c) applied this to trends and suggested that the trend performance of categorical variable may be checked by high correlation between predicted and true facies. Deutsch and Journel (1992) and McLennan (2008) have suggested a method for checking the fairness of the trend model based on the expectation of the data matching the trend.

$$\text{Actual fraction} = E[I(\mathbf{u}_\alpha; k) \mid p(\mathbf{u}_\alpha; k) = p], \quad \forall p, k = 1, \ldots, K$$

where $I(\mathbf{u}_\alpha)$ is the indicator of a category at location $\mathbf{u}_\alpha$ and $p(\mathbf{u}_\alpha; k)$ is the collocated trend proportion. Given the continuous nature of trend model probabilities, a binning approach is suggested. Trend model probabilities at each location, $\mathbf{u}$, are binned into K classes. For each of these classes, the proportion, p, of data at the trend model locations with an indicator I_k is determined. The actual proportion of data with an indicator k in each trend bin can be cross plotted against the expected fraction from the trend model (see Figure 4.13).

The points shown in Figure 4.13 should lie on the 45° line within tolerances if the trend model is completely fair. The tolerances depends on the number of data used for each point. The trend model proportions follow a multinomial distribution, so classical statistics can be used to determine a 99% confidence interval. The probability of observing an actual proportion p based on n data points, given that the true trend model proportion is p_k, is given by the binomial probability distribution (Montgomery and Runger, 2007):

$$\text{Prob}(P = p) = \binom{n}{\text{np}} \frac{p^{\text{np}}}{n} (1 - p_k)^{n(1-p)}$$

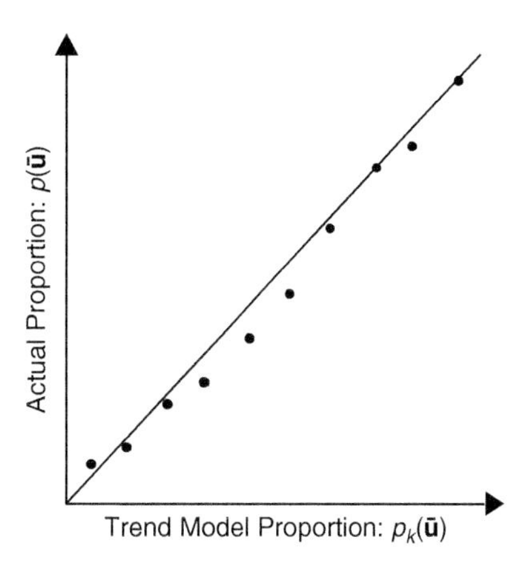

FIGURE 4.13: Fairness Plot for a Categorical Trend Model. On the line indicates a fair trend model, above the line suggests the trend model is too general, and below the line suggests that the trend model is overfit/too specific (Deutsch and Deutsch, 2009).

This can be approximated using the Gaussian distribution with $B(n, p_k) \approx N(n_{pk}, n_{pk}(1 - p_k))$. The cumulative probabilities (0.005 and 0.995) necessary for constructing a 99% confidence interval can be calculated using the binomial probability distribution or, where appropriate, the Gaussian approximation (Deutsch and Deutsch, 2009).

Other Trend Considerations

Once again, it is important to balance variability between the deterministic trend and the stochastic residual. We should not overfit the data, because this will result in a very small stochastic residual and an unfairly small uncertainty model. Conversely, we should not ignore available data on trends supported by data and geological concepts because this may result in an unfairly large uncertainty model. Some trend calculation methods have parameters that directly constrain the level of fit. For others, filters may be applied to smooth trend model results.

Extrapolation is always challenging. Care must be taken to ensure a defendable assumption for trend modeling outside the data and that the distance of extrapolation is reasonable. One model is to assume a constant trend away from the peripheral data. Another model would be a trend extrapolation with the trend values increasing or decreasing with distance away from the data. The latter requires strong support and may result in unreasonable trend values. Moderate extrapolation is unavoidable because the margins of the volume of interest are rarely sampled, but care should be taken with extreme extrapolation.

4.1.4 Multivariate Mapping

Conventional geostatistical techniques have been designed to simulate models of heterogeneity and uncertainty in static rock properties. This is appropriate for input to a transfer function. There are times, especially at exploration scales (and large-scale reservoir framework modeling), when the goal is data integration and uncertainty assessment and detailed realizations are not necessarily required (refer to Section 3.3, Problem Formulation, for these and other work flows). In fact, these models are often 2-D uncertainty maps that integrate all available dense local information and sparsely sampled hard data to aid exploration decisions [for example, Ren et al. (2008)]. Such models may be useful in developing the large-scale framework—for example, defining domains, regions, and trends.

In these settings there may be many different variables: direct measurement from well logs, well cores and field surveys, large-scale remotely sensed variables, interpreted trend-like variables, and other response variables. This problem context is referred to as *massive multivariate*, with typically 10–30 or even more variables of interest. These data often have distinct coverage, scales, and vintages and are variably correlated together and redundant.

Among all the available variables, it is common to have two types of data variables. *Primary* data variables are to be predicted with uncertainty and are available at relatively few locations (i.e., reservoir properties measured directly at well locations from core and/or logs). In Section 3.1, Data Inventory, these are identified as primary hard data and are denoted as y_p, for $y_{p,i}, i = 1, \ldots, N_p$ primary variables. *Secondary* data variables aid with the prediction of the primary variables and are often available more extensively. Examples of secondary variables include geophysical measurements and geological trend and conceptual maps, these may include various derivative maps such as inferred isopachs, distance from source, water depth during deposition, and so on. Consider N_s secondary data variables $(y_{s,i}, i = 1, \ldots, N_s)$. The location is often denoted as $\mathbf{u}_\alpha$, where α is the data index.

Statistical techniques such as principal components, factor analysis, ACE, and cluster analysis could be used to summarize the relationships between the variables, but they do not account for spatial correlation. Conventional geostatistical techniques incorporate the spatial structure, but these techniques are cumbersome in the presence of many secondary variables.

There are a number of geostatistical techniques designed to work with multiple variables (see Section 2.4, Preliminary Mapping Concepts). These techniques account for the spatial relationships between the variables and provide a measure of uncertainty at every estimated location. The main technique is cokriging that can be applied in a multivariate Gaussian or an indicator framework. There are simplifying assumptions such as collocated cokriging and the Markov–Bayes approach. A concern with all these techniques is the inference of the direct and cross variogram measures of correlation, which requires a large number of data. For K variables, they require a total of $(K + 1)K/2$ variogram models, which is difficult in practice. Automatic fitting algorithms have helped; however, the problem of inference remains when we have $K = 10$ or more variables with relatively few data.

Collocated cokriging, in the Gaussian or Bayesian form, simplifies the process to consider only the collocated secondary variables. This also removes the need to model the large number of variograms mentioned above. There are implementation problems associated with this simplification such as variance inflation, but the method has proved very practical. These geostatistical methods for considering multiple variables really only consider 1 to 3 secondary variables; there is no simple way to consider 10 to 30 secondary variables simultaneously common in massive multivariate settings. We must tailor the multivariate statistical and geostatistical tools to the problem of a large number of variables.

Deutsch et al. (2005) propose a work flow for these massive multivariate settings that proceeds as follows;

1. Univariate transformations to Gaussian and assumption of multivariate Gaussianity.
   - Merge secondary data into local likelihood distributions associated with each primary variable with the normal equations.
   - Map each primary data to form a prior distribution from local data only.
   - Calculate a posterior distribution from the prior and likelihood distributions for each primary data.
2. Assess the joint uncertainty with a combined LU/p-field approach/draw from local distributions with correct spatial and multivariate structure.
   - Simulate from these local distributions with P-field simulation to ensure the correct spatial continuity.
   - Apply LU simulation to impose the correct correlation structure between all variables simulated at each location.
3. Back transform from Gaussian back to original data space.

The result of this work flow is a joint uncertainty model in the form of a set of realizations for each primary data at all locations in the volume of interest that integrate all available primary and secondary information and honor the multivariate and spatial relations. Note that the use of the words Prior, Likelihood and Posterior in this context is somewhat arbitrary. We are, in essence, computing two conditional distributions, then merging them into a distribution that is conditional to both data sources.

Likelihood Functions from Secondary Data

All secondary variables are merged into a single *likelihood* distribution for each primary variable at each location. Of course, the number of secondary variables available at each location could vary; the notation $N_S(\mathbf{u})$ denotes the number of secondary data available at location $\mathbf{u}$.

The mean and variance of the likelihood distribution are calculated directly from the normal equations or simple cokriging system of equations discussed in Section 2.4, Preliminary Mapping Concepts. Redundancy between multiple variables or multiple data at different locations is a critical concept in estimation. The normal equations (kriging) resolve redundancy with the correlation between the variables and what we are predicting in an optimal fashion. The system of equations to calculate the weights and subsequently the kriging estimate and kriging variance has been described in Section 2.4, Preliminary Mapping Concepts. Note that the correlations of the secondary data with themselves are on the left-hand side and that the correlation of each secondary variable with the specific primary variable are on the right-hand side.

The correlation coefficients between all pairs of secondary data and all secondary and primary data are required. These should be calculated from the Gaussian-transformed primary and secondary data. The likelihood distributions (denoted as "like") are summarized by a set of mean and variance values for all locations and primary variables:

$$\bar{y}_{\text{like},p}(\mathbf{u}), \sigma^2_{\text{like},p}(\mathbf{u}), \qquad p = 1, \ldots, N_p, \forall \mathbf{u} \in A$$

These distributions are a "collapsed" version of all available secondary variables at location $\mathbf{u}$. The final likelihood distributions account for the correlations and redundancy between the secondary variables and the correlation between the secondary data and the primary variable of interest.

The resulting weights can become unstable and nonintuitive, recognized as large positive or negative weights. Manchuk and Deutsch (2005) suggested sequential removal of the most redundant secondary variable or the secondary variable least correlated with the primary variable. They proposed a metric to measure the importance of each secondary variable for sensitivity analysis and also to ensure that important variables are not removed to improve solution stability. Their method is to sequentially remove variables and calculate the importance metric based on the change in the kriging variance.

$$\text{Importance} = \frac{\sigma_{k,i}^2 - \sigma_k^2}{\sigma_k^2}$$

where i is the variable removed, $\sigma_{k,i}^2$ is the kriging variance for that configuration, and σ_k^2 is the kriging variance when all variables are considered. This relation was developed with the premise that the more a variable contributes to reducing the kriging variance, the more important it is. To gain insight into the mutual interactions of all secondary variables, the full combinatorial is explored—for example, removal of variable 1 in the presence of all other variables, all variables except for variable 2, . . ., only variable 2, and all other possible combinations.

In addition, solution instability may be caused by differences in scale between secondary data and primary data and the use of exhaustive secondary data for the direct secondary data correlations and scaled primary data for the direct primary and cross primary and secondary correlations. This may not result in a positive definite system of equations. Post-processing the system for positive definiteness will improve stability (Deutsch, 2011).

Given a stable solution to the normal equations, the result is an estimate of the likelihood distribution at each location based only on the collocated secondary information; beyond this, no spatial information is accounted for; these values summarize all of the information available in the secondary data related to the primary variables of interest. The spatial information for the primary variable comes in through the prior distributions.

Prior Distributions from Local Data

The distribution of uncertainty in each primary variable is predicted from surrounding data (in a spatial sense) using simple kriging. These estimates are called prior distributions and are denoted with by "prior." Similar to the likelihood distribution, the parameters of the prior distribution are obtained from the normal equation, this time with correlations representing the closeness and redundancy of primary data at other spatial locations (i.e., standard kriging).

The correlation coefficients between all primary data at different locations come directly from a semivariogram or correlogram model. The prior distributions are summarized by a set of mean and variance values for all locations and primary variables:

$$\bar{y}_{\text{prior},p}(\mathbf{u}), \sigma^2_{\text{prior},p}(\mathbf{u}), \qquad p = 1, \ldots, N_p, \quad \forall \mathbf{u} \in A$$

These distributions summarize the spatial information of surrounding data of the same variable type. The likelihood and prior distributions are then combined to get the final updated distribution.

Bayesian Updating

Since the two input distributions are Gaussian in shape, the resulting updated distribution will also be Gaussian. The updated distribution is defined by the updated mean and variance, $\bar{y}_{U,p}(\mathbf{u})$ and $\sigma^2_{U,p}(\mathbf{u})$ respectively. This may be accomplished by utilizing the simplified form for Bayesian updating for Gaussian distributions presented in Section 2.2, Preliminary Statistical Concepts.

The updated distributions defined above must be back-transformed to return the primary variables to their original distributions. The elements of this technique are not new; however, this is a novel way of putting everything together for reliable and simple estimation. A Markov screening assumption is made whereby collocated secondary data screen the influence of nearby secondary data. There is a further assumption that primary data of different types at different locations are also screened. The consequences of these assumptions are not considered severe in most cases. Full cokriging could be implemented to judge their importance.

The percentiles, or arbitrary number of quantiles, could be back-transformed from the local updated distributions for each primary variable. Any summary statistics of the local distributions could then be calculated including the expected value, the local variance, P10, P50, and P90 values, and so on. These summaries could be used to characterize local uncertainty and to assist with well placement and data collection decisions. Local uncertainty in each of the N_p variables at each location $\forall \mathbf{u} \in A$ does not permit multivariate calculations or uncertainty over larger volumes. The local uncertainty models are for each location and each primary variable separately. A simulation approach is required to provide a joint uncertainty model accounting for multivariate and spatial relationships.

Joint Uncertainty

We are often interested in derived variables such as economic value or net calculations. Multiple variables must be combined together. The distributions of uncertainty in the input variables can sometimes be combined analytically, but only when the calculations are simple. In general, a simulation approach is required. Multiple realizations are drawn, each realization is processed to establish the derived variables, and distributions of uncertainty in the derived variables are assembled.

p-field simulation and LU simulation are convenient methods to impose spatial and multivariate correlation, respectively. p-field simulation can be applied to directly sample from the updated local distributions of uncertainty for the primary data while imposing the correct spatial continuity, modeled from the primary data samples and other analog information.

Subsequently LU simulation is applied locally to impose the correct multivariate structure between the primary variable realizations at each location $\mathbf{u}$. LU simulation is selected as a fast simulation method for small systems, as is the case with the local multivariate problem with N_P variables. This is accomplished by the multiplication of the p-field simulated values for each primary variable at each location with the lower triangular matrix, $\mathbf{L}$ of the multivariate correlation matrix $y_p^{\text{LU}}(\mathbf{u}) = \mathbf{L}\mathbf{y}^{p\text{-field}}(\mathbf{u})$. The results from LU simulation are locally standardized to match the local updated mean and variance.

$$Y_{c,p}(\mathbf{u}) = y_p^{\text{LU}}(\mathbf{u}) \cdot \sigma_{U,p}(\mathbf{u}) + \bar{y}_{U,p}(\mathbf{u}),$$
$$p = 1, \ldots, N_p, \forall \mathbf{u} \in A$$

Prediction of uncertainty with multiple primary and secondary variables is an important area of geostatistics. Bayesian updating under a multivariate Gaussian model provides a simple and robust solution to this inference problem. LU and p-field simulation permit calculation of complex derived variables and uncertainty over large areas. This procedure may appear like a hodgepodge of techniques. Each constituent technique is required for a specific purpose of data integration or accounting for multivariate or spatial structure. Simpler techniques would necessarily leave out some aspect of data structure.

Assumptions and Limitations

There are, of course, limitations and assumptions such as representative data, statistical homogeneity, and multivariate Gaussianity. This methodology is Gaussian; that is, all data variables must be transformed to univariate Gaussian distributions prior to analysis, and results must be back-transformed. A parametric distribution model for each variable or a nonparametric normal-scores transformation could be used. Care should be taken to decluster/debias the original data histograms. The variables are assumed to be multivariate Gaussian after univariate transformation of each variable. A check should be performed for clear indications of non-Gaussian behavior: nonlinear relationships, proportional effect (dependency of the variance on the mean), and multivariate constraints such as constant sum constraint of saturations. If this is the case, a special transformation may need to be considered. Log-ratios and the stepwise conditional transformation are two alternatives (see Section 2.4, Preliminary Mapping Concepts). Details of these assumptions are available in Deutsch and Zanon (2007).

While this method will integrate the spatial and multivariate structures, it is possible that this additional integration of multivariate relationships at each location may disrupt spatial heterogeneities that may be important to some transfer functions. For this reason, this method is preferred for settings that are focused on large-scale mapping of a joint uncertainty model, as opposed to cosimulation of reservoir scale realizations to apply to typical flow-based transfer functions.

4.1.5 Summarization and Visualization

Framework modeling can benefit from some specialized summarization and visualization techniques, due to the need to check and communicate associated decisions and given their first-order impact on the subsequent reservoir models. Also, the explicit data integration in multivariate mapping results in unique opportunities to jointly visualize the data sources and their relationships.

With regard to domains or regions, various methods may be applied. For example, statistics by domain or region may be visualized and checked against global input statistics. For example, porosity distributions for each region in each domain may be compared to pooled global porosity distribution over all domains and region or any associated subset. Statistical comparisons or multivariate visualizations may be applied to demonstrate the significant difference in reservoir properties.

Domain top and base surfaces and isopach maps can be utilized to check and communicate volumetric changes with bounding surface, region, and trend uncertainty and to check against the structural model in the conceptual model and any structural constraints mapped from seismic data. This is essential because mistakes at this stage will likely have a significant impact on volumetrics and even perhaps on subsequent exploitation decisions. For example, moving pinch-outs or oil–water contacts may change optimal well locations.

Proportion maps with posted data are important to confirm trend plausibility and agreement with the geological conceptual model. For categorical variables, all proportion models should be viewed jointly to ensure agreement between them. This visualization may be further augmented with trend fairness plots (previously discussed).

Contact profiles (mentioned previously) may be applied to test the type of boundaries that exist between regions. Hard boundaries are commonly assumed and often have significant impact on reservoir connectivity. It may also be useful to calculate contact profiles on the final reservoir property models as a check (see Section 5.1, Model Checking).

Care must be taken when summarizing uncertainty across domains or regions. If correlation exists between domains, then any composite uncertainty model across domains must account for this. For example, if global reservoir quality is related between adjacent domains, failure to account for this correlation in the summarization will result in an underestimation of uncertainty.

Expectation models are valuable to check the reproduction of input trend models and the influence of local data and secondary information. Methods to check trend models with comparison to hard data are presented in this section.

Finally the massive multivariate work flows allow for various visualizations and summarizations that may be useful to investigate and communicate local and global uncertainty. For example, McLennan et al. (2009) present a fracture modeling application with visualization of the exhaustive secondary data, the prior mean and variance maps, the likelihood mean and variance maps, and specific percentiles of the updated local distributions of uncertainty. This presents a good opportunity to view the separate contributions of the sparse primary data and the exhaustive secondary data to the updated uncertainty model. In addition, complete multivariate uncertainty models allow for very interesting model query opportunities. For example, any joint probability map is available, such as the probability of porosity, permeability, saturations and so on, between any bins (defined by their associated upper and lower thresholds). In addition, any lower-order conditional distribution is available.

Further discussion on model checking is available in Section 5.1, Model Checking; model summarization is available in Section 5.2, Model Postprocessing; and uncertainty management is available in Section 5.3, Uncertainty Management.

4.1.6 Section Summary

Large-scale modeling refers to the modeling and mapping at the scales above and as a constraint to geostatistical reservoir property simulation. This includes the modeling of bounding surfaces to characterize the reservoir domain, region identification for separation of distinct subsets within the domain, trend construction for within-region changes, and quantitative mapping techniques that integrate various types of data. These are all important initial work flows and decisions in reservoir modeling.

With the establishment of framework, regions, trends, and all variables integrated, we are ready to apply geostatistical algorithms to estimate and simulate properties within this framework. The remaining sections (4.2–4.7) in this chapter cover the various variogram-based, multiple-point-based, object-based, process mimicking, and optimization-based algorithms that fill in this framework geostatistical reservoir models.

4.2 VARIOGRAM-BASED FACIES MODELING

Facies are often important in reservoir modeling because the petrophysical properties of interest are highly correlated with facies type. Knowledge of facies constrains the range of variability in porosity and permeability. Moreover, saturation functions depend on the facies even when the distributions of porosity and permeability do not. This section presents variogram-based approaches to modeling facies.

Section 3.2, Conceptual Model, provides guidance on choosing the appropriate facies modeling method, given available data, statistical inputs, the conceptual geological model, and project goals. Section 4.1, Large-Scale Modeling, commented on facies modeling in the context of modeling regions with distinct statistics. The *Comments on Facies Modeling* subsection provides some detailed guidance on when to apply facies modeling and when to choose variogram-based, MPS, object-based, and processing-mimicking techniques, which are presented in detail in Sections 4.2–4.5.

The *Sequential Indicator Simulation* subsection presents sequential indicator simulation (SIS), which is a widely used cell-based facies modeling technique. The background of SIS was discussed in Section 2.4, Preliminary Mapping Concepts. Implementation details including the integration of seismic data are discussed more thoroughly in this subsection.

In Section 2.4, Preliminary Mapping Concepts, truncated Gaussian simulation was very briefly introduced as an alternative to indicator simulation. The *Truncated Gaussian Simulation* subsection presents more complete details on the truncated Gaussian simulation method. As mentioned previously, a categorical facies realization is created by truncating a continuous Gaussian simulation. The implementation details of determining the facies ordering, the continuous conditioning data, the right variogram, and integration of seismic data are presented.

Cell-based facies models often show unrealistic short-scale variations that can be removed with image processing techniques; *Cleaning Cell-Based Facies Realizations* subsection presents a simple and yet powerful algorithm for image cleaning. While these methods are presented in this section, they may also be relevant to any cell-based categorical modeling method such as MPS simulation discussed in Section 4.3, Multiple-Point Facies Modeling.

> For the novice modeler this section will be a good introduction to variogram-based facies modeling methods. The comparisons and implementation details should be helpful to select and get started with these methods. For the intermediate modeler the formulations for trend and secondary information integration and practical guidance should be useful to improve their knowledge. For the expert modeler this should all be familiar, but this section may be useful for communication and educating other members of the project team.

4.2.1 Comments on Facies Modeling

In Section 2.1, Preliminary Geological Modeling Concepts, facies were introduced as a classification of rock that distinguishes between reservoir quality. Commonly, facies are based on different grain size or different diagenetic alteration. That is, there are clear distinctions between reservoir quality of sandstone and shale and, often, between limestone and dolomite. Yet, in this book, facies are left ambiguous and may represent separate regions (as discussed in the last section) of the reservoir that have distinct reservoir property input statistics and trends. Facies is not tied to any definitions in any architectural hierarchy and may include lithofacies, facies associations, or even architectural elements. Our concern is not how the facies types are defined, but how to construct realistic 3-D distributions of the facies that may be used in subsequent reservoir decision making.

The first question is whether or not to bother with facies before porosity and permeability modeling. Some considerations are as follows:

1. The facies must have a significant control on the porosity, permeability, or saturation functions; otherwise, modeling the 3-D distribution of facies will be of little benefit since uncertainty will not be reduced and the resulting models will have no more predictive power. There is unavoidable ambiguity in the definition of *significant control*: The histograms of porosity or permeability within the different facies should have significantly different mean, variance, and shape, for example, by more than 30%. The saturation function measurements from different facies should not overlap.
2. The facies must be identifiable at the well locations—that is, from well log data, as well as from core data. Careful examination of core data permits identification of many facies; however, there would be large uncertainty in the 3-D modeling of those detailed facies unless they can also be determined from the more abundant well log data. This consideration is easy to check; one must resist the temptation of taking geological detail available from visual examination of core data unless all wells are cored.

3. An additional constraint on the choice of facies is that they must have straightforward spatial variation patterns. The distribution of facies should be at least as easy to model as the direct prediction of porosity and permeability.

The choice of facies is a difficult case-dependent problem for which there may be no clear solution. Simplicity is preferred. Two facies, a "net" reservoir rock type and a "non-net" rock type, are often sufficient. At times there are two different "net" rock types of different quality, which have clearly different spatial variation patterns. Rarely should four or more be considered. At times there are two different "non-net" rock types, such as a depositional shale and a diagenetic-controlled cemented rock type, which have different spatial variation patterns. It is extremely rare to have to consider three or more "non-net" rock types. Thus, there are up to a maximum of five different facies types. Once the facies are defined, relevant data must be assembled and a 3-D modeling technique selected

Section 3.2, Conceptual Model, includes detailed discussion on the selection of modeling method based on available input statistics and conceptual geological model. The following is a brief discussion related specifically to facies modeling. The alternatives for facies modeling are: (1) variogram-based geostatistical modeling, (2) multiple-point statistics, (3) object-based stochastic modeling, (4) process mimicking, or (5) deterministic mapping. Deterministic mapping is always preferred when there is sufficient data and evidence of the facies distribution to remove any doubt of the 3-D distribution. In many cases, there is evidence of geological trends, which should be included in stochastic facies modeling.

The five main steps for stochastic facies modeling are as follows: (1) Relevant well data together with areal and vertical geological trend information must be assembled within the stratigraphic layer being modeled; (2) seismic attributes are often related to facies in an approximative way; if available, the seismic data must be tested to see if they calibrate to facies probabilities; (3) declustered, spatially representative, global proportions of facies must be determined for any facies modeling; (4) spatial statistics such as variograms, training image, or size distributions must be assembled as input to stochastic modeling; and finally, (5) 3-D facies models must be constructed and validated. Each of these steps will be discussed with the different facies modeling techniques.

Variogram-based techniques are commonly applied to create facies models prior to porosity and permeability modeling (Dubrule, 1989, 1993; Langlais and Doyle, 1993; Murray, 1995; Xu, 1995). The popularity of variogram-based techniques is understandable: (1) Local data are reproduced by construction, (2) the required statistical controls (variograms) may be inferred from limited well data, (3) soft seismic data and large-scale geological trends are handled straightforwardly, and (4) the results appear realistic for geological settings where there are no clear geological facies geometries—that is, when the facies are diagenetically controlled or where the original depositional facies have less structured, short range or amalgamated variation patterns. Of course, when the facies appear to follow clear geometrical patterns, such as sand-filled abandoned channels or lithified dunes, object-based facies algorithms (Section 4.4, Object-Based Facies Modeling) should be considered; or if even greater complexity exists, related temporal sequence process-mimicking facies algorithms (Section 4.5, Process-Mimicking Facies Modeling) may be considered.

All cell-based facies modeling methods (including MPS simulation) can easily reproduce local hard data and trends. As described in Section 2.4, Preliminary Mapping Concepts, the hard data are assigned directly to the model grid cells prior to sequential simulation along a random simulation path through the model grid. Along the sequential path detailed trend, information may be integrated at each cell, resulting in ease in integrating trend and secondary information. In contrast the non-cell-based methods (object-based and process-mimicking) are challenged in dense data settings or in the presence of detailed trends and secondary information.

This flexibility of straightforward data reproduction comes at a cost. Because cell-based facies models are the result of simulation along a random path of cell by cell, there is no concept of geobodies or geometric architecture. For example, while-cell-based simulation may result in connected cells with the same or genetically related facies, there is no rigorous method available to identify these geometries. This limits the ability to provide continuous trends for subsequent continuous reservoir property simulation. While attempts have been made to "detect" the geometries from the cells and to fit trend models to these geometries, these methods break down in the face of geometric complexity and imbrication of the geometries (Pyrcz, 2004). For non-cell-based facies modeling methods these geometries are known and it is straightforward to calculate continuous property trend models within the modeled geometries (see Section 4.4, Object-Based Facies Modeling, and Section 4.5, Process-Mimicking Facies Modeling).

Multiple Depositional Systems

As discussed in Section 2.1, Preliminary Geological Modeling Concepts, a depositional system is a component of the subsurface that is derived from a specific depositional setting—for example, fluvial, deltaic, shoreface systems, or deepwater depositional systems. Genetically linked depositional systems are grouped into systems tracts. An example would include a combined fluvial and coastal systems, or deltaic and deepwater system. Many reservoirs fall into a single depositional setting. As a result, the overall facies heterogeneity remains consistent; therefore, the modeling choices are applicable over the entire reservoir.

Nevertheless, the geological depositional system may change across the areal extent of the model, because depositional system transitions may be transitional or the reservoir may include multiple depositional settings. For example, fluvial systems may express a gradual increase in estuarine influence basinward, or fluvial and related terrestrial facies may sharply transition to deltaic and other marine facies. Such large-scale changes in facies must be accounted for in facies modeling, regardless of whether variogram-based or any other modeling method is used.

Most facies methods could use nonstationary statistics to account for changes in depositional setting. Specific facies modeling methods offer opportunities for customization to account for these transitions. Multiple-point-based methods may work with multiple training images that are applied within each depositional setting. Object-based modeling may be customized to handle such trends; that is, the objects are constructed to progressively change across the reservoir. Fluvial channels are related to their shoreface or marine counterparts; turbidite channels are related to their deepwater sheet and lobe counterparts.

Another approach to account for large-scale changes in the depositional system is to deterministically divide the area into large-scale regions or domains (as discussed in Section 4.1, Large-Scale Modeling). Then, the facies are modeled separately within each region. The regions are modeled sequentially, say from most proximal to most distal, with data conditioning overlapping from one region to the next. This reproduces gradational changes in the facies at the borders between the regions. Separate modeling of different areas could cause artificial discontinuities at the boundaries.

4.2.2 Sequential Indicator Simulation

The building blocks of sequential indicator simulation (SIS) have been discussed. Section 2.2, Preliminary Statistical Concepts, discussed the calculation of representative global proportions and the indicator formalism. Section 4.1, Large-Scale Modeling, discussed construction of locally varying proportion models. Section 2.3, Quantifying Spatial Correlation, covered all the details of variogram calculation. There remain some implementation details, particularly with respect to seismic data, that should be discussed.

SIS is widely used for diagenetically controlled facies because the results have high variability and may be well characterized with anisotropy and variogram measures of spatial correlation (Deutsch, 2006). We consider K mutually exclusive facies categories $s_k, k = 1, \ldots, K$. The indicator transform at a particular location $\mathbf{u}_j$ for a particular facies s_k is the probability of facies s_k prevailing at that location: 1 if it is present, and 0 otherwise.

There are K indicator transforms at each data location. SIS consists of visiting each grid node in a random order. At each grid node a facies code will be assigned by the following steps:

1. Find nearby data and previously simulated grid nodes.
2. Construct the conditional distribution by kriging; that is, calculate the probability of each facies being present at the current location, $p_k^*, k = 1, \ldots, K$.
3. Draw a simulated facies from the set of probabilities.

This entire procedure is repeated with different random number seeds to generate multiple realizations. A random path is considered to avoid any artifacts that can result from a combination of a regular path and a restricted search, which is needed to limit the kriging matrix size. Original data and previously

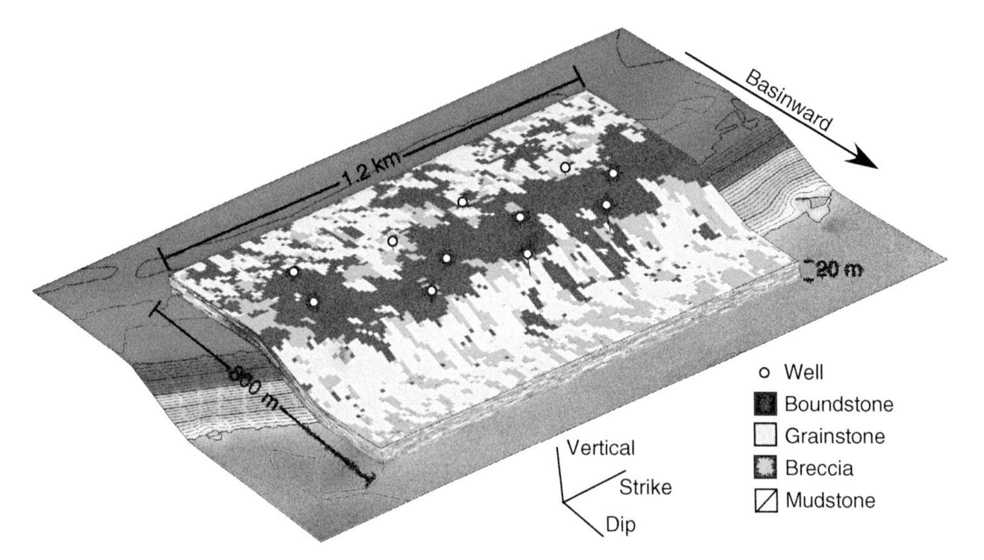

FIGURE 4.14: Oblique View with Strike Sections, Dip Section, and Horizon Slice of a SIS Realization (above) based on Locally Variable Proportions for Boundstone, Grainstone, Breccia, and Mudstone Applied to Impose Facies Trends in a Carbonate Platform.

simulated values are both used so that the realization has correct spatial structure between original data, between original data and simulated values, and between pairs of simulated values.

Figure 4.14 is an example of a SIS from the Carbonate Reservoir from the Synthetic Paleo-basin. The model includes four facies: boundstone, grainstone, breccia, and mudstone. The greater continuity for the boundstone and grainstone is along strike, and that for the breccia is along dip; they are expected geologically and were informed and modeled with directional indicator variograms. Locally variable proportions are applied to impose trends in the facies simulation with boundstone dominant on the reef crest, grainstone and breccia on the flanks, and mudstone becoming more prevalent landward from the reef crest.

Figure 4.15 is an example application of SIS for the Deltaic Reservoir from the Synthetic Paleo-basin. In this model, three indicator variograms were modeled for sand, interbedded sand, and shale and shale and locally variable proportions for each facies to impose facies trends with sand dominant in the proximal delta, interbedded sand and shale dominant in the medial delta, and shale dominant in the distal delta. The lower continuity in sand and shale than in interbedded sand was imposed by the indicator variogram models. Also, note that there is no clear facies ordering in this realization as expected with SIS. Even with a very short vertical range for each facies, the facies trend model concentrates each facies in distinct parts of the model, and this increases vertical connectivity of facies in the realization.

Seismic Data in SIS

Seismic data are often of great value in constraining facies models; it is areally extensive over the reservoir and can be sensitive to facies variations. Of course, facies vary over scales smaller than the seismic measurement volume and there is variability of acoustic properties within the same facies. A first step will be to calibrate the seismic data to facies proportions (see Section 3.1, Data Inventory).

There are typically few well data and an abundance of seismic data. The horizontal indicator variograms for SIS are poorly defined from the well data alone. A solution consists of using the seismic data to help define the 3-D variogram models.

At the heart of indicator simulation is the use of kriging to determine the conditional probability of each facies type. In the presence of seismic data, some form of cokriging or block kriging must be considered to account for the seismic data. The third step will be to choose the form of cokriging and assemble the required input parameters.

Calibration of Seismic Data to Facies

Cokriging does not require the seismic data to be explicitly calibrated to facies proportions; the original

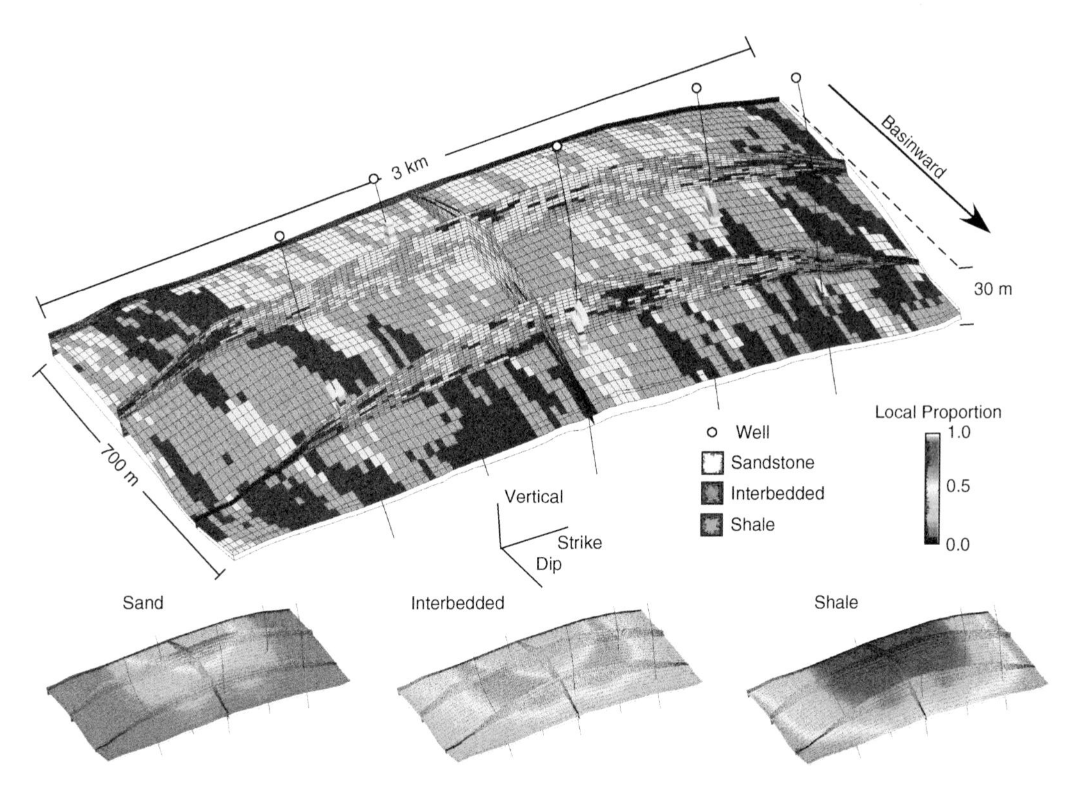

FIGURE 4.15: Oblique View with Strike Sections, Dip Section, and Horizon Slice of a SIS Realization (above) and the Locally Variable Proportions for Sand, Interbedded Sand and Shale, and Shale Applied to Impose Facies Trends with Sand Dominant in the Proximal Delta, Interbedded Sand and Shale Dominant in the Medial Delta, and Shale Dominant in the Distal Delta.

seismic units could be retained and the calibration enters through the cross covariances between the hard indicator data and soft seismic data. Notwithstanding this flexibility of cokriging, calibrated probability values are preferred, because: (1) the calibrated probability values are in units we understand, (2) cross variograms or covariances are often difficult to infer in presence of sparse well data, and (3) calibration allows for integration of other information and constraints.

Seismic data are called acoustic impedance (AI) in the following, but it could be any deemed relevant attribute. There is no need to work with acoustic impedance; the same procedures could be used for any particular seismic attribute or derivative. The range of *AI*-variability is divided into a series of classes. Too few classes and important relationships are lost. Too many classes and there are insufficient data in each class to reliably infer its relation to facies. In general, 10 classes based on the deciles of the *AI* histogram is sufficient.

Figure 4.16 shows a histogram of a seismic attribute at well locations. The upper plot is the cumulative histogram with the deciles specified by the horizontal gray lines. The 10 classes of seismic are identified by the nine vertical gray lines that go from the cumulative histogram through the lower histogram. The classes are numbered from 1 through 10 in the center. Determining these classes is essentially automatic.

The calibration procedure consists of determining the probability of each facies for each class of seismic, that is,

$$p(k|\mathrm{ai}_j), \qquad k = 1, \ldots, K, \quad j = 1, \ldots, N_{\mathrm{ai}}$$

where $p(k|\mathrm{ai}_j)$ is the probability of facies type k for the jth seismic class ai_j. These probabilities are evaluated at the well locations. For each ai datum, at a well location, there are corresponding actual proportions of each facies. The result of this calibration exercise is a table of prior probabilities. Table 4.1 shows an empty calibration table for $K = 3$ facies types and 10 ai classes.

Of particular concern in constructing such a calibration table is the vertical resolution of the seismic

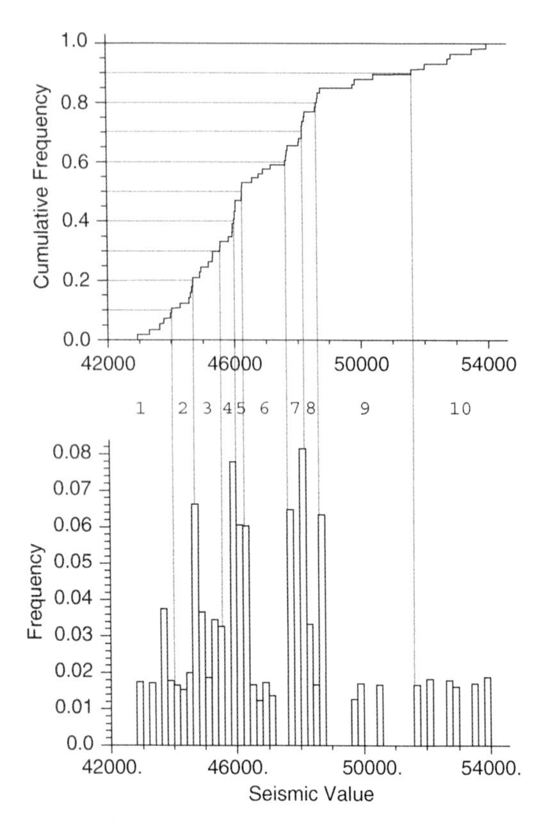

FIGURE 4.16: Cumulative Histogram and Histogram of Seismic Attribute with 10 Classes Defined by Decile Thresholds; There Is the Same Number of Data in Each Class.

TABLE 4.1. AN ILLUSTRATION OF THE CALIBRATION OF FACIES PROPORTIONS TO SEISMIC ACOUSTIC IMPEDANCE (AI)

ai class	$p(k = 1, ai)$	$p(k = 2, ai)$	$p(k = 3, ai)$
$-\infty$–ai_1	—	—	—
ai_1–ai_2	—	—	—
ai_2–ai_3	—	—	—
...	—	—	—
ai_9–∞	—	—	—

data. The seismic data (ai values) may be recorded at a small sample rate (2 ms or less); however, the "real" vertical resolution may be much coarser. It is necessary to consider a vertical window of realistic size to calculate the prior probabilities. Otherwise, the calibration will not be representative of the true seismic resolution and short-scale noise may mask the information value of the seismic data.

The seismic data carries *information* when the calibration probabilities $p(k|\text{ai}_j)$ depart from the global probabilities of each facies, p_k. Ideally, the calibration probabilities will be close to 1 and 0; however, that is never the case due to small-scale variations of facies and the variability of acoustic properties within each facies.

The subject of indicator variogram inference will be discussed before describing the use of these prior probabilities in cokriging.

Indicator Variogram Inference

Regardless of which variant of indicator simulation is used, we must infer a set of K indicator variograms that describe the spatial correlation structure of each facies type. The main challenge in variogram inference is the horizontal direction; there are often too few wells to calculate a reliable horizontal variogram. The vertical resolution, however, usually permits a reliable vertical variogram to be calculated. The seismic data provide an excellent source of information to guide the selection of horizontal parameters for the indicator variograms.

Each facies indicator variogram will be calculated in turn. The first step is to calculate the vertical indicator variogram $\gamma_k(\mathbf{h})$ from the well data and standardize it to unit sill by the variance $p_k(1-p_k)$. Secondly, we calculate the horizontal variogram from the corresponding seismic proportions, that is, the $p(k|\text{ai}_j)$ values. The seismic proportion variograms depend on the facies type $k = 1, \ldots, K$; however, there is unavoidable overlap since the proportions come from the same underlying acoustic rock properties. The seismic variogram is also standardized to unit sill. Finally, the vertical and horizontal variograms are fit with the shape of the vertical indicator variogram; only the length scales or ranges and any zonal anisotropy is taken from the seismic proportion variogram; see Section 2.3, Quantifying Spatial Correlation, for more detail on variogram modeling.

There are a number of assumptions in this hybrid approach to determine indicator variograms. The most important assumption is that the seismic proportions variogram provides a reasonable approximation to the horizontal indicator, i, range. This is reasonable if the seismic is highly correlated to the facies proportions and if the vertical averaging of the seismic is not too pronounced. One must consider slightly reducing the horizontal range by the horizontal support of the seismic data since the seismic variogram range is often longer than the indicator horizontal range (see discussion on Volume Variance in Section 2.3, Quantifying Spatial Correlation).

Locally Varying Mean

Rather than cokriging, the calibrated seismic data probabilities can be used as locally varying means for kriging; that is, the probability of facies k is estimated by kriging as

$$\hat{i}(\mathbf{u};k) = \sum_{\alpha=1}^{n} \lambda_\alpha \cdot i(\mathbf{u}_\alpha;k) + \left[1 - \sum_{\alpha=1}^{n} \lambda_\alpha\right] \cdot p(k|\text{ai}(\mathbf{u}))$$

where $\hat{i}(\mathbf{u};k), k = 1,\ldots,K$, are the estimated local probabilities to be used for simulation, n is the number of local data, $\lambda_\alpha, \alpha = 1,\ldots,n$ are the weights, $i(\mathbf{u}_\alpha;k)$ are the local indicator data, and $p(k|\text{ai}(\mathbf{u}))$ is the seismic-derived probability of facies k at location $\mathbf{u}$ being considered. When there are few local data, the sum of the weight to the data is small and the seismic-derived probabilities receive a large weight. When the weight to local data is high, the seismic-derived probabilities are given low weight.

The set of probabilities $\hat{i}(\mathbf{u};k), k = 1,\ldots,K$, must be corrected for order relations (see Section 2.4, Preliminary Mapping Concepts); otherwise, the SIS procedure remains unchanged. This procedure is simple and quite effective. There is no explicit variogram control on the influence of the seismic data. Some form of cokriging would be required for such control.

The Bayesian Updating Approach

The simplest form of cokriging is collocated cokriging or the *Bayesian Updating Approach* (Doyen et al., 1994, 1996, 1997). This method is popular because of its simplicity and ease with which seismic data are accounted for. Recall the short introduction to Bayesian statistics in Section 2.4, Preliminary Mapping Concepts.

At each location along the random path (recall the procedure described in Section 2.4, Preliminary Mapping Concepts), indicator kriging is used to estimate the conditional probability of each facies, $i^*(\mathbf{u};k), k = 1,\ldots,K$, from hard i data alone, and then Bayesian updating modifies or updates the probabilities as follows:

$$i^{**}(\mathbf{u};k) = i^*(\mathbf{u};k) \cdot \frac{p(k|\text{ai}(\mathbf{u}))}{p_k} \cdot C, \qquad k = 1,\ldots,K$$

where $i^{**}(\mathbf{u};k), k = 1,\ldots,K$ are the updated probabilities for simulation, $p(k|\text{ai}(\mathbf{u}))$ is the seismic-derived probability of facies k at location $\mathbf{u}$ being considered, p_k is the overall proportion of facies k, and C is a normalization constant to ensure that the sum of the final probabilities is 1.0. The factor $p(k|\text{ai}(\mathbf{u}))/p_k$ operates to increase or decrease the probability, depending on the difference of the calibrated facies proportion from the global proportion. No change is considered if $p(k|\text{ai}(\mathbf{u})) = p_k$, which is the case where the seismic value $\text{ai}(\mathbf{u})$ contains no new information beyond the global proportion. An additional step is required to ensure exact reproduction of facies observations at well locations.

The simplicity and utility of this approach is appealing. There are two implicit assumptions behind Bayesian updating that may be important: (1) The collocated seismic data screen nearby seismic data, and (2) the scale of the seismic data is implicitly assumed to be the same as the geological cell size. A relatively simple check on the simulated realizations will judge whether these assumptions are causing artifacts. The calibration table (just like Table 4.1) can be checked using the seismic data and the SIS realization; a close match to the probabilities calculated from the original data indicates the approach is working well. Problems would be revealed by high probabilities (large $p(k|\text{ai})$'s) becoming exaggerated to even higher values in the final model; low values would be even lower. More discussion on model checking is available in Section 5.1, Model Checking.

Markov–Bayes Soft Indicator Kriging

Once the seismic data have been calibrated to probability units, indicator kriging can be used to integrate that information into improved conditional probability values for SIS (Alabert, 1987; Journel, 1986a; Zhu and Journel, 1993).

The Bayesian approach described above is one approach to update the results of indicator kriging using well data alone. Different sources of information such as well data and seismic data are available. With enough data, one could directly calculate and model the matrix of covariance and cross-covariance functions (one cross covariance for each facies k). An alternative to this difficult exercise is provided by the Markov–Bayes model (Zhu and Journel, 1993). For simplicity, we write this model in terms of covariance. The three covariances needed are $C_I(\mathbf{h};k)$, the covariance of the *hard* facies indicator data for facies type k, which is calculated from the well data or a combination of well and seismic data (see Section 2.4, Preliminary Mapping Concepts and above), $C_{IS}(\mathbf{h}; k)$, the cross covariance between the hard facies indicator data and the seismic probability values (the $p(k|\text{ai}(\mathbf{u}))$ values); and $C_S(\mathbf{h}; k)$, the covariance of the seismic probability values. The latter two covariances are given by the following model:

$$\begin{aligned} C_{IS}(\mathbf{u}; k) &= B_k C_I(\mathbf{h}; k), & \forall\, \mathbf{h} \\ C_S(\mathbf{u}; k) &= B_k^2 C_I(\mathbf{h}; k), & \forall\, \mathbf{h} > 0 \\ &= |B_k| C_I(\mathbf{h}; k), & \mathbf{h} = 0 \end{aligned}$$

The coefficients B_k are obtained by simple manipulation of the calibrated seismic probabilities:

$$\begin{aligned} B_k = & E\{P(k|ai(\mathbf{u})) \mid I(\mathbf{u}; k) = 1\} \\ & - E\{P(k|ai(\mathbf{u})) \mid I(\mathbf{u}; k) = 0\} \quad \in \; [-1, +1] \end{aligned}$$

$E\{\cdot\}$ is the expected value operator or simply the arithmetic average. The term $E\{P(k|\text{ai}(\mathbf{u})) \mid I(\mathbf{u}; k) = 1\}$ is close to 1 if the seismic data are good; that is, the seismic-predicted probability of the facies being present is very high if the facies is present. The term $E\{P(k|\text{ai}(\mathbf{u})) \mid I(\mathbf{u}; k) = 0\}$ is close to 0 if the seismic data are good; that is, the seismic-predicted probability of the facies being present is very low if the facies is *not* present.

The K parameters, $B_k, k = 1, \ldots, K$, measure how well the soft seismic probabilities distinguish the different facies. The best case is when $B_k \approx 1$, and the worst case is when $B_k = 0$. When $B_k = 1$, the seismic-probability data $p(k|\text{ai}(\mathbf{u}))$ are treated as hard indicator data. Conversely, when $B_k = 0$, the seismic data are ignored; that is, their kriging weights will be zero.

Retaining only the collocated seismic datum amounts to the collocated cokriging option (Xu et al., 1992). This model and alternatives are covered in Section 2.4.

Block Cokriging

The versions of cokriging presented above are all approximations that try to avoid (1) calculation of cross variograms or covariances, (2) fitting a full model of coregionalization, and (3) handling the large-scale nature of seismic data. We could do the job right. All variograms and cross variograms could be calculated (see Section 2.3, Quantifying Spatial Correlation). These variograms could be downscaled to point scale, assuming that the seismic data provide information on a linear average of facies indicator data. A model of coregionalization could be fit (see Section 2.3, Quantifying Spatial Correlation). Finally, a block cokriging could be performed in SIS that would account for the actual scale of the seismic data *and* the "softness" of the calibration from acoustic impedance to facies proportions. The cokriging formalism is classical.

Most geostatistical modeling software consider the different data types to be at the same spatial scale. Nevertheless, once a small-scale model of coregionalization has been established, linear spatial averages of the variogram can be easily used in the cokriging equations. Implementation of block cokriging in a conventional SIS program (such as `sisim` in GSLIB (Deutsch, 2006)) is straightforward.

Comments on Implementation

SIS is applicable to heterogeneous facies that have been diagenetically altered or that have no clear geometric shapes associated with discontinuities in petrophysical properties. The steps for SIS have been described above and are summarized in the work flows at the end of this chapter. Some comments are as follows:

- There is a need for representative global proportions calculated via declustering or reliable local proportions determined through trend modeling.
- An indicator variogram model is needed for each facies type. In fact, there are $K - 1$ degrees of freedom in the required indicator variograms. Only one variogram is needed for a binary system since the continuity of one facies completely specifies that of the other. We calculate and fit all K variograms in the presence of $K \geq 3$; inconsistency between the K variograms will be revealed in order relations problems.
- The kriging type used in the sequential simulation framework is adapted to handle seismic data.

The lengthy discussion on the type of cokriging to use for seismic data could be confusing. The methods increase in complexity and data requirements. The simplest is best, provided that it does not contradict available data.

Nested Indicators

An alternative implementation of SIS is to apply it in a nested or hierarchical manner. For example,

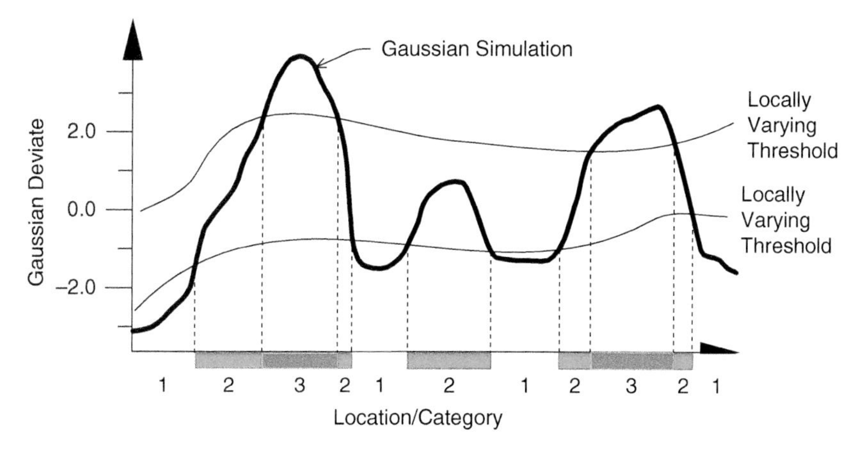

FIGURE 4.17: Schematic Illustration of How Truncated Gaussian Simulation Works: A continuous Gaussian variable is truncated at a series of thresholds to create a categorical variable realization.

consider a first SIS to separate net and non-net facies. Then, a second SIS could be applied within the net facies to distinguish different types of net facies. This procedure allows the "nested" relationships between the facies to be effectively captured. The truncated Gaussian simulation also considers transitions between facies.

4.2.3 Truncated Gaussian Simulation

The key idea with Truncated Gaussian simulation (Armstrong et al., 2003; Beucher et al., 1993; Emery, 2007; Matheron et al., 1987; Xu and Journel, 1993) is to generate realizations of a continuous Gaussian variable and then truncate them at a series of thresholds to create a categorical facies realizations (see short introduction in Section 2.4, Preliminary Mapping Concepts).

A 1-D schematic example is shown in Figure 4.17. Although this example is only 1-D, it illustrates many of the concepts behind truncated Gaussian simulation. The categorical simulation, shown along the lower axis, is derived from a continuous Gaussian simulation shown by the thick black curve. The thresholds for truncating the Gaussian variable to facies need not be constant; we see a greater proportion of facies 1 toward the right side where the first threshold is higher.

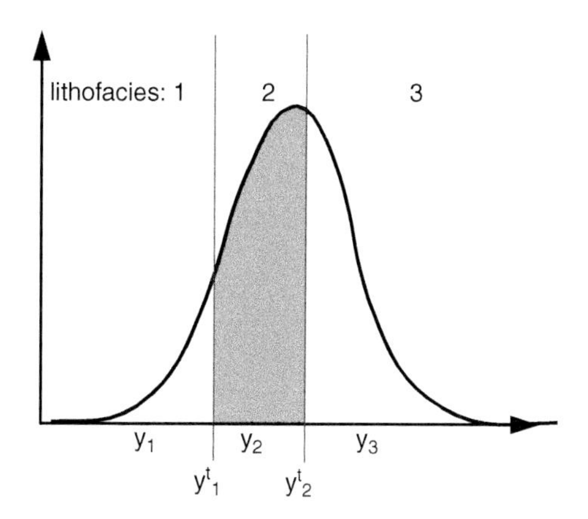

FIGURE 4.18: Standard Normal Distribution with Three Facies codes. There are two thresholds ($y^{t}1$ and $y^{t}2$) and three continuous y values for each facies (y_1, y_2, and y_3). The proportion of each facies is the area under the standard normal curve.

Ordering of Facies

An important feature of truncated Gaussian simulation is the ordering of the resultant facies models. The facies codes are generated from an underlying continuous variable. In the three-facies example of Figure 4.18 we would most often see facies 2 between 1 and 3. Only rarely would code 1 be next to code 3, when the continuous variable changes very quickly from low to high values.

This ordering is a significant advantage of truncated Gaussian simulation when dealing with facies that *are* ordered. For example, the facies may be defined by the amount of shale: 1 = shale, 2 = interbedded sand and shale, and 3 = clean sand. Truncated Gaussian simulation may work well for this case. In another case, facies may be genetically ordered due to depositional process, such as proximal upper shore face, distal upper shore face, and lower shore face in shore face reservoirs. SIS will work better when there is no clear ordering due to explicit constraint on spatial continuity of each facies. The correct ordering for the facies is usually evident from the geological context of the problem.

Proportions, Thresholds, and Conditioning Data

The facies proportions are taken as constant, or they have been modeled as locally variable with the trend modeling techniques of Section 4.1, Large-Scale Modeling. In any case, assume that the proportion of each ordered facies is known at each location $\mathbf{u}$ in the layer A, that is, $p_k(\mathbf{u}), k = 1, \ldots, K, \mathbf{u} \in A$. These proportions can be turned into cumulative proportions:

$$\mathrm{cp}_k(\mathbf{u}) = \sum_{j=1}^{k} p_j(\mathbf{u}), \qquad k = 1, \ldots, K, \quad \forall \mathbf{u} \in A \tag{4.2}$$

where $\mathrm{cp}_0 = 0$ by definition and $\mathrm{cp}_K = 1.0$ since the proportions sum to 1.0 at all locations. The $K - 1$ thresholds for transforming the continuous Gaussian variable to facies are given by

$$y_k^t(\mathbf{u}) = G^{-1}\left(cp_k(\mathbf{u})\right), \qquad k = 1, \ldots, K-1, \quad \forall \mathbf{u} \in A \tag{4.3}$$

where $y_k^t(\mathbf{u}), k = 1, \ldots, K-1, \forall \mathbf{u} \in A$, are the thresholds for the truncated Gaussian simulation, $y_0^t = -\infty$, $y_K^t = +\infty$, $G^{-1}(\cdot)$ is the inverse cumulative distribution function for the standard normal distribution,[1] and $\mathrm{cp}_k(\mathbf{u}), k = 1, \ldots, K-1$ are the cumulative probabilities for location $\mathbf{u}$ [see Eq. (4.2)]. Given a normal deviate or variable, these thresholds may be used to assign a facies code:

$$\text{facies at } \mathbf{u} = k \qquad \text{if} \, y_{k-1}^t(\mathbf{u}) < y(\mathbf{u}) \le y_k^t(\mathbf{u})$$

The locally variable thresholds [see Eq. (4.3)] introduce the trend information. The Gaussian values $y(\mathbf{u})$ are considered stationary, that is, $E\{Y(\mathbf{u})\} = 0, \forall \mathbf{u} \in A$.

The categorical facies data must be transformed into continuous Gaussian conditioning data for conditional simulation of the stationary Gaussian variable, Y. There are two considerations. First, the local proportions and corresponding thresholds must be considered to ensure that the correct facies are obtained on back transformation. Second, the facies are categorical and some decision must be made about the "spikes" caused by categorical data. Handling locally varying proportions and thresholds is straightforward: A local transformation is used. Despiking the categorical data requires additional decisions.

Random despiking (used in the `nscore` program of GSLIB) would introduce unreasonable short-scale randomness. A more sophisticated despiking (see Section 2.2, Preliminary Statistical Concepts) could be used, but with little advantage.

Decisions regarding the parameters for despiking may introduce other artifacts. The simplest solution is to set the normal score transform of each facies to the center of the class in the normal distribution—that is, just leave the spikes unchanged[2]:

$$y(\mathbf{u}) = G^{-1}\left(\frac{\mathrm{cp}_{k-1}(\mathbf{u}) + \mathrm{cp}_k(\mathbf{u})}{2}\right) \tag{4.4}$$

where $y(\mathbf{u})$ is the normal score transform at location $\mathbf{u}$, k is the facies code at location $\mathbf{u}$, and the cp_ks are the cumulative proportions as in Eq. (4.2). All facies measurements are transformed and used in conditional Gaussian simulation (for example, `sgsim` from GSLIB).

Variogram

Truncated Gaussian simulation requires a single variogram for the y variable. This is a convenience in the sense that K indicator variograms need not be calculated and modeled in a consistent manner. This is also a significant disadvantage since there is no ability to control the different patterns of variability for different facies. In general, it is not a good idea to directly calculate and fit the y variogram with the y data defined above in Eq. (4.4) due to the spikes. In the case of two facies (a single threshold y_1^t), the normal score covariance is directly linked, through a one-to-one relation, to the facies indicator covariance (Journel and Isaaks, 1984). We could exploit this analytical relation, but it only exists for the case of two facies.

One way around this limitation would be to consider a series of Gaussian RFs $Y_k(\mathbf{u}), k = 1, \ldots, K,$

[1] Although there are no closed-form analytical expressions for this function, there are excellent polynomial approximations.

[2] The reduced variance (less than 1.0) of the input y data may cause an artifact in the y realizations, but this will not be seen in the categorical variable realization.

each used to simulate (after truncation) a nested set of only two facies; that is, simulate each set nested into a previously simulated set. This would be a more complex version of the nested indicator approach mentioned in the previous discussion on sequential indicator simulation. This approach is not developed further since the facies rarely have such an arbitrary pattern of "nested" variability.

Another approximation consists of inverting the single covariance model $C_Y(\mathbf{h})$ from some average of the K indicator variograms. It is this single average indicator variogram that would be reproduced through simulation, not any particular facies indicator variogram model.

Pluri-Gaussian Simulation (PGS) extends the truncated Gaussian method for simulation of multiple ($K > 2$), non-nested, facies call for multiple Gaussian RFs. Contrary to the case $K = 2$, simulation of multiple, non-nested, facies involves a highly non-unique inversion. Iterative procedures based on trial-and-error selection of the input normal score covariances and cross covariances between the multiple Gaussian RFs have been considered, yet with limited success (Loc'h and Galli, 1996).

Recent applications of PGS (Armstrong et al., 2003; Galli et al., 2006; Mariethoz et al., 2009) have demonstrated the flexibility and utility in being able to honor higher-order connectivity relations. In practice, experience with PGS is required to characterize (a) the facies rules through decisions for the correlation between the multiple Gaussian RFs and (b) the truncation of this higher-order space into K facies.

The recommended approach is to take the most important indicator variogram and "invert" it numerically to the corresponding normal score variogram. This can be done numerically with code available with the paper by Kyriakidis et al. (1999). The normal score variogram has a similar range and anisotropy; the shape is smoother—that is, more parabolic or "Gaussian" at short distances. Figure 4.19 illustrates the relationship between the indicator and

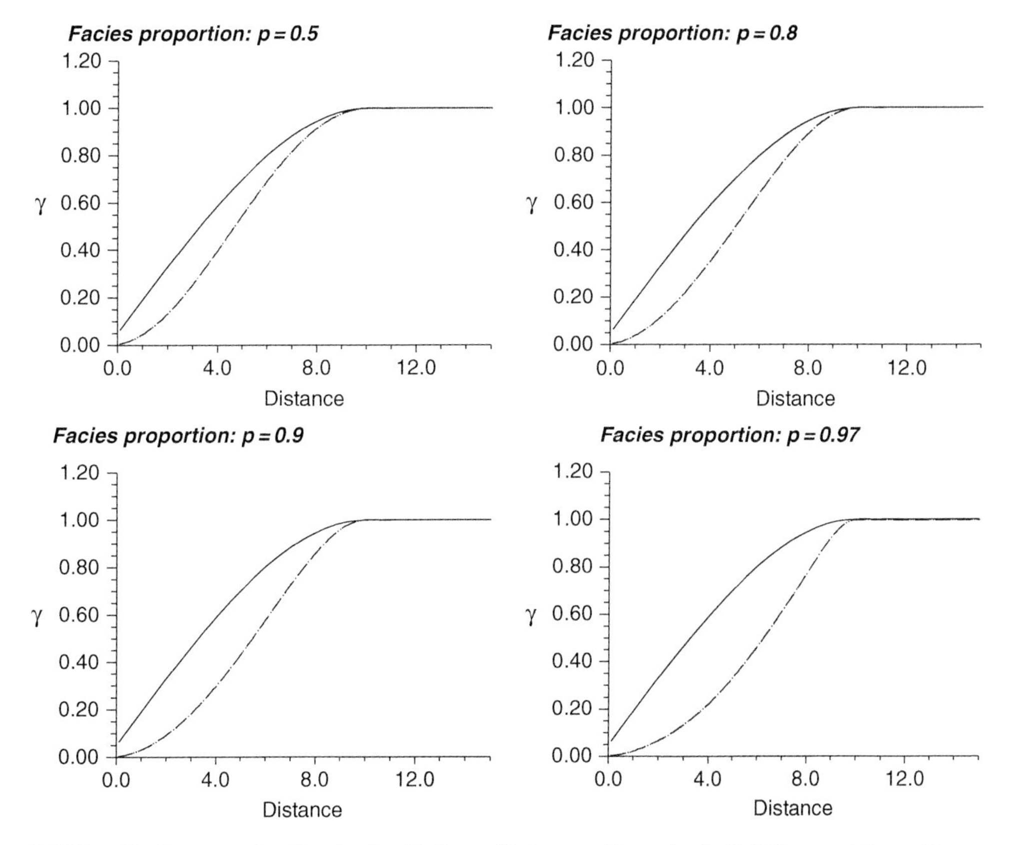

FIGURE 4.19: Corresponding Standardized Indicator Variograms Shown by the Solid lines and Normal Score Semivariograms Shown by the Dashed Lines for Different Univariate Proportions.

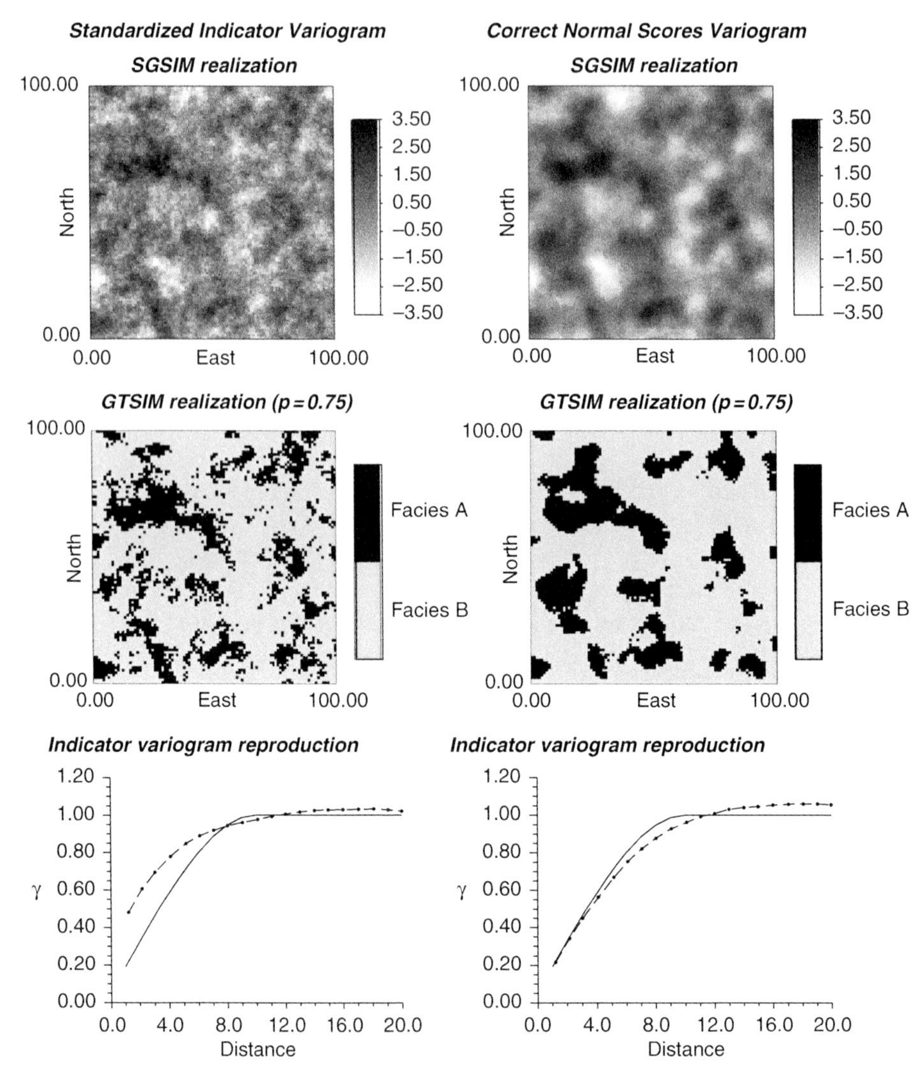

FIGURE 4.20: Illustration of the Consequence of Using the Wrong Variogram for Truncated Gaussian Simulation. The Gaussian realization, truncated indicator realization, and indicator variograms on the left side correspond to using the standardized indicator variogram for the normal score (Gaussian) realization. Note that the resulting indicator variogram has a too-high nugget effect. The realization on the right side corresponds to using the theoretically correct normal score variogram.

Gaussian variogram for a series of different facies proportions.

Figure 4.20 illustrates the consequence of using the wrong variogram for truncated Gaussian simulation. The Gaussian realization, truncated indicator realization, and indicator variograms on the left side correspond to using the standardized indicator variogram for the normal score (Gaussian) realization. Note that the resulting indicator variogram has a too-high nugget effect. The realization on the right side corresponds to using the theoretically correct normal score variogram.

Comments on Implementation

Truncated Gaussian simulation is applicable to heterogeneous facies that have been diagenetically altered, that have no clear geometric shapes, and that are ordered in some predictable manner. The steps for truncated Gaussian simulation have been described above and are summarized in the work flows at the end of this section. Some comments are as follows:

- As with SIS, there is a need for representative global proportions calculated via declustering

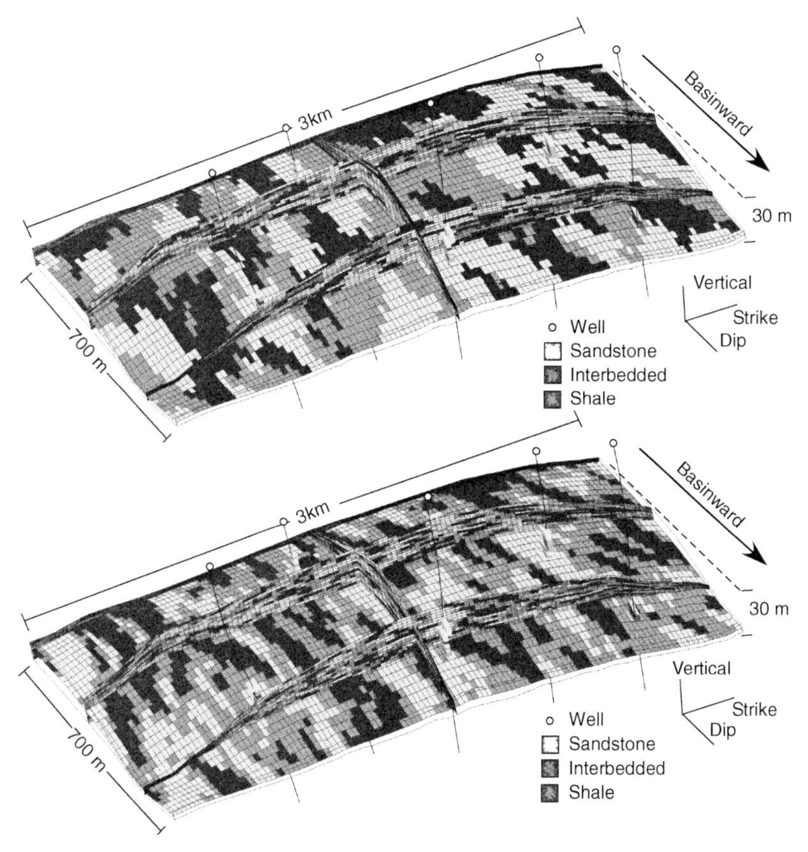

FIGURE 4.21: Oblique View with Strike Sections, Dip Section, and Horizon Slice of a SIS Realization (above) and Truncated Gaussian Simulation (below) Realization. For both, a locally variable azimuth map was applied to rotate the primary direction of the variogram locally to produce the distributary pattern.

or reliable local proportions determined through trend modeling. These proportions determine the locally varying thresholds and conditioning data.

- An indicator variogram model is needed for the most important facies type. This variogram is converted to a variogram for simulation of the Gaussian variable; in practice, the shape is changed to a Gaussian shape.
- Realizations of conditional Gaussian simulation are truncated by the locally varying thresholds to create facies realizations.

Truncated Gaussian simulation requires essentially the same professional involvement and CPU effort as SIS. The consideration to choose one or the other is based on the ordering or nesting of the facies. Truncated Gaussian simulation should be used when there is a clear ordering and no significant difference in anisotropy of the various facies geometries.

Figure 4.21 is an example of truncated Gaussian simulation from the Deltaic Reservoir from the Synthetic Paleo-basin. In this model a locally variable azimuth model was applied to the continuity model to impose the distributary correlation (expanding paleo-flow) commonly observed in deltas. The facies include sandstone, interbedded sand and shale, and shale. Note the natural ordering relationships reproduced with truncated Gaussian with sand to interbedded sand and shale-to-shale transitions.

4.2.4 Cleaning Cell-Based Facies Realizations

One concern with cell-based facies realizations is the presence of unavoidable short-scale variations, which appear geologically unrealistic. In some cases, such variations affect flow simulation and predicted reserves, a more justifiable reason to consider realization cleaning algorithms. A second concern is that the facies proportions often depart from their target input proportions. In particular, facies types with relatively small proportions (5–10%) may be poorly matched. In indicator simulation, the main source of this discrepancy is the order relations correction (the estimated probabilities are corrected to be non-negative and sum to 1.0). There is no evident alternative to the commonly used order relations correction algorithms; post-processing the realizations to honor target proportions is a convenient and attractive solution. An unavoidable consequence of this post-processing is a reduction in the space of uncertainty. The uncertainty in the global proportions could be established ahead of time by, for example, a bootstrap and then used in the cleaning algorithm described below to set the proportion for each scenario (see Section 5.3, Uncertainty Management, for more information).

The general problem of image cleaning has been tackled by a number of workers in the area of image analysis and statistics (Andrews and Hunt, 1989; Besag, 1986; Geman and Geman, 1984; Gull and Skilling, 1985). One proposal to clean facies realizations, closely related to some of these image analysis methods, is based on the concepts of dilation and erosion (Schnetzler, 1994). This approach is well suited to cleaning binary (only two facies) images; however, there is no general extension to more than two facies types. One could consider a nested application two facies at a time.

Iterative or simulated annealing-type algorithms can be designed to clean facies realizations (see Section 4.7, Optimization for Model Construction). These methods can be very powerful; however, there are a number of practical problems, (1) they tend to be CPU-intensive, (2) it is difficult to determine the appropriate values for a number of tuning parameters, and (3) often, a training image is required.

The *maximum a posteriori selection* or MAPS technique replaces the facies type at each location **u** by the most probable facies type based on a local neighborhood (Deutsch, 2005c). The probability of each facies type, in the local neighborhood, is based on (1) closeness to the location **u**, (2) whether or not the value is a conditioning datum, and (3) mismatch from the target proportion.

Consider an indicator realization $i_k^{(0)}(\mathbf{u})$, $k = 1,\ldots,K$, $\mathbf{u} \in A$, of N locations taking one of K facies types, $s_k, k = 1,\ldots,K$. There is no requirement for the facies $s_k, k = 1,\ldots,K$, to be in any particular order. The proportions of each facies type in the realization are $p_k^{(0)} = \text{Prob}\{I_k^{(0)} = 1\} \in [0,\ldots,1], k = 1,\ldots,K$ with $\sum_k p_k^{(0)} = 1$. The target proportions of each facies type are $p_k, \in [0,\ldots,1], k = 1,\ldots,K$ with $\sum_k p_k = 1$.

Consider the following steps to clean the realization $i_k^{(0)}(\mathbf{u})$, $k = 1,\ldots,K$, $\mathbf{u} \in A$, and to bring the proportions $p_k^{(0)}, k = 1,\ldots,K$, closer to the to target probabilities $(p_k, k = 1,\ldots,K)$. At each of the N locations $\mathbf{u} \in A$, calculate a local probability $q_k(\mathbf{u})$ based on a weighted combination of surrounding indicator values:

$$q_k(\mathbf{u}) = \frac{1}{S} \sum_{\mathbf{u}' \in W(\mathbf{u})} w(\mathbf{u}') \cdot c(\mathbf{u}') \cdot g_k \cdot i_k^{(0)}(\mathbf{u}'),$$
$$k = 1,\ldots,K \qquad (4.5)$$

where S is a standardization constant to enforce $\sum_k q_k(\mathbf{u}) = 1.0$, $W(\mathbf{u})$ is a template of points centered at location **u**, and $w(\mathbf{u}')$, $c(\mathbf{u}')$, and $g_k, k = 1,\ldots,K$ are weights to account for closeness to **u**, conditioning data, and mismatch from the global proportions. More precisely:

- $w(\mathbf{u}')$ = weights that define a neighborhood or template to achieve realization cleaning. The definition of the local neighborhood and the weights $w(\mathbf{u}')$ within the neighborhood control the extent of realization cleaning. The appearance of the results is used to determine the optimal neighborhood (see examples below).
- $c(\mathbf{u}')$ = weight to ensure reproduction of conditioning data. $c(\mathbf{u}') = 1.0$ at all non-data locations and is equal to C; $C \geq 10$ at conditioning data locations, $\mathbf{u}' = \mathbf{u}_\alpha, \alpha = 1,\ldots,n$. This discontinuous function ensures that the conditioning data are given priority.
- g_k = weight to ensure that the facies proportions in the cleaned image are closer to the target global proportions: The probability of a particular facies q_k will be decreased if the original realization has a too-high proportion of that facies and will be increased if the original realization has a too-low proportion, specifically,

$$g_k = \frac{p_k}{p_k^{(0)}}, \qquad k = 1,\ldots,K$$

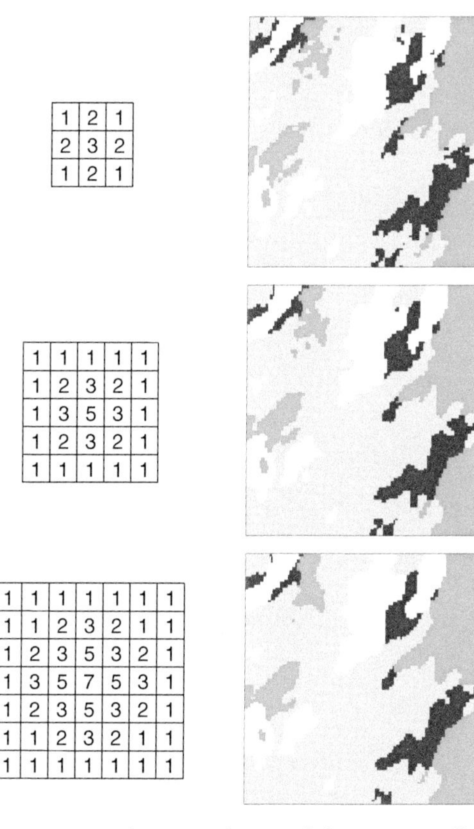

FIGURE 4.22: Three Templates and the Corresponding Cleaned Images Starting from an SIS Realization.

where $p_k, k = 1, \ldots, K$, are the target proportions and $p_k^{(0)}, k = 1, \ldots, K$, are the proportions in the initial realization. Further scaling factors ($f_k, k = 1, \ldots, K$ multiplying the g_k values) could be determined by trial and error to force a closer match.

The size of the local neighborhood and the nature of the weights $w(\mathbf{u}')$ have a significant impact on the "cleanliness" of the results. In general, the image will appear cleaner when a larger window is considered. Figure 4.22 illustrates how an SIS realization is cleaned with three different templates W and weights $w(\mathbf{u}')$. Anisotropy can be introduced to the template W and weights $w(\mathbf{u}')$ to avoid "smoothing" of anisotropic features. Trial and error must be used to determine the weights for any specific case.

Conditioning data are enforced by the weights $c(\mathbf{u}')$ within the local neighborhood; $c(\mathbf{u}')$ is equal to 1.0 at all non-data locations and is equal to C, a larger number, at conditioning data locations, $\mathbf{u}' = \mathbf{u}_\alpha, \alpha = 1, \ldots, n$. Figure 4.23 shows two SIS realizations before and after cleaning. The well data are reproduced in both cases. The facies observed at the vertical well are propagated continuously away from the well. The extent of the propagation depends on the magnitude of C and the size of the cleaning template W; a conditioning datum can affect only those nodes within the template W.

The weights $g_k, k = 1, \ldots, K$, lead the cleaned image toward the global proportions: The probability of a particular facies is decreased if the original realization has a too-high proportion of that facies and will be increased if the original realization has a too-low proportion. As expressed in (4.5), the proportions are not imposed; the probabilities are simply adjusted to be closer. In general, variability in facies proportions is an inherent aspect of uncertainty, and perfect reproduction of target proportions is not a critical goal.

The motivation for cleaning categorical facies realizations is to correct noisy facies realizations that are often geologically unrealistic and have a different flow character than clean images. A drawback of realization cleaning algorithms is that *real* short-scale variations in facies may be removed for the sake of nice, clean, pretty pictures.

4.2.5 Work Flow

Figure 4.24 illustrates a work flow for the calibration of facies proportions from seismic data. The results can be used with SIS or truncated Gaussian facies simulation.

Figure 4.25 illustrates the work flow for facies modeling using sequential indicator simulation (SIS) with seismic data.

Figure 4.26 illustrates the work flow for the preprocessing steps required prior to truncated Gaussian simulation. The locally varying thresholds and proportions must be established. The normal score transformation of the facies is passed to SGS.

Figure 4.27 illustrates the work flow for facies modeling using truncated Gaussian simulation. Stationary SGS is described in Section 4.6.

Figure 4.28 illustrates the work flow for cleaning cell-based facies realizations. The work flow is iterative to achieve the *correct* level of cleaning. Care should be taken to avoid cleaning real consequential short-scale facies variations.

4.2.6 Section Summary

Facies are often an important constraint on reservoir heterogeneity. For this reason we dedicate a

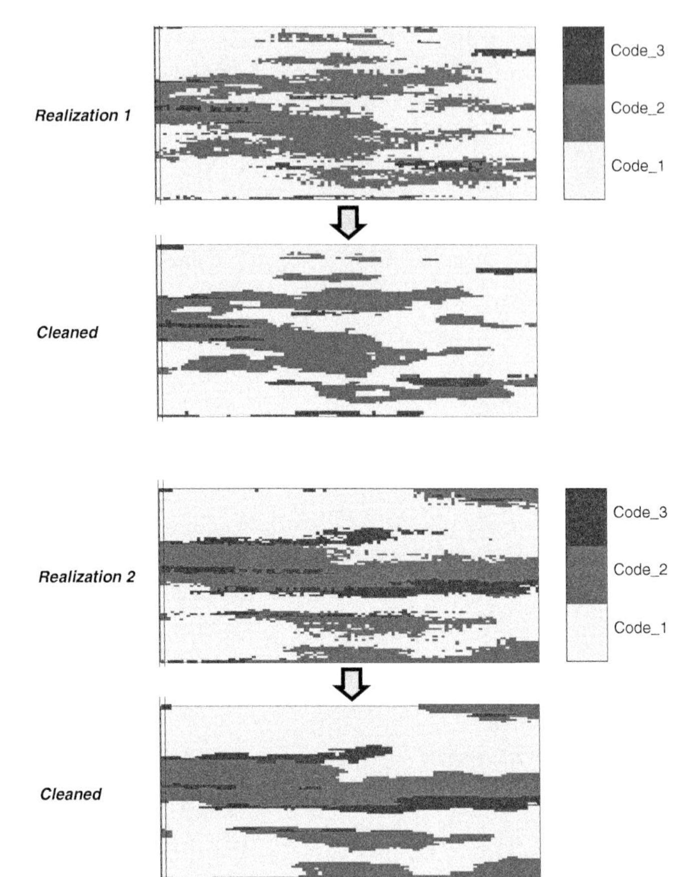

FIGURE 4.23: Two SISIM Realizations Before and After Cleaning. Note the reproduction of the conditioning data at the vertical string of well data at the left.

large part of this chapter to facies modeling (see Sections 4.2–4.5). This section started with the criteria for deciding to model facies and choosing the number of facies. These criteria will be pertinent to any of the other facies modeling methods discussed in Section 4.2–4.5.

This first facies modeling section has focused on the variogram-based methodologies with the traditional workhorse of facies, SIS, providing a robust framework and the flexibility to account for spatial continuity in each facies. For settings in which facies ordering relationships are essential to meeting project goals, truncated Gaussian simulation has been presented as a viable alternative. If more complexity and curvilinear features are required, then multiple-point-based methods may be considered (see the next section); if crisp geometries are required, then consider object-based methods (Section 4.4, Object-Based Facies Modeling); and, finally, if complicated interactions from depositional process are required, then consider process–mimicking methods (Section 4.5, Process-Mimicking Facies Modeling).

For cell-based methods (i.e., variogram- and multiple point-based), short-scale noise may be introduced into the simulations due to algorithm limitations such as order relations discussed for indicator simulation. A simple method for image cleaning is presented to correct this artifact. There is some trial and error to get the right level of smoothing.

With the facies modeling concept and traditional variogram-based facies modeling methods discussed, each of the additional facies modeling methods will be discussed and then followed by a section on the simulation of continuous properties such as porosity, permeability, and saturation (Section 4.6, Porosity and Permeability Modeling).

4.3 MULTIPLE-POINT FACIES MODELING

The widespread use of multiple-point simulation (MPS) for reservoir facies modeling has been a major advancement in geostatistics since the first edition of this book. In Section 2.3, Quantifying Spatial Correlation, the concept of multiple-point statistics

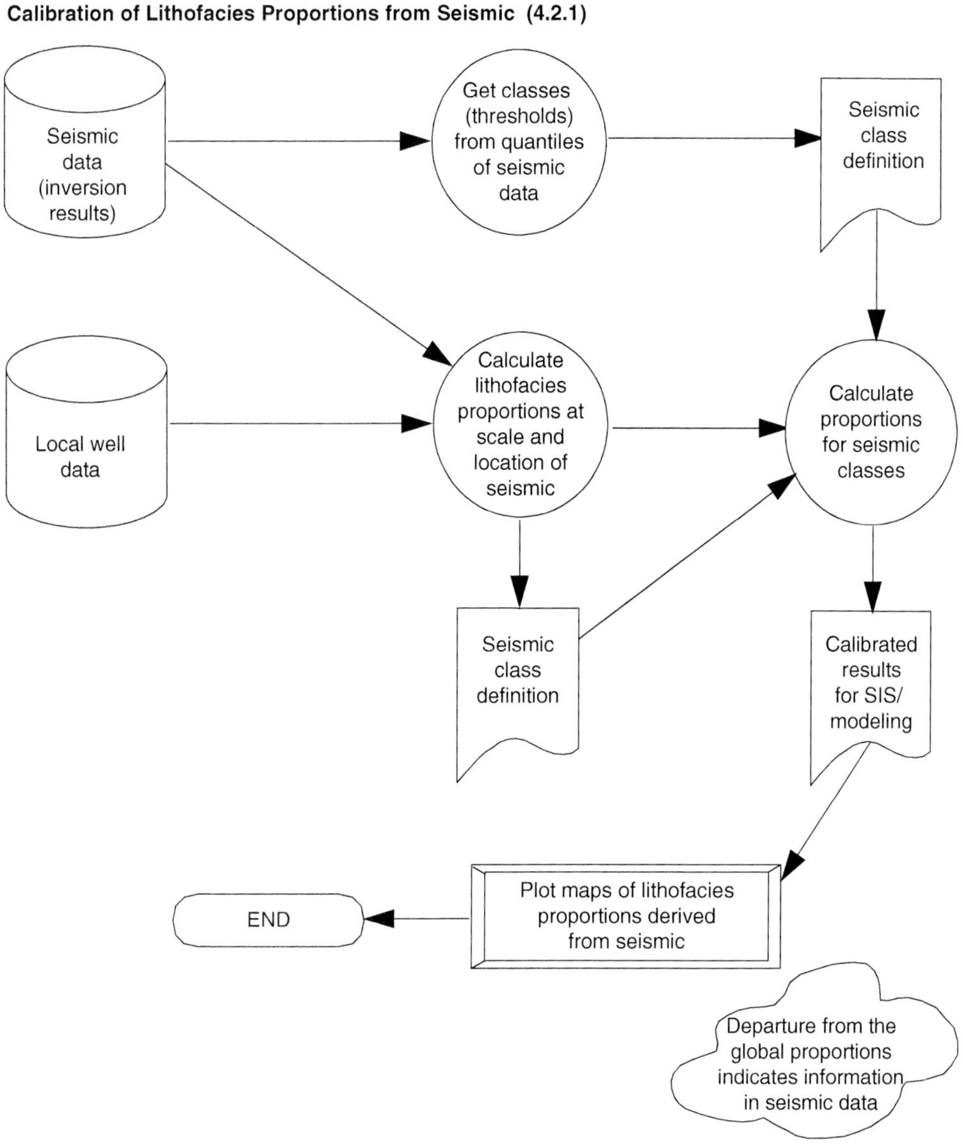

FIGURE 4.24: Work Flow for Calibration of Facies Proportions from Seismic Data.

exported from a training image is introduced; and in Section 2.4, Preliminary Mapping Concepts, the algorithm steps for the inclusion of these statistics is covered as an alternative to kriging in the sequential simulation framework. This section covers facies modeling with MPS.

In the *Multiple-Point Simulation* subsection, the background and methods of MPS are reviewed. Recent developments have resulted in practical MPS methods for facies modeling. The resulting models improve geological input through the utilization of training images that represent reservoir heterogeneity. MPS is still experiencing ongoing development. Throughout this section we refer to these developments, but we base our discussion and demonstrations on the commonly applied method by Strebelle (2002).

Sequential Simulation of MPS results in an efficient, practical methodology that reduces artifacts, while readily honoring constraints such as those related to trend constraints and data conditioning.

In the *Input Statistics* subsection the various statistical inputs are discussed, including global proportions, the model of spatial continuity informed from training images, and locally variable scale, azimuth, and facies proportions.

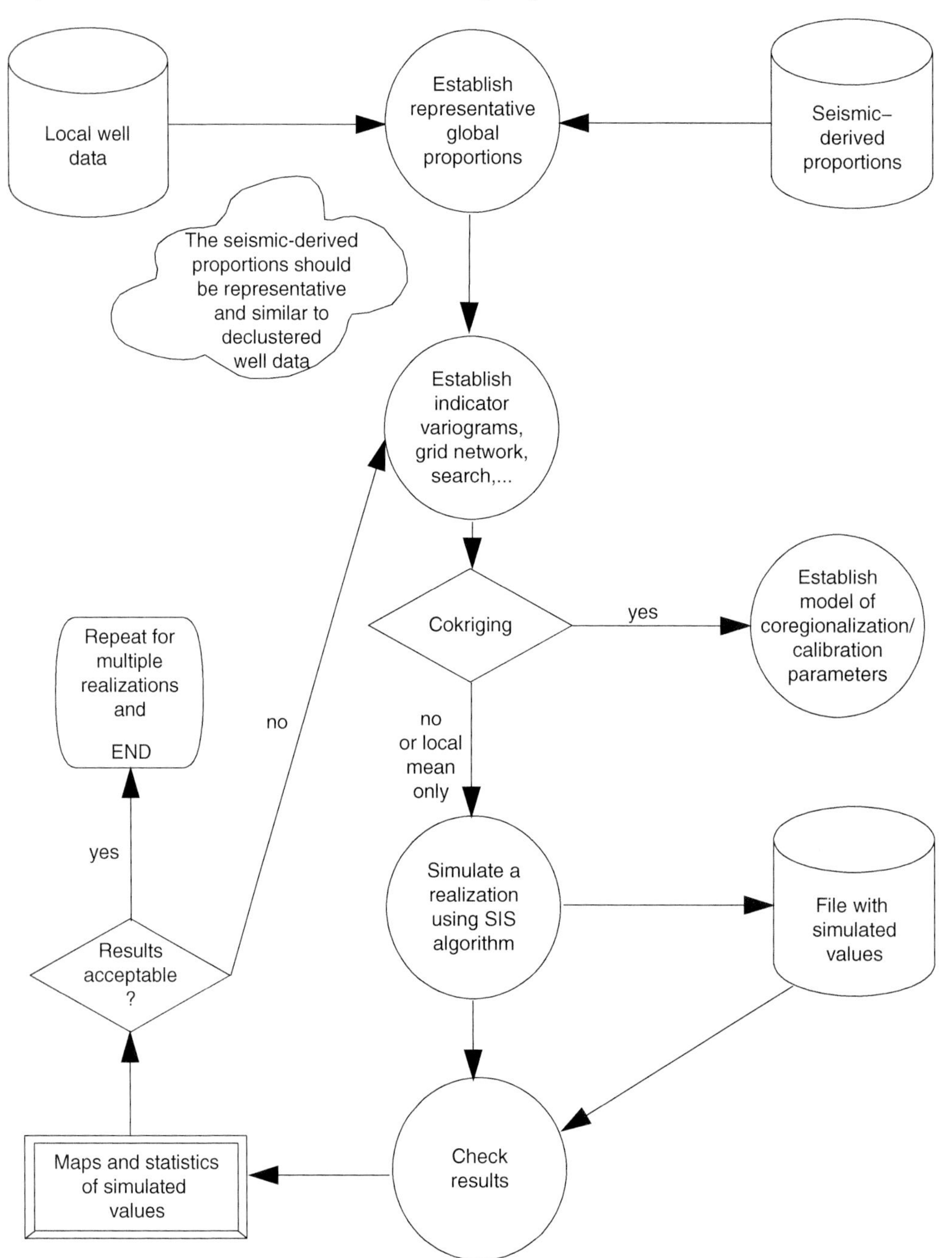

FIGURE 4.25: Work Flow for Facies Modeling Using Sequential Indicator Simulation (SIS) with Seismic Data.

The *Implementation Details* subsection contains specific details related to search trees, multiple-point template design, and continuous trends. Efficient calculation and storage of stationary multiple-point statistics is required for practical reproduction of reservoir heterogeneity. Partitioning search trees (Boucher, 2009) or more efficient methods, data structures may improve MPS's capability to handle complicated spatial heterogeneities. Also, smart multiple-point template design may maximize the impact of the available points in the template. Finally, as with other cell-based facies modeling

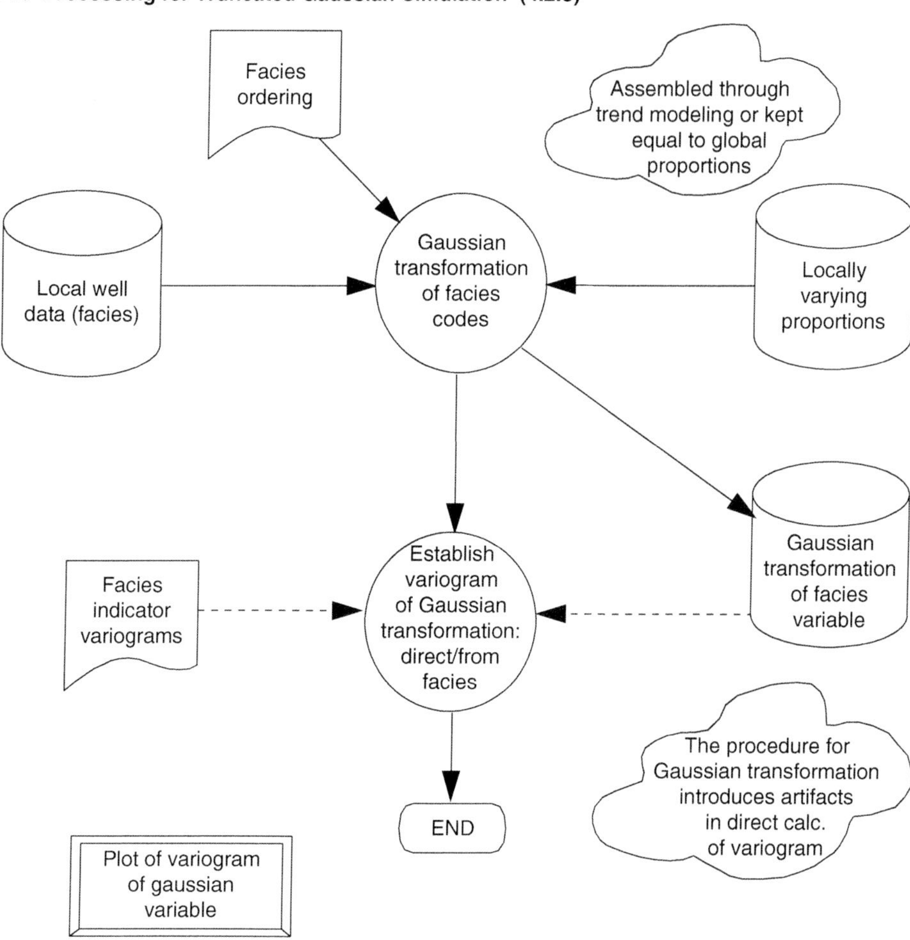

FIGURE 4.26: Work Flow for Pre-processing Steps Required Prior to Truncated Gaussian Simulation.

methods, it is not straightforward to impose trends on the continuous reservoir properties within the facies model. Some potential methods are mentioned.

> The novice modeler will appreciate this introduction to MPS and the very visual and intuitive nature of its inputs and work flows. The intermediate modeler will benefit from the addition of some theory and knowledge of implementation details present in this section. The expert modeler may be an early adopter of new and emerging methods to further improve the reproduction of greater geologic complexity and nonstationarity.

4.3.1 Multiple-Point Simulation

A thorough review of the development of MPS is available in Strebelle (2002). Early MPS methods were based on iterative schemes and were impractical due to computational time and the related difficulty to establish acceptable convergence. Farmer (1988) and Deutsch (1992b) applied MPS within simulated annealing as components within the objective function. Caers et al. (1999) applied Markov chain Monte Carlo simulation and Srivastava (1992) applied the Gibbs sampler. These methods were not practical for dealing with a significant number of multiple-point events.

Guardiano and Srivastava (1993) introduced a noniterative method based on sequential simulation. This is the basis for the algorithm steps listed in Section 2.4, Preliminary Mapping Concepts. The conditional probabilities were calculated directly from the training image, but this was very computationally expensive due to the need to scan the entire training image for each new simulated node within the realization. Strebelle (2002) calculated all conditional

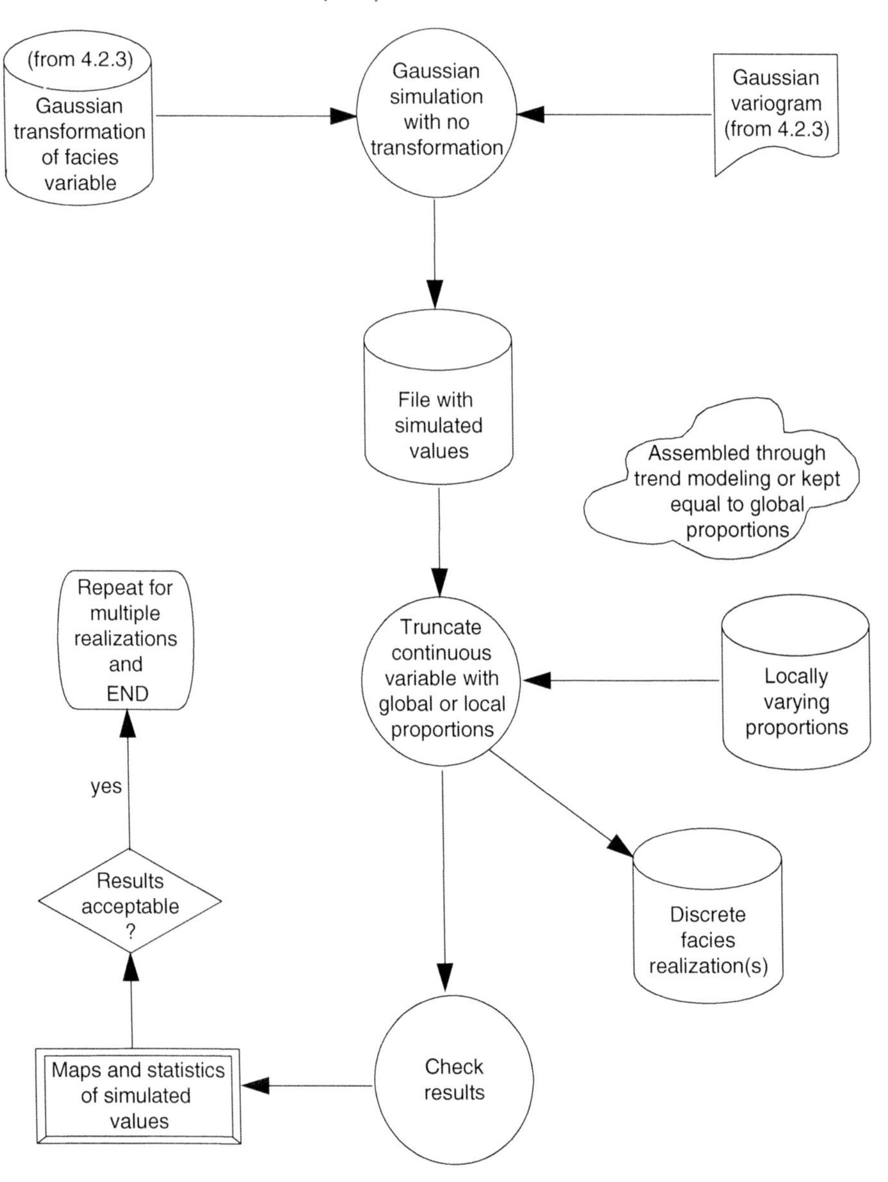

FIGURE 4.27: Work Flow for Facies Modeling Using Truncated Gaussian Simulation.

probabilities prior to simulation and applied dynamic data storage search trees for efficient storage and retrieval and limited consideration to only those conditional probabilities available from a limited data event template applied to the training image. Once calculated, this search tree may be applied to look up required conditional probabilities for all simulated nodes during simulation. The MPS examples shown in this section are based on a recent version of this method discussed in Strebelle (2006). Iterative methods have been recently readdressed by Lyster and Deutsch (2006) with a MPS method based on the Gibbs sampler (see Section 4.7, Optimization for Model Construction).

Design of training images with the right features at the scale of the reservoir grid is essential for MPS facies modeling. Major challenges include the *curse of dimensionality*; that is, a very large number of conditional probabilities are required to characterize multiple-point events (recall K^{n+1} from Section 2.3, Quantifying Spatial Correlation, where K is the number of facies and n is the number of known

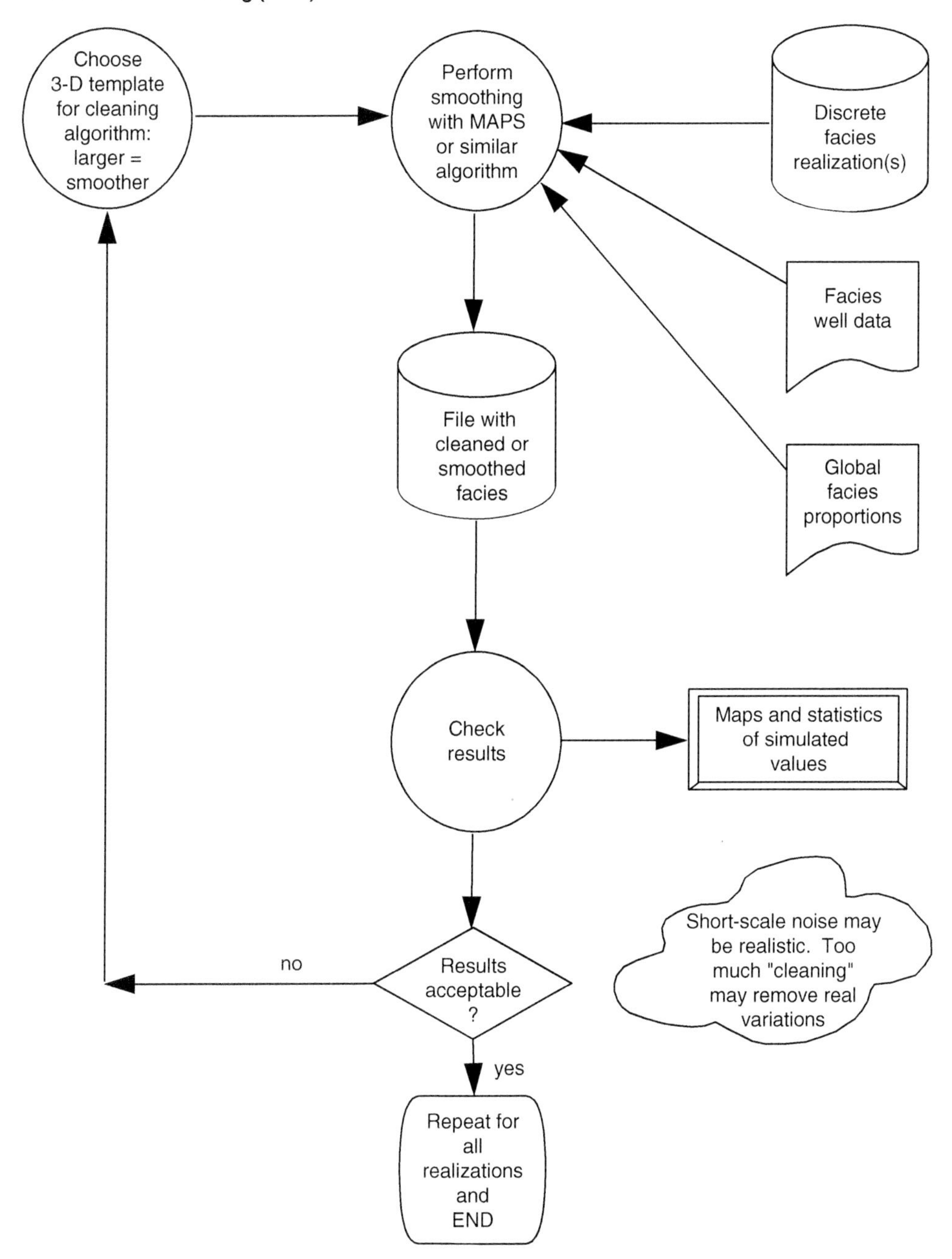

FIGURE 4.28: Work Flow for Cleaning Cell-Based Facies Realizations.

points with the addition of the unknown location, +1). We recognize the recent and emerging research to improve speed and ability of MPS to handle nonstationary features at various scales (see discussion below on Chugunova and Hu (2008), de Vries et al. (2008), and Straubhaar et al. (2010)). Mariethoz et al. (2010) have suggested a direct sampling method that does not rely on data storage; instead they sample the training image for a specific data event with the requirement of a tolerance applied to a closeness measure to data events from model and training image.

MPS-based facies modeling techniques are becoming popular since they retain the ease of conditioning to hard and soft data and trends of cell-based simulation, but integrate some geometric and curvilinear features of object-based models. The utility of MPS methods for modeling reservoirs has been demonstrated in various publications (Caers et al., 2003; Harding et al., 2005; Liu et al., 2004).

Use of MPS is motivated by the observations that (1) local data are reproduced by construction, (2) the required statistical controls (training image) are intuitive and encourage communication between earth scientists and reservoir modelers, (3) soft seismic data and large-scale geological trends are handled straightforwardly, and (4) the results appear realistic for geological settings where geologic facies have characteristic geometries, but lack crisp boundaries.

This type of heterogeneity is common in many depositional settings—for example, when the facies are controlled by the original depositional facies, but have been overprinted with diagenetic alteration or complicated preservation results from reworking of facies. Of course, when the facies appear to follow clear geometric patterns, such as sand-filled abandoned channels or lithified dunes, object-based facies algorithms (Section 4.4, Object-Based Facies Modeling) should be considered; or if even greater complexity exists related to the temporal sequence, then process-mimicking facies algorithms (Section 4.5, Process-Mimicking Facies Modeling) may be considered. See Figure 4.29 for an example MPS training image and simulated realization from the Deltaic Reservoir and refer back to Figure 4.21 for SIS and truncated Gaussian realizations for comparison.

The main steps for multiple-point facies modeling that are common to all facies modeling include the following: (1) Relevant well data together with areal and vertical geological trend information must be assembled within the stratigraphic layer being modeled. (2) Seismic attributes are often related to facies in an approximative way; if available, the seismic data must be tested to see if they calibrate to facies probabilities. (3) Declustered, spatially representative, global proportions of the facies must be determined. The following steps are unique to multiple-point facies modeling: (4) A training image or set of training images must be constructed with the salient features and at the same scale as the reservoir model and with sufficient size to provide reliable conditional probabilities for the applied data event template. (5) Conditional probabilities are calculated from the training image(s) and stored in a search tree and then common to other facies methods. (6) Sequential simulation of a 3-D facies models must be constructed and validated. It is important to establish up front that these conditional probabilities, assessed from frequencies of data event in the training image, are valid. This is discussed next along with unique steps and considerations of MPS.

4.3.2 Sequential Simulation with MPS

The following are some further details on the common sequential MPS method. In Section 2.3, a multiple point event was defined as a set of data values at locations denoted by a lag vector, $z(\mathbf{u}+\mathbf{h}_\alpha), \alpha = 1,\ldots,n$, with respect to an unknown location, $z(\mathbf{u})$. Strebelle (2002) demonstrated that if one solves the single normal equation for an outcome at an unknown location $\mathbf{u}$, given a data event, the exact solution is the conditional probability, also known as Bayes' relation.

$$P\{Z(\mathbf{u}) = z_k | d_n\} = \frac{P\{Z(\mathbf{u}) = z_k, Z(\mathbf{u}+\mathbf{h}_\alpha) = z_{k_\alpha}\}}{P\{Z(\mathbf{u}+\mathbf{h}_\alpha) = z_{k_\alpha}\}} \quad (4.6)$$

that is the conditional probabilities assessed as frequencies of events from a training image are an exact solution to the single normal equation.

Thus the scanning of the training image provides the required conditional probability function required by sequential simulation, can be calculated directly from the frequency of data events associated with a specific occurrence at location $\mathbf{u}$ divided by the total number of data events; therefore, the conditional probabilities calculated from the training image are valid for simulation. An example of this calculation is shown in Figure 4.30 with a simple four-point data event and three replicates identified in the training image. Since two-thirds of the events have a light gray value at the unknown location, $\mathbf{u}$, this proportion is assigned as the probability of light gray and one-third is assigned as the probability of dark gray.

If a specific n-point data event does not have sufficient replicates to calculate conditional probabilities, it is common practice to drop points from the data event until replicates are found. For example, if a specific 10-point statistic is not available in the training image (or there are too few replicates), then the furthest away point, $\mathbf{u}+\mathbf{h}_i$, from the location of interest $\mathbf{u}$ is removed, resulting in a 9-point

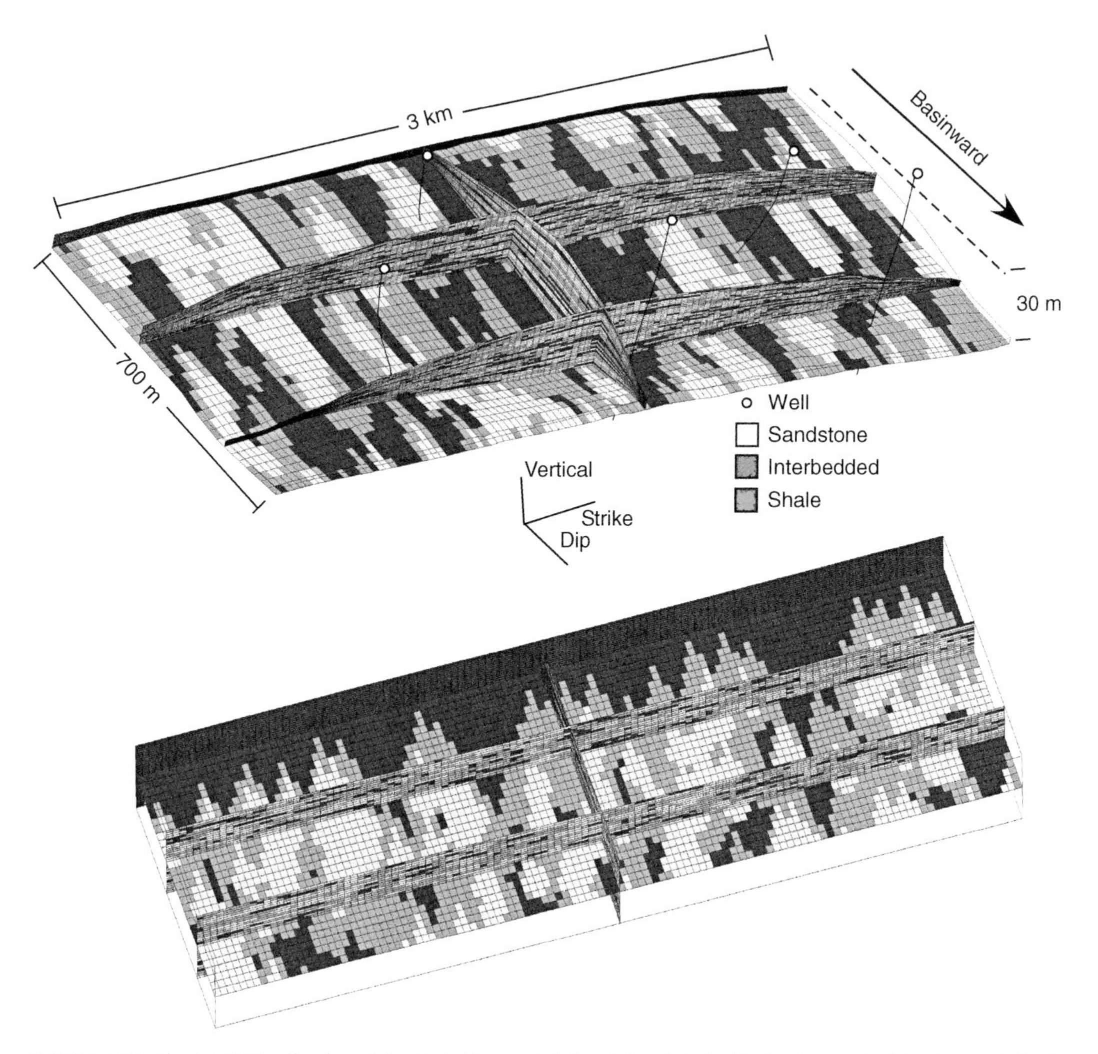

FIGURE 4.29: Single MPS Realization of the Delta Reservoir (above) Simulated with the Shown Training Image (below) and Well Constraints. Note the distinct lobe geometry and facies nesting in the realization. Compare to the same model constructed with SIS and truncated Gaussian simulation (see Figure 4.21 in Section 4.2, Variogram-Based Facies Modeling). Note that the training image is in a regular, flat grid with the average cell size from the reservoir grid above.

statistic. Points are dropped until the conditional probabilities are available.

As mentioned in Section 2.4, Preliminary Mapping Concepts, the common MPS methodology simulates sequentially along a random path. While a sequential random path reduces artifacts, improves data conditioning, and results in efficient simulation, this method may make it difficult to reproduce long-range contiguous features such as channels.

There are good reasons to use a random path. An ordered path applied in sequential simulation may spread features unreasonably along the simulation path. Also, with a random path, any correction to the global statistics (discussed in the next section) should not produce a local bias, while an ordered path could result in bias low or high regions as the simulation attempts to balance the global statistics. In addition, conditioning artifacts are limited with a random path, while an ordered path may result in discontinuities near data due to the strong influence of the adjacent previously simulated nodes.

Yet, a random path results in relatively few multiple points (previously simulated nodes and data) in each data event being informed in the early stage of the simulation and many points informed for the data events in the latter stage of the simulation. As a result, early simulated values are weakly informed

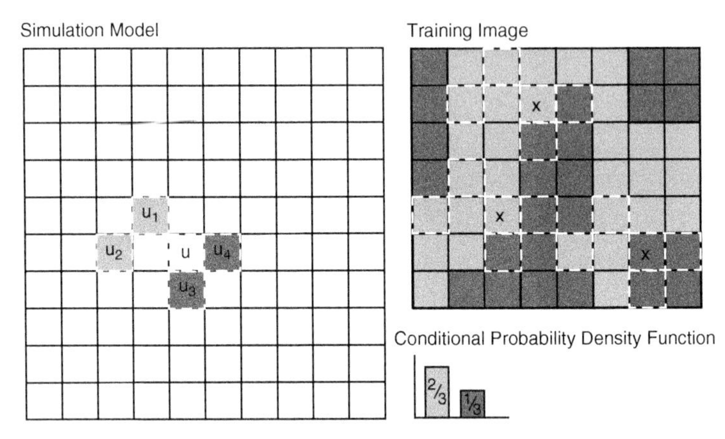

FIGURE 4.30: Simulation at Location **u** Given Four Points in a Local Search Neighborhood. The conditional probability is sampled by scanning the training image for the frequencies of facies at location **u** divided by the total number of occurrences. In this case, $\frac{2}{3}$ for light gray and $\frac{1}{3}$ for dark gray.

from the training image (based on only the few previously simulated nodes in the template and data) and latter simulated nodes are overconstrained. As a result of this effect, MPS may fail to reproduce features with long-range continuity. Imposing long-range features with a random path is analogous to building a bridge from both sides of a river without surveying. The features are seeded randomly, but then fail to meet up correctly; and later nodes on the random path, filling in the middle of bridge, may result in data events not present in the training image. Recall, when a specific data event is not found in the training image, the point furthest from the simulation location is dropped to reduce the number of points in the multiple-point statistic until the event is found in the training image. This allows the simulation to proceed, but issues arise, such breaking up the long-range features (i.e., the bridges that do not meet).

A multiple-grid approach may be applied with simulation proceeding from every n_1^{th} node to every $n_2^{\text{th}}, \ldots, n_m^{\text{th}}$, where n is decreasing to every node at the m step (Tran, 1994) (see an illustration in Figure 4.31). This helps inform large-scale features first and then fill in the short-scale continuity model, and it is successful in improving long-range continuity in variogram-based methods. However, with MPS the spatial model is much more specific; and even with multiple-grid, long-scale features are often an issue.

In Figure 4.32 a complicated training image is shown along with a resulting MPS realization without any local constraints. Note that while many of the short-range features are present, the long-range

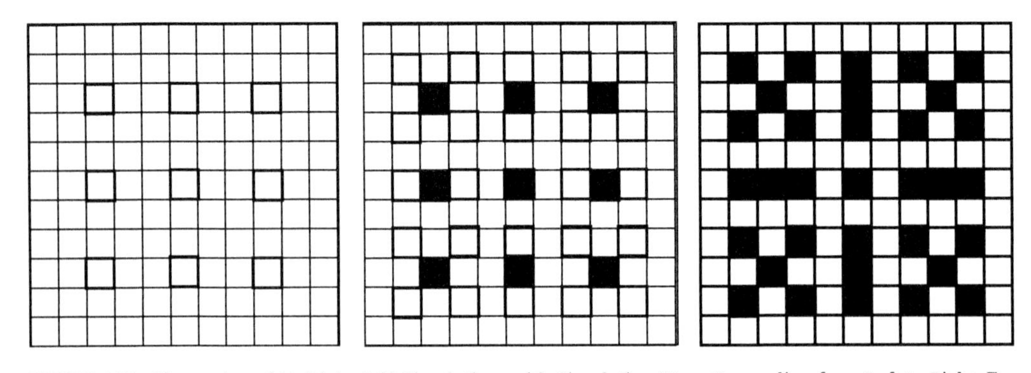

FIGURE 4.31: Illustration of Multiple-Grid Simulation, with Simulation Steps Proceeding from Left to Right. For the first step, every third grid node is simulated in a random order (on the left). For the second step, every second grid node is simulated in a random order (in the center). For the final step, all remaining grid nodes are simulated in a random order.

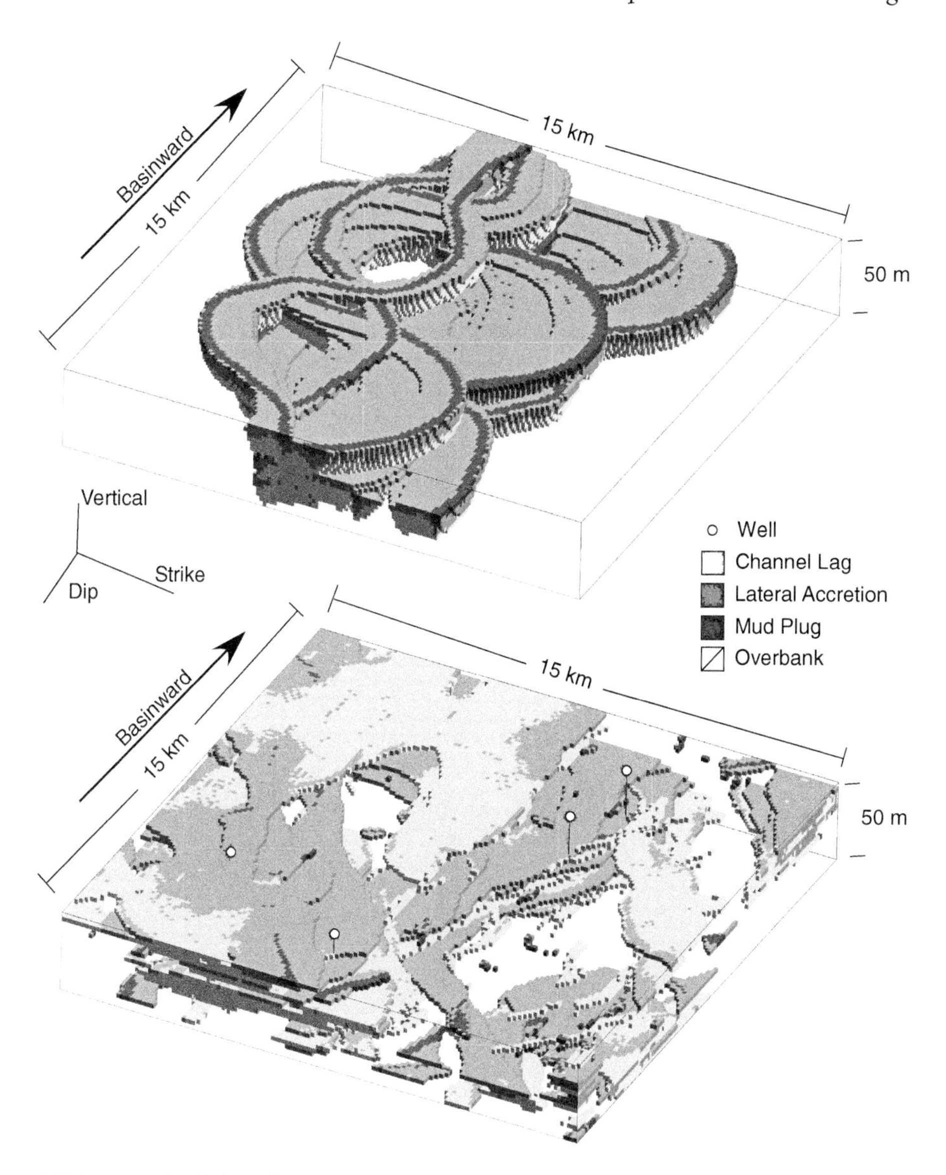

FIGURE 4.32: Simulation of a 15-km × 15-km Section of the Fluvial Reservoir Model with the Very Complicated, Nonstationary Model Applied as the Training Image and No Constraints beyond the Shown Wells. While general geometries are reproduced, the long-range contiguity is disrupted (e.g., see mud plugs segments) and the features are expanded to fill the reservoir in a stationary manner.

features and associated continuity is not well reproduced. Also, the simulation distributes all features in a stationary manner. It is important to reiterate that these issues with long-range features are also present in all other cell-based methods. The difference with MPS is that the spatial continuity model is more specific; therefore, the loss of long-range features is more apparent.

Methods to improve reproduction of training image features include: (1) Use a large enough training image to ensure that more replicates are available for as many data events as possible, (2) Use post-processing methods to locally correct the model, (Strebelle and Remy, 2005) (3) Use application of an order path, or (4) Impose nonstationary constraints, as discussed in the next section.

Strebelle and Remy (2005) have proposed post-processing techniques that track inconsistencies between the simulated realization and the training image as the count of nodes that are dropped from the data event to find the required conditional probability. This measure provides an indication of inconsistency between the emerging simulated realization and the training image at each simulated node. Post-processing is applied through the resimulation of nodes with high levels of inconsistency.

Suzuki and Strebelle (2007) demonstrated that the issues with long-range continuity are seeded with the initial coarse level of multigrid simulation (Tran, 1994). They demonstrated that issues with reproducing long-range training images features are tied to the random path and can be corrected by an ordered path (Daly, 2005). As mentioned before, ordered paths may have issues with artifacts and data conditioning. Suzuki and Strebelle (2007) suggested the use of an ordered path for the first coarse multiple-grid simulation and then switching to a random path for each subsequently finer multiple grid.

The inclusion of nonstationary constraints is very useful to constrain general long-range features. This is discussed in the next subsection.

4.3.3 Input Statistics

As with all cell-based methods, MPS can easily integrate various information sources. Typical statistical inputs include global proportions, training image(s), locally variable proportions, and azimuth and soft secondary data.

Global Proportions

The multiple-point method does not explicitly constrain the global proportions of each facies ($p_k, k = 1, \ldots, K$). With sequential Gaussian simulation, forward and back transform ensures reproduction of the input distribution within statistical fluctuations, and with sequential indicator simulation the local PDF are fundamentally unbiased local estimators based on the hard and soft data. In fact, departures in the global proportions are often seen as a part of the measure of uncertainty, given the input statistics (see Section 5.3, Uncertainty Management).

With MPS there is no such expectation of unbiasedness. The resulting realization global proportions are sensitive to the proportions within the training image; therefore, the spatial continuity measure carries information on global proportions. To minimize this issue, the common practice is to set the global proportions of the training image to the target global proportions of the simulated realizations.

This is complicated by the fact that the simulation is only influenced by conditional probabilities that are applied in the simulation. That is, if only the conditional probabilities from a subset of the training image are used, or specific conditional probabilities are applied more often, then the locations in the training image from which these were borrowed will influence the global proportions in the simulated realization. Given these issues, departures from the target global proportions are expected and tied to arbitrary implementation choices. The variations are not considered a meaningful aspect of uncertainty assessment.

Some type of correction is required to ensure that the realizations match the target global proportions. One approach is to use the *Trans* algorithm from GSLIB to correct the global proportion of the realizations without disrupting conditioning (accomplished by limiting the changes near hard data) (Deutsch and Journel, 1998). Strebelle (2000) has suggested a servosystem that applied a dynamic weighting to the global proportions during simulation of local realizations along the random path.

Nonstationary Statistics

It is common to impose various locally variable statistics on MPS. It is difficult and inefficient to infer very-high-order (very large number of points) statistics to capture large-scale features in the presence of nonstationarities (refer back to Figure 4.32). In fact, departure from simple stationary features in the training image usually results in mixing of features in the resulting simulated realizations. When these large-scale features are mappable, they should be explicitly described with models of nonstationary statistics, such as locally variable facies proportions, azimuths, and scales or even variable training images. When these features are significantly uncertain, multiple realizations or scenarios may be required to account for their associated uncertainty.

Strebelle (2002) proposed the integration of secondary data by the extension of Eq. (4.6). If more than the collocated secondary data are considered, then a training image of the secondary data is required.

$$P\{Z(\mathbf{u}) = z_k | d_n, y\} = \frac{P\{Z(\mathbf{u}) = z_k, Z(\mathbf{u} + \mathbf{h}_\alpha) = z_{k_\alpha}, y\}}{P\{Z(\mathbf{u} + \mathbf{h}_\alpha) = z_{k_\alpha}, y\}}$$

where y is the addition of secondary information available at all locations.

Commonly, locally variable proportions are imposed on the local conditional probabilities calculated from the training image by updating with either a simple weighting scheme (where the weights ensure no interruption of data conditioning) or more complicated probability combination schemes such as the Tau-model presented in Section 2.4, Preliminary Mapping Concepts (Harding et al., 2005). This approach may cause the probabilities to depart significantly from those imposed by the training image, resulting in features not informed from the training image.

Locally variable azimuths and scale may be integrated through simple rotation and scaling of the template (Strebelle, 2006). This is efficient because it allows for the utilization of the conditional probabilities calculated form the initial scan of the training image. Yet, it should be recognized that the multiple-point statistics represented in a search tree provide a discrete spatial continuity model that is linked directly to the model cell scale. This is contrary to a variogram model that provides a consistent spatial continuity model that is valid for all distances and directions and may be scaled to explicitly account for variable support size of the data and model cells. Therefore, any scaling or rotation of multiple-point statistics is somewhat less rigorous and likely relies on (a) a limited number of binned rotational angles and size scale and (b) an approximation with the nearest data and previously simulated nodes being assigned to data event template without explicit accounting for scale. Example realizations from the Delta Reservoir with locally variable azimuth and locally variable scale are shown in Figure 4.33. The locally variable azimuth imposes a distributary pattern, while the locally variable scale results in decreasing element size distal. Another example is included in Figure 4.34 with the addition of locally variable facies proportions to impose a transition from sand to interbedded sand and shale to shale from proximal to distal.

Another method to deal with these nonstationarities may be to divide the simulation model into regions and to use training images designed for the local statistics in each region (de Vries et al., 2008; Wu et al., 2008). Yet, this would require the modeling of transitional training images (5 or more training images) and careful mapping of regions to ensure smooth and consistent transitions at boundaries. de Vries et al. (2008) applied inverse distance weighting to weight conditional probabilities from each tree based on centroids on distance from the regions in the simulated reservoir. These methods add to the difficulty in modeling, reducing the intuitive flow of information from geological concepts to training image to model (Boucher, 2011).

Boucher (2009) suggested a more automated methodology of a computer-based partitioning of the search tree based on filter scores. In this manner the algorithm automatically separates stationary segments of conditional probabilities calculated from the training image and then applies these search tree segments locally. This requires some method of mapping of the regions to identify volumes to apply each search tree segment that may be done by hand or through a simulation of the partition classes with the same autocorrelation identified in the training image.

Chugunova and Hu (2008) suggests the application of auxiliary information such as a model of depositional energy or distance from source mapped continuously over the training image and the reservoir model space. This auxiliary information is then applied during simulation to constrain the regions of the training image that are assessed to borrow stationary statistics for the specific location within the reservoir model. This method is effectively relaxing the assumption of stationarity of statistics in the training image; stationarity is assumed over volumes with similar auxiliary information.

For these methods, multiple secondary data and trends may be integrated with the probability combination schemes mentioned above. In the presence of more than a couple secondary variables and trends, the massive multivariate work flow for Section 4.1, Large-Scale Modeling, may be applied first to combine multiple secondary data together into a single variable to apply as a secondary variable for simulation.

Training Image Construction

The multiple-point model of spatial continuity is the training image: a large enough, exhaustive 3-D representation of the inferred heterogeneity of facies categories at the scale of the model without any *locality*. The training images do not need to be the same size as the model, it should be large enough so that sufficient replicates are available to calculate the multiple-point conditional probabilities and not be so large that it slows down the simulation needlessly. The training image should be limited to the categories considered in the multiple-point statistic, and it should have features that are judged to have a significant impact on the transfer function and that

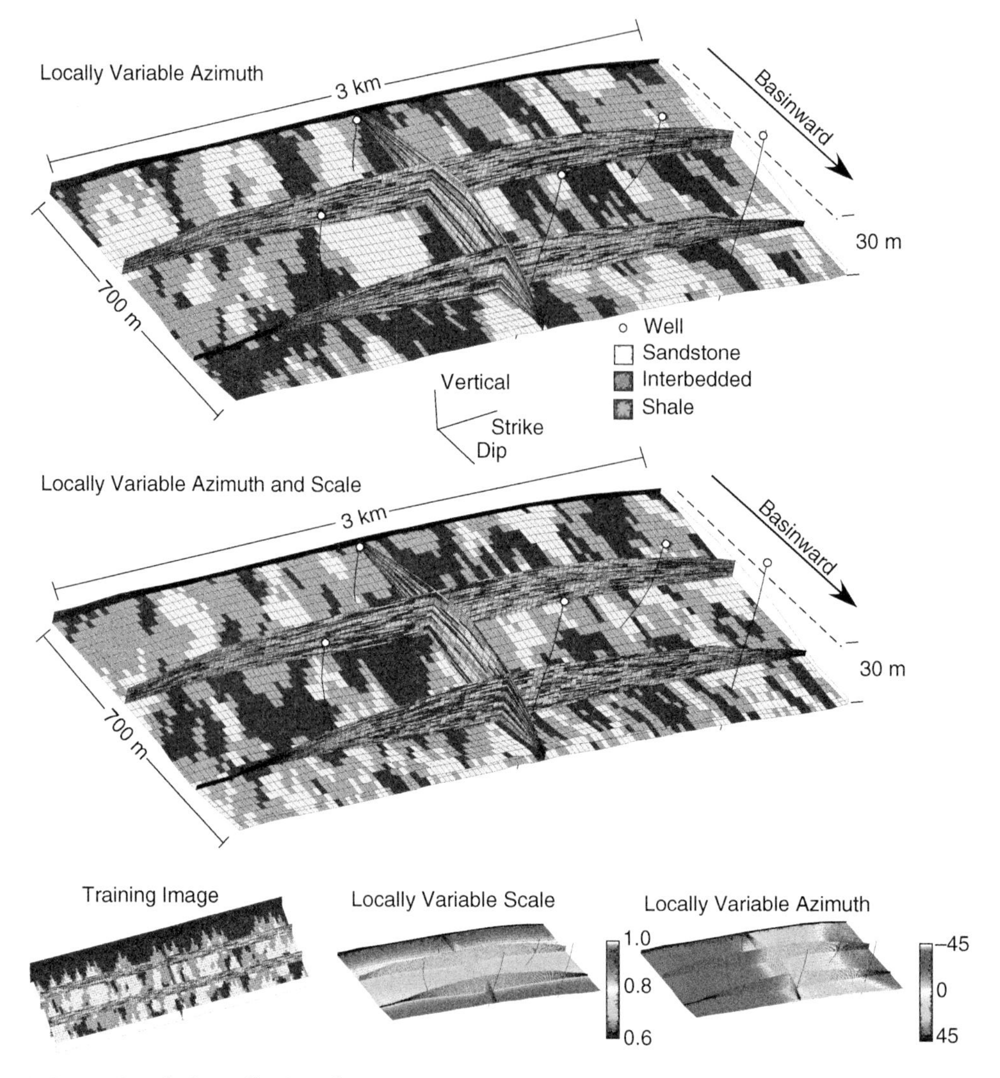

FIGURE 4.33: Single Realization of the Delta Reservoir Simulated with MPS with the Indicated Training Image, Wells, and Locally Variable Azimuth (Upper) and Locally Variable Azimuth and Scale (Lower). Note the distributary pattern and the decrease in lobes size distally. Compare to the same model constructed with MPS and without locally variable constraints in Figure 4.29.

have a good likelihood to be reproduced with the selected data event template. The training image need not be locally accurate or conditional. Spatial statistics are extracted to be applied over the simulated realization under the stationary constraints imposed.

There are a variety of methods to construct these training images. Pyrcz et al. (2007) constructed a training image library of reservoir features typical in common depositional settings. This would allow the user to simply select their spatial continuity model. Greater flexibility may be possible through a training image generator that allows for the drafting of specific heterogeneities for any number of categories and proportions in real time. Maharaja (2008) and Boucher et al. (2010) designed unconditional object-based simulation programs for building training images. In addition, object-based programs like the method developed by Deutsch and Tran (2002) or even process-mimicking methods may be applied to build training images (Pyrcz et al., 2009).

Yet, a key limitation with training images is that they must be assumed stationary and, for best results, be as simple as possible. Complexity in training images often results in local nuance and nonstationarity

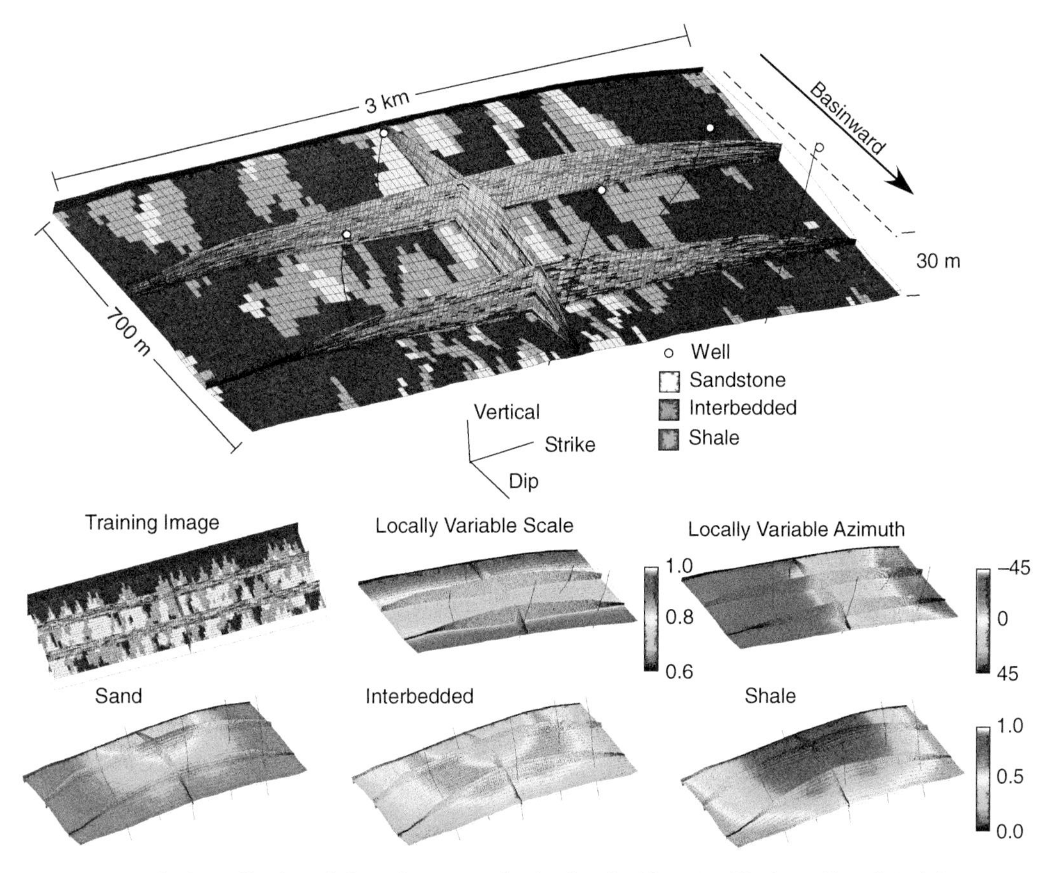

FIGURE 4.34: Single Realization of the Delta Reservoir Simulated with MPS with the Indicated Training Image, Wells, and Locally Variable Azimuth (upper) and Locally Variable Azimuth and Scale (Lower) and Locally Variable Proportions for Each Facies. Note distributary pattern and the decrease in lobes size distally and the transitions from sand to interbedded sand and shale and shale from proximal to distal. Compare to the same model constructed with MPS and without locally variable facies proportions in Figure 4.33.

that are mixed in the search tree and result in ambiguous features in the simulated realizations. For this reason, the training images should include simple features (refer back to the results in Figure 4.32 with a complicated training image and absence of a nonstationarity model). The local complexity is best imposed with nonstationary features in the simulation, discussed in the previous subsection.

MPS (like variogram-based methods from Section 4.2, Variogram-Based Facies Modeling) reproduces local hard data by construction and easily account for local trends. As described in Section 2.4, Preliminary Mapping Concepts, the hard data are assigned directly to the model grid cells prior to sequential simulation along a random path. This allows for conditioning to any density of available local data, but this introduces some unique challenges in dense data settings because the spatial statistics are derived from a training image and not from the data. This introduces the possibility of contradiction between data and multiple-point spatial continuity model that may result in artifacts, typically recognized as discontinuities near data.

In fact, there is no guarantee that the training image features will be reproduced in the realizations. This poses some additional challenges in training image construction and MPS model checking. Boisvert et al. (2007b) have suggested checking training image compatibility with data by comparing various data and training images statistics, such as distributions of runs, thickness distributions, and 1-D multiple-point statistics to avoid data and training image contradictions. Boisvert et al. (2010) suggested statistical comparisons of multiple-point histograms and scaling relations that may be applied to check the reproduction of the multiple-point statistics from the training image in the reservoir realizations. Yet, this is a challenging topic because the multiple-point statistics are generally much more difficult to summarize and visualize than variogram-based spatial continuity measures.

4.3.4 Implementation Details

The search tree and carefully chosen data event templates are important implementation details that impact the efficiency of MPS methods and the quality of the MPS realizations. Also, there may be a need to model continuous trends within the simulated facies model.

Search Tree

The efficient storage and retrieval of conditional probabilities is important for fast MPS applications. It would be computationally expensive to search the training image for the conditional probabilities required at every simulation node during sequential simulation. Strebelle (2002) proposed the use of a dynamic storage method, a search tree, to efficiently store and retrieve conditional probabilities. As mentioned previously, Boucher (2009) developed the concept of partitioned search trees to address training image non-stationarity.

Others have suggested more efficient storage methodologies. Straubhaar et al. (2010) suggest a list structure that requires less RAM for storage and allows for parallelization of the calculation of the conditional probabilities. With more efficient calculation and storage of conditional probabilities, there is an increase in the number of points and number of facies that may be practically applied within MPS, resulting in more complicated reservoir models.

MPS Template

MPS template selection may significantly impact the resulting MPS realizations. This includes template extents, number of points in the template, and orientation. Commonly, the template is assumed to be a simple ellipsoidal shape with user-defined or automatically inferred extent, orientation, and anisotropy, and with all nodes considered within the ellipsoid. Lyster and Deutsch (2006) suggested two-point entropy as a fast proxy for the information content of each node within the multiple-point template. The procedure requires a prior scanning of the training image with the largest practical template and then removal of nodes based on a minimum entropy threshold.

$$H_{\max} = -\sum_{k=1}^{K}\sum_{k'=1}^{K} P_{kk'} \ln P_{kk'}$$

where K is the number of possible facies, and $P_{kk'}$ is the probability of facies k occurring at the central point $\mathbf{u}$ and facies k' occurring at point $\mathbf{u} + \mathbf{h}_\alpha$ of the template. Higher entropy suggests more randomness and therefore less correlation between the central, estimated point, $\mathbf{u}$ and point $\mathbf{u} + \mathbf{h}_\alpha$. Therefore, the points with the lowest entropy should be considered to contain the most information.

There are various considerations in template design. Firstly, the template should be large enough to characterize the heterogeneities of interest. Yet, if the template is too large relative to the training image size, then there may be too few replicates for reasonable statistical inference and a large part of the template, the edges, will not be used frequently. Also, the template size has a first-order control on simulation time and storage requirements. As mentioned in Section 2.3, Quantifying Spatial Correlation, the number of conditional probabilities required is K^{n+1}, where K is the number of categories and n is the number of known nodes in the template and $+1$ is the unknown node in the template. Also if there are secondary data, this increases to $K^{n+1} \cdot S^{n_s}$, with the extension of S as the number of possible categories for the secondary data and n_s is the number of secondary data (Strebelle, 2002).

Continuous Trends

As with other cell-based facies models, MPS proceeds sequentially from cell to cell without any explicit assignment of geobodies or geometric architecture. For example, while MPS may result in connected cells forming recognizable geometries, there is no a priori specification of these geometries. For example, an MPS model may simulate lobes, but these lobes are a spatially clustered set of simulated nodes and assigned conditioning to grid that cannot be readily identified as a lobe; therefore, the simulation cannot label each lobe with a unique lobe index. While this problem may seem trivial when these cell-based geometries are isolated, this is often greatly complicated by imbricated and/or complicated geometries that cannot be readily separated and identified.

This limits the ability to provide continuous trends for subsequent continuous reservoir property simulation. For example, if a channel model is simulated with MPS, it may be expected that the continuous reservoir properties within the channel align with the channel. In some cases the continuity follows the channel path and the reservoir properties fine to the channel margins and from the base of the channel to the top. While geometries emerge, specific relations to continuous properties within the facies regions cannot be readily integrated with MPS work flows.

While attempts have been made to detect the geometries from the cell-based models and to fit trend models to these geometries, these methods break down in the face of geometric complexity and imbrication of the geometries (Cavelius et al., 2012; Pyrcz, 2004). For non-cell-based facies modeling methods these geometries are known and it is straightforward to calculate continuous property trend models within the modeled geometries (see Section 4.4, Object-Based Facies Modeling, and Section 4.5, Process-Mimicking Facies Modeling).

4.3.5 Work Flow

The following are the main work flows related to multiple-point statistics simulation. Figure 4.35 illustrates the steps in building a training image. The first step is to integrate all available information, including well data, seismic data, and conceptual information and experience into a facies framework. This framework includes the number, geometries, and interrelationships of facies. Then a 3-D representation of this facies framework is constructed and checked with the available data and concepts. If satisfactory, then a consistent trend model is constructed to locally adjust the training image features within the conditional simulation. The interplay of trend model and stationary training image should be considered and checked.

Figure 4.36 illustrates the steps in multiple-point simulation. From all available information, including well data, seismic data, and conceptual information and experience, a consistent training image and trend model are constructed. The well data are then assigned as data in the model grid. The simulation proceeds by visiting all grid nodes in a random order and then applying the following steps: pool local data, sample from a conditional distribution constructed from the training image, impose the local trend information on the conditional distribution, apply Monte Carlo simulation, and assign the simulated value as data to the grid. This is repeated along the random path until all grid nodes are visited.

4.3.6 Section Summary

This section has introduced multiple-point-based facies models as a methodology to further introduce geological complexity into cell-based facies models with the addition of curvilinear features and ordering relations. This method has become common practice due to the intuitive work flows, flexible data, trend and secondary information integration, and opportunities to directly integrate geological concepts through training images.

This is a very active area of research. New methods are improving the efficiency in representing geological heterogeneity with multiple-point statistics and the ability to account for nonstationary information in reservoir with location specific and nonstationary training images.

To some extent, the concepts of MPS have led to a change the paradigm of geostatistics from reproduction of statistical descriptions to image processing and image reproductions. In this theme, additional research methods such as FILTERSIM (Wu et al., 2008; Zhang et al., 2006) and SIMPAT (Arpat and Caers, 2005) have been developed. With these methods, patterns are extracted, classified from a set of training images, and then placed in the reservoir model. These methods are currently not practical due to conditioning issues, high level of computational effort, and artifacts.

If crisp geometries are expected in the reservoir, then object-based facies modeling may be the best method (discussed in the next section). If the reservoir heterogeneity includes complicated stacking patterns and interrelationships related to geological process, then process-mimicking methods may be considered (see Section 4.5, Process-Mimicking Facies Modeling).

4.4 OBJECT-BASED FACIES MODELING

In Section 2.4, object-based facies modeling is introduced as an alternative to cell-based simulation (variogram and multiple-point methods discussed in the previous couple of sections). With this method, parameterized architectural geometries are placed sequentially into a reservoir model initialized with a background facies until some criteria is met, such as target global proportion and data conditioning. Object-based facies models are visually attractive because the resulting facies realizations mimic idealized geometries interpreted in outcrops, high-resolution

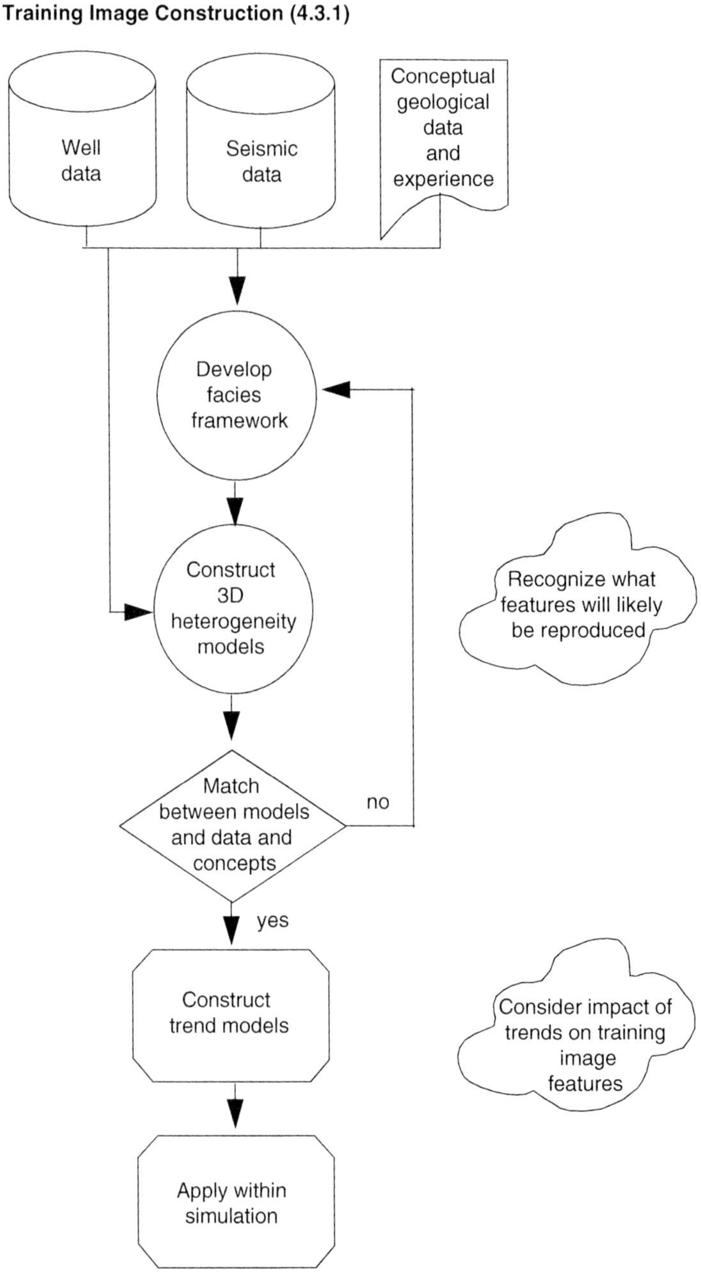

FIGURE 4.35: Work Flow for Training Image Construction.

seismic analogs, and modern analogs. These models result in facies represented by crisp geological shapes and realistic idealized nonlinear continuity and within object continuous property trends that cannot typically be modeled with cell-based approaches.

A discussion is provided in the *Background* subsection on object-based techniques including an overview of some geological shapes that could be considered, the relevant data that object-based models must reproduce, and algorithms for object placement.

The *Stochastic Shales* subsection presents the concepts of object-based modeling in the context of stochastic shales. Shale objects are placed within a net reservoir background. Object-based modeling gained wide popularity in the context of modeling fluvial facies. The *Fluvial Modeling* subsection

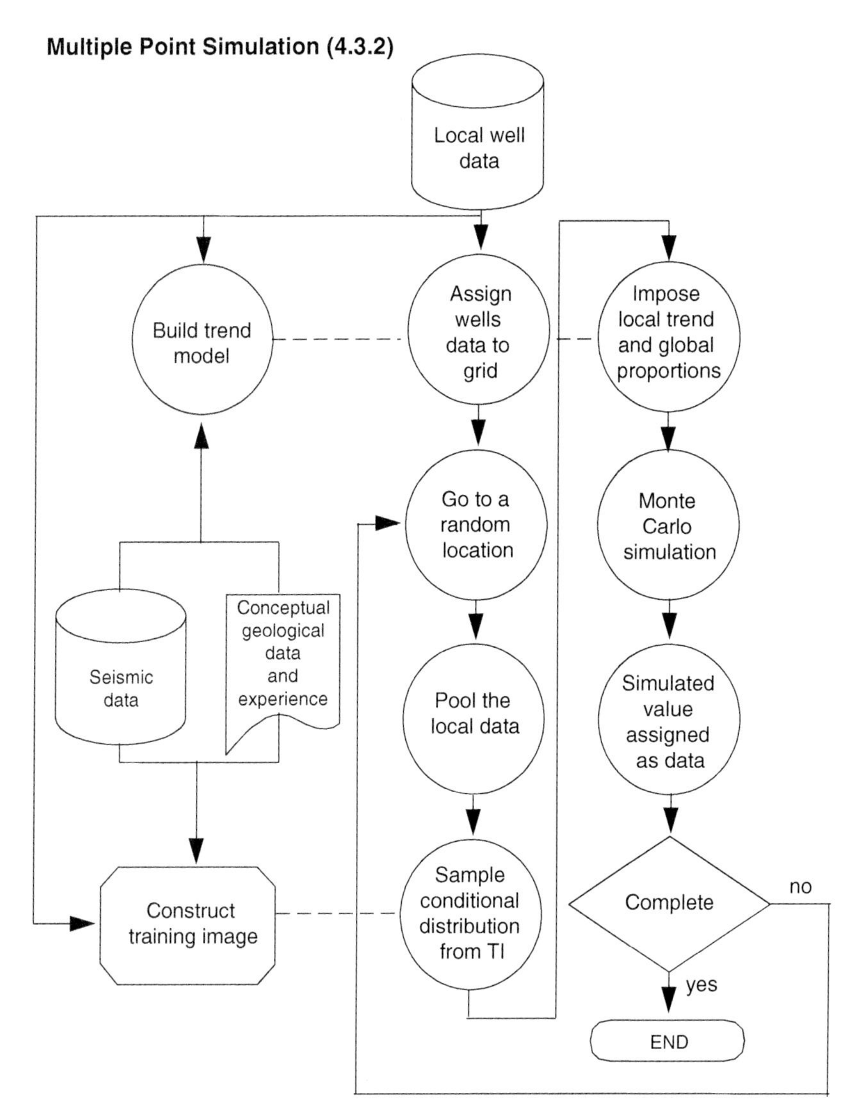

FIGURE 4.36: Work Flow for Multiple Point Simulation.

presents some details related to modeling abandoned sand-filled fluvial sand channels and related overbank facies (levee and crevasse sands). Implementation details including the integration of well and seismic data are discussed.

Non-Fluvial Depositional Systems presents considerations for other depositional systems. In particular, the application to deepwater distributary lobe depositional systems is demonstrated. Finally, the *Work Flow* subsection presents work flow diagrams for some operations related to object-based facies modeling.

The novice and intermediate modeler may benefit from details on specific object-based object parameterization and the associated placement algorithms. This will help improve implementation and explain observed features and potential conditioning issues in the face of dense well data and detailed trends. The expert modeler may be able to expand these methods with new parametric geometries and placement methods tailored to their specific reservoir depositional setting and project goals.

4.4.1 Background

In Section 2.1, Preliminary Geological Modeling Concepts, the concept of architectural hierarchy was introduced. It is convenient to view reservoirs from the perspective of nested building blocks. Architectural elements are typically the basic reservoir building block, organized into architectural complexes and complexes are organized into architectural complex sets. At times it is useful to subdivide elements into stories, beds, and lamina. Recall that we leave the definition of facies ambiguous (see Section 4.2, Variogram-Based Facies Modeling). Facies describe any category at any scale that separates the reservoir into mappable regions with distinct reservoir property statistics.

From this perspective, facies modeling—and by extension the objects in the object-based models—may be concerned with elements, stories within elements, or even beds within stories. For example, an object-based model in a fluvial setting may characterize the channel, levee, and splay architectural elements within a channel complex. In another case, the object-based model may characterize overbank beds or even shale drapes that impeded fluid flow. In microscale modeling the objects may actually be lamina. Regardless of the scale of the objects, the objects should represent significant components of the reservoir. Significance is determined by volumetric contribution and/or contribution to the reservoir flow response (flow conduits, barriers, and/or baffles).

A set of object-based models may be applied to span various hierarchical scales. For example and once again in the fluvial setting, reservoir architecture may be modeled firstly by object-based models of channel complexes, and then subsequently by object-based models of elements within the channel complexes, object-based models of stories within elements, and on through additional smaller-scale features. We consider modeling this genetic hierarchy of heterogeneities by successive coordinate transformations and geometric objects representing each scale. Porosity and permeability models are then constructed at the appropriate scale using coordinate systems aligned with depositional continuity at a specific scale or even with influence from a variety of scales. In general, the number of scales is limited, because large scale features such as complex sets and even complexes may be mappable with available information and, therefore, better modeled by sound deterministic mapping. Also, it is likely that very-fine-scale features (e.g., beds and lamina) do not have a significant impact of volumetrics or flow and/or could not be characterized under the scale constraints of the flow grid (see Section 4.1, Large-Scale Modeling).

As discussed in Section 2.1, significant effort has been focused on sedimentology and stratigraphy of reservoir systems. There are comprehensive texts that cover a broad range of depositional settings (Einsele, 2000; Reading, 1996; Walker and James, 1992), while Galloway and Hobday (1996) provide a convenient synthesis for resource investigation. We provide some key information for designing object-based models, but it is not possible to document a comprehensive approach. Consider the fluvial depositional setting. The book by Miall (1996) provides a well-illustrated description of fluvial sedimentary facies, basin analysis, and petroleum geology with 16 fluvial styles demonstrated with more than 500 figures and 1000 references. Given that the fundamental definition of architectural element relates to describable external and internal geometries, these architectural hierarchical studies are well suited to informing object-based models.

The literature describing fluvial deposits is rich and varied. The history of quantitative computer models for fluvial systems is also extensive. Allen's early qualitative work in the 1960s and 1970s (Allen, 1965) led to quantitative computer simulations (Allen, 1978). Leeder, at about the same time, was also building quantitative models (Leeder, 1978). Bridge published in this area with Leeder (Bridge and Leeder, 1979) and also published computer code (Bridge, 1979) that was updated recently (Mackey and Bridge, 1992).

Although not specifically designed for fluvial facies, object-based models became popular in petroleum reservoir modeling in the mid-1980s due to the work of Haldorsen and others (Haldorsen and Chang, 1986; Haldorsen and Lake, 1984; Stoyan et al., 1987). The importance of fluvial reservoirs in the Norwegian North Sea soon prompted the development of these object-based methods for fluvial facies (Clemensten et al., 1990; Damsleth et al., 1992a; Fælt et al., 1991; Gundesø and Egeland, 1990; Henriquez et al., 1990; Omre, 1992; Stanley et al., 1990). The theory and implementation was refined over a number of years (Georgsen and Omre, 1992; Hatløy, 1995; Hove et al., 1992; Tjelmeland and Omre, 1993; Tyler et al., 1992a,b,c) with increasing practical application of these methods to Norwegian North Sea reservoirs (Bratvold et al., 1994; Tyler et al., 1995). Such applications have set the standard for other oil producing regions of fluvial depositional setting. Other non-Norwegian oil companies also developed object-based modeling capability (Alabert and Massonnat, 1990; Khan et al., 1996), and the object-based model is common in reservoir modeling software.

Recently, there has been an increase in architectural hierarchy studies for deepwater reservoirs with work, including Beaubouef et al. (1999), Sprague et al. (2002), and Sullivan et al. (2004). Similar to the fluvial counterpart, these studies provide detailed descriptions of the characteristic geometries across scales for each distinct type of deepwater system. For example, for the distributary lobe setting common at the slope break and in ponded systems, distributary lobe complexes are composed of lobe elements with internal fill characteristics related to depositional energy. This effort to characterize the deepwater architectural hierarchy has been matched by efforts to integrate these features into object-based models [for example, see Shmaryan and Deutsch (1999)].

Object-based models are now created routinely in reservoir characterization. The three key issues to be addressed in setting up an object-based model are the (1) geological shapes and their parameter distributions, (2) algorithm for object placement modification, and (3) relevant data to constrain the resulting realizations.

Object-based models are able to generate heterogeneity models with crisp geometries that are not possible with cell-based methods. A fundamental assumption with object-based modeling is that the shapes of the specific depositional units are crisp and can be described by a parameterized geometry. While the sedimentological process may generate identifiable geometries, geological complexity due to re-working, imbrication, and diagenesis may result in features that are more broken up or "stochastic" in form, lacking in clear geometries. In these cases, cell-based methods may be favored for their improved conditioning capabilities. Although object-based models may be post-processed to reduce geometric crispness (with the addition of noise), it is not generally possible to add geometric information to cell-based models.

As mentioned in Section 2.1, Preliminary Geological Modeling Concepts, it is important to not confuse geomorphology and sedimentology. It is tempting to apply object-based models to model modern features expressed on the Earth's surface. For example, objects may be designed to represent fluvial channels, eolian dunes, carbonate reefs, and so on. Yet, it is important to ensure that these are the preserved features represented in the reservoir sediments and not just temporary geomorphic expressions. For example, meandering fluvial channels may rework the entire fluvial valley and eolian dunes may climb, and both be pressures as sandstone sheets unrelated to the channel or dune geometries, respectively. Aggradation, denudation, progradation, and retrogradation as a result of eustasy and carbonate growth rates dramatically impact the preserved reef geometry, and the final geometry may vary greatly from a typical idealized reef. In addition, even when architecture is preserved in a simple manner, the preserved geometry may be altered. Consider the case of differential subsidence associated with sand filled channels in muddy overbank. Because the overbank is highly susceptible to compaction while the sand-filled channel will likely be more resistant to compaction, the deposited concave upward and flat top geometry may be transformed into a convex upward mound.

The discussion here is *not* a rigorous presentation of geological concepts for different depositional systems. A pragmatic approach is considered—that is, pragmatic from a geological, flow modeling, and geostatistical perspective. The idea of modeling reservoirs by genetic forward modeling is attractive, but not considered here. Object-based models are *pseudo*-genetic in the sense that erosional rules (younger rocks on top) and other geological principles are used as much as possible. This may even include integrating statistics that constrain the overlap and clustering on objects.

Most process-mimicking facies models (see Section 4.5, Process-Mimicking Facies Modeling) methods rely on objects and surfaces. In separating the two methods, we have drawn the line with consideration of forward models and process rules. If an algorithm places objects without a concept of time sequence, then we call this an object-based model. If an algorithm places objects in sequence and without rules that relate one object to the next, we consider this to still be object-based modeling. If an algorithm places objects in time sequence with rules that mimic depositional process, then we consider this as a process-mimicking model.

There is the inevitable question of semantics and word choice. We choose to use "object-based" modeling. Of course, these object-based models end up as a collection of cells, but they are not called cell-based. The reference to object-based is in the method of creation and not the format of the result. Some prefer "Boolean" models, which has a connotation of greater statistical rigor. Others prefer "marked point process" connoting a statistical point process for object centroids, and then object properties such as size, orientation, and facies type are attributed to the point process.

Geological Shapes

There is no inherent limitation to the shapes that can be modeled with object-based techniques. The shapes can be specified by equations, a raster template, or a combination of the two. They may be placed with a single centroid or fit to a centerline (Wietzerbin and Mallet, 1993). For example, a channel fill cross section fit to a sinuous curve, indicating the position of the channel and associated planform morphology. These shapes may be combined from a set of predefined parametric shapes or even specified or drawn freely prior to modeling. The geological shapes can be modeled hierarchically; that is, one object shape can be used at large scale, and then different shapes can be used for internal small-scale geological shapes. The shapes may be genetically linked to each other. Some evident shapes are as follows:

- Abandoned sand-filled fluvial channels within a matrix of floodplain shales and fine-grained sediments. The sinuous channel shapes are modeled by a 1-D centerline and a variable cross section along the centerline. Levee and crevasse objects can be attached to the channels (Deutsch and Wang, 1996). Shale plugs, cemented concretions, shale clasts, and other non-net facies can be positioned within the channels along with trends and lags in the net facies. Clustering of the channels into channel complexes or belts can be handled by large-scale objects or as part of the object placement algorithm.
- Lower-energy meandering fluvial systems can be modeled as sand lenses or point bars within a non-net background (Caers, 2005; Hassanpour et al., 2013; Pyrcz et al., 2009). We sometimes consider modeling the entire channel (as above) and then assigning sand and shale within the channel in some realistic manner.
- Other channelized depositional systems including deepwater and estuarine systems are often modeled by adapting fluvial channel modeling techniques to system-specific considerations such as channel size, width-to-thickness ratios, and internal heterogeneities (Shmaryan and Deutsch, 1999). The transition from channels to other facies types can also be handled by linked geometries; that is, a channel may evolve into a more lobe-like geometry at some distance into the modeling domain (as energy is lost and the sediments disperse). Channel-type systems are discussed in detail below.
- Shelf edge carbonate reefs may be modeled with a mounded reef core and breccia lobes draping the fore reef and patch reef scatter in the back reef area. Grain stones may form aprons around the reefs. In this setting, it is possible that diagenetic alterations may significantly disrupt the importance the object-based facies model.
- Eolian dune shapes are candidates for object-based modeling. Although the 3-D geometry of such object is not trivial to define analytically, we can make the necessary assumptions for practical modeling.
- Remnant shales may be modeled as disk or ellipsoid objects within a matrix of sand. This may be appropriate in high net-to-gross reservoirs. Although such shales may have a low proportion, they will significantly affect the vertical permeability—hence, horizontal well production and coning.
- Deltaic reservoirs may be modeled as fan-shaped sand units within a matrix of poorer-quality sediments. Historically, cell-based techniques have been used more in such systems.
- Shoreface environments have a systematic progression from proximal or near-shore facies through distal or offshore facies. The position of the shoreline could be modeled as a function of geological time (position in the stratigraphic layer) and the particular ordering of facies linked to this position. A hybrid cell- and object-based modeling scheme would be appropriate.

Another form of object-based modeling is "surface-based modeling" that is described briefly in Section 4.5, Process-Mimicking Facies Modeling, and its future importance is discussed in Chapter 6, Special Topics. Stratigraphic surfaces are viewed as geological objects and modeled with a Boolean object-based formalism. We group these with process-mimicking facies modeling methods as the surfaces are placed in time sequence of events that aggrade and erode the evolving composite surface, and in general the surface placement is determined by process-mimicking rules. However, initial surface-based methods did utilize simple stochastic placement of surfaces (Xie et al., 2000) and could be considered object-based.

Relevant Data

A realistic facies model should reproduce all available data within the volumetric support and precision of each data source. The relevant data include local well data, seismic data, production data, and geological interpretations including deterministic objects, connections, and trends.

Local well data consist of identified facies codes at arbitrary locations in the 3-D reservoir layer. The facies may come from direct core observation or inference using a variety of well logs. Local well data take the form of intersections of facies types at arbitrary orientation and location.

Seismic and production data are most often imprecise large-scale information. There are different ways of representing such information. One way is to calculate the seismic-derived probability of each facies:

$$p_k(\mathbf{u}_V), \qquad k = 1, \ldots, K, \ \ \forall \mathbf{u}_V \in A$$

where $p_k(\mathbf{u})$ is the probability or proportion of facies k at location $\mathbf{u}_V$, there are $k = 1, \ldots, K$ facies codes, and the location $\mathbf{u}_V$ is defined everywhere in the reservoir A. The V subscript is the volume over which the facies proportions are defined. Note that a slightly different notation was considered for the cokriging formalism considered in Section 4.2, Variogram-Based Facies Modeling.

Locally varying facies proportion information may also be obtained from geological interpretation and mapping. Knowledge of the depositional system often allows specification of areas of greater and lesser proportions of different facies.

Seismic, production, and geological data may not be able to distinguish between some of the facies. For example, crevasse splay and levee sands may be indistinguishable from each other in the data. An important constraint in object-based modeling will be to reproduce proportions of "grouped" facies.

The location of some geological objects may be known exactly from seismic observation, interpretation of well test or production data, or geological mapping. The positioning of such objects must be fixed and not left to an algorithm for stochastic placement. There are other types of information regarding object placement: (1) knowledge of the exact position of some objects, (2) knowledge that two (or more) well intersections of the same facies *cannot* relate to the same object, or (3) a probabilistic measure of two (or more) well intersections of the same facies belonging to the same object. These "connections" come from geological interpretation of the particular cored facies and from engineering data that relate pressure of fluid continuity between intersections in different wells. Many implementations of object-based modeling make allowance for such interpreted "connections."

Whereas cell-based models easily reproduce dense well data, locally variable proportions and secondary data, object-based models have some limitations. Dense well data may result in either long run times or artifacts as the specific object-placement scheme attempts to honor the conditioning data. Long run times result from iterations to simultaneously match these difficult constraints. Artifacts generally include: (1) distinctly different model behavior at and near the wells, (2) unrealistic geometries, and/or (3) unrealistic changes in geometry.

A basic check for object-based models is that the location of conditioning data cannot be determined from the model itself and the geometries are stationary, unless nonstationarity is imposed explicitly. Dense conditioning is defined as data spacing that approaches the dimension of the geometry and/or nonsmooth trend and secondary data with significant changes at or below the scale of the objects. For example, conditioning to wells with spacing that is shorter than the width of a channel is challenging due to the need to "fit" channel cross sections to multiple wells while forming realistic planform morphologies across the model. Also, care must be taken not to introduce contradictions with trend and secondary data. For example, high-resolution vertical details cannot be reproduced at scales finer than a channel thickness, and abrupt changes in areal trends cannot be reproduced with contiguous channels. More information on conditioning bias is provided for object-based conditioning in the object placement discussion below.

Lastly, the set of relevant data includes distributions of uncertainty related to the object sizes, shapes, orientations, and interactions. These parameters are specific to a particular problem, but are typically associated with some uncertainty. Hong and Deutsch (2010) presented methods to extract object size and proportions from indicator variograms from well data. It is very difficult to establish reliable distributions for these parameters from sparse well data. In fact, it is hard to get reliable distributions of these parameters from well-exposed outcrops. Recent examples of work to harvest outcrop information for the construction of object-based reservoir models include Novakovic et al. (2002), Willis and White (2000), and White and Willis (2000).

Improvements in seismic imaging and the use of shallow analogs have recently improved this difficult inference problem. Nevertheless, the inference of these distributions remains one of the biggest limitation of object-based modeling.

Grid-Free Methods

The objects in object-based modeling may be represented as a continuous mathematical model. For example, this may be composed of a central flow axis and then of the geometric parameters associated with the attached architecture (Georgsen and Omre, 1992; Wietzerbin and Mallet, 1993). Implicit to this method is the independence from grid. Utilization of this feature varies by implementation. Some methods utilize immediate rasterization of the geometry model before object placement to increase algorithm speed (Deutsch and Wang, 1996), so these grid-free geometries are not readily available and most object-based model immediately rasterize their parametric geometries to the grid after object placement. Other methods include a set of depositional coordinates that may be applied to inform hierarchical trends across scales (Pyrcz et al., 2005b). Furthermore, some recent methods of object-based and process-mimicking methods, known as "grid-free," retain and store the precise object geometry functions (Hassanpour and Deutsch, 2010; Hassanpour et al., 2013; Pyrcz et al., 2011, 2012).

General discussion on grid-free or mesh-free methods is provided in Liu (2002). This geometrical architectural model may be preserved and edited graphically in the geomodeling software and/or stored and edited as an ASCII file. This results in the ability to interact with the model in a very flexible manner. For example, one may add a well intercept by moving a channel to a well to test the impact on flow response. In addition, exact downscaling is possible to any resolution (Pyrcz et al., 2012). An example of exact downscaling is shown in Section 4.5, Process-Mimicking Facies Modeling, with a process-mimicking facies model.

Algorithms for Object Placement

The basic algorithm underlying object-based facies modeling is the Boolean placement of objects. The objects may accumulate or build up from the stratigraphic base (Shmaryan and Deutsch, 1999) or from the stratigraphic top (Viseur et al., 1998). Alternatively, objects may be embedded within a matrix facies according to some stochastic process and erosional rules (stratigraphically higher objects erode older lower objects) imposed afterwards (Deutsch and Wang, 1996). Unconditional simulation is straightforward; objects are placed randomly until the global proportions of the different facies are reproduced. Reproduction of dense well conditioning has been considered difficult; however, many algorithms have been developed to address this challenge.

The objects must be placed so that they appear realistic and honor the available data. This is not trivial. Simplistic algorithms can lead to artifacts near conditioning data. The algorithms for conditioning include: analytical algorithms that force data reproduction, two-step positioning where conditioning data are reproduced first and then the remainder of the domain is filled, and iterative simulated optimization-type algorithms (discussed in Section 4.7, Optimization for Model Construction).

Direct conditioning algorithms modify the size, shape, or position of objects to honor local conditioning data by construction. The procedure may be implemented for a variety of geological shapes, but fluvial channels are the most common. Figure 4.37 gives an illustration of a 1-D process that models that centerline of a channel object. The departure from the main direction line is modeled with geostatistical (typically Gaussian) simulation. The departure from the centerline can be used for conditioning to local well data. Figure 4.38 gives two simplistic examples.

More complicated schemes for centerline definition include control nodes and splined centerlines that may realistically reproduce high sinuosity features. These centerlines may be fit to conditioning data through analytical methods (Oliver, 2002).

Often, direct conditioning algorithms are applied in two steps (Shmaryan and Deutsch, 1999; Viseur et al., 1998). The first step is to add objects to "cover" local facies intersections—that is, match all

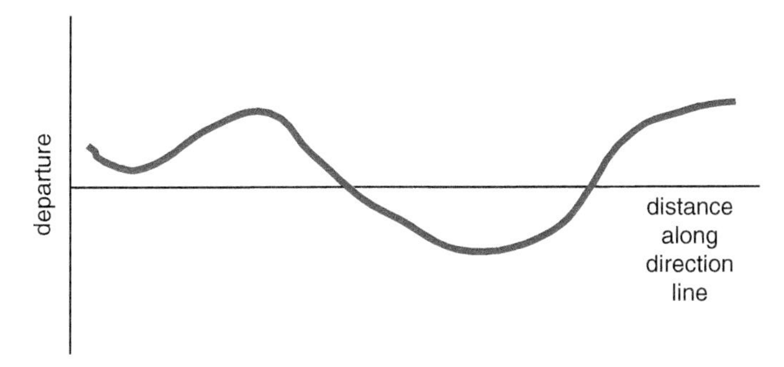

FIGURE 4.37: Illustration of 1-D Variable That Models That Centerline of a Channel Object. The departure from the main direction line is modeled analytically or with geostatistical (typically Gaussian) simulation.

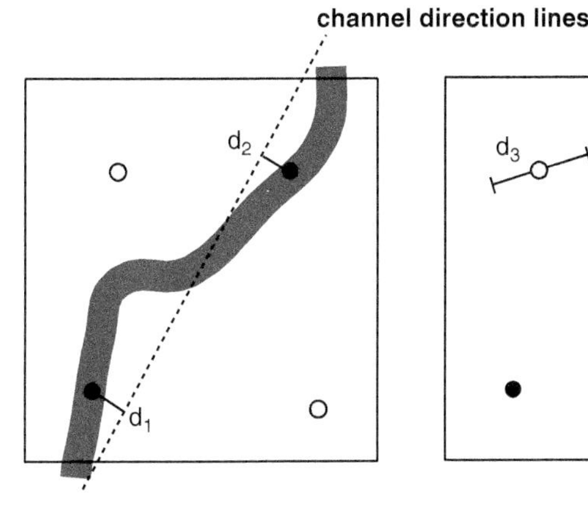

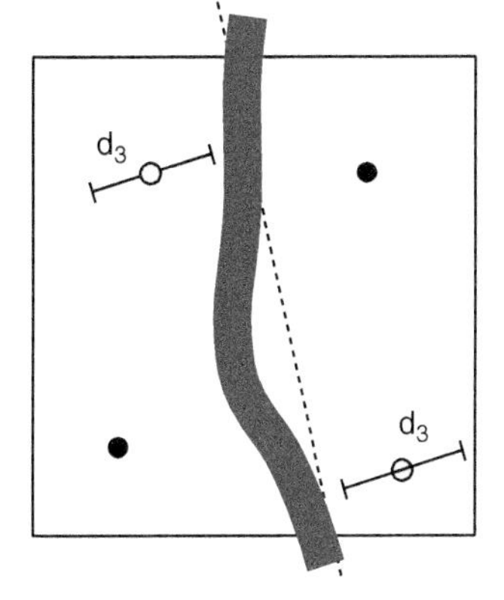

FIGURE 4.38: Two Examples of Direct Conditioning of Channel Centerlines to Honor Local Data in Object-Based Modeling. In the left case, the distances d_1 and d_2 are used to force the channel to go through the two wells. In the right case, the channel is forced away from the wells where it does not exist.

conditioning data that are *not* background facies. The second step is to add additional objects that do not violate known intersections of background facies, until the right proportion of each facies type is present.

A different approach to object placement is iterative. An initial set of objects are placed within the reservoir volume to match global proportions of the different facies. The placement may violate local well data and locally varying proportions. An objective function is calculated to measure the mismatch from the measured data. Then, the model is iteratively perturbed until the objective function is lowered sufficiently close to zero. Details of this method will be discussed in Section 4.7, Optimization for Model Construction.

Finally, Alapetite et al. (2005) introduced a novel method for generating channel forms that honor well data and trends. The input data are applied to condition a continuous random function that is then truncated to calculate the channel object. While this method is notable for novelty, it is not considered to be an object-based model here because it relies on the truncation of a cell-based simulation.

Conditioning routines may introduce bias into the model. Local corrections may result in distinct changes in reservoir architecture near the wells or objects. Two-step processes may result in significant changes in object density intersecting wells relative to those that do not intersect wells. Hauge and Syversveen (2007) have proposed methods to check object-based models for bias.

In addition, conditioning methods may fail to honor all conditioning inputs precisely, unlike cell-based models that ensure reproduction of conditioning (data are generally assigned to the grid nodes as a first step). All conditioning methods are likely to have some amount of mismatch with the wells and trend information in dense data settings. This becomes more of a problem as the density and complexity of conditioning increases and especially in the presence of contradiction between conditioning information (for example, a high fraction of channel elements in wells, while local areal trends indicate low fraction of channel elements). When mismatch occurs, the data and statistical inputs should be first checked for contradictions and then post-processing may be applied to clean up the conditioning at the well locations and/or to improve trend match (Pyrcz et al., 2009).

At times, data conditioning may not be a concern. As MPS (see Section 4.3, Multiple-Point Facies

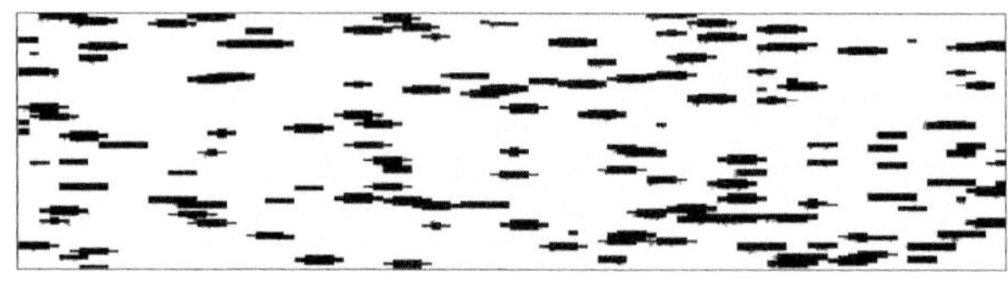

FIGURE 4.39: Illustration of Three Object-Based Stochastic Shale Models. The thickness distribution is the same for all three, but the lateral dimensions is clearly different.

Modeling) has become popular, there has been a growing need for training images. These training images often come from unconditional object-based models (Maharaja, 2008; Pyrcz et al., 2007). Also, Hovadik and Larue (2010, 2011) have demonstrated work flows for studying static and dynamic connectivity with unconditional object-based models. This work with unconditional object-based models has resulted in new insights into the relationships between object geometries, global facies proportions, and connectivity and flow response.

4.4.2 Stochastic Shales

When the net-to-gross ratio is fairly high (greater than 80% or so) it is reasonable to model the remnant shales (or relevant non-net facies) as objects within a matrix of sand. These remnant shales may not affect horizontal permeability to any great extent, but could have a significant effect on vertical permeability, coning, cusping, and performance predictions of horizontal wells. Concretions or stylolites may also be modeled with the same procedure.

The required input to stochastic shale modeling is a geometric specification of the shale geometry, size and orientation parameters, well data, and perhaps information on locally varying shale proportions. The shape is often taken as ellipsoidal. The thickness distribution may be inferred from well locations. The length-to-thickness ratio must be established from geological analog information or matching to some other data type such as well test or production data. Some orientation information may be available through knowledge of the depositional system. It is unlikely that there is any information related to locally varying proportions if the fraction of shale is small (say, less than 20%).

Figure 4.39 shows three images of stochastic shales. Each cross section has the same percentage of shale (15%) and the same thickness distribution. The lateral extent of the ellipsoidal shales was changed systematically.

Conditional realizations are more interesting. Figure 4.40 shows three images that honor the succession of sand and shale at a well location. In this case, a two-step procedure was followed whereby shales were first added to honor the well intersections, and then other shales were added (that do not violate the sand intersections at the well locations) to make up the proportion of 15% shales.

The Delta Reservoir provides a good example to demonstrate object-based shale modeling because it has a high net to gross with a limited fraction of shales that may have a significant constraint on reservoir flow response. The reservoir facies were grouped and set as background, and shales were simulated conditional to well data (see Figure 4.41). The shales are represented as ellipsoids using the `ellipsim` program from GSLIB by Deutsch and Journel (1998). The shales form 10% of the volume and have dimensions of 300 m × 400 m on strike and dip, respectively.

SIS could be applied to build these shale drape models, but would not reproduce the crisp geometries. The object parameterization chosen for the

FIGURE 4.40: Three Object-Based Stochastic Shale Models That Reproduce the Succession of Sand and Shale at a Vertical Well Location.

shales is simplistic and linear. A more complicated application of object-based techniques is for sinuous fluvial channels.

4.4.3 Fluvial Modeling

Historically, object-based modeling has been devoted to abandoned sand-filled channel forms with a matrix of floodplain shales (Deutsch and Wang, 1996). These are classified as "shoestring fluvial reservoirs" by Galloway and Hobday (1996).

More complex fluvial sand geometries such as channel bar forms, laterally extensive sand regions due to lateral accretion, and more discontinuous point-bar sand bodies are considered using the same scheme with more sophisticated sedimentological models (Hassanpour et al., 2013) or process mimicking methods (see Section 4.5, Process-Mimicking Facies Modeling). Also, associated facies such as levee sands, crevasse sands, and cemented zones within the channel sands are modeled at the same time.

Figure 4.42 illustrates the hierarchical approach to modeling. The reservoir is first separated into a number of *major reservoir layers*. Figure 4.42 shows

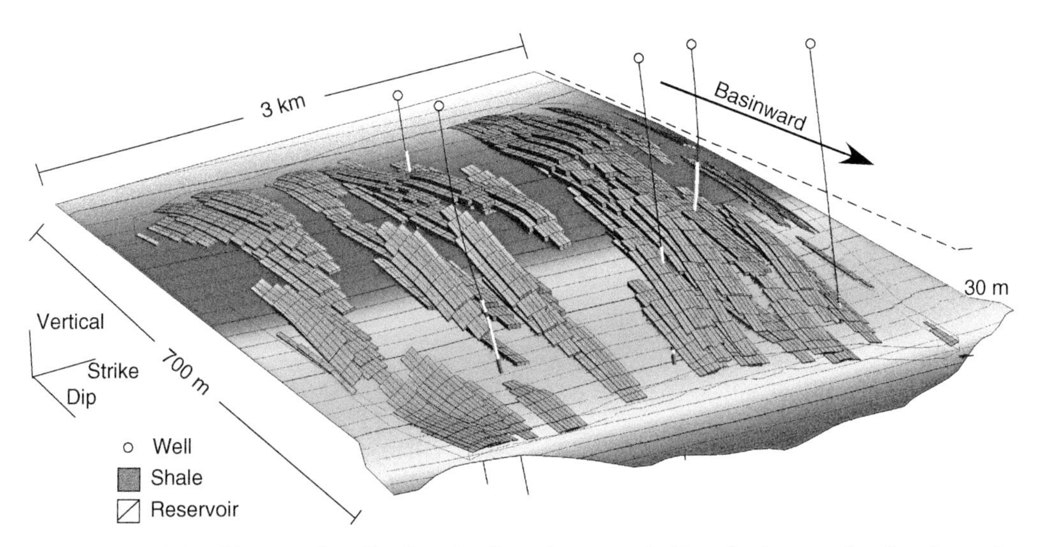

FIGURE 4.41: Shale Object-Based Realizations for the Delta Reservoir. Note the dip axis is lengthen due to the oblique projection. The reservoir facies is hidden to expose the shale objects.

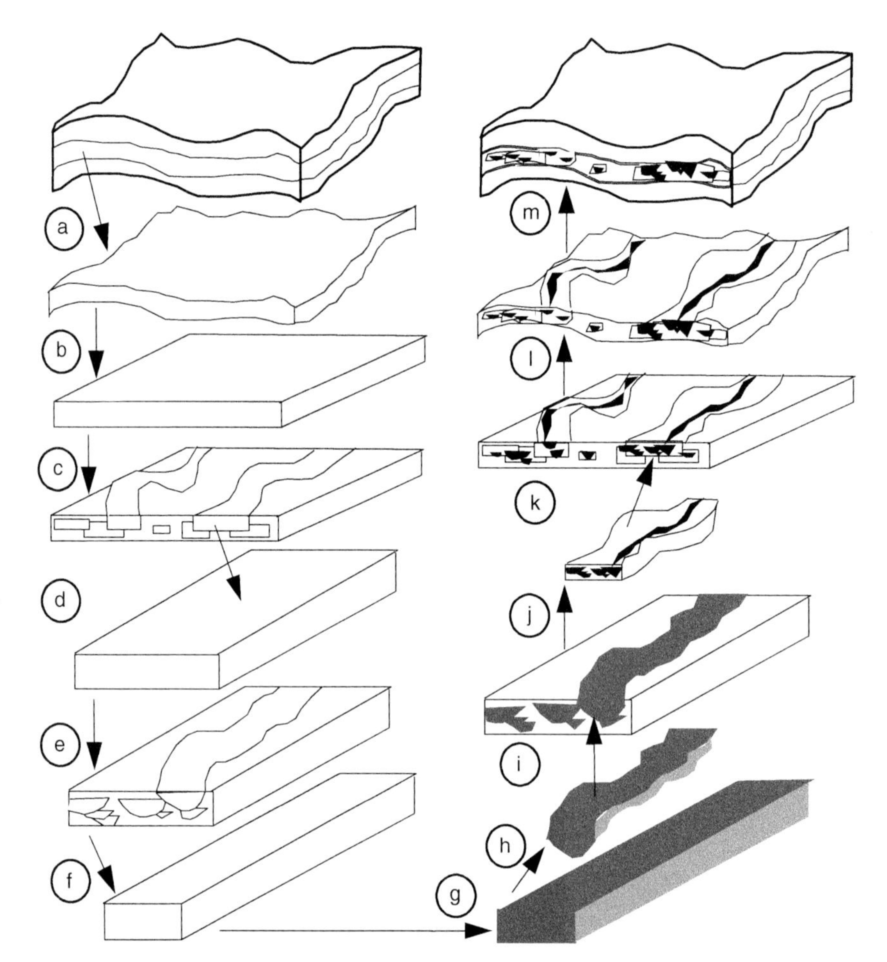

FIGURE 4.42: The conceptual Approach to Hierarchical Object-Based Modeling (a) Each reservoir layer is extracted, (b) stratigraphic reservoir coordinates are calculated, (c) each channel complex is extracted from the layer, (d) a local stratigraphic channel complex coordinate system is established, (e) each channel is extracted from the channel complex, (f) a local stratigraphic channel coordinate system is established, (g) the petrophysical properties are modeled within the appropriate channel coordinate system, (h) the channel coordinates are restored, (i) the channel is placed back in the channel complex, (j) the channel complex coordinates are restored, (k) the channel complex is placed back in the layer, (l) the layer coordinates are restored, and (m) the layer is placed back in the reservoir.

three major reservoir layers. Each of the three layers is modeled independently and then reassembled back into a complete reservoir model. Step (b) indicates a stratigraphic transformation transforming the layer from a vertically irregular volume into a regular 3-D volume. In step (c), a number of channel complexes are stochastically simulated. Then each of these channel complexes is transformed from an irregular shape into a regular 3-D volume, and a number of channels are stochastically simulated, as illustrated by steps (d) and (e). Each channel is in turn transformed from an irregular volume into a regular 3-D volume and filled by cell-based geostatistical algorithms with porosity and permeability, as illustrated in steps (f) and (g). In steps (h) and (i), the property-filled 3-D boxes representing channels are back-transformed to their original geometry and position in the corresponding regular channel complex 3-D volume. When all channel complexes are filled with property-filled channels, they are back-transformed to their original geometries and positions in the corresponding regular reservoir layer volume, as shown in steps (j) and (k). The last steps consist of back-transforming the reservoir layers and restoring them to their original reservoir positions, as sketched in steps (l) and (m).

Coordinate Transformations

One view of object-based modeling is that the coordinate system is adapted to the appropriate principal directions of continuity. The direction of continuity depends on the scale of observation and the specific geological feature being modeled. The different coordinate transformations have all been described in Section 4.1, Large-Scale Modeling; here, they are applied successively and at different scales. For example:

- Each major stratigraphic *layer* or *sequence* is modeled independently. A layer-specific relative stratigraphic coordinate is defined from the existing top, existing base, restored top, and restored base.
- An areal translation and rotation is carried out to obtain an areal coordinate system aligned with the principal paleoslope direction, that is, the direction with the greatest continuity on average.
- Fluvial channels often cluster together in channel complexes or *channel belts*. The channel complexes may have large-scale undulations that are straightened. The channels within the channel complexes are also straightened for modeling of petrophysical properties within them.
- The "straightened" channel complexes and channels are not equally wide; a relative horizontal coordinate makes the boundaries parallel and simplifies subsequent petrophysical property modeling.

Each transform is reversible and the geological objects can be viewed and saved in any coordinate system. Each channel is filled with porosity and permeability using the appropriate within-channel coordinate system. The vertical coordinate system could be taken parallel to the top channel surface or made proportional between the top and base surfaces. The base of each channel may have shale clasts or basal deposits that lower, or coarse lags that raise, porosity and, more importantly, permeability. Furthermore, there may be a systematic trend in porosity from the base to the top of each channel. These features are handled easily, since the position of each channel base is known, by a trend in the porosity modeling program.

Fluvial Facies Objects

Figure 4.43 illustrates one possible conceptual model for fluvial facies for a shoestring fluvial reservoir. There are four facies types where the geometric specification of each is chosen to mimic shapes idealized from observation.

1. The first facies type is impermeable background floodplain shale, which is viewed as the matrix within which the reservoir quality or *sand* objects are embedded.
2. The second facies type is *channel sand* that fills sinuous abandoned channels. This facies is viewed as the best reservoir quality due to the relatively high energy of deposition and consequent coarse grain size. There may be special features within the channel sands such as (1) heterogeneous channel fill, perhaps containing some fine-grained non-net material, (2) a channel lag deposit at the base, and (3) fining-upward trends within the channel fill.
3. The third facies type is *levee sand* formed along the channel margins. These sands are considered to be of poorer quality than the channel fill.
4. The fourth and final facies type is *crevasse splay sand* formed during flooding when the levee is breached and sand is deposited away from the main channel. These sands are also considered to be of poorer quality than the channel fill. As illustrated in Figure 4.43, crevasses often form where the channel curvature is high.

An "object" for fluvial modeling is a channel and all related levee and crevasse sands. More specifically, for computational efficiency the object is a template of cells that would be coded as channel sand, levee sand, and crevasse sand. A combination of an analytical cross section (rectangle or half ellipse) with a 1-D function for the centerline also could be considered. The template provides significant CPU advantages. The connectivity of the resulting realizations is sensitive to the choice of an underlying grid size. The grid size must be chosen small enough to preserve the geological shapes represented by the templates.

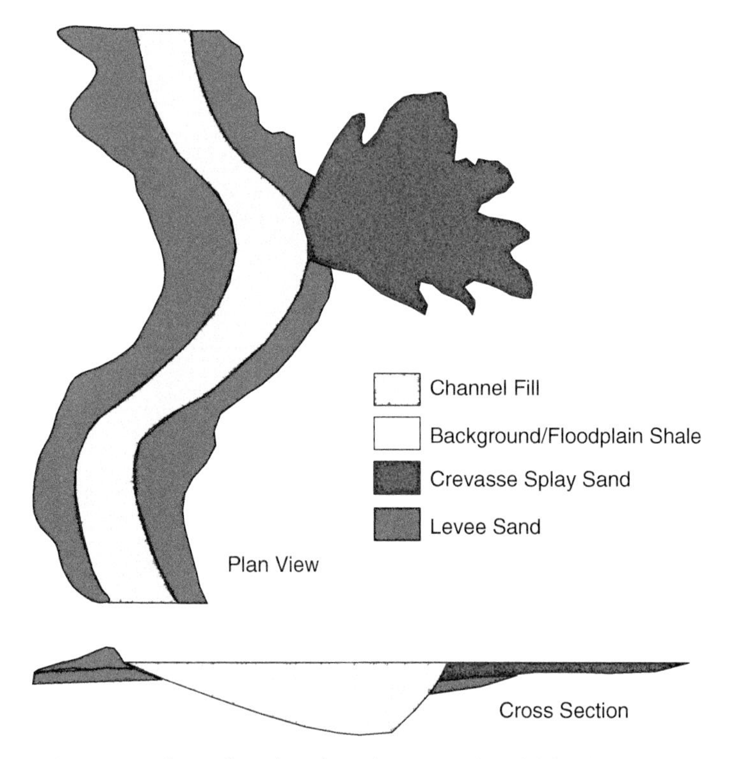

FIGURE 4.43: Plan and Section View of Conceptual Model for Fluvial Facies: Background of Floodplain Shales, Sand-Filled Abandoned Channel, Levee Border Sands, and Crevasse Splay Sands.

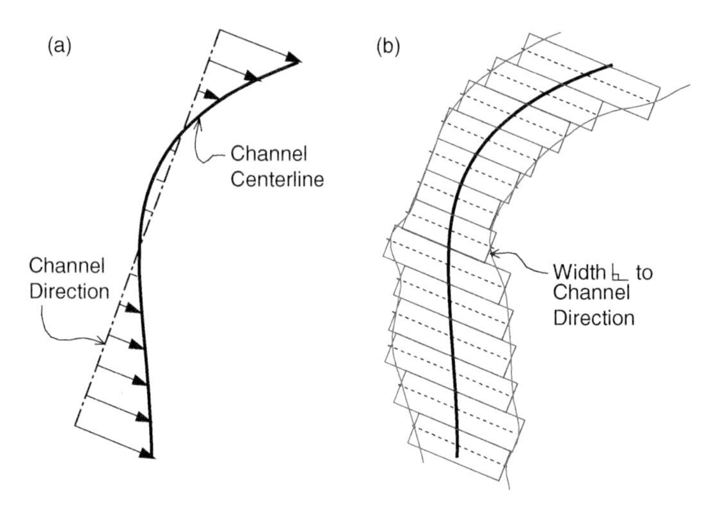

FIGURE 4.44: Areal View of Some Parameters Used to Define Channel Object: (a) Angle for Channel Direction and Deviation for Actual Channel Centerline and (b) Variable Channel Width (and Thickness) with "Blocky" Connection between Channel Cross Section Slices.

The parameters used to define an abandoned sand-filled channel are illustrated in Figures 4.44 and 4.45. The channels are defined by an orientation angle, the average departure from the channel direction, the "wavelength" or correlation length of that average departure, thickness, thickness undulation (and correlation length), width/thickness ratio, and width undulation (and correlation length). Each parameter may take a range of possible values according to a probability distribution.

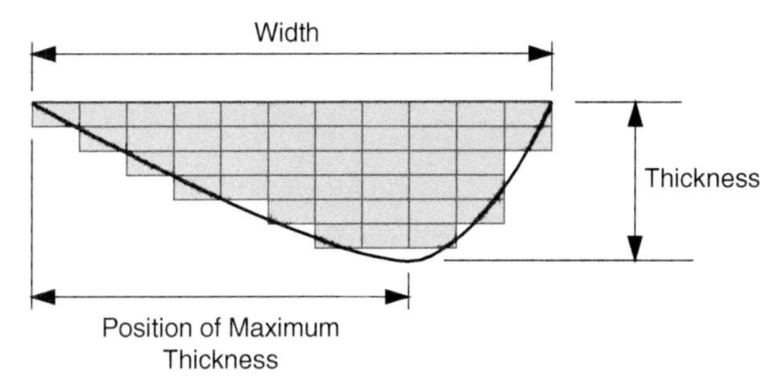

FIGURE 4.45: Cross Section View of a Channel Object Defined by Width, Thickness Position of Maximum Thickness.

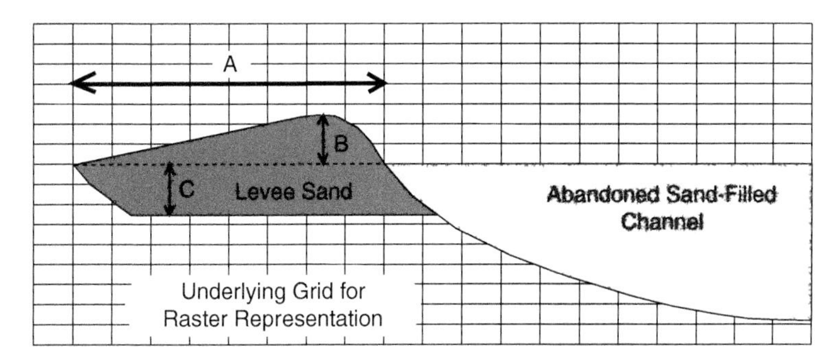

FIGURE 4.46: Cross Section through an Abandoned Sand-Filled Channel and Levee Sand. Three distance parameters (A), (B), and (C) are used to define the size of the levee sand.

Figure 4.46 shows a geometrical form that could be adopted for the levee sand. The three distance parameters (A) lateral extent of the levee, (B) height above the channel datum elevation, and (C) depth below the channel datum are used to define the size. For simplicity, the geometrical shape will remain fixed and only the size will vary. In general, the levee size parameters should depend on the size of the channel; a large channel has larger levees. The sizes of the left and right levee may be different.

A random walker procedure may be used to establish the crevasse geometry (see Figure 4.47). The location of a crevasse along the channel axis is chosen with probability increasing in proportion to the curvature. A number of random walkers are "released" from the location of the crevasse to establish its areal extent (see Figure 4.47). The four control parameters are (1) the *average* distance the crevasse sand reaches from the channel bank, (2) the *average* along-channel distance, and (3) the irregularity of the crevasse sand or the number of random walkers used; more walkers leads to a smoother outline. The thickness of the crevasse sand decreases linearly from a maximum thickness next to the channel.

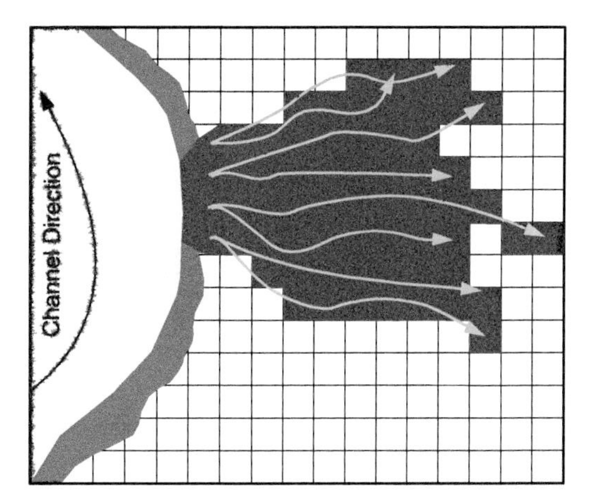

FIGURE 4.47: Areal View of Channel and Crevasse Formed by Breaching Levee. A number of random walkers are released from breach to establish the crevasse geometry. The number, length, and lateral diffusivity control the geometry and size of the crevasse.

Geological Conditioning Data

Many of the object size parameters for object-based models are difficult to infer from available well and seismic data. Wells provide information on the thickness distributions and proportions of the different facies; however, it is often necessary to adapt measurements from analog outcrops, modern depositional systems, and similar densely drilled fields.

The proportion of each facies type (channel sand, floodplain shale, crevasse splay sand, and levee sand) may be specified by a vertical proportion curve, areal proportion maps, and reference global proportions. The reference global proportions are denoted as $P_g^k, k = 1, \ldots, K$, where K is the number of facies. A vertical proportion curve specifies the proportion of a facies k as a function of vertical elevation or time and is denoted as $P_v^k(z)$, where $z \in (0, 1]$. Areal proportion maps specify the facies proportion as a function of areal location (x, y) and are denoted as $P_a^k(x, y)$. These three types of proportions can be obtained through a combination of well and seismic data.

The vertical proportion curve and areal proportion map may have implicit global proportions that may be either inconsistent with each other or with the reference global proportions. The proportions can be scaled. Two characteristics of the geological or geometrical parameters are that (1) each parameter may take a range of possible values according to a probability distribution and (2) the range of values changes with the stratigraphic position/time z. Therefore, the parameters that define each channel are specified with a series of conditional distributions for a discrete set of z values between the stratigraphic base and the stratigraphic top.

As mentioned previously, a collection of closely spaced and genetically related channels is referred to as a channel complex. As described above, the orientation, shape, and size of a channel complex serve as a basic modeling unit. The following input distributions are required to specify the channel complex geometry: (1) angles α, (2) departure from channel complex direction and the correlation length of this departure along the axis of the channel complex, (3) thickness, (4) width-to-thickness ratio, and (5) net-to-gross ratio within the channel complex. A greater number of parameters are required to specify the channel geometry because the thickness and width of each channel is not constant: (1) average thickness, (2) thickness undulation and correlation length of thickness undulation, (3) width-to-thickness ratio, (4) width undulation and correlation length of width undulation, and (5) channel base roughness. There are inevitable problems to reliably infer these parameters from available data.

The facies are known along each well. The data may come from any arbitrary number of wells. The well data in original coordinate space are translated to the appropriate coordinates at each stage of modeling.

Simulation Procedure

All of the geological inputs described above must be specified prior to modeling. Many of these parameters will be difficult to infer from available data. Some parameters, such as width-to-thickness ratios, may be kept constant at some realistic value. Sensitivity studies can be considered on other parameters to judge the importance and to assess the visual acceptability of the resulting realizations.

The simulation procedure is sequential. First, the channel complex distribution is established. Second, the distribution of channels within each channel complex. Third, the porosity is assigned using appropriate channel coordinates. Finally, the permeability is assigned conditional to the facies and porosity. Conventional variogram-based algorithms can be used for porosity and permeability modeling. Conditioning to the facies succession along wells and to facies proportion data should be accomplished directly if possible. An iterative procedure can be used to enforce the facies at wells and proportion curves and maps (see previous discussion on coordinate transformation).

For another example we utilize the paleo-basin. The Fluvial Reservoir is composed of extensive lateral accretion with mud plug channel fills; therefore, it would not be a good fit for the "shoestring" fluvial model described above. Although, the Deepwater Reservoir is composed of a set of stacked channels similar to the fluvial channel based model described above. Deepwater levees are not generally considered reservoir and are omitted from the model and while sediment waves may form outside deepwater channels, they are not considered reservoir in this case. For a demonstration this methodology was applied to model a single channel complex, conditional to available well data (see Figure 4.48). The resulting model may be an overly simplified set of stationary low-sinuosity channels, but it is extremely straightforward to build and reproduces the long-range continuity of the preserved channels that would not be possible with cell-based

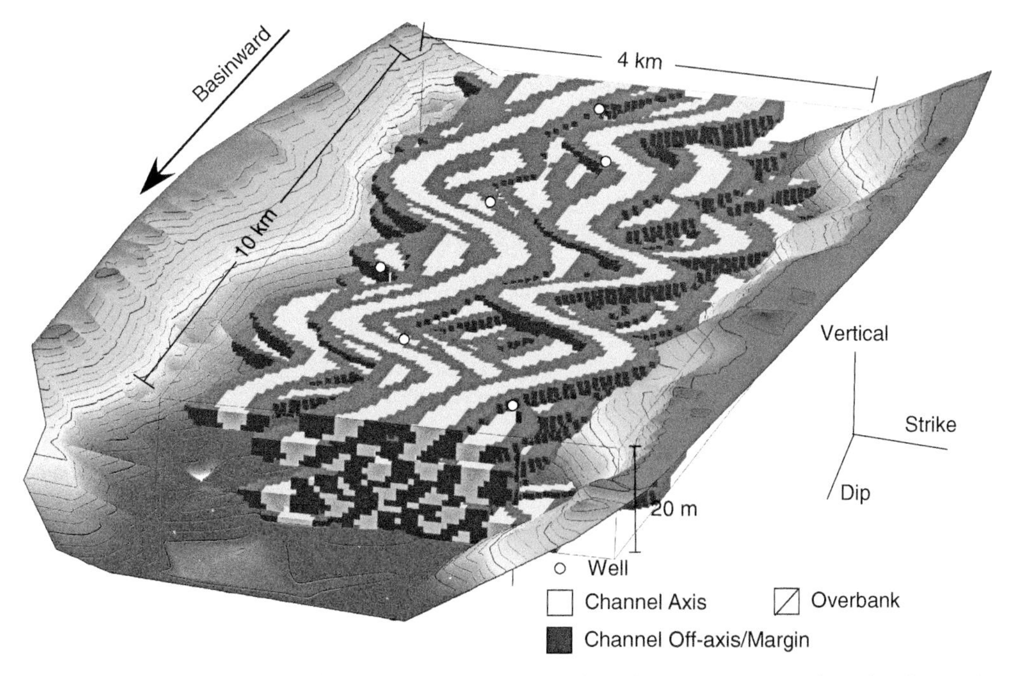

FIGURE 4.48: Deepwater Reservoir Constructed with the Fluvial Object-Based Model. Channels only include two facies: axis and grouped off-axis and margin.

modeling methods. For more complexity, such as distinct stacking patterns, detailed locally proportion models or process-mimicking methods are required (see the next section).

4.4.4 Nonfluvial Depositional Systems

Object-based facies modeling is applicable to a variety of depositional settings. The main limitation is coming up with a suitable parameterization for the geological objects. Deltaic or deepwater lobes are examples of objects that could be defined. Parameters for a "simple" lobe geometry are illustrated in Figure 4.49. This set of parameters provides a balance between overly simplistic geometry (too few parameters) and flexibility with the associated difficult inference (too many parameters). The seven parameters illustrated in Figure 4.49 are as follows:

1. *sw* = starting width; typically set to the final channel width. The user, however, could set this arbitrarily.
2. *ll* = lobe length; total length of the lobe from the channel to terminus.
3. *rm* = relative position of maximum width; the lobe reaches maximum width a distance of rm · *ll* from the start of the lobe.
4. *lwlr* = lobe width/length ratio; maximum width of the lobe is given by lwlr · *ll* at location specified by relative position rm.
5. *st* = start thickness; start thickness of the lobe next to channel (could be associated to the channel thickness at that point). This is the thickness at the lobe centerline.
6. *ft* = final thickness; thickness of the lobe at the terminus (at the lobe centerline) relative to starting thickness.
7. *ct* = cross section thickness correction; amount that the thickness reduces from the centerline to edge of lobe.

The parameterization could easily get more elaborate at the cost of additional parameters that must be inferred from limited observational data. One natural extension is the possibility to keep the base flat for some distance before tapering to zero thickness at the lobe terminus.

In plan view, the constraints on the geometry include the following: (1) The width is equal to starting width at transition point from channel $-y = w$ at $x = 0$, (2) the width is a maximum at relative position $l - y = W$ at $x = l$, (3) the width is zero at maximum lobe length $L - y = 0$ at $x = L$, (4) the tangent to the

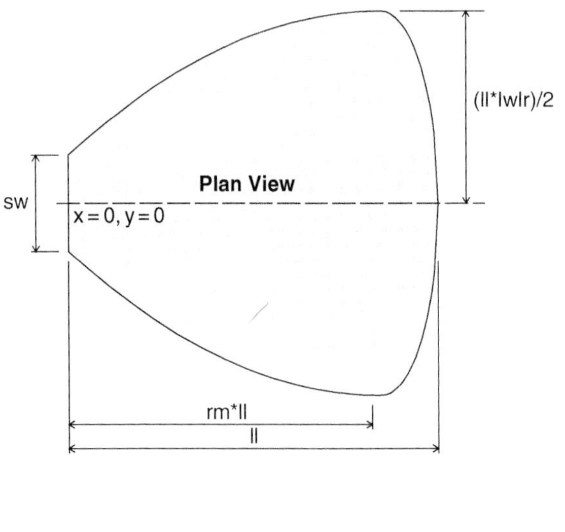

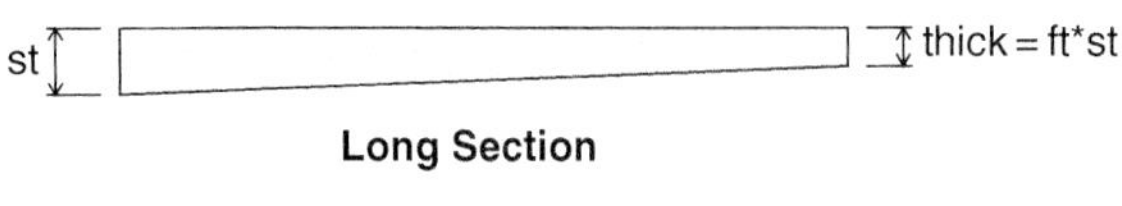

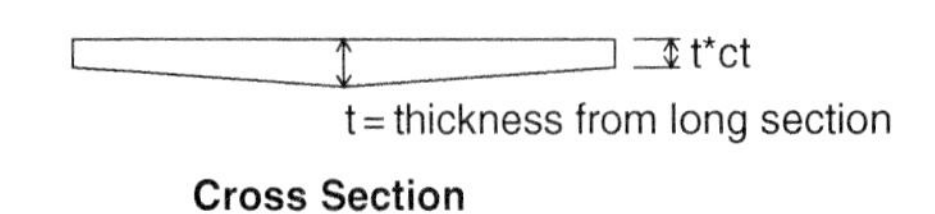

FIGURE 4.49: Parameters Needed to Describe the 3-D Lobe Geometry.

lobe shape has a zero slope at $x = l$, and (5) the tangent to the lobe shape has an infinite slope at $x = L$. The following equation satisfies these constraints:

$$y = \begin{cases} w + 4 \cdot (W - w) \cdot \left[\frac{x}{2l}\left(1 - \frac{x}{2l}\right)\right], & 0 \leq x \leq l \\ W \cdot \sqrt{1 - \left(\frac{x-l}{L-l}\right)^2}, & l \leq x \leq L \end{cases}$$

y is the distance from the centerline, x is the distance along the centerline, w = sw/2.0, l = rm $\cdot$ ll, L = ll, and W = (ll $\cdot$ lwlr)/2. The shape is based on the "$p(1-p)$" shape for the first part (closest to the connection with the channel) and an elliptical shape for the second part. The function and the first derivative are both continuous at all locations around the lobe outline.

Figure 4.50(a) shows a series of lobe shapes for different parameters. A combination of the width/length ratio and relative position parameter provides flexibility in the lobe shape. There are some natural extensions to this areal shape including (1) addition of stochastic variations to the lobe shape for more realism (Figure 4.50(b) shows six examples) and (2) consideration of asymmetric lobe geometries, that is, the W, w, and l parameters could be different on the "top/bottom" of the lobe.

The lobes may be positioned at the end of channels. Figure 4.51 illustrates this relationship together with the channel and lobe template. The final configuration of channels and lobes in any model will ultimately be determined by the simulation algorithm and conditioning data. Areally varying facies proportions and well data could lead to lobes in preferred locations.

4.4.5 Work Flow

Following are the main work flows related to object-based facies modeling. Figure 4.52 provides the steps in characterizing objects for object-based modeling. The first step is to apply all available information, including well data, seismic data and conceptual information to determine if distinct geometries are present in the reservoir. If not, then object-based modeling should not be applied. If distinct geometries are present then determine the placement rules of the objects and the parameters and associated parameter distributions for the objects.

Figure 4.53 illustrates the work flow for fluvial modeling. The main operation is the step to position the facies objects in a manner that reproduces all available data.

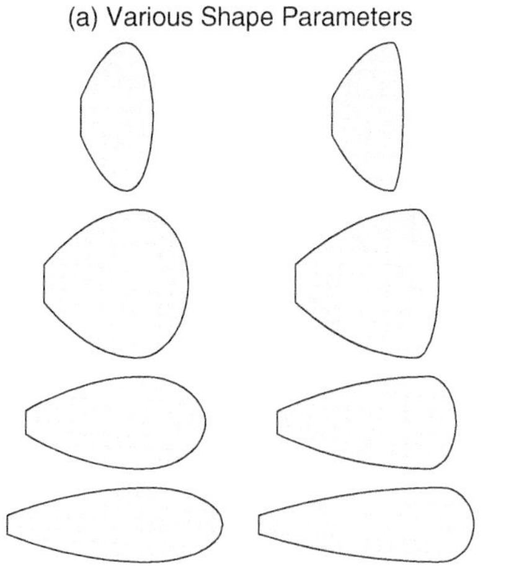

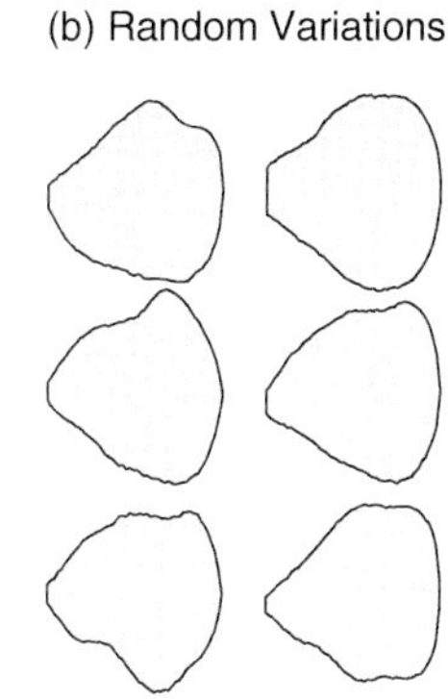

FIGURE 4.50: (a) Example Lobe Geometries for Various Parameters. From top to bottom the width/length ratio decreases from 2.0:1 to 1:1 to 0.5:1 to 0.33:1. The relative position of the maximum width is 0.65 on the left and 0.85 on the right. (b) The basic geometry $W = 0.5, w = 0.01, L = 1.0$, and $l = 0.6$ with six different Gaussian simulations added.

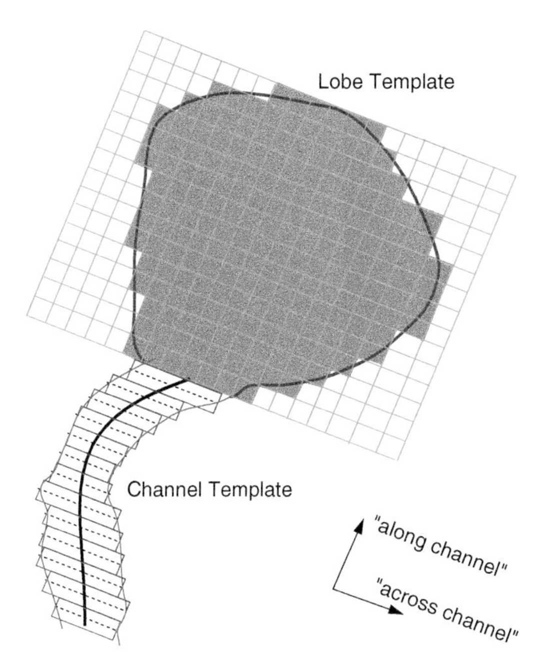

FIGURE 4.51: Channels and Lobes Are Represented by Templates. In the case of lobes, the gray-shaded area is kept in memory together with the depth under each cell.

Figure 4.54 shows more details of the important object-positioning work flow. In practice, there are a number of heuristic schemes to ensure that local conditioning data are reproduced quickly and exactly at the well locations.

4.4.6 Section Summary

Object-based facies modeling is a viable alternative to cell-based facies models, when the reservoir architecture may be represented with crisp parameterized architectural geometries. Objects are placed sequentially into a reservoir model initialized with a background facies to honor conditioning, global proportions, trends, and secondary information.

These methods have been applied to various settings including stochastic shales, fluvial channels, and deepwater channels and lobes. Challenges remain in efficient construction of models that honor dense conditioning data, trends, and secondary data. In addition, formulating characteristic parametric geometries and representative parameter distributions is often difficult.

If even greater geological complexity in the form of relationships between reservoir architectural units is required to meet the project goals, then process-mimicking methods, described in the next section, may be an option.

4.5 PROCESS-MIMICKING FACIES MODELING

Process-mimicking facies models attempt to increase the level of integration of the geological conceptual model (as discussed in Section 3.2, Conceptual Model) by integrating rules, based on the geological process. These rules constrain the sequential construction of reservoir architecture represented by objects and/or surfaces. While this framework results in reservoir models with improved detailed heterogeneity, conditioning to dense well data and seismic remains a challenge (similar to object-based modeling discussed in the last section). This section presents some details on this emerging area and some examples of the method. Admittedly, this methodology is not commonly available in commercial software, with the current exception of the work of Wen (2005). Yet, various academic and industrial

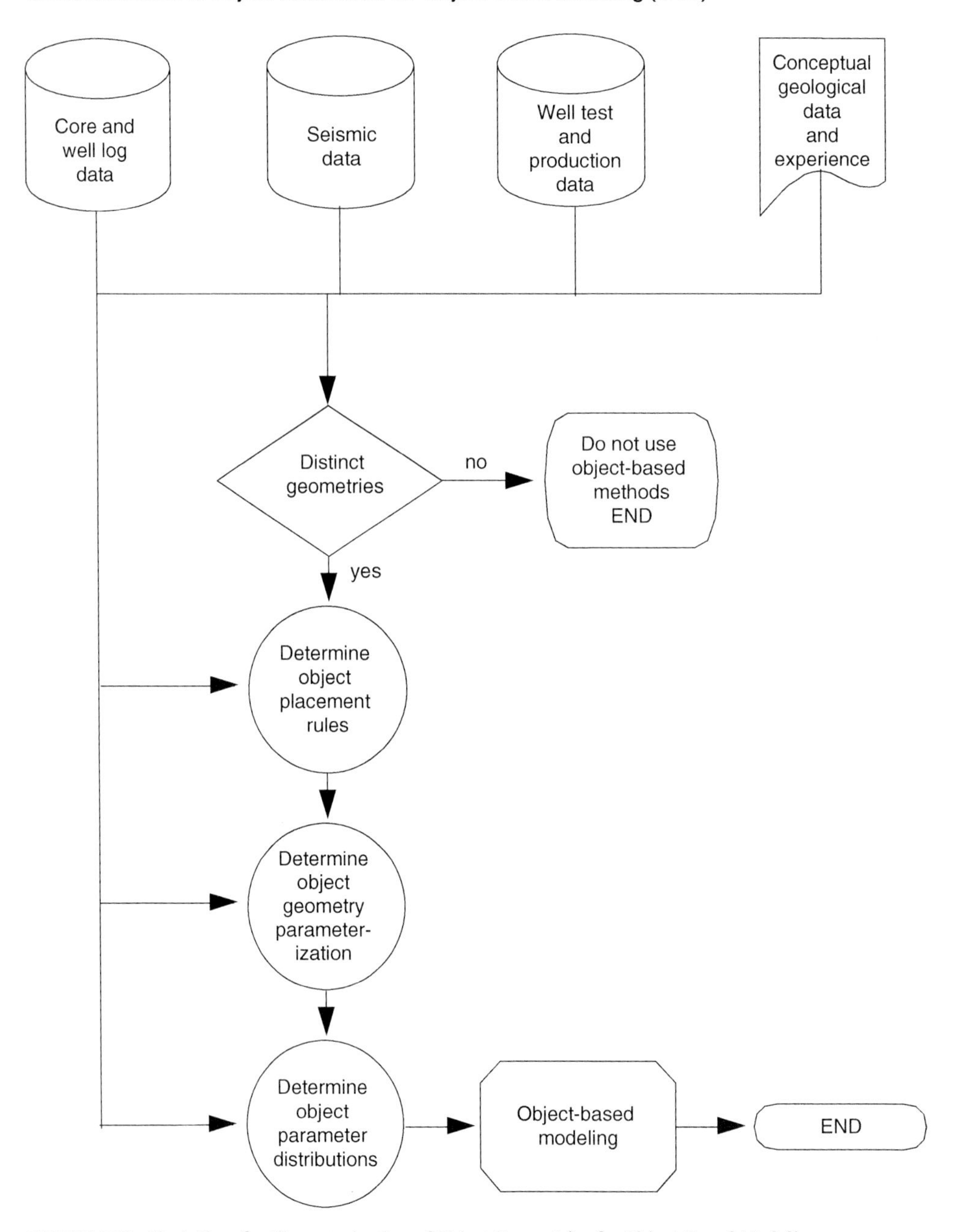

FIGURE 4.52: Work Flow for Characterization of Object Geometries for Object-Based Modeling.

research efforts are focused on this topic with encouraging results and novel applications.

The *Background* subsection covers the development of process-mimicking facies modeling including a discussion on motivation and the linkages between process modeling and more conventional geostatistics.

The *Process-Mimicking Methods* subsection presents the building blocks of process-mimicking models with discussion on forward modeling, types of rules, and comments on information content and conditioning. The *Example Implementations* demonstrates this methodology with a couple example models, including discussion on methods and applications.

Process-mimicking methods are new and most applications are in research code. Novice

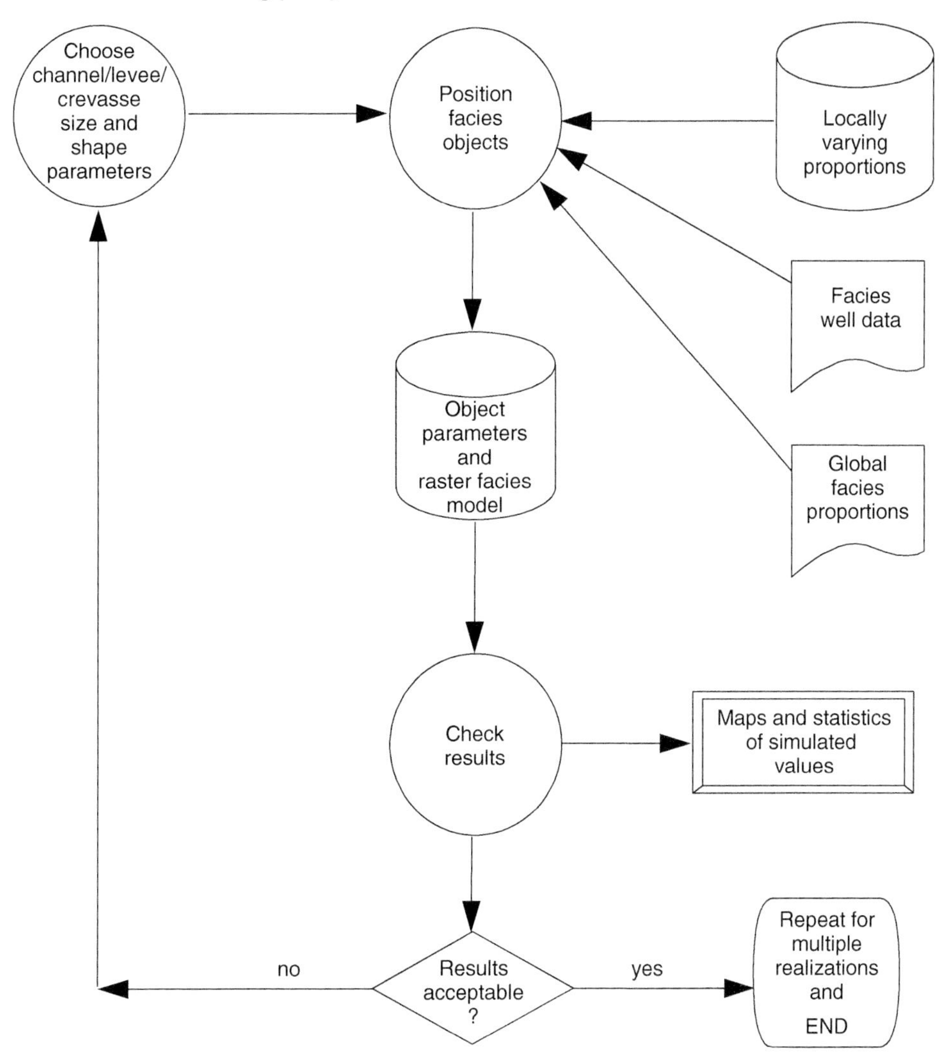

FIGURE 4.53: Work Flow for Fluvial Object-Based Modeling.

and intermediate modelers may have the opportunity to apply these methods within the software developed by Wen (2005) or as they become more broadly available. Expert modelers may look for opportunities to further integrate geological expert knowledge into the reservoir models with research code to meet project goals or to utilize the advanced applications for geologically realistic models presented in this section.

4.5.1 Background

The developments of universal kriging, sequential indicator simulation, Gaussian truncated simulation, object-based modeling, simulated annealing/optimization-based simulation, and multiple-point simulation have all been motivated by the need for improved reproduction of geological concepts. This is the primary focus of process-mimicking methods. While acknowledging the value and practical success of this effort, there remains a niche for even more improved geologic realism.

In this context, process-mimicking facies modeling methods is an effort to further improve the reproduction of geological concepts by integrating geological rules into a geostatistical framework. Current geostatistical algorithms, cell-based or object-based, using variograms, training images, or geometric parameters, enable the reproduction of spatial statistics inferred from available conditioning data and analogs, but rarely integrate information related

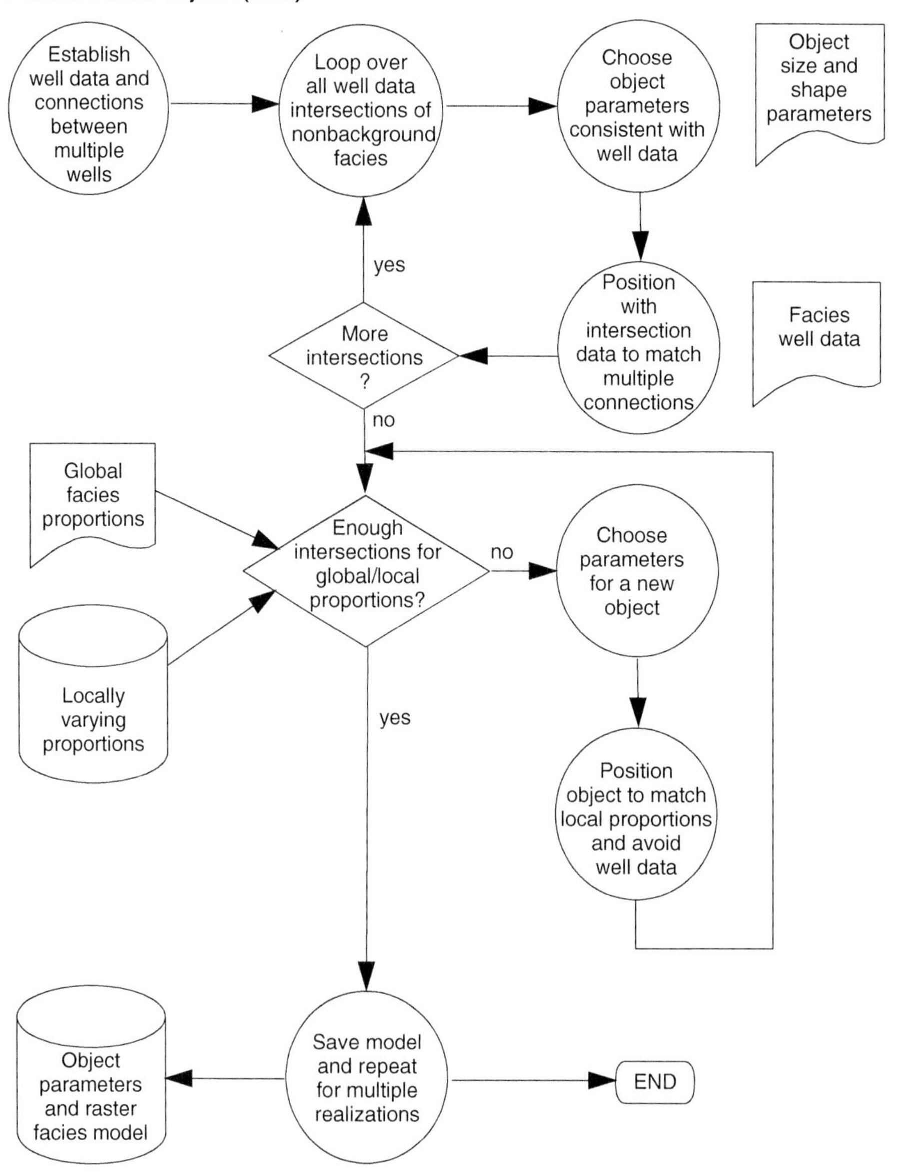

FIGURE 4.54: Work Flow for Positioning of Facies Objects in Object-Based Modeling.

to depositional processes. Indeed, because conventional geostatistical models are constructed without any concept of time or depositional sequence, their ability to incorporate sedimentological rules, which explain facies geobodies interactions and intrabody porosity/permeability heterogeneity, is quite limited. Consider the architecture for the Deepwater Reservoir in Figure 4.55 generated with a process-mimicking model based on a set of expert-derived deepwater channel rules (Pyrcz et al., 2012). It would not be possible to reproduce these details with standard geostatistical method, without a very high level of constraint through nonstationary trend and azimuth models.

Throughout this section, we describe process-mimicking modeling and describe a specific approach known as event-based modeling. There have been various names and associated research efforts in this area. The names include event-based, hybrid, process-inspired, process-oriented and surface-based modeling.

Early work in this area includes parallel efforts with objects and surfaces building blocks. As mentioned in the previous section, surface-based

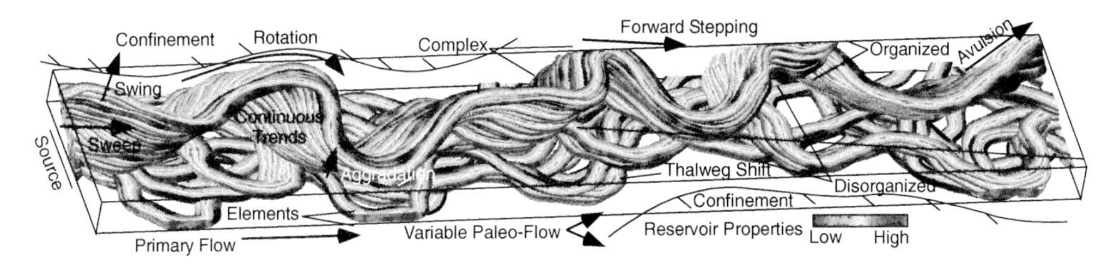

FIGURE 4.55: Complicated Model for the Deepwater Reservoir Constructed with a Process Mimicking Method. Note the identified features that indicated realistic channel behavior and evolution (Pyrcz et al., 2012).

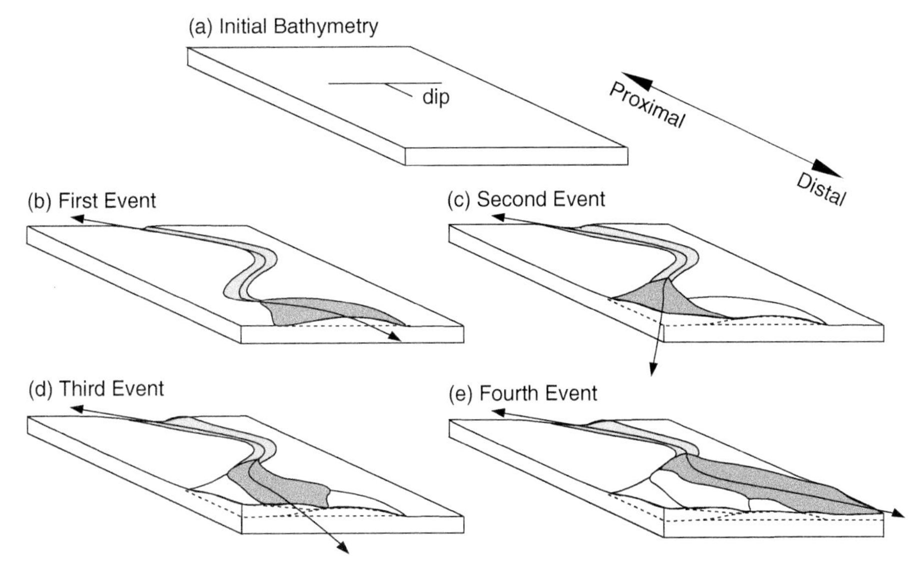

FIGURE 4.56: Surface-Based Model with Sequential Application of Depositional and Erosional Rules That Constrain Object-Based Templates to Aggrade and Erode Surfaces (Pyrcz et al., 2005a).

methods are a variant of object-based methods that track surfaces that bound the objects. In the process-mimicking framework, surface-based methods describe the aggradation and erosion of surfaces based on geometric templates. A schematic of a surface-based process-mimicking method is shown in Figure 4.56, this may be contrasted with object-based process-mimicking methods such as shown in Figure 4.55. However, the line between these object-based and surface-based process-mimicking methods blur because often the object-based process-mimicking methods utilize surfaces to defined local gradients, confinement and surface evolution and objects are used to define the modification of the surfaces (i.e., both objects and surfaces in some fashion are implemented).

Initial work on geostatistical modeling of architectural centerlines and surfaces includes the work of Georgsen and Omre (1992) and Hektoen and Holden (1997), respectively, which played an important role forming the framework for process-mimicking methods. On the object-based process-mimicking front, Sun et al. (1996) applied process extension of the fluvial 1-D process-based bank retreat model by Howard (1992) to construct fluvial models. Lopez et al. (2001) moved this work soundly into the geostatistics realm with their efforts to impose conditioning of this model to well data and trends. On the surface-based front, Xie et al. (2000) and Deutsch et al. (2001) introduced various surface-based simulation methods with hierarchical trend models for reservoir properties. Pyrcz et al. (2005a) demonstrated process-mimicking methods with surface-based models and Abrahamsen et al. (2008) developed new methods to improve conditioning and constrained the geometries attached to depositional event centerline lines with 1-D process models.

As will be discussed later in this section, additional developments have been focused on process rule development (Cojan et al., 2005; Jerolmack and Paola, 2007; McHargue et al., 2011; Pyrcz, 2004; Pyrcz et al., 2011), improved conditioning (Michael et al., 2010; Miller et al., 2008; Pyrcz and Deutsch, 2005; Pyrcz et al., 2012; Zhang et al., 2009), and improved reservoir modeling applications (Pyrcz et al., 2012; Sylvester et al., 2010; Wen, 2005). In this section the brief description of methods and examples models are based on the event-based method summarized by Pyrcz et al. (2012). For greater details on this method the reader may refer to various publications including Pyrcz (2004), Pyrcz and Deutsch (2005), Pyrcz and Strebelle (2006), and Pyrcz et al. (2009).

The authors believe that numerical models directly accounting for geological rules and the processes of deposition are more realistic and suitable for accurate prediction of future reservoir performance. A legitimate criticism of these process-mimicking models is that they are tuned for specific situations and not generally applicable to many different environments. Many different modeling algorithms and programs would be required to address the breadth of geological environments that form the reservoirs in production today. Nevertheless, what is lost in general applicability is gained in site-specific accuracy.

Process-Based Modeling

Process-based modeling is introduced in Section 3.1, Data Inventory, as a possible source of analog data. There are a wide variety of efforts focused on the modeling of sedimentological and stratigraphic process [for example, SEDSIM for siliciclastics (Tetzlaff, 1990) and DIONISOS for carbonates (Granjeon, 2009)]. While there is great variability in the methods and assumptions in process-modeling, they share the following steps: Define initial and boundary conditions, model the process associated with transport, deposition, preservation, and post-depositional processes such as digenesis and faulting, and so on. Useful summaries or methods and discussion on the philosophy of process-based modeling are provided by Paola (2000) and Slingerland and Kump (2011), and an online book is provided by Parker (2012).

These models are valuable because they help develop theory concerning the complicated interactions of process and the resulting architectures and in some well-constrained cases provide predictive models. In the geostatistical work flows described in this book, these modeling methods may aid in the development of conceptual models (Section 3.2, Conceptual Model) and reservoir framework (Section 4.1, Large-Scale Modeling). More direct applications of process-based models to geostatistical work flows include: the development of models for property trends by converting the process model to a locally variable proportion model, or the extraction of spatial statistics, such as variogram parameters, multiple-point statistics, and object geometries for use in geostatistical modeling (Michael et al., 2010; Pyrcz and Strebelle, 2006; Pyrcz et al., 2006).

The application of process models requires extensive characterization of the paleo-conditions during the reservoir formation that is usually difficult to infer. For example, assessment of paleo-physiography and sediment inputs are problematic due to complicated basin history and limited data.

Most importantly, process-based methods are not amenable to the integration of local observations. This is a critical concern because reproducing well and seismic data is central to reservoir modeling. Current conditioning methods typically employ significant supervision and modification of the model to approximately match well and seismic observations. These methods are impractical in the presence of significant levels of conditioning.

In addition, reservoir heterogeneity is an output for process-based models, and these models may have great sensitivity to inputs (even chaotic behavior), resulting in overly broad and potentially poorly understood uncertainty models. Often pseudo-inverse modeling is employed, with iteration of input parameters to attempt to match expected architecture (Tetzlaff, 1990). While this results in lessons concerning the complicated interactions of inputs to output architectures and the resulting model should be internally consistent, these methods are laborious to apply in practice for multiple realizations for uncertainty modeling.

Also, process-based methods are generally significantly slower than conventional geostatistical simulation and many methods are not able to produce reservoir scale architecture. The resulting architectures are often at a larger than reservoir scale, representing an averaged behavior; they are not at

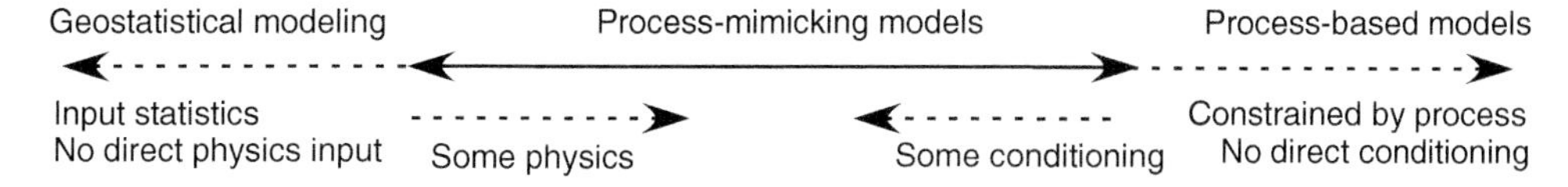

FIGURE 4.57: Simple Comparison of Geostatistical and Process-Based Modeling. There is an opportunity to move into the space between the two with process mimicking models.

a reservoir scale, element, or complex architecture that is important to reservoir flow prediction. Those methods that produce reservoir scale architecture often do so practically over short spatial extents and require strong initial constraints (e.g., the initial channel shape is mapped and the process model fills the channel with high-resolution heterogeneity).

At this time, process-based modeling is not likely the answer to improving geological realism in specific reservoir models; the answer may be to move into the space between geostatistical and process modeling methods. This is illustrated in Figure 4.57. In this section, we focus on the integration of some physical control in the form of rules. The efforts to speed up the process with fast proxies and to improve conditioning is not discussed here, although the reader may be interested in examples such as FUZZIM (Nordlund, 1999) and cellular automata (Salles et al., 2008).

While subsequent discussion focuses on improving geological complexity in geostatistics with process information, there are opportunities to integrate geostatistical concepts into process modeling. Quantification of the architectural heterogeneities that result from process models is important to developing new concepts with regard to spatial and scaling relations of process and products. These new insights may be useful for formulating secondary information to be integrated into the model with methods described in Section 4.1.

Improved Geological Complexity

For some depositional settings and project goals, reservoir significant architectures may not be adequately characterized with the cell-based or object-based methods described in Sections 4.2 through 4.4. This limitation is due to features common to geostatistical methods, such as simulation along random paths and heterogeneity limited to those characterized by stationary input statistics and nonstationary trends. These features are important because they ensure robust reproduction of input statistics calculated from local and/or analog data, and importantly, ensure efficient data conditioning. This has resulted in a set of geostatistical tools that have served well for reservoir modeling. There remains a place for conventional geostatistical methods, often in combination with the process-mimicking methods described here.

With random paths there is no consideration of temporal sequence in model construction. This is common to cell-based and object-based methods. In Sections 4.2 and 4.3 random paths are described in the sequential simulation in cell-based methods, and in Section 4.4 the random placement of objects are discussed in object-based simulation. This random sequence (as opposed to temporal sequence) is critical for conditioning and to ensure reproduction of global and local statistics and secondly to avoid artifacts. Yet, temporal considerations are central to geological concepts, and random path geostatistical methods have no concept of time. With adoption of the forward modeling (i.e., time sequence) framework, event-based models readily integrate these concepts. Time is integrated as a sequence of events. While it is possible to impose a hierarchy of scales, no effort is made to rigorously account for the exact lapse of time; instead the method accounts for the sequence of placement (and erosion) of significant reservoir elements (see a simple strike section illustration of an event-based model in Figure 4.58).

Another feature of geostatistics is the use of heterogeneity models limited to statistics that may be directly calculated from data or analogs. Geostatistical models are not predictive of heterogeneity; they reproduce the stationary statistical description of heterogeneity, and any statistics beyond these are determined by the implicit random function model. Some may argue that this reproduction of statistics from data improves objectivity by limiting unwarranted features beyond those inferred from the available information. Certainty this feature does provide a simple and defendable modeling assumption.

This is an inefficient framework for generating high levels of geological complexity. A more efficient method is to characterize the rules for interactions

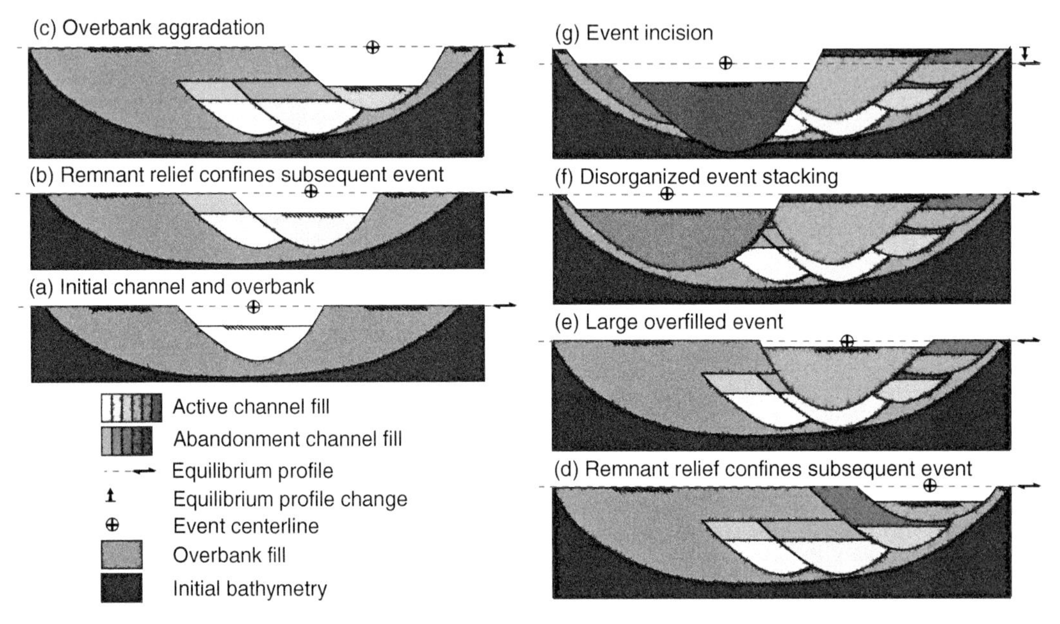

FIGURE 4.58: Time Sequence in a Single Strike Section Demonstrating the Architecture Resulting from Seven Events in Response to Varying Allogenic Parameters in Event-Based Modeling. Note the impact of under-filled versus filled channels on stacking pattern. [from Pyrcz et al. (2011).]

of depositional events and to allow the heterogeneity some flexibility to emerge as a product of coupled rules. As a result, it may be difficult to honor very specific heterogeneity concepts informed from local data and precise trend information. A practical method is to simplify the rule set for practical conditional reservoir models to enable the reproduction of the basic required architectural concepts. Then, for a numerical laboratory approach, advanced rules are enabled, resulting in complicated emergent behaviors.

Two components of emergent behaviors are new heterogeneities that are not imposed directly and drift from initial seed architectures. This may be a concern as it may be considered an artifact of mismatch of the seed form and the rules. This may be dealt with through inversion for the required input parameters given the seed form or vice versa by running the rules until the form stabilizes and removing the transitional products (Howard, 1992).

The emergence of low entropy features not explicitly constrained in the model inputs is considered part of the predictive nature of these models. This is a challenging area of research and admittedly the extent of the predictive value and reasonable applications need to be explored. The efficient reproduction of emergent reservoir scale features has resulted in new concepts and directions for investigation (McHargue et al., 2010; Pyrcz et al., 2011; Sylvester et al., 2010).

Direct reproduction of input statistics and strong conditioning are important strengths of geostatistics and must be maintained. These are critical components for data driven methods. Yet some heterogeneities do not lend themselves to such random path methods and statistical descriptions. These features may be more readily explained in sedimentological and stratigraphic terms. Central to sedimentology and stratigraphy are concepts of temporal sequence of coupled deposition and erosion process driven by mass and energy balance and constrained by topography and sediment sources (Boggs, 2001; Galloway and Hobday, 1996; Reading, 1996; Walker and James, 1992).

Process-mimicking methods retain many of the paradigms of geostatistics while improving the integration of geological information. They remain fast, conditional to sparse data, and can impose some statistical features and trends. Process-mimicking requires a shift to methods that sacrifice rigor in data conditioning and reproduction of input statistics for efficient reproduction of architecture or conditioning to architectural information. These types of compromises between disparate inputs are common in geostatistical methods. For example, a trend model may reduce the reproduction of a specified variogram model, and a cloud transform simulation will reproduce the detailed bivariate relationship, but may not precisely honor the input histogram and spatial continuity.

4.5.2 Process-Mimicking Modeling

When specific architectures such as interelement stacking patterns and associated heterogeneities are relevant to model response, it is important to model them. These may be reproduced by a set of simple rules that describe reservoir elements and their relationships spatially and temporally. For example, one may characterize a distributary lobe complex with a rule that subsequent events are attracted to erosional relief, repulsed from depositional relief and transition from channel to lobe element due to change in gradient (see Figure 4.59).

While this set of rules are intuitive and straightforward, it would not be feasible to impose these features in geostatistical modeling without temporal sequence. In cell-based methods, this would require the development of a statistic that would characterize the complicated, nonstationary product of this rule. In object-based models this would require a constraint that randomly placed channels and lobes form this complicated stacking pattern. In the least, this would require a statistics that describe this complicated stacking pattern and an object placement method that can impose this statistical description. While such a method may be possible, it would likely miss many of the features readily reproduced with process-mimicking methods. The ultimate decision to consider process-mimicking methods will depend on the geological accuracy required to meet the project goals.

The following discussion covers the principles of process-mimicking methods, including the building blocks, temporal sequence, types of rules, and some general comments on predictive power.

The Building Blocks

The fundamental building blocks of process-mimicking models include: definition of model state properties, definition of events, placement of events, and updating the model state properties.

State properties are defined and quantified over the model domain. These are properties that provide local information to interact with the rules and may include properties that describe attributes at any scale, including morphometric properties such as elevation, erodibility, gradient, gradient change, and derived properties such as event frequency, event proximity, available accommodation, isopach thickness of deposits, and process properties such as approximations for velocity and energy. These properties are initialized and then updated during model construction.

Events and associated architecture represent the basic reservoir modeling units. As with the facies objects in object-based facies modeling, events are

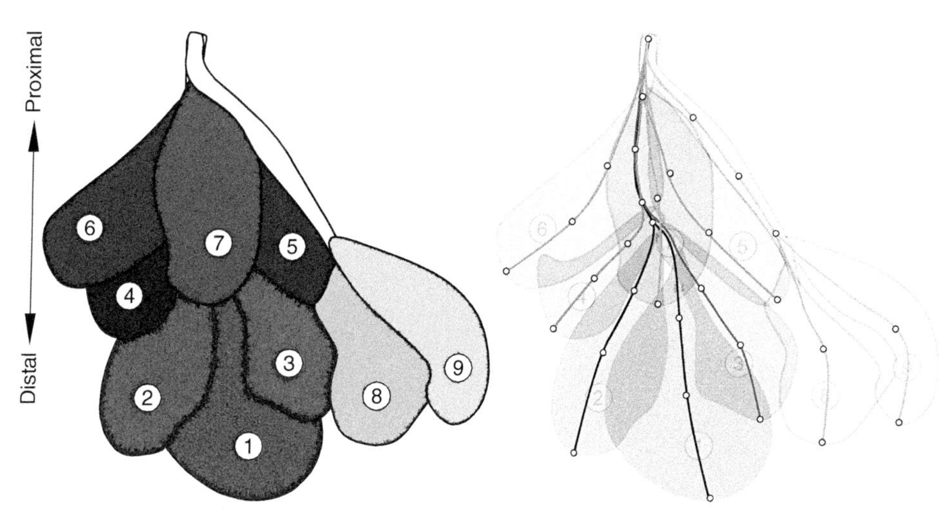

FIGURE 4.59: This Pattern is Reproduced Simple Rules from Depositional Process, Lobes Attach to the Ends of Channels, Channels Attract Architecture Due to Erosional Relief, Lobes Repel Architecture Due to Aggradational Relief, and Lobes Form in Locations with Rapid Decrease in Gradient (Pyrcz et al., 2006). The event geometries are shown on the left, and the event control nodes and centerlines are shown on the right.

defined at a specific scale to meet the project goals—that is, the fundamental building blocks that influence reservoir volumetrics and flow character. Scale choices will depend on required resolution for the model purposes, available information, and time. Hierarchical modeling methods may be applied to model salient heterogeneities across reservoir scales from complex sets to complexes and to elements and within elements (Deutsch and Wang, 1996; Pyrcz and Deutsch, 2005). Events may include objects defined by centroid or centerline and with associated parametric geometries and may include composite objects with embedded or transitional patterns. Rules may be applied to determine the geometry or changes in geometry as a function of model state.

Event placement includes orientation for centroid-based elements and route finding for centerline-based elements. Route finding may include rules for the response of events to confinement and ponding (Pyrcz and Strebelle, 2006; Pyrcz et al., 2006). In addition, the route may impose constraints on the event because the geometry may change as events are confined by local topography. Rules are required for areal and vertical placement as a function of model state. Areal placement rules may include a spectrum from simple rules that are only based on the previous element to more complicated rules that consider the cumulative state properties of the model. Vertical placement may be based on simple additive or subtractive methods or based on an assumption for local datum (Pyrcz, 2004). Refer back to Figure 4.56 for an illustration of event placement based on route finding and geometry and geometry transitions that respond to the current state of the model.

After an event, the *local state properties are updated.* Updated state variables then constrain subsequent events. There is a continuum of possible implementations depending on the complexity of the rules. For example, the topography may be raised by deposition from a previous event, and this may repulse a subsequent event. In a more simple stochastic approach a derived map of local frequencies may be used to force attraction or repulsion with a subsequent event (Xie et al., 2000).

Temporal Sequence

As previously discussed, the utilization of temporal sequence enables the integration of intuitive confinement and stacking rules (e.g., consider tracking topographic evolution over time). In addition, the framework allows for the integration of concepts related to allogenic cycles (see Section 2.1, Preliminary Geological Modeling Concepts) with systematic changes in architecture over time and results in models that directly integrate scale and hierarchy. For example, the input constraints of the model may cycle with a user-defined allogenic curve (Pyrcz et al., 2011).

Types of Rules

There is a continuum of rule-based approaches within process-mimicking methods (see Figure 4.60). One could formulate simple rules that predictably result in a desired stacking pattern or other heterogeneity. Conversely, one could formulate rules concerning flow routing within topography and other state properties and allow a stacking pattern to emerge from the model. At one end of the continuum there are independent stochastic rules, in which objects are randomly placed without consideration of previous events as in object-based simulation (Section 4.4). At most, these rules may include statistics to constrain a general stacking pattern with clustering or repulsion. At the other end of the continuum there are fully coupled process constructs that represent first principles, conservation laws, and coupled processes of reservoir formation. Under process-mimicking applications, three types of rules are identified.

Drafted rules are rules implementations in which the user specifies detailed architectures and the rules fill in heterogeneity in a completely predictable and

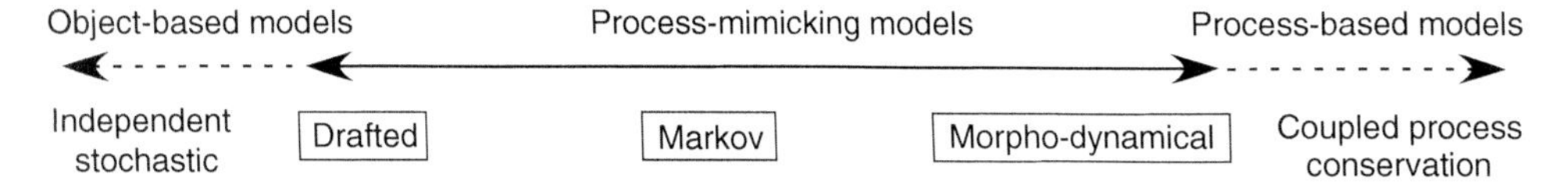

FIGURE 4.60: Continuum of Rules Implementations within Process Mimicking Methods. Within object-based models the rules are typically based on independent, random placement of objects with perhaps limited constraint on overall stacking while with process-based models the rules or rigorous coupled process and conservation laws. Three distinct levels of process-mimicking rules are identified.

consistent manner (Wen, 2005). These methods provide excellent control on heterogeneity, but require significant user inputs and constraints. *Markov rules* are only dependent on the previous event (e.g., the previously mentioned bank retreat model). These methods provide complicated, realistic architecture with limited user constraint required. The general features should be predictable, but there may be some interesting emergent specific features. A set of possible Markov rules for channels are shown in Figure 4.61. *Morpho-dynamical rules* consider the cumulative state properties of the model to constrain the subsequent event (e.g., the previously discussed distributary lobe model in Figure 4.56).

In general, for simple architecture, the independent stochastic rules with typical object-based methods are often appropriate. For more complicated architectures, drafted and Markov rules are applied to reproduce inferred architectures. Morpho-dynamical rules may be useful in practical terms to impose confinement constraints—for example, the model of turbidite channels confined by a slope valley in the Deepwater Reservoir Model. Yet, efforts to integrate detailed morpho-dynamical rules have resulted in models that exhibit numerical chaos intrinsic to process-based models (Pyrcz et al., 2012). While conditioning methods may be efficiently imposed, the overall architectural heterogeneity is strongly emergent and difficult to constrain for practical reservoir models. For example, areal trends may significantly depart from uniform even in the absence of a trend model. The use of morpho-dynamical rules will likely remain in the short term as useful tools for experimentation and research, but

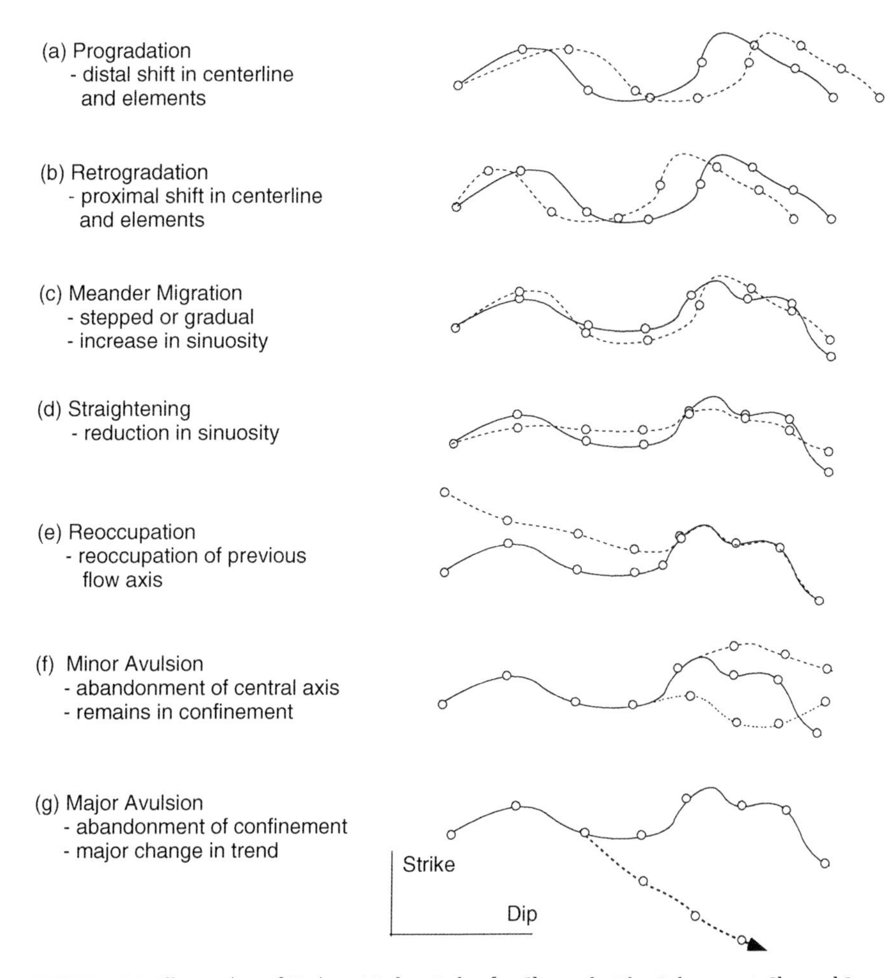

FIGURE 4.61: Illustration of Various Markov Rules for Channels. The Subsequent Channel Is a Function of Only the Previous Channel. For example, progradation results in a forward step of the channel features, while minor avulsion results in branching from the course of the previous channel.

not for practical reservoir models (McHargue et al., 2010; Pyrcz et al., 2011; Sylvester et al., 2010).

Rules may be further subdivided by the associated method of rule inference and rule expression. *Empirical rules* are observed from data or analog and coded as geometries and probability distributions (e.g., element geometries, avulsion frequency etc.). *Predictive rules* require a level of interpretation concerning sedimentological process and allogenic forcing (e.g. avulsion location rules, flow response to gradient). Also, rules may be formatted deterministically as constants, functions or relationships and thresholds, and stochastically as conditional probability density function or even as fuzzy rules informed by an expert system.

Furthermore, rules may be applied to various components of the model. *Geometric rules* constrain the internal and external geometry of elements and the transitions between elements. For example, a geometric rule may relate the cross section of a channel to the local channel centerline curvature. *Placement rules* constrain the sequential placement of elements. Examples of placement rules include a depositional probability map based on accommodation and substrate erodibility or a routing algorithm that places channels from sources through a state model. *Initiation rules* constrain the source and type of event and may be tied to allogenic forces and current topography. *Termination rules* constrain the event termination, often at a sink location along the model periphery or potentially in the model if event ponding is considered.

Information Content

During model construction all spatial and temporal information is preserved and can be applied to constrain subsequent modeled events or reservoir properties within and between elements at any scale. This information links all spatial and temporal positions within the model with all other positions. For example, positions on channels and levees may be related by updip avulsions, overbank locations may be related to adjacent channels, and potential for course overbank flows is due to channel curvature. All locations may be related to an allogenic scheme, relating energy, discharge, sediment supply, water depth, and so on (see Figure 4.62).

All of this is grid-free or independent of a grid (see discussion in Section 4.4, Object-Based Facies Modeling). While there are some similarities

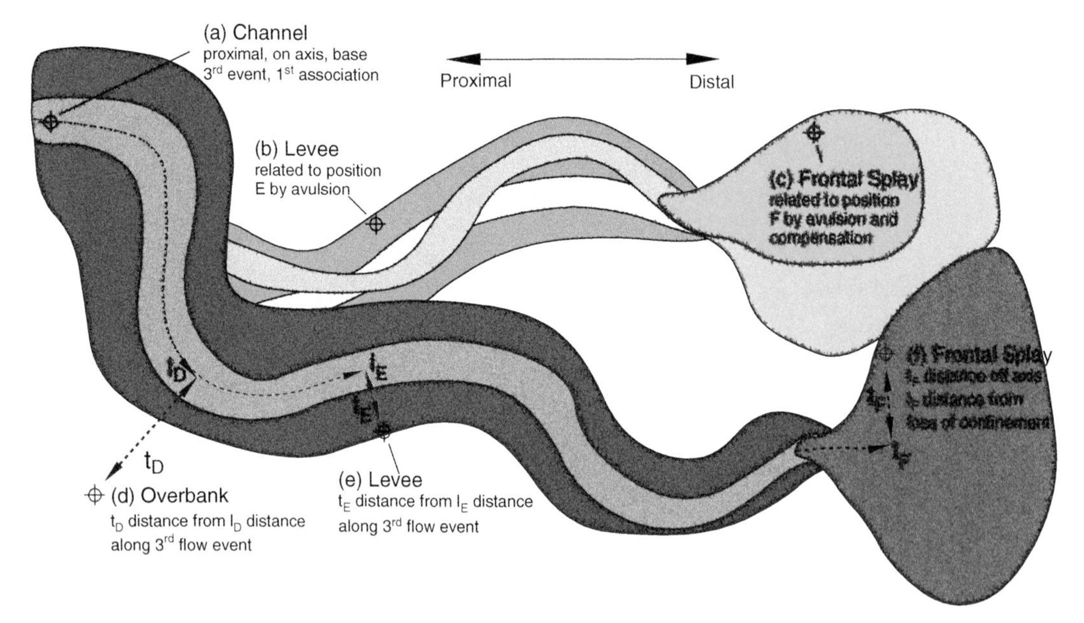

FIGURE 4.62: Illustration of the Information Content within an Event-Based Model. For any location within an event-based model the story of the genesis of the location may be extracted: (a) is on the axis, near the base of the main channel, near the source, (b) is on a levee related to position (e) by an updip avulsion, (c) is on a marginal frontal splay related to (f) by an updip avulsion, and (d) is overbank with some potential for courser grained overbank due to the curvature of the adjacent channel at I_D.

From Pyrcz and Strebelle (2008).

with grid-free approaches in object-based modeling, the level of information, including state and process proxies and the ability for these to interact with model construction, is unique. This level of information preservation aids the actual rules-based implementations.

In addition, this information may be utilized for detailed trend and property models to fill in the event geometries. This information can be applied to calculate depositional coordinate systems, properties, and locally variable azimuth fields that may be combined to formulate property trend models and constrain subsequent reservoir property simulation. These coordinates may include distance from the thalweg of a channel fill, the curvature of the channel at a location in a channel fill, or the velocity profile along strike of a channel fill (Pyrcz, 2004; Pyrcz et al., 2005b; Stright, 2006).

As mentioned in Section 4.4, Object-Based Facies Modeling, grid-free work flows enable model refinement. A sector model was extracted between two wells from an event-based model of the Deepwater Reservoir Model to demonstrate this capability (see Figure 4.63). The facies associations and reservoir property trend models are rescaled from a low to medium and high resolution without (a) loss of information or (b) need to regenerate the model by building another stochastic realization.

Other Considerations

As mentioned above, in the case of morphodynamical and to a lesser extent Markov rules implementations, the models may exhibit emergent, unanticipated features. In this case some components of architecture are not explicitly defined by the inputs. Care must be taken to ensure that the architecture is reasonable for the modeling needs and honors local information. In extreme cases, mismatch between the initial state and the rules may result in architectural transitions. For example, a seeded initial low-sinuosity channel becomes high sinuosity over multiple events with Markov rules. This is analogous to spatial shifts that occur in geostatistical modeling away from data when data statistics do not match the input simulation statistics. In another example, the combination of slope valley confinement, path routing, and data constraints will result in shadows in the expected channel density maps calculated over many realizations, while object-based and multiple-point facies models would produce more naïve, uniform expected channel density maps. This added information and constraint must be understood and geologically defendable.

Conditioning to dense well data and specific soft information and trends remains a challenge. Interactions between the event objects, along with the utilization of a forward framework, increase the difficulty

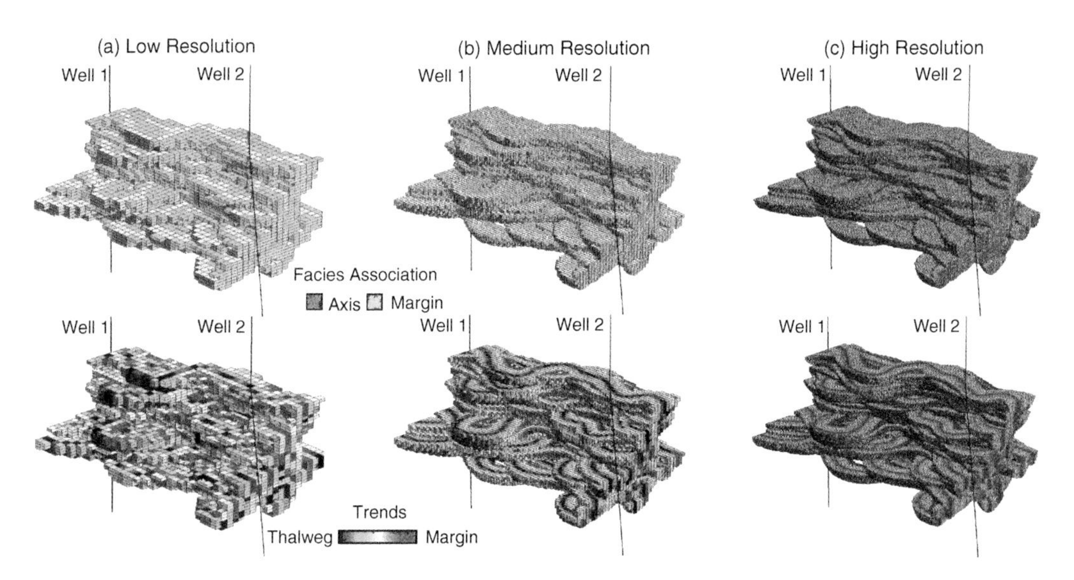

FIGURE 4.63: Sector Model between Two Wells from the Deepwater Reservoir at Three Scales. The model was constructed from the truth model by "reloading" the architecture to three different grids with low, medium, and high resolution. The within channel trend model is based on an internal coordinate based on distance relative to channel thalweg and margin. The truth model was constructed from a process-mimicking model for channel stacking in deepwater slope valley channels.

of conditioning over object-based simulation. It may not be possible to freely move an element to match conditioning data, given the fact that rule constraints between objects and forward modeling may move the model into positions that may prevent subsequent conditioning. Also, the forward framework increases the difficulty in honoring global proportions and trends because it is not possible to simply place another event in any position to improve statistical reproduction. There are a variety of methods that may be applied to condition these models, including (1) dynamically constrain model parameters during model construction to improve data match (Abrahamsen et al., 2008; Lopez et al., 2001), (2) a posteriori correction with kriging for conditioning (Ren et al., 2004), (3) pseudo-reverse modeling (Tetzlaff, 1990), (4) apply as a training image for multiple-point geostatistics (Strebelle, 2002), and (5) direct fitting of geometries to data (Shmaryan and Deutsch, 1999; Viseur et al., 1998), (6) extracting features from the model and associated statistics (Miller et al., 2008; Pyrcz et al., 2011), and (7) postprocessing with image cleaning (Pyrcz et al., 2009).

Each of these methods may demonstrate success in distinct settings. Any current process-mimicking method will have some variant of these methods integrated for conditioning. It is important to check the resulting models for conditioning mismatch and artifacts as discussed in Section 4.4, Object-Based Facies Modeling.

Example Implementations

The following examples demonstrate a variety of rule-base implementations.

Drafted Rules Approaches

The drafted approach allows for the precise reproduction of complicated channel architecture. These models are strongly constrained by the user with drafting of the critical elements and rule-based interpolation. For example, one method included drawing of the initial and final channel in a fluvial meander channel belt combined with geometric parameterization of the infill architectures to construct a detailed model (Hassanpour et al., 2013; Wen, 2005). This method is useful for fast generation of architectural complexes. Complex sets required precisely parameterized architectures based on the nesting of multiple drafted complexes. These methods have been be applied to test architectural flow response and scale-up relations (Wen, 2005) or constrained to condition in very sparse data settings.

Markov Rules Approaches

With the Markov model approach, the subsequent event is determined only by the previous event. This allows for these models to evolve over time, yet most long-range and all short-range features are known a priori. For example, the placement of a channel is determined by the location of the previous channel and rules with respect to avulsion and/or lateral accretion (Howard, 1992; Lopez et al., 2001; Pyrcz and Deutsch, 2005; Pyrcz et al., 2009; Sun et al., 1996; Sylvester et al., 2010).

The Fluvial Reservoir Model truth model was constructed with Markov rules (see Figure 4.64). These rules included major and minor avulsion and meander migration with associated probabilities of occurrence (10%, 10%, and 80%, respectively). For

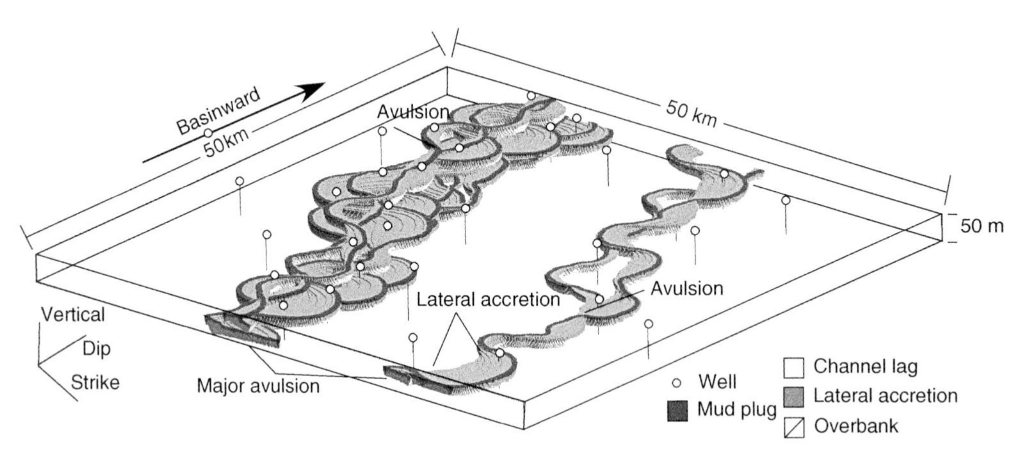

FIGURE 4.64: An Event-Based Model Driven by Markov Rules for Major Avulsion and Minor Avulsion and Meander Migration, Resulting in the Identified Features.

each event, the type of event was first drawn from the event type PDF and then the rules for the drawn event were applied. Major avulsions seed a new channel centerline based on initiation rules for channel source constrained by seismic information. Minor avulsions find an avulsion node as a function of curvature and near bank velocity of the previous channel centerline and seed a new channel at that location. Meander migration utilized a rules-based simplified form of the bank retreat model to produce lateral accretion features (Howard, 1992, modified from).

Rules-Based Morpho-dynamic Models

With morpho-dynamic models, the subsequent event is determined by the current accumulated state of the model. This allows for complicated evolution of emergent behavior—for example, the generation of a distributary lobe complex through the erosion and deposition of channels and lobes along a steepest gradient path with transition from erosion to aggradation dependent on a flow energy proxy (Abrahamsen et al., 2008; Pyrcz et al., 2005a, 2006; Zhang et al., 2009). See Figure 4.65 for an example of this type of model. Note the initial feature, a single channel and lobe, is directly determined by the rules, but the subsequent features, including compensation stacking, progradation, avulsion, and retrogradation, are all emergent features that could not readily be determined a priori by interrogating the rules. A process-mimicking model was required to observe their interactions. Another example of this approach with a deepwater slope valley model that integrates allogenic forcing, substrate erodibility, and flow confinement to dynamically model the valley erosion and fill is provided by Pyrcz and Strebelle (2008) and McHargue et al. (2011).

Additional Applications

Some further comment is provided on applications of process-mimicking models. Detailed, geologically realistic reservoir models allow for various conventional and new applications.

Conventional Geostatistics

The ability to very quickly and easily build a large suite of scale-refineable, information rich, detailed numerical representations of reservoir architecture is useful in conventional reservoir geostatistics. The

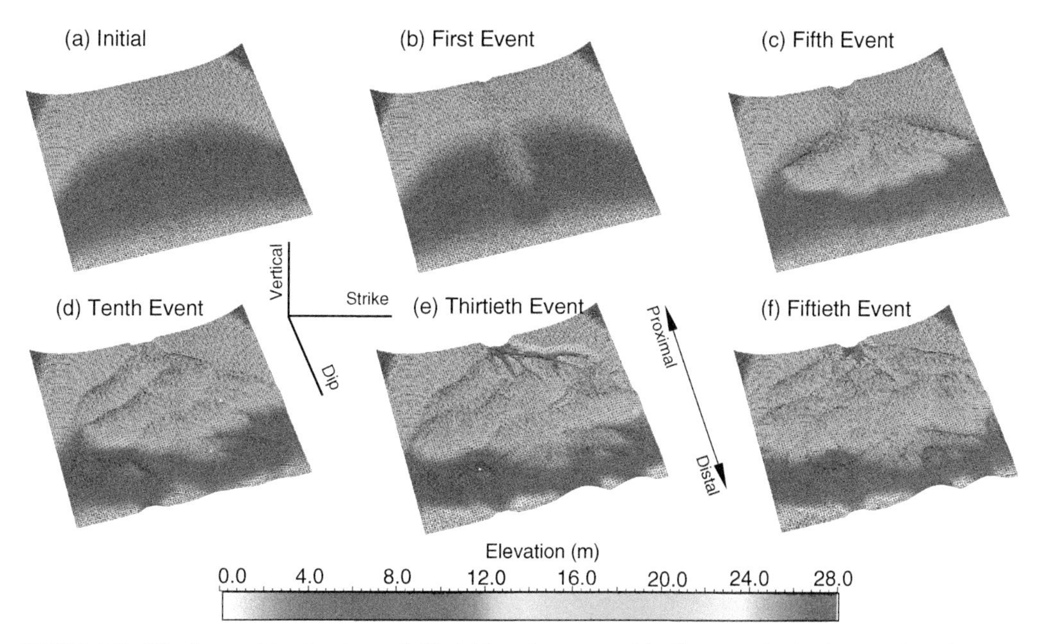

FIGURE 4.65: Distributary Morpho-Dynamic Model for a Deepwater Distributary Lobe Setting Based on the Coupling of Simple Rules. (a) Simple initial bathymetry model, (b) Initial event follows steepest descent and switches from erosion to deposition with a decrease in gradient. (c) Multiple event emergent compensates due to remnant relief of previous events. (d–f) Event exhibit emergent events avulse, prograde, and retrograde with various frequencies.

Taken from Pyrcz et al. (2006).

potential applications include training images and nonstationary statistics.

Exhaustive training images are required by MPS to calculate the large number of required conditional probabilities for specified data events (see Section 4.3, Multiple-Point Facies Modeling). It is natural to consider the use of process-mimicking methods to furnish such images. Yet, experience has shown that nonstationarity and complexity in training images may degrade the quality of MPS realizations in methods that do not account for nonstationarity in the training image (see Section 4.3). Often, the best practice is to construct simplified training images and impose nonstationarity in the MPS simulation (Harding et al., 2005; Strebelle, 2002, 2012). Given this consideration, care must be taken when using process-mimicking models for training image generation. Often, the more simplified methods such as the training image libraries and generators (Maharaja, 2008; Pyrcz et al., 2007) discussed in Section 4.3 are sufficient. There may also be opportunities to utilize process-mimicking models to extract traditional spatial continuity models (that are less sensitive to nonstationarity) such as semivariogram models and indicator semivariogram models and transitional probabilities.

Another application is to calculate nonstationary statistics such as trends and locally variable azimuths from process-mimicking models and apply them to guide conventional geostatistical models. For example, event-based methods may be useful for inferring consistent, detailed locally variable azimuth models, given local constraints. This is straightforward, given the preservation of information in these models. The design of unstructured grids from these detailed architectures could also be useful.

Numerical Analog Models

In a more advanced application, the event-based models are treated as numerical analog models for the reservoir of interest. Once this decision is made, a variety of applications are possible, including determining architectural relationships, well risk analysis, and value of well data. Admittedly the following applications rely on strong assumptions of the selected models being analogous and representing appropriate models of uncertainty.

It may be important to understand the relationships between model architectural parameters and/or between model architectural parameters, reservoir parameters such as fluid volumes, connectivity, and flow response. With a detailed set of realistic architectural models, this is possible. For example, in a simple channel setting, the influence of the channel stacking and system aggradation rate can be directly related to reservoir volume. Process-mimicking methods enable this experiment, because this volume is not the simple sum of the volume of individual channels, nor constrained by the model as an input, but the result of a complicated preservation operator inherent to the rules (see Figure 4.66). In another case, the preservation potential of components of the element fill may be quantified for representative statistics—for example, the fraction of axial channel lag preserved in an aggrading and meandering channel model. There are much more complicated experiments possible, such as quantifying reservoir connectivity for various types and frequencies of channel avulsion.

Journel and Bitanov (2004) and Maharaja (2007) developed methodologies for exploring NTG uncertainty through spatial bootstrap from reservoir models. In a similar manner, a proposed well design may be applied to a suite of event-based models to calculate the resulting probability distribution of any associated well result of interest, such as the net pay length, average NTG, proportion the well of above specific thresholds, number of isolated units, and so on. If information is available concerning well site selectivity in the analog models, this may be imposed with a selectivity bias surface that adjusts spatial bootstrap sampling rates (see a suite of well samples and well-based statistics from an event-based model in Figure 4.67). This work flow is useful, because it may be difficult to transfer 3-D concepts of architecture to a set of 1-D concepts of well outcomes (along well trajectory). In addition, this method allows for the comparison of multiple well plans and their associated risks. For example, the distribution of possible NTG can be compared for a single well and averaged over a multiple well template, providing a direct indication of the mitigation of well risk through multiple wells. Also, vertical, horizontal, and deviated wells may be compared to assess the value of the added cost of directional drilling to mitigating well risk.

As an extension of well risking, the value of well information may be analyzed. It is straightforward to sample well templates through a suite of event-based numerical analogs to assess the probability of a well result, given the architectural model or

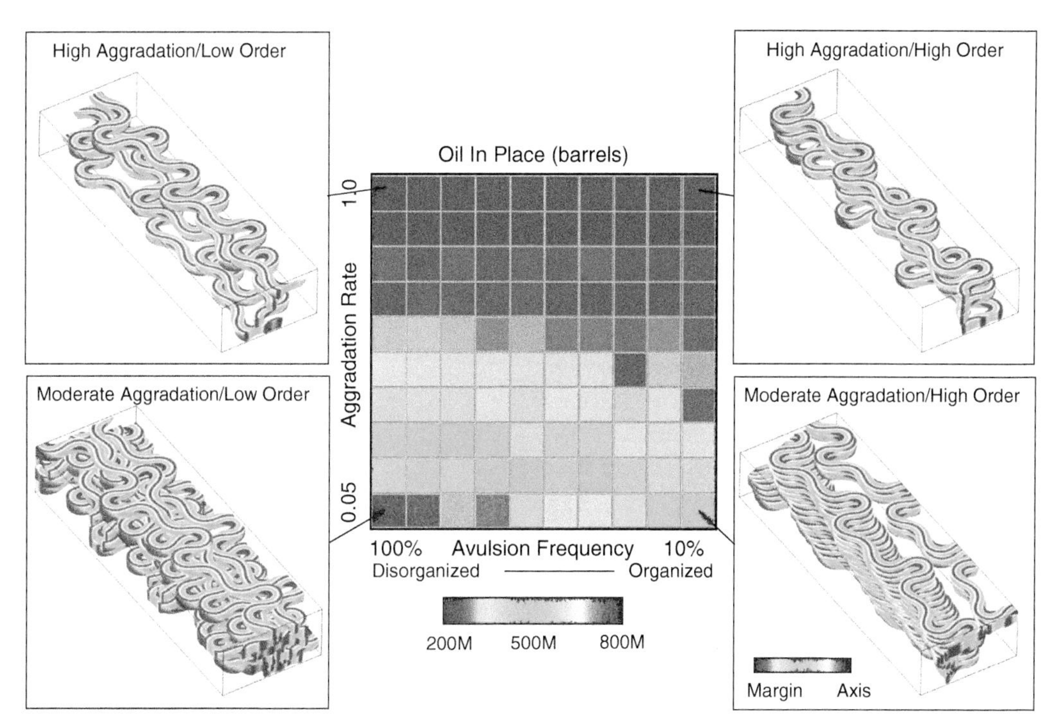

FIGURE 4.66: Response Surface for Oil in Place from a Spectrum of Event-Based Channelized Models with Variable Aggradation Rate and Degree of Channel Organization as Constrained by the Frequency of Avulsions. Assuming that the sediment-fill composition of channel elements remains constant, the volume of oil in place is constrained primarily by high aggradation rates and secondly by disorganized stacking of channel elements.

From Pyrcz et al. (2011).

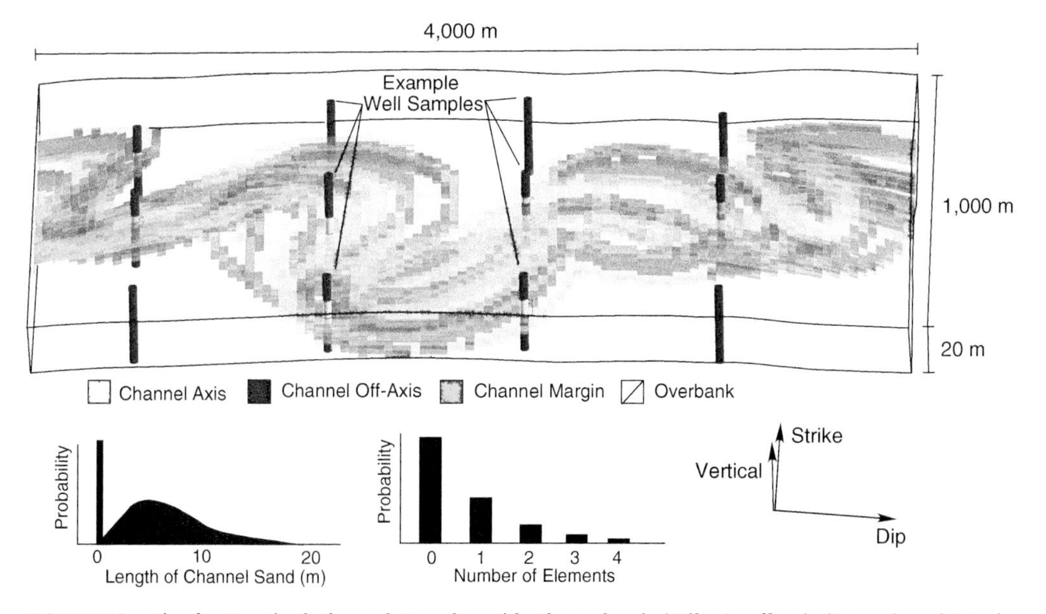

FIGURE 4.67: Simple Organized Channel Complex with Channel Axis (Yellow), Off-Axis (Orange), and Margin (Green) and a Subset of Regularly Spaced Spatial Bootstrap Samples and Two-Example Well-Based Statistics from Exhaustive Sampling.

any description or summary statistic for the model. Bayesian inversion allows for the assessment of the probability of a specific architecture given a well result (Pyrcz et al., 2012). The concept of inferring architecture from well statistics has been applied with bed thickness from indicator variograms by Hong and Deutsch (2010).

A simple numerical approach to this inversion is possible. A large suite of models are generated that are deemed to represent reservoir uncertainty. Then a variety of well configurations are sampled from all of these models, and statistics of interest are binned and tabulated. A simple example for a several-well template and from 4000 architectural models is shown in Figure 4.68. Note the difference in the probability contours between the 1 well versus 7 well cases. Changes in the number of wells, well rules, and type of wells also impact these conditional distributions and provide information on the value of a well plan (measured by separation of conditional probabilities $P\{\text{model}|\text{well}\}$).

4.5.3 Work Flow

This section presents the methods and considerations for the application of process-mimicking simulation. Figure 4.69 illustrates the steps in building a process mimicking simulation. The first step is to integration all available information, including well data, seismic data, and conceptual information and experience into a facies framework. This framework includes the conceptual facies model with number, geometries, and interrelationships of facies. From this framework, geometric parameters of architectural elements and rules for sequential construction of the reservoir are inferred. Model construction includes the sequential addition of elements to match local and global proportions and conditioning data. When all of these constraints are met, the results are checked. There may be a need to iterate and change the geometries and rules to match the concepts or to improve conditioning.

4.5.4 Section Summary

Process-mimicking methods are becoming increasingly common geostatistical tools that utilize a geological process coded into rules in a forward context to further improve geological realism. This methodology is useful when geological complexity is essential to the project goals and local data constraints are sparse.

This method requires a retooling of geostatistics, with the integration of forward modeling and rules-based inputs. In turn, these methods may result in more challenging conditioning and emergent behaviors that may be deemed useful and accurate or conversely may be considered artifacts. Simple drafted and Markov rules methods are practical in reservoir modeling. Morpho-dynamic rules-based methods are very interesting research tools, but may not be practical for direct modeling of reservoirs in the short term. New work flows with these types of models require thorough model checking (see Section 5.1).

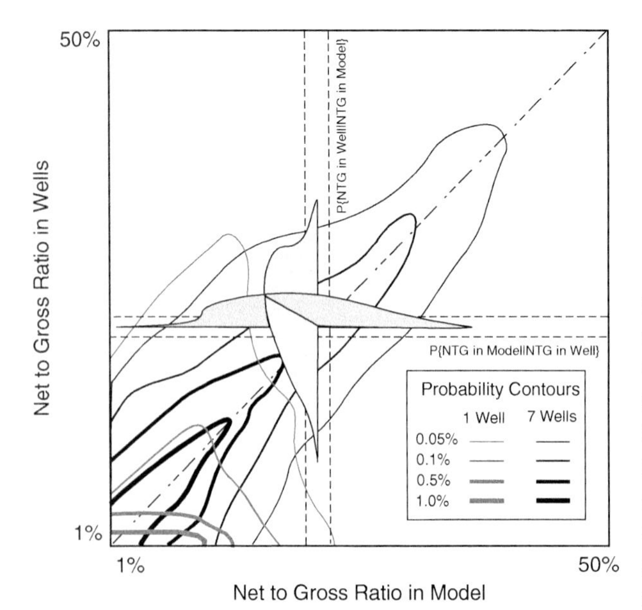

FIGURE 4.68: Net to Gross in Model versus Net to Gross Observed in One or Several Wells Bivariate Relationship Calculated by Scanning 4000 Channel Models with Both Well Templates, Allowing for Inversion for the Probability Distribution for NTG in Model Given Well NTG. The degree of separation in the conditional distributions (net to gross distribution in model given net to gross observed in wells) indicates information content for wells to inform reservoir.

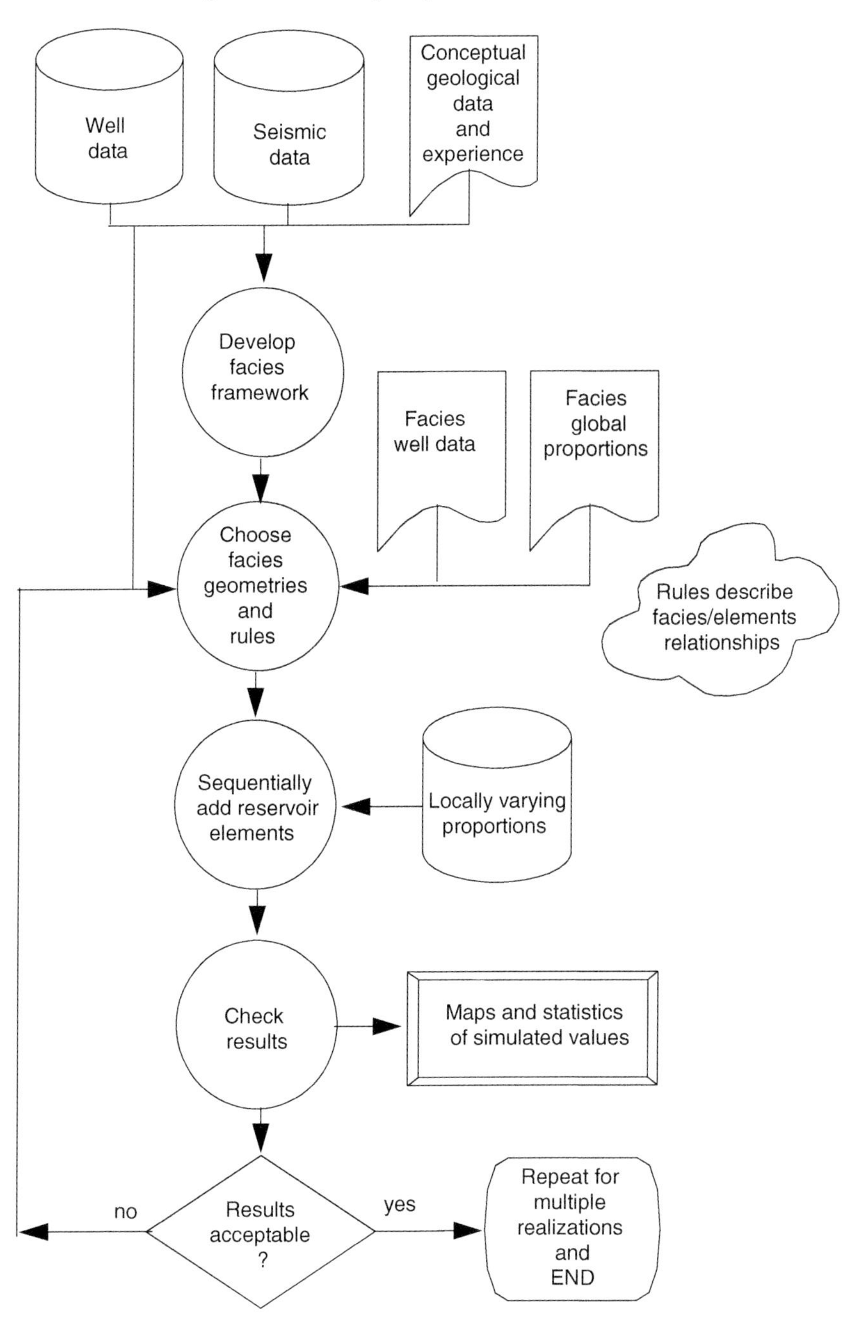

FIGURE 4.69: Work Flow for Process-Mimicking Facies Models.

Constructing reasonable facies models to meet project goals is an important aspect of geostatistical reservoir modeling. The next step, as presented in the next section, is the modeling of continuous reservoir properties such as porosity, permeability, and saturations within the facies framework.

4.6 POROSITY AND PERMEABILITY MODELING

Petrophysical properties such as porosity and permeability are modeled within each facies and reservoir layer. Shale and non-net facies may be assigned arbitrary low values; however, the petrophysical properties within most facies must be assigned to reproduce the representative histogram, variogram, and correlation with related secondary variables. The Gaussian and indicator techniques introduced in Section 2.4, Preliminary Mapping Concepts, are used for this purpose. This chapter discusses practical considerations and implementation details.

Multiple-point methods for continuous variables are an active area of research, for example, see the work of Honarkhah and Caers (2010); Mariethoz and Kelly (2011); Mariethoz et al. (2010); Zhang et al. (2006). This promising area of study is still early and is not currently available. For this reason, we focus on traditional continuous variable modeling methods. Some discussion on continuous MPS is reserved for Chapter 6, Special Topics.

In many reservoir modeling problems the majority of reservoir heterogeneity is constrained by the facies models. In these cases, continuous reservoir property variability between facies is greater than variability within facies, and spatial heterogeneity of facies is a more significant constraint on fluid flow than spatial continuity of the continuous reservoir properties. In these cases, focus should be on facies modeling and then reasonable continuous property modeling within the facies framework. However, sometimes discrete, mappable facies are not available; therefore, the modeling work flow omits facies and proceeds directly to continuous properties to characterize reservoir heterogeneity.

The *Background* subsection describes the characteristics of porosity and permeability and the considerations for modeling. Porosity is relatively straightforward since it averages linearly and has low variability. Permeability is more difficult. It is more variable, has a highly skewed histogram, does not average linearly, and must reproduce an imperfect relationship with porosity.

The *Gaussian Techniques for Porosity* subsection describes how sequential Gaussian simulation can be used to create porosity models. Porosity is simulated within each facies type, defined within the correct layer-specific coordinate system. This Gaussian technique is used extensively because of its simplicity of application and flexibility to create realistic heterogeneities.

The *Porosity/Permeability Transforms* describes basic methods to get permeability with porosity/permeability transforms such as regression, conditional expectations, and simple Monte Carlo simulation from porosity/permeability cross plot. These simplistic techniques only work when (1) there is excellent correlation between porosity and permeability or (2) the permeability histogram has very low variance. In general, some technique that reproduces the variogram of permeability is required. For a more general applications, conditional cloud transform simulation is an option.

The *Gaussian Techniques for Permeability* subsection presents Gaussian techniques to cosimulate permeability using porosity as a secondary variable. Gaussian cosimulation is easy to apply and reproduces the basic characteristics of permeability variation. Indicator techniques must be considered when permeability shows complex patterns of spatial variation such as exceptional continuity of extreme values. The *Indicator Techniques for Permeability* subsection gives details of how indicator techniques are applied to permeability. The final subsection *Work Flow*, presents work flows for porosity and permeability modeling.

The novice modeler should benefit from the detailed descriptions and examples for porosity and permeability modeling. The intermediate modeler may benefit from the comparison of various methods to model permeability. This material may be skipped by the expert modeler.

4.6.1 Background Variables

Porosity and permeability are the main variables needed for reservoir characterization. Other variables such as the volume fraction of shale, impedance, velocity, and residual water saturation may also be of interest. In the following we refer to these variables as petrophysical properties, reservoir properties, or just properties.

Porosity is the volumetric concentration of pore space. It is a well-behaved variable since it typically has small variability and averages according to a simple arithmetic average. There are different definitions for porosity. It is common to model the "effective" porosity. The "total" porosity includes volume that may not be accessed through practical recovery mechanisms.

Permeability and some other variables are not intrinsic rock properties; they depend on boundary conditions outside of the volume of measurement or specification. These variables can vary over several orders of magnitude. Permeability is a rate constant or effective property that is not a simple average of smaller-scale values. There are some considerations in modeling permeability: (1) The data should be corrected to reservoir fluid and pressure conditions; these conditions are different from the laboratory where the measurements were made and a prior correction using well test-derived permeability data must be considered. (2) Horizontal to vertical anisotropy is an important factor; the principal directional components of permeability can be modeled or global vertical to horizontal anisotropy ratios can be applied on a by-facies basis.

As discussed in Section 2.4, Preliminary Mapping Concepts, vertical and areal trends in the mean and direction of continuity are common. Facies modeling may have accounted for some of the trend. Deterministic trends must be modeled, and some decision must be made about how to account for the trend—that is, a cokriging or a mean plus residual approach. Details are presented in Section 2.4. The "variable" to model may be residuals instead of original units.

Scale Differences

Petrophysical properties must be assigned to every cell in the geological model. As discussed in Chapter 1, the volume difference between different core measurements, well log-derived properties, and the geological modeling cell size is not considered. Therefore, we are essentially assigning *point* scale values on a regular grid in stratigraphic coordinate space. These values are associated to the entire geological modeling cell for calculation of fluid volumes and properties for flow simulation. Reservoir modelers will increasingly consider the vast volume difference between well data and the geological modeling cells (see the "missing scale" discussion of Section 2.3, Quantifying Spatial Correlation); however, the conventional practice presented here has led to reservoir models with significant predictive ability. In Section 2.4, Preliminary Mapping Concepts, alternative work flows are presented that assume point scale properties at model cell centroids, this may aid with the scale issue. However, eventually some type of scale-up will be required to calculate effective flow grid properties for flow simulation. More details on scaling are available in Section 5.2, Model Post-processing.

Facies

Porosity and permeability are modeled within each facies and reservoir layer or region (see Section 4.1). The reservoir layers are distinct due to deposition at different times. Reservoir properties within the *same* facies and *different* layers may have similar characteristics. Nevertheless, they have to be modeled separately because of different layer-specific stratigraphic coordinate systems. The properties within *different* facies in the *same* layer often are significantly different and unrelated.

The properties within the different facies can be modeled independently when the facies are unrelated. The "cookie-cutter" approach is sometimes used: (1) Complete 3-D models of the properties are constructed for each facies, and (2) the models are merged using the facies model, that is, the porosity and permeability are taken from the appropriate facies-dependent model. This is wasteful. A faster approach is to use the facies model as a template during simulation of the properties. The predictions are limited to where they are needed. This does not introduce any biases or other problems.

Properties in different facies may be related to each other; that is, there is a correlation in petrophysical properties across facies boundaries (see discussion on boundary modeling and contact profile method to check for boundary types in Section 4.1, Large-Scale Modeling). There are a number of approaches to deal with this correlation:

- Ignore the correlation and model the properties independently in the different facies. The correlation across facies boundaries likely has little consequence on volumetrics and flow performance predictions. We could argue that such correlation is not a first-order effect.
- Consider a form of cokriging with the properties in different facies as secondary variables. Calculation and modeling of the necessary models of coregionalization is difficult. This approach is too awkward for practical application.
- Create the models sequentially; that is, construct the model of porosity and permeability in one facies and then use the result to assist in modeling the other facies. This is similar to the standard practice of using a *logic matrix* in modeling mineral deposits.

A logic matrix specifies whether to use data from another facies, i, when assigning the current facies, j. Figure 4.70 gives an example. Note that this matrix

Data in Facies: \ Modeling Cell in Facies:	shale	channel	crev.	conc.
shale	Yes	No	No	No
channel	No	Yes	No	No
crev.	No	Yes	Yes	No
conc	No	No	No	Yes

FIGURE 4.70: Illustration of a Logic Matrix of Using Data from Different Facies (the Rows) to Simulate the Petrophysical Properties in Other Facies (the Columns).

may not be symmetric. In the example of Figure 4.70, measurements from crevasse sand are to be used in the simulation of channel facies, but measurements from channel sand are not to be used to predict crevasse. The rationale for this may be due to geological considerations or data acquisition considerations. Once data from another facies are used, they are considered the same as data from the same facies as the cell being modeled. One could imagine some kind of weighting scheme.

Secondary Data

In many cases there are secondary data to be considered: Seismic impedance is negatively correlated with porosity, permeability is positively correlated with porosity, and irreducible water saturation is correlated with permeability. Often, the properties we are modeling are correlated. Seismic or inversion of production data provides an external source of information.

Typically, modeling proceeds sequentially. The secondary data available at each stage may be of multiple types: Either they are (1) redundant or (2) one data type brings the most information. Therefore, most often only one secondary variable is considered for each property. Of course, this is not a theoretical limitation; it is merely practical. Methods for dealing with multiple secondary data and merging to form super variables is discussed in Section 4.1, Large-Scale Modeling.

Typically, porosity is modeled first using seismic as the secondary variable and then permeability is modeled using the previous simulation of porosity as a secondary variable. Seismic data are indirectly used for permeability through the porosity simulation.

Due to geological complications and inherent limitations in seismic data acquisition, seismic inexactly measures average porosity. Seismic-derived porosity may be correlated with the true porosity with a correlation of 0.5 to 0.7. The specific seismic attribute and degree of correlation must be calibrated for each reservoir. We account for this precision. Seismic-derived porosity represents a volume significantly larger than the typical geological modeling cell. The areal resolution is often comparable. The vertical resolution, however, is 10 to 100 times the resolution of the geological modeling cells. Current geostatistical models are built at a vertical resolution of 0.3–1 m and current seismic data informs a 10 to 30-m vertical average. We often account for this scale difference by considering the seismic data in block cokriging or as a local mean (Section 2.4, Preliminary Mapping Concepts).

Number of Realizations

A single realization of porosity and permeability should be constructed for every facies realization. As mentioned above, the facies geometry is typically much more consequential for flow. Therefore, there may be no need to generate too many realizations of porosity and permeability. A few realizations may be generated for validation and the uncertainty from the facies realizations accounted for in subsequent ranking and flow simulation. In many cases, for each facies realization a porosity and permeability realization is simulated. Section 5.3, Uncertainty Management, gives more details related to uncertainty management and the number of realizations to generate.

4.6.2 Gaussian Techniques for Porosity

There are many methods for simulating the spatial distribution of a regionalized variable. The Gaussian formalism, however, is the simplest way to create a realization that honors local data, a histogram, and a variogram. Major abrupt discontinuities in the reservoir are captured by the structural and facies model. Considering a stationary histogram and variogram to quantify the properties within reasonably homogeneous facies is deemed adequate in most cases.

The building blocks of sequential Gaussian simulation (SGS) have been discussed. Section 2.2,

Preliminary Statistical Concepts, discussed the calculation of a representative histogram. Section 2.4, Preliminary Mapping Concepts, discussed construction of locally varying mean models. Section 2.3, Quantifying Spatial Correlation, covered all the details of variogram calculation. Section 2.4, also covered SGS. There remain some implementation details, particularly with respect to seismic data, that should be discussed.

The porosity data or residual porosity values (obtained by subtracting locally varying mean porosity values) must be transformed to a normal distribution. A variogram must be established. Sequential Gaussian simulation consists of visiting each grid node in a random order and doing the following:

1. Finding nearby data and previously simulated grid nodes,
2. Constructing the conditional distribution by kriging, that is, calculating a mean and estimation variance by simple kriging, and
3. Drawing a simulated value from the conditional distribution (normal with mean and variance from step 2).

This entire procedure is repeated with different random number seeds to generate multiple realizations. There are different ways of understanding this procedure from a conceptual or theoretical framework. The authors prefer to view the multivariate distribution of porosity as following a multivariate Gaussian distribution, then this distribution is decomposed into a series of conditional distributions by recursive application of Bayes' Law. A Markov screening assumption is invoked in the calculation of each conditional distribution whereby only nearby values are used for conditioning. There are few assumptions and the procedure is robust.

4.6.3 Seismic Data in SGS for Porosity

Seismic data are areally extensive over the reservoir and are sensitive to porosity variations. The large-scale information provided by seismic will be accounted for first in the facies model. The only additional information to gain is porosity variations within the facies, which may be minor compared to the differences between facies. The calibration procedure will reveal whether seismic should be used for SGS of porosity.

Calibration of Seismic Data to Porosity

Calibration is done on a by-facies basis. The calibration depends on modeling procedures adopted later: block (co)kriging requires the vertical average of porosity to be calibrated against seismic, collocated cokriging requires the small-scale data to be calibrated against seismic data.

There are many seismic attributes that could be considered. In general, the most reliable attribute is acoustic impedance. The seismic data should be processed to provide a single seismic attribute for use as a secondary variable for geostatistical modeling. Neural networks, discriminant analysis, rule induction, or many regression-type procedures (not covered here) can be used to arrive at a single seismic-derived porosity. The calibration can appear unrealistically good when the well data used for calibration are the same as used to arrive at the single seismic attribute. Independent well data (held back from the derivation of seismic attribute) should be used for the calibration. Then, after the calibration statistics have been extracted, all of the well data can be used for the ultimate calculation of a seismic-derived porosity.

Calibration considers paired observations of porosity (at the right scale), and seismic attributes have been assembled: $Z_S(\mathbf{u}_\alpha), Z_\phi(\mathbf{u}_\alpha), \alpha = 1, \ldots, n_c$. Both variables must be transformed to standard normal distributions (see Section 2.2, Preliminary Statistical Concepts). The paired values for well averages are shown from the Fluvial Reservoir Model for the 19 wells that penetrate the reservoir facies (including lateral accretion and lag facies) are shown on a scatter plot with the correlation coefficient. Figure 4.71 shows an example cross plot. Only the correlation coefficient is used for Gaussian techniques; the bivariate distribution is assumed to follow a bivariate Gaussian distribution with elliptical probability contours.

The correlation coefficient is a two-point measure of correlation. Any lateral shift or mismatch in the seismic data could significantly affect the correlation coefficient. In general, the correlation coefficient will be reduced if the well data are being compared to a nearby, but not collocated, seismic data value. Careful data cleaning and some repositioning may be required.

Horizontal Variogram Inference

Regardless of which variant of simulation is used, a full 3-D variogram model is needed to describe the

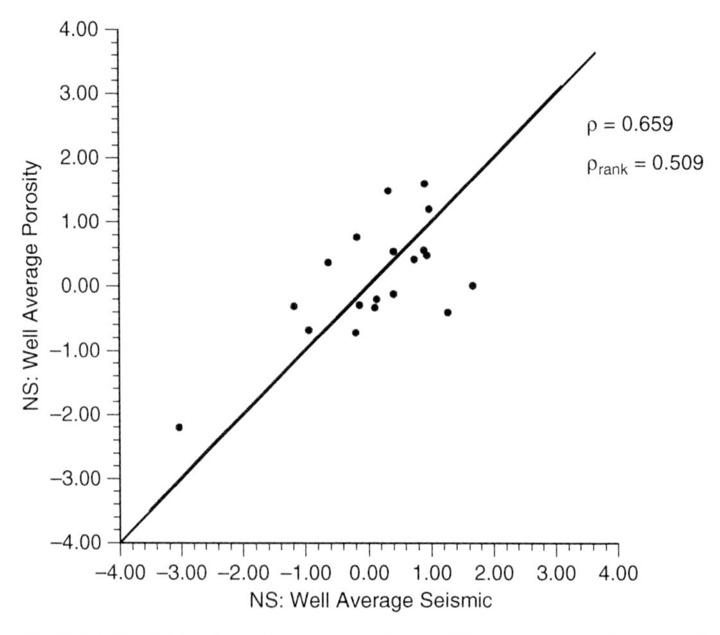

FIGURE 4.71: Calibration of the Normal Score of Seismic versus the Normal Score of Porosity. There are 19 well data from the Fluvial Reservoir within the reservoir region, showing an excellent correlation of 0.659 with a lower rank correlation of 0.509, suggesting that the correlation is strongly influenced by the single data point.

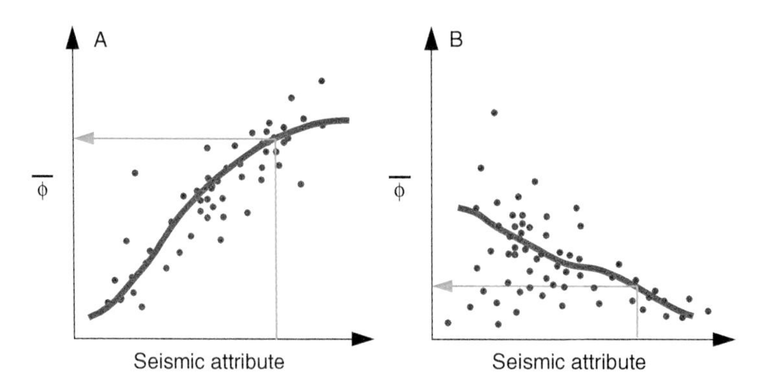

FIGURE 4.72: Two Examples of Converting a Seismic Attribute to Average Porosity. There can be positive or negative correlation. An example of converting the seismic attribute to average porosity is shown on each cross plot. The correlation can be positive (A) or negative (B).

spatial correlation structure of porosity. The main challenge in variogram inference is the horizontal direction. The seismic data provides an excellent source of information to guide the selection of horizontal range parameters. As discussed in Section 2.3, Quantifying Spatial Correlation, we must assess the presence of any zonal anisotropy in addition to adopting the horizontal range from the seismic data.

Locally Varying Mean

Rather than cokriging, we can convert the seismic data to the units of porosity and then use them as a local mean. The first step is to calculate the average porosity at all locations from the seismic attribute; see Figure 4.72 for two schematic illustrations. Then, simple kriging is used in the Gaussian simulation using the seismic-derived mean values. The kriged estimate is calculated as

$$y^*(\mathbf{u}) - m(\mathbf{u}) = \sum_{\alpha=1}^{n} \lambda_\alpha \cdot [y(\mathbf{u}_\alpha) - m(\mathbf{u}_\alpha)]$$

$$y^*(\mathbf{u}) = \sum_{\alpha=1}^{n} \lambda_\alpha \cdot y(\mathbf{u}_\alpha) + \left[1 - \sum_{\alpha=1}^{n} \lambda_\alpha\right] \cdot m(\mathbf{u})$$

where $y^*(\mathbf{u})$ is the kriged estimator or the mean of the conditional distribution for Gaussian simulation, n is the number of local data, $\lambda_\alpha, \alpha = 1, \ldots, n$, are the kriging weights, $y(\mathbf{u}_\alpha)$ are the local transformed porosity data, and $m(\mathbf{u})$ is the seismic-derived mean porosity (transformed to Gaussian units) at location $\mathbf{u}$ being considered. When there are few local data, the sum of the weight to the data is small and the seismic-derived mean porosity receives a large weight. When the weight to local data is high, the seismic-derived mean is given low weight.

Collocated Cokriging

Some type of cokriging must be applied to weight the seismic data in a way that explicitly accounts for the "softness" of the seismic data. Doyen (1988) showed one of the first applications of cokriging porosity with seismic data. A large number of refinements and simplifications have followed (Almeida and Journel, 1994; Almeida, 1993; Xu et al., 1992).

The collocated cokriging formalism was developed in Section 2.4, Preliminary Mapping Concepts, and further explained, in the context of SIS, in Section 4.2, Variogram-Based Facies Modeling. The sole input required to use the seismic data is the correlation coefficient between the *hard* Gaussian transform of porosity and the *soft* Gaussian transform of seismic.

As mentioned in Section 2.4, Preliminary Mapping Concepts, keeping only one collocated secondary datum does not affect the estimate (close-by secondary data are typically very similar in value), but it may affect the resulting cokriging estimation variance: That variance is overestimated, sometimes significantly. Since the kriging variance defines the spread of the conditional distribution from which simulated values are drawn, this may be a problem. The collocated cokriging variance should then be reduced by a factor (assumed constant $\forall \mathbf{u}$) to be determined by trial and error.

The variance of the simulated values should be close to 1.0 in Gaussian units. We start with the variance reduction factor *varred* = 1.0 and run one simulation to see the variance of the resulting simulated values. The variance reduction is systematically lowered *varred* < 1.0 until this variance is reduced to 1.0. There may be a highly nonlinear relation between the variance reduction factor *varred* and the variance of the simulated values. There is more likely to be a problem when the seismic data are much smoother than the porosity data.

Block Cokriging

Collocated cokriging is an approximation that avoids (1) calculation of cross variograms or covariances, (2) fitting a model of coregionalization, and (3) handling the large-scale nature of seismic data. As with SIS, we could do the job right. All variograms and cross variograms could be calculated (see Section 2.3, Quantifying Spatial Correlation). The corresponding point-scale variograms could be determined with analytical models. A model of coregionalization could be fit (see Section 2.3, Quantifying Spatial Correlation). Finally, a block cokriging could be performed in SGS that would account for the scale of the seismic data and the "softness" of the calibration from seismic to porosity. The cokriging formalism is classical. Implementation of block cokriging in a conventional SGS program [such as `sgsim` in GSLIB (Deutsch and Journel, 1998)] is straightforward.

An implicit assumption behind this block cokriging formalism is that all variables average linearly. This may be the case for porosity; however, the seismic variable certainly does not average linearly. Notwithstanding this assumption, it may be better than ignoring the measurement scale (Behrens, 1998).

Stochastic Inversion

An alternative to the cokriging approach to integrating seismic data is to consider a form of "stochastic inversion." The idea is to directly account for seismic data was launched in the early 1990s by workers at Elf (Bortoli et al., 1993). The basic idea is to simulate acoustic impedance, process the impedance through a forward seismic model, and choose the impedance model by rejection sampling—that is, the impedance model that reproduces the original seismic data. Porosity is then linked to acoustic impedance.

The original idea proposed by Haas was to generate a number of cross sectional models, say 10 to 100, and perform forward seismic modeling on each cross section. The difference between the forward simulated seismic and the real seismic data is computed for each realization. Then, the simulated cross section with seismic traces closest to the original seismic data is retained. Each cross section is simulated conditional to all well data and previously simulated cross sections. Dubrule, Haas, and coworkers extended the original approach to work on each 1-D column one at a time (Dubrule et al., 1998; Haas and Dubrule, 1994). This speeds convergence.

There are different versions of geostatistical or stochastic inversion appearing in the literature (Francis, 2005; Helgesen et al., 2000; Lo and Bashore, 1999; Sams et al., 1999; Torres-Verdin et al., 1999). Many of the more recent approaches start with an initial model and loop over every cell in the entire model some number of times (10–15). At each iteration the facies or acoustic impedance is retained according to some decision rule. The decision rule is based on either simulated annealing (see next section) or the result that matches the seismic data the closest.

The geostatistical inputs to stochastic inversion are conventional—that is, a global histogram and variogram. These must be supplemented by petrophysical relationships that link acoustic impendence to porosity and the wavelet that links acoustic impendence to the original seismic response.

Small Example

The following example is based on the synthetic Fluvial Reservoir Model introduced in Section 3.3, Problem Formulation. For this example, porosity and permeability are simulated within the lateral accretion and channel lag facies combined together. Figure 4.73 shows a color-scale image of the 2-D grid seismic data and a scatter plot between the normal score transform of porosity and the normal score transform of the seismic variable. The correlation coefficient is 0.4, which is high given the small scale of the porosity data. As shown on Figure 4.71, the correlation increases with porosity averaging over wells to almost 0.6. The correlation coefficient can increase or decrease with scale-up, depending on the direct and cross variogram structure.

There are 19 wells in the area of interest; see the porosity and permeability logs in Figure 4.74. No explicit trend modeling is considered since the areal trends are captured by the seismic data. The vertical and horizontal variograms of the normal score transform of porosity are shown in Figure 4.75. The stratigraphic coordinate was taken proportional between the upper and lower bounding surfaces of the reservoir layer.

Sequential Gaussian simulation with collocated cokriging based cosimulation using the seismic data was performed. An oblique view of a single porosity realization is shown in Figure 4.76. The correlation between the simulated porosity and secondary seismic data is apparent with comparison to Figure 4.73. This illustrates a simple and effective procedure for property modeling. Permeability can be modeled using porosity as a secondary variable, multiple realizations can be constructed, and the models can be scaled for flow simulation.

4.6.4 Porosity/Permeability Transforms

Porosity is modeled on a by-facies basis. The porosity models are used directly to assess pore volume, pore volume uncertainty, and connectivity based on porosity. Multiple realizations can be ranked and selected on the basis of hydrocarbon pore volume; see Section 5.3, Uncertainty Management. However, permeability is required in addition to porosity for flow simulation.

Three approaches will be developed for permeability modeling. This subsection *Porosity/permeability Transforms* covers porosity/permeability transforms, subsection *Gaussian Techniques for Permeability* discusses Gaussian techniques, and subsection *Indicator Technique for Permeability* presents indicator techniques. These are in order of increasing complexity and flexibility. Most of the porosity/permeability transforms described below are only rarely used because they fail to reproduce permeability univariate and spatial statistics. Although cloud transform provides a practical method that may produce reasonable permeability pattern, Gaussian techniques are widely used. Indicator techniques are occasionally used when there are sufficient data to infer the complex required statistics.

Regression

Classical regression can be applied to develop a relationship between porosity and permeability. Typically, the logarithm of permeability is used because of the skewed and approximately lognormal character of many permeability histograms. The regression equation takes the form

$$\log(K)^* = a_0 + a_1 \cdot \phi + a_2 \cdot \phi^2 + a_3 \cdot \phi^3 \quad (4.7)$$

where $\log(K)^*$ is the predicted logarithm of permeability; $a_i, i = 0, \ldots, n_R$, are regression coefficients;

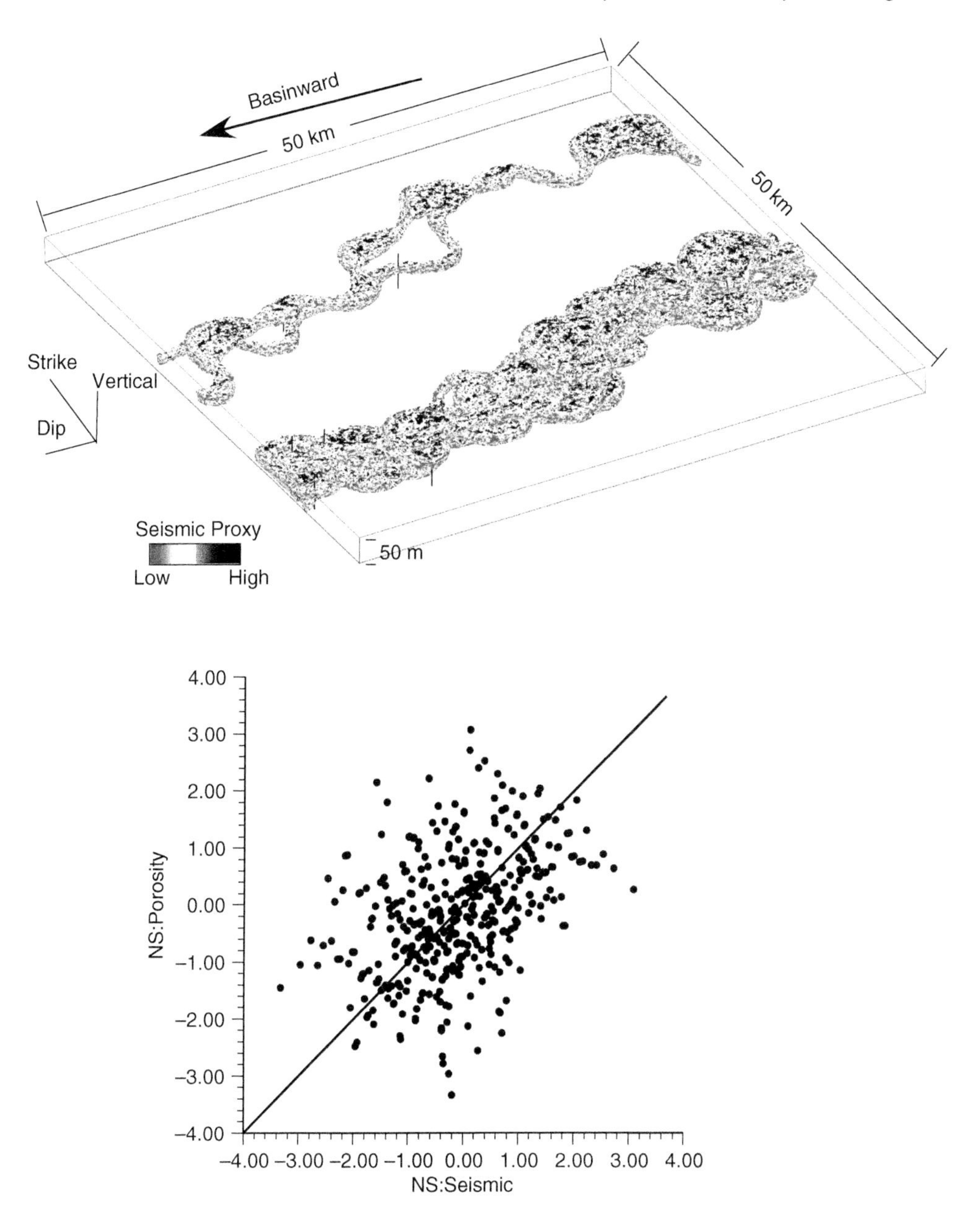

FIGURE 4.73: Oblique View of the Seismic Variable on the Reservoir Facies Model of the Fluvial Reservoir Model (above) and the Scatter Plot of the Well-Based Porosity and Seismic Information Projected to the Wells. The correlation between the normal score of the seismic and normal score of porosity is 0.41, which is very good given the scale of the porosity data. In this case, the correlation increases with porosity averaging, shown previously in Figure 4.71.

and ϕ is porosity. Log-linear regression is common; that is, only the first two terms are used. The second-order term $a_2 \cdot \phi^2$ may be used to capture nonlinear features. Only rarely are higher-order terms ≥ 3 needed or recommended. The calculation of the regression coefficients $a_i, i = 0, \ldots, n_R$, is automatic in most software; details can be found in most statistics books.

Figure 4.77 gives four real examples of porosity/permeability cross plots and second-order regression curves. These equations can be used directly to predict permeability given porosity. The

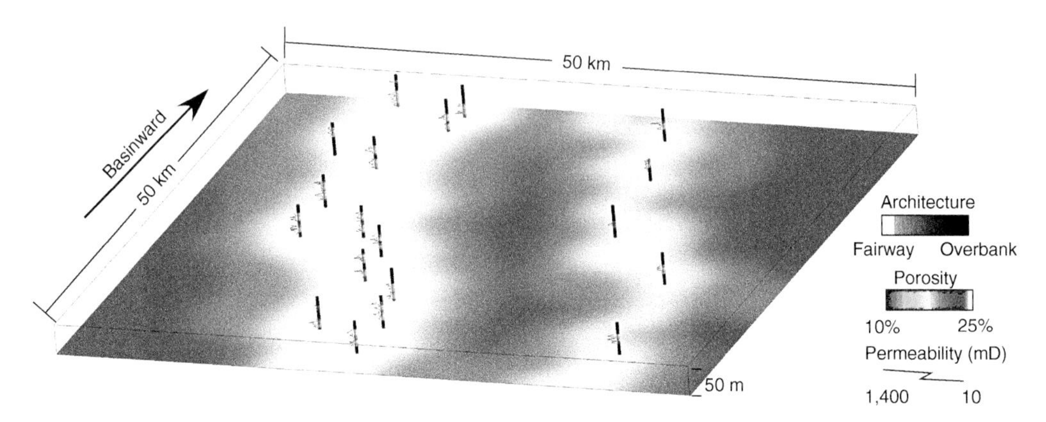

FIGURE 4.74: Oblique View of the 19 Wells with Porosity Logs Shown in Color and Permeability Logs as Lines. A smoothed map of the average reservoir thickness is shown in gray scale to indicate the locations of the channel complexes.

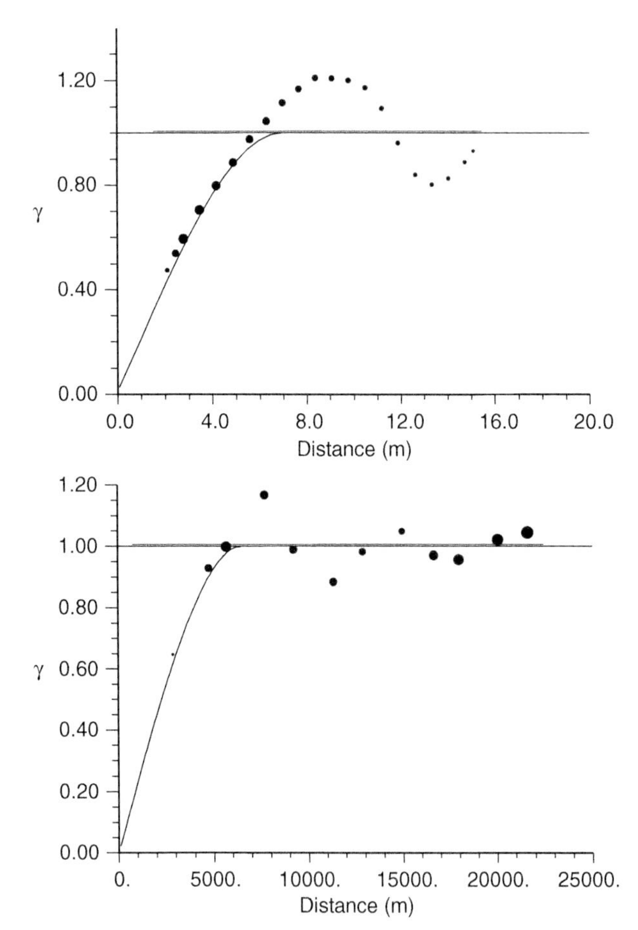

FIGURE 4.75: Vertical and Horizontal Variograms with Fitted Models for 3-D Porosity Data of Figure 4.74. Given the spacing and limited number of wells, inference of the horizontal variogram is challenging. Directional variograms did not result in any clear features.

advantages of this approach include: (1) The approach is simple and almost automatic and (2) the relation between porosity and permeability is approximately reproduced.

There are disadvantages to regression. The low and high permeability values are smoothed out; that is, it is unlikely to encounter predicted values as extreme as found in the data. The variogram or spatial variability of the predicted permeability values is borrowed from porosity; however, permeability is usually more variable than porosity. Finally, regression-based permeability values do not

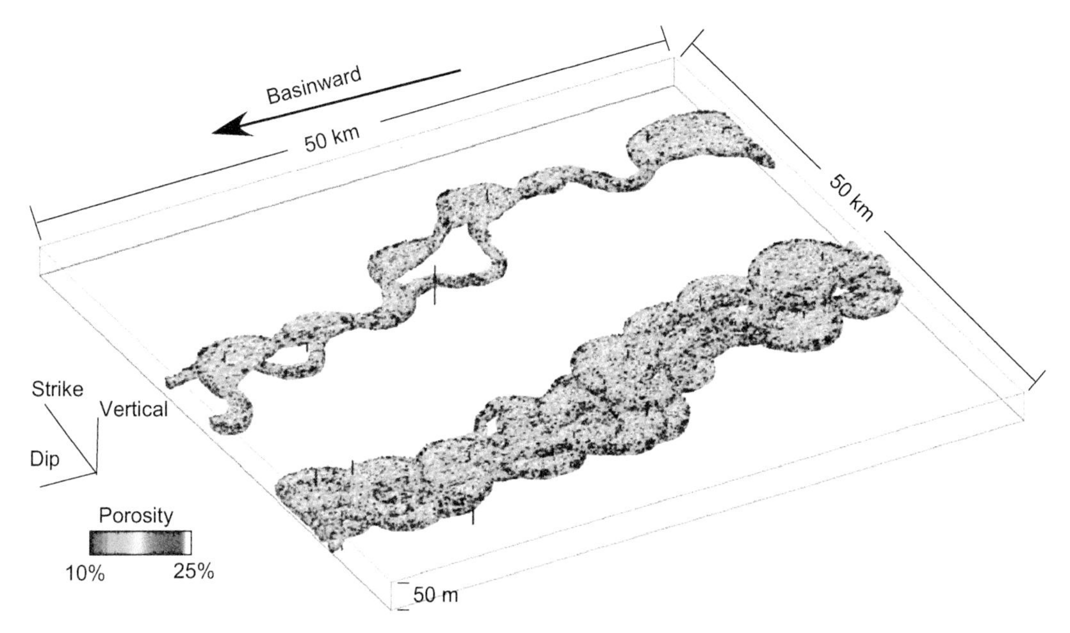

FIGURE 4.76: Oblique View of a Porosity Model Generated by Sequential Gaussian Simulation. The seismic variable was considered as a secondary variable using collocated cokriging with a correlation coefficient of 0.4; see the seismic data in Figure 4.73.

account for uncertainty specific to permeability, beyond that of porosity. Geostatistical simulation techniques overcome these limitations.

Conditional Expectation

Assuming a parametric relation between (log) permeability and porosity can be limiting. Often, the relation is more complex than the simple polynomial relations used in Eq. (4.7). For that reason, the conditional expectation of permeability given porosity is more flexible to capture complex nonlinear relationships and the conditional variance of permeability given porosity. The concept of a conditional expectation transform was illustrated in Figure 4.72. Here is the procedure to construct a conditional expectation curve:

- Sort the N porosity/permeability pairs in order of ascending porosity.
- Choose a moving window filter size M on porosity. This number will depend on the number of paired observations available. M must be greater than 10 to avoid erratic fluctuations. M is usually less than $N/10$ to avoid oversmoothing.
- Average the porosity/permeability values for moving window averages of M values at a time. This generates $N - M$ pairs of smoothed porosity/permeability values.

Permeability is then predicted by interpolating between the smoothed porosity–permeability pairs. The permeability can be kept constant for each porosity class or linear interpolation can be used. The most critical decision is how to handle the lower tail (below the lowest averaged porosity value) and the upper tail (above the highest porosity value). Figure 4.78 gives the four conditional expectation curves that correspond to the paired porosity/permeability data introduced in Figure 4.77.

There may be too few data points to fully specify the bivariate relationship between porosity and permeability. A bivariate smoothing algorithm could be used to fill in the cross plot (Deutsch, 1996a; Scott, 1992).

Monte Carlo Simulation from Conditional Distributions

The concept of conditional expectation curves can be taken one step further. The permeability value at a location $\mathbf{u}$ can be drawn by Monte Carlo simulation from the conditional distribution of permeability given the porosity at that location $f(k|\phi(\mathbf{u}))$. A series of conditional distributions are constructed;

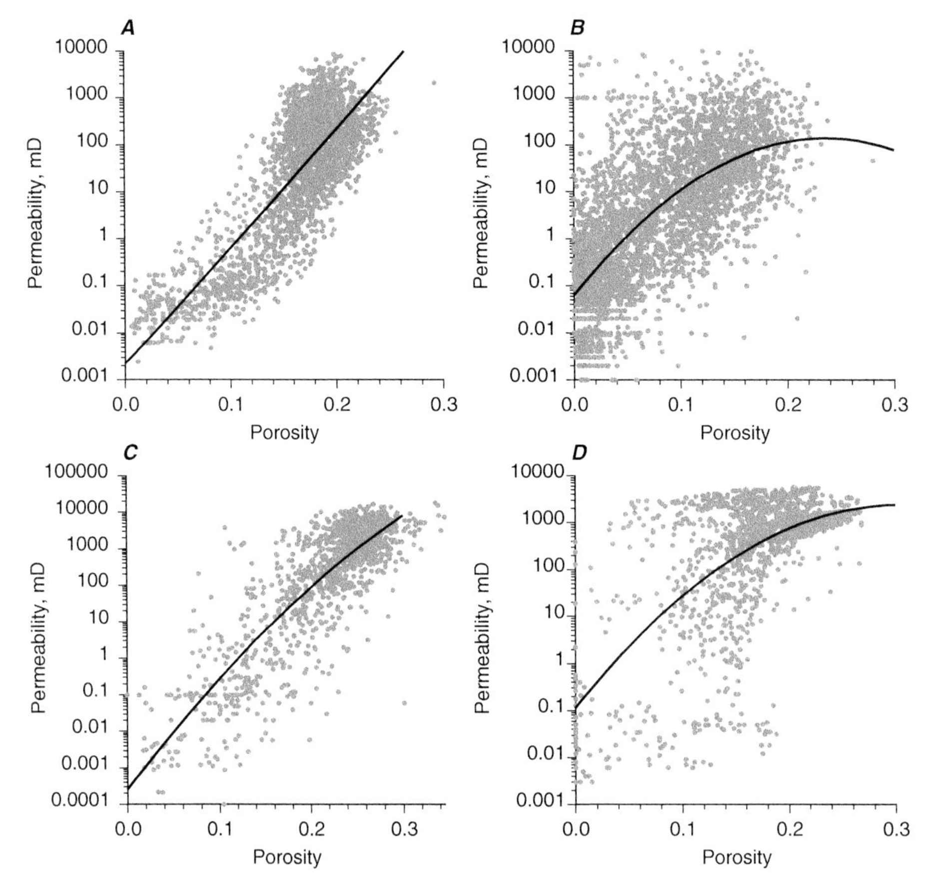

FIGURE 4.77: Four Examples of Porosity/Permeability Cross Plots with Second-Order Regression Curves.

see the three in Figure 4.79. In general, 10 or more conditional distributions are used. The porosity "windows" used to construct the conditional distributions can overlap. The details of Monte Carlo simulation were covered in Section 2.2, Preliminary Statistical Concepts.

The histogram of permeability and the full scatter between porosity and permeability is reproduced with this approach. Some spatial correlation is imparted to the predicted permeability values because of correlation with porosity; however, the spatial variability of the predicted permeability values will be too random since previously drawn permeability values are not considered. A geostatistical approach is required to impart the correct spatial correlation to permeability.

Cloud Transform

The cloud transform is a method that imposes spatial correlation for permeability in Monte Carlo simulation from conditional distribution (permeability conditional to porosity). The fundamental addition to the Monte Carlo method described above is to apply a conditional and correlated p-field to draw from the conditional distributions. As mentioned in Section 2.4, Preliminary Mapping Concepts, a p-field is a uniformly distributed ($U[0, 1]$) random function with a specified correlation structure, typically defined by a stationary covariance function ($C(\mathbf{h})$). For cloud transform, the p-field is conditioned so that the hard data values are drawn at the data locations from the conditional bins.

As with the previous method Monte Carlo Simulation from Conditional Distributions, the full scatter between porosity and permeability are reproduced and in addition there is some level of spatial continuity imposed with the p-field. However, there is no guarantee that the precise histogram and spatial continuity models for permeability are reproduced. Although, some view cloud transform as an opportunity to infer the often poorly informed permeability

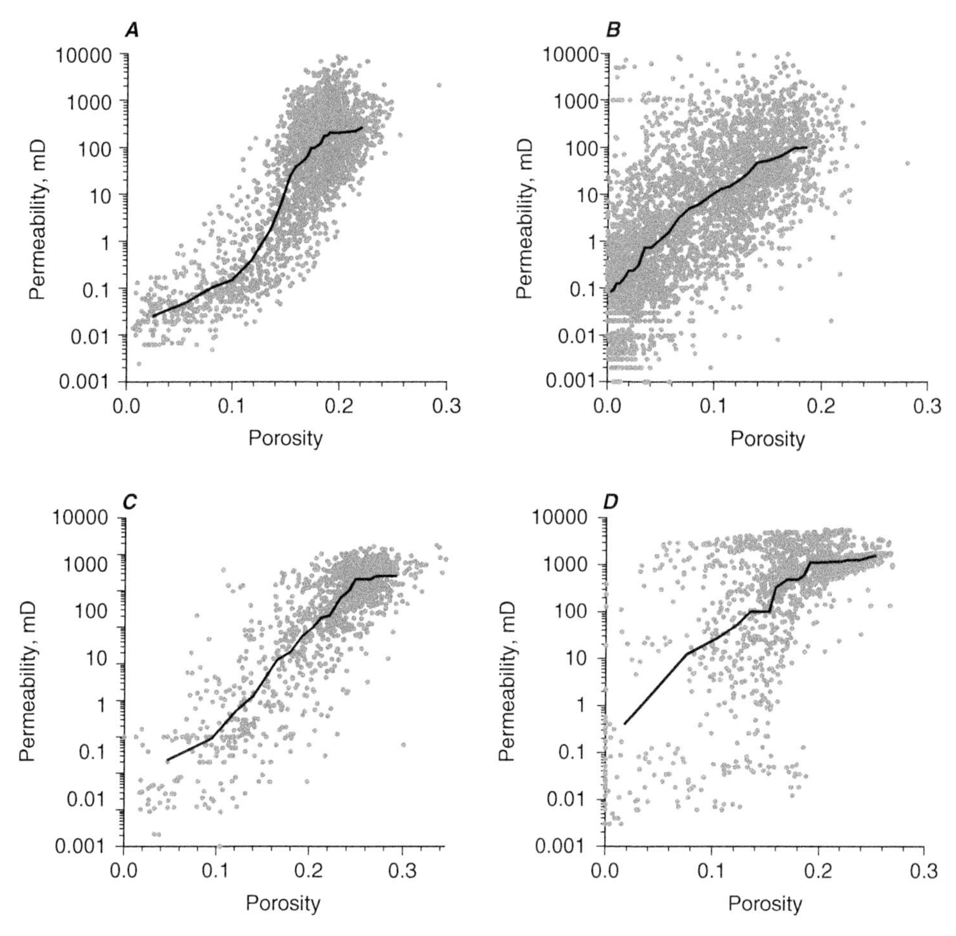

FIGURE 4.78: Four Examples of Porosity/Permeability Cross Plots with Conditional Expectation Curves; Second-Order Regression Equations Are Shown in Figure 4.77.

distribution from the better informed porosity distribution and the porosity and permeability relationship (i.e. a form of debiasing). Post-processing is often applied to correct the histogram. Spatial continuity should be checked.

Cloud transform was applied to the Fluvial Reservoir Model constrained by the previously simulated porosity realization and a modeled bivariate relationship between porosity and permeability (see Figure 4.80). This model was fit based on a densification and extrapolation of the paired well porosity and permeability data. Ten bins were applied to calculate the associated conditional distributions.

4.6.5 Gaussian Techniques for Permeability

The cokriging procedure described above in subsection *Gaussian Techniques for Porosity* to simulate porosity using seismic data works very well for permeability correlated to porosity. The implicit assumption is that porosity, permeability, and the bivariate relationship of porosity and permeability follow a (bi)Gaussian distribution after univariate Gaussian transforms of each. The procedure is as follows:

- Normal score transform porosity and permeability. The normal score variogram of permeability and the correlation coefficient between the normal score transforms of porosity and permeability are needed. Full cokriging would require a linear model of coregionalization between porosity and permeability (see Section 2.3, Quantifying Spatial Correlation).
- Simulate the normal score transform of permeability using porosity as a covariate. The results are checked and back-transformed.

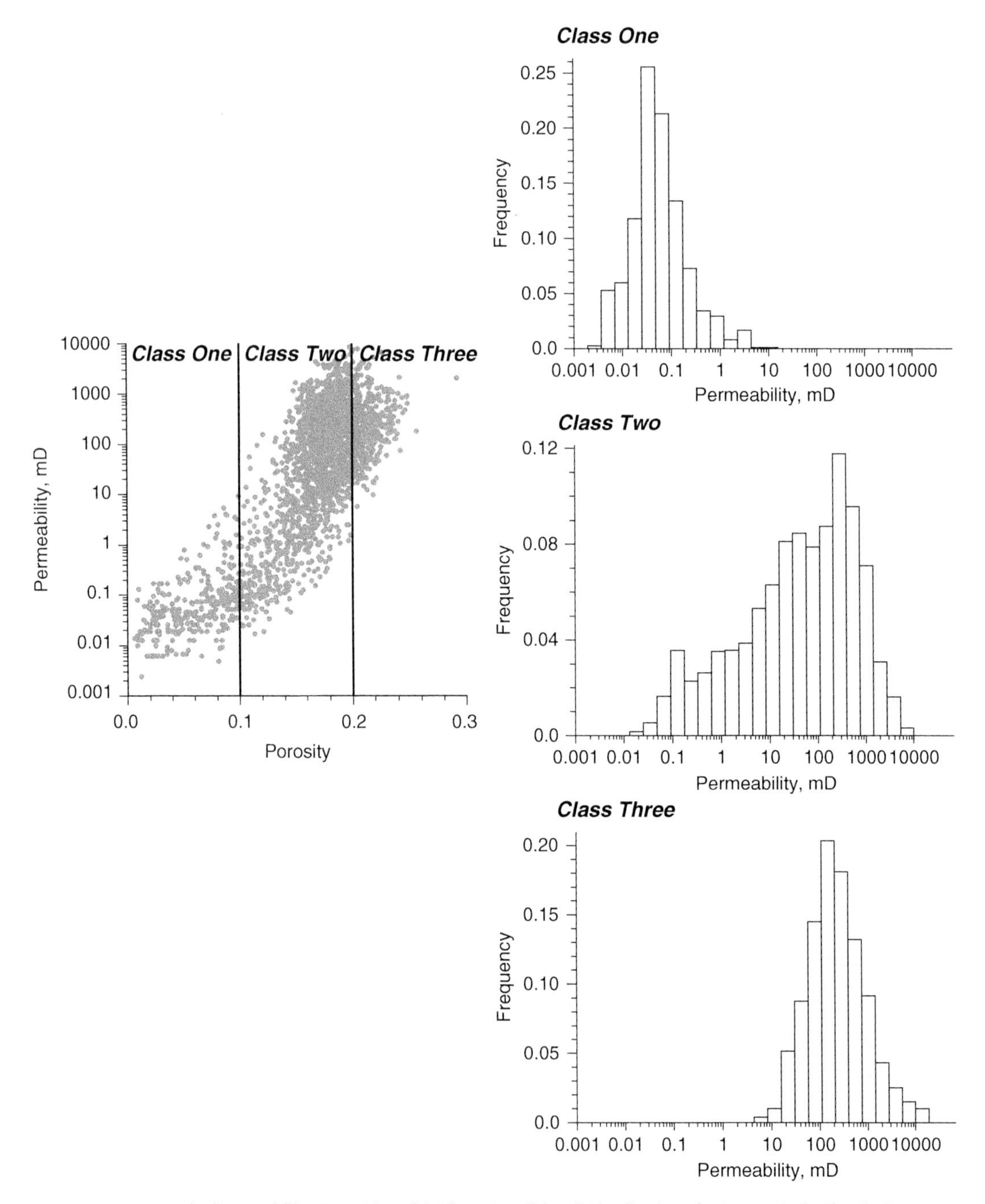

FIGURE 4.79: Porosity/Permeability Cross Plot with Three Conditional Distributions for Monte Carlo Simulation. More conditional distributions would be used in practice; three are shown for illustration.

Collocated cokriging works well since previously simulated porosity is present at all locations. Figure 4.81 shows the four examples of normal score porosity permeability cross plots corresponding to previous cross plots in Figures 4.77 and 4.78. The bivariate distributions shown in all four cases do not appear to have the elliptical probability contours required of Gaussian techniques because they are not bivariate Gaussian.

In general, such Gaussian techniques are widely used because they are simple to apply and reproduce most required features. A characteristic feature of permeability, however, is that the continuity of extreme low values (flow barriers) and extreme high

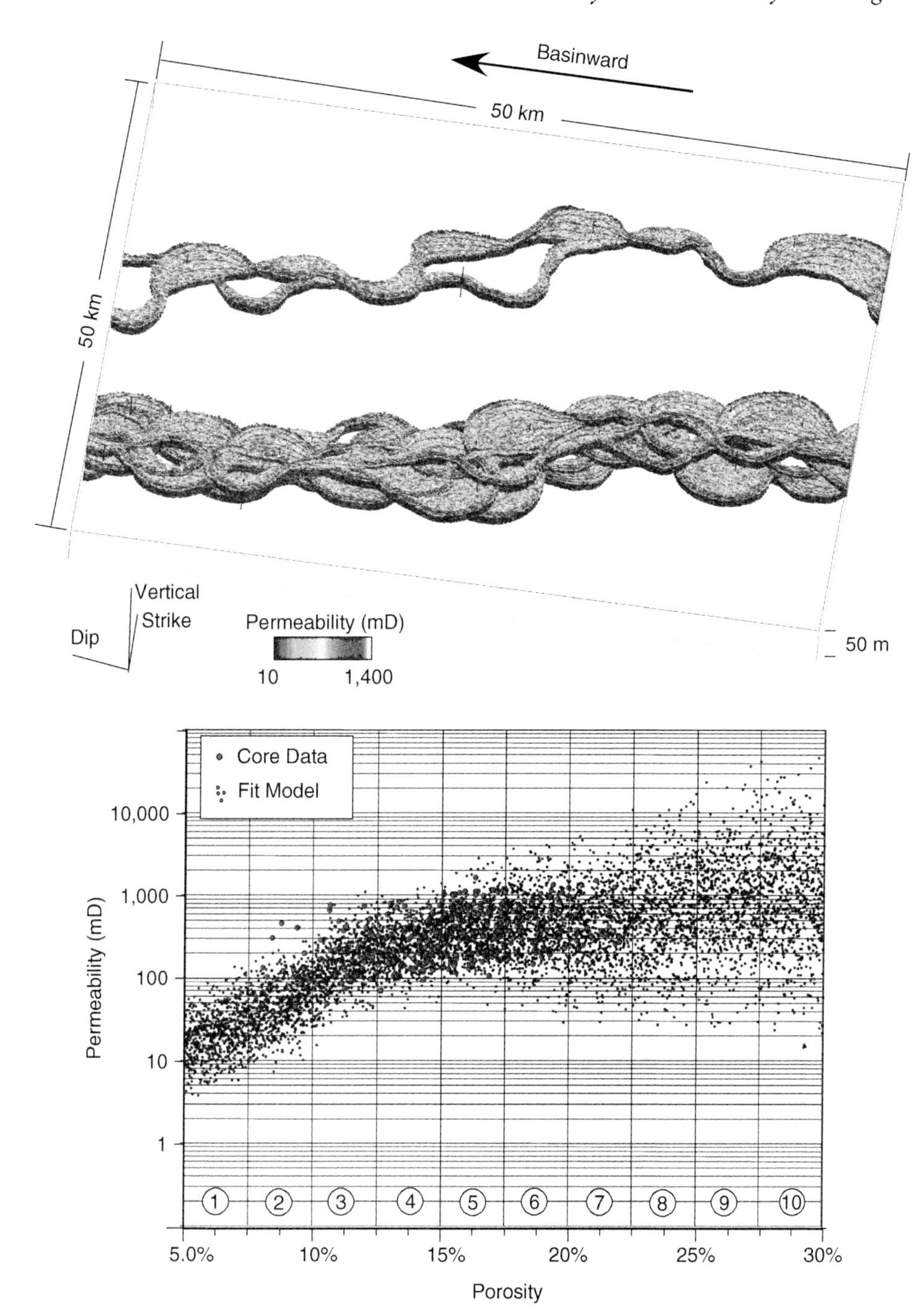

FIGURE 4.80: Permeability Realization Simulated with Cloud Transform. The well-based and fit bivariate relationships are shown below.

values (flow conduits) are very important in subsequent flow modeling studies.

As mentioned in Section 2.4, Preliminary Mapping Concepts, a characteristic feature of Gaussian techniques is that the extreme values are disconnected. For this reason, Gaussian techniques are not always recommended for permeability. Prior facies models would capture large-scale continuity. In the presence of a "good" facies model it is unlikely that a more sophisticated permeability model is warranted. Nevertheless, indicator methods, described in the next section, may be used to account

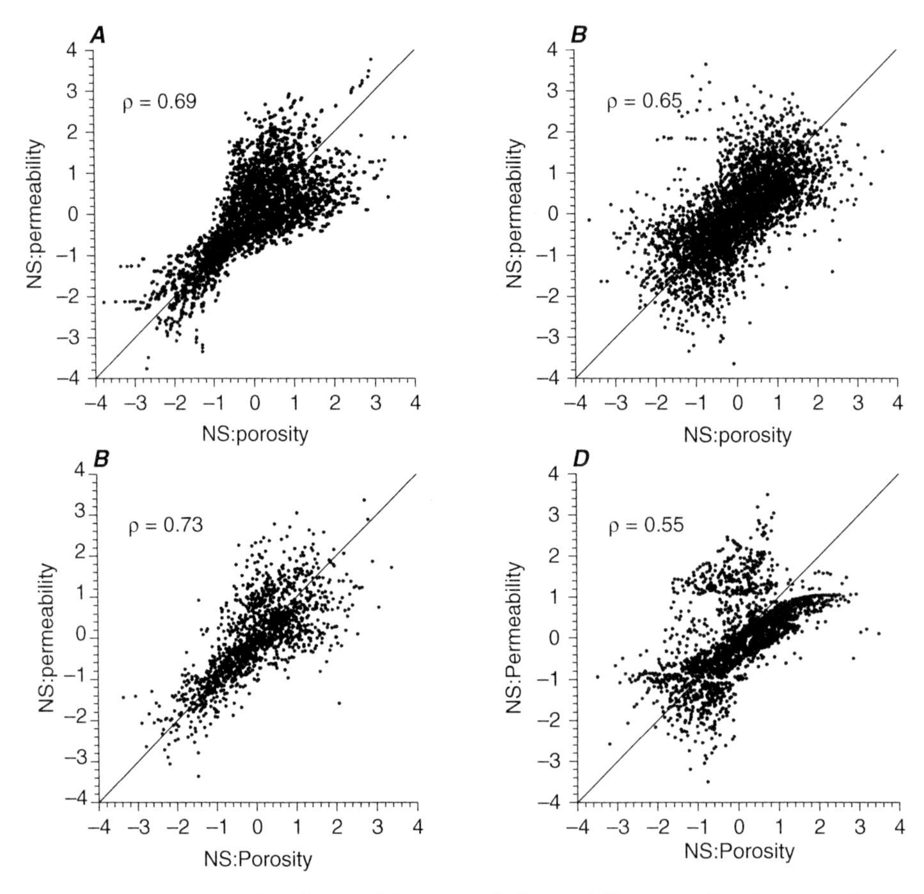

FIGURE 4.81: Four Examples of Normal Score Porosity/Permeability Cross Plots Corresponding to Previous Cross Plots in Figures 4.77 and 4.78. Note that the bivariate distributions shown in all four cases do not appear to have the elliptical probability contours required of Gaussian techniques.

for greater or lesser continuity of extreme values. A potentially larger problem is the limitations in the Gaussian methods to capture the potentially complicated porosity and permeability bivariate relationships, including heteroscedasticity, nonlinear features, and constraints. The previously discussed transform methods are required to capture these relationships. For Gaussian methods these should be checked for reproduction by comparing the scatter plots of porosity and permeability data and realizations.

Small Example

Figure 4.76 shows the covariate realization of porosity. The cross plot between porosity and permeability for this example is given on lower part of Figure 4.82. Gaussian simulation was used to generate permeability realizations (after calculating the variogram of the normal scores of permeability). A realization of permeability generated by Gaussian simulation using porosity is shown in the upper part of Figure 4.82. Note the good correlation between porosity and permeability. The sample correlation coefficient is reproduced exactly, yet specific features of the bivariate porosity and permeability relationship are likely not reproduced.

4.6.6 Indicator Technique for Permeability

The indicator formalism for kriging and simulation is described in Section 2.4, Preliminary Mapping Concepts. Details on indicator cokriging for facies modeling is given in Section 4.2, Variogram-Based Facies Modeling. Implementation details of indicator methods for permeability are described here. Indicator methods for permeability would be used when the spatial continuity at low and high permeability thresholds does not fit the Gaussian model.

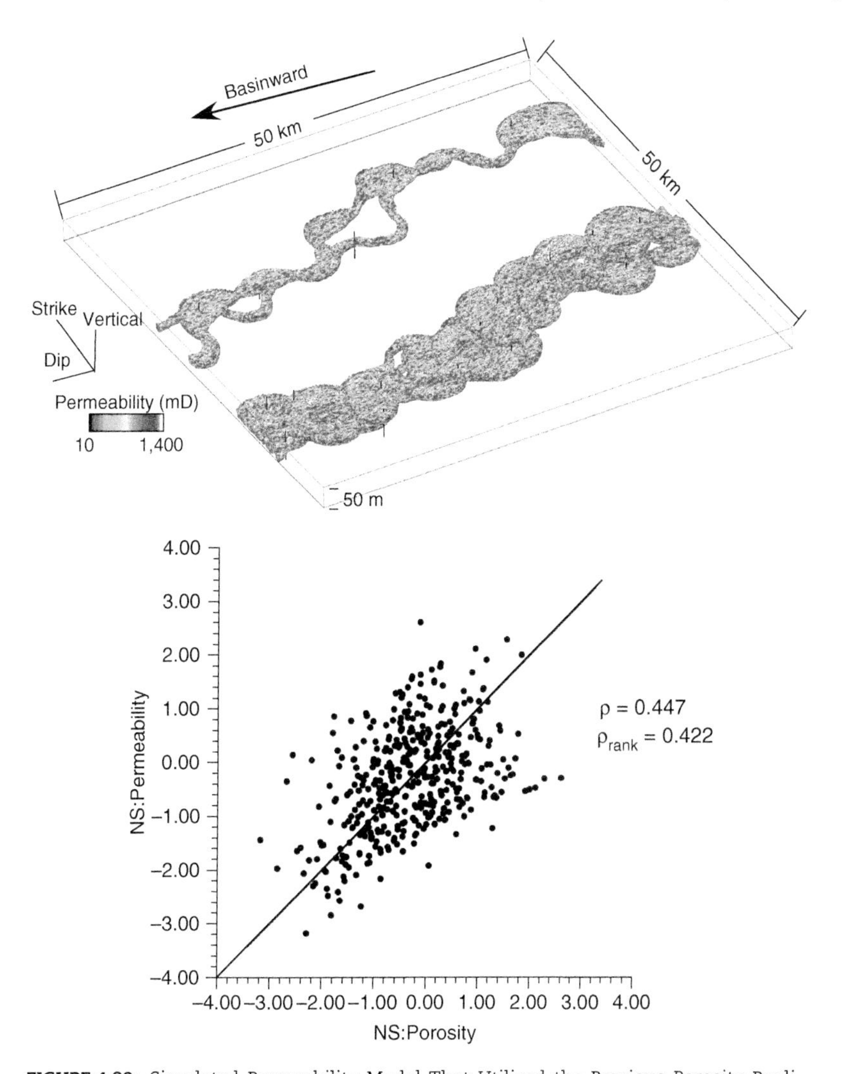

FIGURE 4.82: Simulated Permeability Model That Utilized the Previous Porosity Realization as a Covariate (See Figure 4.76) Is Shown Above. The normal score porosity and permeability scatter plots and applied correlation coefficient of 0.45 is shown below.

The price of this greater flexibility is more professional and CPU time for model building. Some form of cokriging is required to account for the correlation with porosity.

Indicator kriging directly estimates the conditional probability distribution at an unsampled location with no explicit Gaussian assumption. The range of permeability variability is divided with a series of threshold values $z_j, j = 1, \ldots, N_J$. The symbols k or K are commonly used for permeability; therefore, j notation is used to avoid confusion. As mentioned in Section 2.4, Preliminary Mapping Concepts, between 5 and 11 thresholds are considered between the 0.1 and 0.9 quantiles. Choosing the nine deciles is a good starting point. Fewer thresholds could be considered to simplify the procedure. Additional thresholds could be added to provide greater resolution. Choosing too many thresholds results in significant order relation violations, which defeats the goal of choosing more thresholds.

A variogram is needed for each threshold. Most software makes this easy. Nevertheless, the same variogram can be used for a number of thresholds; for example, indicator variogram at the 0.1 quantile can be used for the 0.1, 0.2, and 0.3 quantiles. The common origin of the indicator data will impart consistency in the indicator variograms; that is, the parameters such as nugget effect and anisotropy directions of the indicator variograms should vary smoothly from one cutoff to another. Drastically different indicator variograms will be difficult to reproduce in indicator simulation because of order relation violations.

The porosity values must be coded or transformed to pre-posterior probabilities of permeability at the thresholds $z_j, j = 1, \ldots, N_J$, considered for permeability modeling:

$$y(\mathbf{u}; z_j) = \text{Prob}\left\{Z(\mathbf{u}) \leq z_j | \phi(\mathbf{u})\right\}, \qquad j = 1, \ldots, N_J$$

A high porosity $\phi(\mathbf{u})$ at a particular location would, in general, entail a high permeability—that is, a small probability $y(\mathbf{u}; z_j)$ to be less than a low threshold z_j. A low porosity $\phi(\mathbf{u})$ would, in general, entail a low permeability—that is, a high probability $y(\mathbf{u}; z_j)$ to be less than a low threshold z_j. These probabilities are calculated using the conditional distributions from a cross plot of collocated values; see details in Subsection Porosity/Permeability Transforms.

A distribution of uncertainty in permeability at each location $\mathbf{u}$ is then directly estimated through kriging the indicator transform at each threshold:

$$\left[i(\mathbf{u}; z_j)\right]^* = \sum_{\alpha=1}^{n} \lambda_\alpha i(\mathbf{u}_\alpha; z_j) + \sum_{\beta=1}^{n'} \lambda'_\beta y(\mathbf{u}_\beta; z_j), \qquad j = 1, \ldots, N_J \tag{4.8}$$

There are n hard nearby permeability data with their indicator transforms $i(\mathbf{u}_\alpha; z_j)$ and n' nearby secondary porosity data (choosing the single collocated porosity value is often adequate) with their indicator transforms $i_{\text{soft}}(\mathbf{u}_\beta; z_j)$. The weights $\lambda_\alpha, \alpha = 1, \ldots, n$, and λ'_β, $\beta = 1, \ldots, n'$, are calculated by indicator cokriging.

A linear model of coregionalization could be built between the i-permeability indicators and the y-porosity indicators at each threshold (see Section 4.6); however, the simpler Markov–Bayes model (Zhu and Journel, 1993) is commonly adopted. This was introduced in Section 2.4 in the context of categorical indicator data.

Calibration parameters $B_j, j = 1, \ldots, N_J$, are required for each threshold. These calibration parameters and the i-permeability indicator variograms fully specify the direct and cross variograms needed for the cokriging (see discussion in Section 2.4). The coefficients B_j are obtained by comparing collocated hard and soft indicator data:

$$\begin{aligned} B_j = {} & E\{Y(\mathbf{u}; z_j) \mid I(\mathbf{u}; z_j) = 1\} - \\ & E\{Y(\mathbf{u}; z_j) \mid I(\mathbf{u}; z_j) = 0\} \quad \in [-1, +1] \end{aligned}$$

$E\{\cdot\}$ is the expected value operator or simply the arithmetic average. The term $E\{y(\mathbf{u}; z_j) \mid I(\mathbf{u}; k) = 1\}$ is close to 1 and $E\{y(\mathbf{u}; z_j) \mid I(\mathbf{u}; z_j) = 0\}$ is close to 0 if the porosity is highly correlated to permeability. The best case is when $B_j \approx 1$. When $B_j = 1$, the porosity indicator data are treated as hard indicator data. Conversely, when $B_j = 0$, the porosity data are ignored; that is, their kriging weights will be zero.

Indicator kriging (IK)—that is, estimation of the probability at each threshold with Eq. (4.8)—leads to a distribution of uncertainty. This IK-derived distribution can then be used for stochastic simulation as part of the sequential indicator simulation (SIS) algorithm. The steps for indicator simulation of permeability may be summarized as follows:

1. Choose permeability thresholds and establish indicator variogram models for each threshold. The same variogram model can be used for multiple thresholds.
2. Transform the porosity values to secondary indicator data using the cross plot between collocated porosity and permeability values (the `bicalib` program of GSLIB can be used for this purpose).
3. Calculate the B calibration parameters for each threshold. The porosity indicator data need not be used if these parameters are all close to zero, which would indicate that there is no correlation between porosity and permeability.
4. Perform sequential indicator simulation to create multiple realizations of permeability that reproduce (1) the permeability conditioning data and histogram, (2) the continuity at different permeability thresholds, and (3) the correlation with porosity.

While we have covered continuous indicator simulation here for completeness, the method has fallen out of practice due to known artifacts with the spatial continuity. The between-class spatial

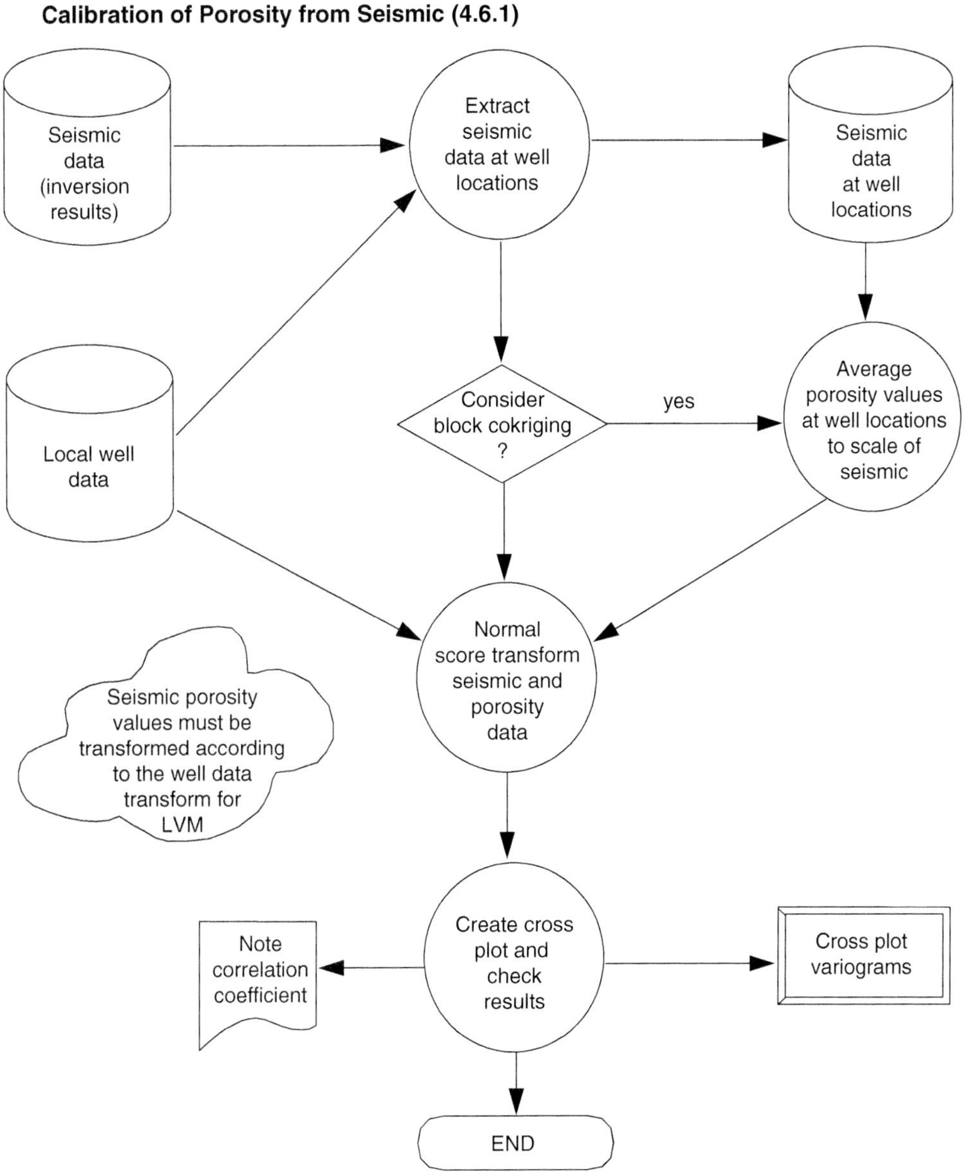

Calibration of Porosity from Seismic (4.6.1)

Seismic data (inversion results)
Extract seismic data at well locations
Seismic data at well locations
Local well data
Consider block cokriging ?
yes
Average porosity values at well locations to scale of seismic
Normal score transform seismic and porosity data
Seismic porosity values must be transformed according to the well data transform for LVM
Note correlation coefficient
Create cross plot and check results
Cross plot variograms
END

FIGURE 4.83: Work Flow for Calibration of Porosity from Collocated Seismic Data. There are times when the search for nearby seismic data must be expanded to include more than the collocated data.

continuity is often poorly reproduced with sharp changes between the continuous bins.

4.6.7 Work Flow

Following are the main geostatistical work flows related to property modeling. Figure 4.83 illustrates the work flow to calibrate seismic data to porosity data at well locations. The main considerations are the scale difference and the normal score transform. Attention should be paid to the fact that averaging of porosity is not linear after normal score transformation.

Figure 4.84 illustrates the work flow for porosity modeling by sequential Gaussian simulation with seismic data. This is very similar to work flow 4.2.2 (Figure 4.25) in Section 4.2; however, there is often seismic data and trends that must be considered; these additional data inputs and implementation considerations are shown.

Figure 4.85 shows the work flow for permeability modeling by porosity-permeability regression-type

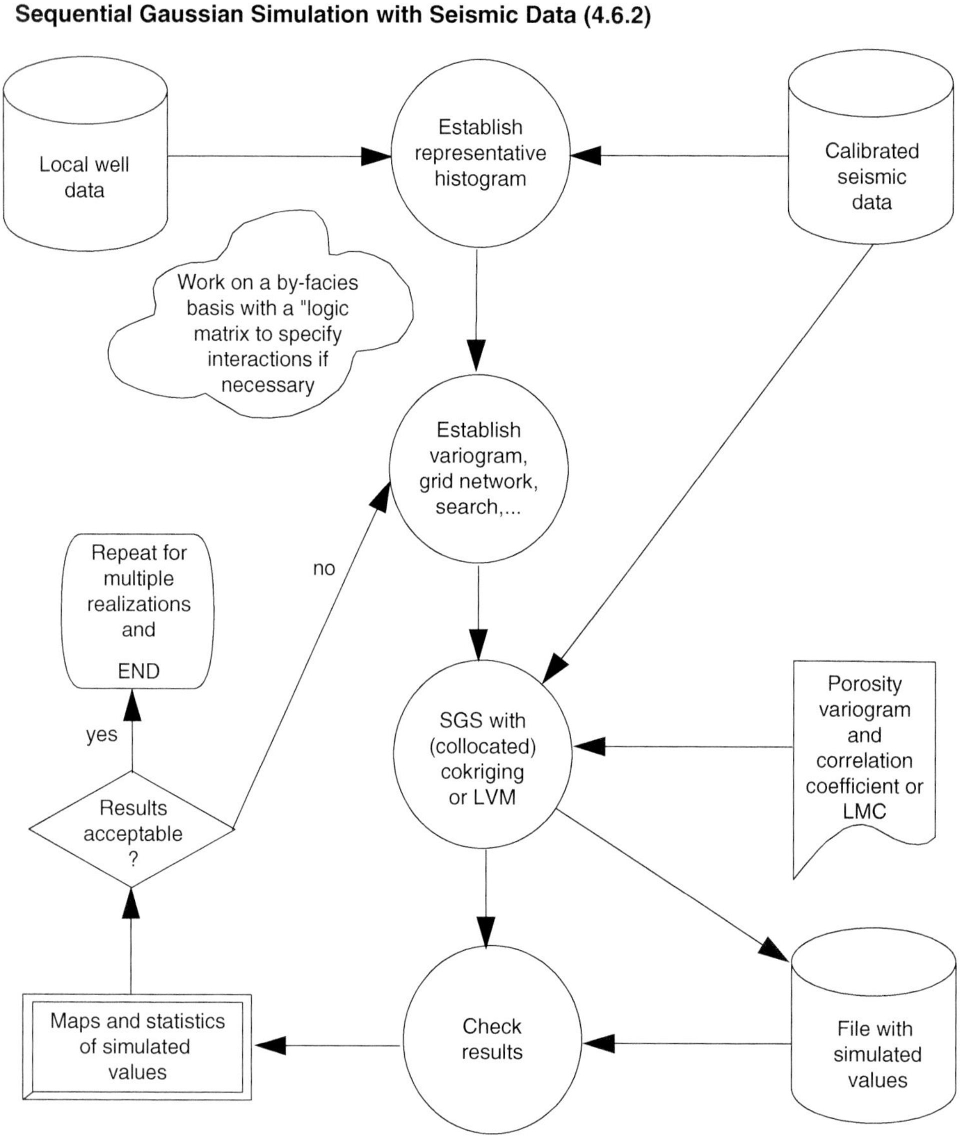

Sequential Gaussian Simulation with Seismic Data (4.6.2)

FIGURE 4.84: Work Flow for Porosity Modeling by Sequential Gaussian Simulation with Seismic Data.

transforms. The work flow is straightforward and requires little geostatistical input, which is why it is commonly used in practice.

Figure 4.86 illustrates the work flow for permeability modeling using a porosity model for conditioning. The collocated approach is commonly used for simplicity; it is relevant since a previously simulated porosity value is available for every grid cell where permeability is being simulated.

The continuity of extreme high and low permeability values can be very important in fluid flow predictions. Sequential indicator simulation is a technique to account for this greater continuity. Figure 4.87 illustrates the work flow for sequential indicator simulation of permeability using a prior porosity model.

4.6.8 Section Summary

Petrophysical properties such as porosity and permeability are modeled within each facies and reservoir layer or may be modeled without consideration of facies. Variogram-based methods remain as the primary tools for accomplishing this, although new developments in multiple-point methods may produce practical continuous property work flows in the

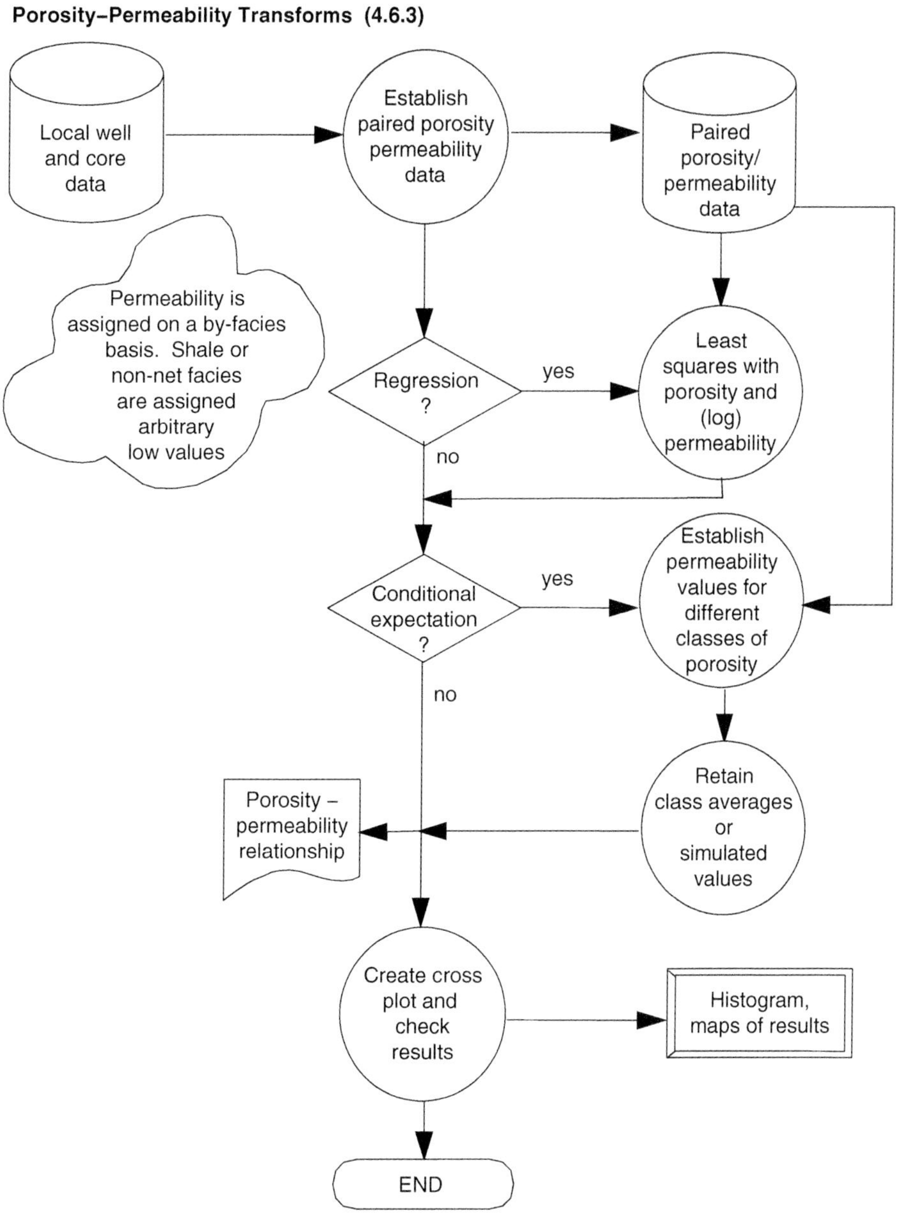

Porosity–Permeability Transforms (4.6.3)

Local well and core data

Establish paired porosity permeability data

Paired porosity/ permeability data

Permeability is assigned on a by-facies basis. Shale or non-net facies are assigned arbitrary low values

Regression ?

yes

no

Least squares with porosity and (log) permeability

Conditional expectation ?

yes

no

Establish permeability values for different classes of porosity

Retain class averages or simulated values

Porosity – permeability relationship

Create cross plot and check results

Histogram, maps of results

END

FIGURE 4.85: Work Flow for Permeability Modeling by Regression-Type Transformation Using Porosity.

short term (see Chapter 6, Special Topics, Continuous MPS discussion).

Porosity is relatively straightforward since it averages linearly, exhibits isotropy and has a well-constrained range often close to 0% to 30% or more, depending on rock type and compaction and cementation history. Sequential Gaussian simulation is often applied with constraints to trends and secondary data if available.

Permeability is more difficult. It is more variable, has a highly skewed histogram, does not average linearly, exhibits anisotropy, and is dependent on fluid and boundary conditions. Various methods are available and their selection depends on the statistical inputs and features that must be reproduced. Porosity/permeability transforms provide an opportunity to honor regression fit or exact relationships between porosity and permeability. Gaussian

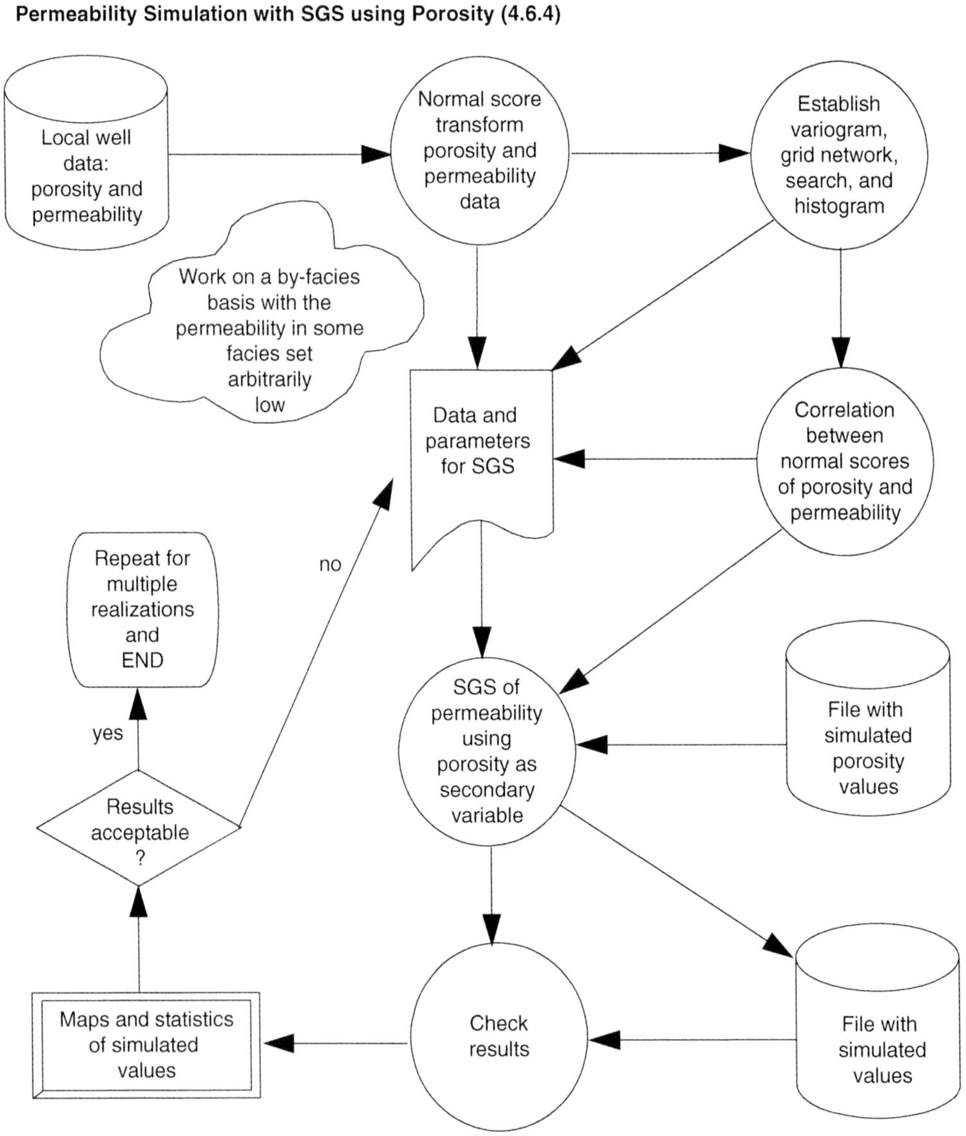

Permeability Simulation with SGS using Porosity (4.6.4)

Local well data: porosity and permeability
Normal score transform porosity and permeability data
Establish variogram, grid network, search, and histogram
Work on a by-facies basis with the permeability in some facies set arbitrarily low
Data and parameters for SGS
Correlation between normal scores of porosity and permeability
Repeat for multiple realizations and END
no
yes
SGS of permeability using porosity as secondary variable
File with simulated porosity values
Results acceptable ?
Maps and statistics of simulated values
Check results
File with simulated values

FIGURE 4.86: Work Flow for Permeability Modeling by Sequential Gaussian Simulation Using Porosity as a Covariate.

techniques reproduce the permeability univariate and spatial distributions approximately, but often fail to capture the complicated porosity/permeability bivariate relationships. Indicator methods provide an opportunity to capture the variable spatial continuity for low and high permeability values, but continuous indicator methods are known to produce artificial spatial discontinuities across continuous bins.

The next section deals with optimization methods applied in model construction. These methods provide opportunities to integrate various data and input statistics into reservoir models.

4.7 OPTIMIZATION FOR MODEL CONSTRUCTION

Optimization provides another approach to construct reservoir models. These are useful when the data and input statistical constraints are not amenable to kriging or training image-based simulation methods. Conventional random function (RF)-based simulation refers to the noniterative, sequential methods like those used in sequential Gaussian simulation, sequential indicator simulation and the MPS method described in Section 2.4, Preliminary Mapping Concepts. In general, the RF-based

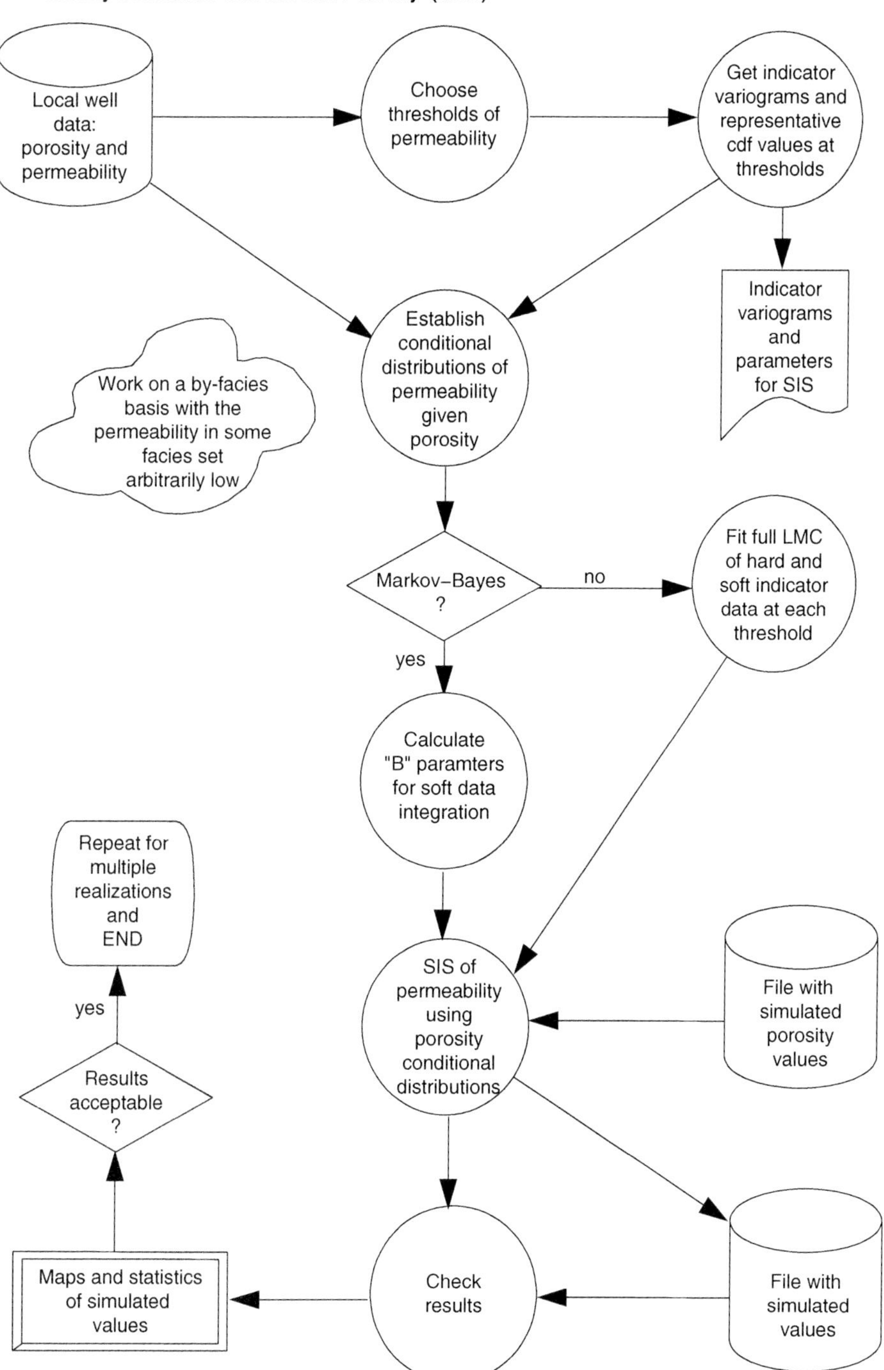

FIGURE 4.87: Work Flow for Permeability Modeling by Sequential Indicator Simulation Using Porosity as a Covariate.

methods are limited to reproducing limited statistical descriptions. Conversely, the optimization methods are able to impose additional constraints that either can be incorporated into an objective function or may be coupled to a forward model. This includes complicated inverse problems such as reproducing connectivity and flow response descriptions that could not be imposed in RF-based methods straightforwardly. While optimization methods are powerful and flexible, their implementation is typically more complicated and CPU intensive. It is generally difficult to set up and tune optimization methods such that the results are reasonable over a manageable calculation time (or number of iterations).

Optimization could also be considered for nonreservoir modeling steps such as well placement optimization. This is discussed in Section 5.3, Uncertainty Management.

The *Background* section provides general motivation for the application of optimization methods in model construction. This includes a discussion of Monte Carlo Markov Chain methods and their general work flow.

Simulated Annealing is a commonly applied optimization methods for model construction. It is flexible at encoding data conditioning and statistical constraint in an objective function and imposing them on final realizations. A discussion is provided on the implementation of simulated annealing, applications, and potential issues with implementation.

The *Other Methods* section presents other optimization methods such as maximum a posteriori selection, the Gibbs sampler, and nature-inspired methods such as evolutionary algorithms and particle swarm optimization. Admittedly, the current applications of these methods for reservoir model construction are limited. The Gibbs sampler has been amenable to iterative MPS from training images, and the other methods have been applied to update models and to optimize input statistics and trends from production and seismic data matching.

> The novice modeler can skip this section because the use of optimization in standard reservoir modeling work flows is very limited. The intermediate modeler may be interested in the simulated annealing section to learn more about the algorithm behind commonly applied object-based facies modeling tools such as `fluvsim` (Deutsch and Tran, 2002). The expert modeler may read the entire section to identify new opportunities to further integrate complex and disparate data sources with novel optimization approaches.

4.7.1 Background

Most of the modeling methods described up to this point have been direct—that is, based on a conventional random function framework. Input statistics, hard data, and secondary data conditioning and trends are imposed within a noniterative, sequential simulation scheme. In the case of variogram-based methods, the kriging system is applied sequentially at each unsampled location to draw from the random function characterized by the histogram and variogram. In the case of multiple-point methods the single normal equation is applied to each unsampled location, which was shown to be equivalent to the conditional probabilities taken directly from frequencies of multiple-point events in a training image. These frameworks are efficient and theoretically sound. Simulation proceeds by visiting each model node once; therefore, the run times are typically fast and known a priori.

Yet, there are times when such a direct sampling of the random function is not possible. This is due to more complex desperate data and input statistical constraints or the need to solve an inverse problem. For example, the placement of objects in object-based methods to honor hard and soft conditioning, and trends is complicated. Consider all the possible geometry parameterizations and placement rules coupled with the requirement to fit these geometries to available conditioning. There is no known analytical solution for conditional distributions that could be sampled sequentially to construct such a model with any form of guarantee of convergence. At best, there are ad hoc methods that sequentially place objects in two steps: first to honor data and then to honor global statistics (see discussion in Section 4.4, Object-Based Facies Modeling). Due to this complexity, the majority of object-based modeling methods are based on optimization. Other complicated settings include seismic data integration (Saussus, 2009) and dynamic data integration (Martínez et al., 2012). These are difficult inverse problems, which typically rely on coupling the forward model for seismic or flow simulation to an optimization engine.

Also, higher-order constraints may add complexity that cannot be reproduced with conventional RF-based methods. For example, consider well connectivity through the model. This constraint is a function of the potentially tortuous and complicated paths between wells through cells of variable reservoir. Even with an efficient statistical description of interwell connectivity, there is no straightforward method to impose such a complicated constraint on a sequential simulation scheme. Once again, any attempt to impose these complicated constraints in a direct framework would likely be ad hoc and not guaranteed to converge free of artifacts.

Finally, geological quantification and heterogeneity characterization are not limited to variograms and multiple-point statistics. *STRATISTICS* is a software package from Professor Plotnick at the University of Illinois that includes various point- and raster-based statistics that describe spatial heterogeneities. For example, lacunarity is a raster-based statistic that describes the space-filling nature of spatial heterogeneity across scales (Plotnick et al., 1996), and Ripley's K function describes the clustering, randomness, or repulsion of points in space (Ripley, 2004). These statistical descriptions are now being actively applied to outcrop to improve quantification and classification of these natural examples and flume experiments to allow for comparison of results and sensitivity analysis of input initial and boundary conditions (Wang et al., 2011). There is no current method to impose such statistical descriptions in a conventional RF-based method. As these statistical descriptions become more accepted and more commonly available from reservoir analogs, optimization methods will provide the opportunity to integrate them into reservoir models (Honeycutt and Plotnick, 2008; Middleton et al., 1995; Perlmutter and Plotnick, 2002; Plotnick, 1986).

Caution must be exercised when applying optimization methods. Consider the following warnings:

- *Convergence is not guaranteed.* The models may not honor the input statistics constraints and may not reproduce the hard data. The optimization may be stuck in a local minimum. These models should be checked carefully.
- *No theoretical foundation exists.* As a result, various ad hoc parameters and tricks may be required to construct models. These may not function well in all settings and may require a high degree of experience with the specific tool.
- *The dimensionality of practical 3-D problems may be too great to solve in reasonable time.* Run times may be very long and results may be suboptimal.
- *The complicated input constraints and conditioning may be contradictory.* While this may occur for all simulation methods, it is more difficult to anticipate these contradictions and to understand their results due to the potentially complicated nature of the constraints.

Most of the optimization methods discussed here fit into the family of Markov chain Monte Carlo (MCMC).[3] MCMC methods may be applied to deal with these complicated settings for which no analytical solution is available for direct sampling (Gelfand and Smith, 1990; Lyster, 2007; Robert and Casella, 2004). The MCMC methods utilize Markov chains with a sequence of variables with the following property:

$$\begin{aligned} P(Z_i = z_i | Z_{i-1} = z_{i-1}, Z_{i-2} = z_{i-2}, \ldots, Z_0 = z_0) \\ = P(Z_i = z_i | Z_{i-1} = z_{i-1}) \end{aligned}$$

where a Markov chain explores a state space such that the current state i is only dependent on the previous state $i - 1$. With the assumption of Markov screening the states $i - 2, i - 3, \ldots,$ are not considered. In the model building context the state Z may represent the local realization at a location $\mathbf{u}$ with limited search neighborhood or any set of locations $\mathbf{u}_\alpha, \alpha = 1, \ldots, n$, where $n \leq m$ where m is the total number of locations within the reservoir model.

Types of MCMC methods include random walk, Metrolpolis–Hastings algorithms, simulated annealing, and the Gibbs sampler. For a random walk, a random component is added to the previous state. While random walk is interesting, for the typical complicated constraints required in reservoir modeling, random walk is not a practical optimization method. Metropolis–Hasting algorithms utilize accept–reject criteria to move the Markov chain towards more probable states (Metropolis et al., 1953). Simulated annealing is a special case of

[3] This is a convenient manner to group and classify optimization methods for our discussion. Other schemes are possible given different criteria.

Metropolis–Hastings algorithms that applies an energy and objective function that provides probabilities of accepting favorable and unfavorable changes (Deutsch, 1992a), while the Gibbs sampler allows for repeated sampling from conditional distributions to represent the full, likely unavailable, joint probability distribution (Casella and George, 1992; Geman and Geman, 1984). This is one perspective on the general framework of optimization methods. Conversely, simulated annealing may be classified as a nature-inspired optimization method (discussed next).

Furthermore, the concept of nature-inspired optimization schemes is quite powerful. As will be discussed in the next section, simulated annealing is analogous to metallurgical annealing. Simulated annealing is an extremely flexible method for integrating a variety of data and statistical constraints. Other nature-inspired methods such as evolutionary algorithms and swarm optimization are commented on after the simulated annealing discussion.

Optimization for model construction includes model construction with inverse modeling, complicated data, and statistical constraints and model updating to impose these constraints. Yet, there are various applications for optimization aside from reservoir model construction. For example, optimization applied to multiple reservoir model realizations for optimum exploitation planning/well placement is a common application (Norrena and Deutsch, 2002). The following discussion focuses on optimization for integration of data and statistics into reservoir models only, while discussion on optimization of well placement is deferred to Section 5.3, Uncertainty Management.

Optimization-based methods are typically more difficult to apply than the conventional RF-based methods. The iteration and convergence schemes typically have parameters that require expert knowledge to ensure inputs that will result in efficient and reasonable results. Undoubtedly, with careful expert application, these tools extended the efficacy of geostatistical reservoir modeling.

Methodology

Regardless of the specific optimization method that is applied, the basic setup of optimization for model construction is as follows:

1. Formulate an objective function.
2. Generate a starting point.
3. Perturb the model.
4. Stop when constraints are satisfied.

These are shown in a work flow in Figure 4.97 in the Work Flow Subsection and are discussed in detail below.

Objective Function

An objective function is an encoding of the constraints to be imposed on the model. In simulated annealing the object function represents the closeness of the current model to the target state. Various possible components and design considerations for simulated annealing object functions are listed in the next section. For the Gibbs sampler the objective function is expressed as a set of conditional probabilities that may be calculated up front. With both of these methods the objective function may be formulated from multiple components, each representing different data and statistical constraints on the model. The weighting of each of these components of the objective function remains as a challenge with these methods.

Initial Realization

The initial realization is the starting point for the iterative optimization. We commonly consider a facies indicator $i(\mathbf{u})$ or a petrophysical property $z(\mathbf{u})$ on a regular grid, that is, a regular Cartesian grid of size $i = 1, \ldots, nx$ by $j = 1, \ldots, ny$ by $k = 1, \ldots, nz$ (typically, $nx \cdot ny \cdot nz \approx 10^{5-7}$) with

$$f_{i,j,k} = \text{facies code taking a value } f = 0, 1, \ldots, n_F$$

$$\phi_{i,j,k} = \text{porosity } \phi \in (0, 1]$$

$$K_{i,j,k} = \text{horizontal permeability } k \in (0, \infty)$$

The use of a regular Cartesian grid is not a condition for optimization; however, a Cartesian grid that conforms to major stratigraphic layering makes updating the objective function more straightforward (see later).

The space of uncertainty or possible configurations is N^K, where N is the number of grid nodes and K is the number of allowable outcomes. This space is larger than we can possibly imagine; however, we are only interested in the configurations that reasonably approximate all available data (within their reliability).

We do not have to consider cell-based representations of geological properties. The "realization" in optimization could be an object-based model of geological facies units such as channels, levees, and crevasse splays. The space of uncertainty would then be the parameters that control the specific positioning and geometry of the geological facies objects.

Consider two options for constructing the initial realization for optimization: (1) a completely random or even constant value realization or (2) the result of some faster (but less flexible) simulation algorithm, which imparts some desirable features. Although the second option provides an initial realization that starts at a lower objective function, it is often better to start with a random or noncommittal realization. There are a number of reasons for initial random realizations. Firstly, there are no initial large-scale features in the wrong place that are hard to undo. In fact, it takes more iterations and time to reconfigure a wrong image than to configure an initially random image. A second reason in support of an initial random realization is the resulting space of uncertainty, which is larger and more realistic with random initial realizations (Deutsch, 1992a).

Perturb the Model

The optimization methods proceeds by randomly visiting all locations in the model ($\mathbf{u}^i_\alpha, \alpha = 1, \ldots, n$, $i = 1, \ldots, m$, where there are n locations in the model and m is the number of iterations) and applying perturbations to the current model ($Z(\mathbf{u}^{i-1}_\alpha)$). For each iteration, changes are proposed at each location. A decision rule is applied to determine if the change is retained discarded. In the case of the Gibbs sampler, a new value is drawn from the updated conditional distribution.

Stopping Criteria

The iterative procedure should be stopped when the input data are reproduced within their required level of certainty. In presence of soft or imprecise data, the objective function should include the imprecision. Thus, each component objective function should get close to zero. How close to zero? Less than 1% of the starting value would normally ensure close reproduction of the input statistic. The results can be checked if there is any question about how closely the input statistics are being reproduced.

In practice, the procedure is also stopped once some reasonable CPU limit is reached. In this case, the reproduction of each component objective function must also be checked. Slow convergence may be due to a large and difficult problem that truly requires many perturbations and conflicting objective function components, which should be resolved by retaining only the most reliable components. Conflicting objective functions may not be known in advance; however, they may become evident by noting the objective functions that do not decrease to zero.

We do not search for the global minimum; the input statistics are not known so precisely and we would not have sufficient CPU power. A realization that reproduces all known data within an acceptable tolerance is a candidate realization for subsequent reservoir management decision making. We can find many realizations that reproduce all known data. Taken together these realizations represent a space of uncertainty implicit to the random function defined by the optimization problem formulation.

Comments on Iterative Methods

While iteration usually implies in greater CPU time than direct modeling methods, there are some advantages to iterative methods. For example, there is no need to account for irregular, variable patterns of samples within a local neighborhood. As a result, search is very straightforward and fast. All local neighbors may be pooled without a need to find nearest neighbors with some search scheme. Also, since all local neighborhood values are informed, the calculations associated with the objective function are simplified. For example, in the example of the Gibbs sampler applied to simulate multiple-point statistics, all nodes in the multiple-point template are populated and there is no need to apply partial templates.

While iterative methods are powerful for simultaneously honoring multiple data and statistical constraints, there may be issues with reproduction of all constraints. Firstly, care must be taken to ensure that there are no contradictions between the constraints. Secondly, as mentioned before, component weighting is a challenge. Even with careful component weighting, some methods are disposed to impose a specific component while sacrificing another.

Lyster (2008) noted this issue with Gibbs sampling-based MPS. It has been observed that when the local PDF is integrated in the usual manner by updating the conditional distributions, this has a tendency to overwhelm the components associated with spatial continuity. Local trends are reproduced with a loss of important spatial continuity features. Lyster proposed the following procedure;

(1) The initial seed at each location should be drawn from the local PDF instead of the global PDF to assist with reproduction of the locally variable PDF, and (2) a servo system should be applied to the local conditional probabilities based on binning of the PDFs for each facies over the entire model. This imposes the global constraint fraction of facies in each trend bin B_k. For example, in the bin 30% to 40% for the k facies, there will be 30% to 40% of facies k.

4.7.2 Simulated Annealing

The following discussion provides details on the methods and implementation of simulated annealing. Many of the considerations such as the initial realization, objective function, perturbation, and stopping criteria are common to most optimization methods.

The *method of simulated annealing* is an optimization technique that has attracted significant attention. For practical purposes, simulated annealing has effectively "solved" the famous *traveling salesman problem*, that is,

> Find the shortest cyclical itinerary for a traveling salesman who must visit each of N cities once and only once in a sequential path.

This is a classic combinatorial minimization problem where the solution space is not simply the N-dimensional space of N parameters. There is a discrete but factorially large configuration space, for example, the number of routes increases as $N!$. Another characteristic feature is that there is no notion of "direction" or "downhill." Thus the minimization cannot rely on the calculation of gradients or derivatives, which are commonly used in optimization problems.

The central idea behind simulated annealing is an analogy with thermodynamics, specifically with the way liquids freeze and crystallize, or metals cool and anneal. At high temperatures the molecules can move freely. As the temperature is slowly lowered, the molecules line up in crystals which represent the minimum energy state for the system. The Boltzmann probability distribution, $P\{E\} \sim \exp(-E/(k_b T))$, expresses the idea that a system in thermal equilibrium at a temperature T has the energies of its component molecules probabilistically distributed among all different energy states E. The Boltzmann constant k_b is a natural constant that relates temperature to energy. Even at a low temperature there is a probability that the energy is quite high; in other words, a system will sometimes give up a low energy state in favor of a higher energy state (Press et al., 1986).

Metropolis et al. (1953) extended these principles to numerically simulate molecular behavior. A system will change from a configuration of energy O_1 to a configuration of energy O_2 with probability $p = \exp(-(O_2 - O_1)/(k_b T))$. The system will always change if O_2 is less than O_1 (that is, a favorable step will always be taken), and sometimes an unfavorable step is taken; this has come to be known as the Metropolis algorithm. More generally, any optimization procedure that draws upon the thermodynamic analogy of annealing is known as simulated annealing.

In the early 1980s Kirkpatrick et al. (1983) and independently Černý (1985) extended these concepts to combinatorial optimization. They formulated an analogy between the objective function and the free energy of a thermodynamical system (Aarts and Korst, 1989; Press et al., 1986). A control parameter, analogous to temperature, is used to control the iterative optimization algorithm until a low energy (objective function) state is reached.

One of the first direct applications to spatial phenomena was published by Geman and Geman (1984) who applied the method to the restoration of degraded images. About the same time Rothman (1985) at Stanford applied the method to nonlinear inversion and residual statics estimation in geophysics. Independent research by C. L. Farmer (Farmer, 1992) led to the publication of a simulated annealing algorithm for the generation of rock-type models. This triggered considerable interest in the method among geostatisticians (Deutsch, 1992a; Srivastava, 1990b).

The essential contribution of simulated annealing is a prescription for when to accept or reject a given perturbation. The acceptance probability distribution is given by

$$P_{\text{accept}} = \begin{cases} 1, & \text{if} \quad O_{\text{new}} \leq O_{\text{old}} \\ \exp\left(\frac{O_{\text{old}} - O_{\text{new}}}{t}\right), & \text{otherwise} \end{cases}$$

This probability distribution is shown in Figure 4.88. All favorable perturbations ($O_{\text{new}} \leq O_{\text{old}}$) are accepted, and unfavorable perturbations are accepted with an exponential probability distribution. The parameter t of the exponential distribution is analogous to the "temperature" in annealing. The higher the temperature, the more likely an unfavorable perturbation will be accepted.

In the physical process of annealing, the temperature must not be lowered too fast or else the

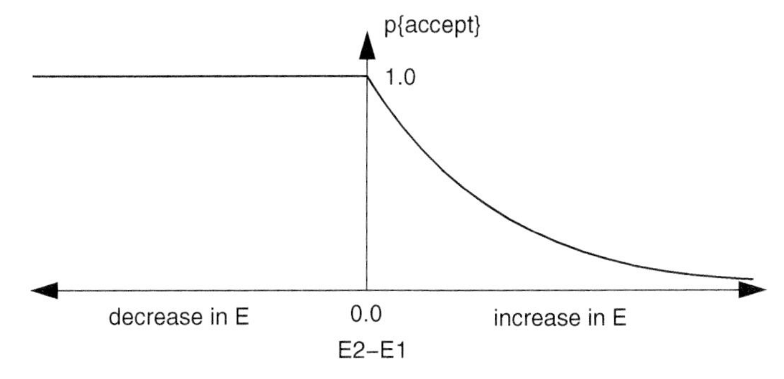

FIGURE 4.88: Probability of Accepting a Change to the System in Simulated Annealing: Probability is 1.0 When the Objective Function Decreases and the Probability Follows an Exponential Distribution when the Objective Function Increases.

material will be "quenched" (frozen in its evolution) and never achieve a low-energy state. The same phenomenon happens in simulated annealing; the image or realization will get trapped in a suboptimal situation and never converge when the t parameter is lowered too quickly. Time in the context of simulated annealing may be considered as the total number of attempted perturbations. The scheme employed to reduce the temperature with time is known as the *annealing schedule*. The determination of a suitable annealing schedule for different problems remains an outstanding challenge in the application of simulated annealing to geostatistical problems.

Steps in Annealing

The steps within simulated anneal are similar to those enumerated above for optimization methods for model construction with some additional considerations. Figure 4.89 illustrates the overall flow chart for simulated annealing applied to geostatistical problems. The steps and considerations unique to annealing are considered in the following sections.

Initial Realization

A randomized initial realization is the most effective to avoid artifacts and fairly represent the full model of uncertainty. Simulated annealing "freezes" large-scale features early, at relatively high temperatures, and then converges on details at lower temperatures. It is necessary to set an initially high temperature to permit large-scale features to be established by all components of the objective function. Such high temperature results in the initial image being randomized, which defeats the goal of using an initially correlated image.

Objective Function

Simulated annealing requires an energy or objective function that measures closeness to data or desired features. In general there are multiple data sources that are coded as separate component objective functions. These components are put together in a weighted sum for the total objective function:

$$O = \sum_{i=1}^{N_c} \omega_i \, O_i$$

where N_c is the number of components, ω_i, $i = 1, \ldots, N_c$, are the weights, and O_i, $i = 1, \ldots, N_c$, are the component objective functions. Some possible component objective functions are listed and then a procedure to determine the weights will be discussed.

Each component objective function O_i is a mathematical expression to enforce some data or impart some desirable spatial character to the realization. All component objective functions are measures of mismatch between a desired property and that of a 3-D model. The $*$ superscript in the following objective functions denotes the property of the 3-D model; no subscript indicates the target property. Some examples:

- All reservoir models must reproduce local facies, porosity, and permeability data coming from core and well log data. These data may be forced into the model by construction—that is, by assigning them to the closest grid node location—or they may be reproduced as part of the objective function. The component objective functions for local data are

$$O_i = \sum_{\alpha=1}^{n_{\text{data}}} \sum_{k=1}^{n_{\text{facies}}} \left[i(\mathbf{u}_\alpha; k) - i^*(\mathbf{u}_\alpha; k) \right]^2$$

$$O_Z = \sum_{\alpha=1}^{n_{\text{data}}} \left[Z(\mathbf{u}_\alpha) - Z^*(\mathbf{u}_\alpha) \right]^2$$

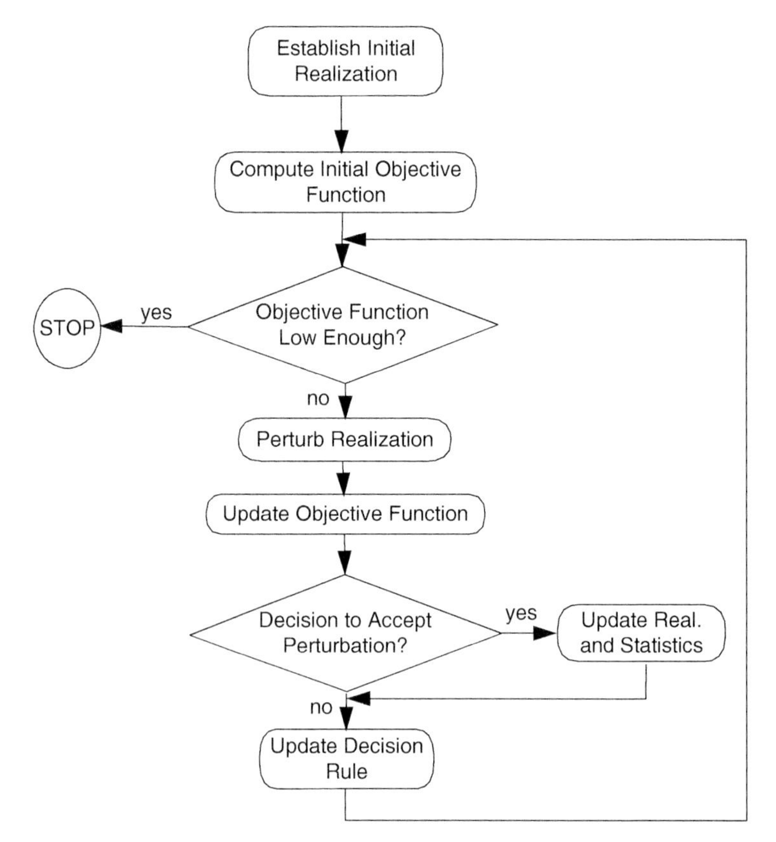

FIGURE 4.89: A Flow Chart with the Steps in Simulated Annealing. Each step is presented in detail.

where $\mathbf{u}_\alpha, \alpha = 1, \ldots, n_{data}$, are data locations, the $*$ superscript denotes the facies, porosity, or permeability in the candidate model, i (in this context) is the facies indicator, and Z is a continuous variable such as porosity, permeability, or water saturation. There may be a different number of data for each variable; the notation has been kept simple.

- The declustered representative histogram of the facies or continuous variable should also be reproduced by the numerical model:

$$O_{p_i} = \sum_{k=1}^{n_{facies}} \left[p_k - p_k^*\right]^2$$

$$O_{F_Z} = \sum_{j=1}^{n_q} \left[F_Z(q_j) - F_Z^*(q_j)\right]^2$$

where p_k is the proportion of facies $k = 1, \ldots, n_{facies}$ and $F_Z(q_j)$ is the cumulative distribution function at a particular quantile $q_j, j = 1, \ldots, n_q$. The number of quantiles n_q would be chosen by the modeler. Once again, the $*$ denotes the property of the candidate model.

- Beyond local data and the correct univariate distribution, reservoir models should reproduce two-point variogram measures of spatial correlation:

$$O_\gamma = \sum_{j=1}^{n_\mathbf{h}} \left[\gamma(\mathbf{h}_j) - \gamma^*(\mathbf{h}_j)\right]^2$$

The number of lags, $n_\mathbf{h}$, is chosen to "inject" all spatial correlation deemed important. It is unnecessary to consider distances and directions where there is no spatial correlation; all features not imposed by a component objective function will show no correlation due to the initial random model.

- Multiple-point measures of spatial correlation can be coded into an objective function with little difficulty. For example, the N point connectivity in a particular $\mathbf{h}$ direction is given by

$$O_C = \sum_N \left[C(\mathbf{h}; N; z_c) - C^*(\mathbf{h}; N; z_c)\right]^2$$

where C is a multiple N-point covariance for unit lag $\mathbf{h}$ at threshold z_c. Different multiple-point statistics of arbitrary configuration and size could be considered. If N is small, there may be sufficient data for reliable inference; however, in most settings these statistics need to be exported from an exhaustive training image (see Section 4.3, Multiple-Point Facies Modeling).

- As mentioned in Section 4.1, Large-Scale Modeling, deterministic trends are important in reservoir modeling. Such trends could enter as component objective functions. For example, in the case of vertically varying facies proportions we have

$$O_v = \sum_{k=1}^{n_{\text{facies}}} \sum_{z=1}^{n_z} \left[p_k(z) - p_k^*(z)\right]^2$$

where $p_k(z)$ is the proportion of facies k at areal slice z, $k = 1, \ldots, n_{\text{facies}}$ the number of facies, and $z = 1, \ldots, nz$ the number of areal slices. Areal variations could be considered in a similar manner. Areal and vertical variations in the average of a continuous variable may also be expressed in a similar manner.

- Multiple petrophysical properties are almost always considered—for example, facies, porosity, and permeability. The relationships between the variables must be reproduced by the reservoir models for reliable predictions. As an example, consider the bivariate distribution between porosity and permeability:

$$O_{\phi/K} = \sum_{i=0}^{n_\phi} \sum_{j=0}^{n_K} \left[f(\phi_i, K_j) - f^*(\phi_i, K_j)\right]^2 \tag{4.9}$$

where the bivariate distribution is discretized into $i = 1, \ldots, n_\phi$ porosity classes and $j = 1, \ldots, n_K$ permeability classes, and $f(\phi_i, K_j)$ is the probability of a porosity/permeability pair falling into a particular class. The correlation coefficient could be considered as a simpler measure of correlation; however, bivariate probabilities are more flexible to capture nonlinear and complex relationships.

- Seismic data are of significant importance when well data are sparse. The bivariate relation between seismic data and average porosity could be captured as bivariate probabilities, as above in Eq. (4.9); however, there are often too few calibration wells to infer the full bivariate relation. In many cases, we can only infer the correlation coefficient between the porosity and permeability with any degree of reliability. This correlation between seismic data and porosity could be imposed at the correct scale:

$$O_\rho = \left[\rho_{\overline{\phi},\text{Impedance}} - \rho^*_{\overline{\phi},\text{Impedance}}\right]^2$$

where $\rho_{\overline{\phi},\text{Impedance}}$ is the classical correlation coefficient.

- Another important source of data is well test and historical production data. Such data must be reproduced by numerical reservoir models for reliable future predictions. There are a large variety of production data that are handled differently; Section 3.1, Data Inventory, discusses this dynamic data in more detail. As an example, we could construct a component objective function for a well-test-derived effective permeability:

$$O_{\text{wt}} = \sum_{i=1}^{n_{\text{well}}} \sum_{t=1}^{n_{\text{time}}} \left[\overline{k}(\mathbf{u}_i, t) - \overline{k}^*(\mathbf{u}_i, t)\right]^2$$

where $i = 1, \ldots, n_{\text{well}}$ is the number of wells with well test intervals, $t = 1, \ldots, n_{\text{time}}$ are the time intervals, and $\overline{k}(\mathbf{u}_i, t)$ is the effective permeability for well i and time t. The link between time and space is commonly made by Oliver's weighting function (Oliver, 1990a).

The component objective functions must be positive and decrease to zero when data or spatial statistics are reproduced by the realization.

In general, the number of required perturbations to achieve convergence is between 10 and 1000 times the number of variables in the system. Each component objective function must be updated or recalculated after each perturbation. As discussed below, we should aim at being able to locally update the objective function after a local perturbation to the 3-D model. Global recalculation of an objective function is avoided because of CPU cost.

The units of each component objective function are different. Moreover, the rate at which each component objective function decreases to zero is different. It is necessary to weight each component to ensure that all components i are considered in the global objective function. Weights are first determined to ensure that each component has an equal importance. Then, we may increase the importance of a particular objective function as a subjective choice or in the presence of conflicting objective function components.

The absolute magnitude of the objective function is not used in making a decision; the difference is considered:

$$\Delta O = O_{\text{new}} - O_{\text{old}}$$

$$\Delta O = \sum_{i=1}^{N_c} \omega_i \Delta O_i = \sum_{i=1}^{N_c} \omega_i \left[O_{i_{\text{new}}} - O_{i_{\text{old}}} \right]$$

Our goal is to establish weights $\omega_i, i = 1, \ldots, N_c$, so that, on average, each component contributes equally to a change to the global change in the objective function ΔO, that is,

$$\omega_i = \frac{1}{\overline{|\Delta O_i|}}, \qquad i = 1, \ldots, N_c$$

The average change of the objective function for each component can be determined by proposing a some number M of changes to the system and calculating the average change. Approximate by

$$\overline{|\Delta O_i|} = \frac{1}{M} \sum_{m=1}^{M} |O_i^{(m)} - O_i|, \qquad i = 1, \ldots, N_c$$

These weights could be updated throughout the annealing process; however, that has not been found necessary (Deutsch, 1992a).

Stop Criteria

As discussed previously, the iterative procedure should be stopped when the objective function is low enough to ensure that all input data are reproduced within their level of certainty. For simulated annealing, the component objective functions should be designed to go to zero.

Also, the procedure is also stopped once some reasonable CPU limit is reached. Simulated annealing has the added consideration of the annealing schedule. Slow convergence may be due to a slow annealing schedule, which could be changed (see below). Also given the flexibility to integrate any data or statistical constraint, extra care is required to avoid a conflicting objective function component (recall previous discussion on conflicting components).

4.7.3 Perturbation Mechanism

Initial applications of simulated annealing considered "swapping" as the perturbation mechanism (Farmer, 1992). This has the advantage that the histogram of the initial image is never changed, so there is no need to include the histogram in the objective function; however, this scheme is inflexible in the presence of complex conditioning data. Secondary data may be available in the form of a component objective function that would lead to a different histogram than was used for the initial realization.

Previous applications consider proposing a change at one location at a time (Deutsch and Journel, 1998). The locations may all be considered before reconsidering a particular location. This scheme is sometimes referred to as the "heat bath" algorithm (Datta-Gupta et al., 1995). Some advantages of the heat bath over random grid node selection have been reported.

After selecting a grid node location to perturb, a new value is drawn from either the global histogram or a local distribution, considering the values at nearby grid node locations. The global distribution is noncommittal and is implemented in the `sasim` program of GSLIB (Deutsch and Journel, 1998). Considering a local distribution, however, can speed convergence.

The local distribution may be determined by kriging weights obtained by some representative variogram and a small local template of grid node locations that excludes the collocated grid node that is being perturbed (see upper left of Figure 4.90). The kriging weights provide a direct estimate of local proportions of categorical variables. For continuous variables, there is a need to provide a continuous cdf model between the available quantile values. A straightforward linear interpolation between the

Template of Kriging Weights

	0.15	0.20
0.15	j′	0.15
0.20	0.15	

$$\gamma(\mathbf{h}) = 0.1 + 0.9\,\text{Sph}\left(\sqrt{\left(\frac{hx}{10}\right)^2 + \left(\frac{hy}{5}\right)^2}\right)$$

x is in +45°direction and
y is in –45°direction

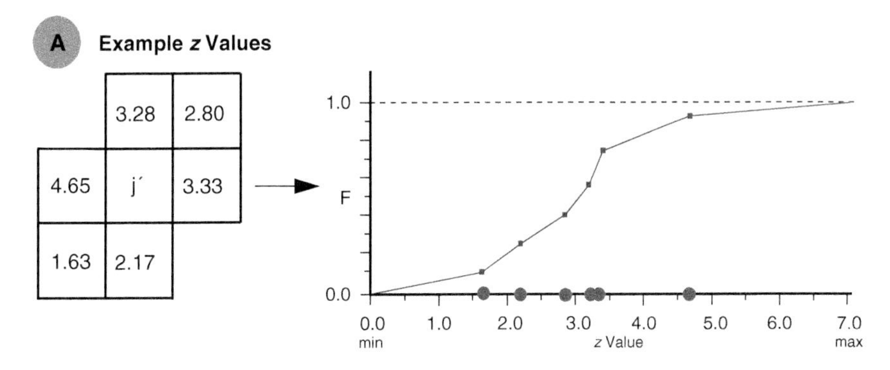

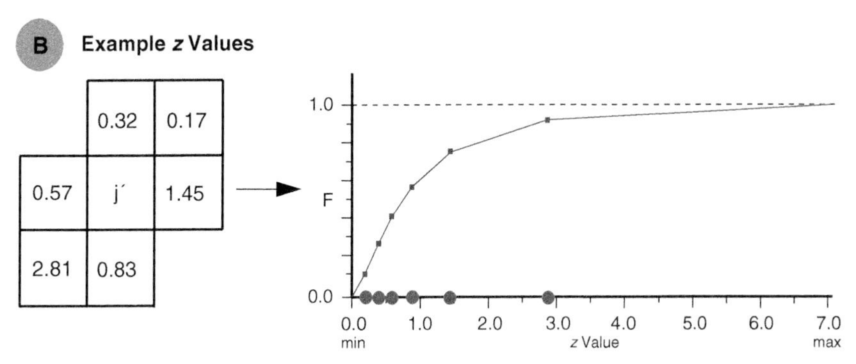

FIGURE 4.90: An Illustration of How Local Distributions at Location j' Are Constructed by Applying Kriging Weights to Local Data. Two examples are shown: (A) with values that are in a relatively high area and (B) with values that are in a relatively low area.

minimum, the available quantiles, and maximum is considered.

In summary, the implementation steps for this perturbation mechanism are illustrated (Figure 4.90) and described below:

1. Establish kriging weights for a template of points (excluding the location being perturbed) using a representative variogram. The median indicator variogram would be a good choice.
2. Sort the data within the template in ascending order, $z_{(1)}, z_{(2)}, z_{(3)}, \ldots, z_{(n)}$, with kriging weights: $w_{(1)}, w_{(2)}, w_{(3)}, \ldots, w_{(n)}$.
3. Calculate the cdf values for each datum

$$cp_{(i)} = \sum_{1}^{i} w_{(i)}, \qquad i = 1, \ldots, n$$

where the weights $w_{(1)}, w_{(2)}, w_{(3)}, \ldots, w_{(n)}$ sum to 1, $cp_{(0)}$ at $z_{\min}$ is 0.0, and $cp_{(n+1)}$ at $z_{\max}$ is 1.0.

4. Establish intermediate ccdf values:

$$F(z_{(i)}) = \frac{cp_{(i-1)} + cp_{(i)}}{2}, \qquad i = 1, \ldots, n$$

5. Linear interpolation allows a complete specification of the relation between $F(z)$

and z. More elaborate "tail" extrapolation methods could be considered for highly skewed data distributions with limited data (Deutsch and Journel, 1998).

6. Lastly, a new candidate value $z^{new}(\mathbf{u}_j)$ is drawn for location $\mathbf{u}_j$ from this local ccdf. This candidate value is more likely to be accepted than a value drawn from the global distribution, since it is consistent with other cell values in the local neighborhood. Moreover, since it is drawn from a local conditional distribution built by kriging, the variogram of the perturbed model will more likely be improved.
Note that a candidate value $z^{new}(\mathbf{u}_j)$ will be drawn from the global distribution if the variance of the local distribution is greater than that of the target histogram.

The weights at the top of Figure 4.90 consider an anisotropic spherical variogram model. Two different sets of local data, hence conditional distributions (A and B), are shown in Figure 4.90.

The only additional parameter required to use this scheme for constructing a local distribution is the size of the template, n_{tem}. There is a small CPU penalty for using a larger template; that is, it takes some CPU time to assemble the local data into a distribution. A perturbation with $n_{tem} = 4$ takes 5% more CPU time to consider than simply drawing from the global distribution. A perturbation with $n_{tem} = 40$ takes 45% more CPU time. From this perspective, a smaller template is to be preferred. A larger template, however, considers more spatial information and requires fewer perturbations to achieve convergence. The CPU time versus template size can be plotted to determine a minimum CPU value. Typical results indicate that templates of size 8–12 yield optimal results. The CPU time is decreased by at least an order of magnitude.

4.7.4 Update Objective Function

The objective function should be updated after every attempted perturbation. There are many attempted perturbations. As mentioned, there are normally between 10 and 1000 times the number of variables or grid nodes. For a 1-million-grid-node problem there are between 10 million and 1 billion attempted perturbations. Updating the objective function only occasionally slows convergence, but is the last resort for difficult objective functions. In general, all components of the objective function must be updated after each perturbation. It would be very CPU-expensive to recalculate the objective function after each perturbation. In almost all cases, however, it is possible to update the underlying spatial statistics without global recalculation.

Consider the traditional variogram as an example. The variogram is defined as

$$\gamma(\mathbf{h}) = \frac{1}{2N(\mathbf{h})} \sum_{i=1}^{N(\mathbf{h})} \left[z(\mathbf{u}) - z(\mathbf{u}+\mathbf{h})\right]^2$$

with the global cost of calculation proportional to $\propto (nx-1)\cdot ny \cdot nz$ operations. The variogram, however, may be updated by few CPU operations:

$$\begin{aligned}\gamma_{\text{new}}(\mathbf{h}) = \gamma_{\text{old}}(\mathbf{h}) & \\ &-\left[z(\mathbf{u}) - z(\mathbf{u}+\mathbf{h})\right]^2 - \left[z(\mathbf{u}-\mathbf{h}) - z(\mathbf{u})\right]^2 \\ &+\left[z'(\mathbf{u}) - z(\mathbf{u}+\mathbf{h})\right]^2 + \left[z(\mathbf{u}-\mathbf{h}) - z'(\mathbf{u})\right]^2\end{aligned}$$

where value $z(\mathbf{u})$ is being perturbed to $z(\mathbf{u}')$. The local cost of updating is two operations. As illustrated in Figure 4.91, the $+\mathbf{h}$ and $-\mathbf{h}$ contribution for a particular lag $\mathbf{h}$ must be updated; there are two operations regardless of grid size.

Virtually all spatial statistics may be updated and not globally recalculated, including the example objective functions given above: (1) local data, (2) global distributions, (3) variogram measures of correlation, (4) multiple-point spatial statistics, (5) vertical and areally varying proportions or averages, (6) bivariate distributions between two variables, and (7) the correlation coefficient between two variables at any scale.

There are some objective functions that cannot be locally updated—in particular, anything that is a function of *all* cells *simultaneously*, for example, dynamic flow data. In this case, it is desirable to replace the complex calculation by a simpler proxy if possible; for example, consider inverted large-scale features or multiple-point measures of connectivity. The inversion or "spatial coding" of dynamic data is discussed in detail in Section 3.1, Data Inventory.

When complex data cannot be represented by a spatial statistic, we can resort to global calculation. Of course, the difficult component objective function could be updated less frequently than the other objective function components. Implementation becomes difficult and must be tackled by trial and error.

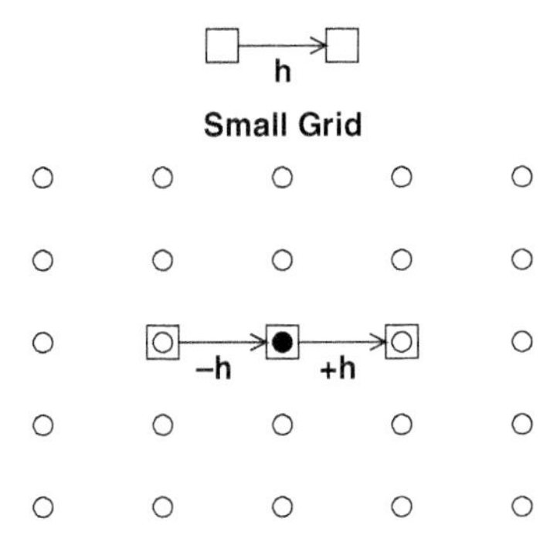

FIGURE 4.91: Illustration of the Lags That Need Updating in the Variogram Calculation.

4.7.5 Decision Rule

The temperature parameter t in the simulated annealing decision rule must not be lowered too fast or else the image may get trapped in a suboptimal situation and never converge. However, if lowered too slowly, then convergence may be unnecessarily slow. The specification of how to lower the temperature t is known as the "annealing schedule." There are some annealing schedules that ensure convergence in very particular cases (Aarts and Korst, 1989; Geman and Geman, 1984); however, they are much too slow for practical application. One form of an annealing schedule that ensures convergence is

$$t(p) = \frac{C}{log(1+p)}, \qquad p = \text{pass through system}$$

Under certain assumptions it is possible to show that, with correct choice of C, the system will converge. This is too slow for practical application.

The following empirical annealing schedule is one practical alternative (Farmer, 1992; Press et al., 1986). The idea is to start with an initially high temperature t_0 and lower it by some multiplicative factor λ whenever enough perturbations have been accepted (K_{accept}) or too many have been tried (K_{max}). The algorithm is stopped when efforts to lower the objective function become sufficiently discouraging. The following parameters describe this annealing schedule (see also the chart of Figure 4.92):

t_0: the initial temperature.

λ: the reduction factor $0 < \lambda < 1$ usually set to 0.1.

K_{max}: the maximum number of attempted perturbations at any one temperature (on the order of 100 times the number of nodes). The temperature is multiplied by λ whenever K_{max} is reached.

K_{accept}: the acceptance target. After K_{accept} perturbations are accepted, the temperature is multiplied by λ (on the order of 10 times the number of nodes).

S: the stopping number. If K_{max} is reached S times, then the algorithm is stopped (usually set at 2 or 3).

ΔO: a low objective function indicating convergence.

Figure 4.93 shows the typical reduction in the global objective function versus the number of attempted perturbations. The rate of decrease in the objective function is nonlinear. The objective function remains quite high until a "critical temperature," then it drops quickly until convergence, and finally the objective function drops slowly. Note the logarithmic scaling on both axes of Figure 4.93.

4.7.6 Problem Areas

There is no equivalent to the metallurgical cooling in applications of simulated annealing. Nevertheless, there is an artifact that appears as "thermodynamic edge effects" (see Figure 4.94). Extreme values (black and white colors in Figure 4.94) are preferentially located at the edge of the grid when the temperature is lowered quickly (Deutsch and Cockerham, 1994b). This is due for example to the variogram component of the objective function. The pure nugget variogram of the initial random realization can be improved quickly by placing high (and

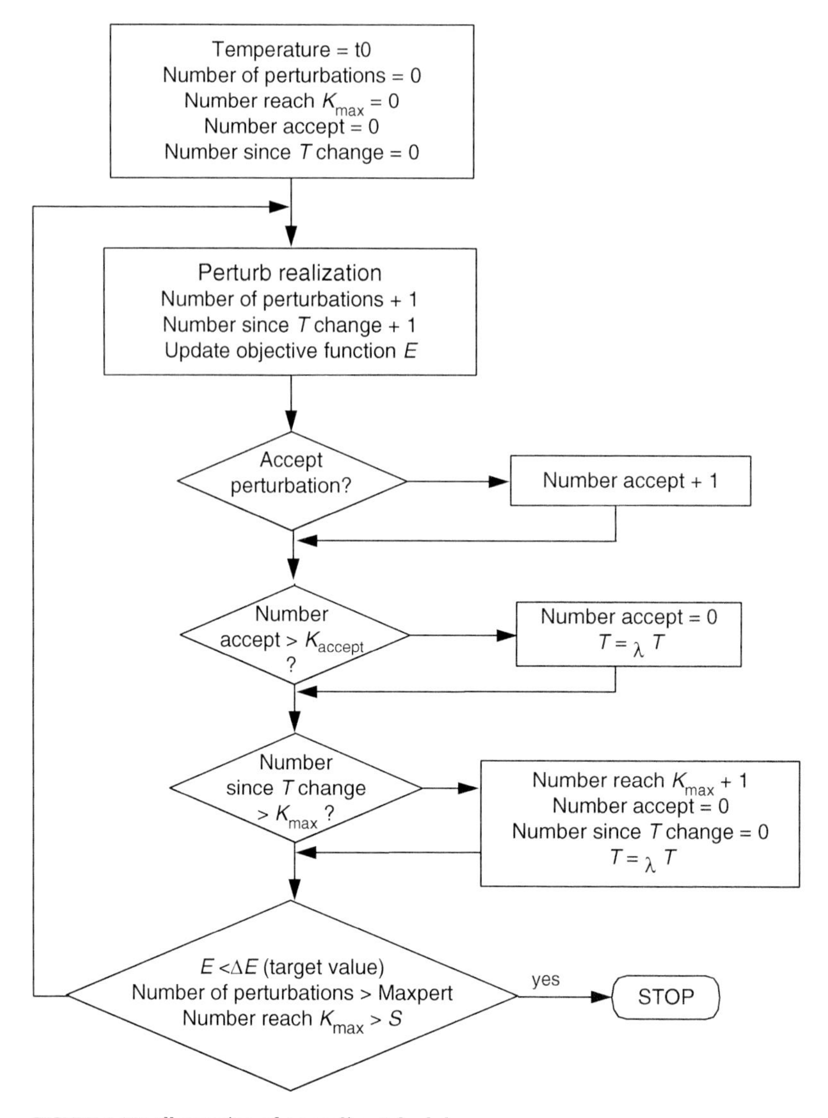

FIGURE 4.92: Illustration of Annealing Schedule.

low) values at the grid edge where they only enter a variogram pair once and not for $+\mathbf{h}$ and $-\mathbf{h}$. The histogram is reproduced and the variogram is improved by such placement. Of course, this is unsatisfactory and a special weighting of edge pairs can be considered in software.

Another artifact of fast annealing is discontinuities near well data; see Figure 4.95 where there is artificial discontinuities near two well locations. The objective function is near zero for such an image: The data are honored, albeit with a discontinuity; the proportion of white and black facies is reproduced; and the variogram is reproduced. The problem is that the variogram pairs involving the conditioning data are overwhelmed by the pairs of the remaining grid nodes. Convergence takes place without particular attention to the conditioning data, which are reproduced by construction. One solution to this problem is to place extra weight on the variogram pairs involving conditioning data (Deutsch and Journel, 1998).

The potential for a quasi-exact reproduction of input statistics (see Figure 4.96), is another problem with simulated annealing. Uncertainty in input statistics must be accounted for explicitly. There are "ergodic" fluctuations in kriging-based techniques, which somehow account for the variability of the input statistics for the region being simulated. There is a concern in simulated annealing that such close reproduction of input statistics artificially reduces the final "space of uncertainty." A number of case studies (Deutsch, 1992a, 1994) show that

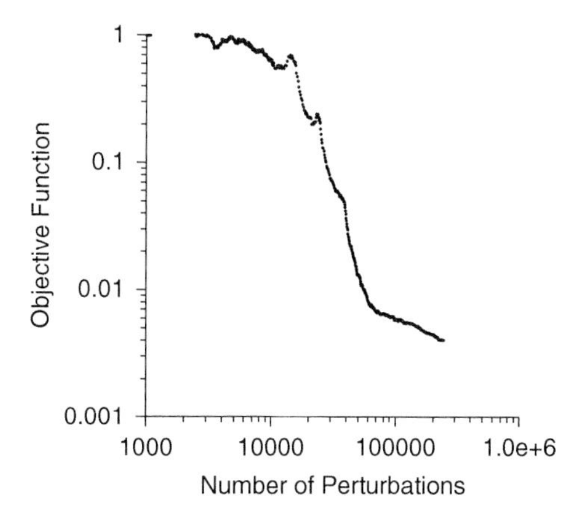

FIGURE 4.93: Illustration of Typical Reduction in Objective Function versus Number of Perturbations.

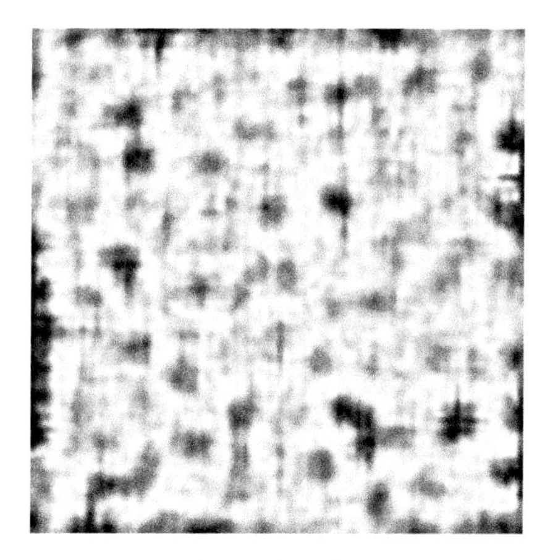

FIGURE 4.94: Illustration of Thermodynamic Edge Effects Caused by Too-Fast Cooling and no Special Weighting of Edge Pairs.

this is not a concern; in fact, the space of uncertainty of simulated annealing is larger than that of kriging-based techniques. In addition, as discussed in Section 5.3, Uncertainty Management, ergodic fluctuations alone are not a sufficient model of uncertainty. Explicit uncertainty in model inputs should be modeled and pass through the modeling method. Given, the quasi-exact reproduction of simulated-annealing this is essential.

The CPU time requirements for simulated annealing methods can become excessive particularly when there are many perturbations required with a complex objective function. In attempts to reduce the CPU requirements, one often embarks upon the time-consuming process of adjusting the tuning parameters that constitute the weighting function and the annealing schedule. The tradecraft required to do this successfully is a significant problem.

Threshold Accepting (TA)

Another published stochastic relaxation technique is a method called threshold accepting (TA) (Dueck and Scheuer, 1990). Threshold accepting differs from annealing and the greedy methods (see maximum a posteriori discussed below) approaches only in the decision rule to accept or not an unfavorable perturbation. An unfavorable perturbation will be accepted if the change in the objective function is less than a specified threshold. The threshold is lowered in much the same way as the temperature parameter t is lowered in simulated annealing.

Many different variants of the same basic annealing algorithm are possible by considering different objective functions, ways to create the initial image, perturbation mechanisms, and decision rules.

4.7.7 Other Methods

There are a variety of other methods for optimization for model construction. These include maximum a posteriori (MAP) method and Gibbs sampler. In addition, brief comments are made on nature-inspired methods such as evolutionary algorithms and swarm optimization.

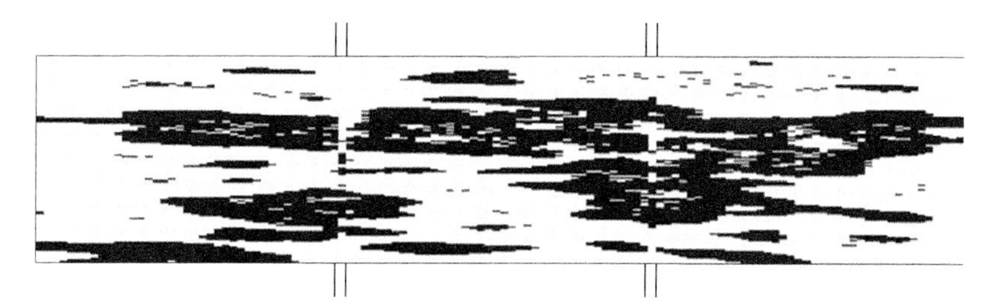

FIGURE 4.95: Illustration of Discontinuities at Conditioning Data Locations Caused by Too-Fast Cooling and No Special Weighting of Well Locations.

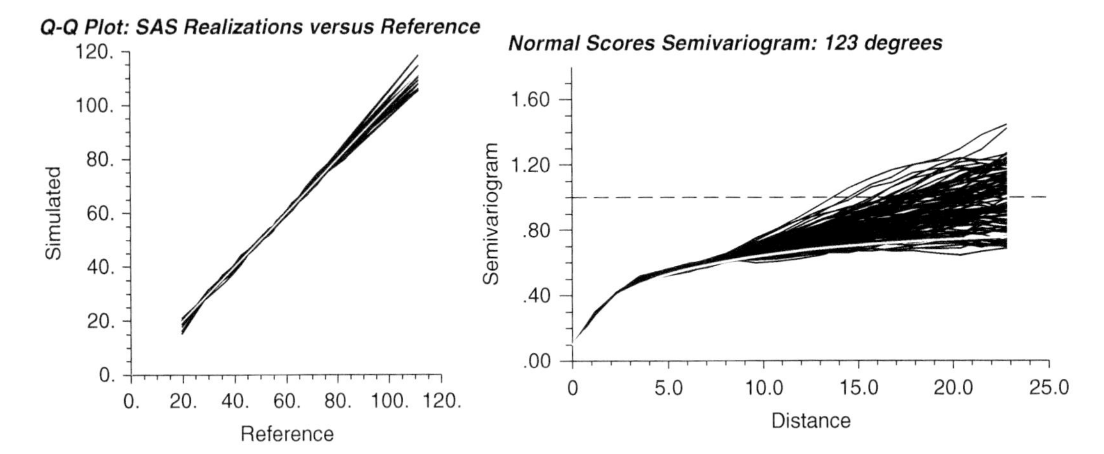

FIGURE 4.96: Illustration of Quasi-Exact Reproduction of Input Statistics, Which Could Lead to a Too-Narrow Space of Uncertainty.

The Maximum A Posteriori (MAP)

Bayesian classification schemes provide a variant to the simulated annealing algorithm (Andrews and Hunt, 1989; Besag, 1986). The classic works of Geman and Geman (1984), Besag (1986), and others have been applied in geostatistics by Doyen and Guidish (1989). The same basic algorithm is followed; however, all favorable perturbations are retained and all unfavorable perturbations are rejected. Such a greedy method is likely to get caught in highly suboptimal local minimums, and convergence is not achieved for difficult problems.

Gibbs Sampler

The Gibbs sampler is a statistical algorithm that was first proposed by Geman and Geman (1984); it is a special case of the Metropolis algorithm (Metropolis et al., 1953). It is a method used for drawing samples from complicated joint or marginal distributions, without requiring the density functions of the distributions (Casella and George, 1992). The basic idea of a Gibbs sampler is that of resampling individual variables conditional to others in the same sample space. Drawing new values conditional to all others, and repeating this process many times, results in an approximation of the joint (and marginal) distribution(s). For example, to determine the density function for $f(X, Y, Z)$, one would start with an initial state (values) for X^0, Y^0, and Z^0 and then sequentially draw values from the conditional distributions $f(X^{i+1}|Y^i, Z^i)$, $f(Y^{i+1}|X^i, Z^i)$, and $f(Z^{i+1}|X^i, Y^i)$. When a number of conditional values have been drawn, the state of (X, Y, Z) approximates a sample from the joint distribution $f(X, Y, Z)$. Repeating this process enough times allows the density (or histogram) of the joint distribution to be constructed empirically.

In the context of geological modeling, a sample from the joint distribution is a simulated realization. In a Gibbs sampler, only the conditional distributions are required to sample from the joint distribution. While the full joint distribution is not practically available in model construction, the conditional probabilities are often available at each location. For example, all necessary conditional probabilities may be determined from a training image for MPS (see Section 4.3, Multiple-Point Facies Modeling).

An advantage of the Gibbs sampler work flow for geospatial modeling is that there is no strict specification for the conditional distributions' need to be determined. Guardiano and Srivastava (1993); Srivastava (1992) developed methods using kriging, two-point statistics, and multiple-point statistics. Lyster (2008) developed a MPS algorithm known as `mpsesim` based on the Gibbs sampler. Besides a multiple-point statistics template, it is possible to integrate any desired statistics into the conditionals, such as locally varying means, secondary data, and lower-order statistics to ensure their proper reproduction (using the probability combination schemes discussed in Section 2.4, Preliminary Mapping Concepts).

Yet the Gibbs sampler is not as flexible as simulated annealing. Any data or statistical constraint may be coded into the simulated annealing objective function, while Gibbs sampler requires a set of conditional probabilities. It would not be straightforward to code complicated constraints such as interwell connectivity into the required conditional probabilities.

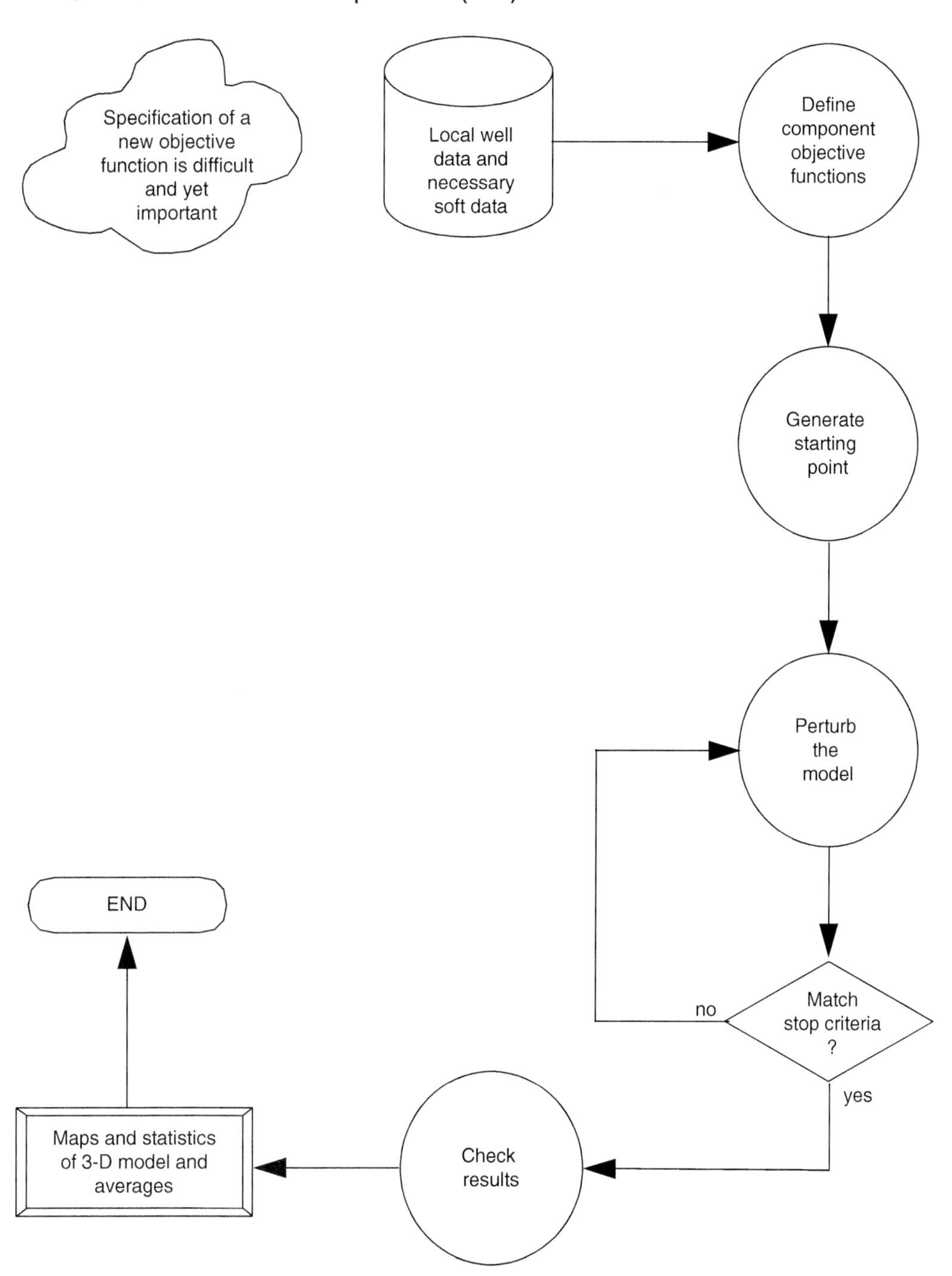

FIGURE 4.97: Work Flow for 3-D Model Construction with Optimization.

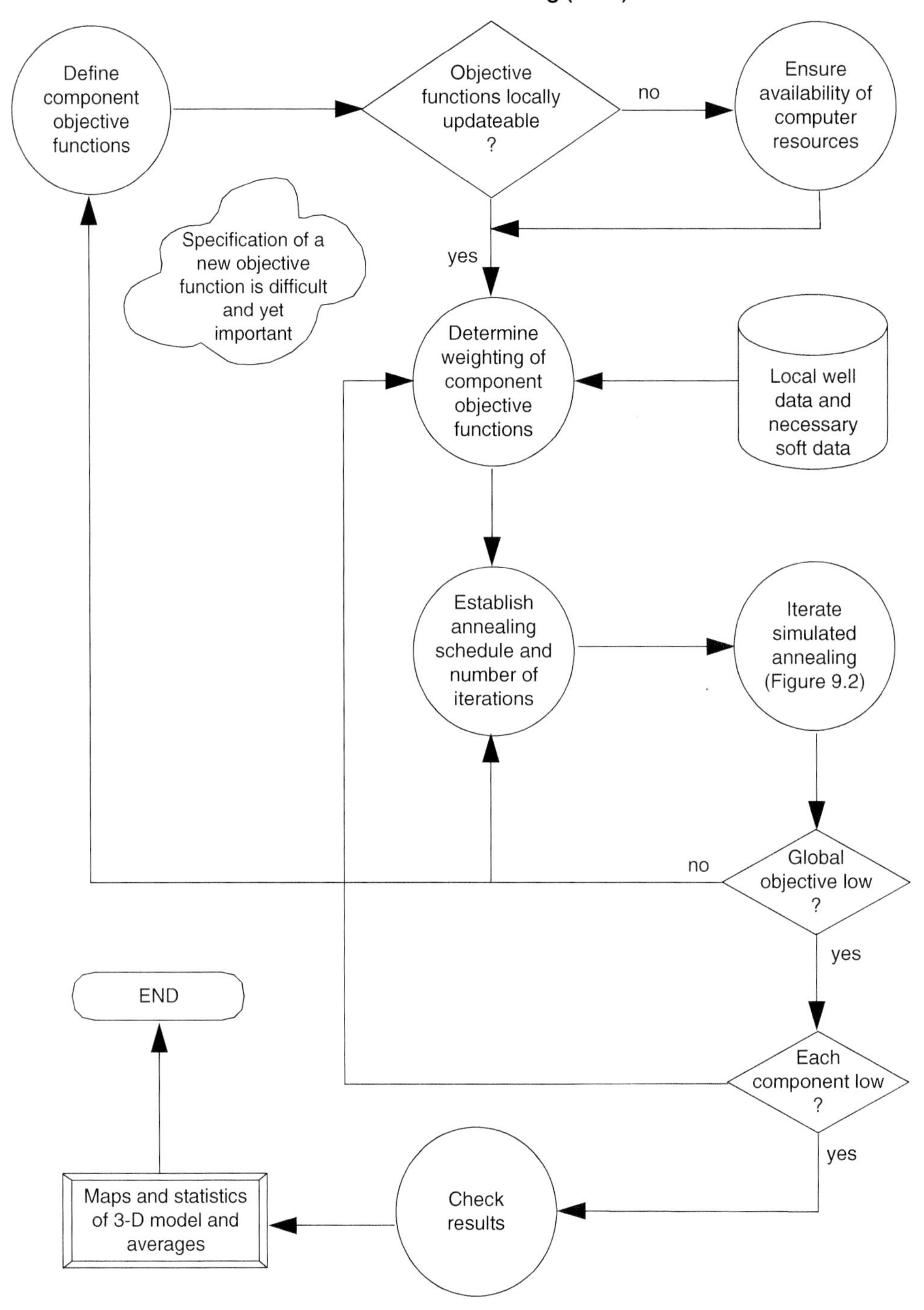

FIGURE 4.98: Work Flow for 3-D Model Construction with Simulated Annealing.

Evolutionary Algorithms

Evolutionary algorithms, and specifically genetic algorithms, have been proposed to aid with history match reservoir models (Romero et al., 2000; Schulze-Riegert and Ghedan, 2007; Schulze-Riegert et al., 2002; Soleng, 1999; Williams et al., 2004). This nature-inspired optimization scheme is analogous to evolution of species. This method requires a genetic representation of the solution set and a fitness function, similar to the object function. It would not be practical to represent every model cell within genetic representation. Not only would this be computationally prohibitive with the typical million to tens of millions of cells in a standard reservoir model (recall the previous discussion of the immense size of the solution space), but the manipulation of each cell independently would destroy spatial continuity. Instead, the typical method is to genetically code the input statistics and parameters and to then apply a standard geostatistical algorithm, such as sequential Gaussian simulation; therefore, the evolutionary algorithm is not applied to construct the reservoir models, but to optimize the input parameters for a set simulated models. For example, Romero and Carter (2003) and Romero et al. (2000) utilized a genetic algorithm to optimize the property trend (defined by a few hundred control points in the model), variogram parameters, and various fault and skin properties to improve match with production history.

With genetic algorithms, multiple solutions are simultaneously preserved, forming a population. The perturbation mechanisms mirror evolutionary processes with combinations of fit parents resulting in children with crossovers and mutations. The initialization and stopping criteria are similar to those discussed for simulated annealing; randomized initial states more effectively explore the uncertainty and space, stopping criteria usually is related to goodness of the solution and/or computational time.

Particle Swarm Optimization

Particle swarm optimization (PSO) is another useful nature-inspired optimization method (Kennedy and Eberhart, 1995). Pedersen and Chipperfield (2010) provide a good summary and discussion on practical set up and application. Like evolutionary algorithms, a population (known as a swarm) of solutions is constructed. In this case, these solutions (known as particles) are seeded into the solution space with a position, a velocity vector, and a best-known solution (the occupied location that minimizes the objective function). Over each iteration, the velocity of each particle is updated with a random component and attraction to the best solution found by other particles. Parameters are assigned to favor search (random movement) versus convergence (moving to the best known solution) of the swarm. Martínez et al. (2012) demonstrated methods to apply PSO to update reservoir models to honor seismic information and to optimize reservoir exploitation.

4.7.8 Work Flow

This section has covered the use of optimization methods to construct reservoir models. Figure 4.97 shows the general work flow for optimization. This includes the specification of an objective function (including match to local well data, soft data, and heterogeneity models), generate an initial model, calculate the objective function on the current model, perturb the model, recalculate the objective function and determine whether to accept or reject the perturbation and if convergence has been reached. Upon completion, the results should be checked carefully with maps and summary statistics.

Figure 4.98 illustrates a work flow for 3-D model construction with simulated annealing. The use of simulated annealing is not standard and defies simple presentation on a work flow. Correct setup of the component objective functions and their updating are a challenge.

4.7.9 Section Summary

Optimization methods provide another set of methods to construct reservoir models. While simulated annealing is commonly applied to construct object-based models, applications of the other methods is less common. While these methods are powerful and flexible, they are typically more complicated and CPU-intensive.

With this short discussion of optimization methods to construct reservoir models, we conclude our discussion on reservoir model construction. Chapter 4 has covered model methods from large-scale to facies modeling with variogram-based, multiple-point, object-based, and process-mimicking, continuous reservoir property modeling and finally this concluding discussion on optimization methods.

In Chapter 5 we discuss model applications—specifically, model checking (Section 5.1), model post-processing (Section 5.2), and uncertainty management (Section 5.3).

5

Model Applications

Geostatistical reservoir models are usually not the final product of a reservoir modeling project. There are modeling applications to ensure that these models provide optimum support for decision making.

These models must be checked for accuracy. The *Model Checking* section reviews the methods for checking models to ensure data that input statistics and concepts are reproduced, the model is predictive away from conditioning, and the uncertainty model is fair.

Modeling Post-processing is often applied to adjust model scale or to learn from the model through a variety of summarizations that provide local and global measures with uncertainty with all available information. While some of these summarizations could be considered transfer functions, the general concept of model transfer functions such as volumetric calculation and flow response to support decision making are not in the scope of this book.

Uncertainty Management provides guidance in developing uncertainty models and using the results with model ranking and decision making in the presence of uncertainty.

5.1 MODEL CHECKING

As discussed in Section 3.3, Problem Formulation, reservoir modeling consists of many interdependent modeling steps, each with many parameters and assumptions. As a result, there are many opportunities for mistakes. It is impossible to completely validate, in an objective sense, stochastic models. Nevertheless, there are some basic checks that can be applied to detect errors in the geostatistical modeling. At the very least, we need to be familiar with the performance and limitations of our modeling work flows.

The *Background* subsection discusses the sources of error and warns that they are common in complicated work flows. Model checking will reveal modeling issues, but cannot replace rigorous data checking and management, as discussed in Section 3.1, Data Inventory. Issues related to ergodic fluctuations, non-stationary models, and other applications for model checking tools are explored.

The *Minimum Acceptance* subsection reviews the work of Leuangthong et al. (2004). They suggest a set of basic models checks to ensure that data and input statistical constraints are reproduced. These include checking data, histogram, variogram, and correlation coefficient reproduction.

The *Higher-Order Checks* subsection reviews the work of Boisvert et al. (2007a) to establish a set of model checks that are applicable to multiple-point methods. While the minimum acceptance checks are useful, they do not account for the multiple-point spatial continuity constraints from the training image. Summaries of multiple-scale histograms, multiple-density functions, and zero bins are demonstrated as metrics to check these multiple-point realizations.

Geostatistical estimation may be checked by setting each well aside and predicting the corresponding reservoir properties with the remaining data. Estimation is different from simulation, which provides an entire distribution of uncertainty rather than a single estimated value. Continuous variables are different from categorical variables. The *Cross Validation and the Jackknife* subsection discusses methods to check model estimation for both continuous and categorical variables. The *Checking Distributions of Uncertainty* subsection describes cross validation of simulation techniques for both continuous and categorical variables.

Model checking is essential and must be integrated into the geostatistical reservoir modeling work flow. The novice modeler will find this section useful for the examples of potential pitfalls in modeling. Both novice and intermediate modelers will benefit from integrating these checks into their work flows. These checks tend to be simple; even if a specific earth modeling software tool does not have these checks available, they may be added with some basic scripting and work flow steps. The expert modeler is likely already familiar with the potential sources of error and the minimum acceptance checks, but may be interested in the higher-order checks to check MPS models and accuracy plots for checking uncertainty models over multiple realizations.

5.1.1 Background

Model checking is very important, but often neglected. Often the available software and work flows lack the required methods and displays for model checking. It would be a mistake to assume that since an input is supplied to an algorithm, it will be reproduced in the resulting numerical models. There are some common causes and comments regarding issues/problems with numerical geological models:

1. *Lack of Documentation* commonly results in future errors within a project. Good documentation as discussed in Section 3.3, Problem Formulation, is a first line of defense against future errors. Rarely do we construct the first model of a reservoir; others have almost certainly constructed some representation of the reservoir that should be reviewed.
2. *Data Interpretation and Data Management Errors* include misclassifying facies, omission of data events or entire wells from the database, or shifted / incorrect locations. These may not be dealt with through model checking. While data visualizing may assist, simulated realizations tend to mask data issues. Robust data documentation (see Section 3.1, Data Inventory) is a prerequisite, but thorough data checking is essential and outside the scope of this book.
3. *Poor Inference* of input statistical constraints is also common. For example, application of a conspicuously biased naïve data distribution will significantly bias the simulated models. Model checking may identify input contradictions due to poor statistical inference, but careful analysis with tools outlined in Section 2.2, Preliminary Statistical Concepts and Section 2.3, Quantifying Spatial Correlation is best practice.
4. *Poor Implementation* includes decisions that prevent the modeling methods from producing reasonable models. For example, overly small search limits will prevent long-range features from being reproduced in the model or too few bins will cause binning artifacts in cloud transform. Model checking will efficiently reveal these issues.
5. *Model Parameter Errors* include errors in data entry of the parameters required for the modeling method. For example, the incorrect range is entered for a variogram model or the wrong minimum value is entered for the tail extrapolation of the input distribution. Model checking will immediately expose these issues.
6. *Misunderstanding the Implementation* is common when geostatistics is applied as a block box. An example of this is applying the same random number seed for the simulation of porosity and the subsequent cosimulation of permeability; the software most likely does not know to change the seed - we have to. The artificial inflation in correlation will be detected when the porosity and permeability bivariate relationship is checked.
7. *Contradictions in Inputs* can easily occur when a variety of data sources are integrated. A common example is the combination of a high continuity primary variable cosimulated with a low continuity secondary data with a high correlation coefficient between primary and secondary variables. Clearly, these statistical inputs are contradictory and cannot be jointly reproduced in the models. This will be discovered in the model checks.

Given the complexity of the methods and large work flows, model checking is important and will save significant cost in project recycle and poor decision quality due to incorrect models. Careful checking of geostatistical models is best practice. Sometimes there are so few data that there is unavoidable reliance on a conceptual model; it may not be possible to perform many quantitative checks. At a minimum, the final model should adhere to the conceptual geological model and the data and chosen statistical parameters must be reproduced and used correctly by the estimated model or simulated realizations. Even if the inputs are highly uncertainty, it is better they are based on an explicit decision rather than poor implementation or blunder.

Data Preparation

Data assignment to grid may be complicated. As discussed in Section 2.3, Quantifying Spatial Correlation, the data should be scaled up to the assumed model support size. While porosity averages linearly and is somewhat straightforward to deal with, facies and other categorical information do not scale up in an intuitive manner and permeability does not average linearly and exhibits anisotropy that complicates scale up. Reproduction of thin beds remains a challenge.

Any comparison between the raw sample data and the values simulated in the collocated grid cell will show some scatter due to (1) the raw sample data may be outside the grid, (2) the value is inside the grid but has been identified as invalid and trimmed, or (3) there are multiple values within a cell and the values assigned to the cell are scaled-up values. Data checking must be performed prior to model checking. While mapping the data may be a valuable data check, as mentioned previously, simulation will likely hide data issues. In fact, it is not likely that data issues will be detected from the simulated realizations themselves.

For any of the following checks it is assumed that the data have been sufficiently prepared, formatted, cleaned, scaled up, and assigned to model grid nodes.

Statistical Fluctuations

Departures of simulated realizations from input statistics are expected. These are statistical or ergodic fluctuations (Srivastava, 1996). Statistical fluctuations result from uncertainty and the limited spatial extent of the reservoir; there is no infinite domain for the random high and low values to completely cancel out. As a practical consequence, the realization statistics should converge to the input statistics as the size of the reservoir becomes large relative the range of spatial continuity.

These statistical or ergodic fluctuations are a component of uncertainty modeling (see discussion in Section 5.3, Uncertainty Management) (Leuangthong et al., 2004). Some modeling methods, such as cloud transform and multiple-point simulation, regularly perform distribution transformation to suppress fluctuations and ensure precise reproduction of input statistics. Regardless of how these fluctuations are interpreted and used, the acceptable level of ergodic fluctuations can be calculated to aid in model checking.

Deutsch and Journel (1998) discussed the effect of domain size and the variogram model on ergodic fluctuations; Goovaerts explored the magnitude of ergodic fluctuations and the space of uncertainty from four different simulation algorithms (Goovaerts, 1999); Srivastava touched on the ability of simulation to fairly sample from the space of uncertainty (Srivastava, 1996, page 60); Chilés and Delfiner mentioned the use of the integral range as a measure of practical ergodicity (Chilès and Delfiner, 2012; Lantuéjoul, 2011).

Dispersion variance, introduced in Section 2.3, Quantifying Spatial Correlation, provides another way to "predict" the amount of ergodic fluctuations one should expect. This is expressed as the dispersion variance of the domain relative to the assumption of an infinite domain, $D^2(A, \infty)$; this can be calculated numerically from $\overline{\gamma}$ values or directed calculated from model realizations with the correct spatial continuity. If realization statistics exceed this expected range of ergodic fluctuations, then the problem formulation and model parameters should be checked.

Preliminary Checking

Clearly, the data assigned to model cells must be reproduced at those cell locations. This is relevant to both estimation and simulation. Estimation methods may reproduce trend models and in the case of kriging provide local distributions of uncertainty that may be checked. Aside from these constraints, estimation methods do not reproduce the histogram or the model of spatial continuity; therefore, checks for estimation models are limited. For simulated realizations, all input statistics should be checked, including variogram, correlation coefficient, and multiple-point statistics.

As discussed above, in some cases the expected fluctuations of statistics in the simulated realizations are known. Either the method applies a correction to enforce reproduction or the expected fluctuations may be calculated. This may be applied as a guide for the acceptability of model realizations.

Tools such as the bootstrap (discussed in Section 2.2, Preliminary Statistical Concepts) may be applied to the data set to calculate uncertainty in the input statistic. This may be applied to provide a standard for judging the acceptability of statistical fluctuations in the models.

Aside from this, good judgment is required. A decision on acceptable fluctuations from the input statistics should be based on knowledge of the certainty of these statistics and the impact of their associated fluctuations on the transfer function.

Finally, the true distribution in the reservoir will remain unknown and it is impossible to completely validate or verify a numerical geological model (Oreskes et al., 1994). Nevertheless, geological models can be subjected to a series of tests that increase their credibility and identify significant bias, rather than verify their correctness; this is deemed model confirmation (Oreskes et al., 1994). This is our goal in this section.

Nonstationarity Models

Reservoirs are nonstationary; many of the statistical parameters used in modeling may be locally variable and accounting for this local variation, or nonstationarity, may be important to achieve models with high local accuracy. Nevertheless, the use of nonstationarity models such as locally variable means, proportions, azimuths, scale, and so on, complicate the task of model checking. For example, in the discussion below, it is proposed that the variogram be checked between the input and realizations. In the presence of a locally variable azimuth model or even trend model, the global variogram will not match the input variogram. In order to check the variogram, the effect of the locally variable azimuth or trend model must be removed; therefore, the variogram should be calculated with the search template aligned to the local azimuth on the residual.

In addition, the reproduction of the nonstationarity model may be checked. This is best accomplished with moving window calculations of statistics such as means, proportions, and correlation coefficients over the realizations. While this is useful, it should be noted that local conditioning will update these nonstationarity models. That is, the expected value over realizations will approach data values near data locations. Data locations should be identified in these nonstationary model checks.

5.1.2 Minimum Acceptance Checks Other Applications

Various model checking methods are provided that quantify and summarize realizations for the purpose of model checking. Yet, these quantifications have other possible applications, including (1) comparing modeling methods with each other to determine the method best able to reproduce specific statistical inputs, (2) simple ranking of realizations (see Section 5.3, Uncertainty Management), (3) determining the fitness of realizations for specific purposes, and (4) helping select input parameters.

It may be challenging to select the specific modeling method that is best matched to meet the project objectives. These tests may provide an indication of performance of various methods to honor statistical inputs that are deemed important. For example, if the porosity and permeability bivariate relationship details are determined to be important to flow simulation, checks on the reproduction of this relationship may determine if Gaussian cosimulation is sufficient or if cloud transform is required.

A simple form of realization ranking could be based on realization summary statistics. For example, the averages of porosity realizations may be related to oil in place, or permeability variance may be related to flow heterogeneity.

It may be possible that individual realizations may be determined to be unfit and discarded from the suite of realizations that represent the uncertainty model. These models may have features and/or statistics that are not plausible. Model checking provides an opportunity to identify these realizations.

It may be difficult to assess the appropriate model parameters to produce reasonable models. For example, competing data and statistical inputs may confound the reproduction of specific statistical inputs. Iterative modeling with modeling checking provides an opportunity to tune the model parameters *to improve model performance.*

The concept of minimal acceptance was developed by Leuangthong et al. (2004). This subsection is a summary of this work. Minimal acceptance criteria are confirmation that a model acceptably reproduces the available data and input statistics. This does not imply that the model is geologically realistic or good for production forecasting. Neither does it suggest the model will perform well for forecasting at locations away from the data (this will be covered in the Cross Validation and the Jackknife subsection). Minimal acceptance only checks that the following inputs are reproduced:

1. Data values at their location
2. Distribution of the variable of interest

3. The spatial continuity characterized by the variogram model
4. The bivariate relationship characterized by correlation coefficient or full scatterplot

As discussed above, statistical fluctuations are expected and input statistics are reproduced in expected value considering that the available data may update or modify the expected value. Also, in the case of multivariate simulation, the multivariate distributions and corresponding summary statistics should be reproduced within acceptable fluctuations.

Specific implementation details may cause problems. In such situations, a careful examination should be conducted to confirm the cause of violation, and whether this is acceptable. Some of these checks can be performed on individual realizations while others require consideration of the full suite of multiple realizations.

Visual Inspection

The first check on a set of realizations should be visualization. This visualization should highlight low- and high-valued areas, trends, and continuity. The project team should be satisfied with these features. The variability and consequent uncertainty should be reasonable and plausible; for example, there should be no high values in clearly low areas and vice versa. The models should be compared to the conceptual model and data constraints. Comparisons against simple geological contours of trends, generated by methods such as hand contouring, inverse distance, and other estimation techniques, provide a level of comfort and confidence in the simulation models.

Fortunately, modern 3-D model software allow for efficient model visualization, along with various operations such as slicing, filtering, and segmenting. It is essential to take advantage of these tools to fully understand the models. An example porosity model is shown in Figure 5.1 for the Carbonate Reservoir. We assumed that diagenetic alteration has significantly disrupted facies control on porosity and directly modeled porosity with a trend model related to the energy associated with the reef. A plan section and five dip sections are shown through the model and trend model.

We need to recognize the limitations in visualization. As mentioned previously, simulated realizations may hide model issues. To understand a geological model, one must actually consider all realizations jointly. Post-processing methods discussed in the next section may be useful to summarize over realizations. Also, due to vertical exaggeration and limited model resolution or large scale of the models, features in the model may not always appear geologically plausible. Finally, we are often expecting more continuity than is reasonable; reservoirs are large, and geological variability will surely exist.

Data Reproduction

The models must honor the data at the data locations. A model that fails to do this will obviously lack credibility, but will also likely encounter significant issues with flow prediction. For example, perforated intervals may not line up with reservoir quality rock. Data reproduction may be related to issues in (1) assignment to grid and (2) conditioning in the simulation method.

As mentioned previously, with model checking, it is assumed that data have been interpreted correctly and have been assigned to the grid appropriately. Checks for data reproduction should be conducted on the data assigned to cells to investigate the performance of the model not the interpretation.

Cell-based modeling methods should not encounter data reproduction issues. The kriging system explained in Section 2.4, Preliminary Mapping Concepts, has the property of exactitude. Data are estimated at the data locations with zero kriging variance. It follows that simulation based on kriging will draw the data values at the data locations. In addition, the sequential simulation methods for Gaussian, indicator, and multiple-point methods initialize the grid by assigning data to data locations and do not resimulate at these locations. If these methods fail to reproduce the cell assigned data at the specific cell locations, then a significant implementation mistake has occurred, such as data mishandling or unintentional use of block kriging.

Object-based, process-mimicking, and optimization methods do not guarantee data conditioning. In fact, given the complexity of honoring potentially dense data with complicated geometries, it is likely that the placement routine (that is often optimization based) will not completely converge and honor all data. With these methods it is essential to check data reproduction. If this is a problem, then adjustment on the simulation routine or a post-process conditioning correction (see Section 5.2, Model Post-processing) may be required.

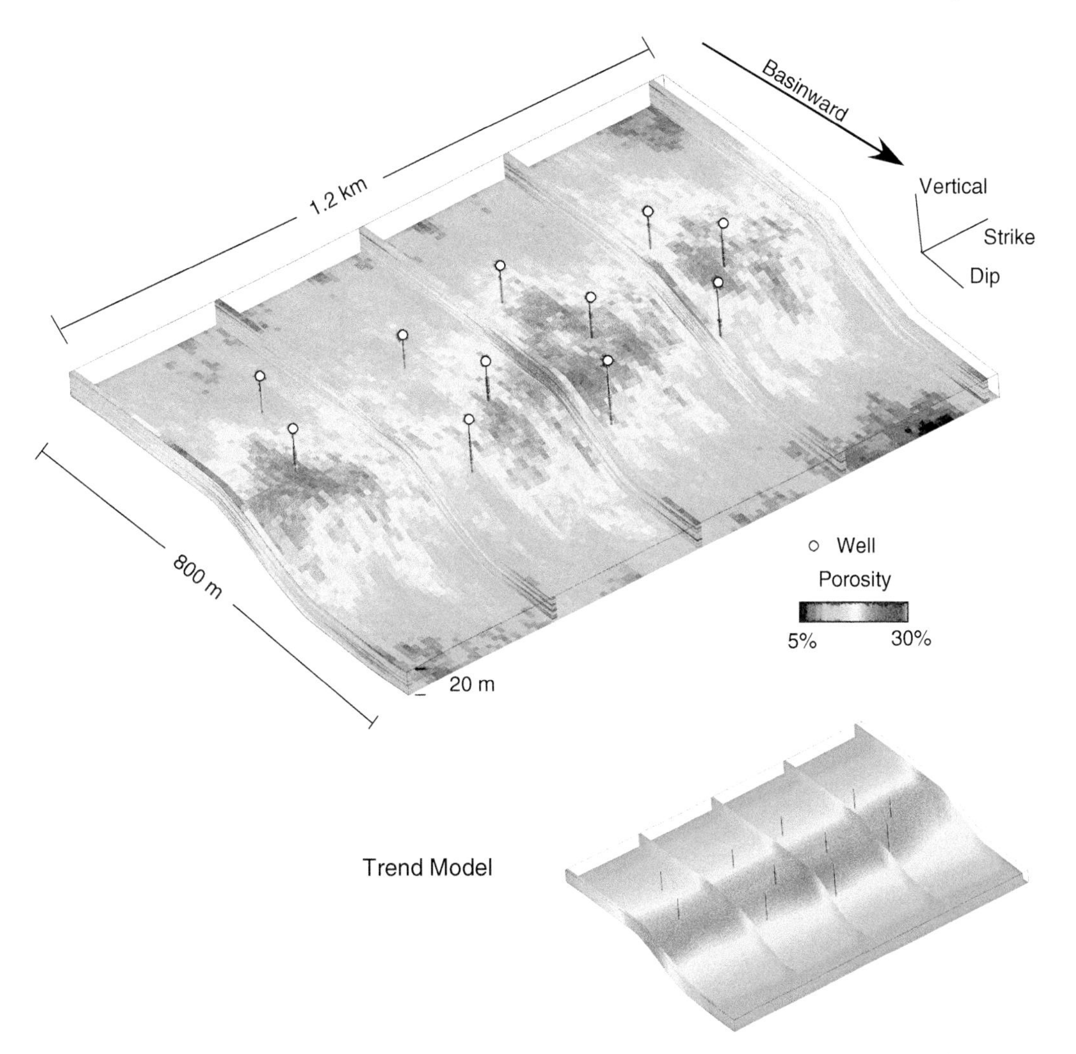

FIGURE 5.1: Oblique View of a Porosity Realization, the Associated Trend Model and Well Data Logs. The sections include a plan section at the base of the model and five equally spaced dip sections.

To verify hard data conditioning, a cross plot of the data assigned to model cells and the simulated values at the data assigned cells should be generated. All points in the resulting cross plot should lie perfectly on the 45° line. Any cell with associated points that fall off of the 45° line should be located and investigated. The cause of this failure to match hard conditioning should be explained and corrected.

Finally, the data should be reproduced without artifacts near the data. In other words, if the data locations were not labeled on the model, it should not be possible to infer their locations. Common artifacts include discontinuities or data being honored as local minima or maxima. The practical check for this issue is visual inspection of various slices through the model that intercept the data.

Histogram Reproduction

Another minimum acceptance check is to verify that the histogram is reproduced. As discussed Section 2.2, Preliminary Statistical Concepts, the histogram is a first-order model control. Significant effort must be conducted to ensure that the input histograms are representative of the volume of interest. With all simulation methods the input histogram is identified. This may be implicit through a histogram from the input data (although this raw histogram is not typically representative) or explicitly through a set of data and declustering weights or a modeled reference histogram. In some cases, such as stepwise conditional transform (see Section 2.4, Preliminary Mapping Concepts) and cloud transform (see Section 4.6, Porosity and Permeability Modeling), the histogram may be constrained through a conditional

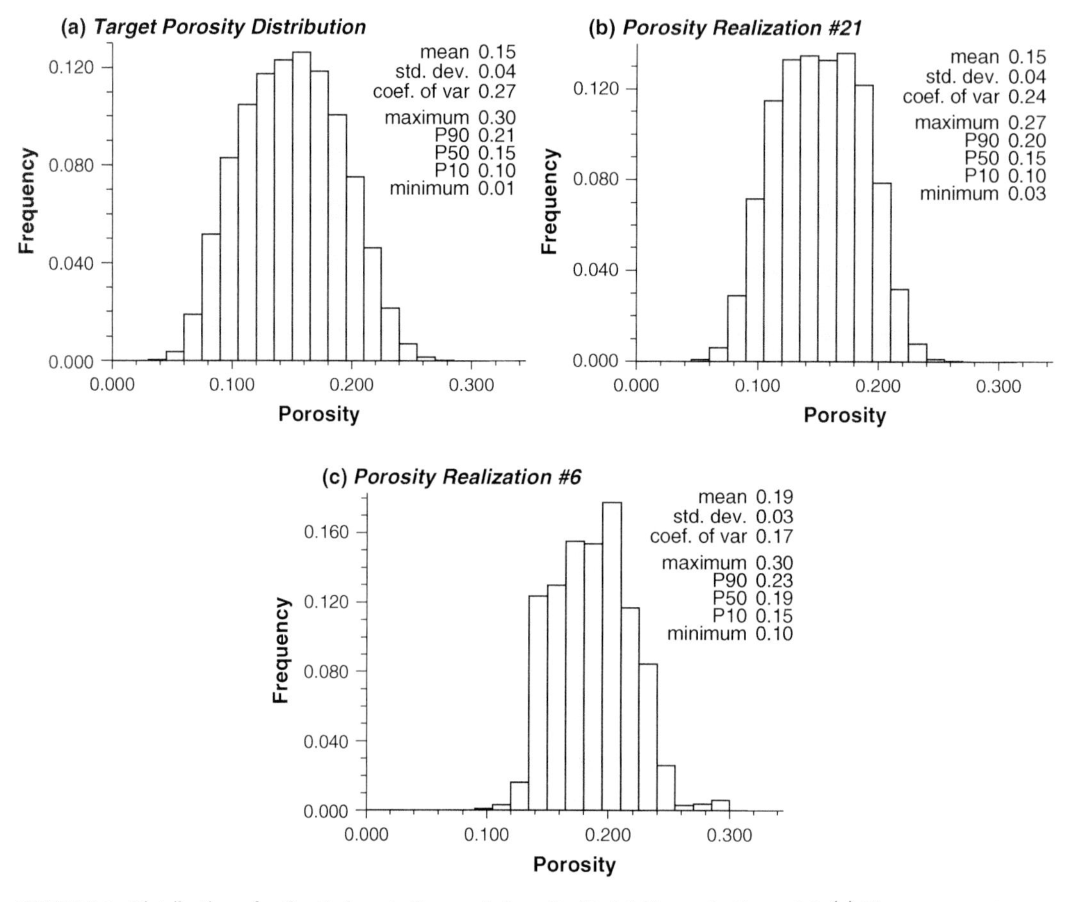

FIGURE 5.2: Distributions for the Carbonate Reservoir Porosity Model Shown in Figure 5.1. (a) The representative porosity distribution. (b) The distribution from a realization that matches the target well. (c) The distribution from a realization that exhibits a poor match due to implementation issues related to trend mishandling and poor tail extrapolation.

transform. In one form or another, the histogram is an input statistic that should be reproduced.

To verify global reproduction of the histogram, the histogram of the model should be examined. Direct inspection of realization histograms is important. While it may be impractical to examine the histograms from all realizations, a few randomly selected realizations should be checked. Checking a couple tabulated or averaged summary statistics could miss significant histogram reproduction issues such as unreasonable values and unexpected histogram shape.

In this type of visual checking, the key features to note include reproduction of (1) the histogram shape, (2) the range of the simulated values, and (3) the summary statistics, such as the mean, median, variance, min, and max. For the Carbonate Reservoir, the input porosity histogram and histograms for two realizations are shown with good and poor histogram reproduction (see Figure 5.2).

A quantile–quantile (Q–Q) plot (see Section 2.2, Preliminary Statistical Concepts) may provide a better indication of histogram reproduction, because binning may hide some features in the histogram. This type of check permits multiple realizations to be visualized at once. These plots may be viewed side-by-side or with multiple realizations posted on the same plot for comparison, to assess whether the suite of distributions honors the input histogram with reasonable statistical fluctuations (see Figure 5.3 for the Q–Q plots for the carbonate porosity distribution in Figure 5.2).

Trend Reproduction

Trend models represent the model of local expected value in a reservoir property. It is straightforward to check trend reproduction. Boisvert (2010) proposed checks for reproduction of the trend: (1) globally for a single realization or (2) locally over a set of realizations. It is often more straightforward to understand

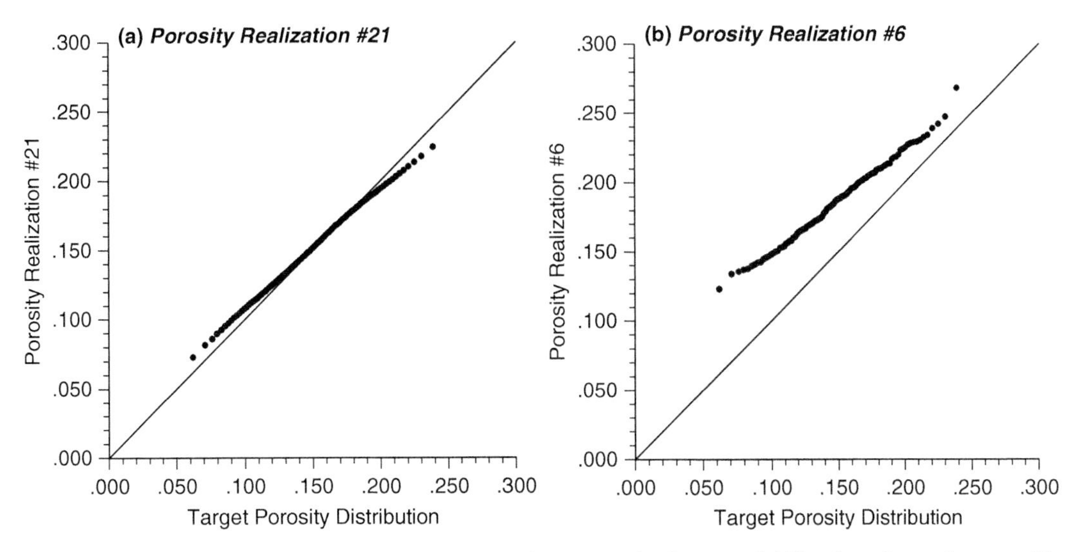

FIGURE 5.3: Q–Q Plots for the Two Realizations Distributions Shown in Figure 5.2. (a) The plot indicates the reasonable match for the realization. (b) The plot indicates various problems including a significant increase in mean, decrease in variance in the realization and issues with the tails.

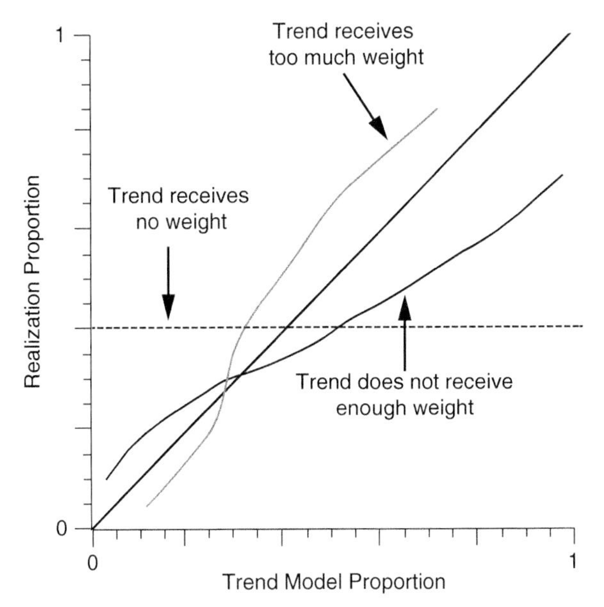

FIGURE 5.4: Illustration of a Global Trend Check and Interpretation of Results. The realization proportions are compared to the trend model proportions to check trend model reproduction.

trends than to assess the validity of the short-scale variability of the reservoir models.

Evaluating the global reproduction of the trend for a single realization ensures that the trend is reproduced in expected terms over the realization. For example, all locations where the categorical trend map indicates a proportion between 0.2 and 0.3 can be examined. For a single realization the proportion in these locations should be about 0.25. The expected trend proportion from the trend model (0.25 in this case) can be compared to the actual proportion in the realization.

An illustration of the results of this test is shown in Figure 5.4. The simulation implementation may place too much emphasis on the trend. As a result, when the trend proportion is high, the realization proportion is even higher and when the trend proportion is low the realization proportion is even lower. Conversely, if the simulation implementation places too little emphasis on the trend, the

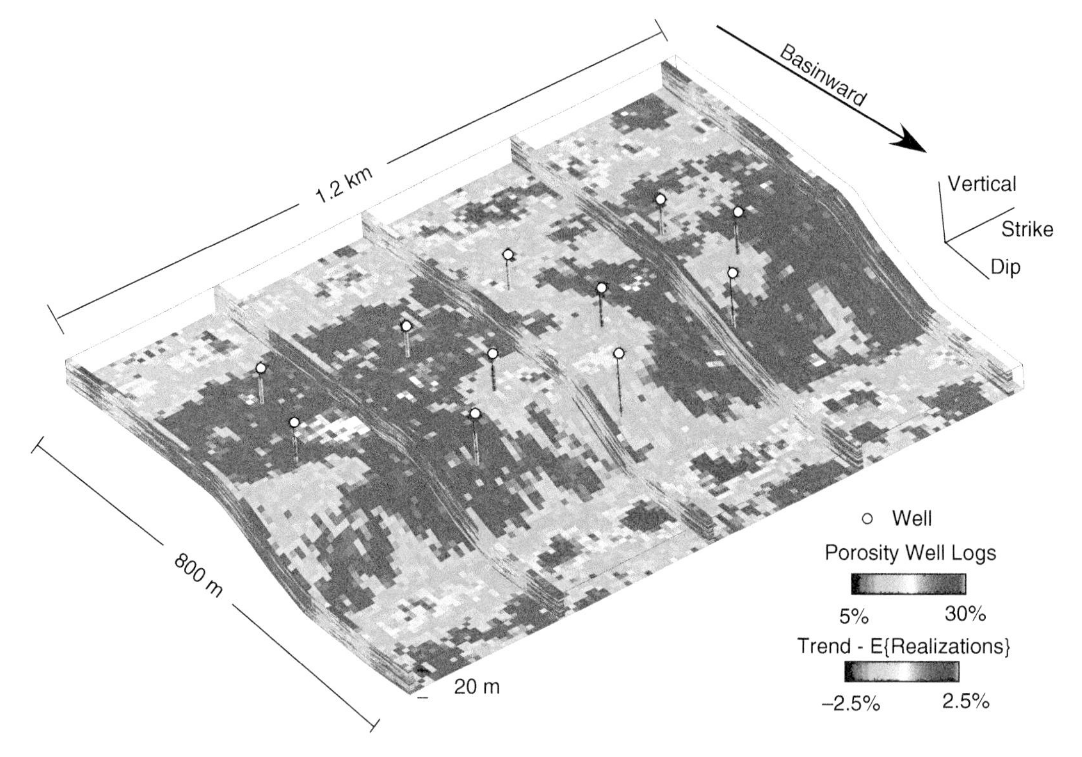

FIGURE 5.5: Oblique View of the Residual of the Average of 10 Realizations Subtracted from the Trend Model. The residual is displayed with an adjusted color scale to show features. The impact of the conditioning data can be seen with generally higher values than the trend near the well data. In general, the trend reproduction is good, but could be improved by better calibration of the trend model to well data.

results all approach the global proportion or mean for all trend bins.

Assessing the local trend ensures that on average the realizations honor a local trend at a specific location. The average proportion over all realizations at each location is compared to the actual trend proportion by calculating the local difference.

$$m(\mathbf{u}) - E\{Z^i(\mathbf{u})\}, \qquad i = 1, \ldots, L$$

where $m(\mathbf{u})$ is the local mean or proportion and there are L realizations. Significant departures may be expected near data or with significant secondary constraint information. These should be investigated and explained. This check was applied to 10 porosity realizations from the Carbonate Reservoir Model (see Figure 5.5).

Variogram Reproduction

Given the data assigned to model cells are reproduced and the histograms are correct, the spatial continuity model is checked. Checking variogram-based spatial continuity models is discussed here, while multiple-point spatial continuity is discussed in the Higher-Order Checks Section.

For Gaussian simulation, it is important to note that this check must be performed in Gaussian space with trend removed, since only the normal scores variogram of the residual is imposed directly. Recall that the variogram of the normal scores of the data is required for Gaussian simulation. This can be accomplished by either suppressing the back transform of the simulation algorithm or transforming the simulated realizations to standard normal prior to checking the variogram. Once spatial continuity is checked in Gaussian space, it may be instructive to compare the variogram of the data with the realizations in real space (not transformed to standard normal). Departure of these results includes issues related to the acceptability of the Gaussian assumption.

The variogram should be calculated for multiple realizations and compared to the input variogram model in the same direction (see Figure 5.6). The model variogram should be reproduced within acceptable ergodic fluctuations (see previous discussion). If greater rigor is required, Ortiz and Deutsch

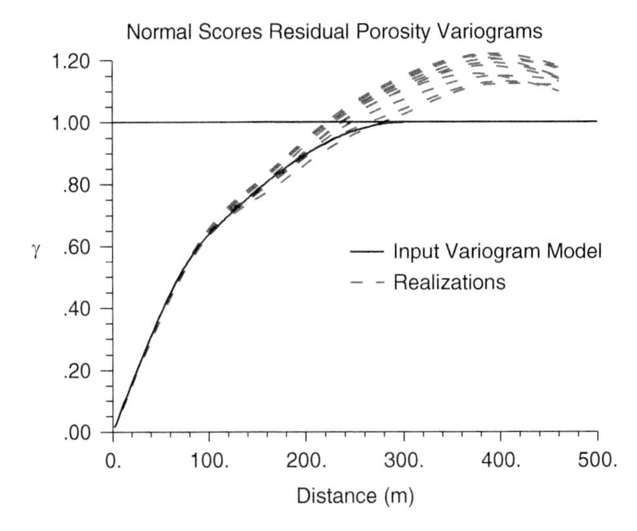

FIGURE 5.6: The Experimental Variograms Calculated from the Normal Score Transform of the Residual of 10 Realizations (Trend Removed) Compared with the Input Model Variogram. Note the fluctuations are limited and reproduction is good. Some of the data-constrained trend remains in the residuals, resulting in the trend feature.

(2002, 2007) and Rahman et al. (2008) provide a review of various methods to quantify acceptable fluctuations in the variogram and proposed a hypothesis test for variogram reproduction.

Reproduction of Bivariate Relationships

Commonly, a problem formulation includes multivariate relationships constrained with a sequence of bivariate cosimulations. For example, the following method is commonly applied, porosity simulated within facies, and permeability cosimulated with porosity. In this case of cosimulation with collocated cokriging, the correlation coefficient of the Gaussian transform of the variables is an input statistic. The paired realizations of each variable should be transformed to standard normal and the correlations coefficients should be checked.

In addition, the real space realizations may be cross plotted. This will explore issues related to the assumption of bivariate Gaussianity after the univariate transform of each variable. If this check revels significant issues that may negatively impact the transfer function, then the choice may be made to change the modeling method from Gaussian cosimulation to a method that reproduces the bivariate relationship, such as cloud transform.

For the cloud transform, the above-mentioned scatter plot of the real space paired realizations should be checked for departures from the modeled scatter plot. Typical issues include binning artifacts and anomalous values near the tails. If binning issues are significant, the choice may be made to densify the cloud model and assign more conditional bins to smooth out the influence of binning. If anomalous values exist near the tails, an effort may be required to ensure that the cloud model sufficiently covers the range of each variable.

5.1.3 High-Order Checks

Higher-order checks may be required. For example, multiple-point methods reproduce features beyond the histogram and the variogram. Given the greater degree of control of spatial heterogeneity available with multiple-point geostatistics, checking these higher-order statistics will be even more important to ensure a reasonable reproduction of input statistics. As mentioned in Section 4.3, Multiple-Point Facies Modeling, there is no guarantee that the MPS implementation will reproduce the features in the input training image. One of the most important checks is visual inspection and comparison of the input data, training image, and individual realizations.

Also, nonstationarity models are commonly applied with MPS to assist with the strong stationarity assumption of the training image or even multiple training images may be applied. When these models are present, they should be considered jointly in the above visual inspection of the input data and individual realizations.

Quantification is more challenging. Boisvert (2010) attempted to extend the minimum acceptance criteria of Leuangthong et al. (2004) to include a set of higher-order checks that may be useful for MPS.

The multiple-point statistic is a difficult criterion for judging the acceptability of a model. It is not possible to visualize or summarize multiple-point

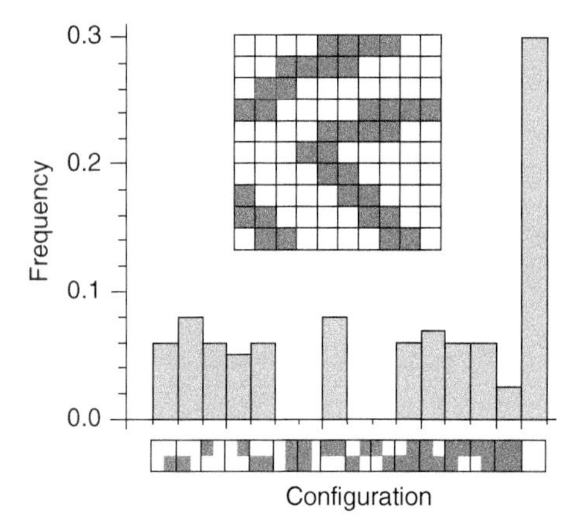

FIGURE 5.7: A Multiple Point Density Function of the Example TI. This example demonstrates a four-point configuration with two categories (Boisvert et al., 2007a).

statistics due to the very high dimensionality (recall that there are K^{n+1} conditional probabilities for a multiple-point statistic, where K is the number of categories, n is the number of points of known locations and +1 represents the unknown location) and there is no straightforward ordering relationship for these conditional probabilities. To demonstrate this, consider the full multiple-point density function for a simple four-point statistic with two categories shown in Figure 5.7. Note each bin in the multiple point density function represents the frequency of a specific multiple point configuration.

Boisvert (2010) presents alternative metrics and summary statistics that can be used to compare the multiple-point statistics of input training images and the associated simulated realizations, including:

1. Multiple-scale histogram
2. Multiple-point density function
3. Missing bins in the multiple point density function

Multiscale Histogram

The multiscale histogram refers to the suite of histograms over various support sizes. Each histogram is calculated by scaling up the original model to a new support size. In the case of categorical models, the property immediately scales to continuous proportions of each category. The behavior of the multiscale histogram is dependent on the underlying spatial structure of the training images and realizations. Short-scale variability and the absence of trends result in rapid decrease in variability with increase in support size.

One method is to visually assess and compare the behavior of the scaled histogram of the training image to that of the realizations. Another is to quantify the change in the proportion distributions over various scales for the training image and the realizations. Figure 5.8 shows such a comparison specified further to proportion in the upper and lower tails for two MPS implementations (one with 4-point and the

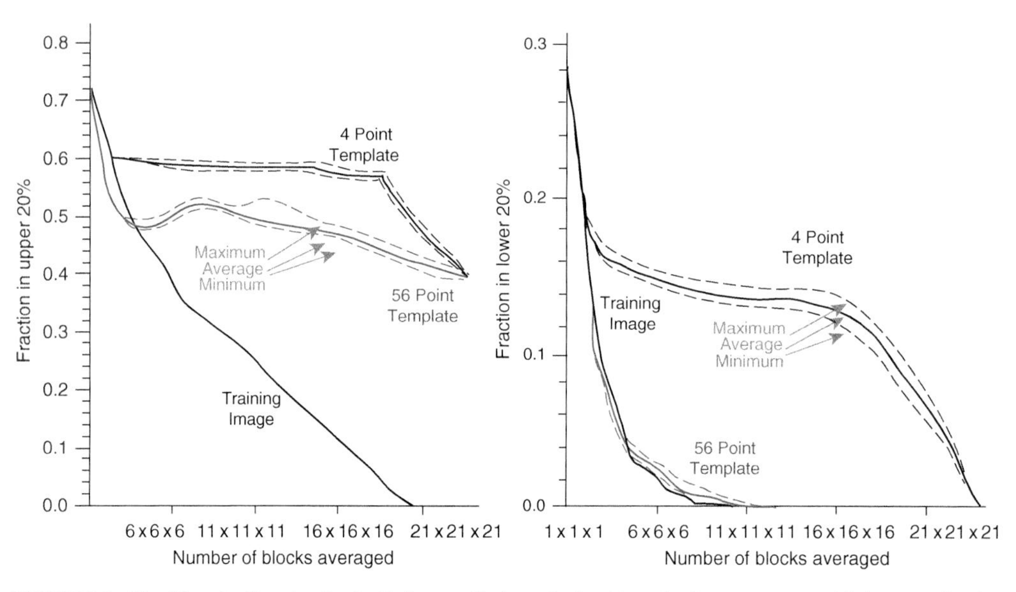

FIGURE 5.8: Checking the Fraction in the Tails over Various Scales. Note the improvement with increase in size of multiple point template from 4 point to 56 points (Boisvert et al. 2007a).

other with 56-point templates). The greater number of points results in scaling behavior closer to the training image.

Multiple-Point Density Function

Checking the multiple-point density function is analogous to checking the variogram for a variogram-based method. As mentioned previously, checking the multiple-point density function is complicated by its high dimensionality and lack of intuitive ordering. Boisvert et al. (2007b) proposed a difference measure between two multiple-point density functions.

$$\delta = \sum_{i=1}^{n} | g_i^{\mathrm{TI}} - g_i^{l} |$$

where δ is the difference between multiple-point distributions, i is the bin index, n is the total number of bins, and g_i^{TI} is the conditional probability from the training image and g_i^l is the conditional probability for the realization, l for the bin, i. The authors noted the strong influence of empty bins, multiple configurations not included in the training image of the realization and suggest a variant that omits bins without values for either the training image or the realization.

While this comparison is useful for assessing multiple-point characteristics, it is important to realize that this difference measure reduces a very-high-order statistic to a single value. Much information contained in the multiple-point density function is lost. An example comparison is shown for various MPS realizations with three different template sizes on the left of Figure 5.9.

Missing or Zero Bins

Various other summaries may be considered to extract more information from the multiple-point density function. For example, the absence of specific multiple-point configurations (a zero bin) may also indicate important information concerning the spatial heterogeneity. However, zero bins in the multiple-point density function may also be the result of a limited training image size. An assumption is made that the training image is representative of the modeling domain; therefore, if a configuration does not exist in the TI, it should not appear in the realizations.

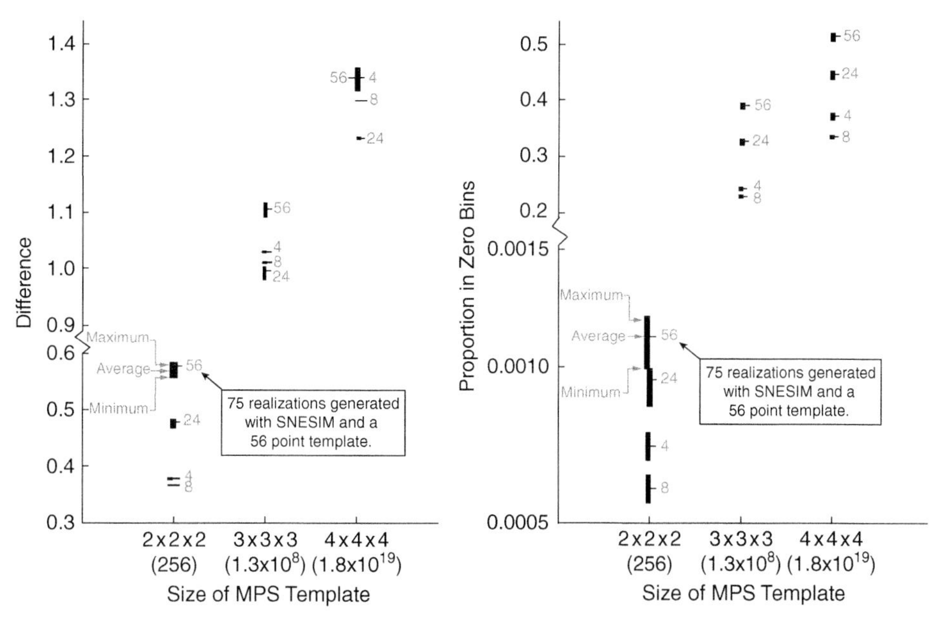

FIGURE 5.9: Multiple-Point Comparisons Based on the Multiple-Point Density Functions. The X axis is the size of the multiple-point template used in the calculation, and the number of nodes used in the MPS realizations are indicated for each set of realizations. On the left, the summed difference between realizations and the training images are shown. On the right the proportion of implausible configurations (those not found in the training image).

On the right-hand side of Figure 5.9, the proportion of zero bins is shown for multiple realizations of MPS with various multiple-point template sizes. These are multiple-point configurations found in the realizations that are not present in the training image.

5.1.4 Cross Validation and the Jackknife

This subsection describes some validation procedures for "estimation". The next subsection describes a complementary set of procedures to validate "simulation" models of uncertainty.

The basic idea is to estimate the attribute value (e.g., porosity) at locations where we know the true value. Analyzing the error values indicates the *goodness* of our modeling parameters (Davis, 1987; Efron, 1982). In cross validation the actual data are deleted one at a time and re-estimated from the remaining neighboring data. The term jackknife applies to resampling without replacement—that is, when an alternative set of data values is re-estimated from another non-overlapping data set. The jackknife is a more stringent check since the non-overlapping data are not used to establish the statistical parameters such as the histogram and variogram. Also, the idea behind the jackknife is that it should be repeated with different non-overlapping data sets to filter out the statistical fluctuations of choosing one data set.

The jackknife requires a significant amount of CPU and professional effort to repeat the estimation procedure, including calculation and fitting of all variograms, with different sets of data; therefore, the jackknife is not commonly used. Implementation of cross validation requires reasonable care to make estimation as difficult as it will be in practice. For example, each well must be removed in turn; removal of each sample individually would lead to unrealistically good results due to very close adjacent samples. Cross validation is only practical with some reasonable number of wells, at least five; otherwise, there are too few data for checking.

In general, the real unsampled locations have the benefit of all data, and the data spacing for estimation on a grid will be less. The distance to the data is quite large in cross validation; consider leaving one data out at a time. When estimating on a grid, some of the locations are right by the well locations and even in between the data, the wells are closer than in cross validation. The disparity between data spacing in checking and real estimation may be even larger in the jackknife. Careful evaluation of the results is still relevant; blunders and problem data could be detected and the overall quality of prediction is assessed.

Visualization of Cross Validation/Jackknife

The most informative display is a cross plot between the pairs of estimates and true values. Any standard cross plotting software could be used, but there are some important considerations in making a good display: (1) The estimate should be the independent variable (X or abscissa axis) because we will have access to the estimates and not the truth, (2) the units of both axes should be the same and the 1:1 line should be shown, (3) the regression of the truth on the estimates should be shown to provide an indication of conditional bias, (4) relevant summary statistics should be shown, and (5) the worst cases of over- and underestimation should be identified with the data event index so that they can be checked. An example of this plot is shown in Figure 5.10.

Checking summary statistics alone would not be as informative as checking the plot. We are looking for patterns or particular values that are anomalous. Some summary statistics are of interest. The means of the true and estimated values tells us of any systematic bias; they should be the same. The standard deviations of the true and estimated values tells us of the smoothing effect. The mean squared error (MSE) is a common summary of prediction performance; the MSE should be small. The covariance between the estimate and truth is another useful measure; it should be as high as possible for a good estimate. The correlation coefficient is also useful, but it is sensitive to smoothing and a low variance of the estimates. A smooth estimator will have a higher correlation with the truth, but that is not necessarily a good thing. The slope of the regression line should be close to one. A slope less than one would indicate conditional bias. This may be acceptable for interim estimates where smoothing is an even larger problem. The practitioner will have to assess these statistics in the context of their problem.

Categorical variables are different. Assuming each category or facies one at a time, the true values are only 0 or 1. The estimates are continuous values between 0 and 1, representing the probability estimated for each category. The plot is somewhat different. Some considerations include: (1) The 1:1 line is not relevant, (2) the regression line is interesting, but not as relevant as with

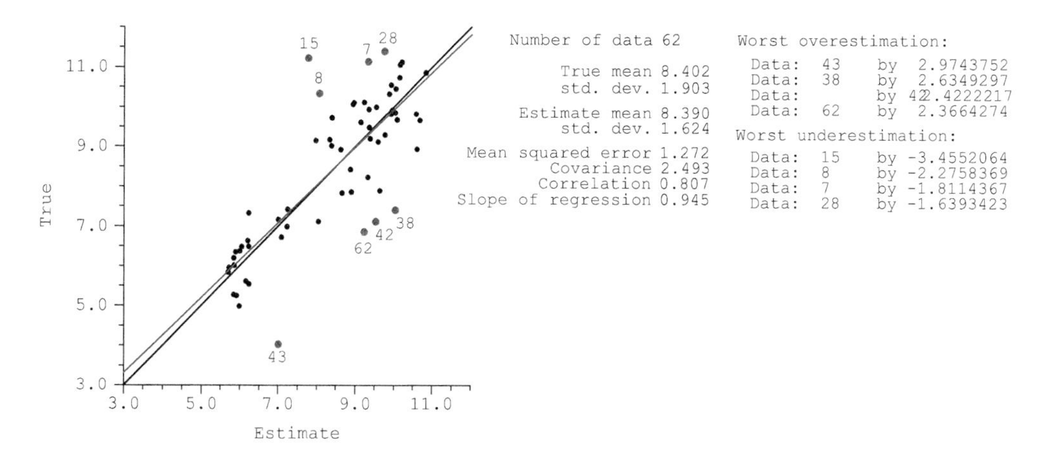

FIGURE 5.10: Plot That Provides Useful Summaries for Cross Validation or Jackknife of a Continuous Reservoir Property from Deutsch and Begg (2001). Regression of truth on estimates is shown as a red line. The outlier data are identified with red dots and data index.

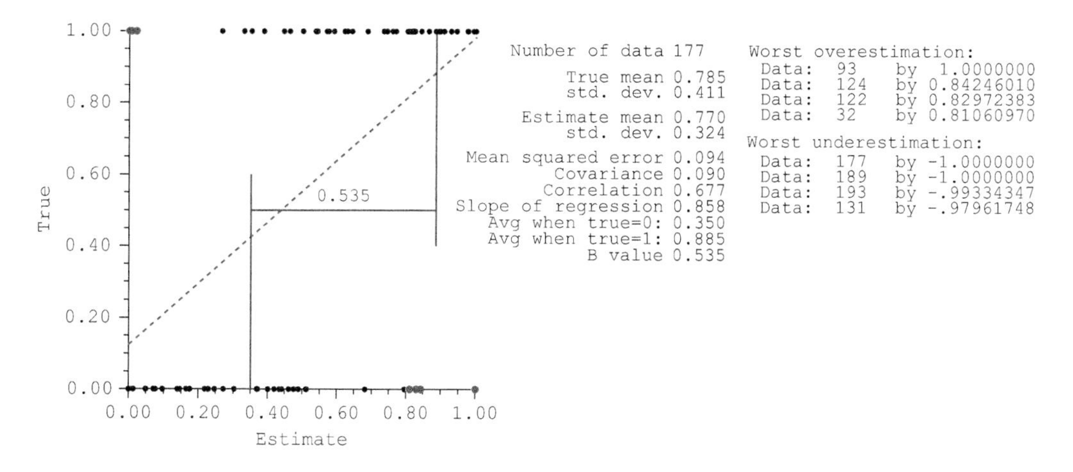

FIGURE 5.11: Plot That Provides Useful Summaries for Cross Validation or Jackknife of a Categorical Reservoir Property (Deutsch, 2010b). The points represent the paired true indicators versus the probability estimated for that indicator. The regression line of truth on estimates is shown as a red dashed line. The mean of estimates for $I(\mathbf{u}) = 0$ and $I(\mathbf{u}) = 1$ are shown with vertical blue lines, and the difference (B value) is shown above the horizontal blue line. Outliers are indicated as red dots with data indices for retrieval and investigation. Note that outlier indice labels are removed from this figure because they were not readable due to overlap and plotting scale.

continuous variables, and (3) the average of the estimated probabilities conditional to the true values are important. The difference between those averages is identified as "B" value reminiscent of the B_j calibration parameter in the Markov–Bayes model presented in the Section 4.6, Porosity and Permeability Modeling, for permeability indicator simulation with soft porosity indicators. Recall that the B_j close to one indicates perfect ability of the soft secondary indicator to inform the primary variable. In this case, B is an important measure of the quality of the estimates. A large value is good, which means that the presence/absence of the category is predicted correctly. An example of this plot is shown in Figure 5.11.

There are other displays that could be useful. An error value can also be calculated for each re-estimated location:

$$e(\mathbf{u}_i) = z^*(\mathbf{u}_i) - z(\mathbf{u}_i)$$

These errors can be analyzed in different ways: (1) The distribution of the errors $\{e(\mathbf{u}_i), i = 1, \ldots, n\}$, should be symmetric, centered on a zero mean, with a minimum spread; (2) the mean-squared error $\text{MSE} = 1/n \sum_{i=1}^{n} [e(\mathbf{u}_i)]^2$ should be minimum; (3) the plot of the error $e(\mathbf{u}_i)$ versus the estimated value $z^*(\mathbf{u}_i)$ should be centered around the zero-error line, a property called "conditional unbiasedness"; and (4) the n errors should be independent from each other. This can be checked by contouring or posting the error values. Areas or zones of systematic over- or underestimation are a problem and, perhaps, the trend model should be looked at more carefully.

The results of cross validation must be used with care. Cross validation does not prove the optimality of a given estimation scheme. The analysis of re-estimation scores does not consider the *multivariate* properties of the estimation algorithm, that is, of the estimates taken all together. Cross-validation results are also insensitive to the variogram at distances less than the sample spacing, which is an important unknown aspect of estimation. Cross validation should be used to check for mistakes and identify problem data. The techniques to check estimation are straightforward. In the context of simulation, however, there is no single estimate; there are a potentially large number of simulated values. A different approach must be considered.

5.1.5 Checking Distributions of Uncertainty

It is impossible to rigorously validate a model when applied to unknown values (Oreskes et al., 1994). The true and unknown reality will certainly not follow our relatively simplistic probabilistic models and may not be reliably informed by the data we have at hand. Nevertheless, we can use cross validation to assess geostatistical models of uncertainty.

The result of cross validation and the jackknife applied in a simulation setting is a set of true values and paired distributions of uncertainty, $\{z(\mathbf{u}_i), F_Z(\mathbf{u}_i; z), i = 1, \ldots, n\}$. Different modeling decisions would lead to different distributions of uncertainty—that is, different $F_Z(\mathbf{u}_i; z)$'s. These local distributions or ccdf models may be (1) derived from a set of L realizations, (2) calculated directly from indicator kriging, or (3) defined by a Gaussian mean, variance, and transformation. Checking distributions of uncertainty or comparing alternative approaches requires an assessment of the *goodness* of distributions of uncertainty (Deutsch, 1996b).

In the context of evaluating the goodness of a probabilistic model, specific definitions of accuracy and precision are proposed. For a probability distribution, accuracy and precision are based on the actual fraction of true values falling within symmetric probability intervals of varying width p:

- A probability distribution is accurate if the fraction of true values falling in the p interval exceeds p for all p in $[0, 1]$.
- The precision of an accurate probability distribution is measured by the closeness of the fraction of true values to p for all p in $[0, 1]$.

A procedure for the direct assessment of local accuracy and precision is now described. In checking the goodness of a probabilistic model, we can compare alternative models and perhaps fine-tune the parameters of a chosen model.

Direct Assessment of Accuracy and Precision

The first step is to calculate the probabilities associated to the true values $z(\mathbf{u}_i), i = 1, \ldots, n$, using the model of uncertainty:

$$F^*(\mathbf{u}_i; z(\mathbf{u}_i)|n(\mathbf{u}_i)), \qquad i = 1, \ldots, n$$

For example, if the true value at location $\mathbf{u}_i$ is at the median of the simulated values, then $F(\mathbf{u}_i; z(\mathbf{u}_i)|n(\mathbf{u}_i))$ would be 0.5.

Then, a range of symmetric p-probability intervals (PIs)—say, the percentiles 0.01 to 0.99 in increments of 0.01—can be considered. The symmetric p-PI is defined by corresponding lower and upper probability values:

$$p_{\text{low}} = \frac{(1-p)}{2} \quad \text{and} \quad p_{\text{upp}} = \frac{(1+p)}{2}$$

For example, for $p = 0.9$ we have $p_{\text{low}} = 0.05$ and $p_{\text{upp}} = 0.95$. Next, define an indicator function $\xi(\mathbf{u}_i; p)$ at each location $\mathbf{u}_i$ as

$$\xi(\mathbf{u}_i; p) = \begin{cases} 1, & \text{if } F(\mathbf{u}_i; z(\mathbf{u}_i)|n(\mathbf{u}_i)) \in (p_{\text{low}}, p_{\text{upp}}] \\ 0, & \text{otherwise} \end{cases}$$

The average of $\xi(\mathbf{u}_i; p)$ over the n locations $\mathbf{u}_i$,

$$\overline{\xi(p)} = \frac{1}{n} \sum_{i=1}^{n} \xi(\mathbf{u}_i; p)$$

is the proportion of locations where the true value falls within the symmetric p-PI.

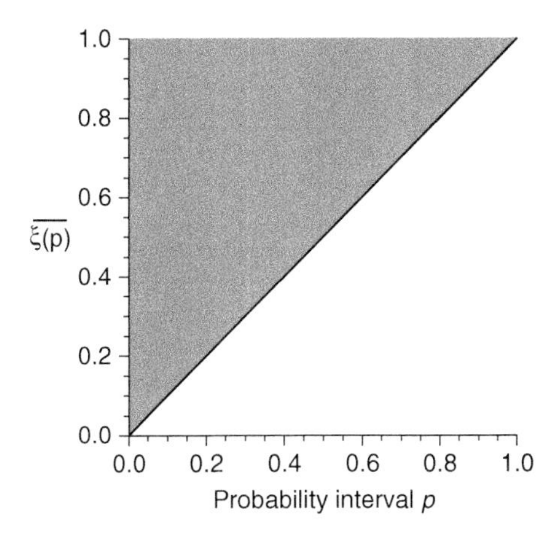

FIGURE 5.12: Blank (Template) Accuracy Plot: The Actual Fraction of True Values Falling within a Probability Interval p, Denoted $\overline{\xi(p)}$, Is Plotted versus the Width of the Probability Interval p.

According to our earlier definition of accuracy, the simulation algorithm used to generate the distributions of uncertainty (ccdfs) is accurate when $\overline{\xi(p)} \geq p,\ \forall\ p$. A graphical way to check this assessment of accuracy is to cross plot $\overline{\xi(p)}$ versus p and see that all of the points fall above or on the 45° line. This plot is referred to as an *accuracy plot*. A blank accuracy plot is shown in Figure 5.12. The ideal case is where points fall on the line; that is, the probability distributions are accurate and precise. Points that fall above the line (in the shaded region) are accurate but not precise. Finally, points that fall below the line are neither accurate nor precise.

A distribution is accurate when $\overline{\xi(p)} \geq p$. Precision is defined as the closeness of $\overline{\xi(p)}$ to p when we have accuracy ($\overline{\xi(p)} \geq p$). Quantitative measures of accuracy and precision could be defined; however, visual inspection of the accuracy plot is adequate. The results are acceptable if the points fall close to the 45-degree line. The distributions of uncertainty are too wide when the points fall above the 45-degree line and too narrow when the points fall below the 45-degree line. There are different ways to increase or decrease the spread of the distributions of uncertainty. One way is to increase the spatial correlation, which will decrease the spread of the local distributions of uncertainty. Decreasing the range of spatial correlation will increase the width of the local distributions of uncertainty.

Good probabilistic models must be both accurate and precise; however, there is another aspect of probabilistic modeling: the spread or uncertainty represented by the distributions. For example, two different probabilistic models may be equally accurate and precise and yet we would prefer the model that has the least uncertainty. Subject to the constraints of accuracy and precision, we want to account for all relevant data to reduce uncertainty. For example, as new data become available, our probabilistic model may remain equally good and yet there is less uncertainty.

The overall uncertainty of a probabilistic model may be defined as the average conditional variance of all locations in the area of interest:

$$U = \frac{1}{n}\sum_{i=1}^{N}\sigma^2(\mathbf{u}_i)$$

where there are n locations of interest $\mathbf{u}_i, i = 1, \ldots, n$, with each variance $\sigma^2(\mathbf{u}_i)$ calculated from the local ccdf $F(\mathbf{u}_i; z|n(\mathbf{u}_i))$.

Reservoir Case Study

To further illustrate the direct assessment of accuracy and precision, consider the "Amoco" data consisting of 74 well data related to a West Texas carbonate reservoir. Figure 5.13 shows a location map of the 74 well data and a histogram of the vertically averaged porosity for the main reservoir layer of interest. The vertically averaged porosity data were transformed to a Gaussian distribution and all subsequent analysis is performed with transformed data. This transform is reversible so the results would be the same were we to consider back transform porosity values.

A number of procedures were checked for building distributions of uncertainty at the 74 well locations. The variable being considered was the vertical average of porosity over the vertical thickness of the reservoir. A "leave-one-out" cross-validation approach was considered. A number of alternative geostatistical methods were considered to build distributions of uncertainty in the vertical average of porosity:

1. Gaussian distribution of uncertainty built by kriging of normal score values with the variogram modeled using the complete 3-D set of normal score porosity data. Figure 5.14 shows the variogram and accuracy plot. The variogram of the vertical average data should have been considered; this variogram applies to the small-scale data and is an incorrect

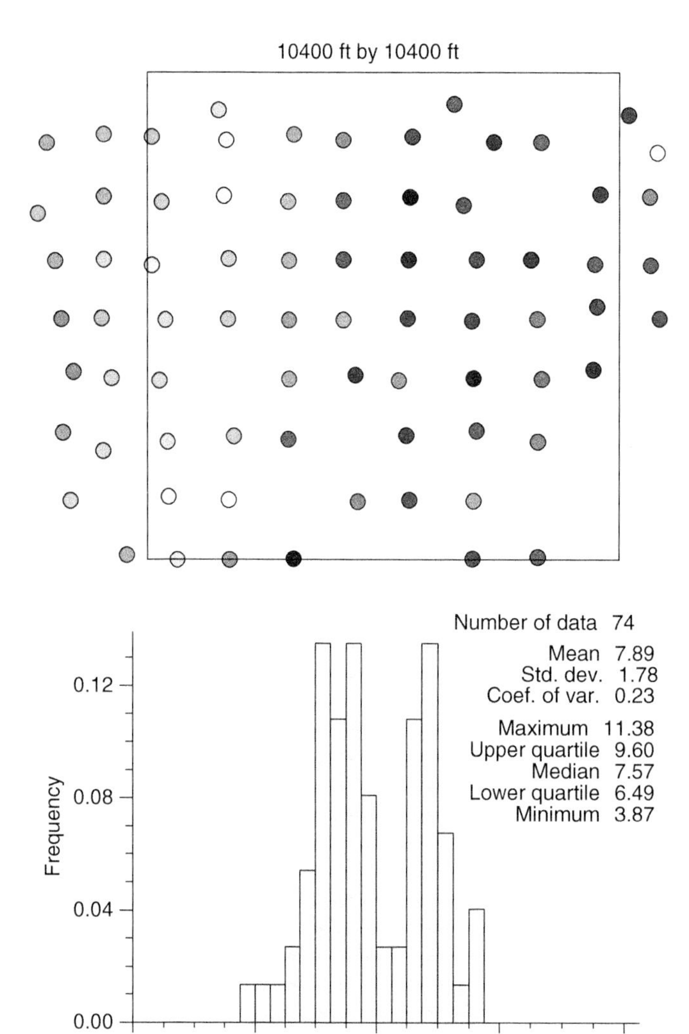

FIGURE 5.13: Location Map and Histogram of 74 Well Data.

choice for the 2-D vertical average. The average uncertainty in this case is 0.757 (relative to 1.0 for no local information).

2. Gaussian distribution of uncertainty built by kriging of normal score values with isotropic variogram from the set of normal score transforms of 2-D average porosity. Figure 5.15 shows the variogram and accuracy plot. The average uncertainty is 0.433, which is significantly better that the incorrect choice of the 3-D variogram.
3. Gaussian distribution of uncertainty built by kriging of normal score values with an anisotropic variogram model built from the 74 vertically averaged values. Figure 5.16 shows the variogram and accuracy plot. The average uncertainty is 0.385, which is slightly better than the less correct choice an isotropic variogram.
4. An indicator-based approach (see Section 4.6, Porosity and Permeability Modeling for details given in the context of permeability) with nine thresholds reduces uncertainty to 0.244 with reasonable accuracy. This improvement in results is at the cost of greater effort: professional time to calculate and model nine anisotropic indicator variograms rather than just one and nine times the CPU effort for the full indicator kriging.

The accuracy plot and uncertainty can be used together to assess local distributions of uncertainty

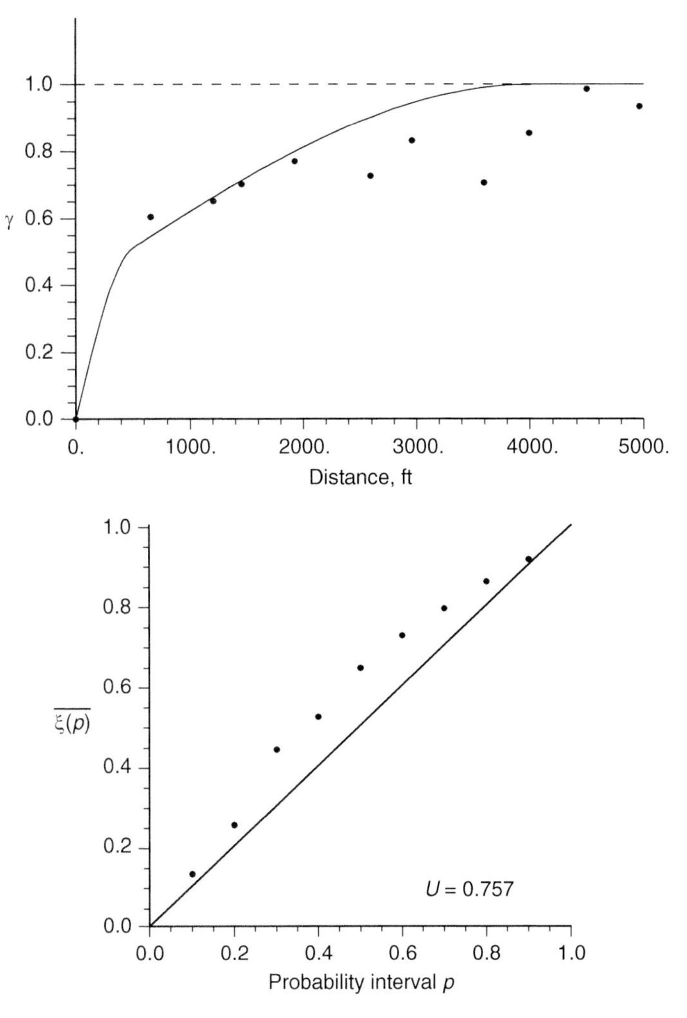

FIGURE 5.14: Horizontal Normal Scores Semivariogram from the 3-D Porosity Data (Calculated Using Stratigraphic Coordinates) and the Corresponding Accuracy Plot Considering Cross Validation with the Normal Scores of the 74 Well Data in Figure 5.13.

and compare alternatives. The indicator approach appears suited to this example (see Figure 5.17).

Checking Categorical Variable Distributions

The following is additional considerations for checking categorical variable distributions. The discrete nature of categorical variables and the lack of an ordering require different cross-validation techniques than continuous variables. The prediction of facies may be checked by (1) closeness of the estimated probabilities to the true facies and (2) accuracy of the local probabilities. Cross validation in a categorical variable setting leads to a set of true facies types and predicted probabilities, that is, $\{s(\mathbf{u}_i), p^*(\mathbf{u}_i), k = 1, \ldots, K\}, i = 1, \ldots, n$, where K is the number of facies and n is the number of data locations.

All variants of sequential indicator simulation provide direct estimates of the facies probabilities at the locations being (re)estimated. The cross validation of object-based methods is quite tedious; the facies modeling must be repeated for every well to be (re)estimated. The distributions of uncertainty must be built up directly by simulation without taking the shortcut of kriging indicator transforms. Although tedious, this procedure permits a quantitative comparison with other methods and an ability to fine-tune the object shape and size parameters.

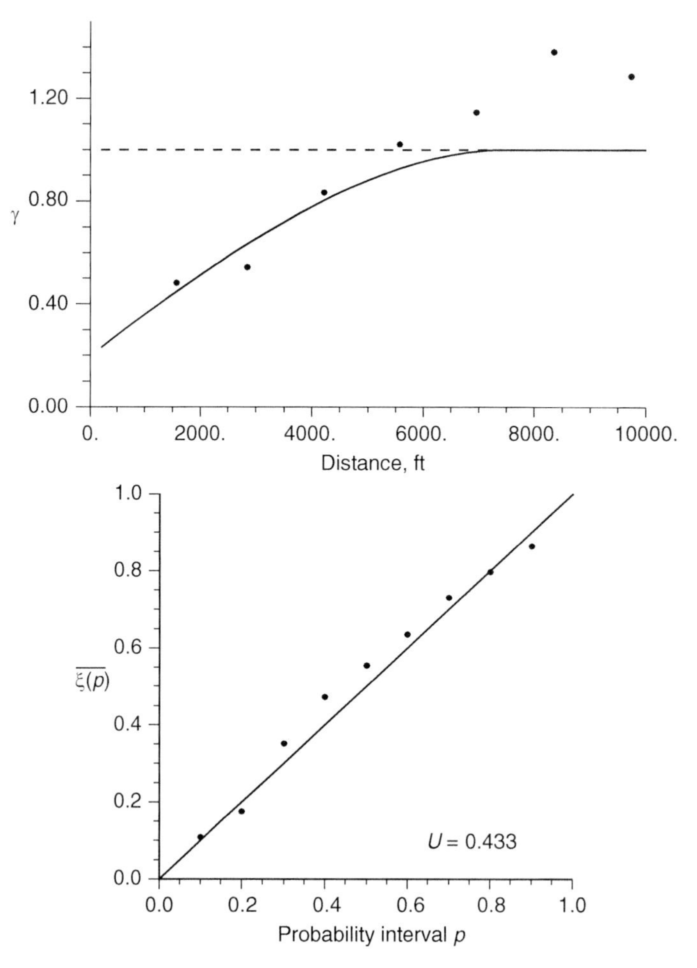

FIGURE 5.15: Horizontal Normal Scores Semivariogram from the Vertically Averaged Porosity and the Corresponding Accuracy Plot Considering Cross Validation with the Normal Scores of the 74 Well Data in Figure 5.13.

The predicted probabilities, $p^*(\mathbf{u}_i), k = 1, \ldots, K$, $i = 1, \ldots, n$, can be directly calculated for truncated Gaussian simulation without repeated simulation. The distribution of uncertainty of the Gaussian variable and knowledge of the locally variable thresholds allow us to calculate these probabilities analytically. The predicted probabilities may be checked in the same way as the direct estimates from indicator kriging methods and the simulation result of object-based methods.

The challenge is to summarize the results in a way that permits easy comparison. Three measures are suggested:

1. *Closeness to the true facies* measures how large the predicted probabilities of the true facies are. Ideally, the probability of the true facies $p^*(\mathbf{u}_i; s(\mathbf{u}_i))$ should be close to one.
2. *Fuzzy closeness to the true facies* accounts for the fact that the consequences of misclassification are different; that is, assigning the wrong type of sand is less consequential than assigning a sand facies as non-net shale.
3. *Accuracy of the local probabilities* measures the reliability of the predicted probabilities; that is, 80% of the locations assigned an 80% chance to be facies s_k *should be* facies s_k.

A quantitative measure of closeness to the true probabilities may be summarized by

$$C = \frac{1}{N} \sum_{i=1}^{N} \{p(\mathbf{u}_i; k) \mid \text{true} = k\}$$

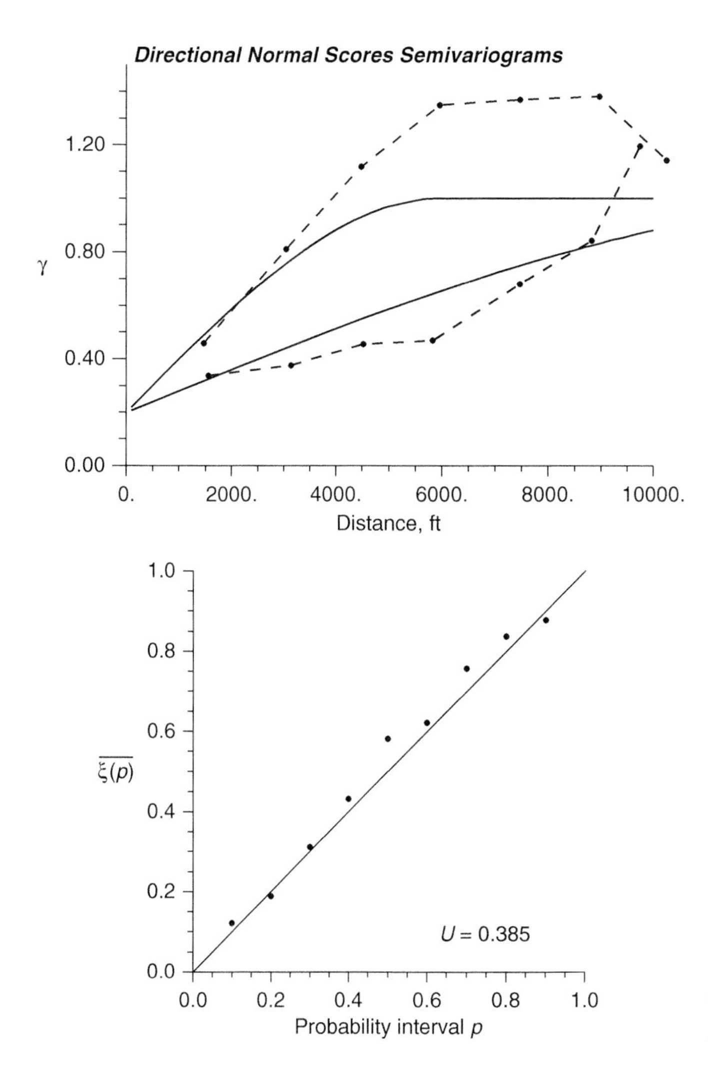

FIGURE 5.16: Directional Variograms from Vertically Averaged Porosity and Accuracy Plot.

This is interpreted as the average predicted probability of the true facies. This probability would be 1.0 in an ideal case of complete information and $1/K$ in the case where no information is available.

The closeness measure C described above quantifies how "right" the probabilities are for the true facies; however, it gives no credit when a similar facies is predicted, nor does it penalize the case when drastically different facies are predicted. A closeness matrix could be defined to account for these consequences. The closeness value $c(i, j)$ specifies how close facies i is to facies j. By definition, the closeness $c(i, j)$ is one for $i = j$, the same facies, and $c(i, j)$ is zero for completely different facies, say i = clean channel sand and j = shale. The exact values of $c(i, j)$ must be determined qualitatively considering the impact of misclassification. The need for a qualitative assignment is a shortcoming; however, some means must be used to account for the fuzzy closeness of facies.

The third check is the accuracy of the local probabilities. The local probabilities are "accurate" or "fair" if they accurately reflect the true fraction of times the predicted facies occurs. For example, consider all locations where we predict a 65% probability of facies 1 ($p_1^* = 0.65$); then 65% of those locations should be in facies 1. There is nothing special about 0.65; we check a range of predicted probabilities. Cases where the actual fraction departs significantly from the predicted probability are a problem. This is similar to the method applied in Section 4.1, Large-Scale Modeling, to check a categorical trend model, except we are checking a distribution of uncertainty against data with a cross-validation work flow as opposed to local trend probabilities against data.

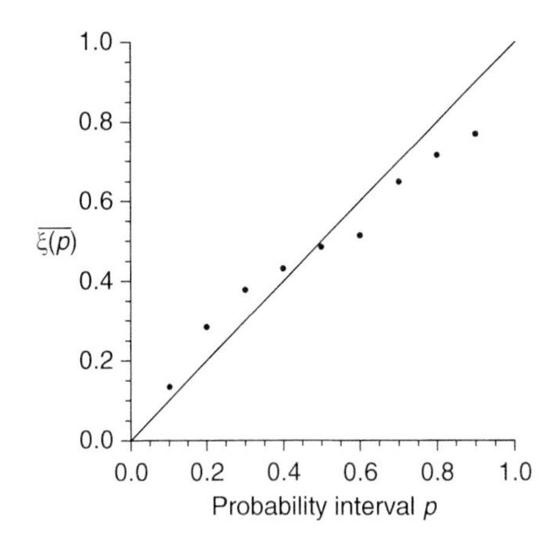

FIGURE 5.17: Accuracy Plot from Indicator Kriging (IK) ccdfs.

Also, the previously discussed accuracy plot method discussed previously and shown in Figure 5.12 may be applied to categorical distributions. A categorical accuracy plot is shown in Figure 5.18. The previously discussed symmetric p-PI are applied from zero to one and the same concepts of fairness, accuracy and precision apply. Instead of overall uncertainty (not possible with our categorical case), we rely on average Shannon entropy, H:

$$H_{\text{avg}} = -\frac{1}{N}\sum_{i=1}^{N}\sum_{k=1}^{K} p_k \ln(p_k)$$

with comparison to the maximum possible entropy given the number of categories, K:

$$H_{\text{max}} = \sum_{k=1}^{K} \frac{1}{K} \ln(K)$$

Probability intervals are useful to interpret the significance of the results (as shown in Figure 5.18 with the vertical lines in each bin). The expected fraction in each interval is equal to the mean of the probability interval; however, with either a low number of data in the interval or wide range of values taken in the interval, deviations are expected. A bootstrap-like approach (refer to Section 2.2 for information on bootstrap) is proposed to quantify this expected deviation by the construction of a 90% probability interval (or for any desired interval size). The algorithm calculates a 90% probability interval for each bin separately:

1. For each bin (probability interval), find the probability values in the bin.
2. Generate a random number [0,1]. If the random number is lower than the probability value, then $i_k^* = 1$, else $i_k^* = 0$.
3. Calculate the fraction of these random indicators in the bin.
4. Repeat this L times (where L is large enough, e.g., 1000 times), get the 5% and 95% quantiles to get the 90% probability interval.

This calculation assumes that the probability values are unbiased estimates of the category and that

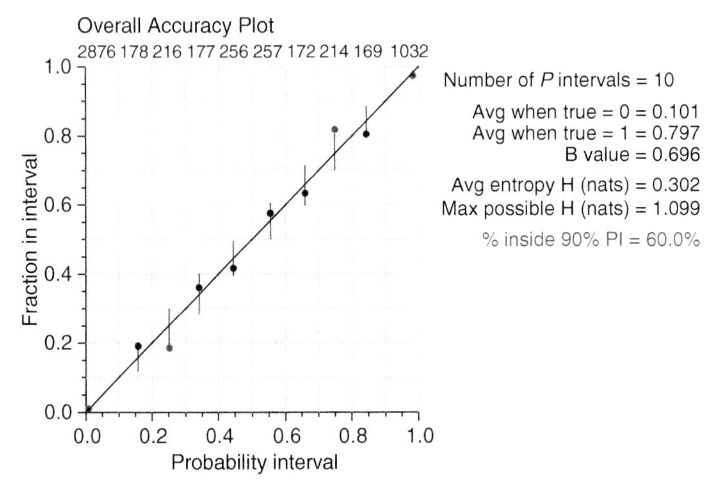

FIGURE 5.18: Example Accuracy Plot for a Categorical Variable Example. These results are quite good, that is, the proportion of times the facies is truly present is very close to the predicted probability. The method and software utilized is from Deutsch and Deutsch (2012).

probability values are independent. For the cross validation of a kriged categorical variable, the probability variables will be conditionally unbiased, given a sufficiently large search; however, the probability estimates will not be independent. The 90% probability interval constructed with this algorithm is therefore a conservative estimate.

For yet another example, five wells are used for a small example to check facies prediction. Seismic data are available to improve the predicted models of uncertainty. Indicator kriging (IK), Bayesian updating (BU), and a full block cokriging (BC) of the facies using the seismic data as a secondary data were considered. The closeness statistic C is 0.320, 0.358, and 0.349 for the three methods (note that a statistic of 1.0 would be ideal). The Bayesian updating approach appears best. We have an indication that the block cokriging approach is more sensitive to the larger number of input parameters. Figure 5.19 shows a cross plot of actual probabilities versus predicted probabilities.

The main uses for the diagnostic tools presented here are: (1) detecting implementation errors, (2) quantifying uncertainty, (3) comparing different simulation algorithms (for example, Gaussian-based algorithms versus indicator-based algorithms versus simulated annealing based algorithms), and (4) fine-tuning the parameters of any particular

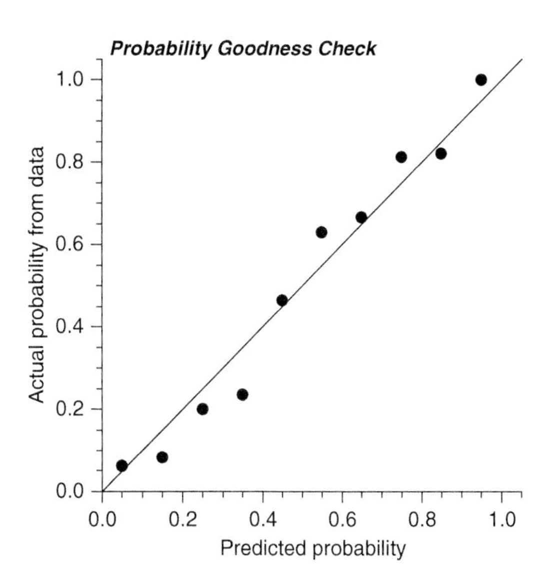

FIGURE 5.19: Example Cross Plot of Actual Probabilities versus the Predicted Probabilities. The closeness of the results to the 45-degree line attests to the goodness of the probabilities. These results are for block cokriging in all facies. We could also look at the results on a by-facies basis.

probabilistic model (for example, the variogram model used). These tools provide basic checks—that is, necessary but not sufficient tests. They do not assess the multivariate properties of the simulation. Care is needed to ensure that features that impact the ultimate prediction and decision making, such as continuity of extreme values, are adequately represented in the model.

5.1.6 Work Flow

Figure 5.20 illustrates the work flow for cross validation of an estimation technique. The cross validation of estimation and simulation techniques is quite different.

Figure 5.21 illustrates the work flow for validation of a Gaussian simulation. Once again, any variable could be considered. The cross validation is done in Gaussian space. The transformation back to real units is straightforward (see Section 2.4). The distributions are represented differently with indicator simulation; the principles of checking a distribution of uncertainty remain the same. Distributions of uncertainty from annealing and object-based modeling require multiple realizations with each well left out, which is a great deal of work.

5.1.7 Section Summary

Model checking is essential, considering typical method and work flow complexity. Many things can go wrong, and project recycles and poor decision quality result in direct and lost opportunity costs. The methods to check models are simple and typically fast.

Minimum acceptance, higher-order check, cross validation and the jackknife, and checking distributions of uncertainty provide a comprehensive set of methods to detect issues with simulated realizations and some basic checks for estimated models.

In the next section, various methods for post-processing reservoir models are discussed.

5.2 MODEL POST-PROCESSING

Model post-processing operates on the *geological model* in order to modify or fix the models, to scale the models, or to summarize the models. These operators may be applied to all or a subset of realizations, to a single realization over the entire area of interest, or to a subset or region of the model.

In the *Background* subsection, the geological model is defined and the need for model post-processing is explained. At times, post-processing is

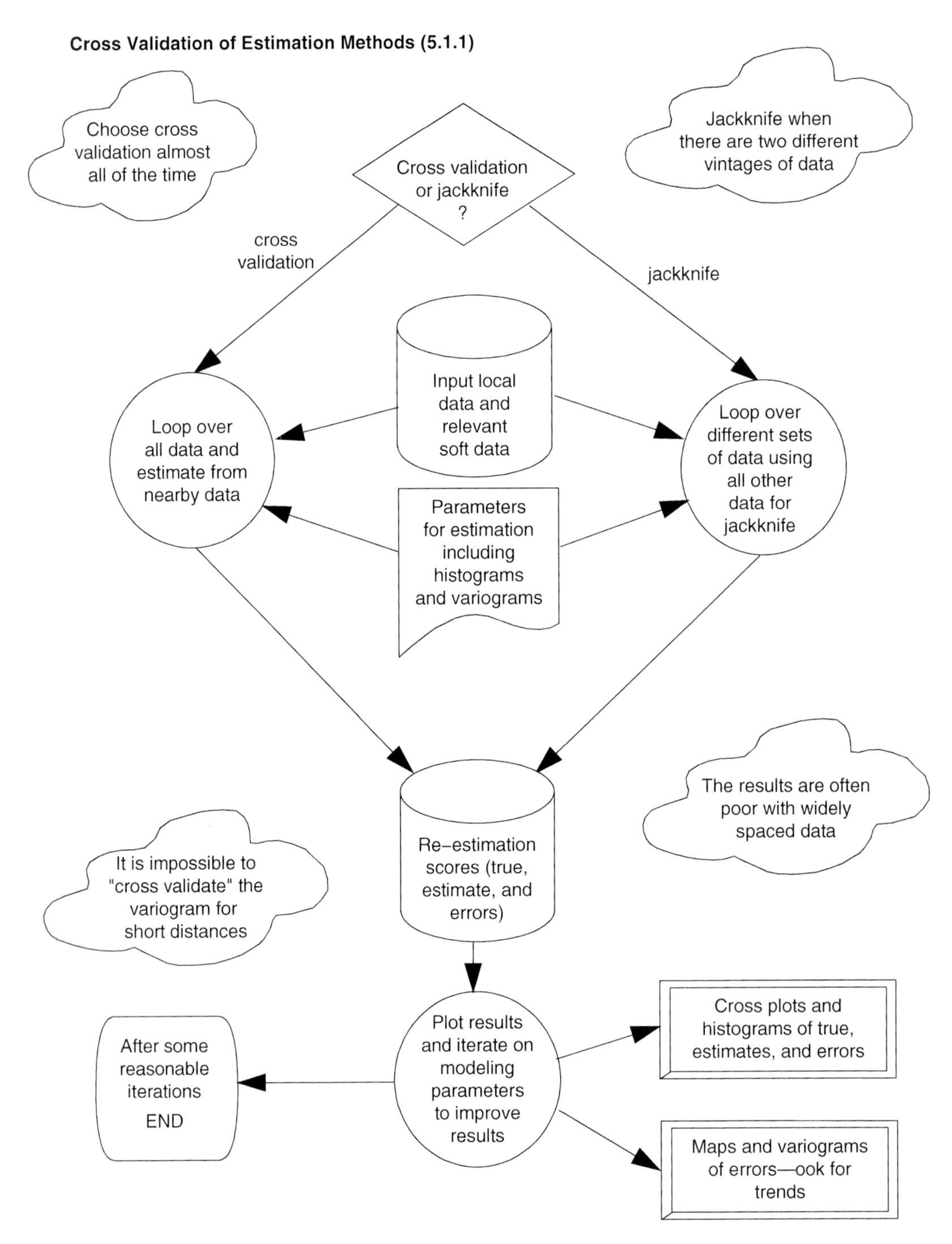

FIGURE 5.20: Work Flow for Cross Validation and Jackknife of an Estimation Technique.

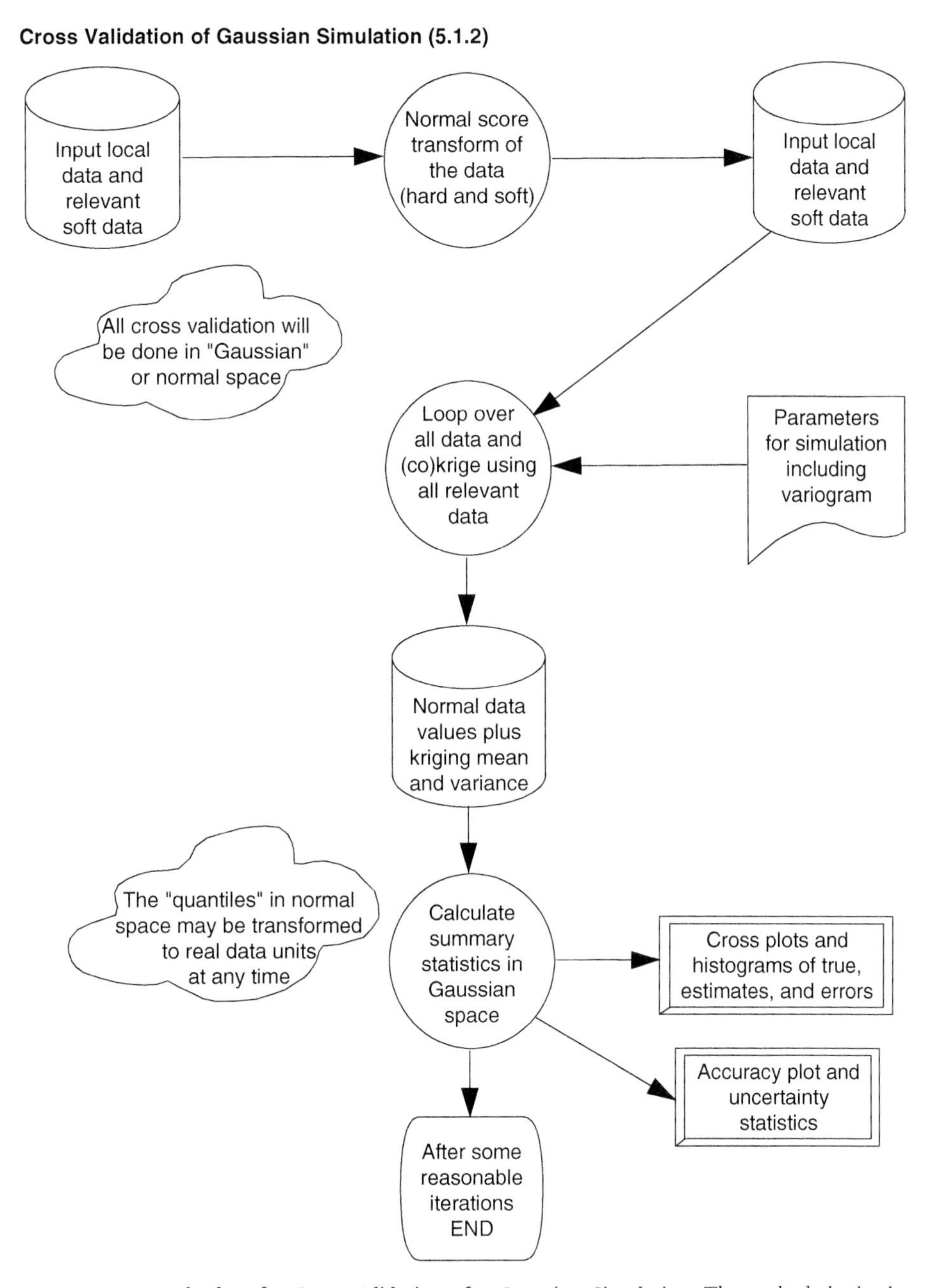

FIGURE 5.21: Work Flow for Cross Validation of a Gaussian Simulation. The method checks the accuracy and precision of the uncertainty model. This can be performed with or without secondary data.

required to honor specific constraints or to work at the correct scale. In other cases, post-processing may be a powerful tool to summarize, explore, and learn from our models.

In the *Model Modification* subsection the case for model fixing is explained. These operators change the model, often only slightly, and are often applied in work flows. At times this type operator is required to honor data and input statistics constraints.

Model Scaling adjusts the model scale or the support volume of the simulated values. Upscaling is typically straightforward and can be accomplished by averaging or flow-based upscaling applied to the fine-scale exhaustive model. Downscaling is an ill-posed inverse problem; there is no unique solution. In the absence of methods for practical exact downscaling, a practical approximate method is proposed. Scaling may be postponed until the necessary construction of a flow grid with effective properties for model cells, with simulation at data support size (see Section 2.4, Preliminary Mapping Concepts).

The *Pointwise Summary Models* subsection presents a variety of simple summaries that may be applied locally over multiple realizations. These operators result in 3-D summaries of the uncertainty model, such as average, variance, and percentile. These operators are useful to identify locations of interest within the model.

The *Joint Probability Models* subsection presents a variety of summaries that account for multivariate and spatial relationships. For example, joint probabilities of meeting specific multivariate criteria may be applied to identify sweet spots, and connected geobodies may provide summaries of local and global heterogeneity.

> These post-processing operators are straightforward. This section is easily understandable and applicable to novice through expert modelers. Surprisingly, post-processing is generally underutilized in current practice.

5.2.1 Background

The numerical representation of the reservoir consists of a suite of realizations that represent the reservoir and associated uncertainty. Each realization is a model of the entire volume of interest with a complete set of reservoir properties at each location, $\mathbf{u}$, within the reservoir model.

$$\text{facies}^{\ell}(\mathbf{u}), \text{porosity}^{\ell}(\mathbf{u}), \text{perm}^{\ell}_{h_1}(\mathbf{u}), \text{perm}^{\ell}_{h_2}(\mathbf{u}), \text{perm}^{\ell}_{v}(\mathbf{u}),$$
$$\text{saturation}^{\ell}_{\text{water}}, \text{saturation}^{\ell}_{\text{oil}}, \ldots, \mathbf{u} \in V, \quad \text{where } \ell = 1, \ldots, L$$

where L is the number of realizations and V is the volume of interest for the reservoir model. Importantly, note that a realization has a single value for each reservoir property at each location. This is the most efficient manner to represent reservoir uncertainty. Other schemes, such as multiple porosity realizations matched to a single facies realization, or multiple permeability realizations matched to a single porosity realizations, are suboptimal. An exception to this rule is scenario-based modeling. Typically, a large number of realizations may be matched to a few scenarios; thus a few facies scenarios would be matched with many realizations of porosity and permeability.

These realizations are calculated such that the appropriate relationships between the reservoir properties are reproduced. For example, the constraints of facies on all subsequent reservoir properties are imposed. Each facies will have unique univariate and spatial distributions and perhaps trends and the bivariate relationships. These are enforced on each continuous reservoir property in sequence. More discussion on models of uncertainty and the number of realizations required is provided in Section 5.3, Uncertainty Management.

This set of realizations and all associated summaries and derivative models is the *geological model.* This section deals with the various summaries and model manipulations that may be applied to correct, interrogate, and understand the geological model. Fundamental to geostatistical *realizations* is the concept that these numerical models help the modeler and project team learn or realize new ideas concerning the reservoir. This is an important concept that is often not considered. Consider these questions; What do we learn from our realizations? How do we learn from our realizations?

Firstly, through the construction of these models that integrate all available data and input statistical constraints, we facilitate communication and visualization and learn the spatial heterogeneities that may exist in the reservoir. This may be qualitative based on visual inspection of the model or may be more quantitative with application of the model checking tools discussed in Section 5.1, Model Checking. Nevertheless, the model is a shared concept that can be supported and challenged.

Secondly, through the application of the realizations to the transfer function, we learn about important reservoir response variables such as OIP, recovery factors, and production rates. The transfer functions may be complicated and often result in new concepts and understanding concerning geological sensitivity and risk. While this is often a straightforward application of the models to volumetric and flow simulation methods, at times model post-processing is required to correct and scale models.

Thirdly, the realizations may be post-processed to further improve the models and extract information from the geologic model. This information and subsequent summaries provide additional important information—for example, information on the impact of distribution assumptions and model scale on flow response, locations in the model that are surely high reservoir quality, and locations that have specified probability of occurrence of the right combination of reservoir properties for a sweet spot. Post-processing includes the following categories:

- Model Modifications
- Model Scaling
- Pointwise Summary Models
- Joint Summary Models

These are discussed and demonstrated in the following subsections.

Model post-processing is applied to the realizations after they have been generated and checked. These methods do not include any algorithms that are applied during the model simulation. For example, methods within a sequential simulation method to impose a specific univariate distribution are not considered post-processing, while an algorithm that fixes the global distribution of a realization after simulation is considered post-processing.

5.2.2 Model Modification

Model modification represents any model post-processing method that changes the model. This could be equivalently expressed as *model fixing, model cleaning,* or even *model tweaking.* This includes: routines that remove artifacts such as isolated facies that are common in sequential indicator simulation, routines that correct model conditioning to be exact, and routines that change or correct the statistics of the model. *Why would simulated models require fixing?* Recall in Chapter 4 that all the simulation tools represent some form of compromise. Inherent to all simulation methods is a hierarchy of priority for honoring data and statistical inputs and implicit assumptions. Common to all simulation methods is the ability to integrate various forms of data and input statistical constraints, along with implicit assumptions. Combinations of trends and secondary data may result in unreasonable values. In addition, they vary in their ability to reproduce these statistics and often the reproduction of these statistics exhibit ergodic fluctuations that may not be considered useful. Post-processing is an opportunity to override this hierarchy of priorities or to address an algorithm limitation to meet project goals.

Model modification should not be applied to *hide* poor implementation. For example, a biased trend model may bias the univariate distributions of simulated realizations. It would be best practice to correct the trend model, rather than to correct the bias that results in the distribution of the realizations. Nevertheless, it may not be possible to formulate a work flow to satisfy all required model constraints. In these cases, careful application of model modification may be best practice.

The following are various types and examples of model modification.

Distributions, Trends, and Uncertainty Model Correction

In some cases there are systematic biases that must be removed in simulated realizations' distribution. These biases may result from (1) variance inflation in cosimulation or simulation with a trend, (2) departures of the data from the implicit multivariate distribution model chosen for the simulation, or (3) implementation considerations such as a limited data neighborhood or numerical precision.

Deutsch and Journel (1998) include a distribution correction method known as `trans` devised to transform any set of values so that the distribution of the transformed values matches the target distribution. An important feature of this program is the ability to *freeze* local conditioning data. When data are frozen, the target distribution is only approximately reproduced; the reproduction is excellent if the number of frozen values is small with regard to the number of values transformed.

The transformation method is a generalization of the quantile transformation used for normal scores. The p quantile of the original distribution is transformed to the p quantile of the target distribution (see Section 2.2, Preliminary Statistical Concepts). This transform preserves the p quantile indicator variograms of the original values. The variogram (standardized by the variance) will also be stable, provided that the target distribution is not too different from the initial distribution.

When freezing the original data values, the quantile transform is applied progressively as the location gets further away from the set of data locations (Journel and Xu, 1994). The distance measure used is proportional to a kriging variance at the location of the value being transformed. That kriging variance is zero at the data locations (hence no transformation) and increases away from the data (the transform is increasingly applied). An input kriging variance file must be provided.

Because not all original values are transformed, reproduction of the target histogram is only approximate. A control parameter, $\omega \in [0, 1]$, allows the desired degree of approximation to be achieved at the cost of generating discontinuities around the data locations. The greater ω, the lesser the discontinuities.

Deutsch (2005b) developed a new method with the addition of (1) accounting for a trend, (2) explicit accounting for uncertainty, and (3) iterative implementation to ensure reproduction of all constraints.

Trends

Section 4.1, Large-Scale Modeling, provides details on trend modeling and construction and Section 2.4, Preliminary Mapping Concepts, discusses methods to integrate trends into simulated realizations. Importantly, the trend model provides important local spatial information and a representative average that assists with declustering/debiasing. The trend is represented as a locally variable mean or proportion value for the property of interest at all locations in the model.

Enforcing reproduction of the trend requires subsetting of the model. If we insist on the trend being reproduced at every grid cell, then we would get back the trend exactly, which is unreasonable. There are three reasonable subsets that we may choose: nz vertical trend values, nx·ny areal trend values, or nq classes of trend values based on the input trend model. Choosing to enforce the vertical and areal trend values is reasonable when the 3-D trend was constructed as a combination of these lower-order trends. Considering classes of the trend model, that may not have a consistent vertical or areal trend over the entire model.

The schematic sketch is Figure 5.22 illustrates how nq = 7 classes would be chosen from a histogram of trend values (the green distributions). The histogram of the data is sketched in red to illustrate how the original data would be more variable.

The approach to enforce reproduction of the trend model in different classes is the same: (1) Calculate the target mean in each subset, (2) calculate the actual mean in each subset, and (3) multiply all data in the subset by the ratio of the target divided by the actual. This will be described below in the context of correct of trends and uncertainty model.

Explicit Accounting for Uncertainty

Local uncertainty is largely insensitive to large-scale uncertainty in parameters such as the global histogram; however, geostatistical realizations are being used increasingly to assess global uncertainty in recoverable reserves. Techniques such as the spatial bootstrap and conditional finite domain (see Section 2.2, Preliminary Statistical Concepts) are being used to assess uncertainty in the input histogram or (at least) the input mean. There is a need to account for this input parameter uncertainty. Global uncertainty would be underestimated if the global histogram/trend were imposed on every realization.

Global uncertainty in the univariate distribution should be transferred through to geostatistical simulation and then onto response uncertainty. We could assemble a database of equiprobable (equally likely to be drawn) target univariate distributions by the spatial bootstrap or some similar technique. Each realization could be transformed to reproduce a different target distribution.

With the Deutsch (2005b) method, this is streamlined with the assumption that the uncertainty in the histogram shape is considered a second-order effect. Uncertainty in the mean is a first-order effect that should be quantified and transferred to each realization. Multiple realizations of the property distribution are calculated by scaling the representative distribution.

$$z_i^t = z_i \cdot \frac{m_{\text{targ}}}{m_{\text{orig}}}, \qquad i = 1, \ldots, N$$

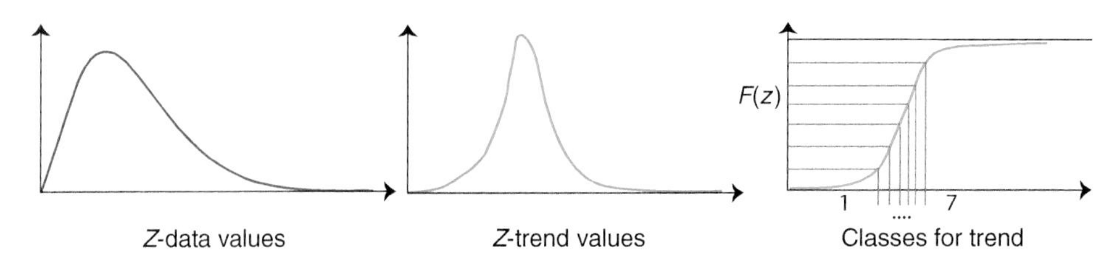

FIGURE 5.22: The Schematic Sketch Illustrates How nq = 7 Classes Would Be Chosen from a Histogram of Trend Values (the Green Distributions). The histogram of the data is sketched in red to illustrate how the original data would be more variable.

where m_{orig} is the representative mean and m_{targ} is the mean for the specific realization determined by the uncertainty model.

This is applicable to the commonly encountered positively skewed distributions of non-negative variables. It is possible to generate too-large values when $m_{tar} > m_{orig}$. The solution is to reject those values and rescale the remaining data:

1. Reject $z_i^t > z_{max}$.
2. Calculate mean of remaining data: m_t.
3. Scale the remaining values: $z_i^{t+1} = z_i^t \cdot \frac{m_{targ}}{m_t}$.

This will have to be applied iteratively. Convergence to a mean within a small tolerance of the target is generally achieved within five iterations. The trend values must also be scaled to be consistent with the target mean. A multiplicative approach is also followed:

$$m^t(\mathbf{u}) = m(\mathbf{u}) \cdot \frac{m_{targ}}{m_{trend}}$$

where m_{trend} is the mean of the original trend values. The trend values should not be highly variable; otherwise, they are not really a trend. For this reason, no iteration is performed to remove high values. To account for the uncertainty model, each realization will be scaled to a different target distribution. Randomly pairing an input realization with a target distribution/mean may lead to large changes for some realizations. Ideally, we might choose to transform each realization to a target mean that is close to the original.

Iterative Algorithm

The distribution of the initial realization is denoted $F_{init}(z)$. The target distribution is denoted $F_{targ}(z)$. All initial values $z_i, i = 1, \ldots, N$, could be transformed to ensure reproduction of the target distribution, with the regular quantile–quantile transformation (see Section 2.2, Preliminary Statistical Concepts) :

$$z_i^{hist} = F_{tar}^{-1}(F_{init}(z_i)), \qquad i = 1, \ldots, N.$$

The initial values could also be transformed to ensure reproduction of the trend within the $k = 1, \ldots, K$ subsets of the trend model:

$$z_i^{trend} = z_i \cdot \frac{m_{targ}^k}{m_{init}^k}, \qquad i = 1, \ldots, N.$$

The k superscript in this equation refers to which class of the mean model the ith data falls. We could attempt to enforce the mean within multiple subsets—for example, the vertical trend and the horizontal trend. We could add an index for the trend subset: $z_i^{trend}, s = 1, \ldots, N_T$.

The transformed values that would lead to reproduction of the histogram and the trend will likely be different. We could choose to average them to approximately reproduce all of the data constraints:

$$Z_i^{ht} = \text{avg}(z_i^{hist}, z_i^{trend,s}, s = 1, \ldots, N_T), \quad i = 1, \ldots, N.$$

Ensuring consistency in the target histogram and the trend model may make it possible to enforce the local data, the target histogram and the trend model in one pass; however, in general there will be trade-offs and the transformed values will not exactly reproduce the input statistics. It seems reasonable to reset the initial realization to the transformed realization and rerun the algorithm. This is easy to implement. Experience shows that this permits convergence to all of the objectives provided that they are consistent with each other. Inconsistent inputs such as the target mean being different than the mean of the trend model would lead to the results not converging. An extreme example of distribution correction is shown for a porosity realization from the Carbonate Reservoir. Note that the values remain the same near the well data, but change significantly away from the well locations (see Figure 5.23).

Categorical Image Cleaning

As discussed in Section 4.2, cell-based categorical simulation methods often result in artificial decrease in short-scale continuity represented by isolated pixels. For example, a single sand facies cell surrounded by mud facies in a facies realization. Also, reproduction of facies proportions can be problematic; in particular, facies types with relatively small proportions may be poorly matched. Image cleaning by MAPS was presented as a method to correct for these artifacts and also to simultaneously correct the global pdf to a target global pdf (see Section 4.2). Because this form of post-processing is almost always required with sequential indicator simulation, it is included with the discussion on variogram-based facies models. To illustrate this operation, Figure 5.24 shows an example of the Delta Reservoir facies model before and after application of the categorical image cleaning.

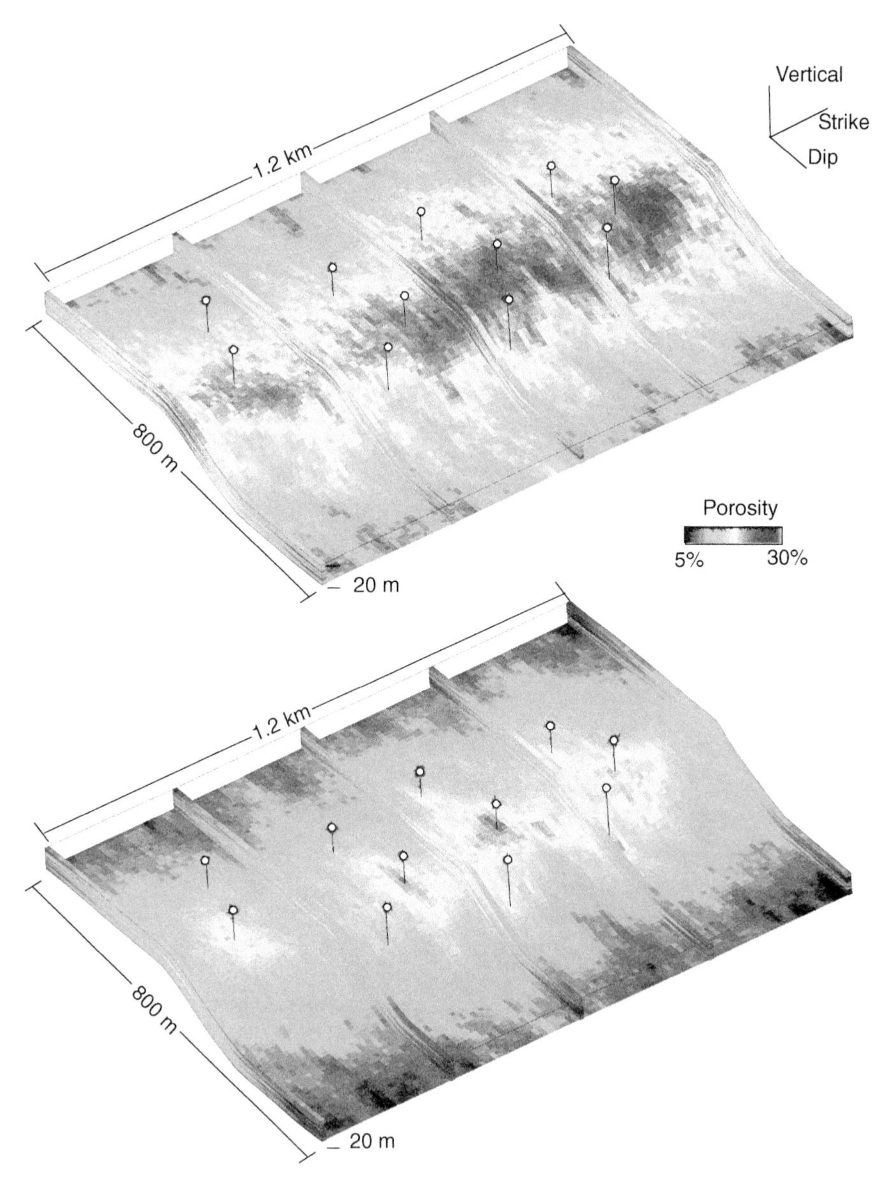

FIGURE 5.23: Porosity Realizations with the Original Porosity Distribution (above) and the Corrected Porosity Distribution (below).

Conditioning Correction

Post-processing may be applied to correct the conditioning within a model. As discussed in Section 5.1, Model Checking, model exactitude or the data assign to model cells reproduced in the model realizations is one of the realization minimum acceptance criteria.

There are two types of conditional correction: (1) conditioning unconditional realizations and (2) correcting conditional realizations for precise exactitude.

There are simulation methods that are inherently unconditional, such as turning bands, that require a second step to impose conditioning. These methods general rely two kriging-based models, one of the data, $y^*_{kc}(\mathbf{u})$, and one of the unconditional simulated values at the data locations, $y^*_{ku}(\mathbf{u})$, to form a correction model:

$$y^{\ell}_{cs}(\mathbf{u}) = y^{\ell}_{uc}(\mathbf{u}) + \left[y^*_{kc}(\mathbf{u}) - y^*_{ku}(\mathbf{u})\right]$$

where $y^{\ell}_{uc}(\mathbf{u})$ is the unconditional realization and $y^{\ell}_{cs}(\mathbf{u})$ is the conditional realization.

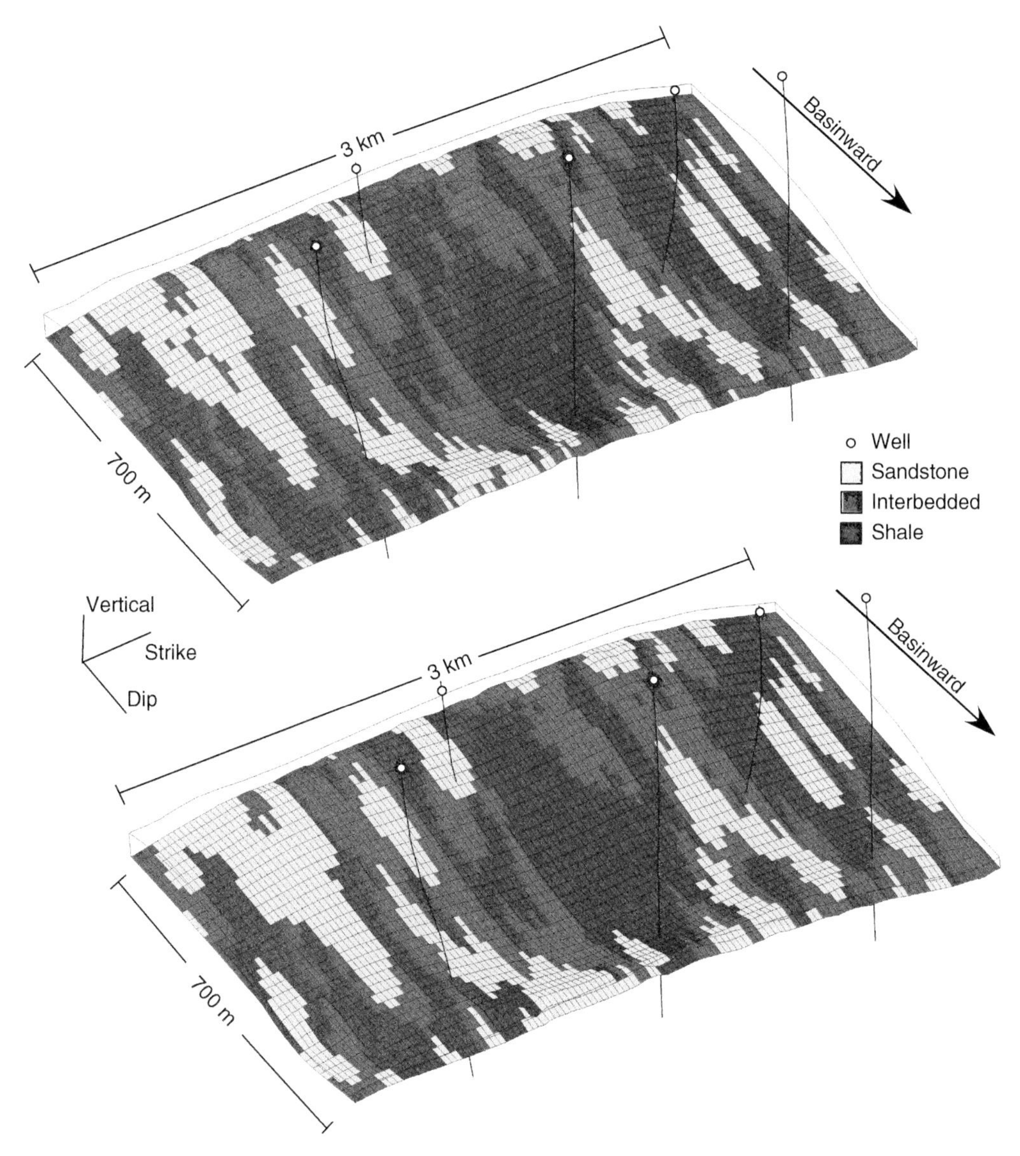

FIGURE 5.24: Facies Model before (above) and after Image Cleaning from the Delta Reservoir.

In the second case, the conditional realizations fail to perfectly match the data. This is typically observed with object-based, process-mimicking, and optimization methods (see Sections 4.5, 4.6, and 4.7, respectively). There is a need for a flexible post-processing algorithm to enforce well data reproduction without introducing artifacts.

Deutsch (2005c) demonstrated a method based on erosion/dilation to account for the following situations. It may be unrealistic to embed the actual channel-like geometry in the algorithm as shown by the schematic black line; however, we would like the final cell-based model to appear realistic. The method is an extension of the MAPS method discussed in Section 4.2 (see illustration in Figure 5.25).

The Deutsch (2005c) methodology works with the cells and their statistical relationship with surrounding cells; there is no explicit manipulation of object geometries. A cell-based statistical procedure has the advantage of (1) simple post-processing and (2) easily extended to multiple facies such as levees and crevasse splays.

The image analysis technique is applied in post-processing mode to enforce well conditioning data with smooth changes. The realization should only be changed near wells where the observed intersections do not match the image to be post-processed. Cells

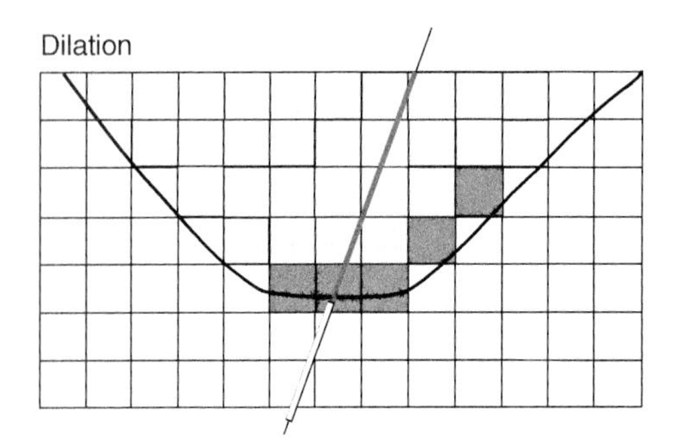

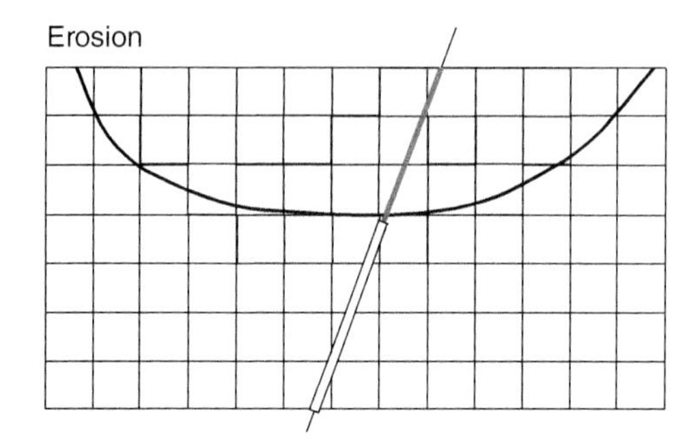

FIGURE 5.25: Illustration of the Use of Erosion and Dilation to Post-process a Cell-Based Model to Enforce Precise Well Conditioning.

From Deutsch (2005c).

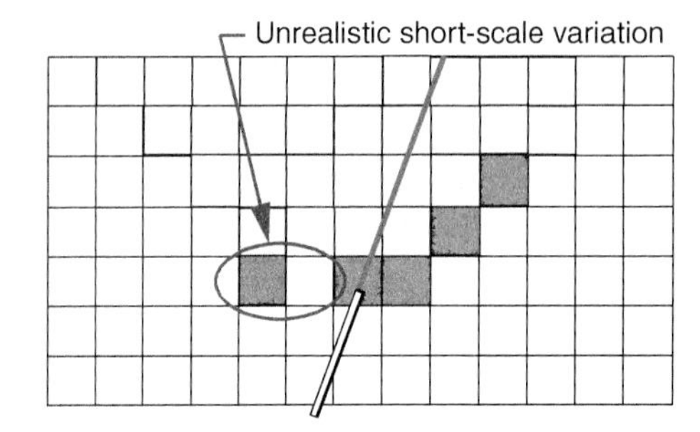

FIGURE 5.26: Illustration of the Unrealistic Short Scale Variability That Is Corrected with the MAPS Post-processing Method.

From (Deutsch, 2005c).

that are candidates for a change in facies are identified as those within an ellipsoidal range from cell-well values that mismatch. The cells at the mismatched locations will be visited first and a spiral search will be used until the ellipsoidal range is reached.

The algorithm is applied sequentially; a change is considered at a grid location by considering the well mismatch and all previous grid node changes. The original MAPS algorithm is not sequential; however, the goal here is for the erosion/dilation to be smooth away from the wells and not have unrealistic short scale variations. The sketch in Figure 5.26 shows a situation we want to avoid. If the cell on the right is unchanged, then the cell on the left should also be left unchanged; the algorithm has to be sequential so that changes are smoothly propagated away from the wells with a mismatch.

The probability of changing the facies assignment at a grid cell location is established by two factors: (1) the facies in nearby grid cells and (2) the new facies that is being assigned at the cell intersected by the well (to correct a mismatch). The effect of the mismatch will decrease as the cell under consideration gets further from the well.

The probability of each facies prevailing at any particular cell location u is calculated based on a weighting function:

$$p(\mathbf{u}, k) = \frac{1}{S} \sum_{\mathbf{u}' \in W(\mathbf{u})} w(\mathbf{u}') \cdot i(\mathbf{u}; k), \qquad k = 1, \ldots, K$$

where S is a standardization constant, $W(\mathbf{u})$ is a template of weights centered at the location under consideration, and $i(\mathbf{u}; k)$ is the indicator of facies k at location $\mathbf{u}$. There is much discussion on the weighting template in Cleaning Cell-Based Facies Realizations subsection in Section 4.2, Variogram-Based Facies Modeling. A reasonably small template $(5 \times 5 \times 5)$ appears to work well. A larger template induces excessive smoothness, and a smaller template does not enforce smooth enough transitions away from the well locations.

All cells under consideration in the sequential path are within a reasonably close distance to a cell-well mismatch. There is some probability that the cell under consideration should also be changed to the observed facies at the well. This probability should be one at the well location and decrease as the distance from the well increases. The distance of the cell to the well is standardized by the ellipsoidal radius, and then the probability to observe the same facies as the well is increased by the following factor:

$$f = (1/d)^{\omega}$$

where f is the increased probability of the same facies as the well, and d is the standardized distance between 0 and 1, where $d = 0$ at the well location and $d = 1$ at the maximum distance from the well. The ω factor controls how quickly this factor decreases with distance; a value of $\omega = 2$ was found reasonable. So, f is added to the $p(\mathbf{u}; k)$ value corresponding to the facies k at the well.

The only other factor change is to slightly modify f by a random number; that is, multiply by a random number between 0.9 and 1.1. This avoids an excessively blocky behavior if the mismatch is in a homogeneous region of another facies. The algorithm runs extremely fast, and wells are honored smoothly with few visual artifacts (see Figure 5.27).

A third potential conditioning correction by post-processing case is model updating to honor new information. These approaches that attempt to update without complete model resimulation typically rely on resimulation of the portion of the model close to the new information and corrections to ensure the global distributions are correct.

5.2.3 Model Scaling

It may be necessary to change the model scale with upscaling to decrease in realization resolution or with downscaling to increase the realization resolution. Upscaling is required to move from data scale to model grid scale and then subsequently to flow grid scale. As discussed in Section 2.4, Preliminary Mapping Concepts, an alternative work flow is to consider the geological model to be a mesh of reservoir properties at cell centroids. This work flow has postponed model upscaling until immediately before the transfer functions that require effective properties over grid cell (such as flow simulation).

Upscaling

Upscaling may be required to reduce the computational time of the transfer function. This is commonly the case with flow simulation. Another purpose of upscaling is to provide a model summary—for example, to convert the 3-D models into 2-D maps or to generate large-scale models. Finally, in conventional work flows that consider simulated models to provide properties at model grid cell support, upscaling from well measurements and downscaling from seismic is typically attempted (although not usually rigorously).

An exhaustive fine-scale realization is available; therefore empirical upscaling is possible; otherwise, analytical tools are available (see Section 2.3, Quantifying Spatial Correlation) to adjust univariate and spatial distributions based on the point-support variogram model.

Scaling Facies

Change in scale for categorical variables is problematic. Clearly, scale-up of categories should transform into a continuous variable, because large scale cells now represent mixtures of categories. Yet, while facies are not directly input for flow simulation, facies must be scaled up, because relative permeability and other saturation functions are often assigned on a by-facies basis in the simulator. The most common facies within each large-scale flow simulation block is often taken as the large-scale facies. This may lead to underrepresentation of facies that have small global proportions. Alternative schemes could be considered, such as fixing the small proportion facies first, based on the maximum probability of that facies, and then moving on.

Scaling Porosity and Saturations

Porosity and saturation are both volumetric concentrations and average linearly. Arithmetic averaging is theoretically correct. The only complication is in low net-to-gross reservoirs where some of the porosity at

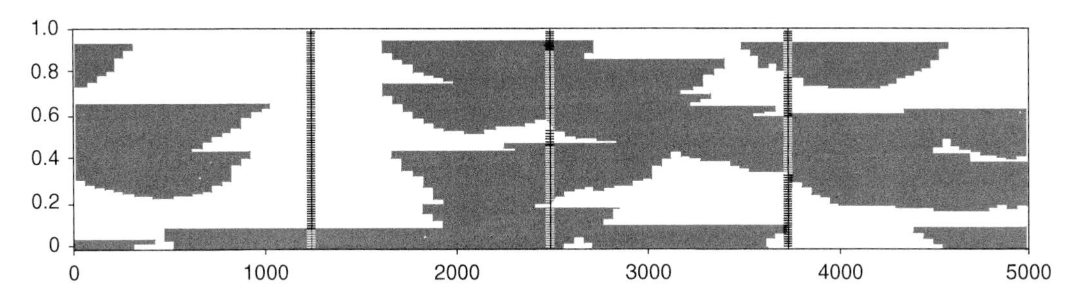

FIGURE 5.27: Example Cross Section from a Object-Based Channel Realization That Has Been Corrected with the MAPS Post-processing Method for Consistent, Precise Conditioning of Well Data.

From Deutsch (2005c).

a small scale may be "ineffective." This complication is not currently considered in scale-up.

Given that there are a sufficient number of high-resolution values in each upscaled cell, arithmetic averaging provides a good upscaled value. Summary statistics such as the univariate distribution and spatial continuity may be calculated from the upscaled porosity or saturation model to directly observe the impact of change in volumetric support on these statistics.

Scaling Permeability

The scale-up of permeability is more problematic (Tran, 1995; Wen and Gomez-Hernandez, (1996). The effective permeability of a flow simulation grid block depends on the spatial arrangement of the constituent geological modeling cells and the flow boundary conditions. The scaling of relative permeability curves and other saturation functions is even more complex than absolute permeability. A very brief discussion on the scale-up of absolute permeability is given here. The reader is referred to the literature (primarily SPE papers) for more details (Renard and de Marsily, 1997).

The three common simple averaging methods are (1) arithmetic averaging, which is correct for linear flow in parallel composites, (2) harmonic averaging, which is correct for linear flow in series composites, and (3) geometric averaging, which is correct for white-noise random media in 2-D with a lognormal distribution of permeability (Matheron, 1966):

$$k_a = \frac{1}{n}\sum k_i\,,\quad k_h = \left[\frac{1}{n}\sum\frac{1}{k_i}\right]^{-1},\quad k_g = \left(\prod k_i\right)^{\frac{1}{n}}$$

The geometric average can be equivalently calculated as the exponentiation of the arithmetic average of the logarithms of the data. The arithmetic average is the upper bound and the harmonic average is the lower bound of the effective permeability. These may not be the bounds if extreme permeability features such as "zero" permeability shales and "infinite" permeability fractures have not been representatively sampled.

Directional averaging (Cardwell and Parsons, 1945) gives much more realistic bounds on the effective permeability. Consider a flow simulation grid block made up of a regular 3-D network of geological modeling cells. The upper bound of the effective permeability in a particular direction can be calculated by a two-step procedure: (1) Calculate the arithmetic averages of the permeability in 2-D slices perpendicular to the flow direction, and (2) calculate the harmonic average of the 2-D slice averages. The lower bound of the effective permeability is given by a similar two-step procedure: (1) Calculate the harmonic averages in each 1-D column of cells in the flow direction, and (2) calculate the arithmetic average of the 1-D column averages. Averaging is problematic when these two bounds are very different. A map of the difference can be used to identify areas where the grid size or orientation could be changed.

The conventional averages can all be written in a general "power" averaging (Deutsch, 1989a; Journel et al., 1986; Korvin, 1981) formalism:

$$k_\omega = \left[\frac{1}{n}\sum k_i^{\omega}\right]^{\frac{1}{\omega}}$$

where the arithmetic average is at $\omega = 1$, the harmonic average is at $\omega = -1$, and (at the limit) the geometric average is at $\omega = 0$. Studies (Deutsch, 1989a) show that the averaging power depends on direction, where it is greater in the horizontal direction and less in the vertical direction. The averaging power depends on the spatial continuity and not on the histogram or relative amounts of high- and low-permeability values. The averaging power is close to 0.6 for horizontal permeability and -0.4 for vertical permeability for a wide range of geostatistical models. Of course, it should be calibrated to the particular heterogeneity present in each reservoir. The calibration procedure requires the "true" effective permeability to be calculated by single-phase steady-state flow simulation on the flow simulation grid blocks.

These small-scale flow simulations are remarkably fast. This leads us to the most common approach to scale-up of permeability; just solve for it directly. Boundary conditions must be chosen. No flow boundary conditions on the four faces perpendicular to flow are easy to apply, but can unrealistically confine flow and lead to low permeability. Periodic boundary conditions are often more realistic. Two recent approaches with reasonable literature reviews of direct permeability upscaling are provided by Razavi and Deutsch (2012) and Manchuk et al. (2012).

Downscaling

Scale-down is an ill-posed inverse problem; there is no unique solution and smaller-scale properties necessarily have more variability. Downscaling has often been used to deal with multiscale data in geostatistical modeling because large-scale data is commonly available for geostatistical modeling. The large-scale data (block data) may come from: (1) a 2-D large-scale prior model, which usually is built over a very large area; (2) inversion from production data and well testing data, which usually have a much larger scale than the core and well log data; (3) inversion from seismic data; (4) geological trend data that are considered reliable; or (5) any other large-scale data from measurement or expert interpretation.

Downscaling is required when building a model from these large-scale data. Downscaling is also used to generate fine-scale 3-D models from large-scale models for detailed flow simulation on small areas, such as pad area. The large-scale model could be a 3-D coarser grid model or a 2-D model where the downscaling is actually extending the 2-D model to 3-D model (Ren and Deutsch, 2005). Most likely the model is a coarse-scale 2-D model because a coarse-scale 3-D model would commonly have low vertical resolution relative to the reservoir layers or domain.

Some of the techniques used for downscaling with multiscale data include cokriging, collocated cokriging, and using the block data as locally varying mean data. Because none of these methods reproduce the block data exactly, Ren and Deutsch (2005) presented a combined block kriging direct sequential simulation method for exact downscaling (in this case, exact means that the downscaled values will scale up to the original coarse-scale values). Given the significant challenges with direct sequential simulation, a simple methodology for inexact down scaling is preferred.

1. Assign coarse-scale values to fine-scale cell nearest to centroid.
2. Apply volume variance relations to calculate fine-scale input statistics.
3. Simulate on fine grid with coarse-scale conditioning and fine-scale statistics.

Assign Coarse Scale to Fine Grid

The coarse-scale values are assigned to the fine-scale cells that are nearest to the course-scale cell centroids. An alternative would be to assign the coarse-scale value to all fine-scale cells within the coarse scale cell, but this would under represent the variability expected with downscaling. The subsequent resimulation on the fine-scale grid will impose the correct level of variability, and the conditioning will ensure that the fine-scale realization will tend toward the large-scale value over the larger volume.

Volume-Variance Relations

Simulation with the coarse-scale input statistics would underestimate the variability at the fine scale. In addition, there are secondary effects to the variogram. In Section 2.3, Quantifying Spatial Correlation, the methods for scaling continuous property distributions and variograms are presented.

While the volume-variance relations assume stationarity and linear averaging, in most cases they should provide a good approximation for the input statistics with change of support.

Simulation on Fine Scale Grid

Once the coarse-scale data are assigned to the fine-scale grid and the input statistics are calculated for the fine-scale support size, simulation may proceed on the fine-scale grid.

Finally, as mentioned previously, the method of simulating a data mesh at data support may remove the need for downscaling. It would be straightforward to augment the mesh with new simulated values at infilled locations conditional to the previous mesh at any spacing required. Because the support size does not change, no scaling is necessary.

Because upscaling is straightforward, model checking may include upscaling of the downscaled realizations and comparison to the coarse-scale realizations. Special attention should be paid to the comparison of the course-scaled values and upscaled downscaled realizations.

5.2.4 Pointwise Summary Models

Pointwise summary models include a suite of statistics that may be easily calculated for each location, $\mathbf{u}_\alpha$, over the set of realizations that represent the geological model. Deutsch and Journel (1998) provided utilities to calculate e-type, conditional variance, and local percentile models. Details on these statistics are provided below.

None of these summary models should be applied independently to represent the geological model. For example, the e-type model is too smooth and the local P10 model is *not* the P10 model (as discussed in the next section). In fact, the local P10 model would be an inconceivably pessimistic model (P10 values at each location) and would significantly overestimate spatial continuity.

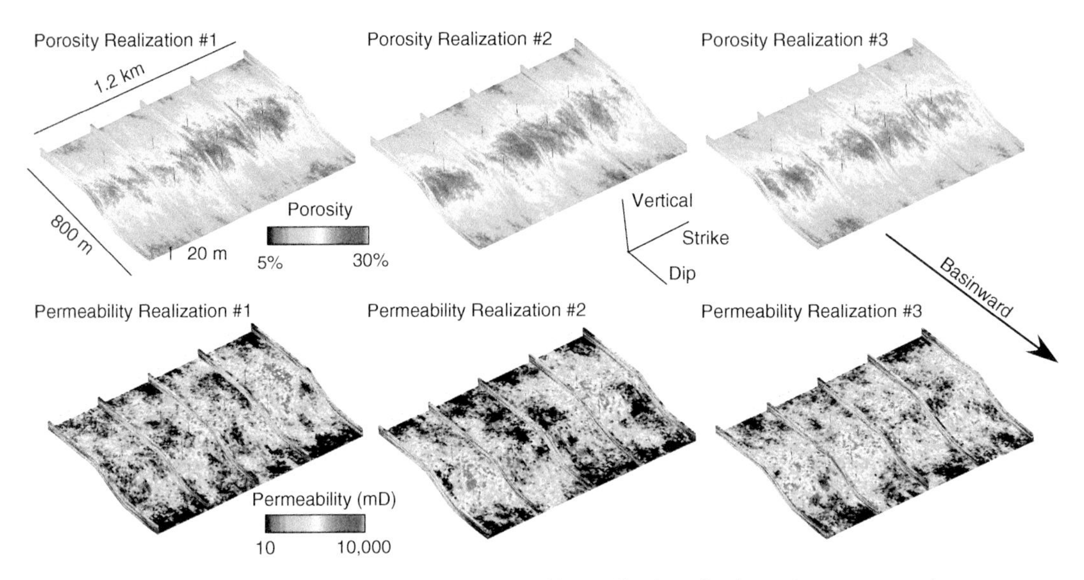

FIGURE 5.28: Oblique View of Three Porosity and Permeability Realizations for the Carbonate Reservoir.

To demonstrate the following statistics, we utilize a set of 20 realizations of porosity from the Carbonate Reservoir model. Three example realizations of porosity and cosimulated permeability are shown in Figure 5.28.

Expected Value Models

An e-type model is the expected value of each local distribution of uncertainty for a specific continuous reservoir property. For categorical reservoir properties, a local proportion may be calculated for each category. Since all realizations are equiprobable, the model is simply the average of all realizations at each location (in the case of scenarios with assign probabilities as discussed in the next section, weighting may be applied).

$$\bar{z}(\mathbf{u}_\alpha) = \frac{1}{L}\sum_{\ell=1}^{L} z^{\ell}(\mathbf{u}_\alpha), \qquad \forall \alpha \in V$$

where there are L realizations of the property z at all locations within the volume of interest, V.

E-type maps are useful for checking the reproduction of trends in the presence of conditioning data. Near the conditioning data the e-type model should approach the conditioning data value, and away from the data the e-type model should approach the trend or local or global mean in the absence in trend. Also, e-type models provide a best local estimate after integration of all data and input statistics. See an example e-type model in Figure 5.29.

Conditional Variance Models

The conditional variance model is the variance of the local realizations, at each location in the model.

$$\sigma^2(\mathbf{u}_\alpha) = \frac{1}{L}\sum_{\ell=1}^{L}[z^{\ell}(\mathbf{u}_\alpha) - \bar{z}(\mathbf{u}_\alpha)]^2, \qquad \forall \alpha \in V$$

where $\bar{z}(\mathbf{u}_\alpha)$ is the e-type value at location $\mathbf{u}_\alpha$, and there are L realizations of the property z at all locations within the volume of interest, V.

A useful representation is the conditional standard deviation:

$$\sigma(\mathbf{u}_\alpha) = \sqrt{\sigma^2(\mathbf{u}_\alpha)}, \qquad \forall \alpha \in V$$

because the units are equivalent to the units of the property of interest. With a distribution assumption, this single value may be used to communicate local uncertainty. For example, with the Gaussian assumption, there is a 95% chance of the true value existing within two standard deviations of the e-type value.

If no distribution assumption can be made, then the local percentiles may be calculated directly from local realizations and used to calculate a measure of spread such as interquartile range.

For categorical variables, entropy may be calculated as an indication of "spread" or local uncertainty.

Conditional variance and expectation maps together provide a visualization of the homoscedastic or heteroscedastic nature of the uncertainty model. For example, homoscedastic models are identified by independence between conditional expectation

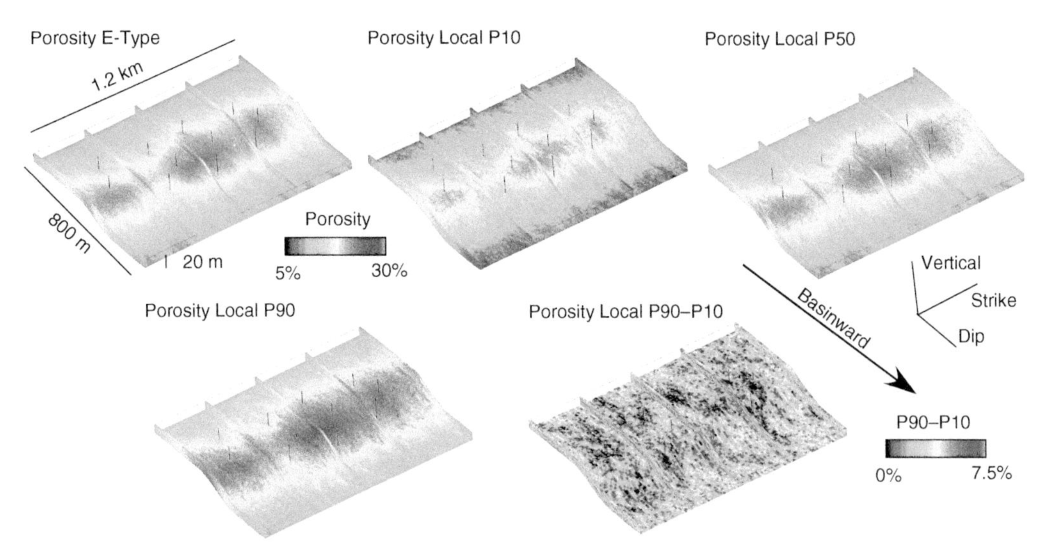

FIGURE 5.29: Oblique View of Post-processed Models Derived from the 20 Realizations of Porosity (See Three Realizations in Figure 5.28). These models include e-type, local P10, P50, and P90 and local P90–P10.

and conditional variance; that is, maximum conditional variance in areas of low conditional mean and high conditional mean are equivalent, and this may occur with the decomposition trend and residual modeling method.

Local Percentile Models

Care must be taken to specify these models as *local* percentile models. These local percentile models are not percentile models, commonly expressed as for example P10, P50 or P90 and so on. Local percentile models represent the specified p value from each local distribution of uncertainty. They are calculated by

$$z^p(\mathbf{u}_\alpha) = F^{-1}(p)(\mathbf{u}_\alpha), \qquad \forall \alpha \in V$$

where F is the local distribution function, typically formed by listing the local realizations at location $\mathbf{u}_\alpha$, $z^\ell(\mathbf{u}_\alpha)$ over the realizations $\ell = 1, \ldots, L$.

The local percentile models are a powerful communication tool. For example, if a specific location is high in a local P10 model, then it is surely high, because there is a 90% probability that the true value will be even higher. Conversely, if a location is low in a local P90 model, then it is surely low, because there is a 90% probability that the true value will be even lower (see local P10, P50, and P90 models in Figure 5.29). As mentioned above, paired local percentile models may be applied to calculate measures of spread such as interquartile range (see the example range model based on local P90–P10 in Figure 5.29).

Indicator Models

A set of indicator maps may also be a valuable method to visualize individual realizations [see examples from Isaaks and Srivastava (1989)]. Recall the indicator transform from Section 2.2, Preliminary Statistical Concepts.

$$i^\ell(\mathbf{u}_\alpha; z_k) = \begin{cases} 1, & \text{if } z^\ell(\mathbf{u}_\alpha) \le z_k \\ 0, & \text{otherwise} \end{cases}$$

These models provide the coding of whether a specific local realization, $z^\ell(\mathbf{u}_\alpha)$, is less than a specific threshold z_k. This may be interesting if z_k represents a specific economic, or otherwise important, threshold.

Perhaps of more interest is the expectation of these indicators over all realizations, L. This represents the probability of the true value being less than (or, if specified, greater than) the threshold.

$$P\{Z(\mathbf{u}_\alpha) \le z_k\} = \frac{1}{L} \cdot \sum_{\ell=1}^{L} i^\ell(\mathbf{u}_\alpha; z_k), \qquad \forall \alpha \in V$$

These models may be applied directly to indicate the risk of failing to match or exceed an economic threshold for a specific reservoir property. Example models based on indictor transforms of the porosity realizations are shown in Figure 5.30 along with an expected indicator model that represents the probability of exceeding a porosity threshold of 15%.

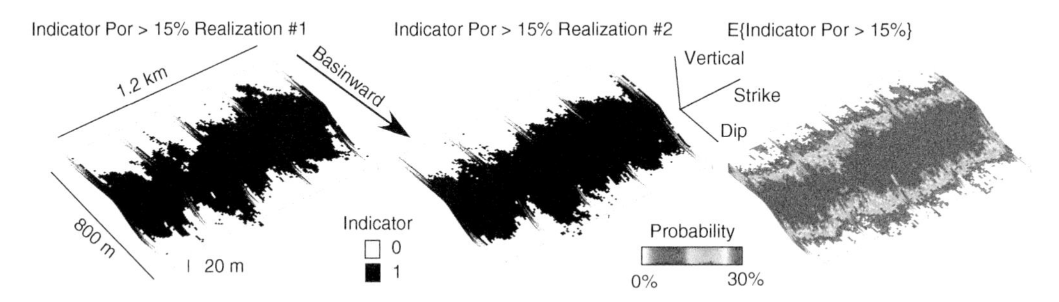

FIGURE 5.30: Oblique View of Indicator Results Derived from the Realizations of Porosity (See Three Example Realizations in Figure 5.28). The indicator models are based on the first two realizations and the expected model is based on the first 10 realizations.

Moving Window Models

There are various moving window statistics that may be applied to investigate reservoir realizations. As discussed in Section 4.1, Large-Scale Modeling, moving window methods are commonly applied to calculate local means and proportions to trend models (Manchuk and Deutsch, 2011). Application of these methods to a single realization provides a method to assess the reproduced trends in a single realization and fluctuation in these trends over multiple realizations.

Moving window methods may be applied to calculate a variety of statistics from individual local realizations and derivatives that compare the results over realizations. These include: locally variable azimuth models (Boisvert, 2011), locally variable connectivity or connected volume (Deutsch, 1998b), locally variable correlation coefficient, spatial continuity, and so on. In general, more complicated statistics will likely produce more noisy local results from moving windows. Careful tuning of the windows size may be helpful (Boisvert, 2011) or the resulting models may be averaged to remove noise.

Beyond model checking and exploration, these locally variable statistics can be applied to constrain subsequent models. For example, a locally variable azimuth model extracted by post-processing a cell-based facies model can be applied to constrain the orientation of the continuous property simulation within the facies model.

5.2.5 Joint Summary Models

The geological model represented by multiple realizations provides a model of joint uncertainty (see Section 4.1.4, Multivariate Mapping), given the imposed multivariate and spatial relationships. These models should provide realistic results in the subsequent transfer function. Yet, there are various other opportunities to explore this joint uncertainty model. This may include: (1) joint probabilities, (2) geobodies, and (3) spatial features.

Joint Probability Models

Since the model realizations represent the correct spatial and multivariate relationships, it is possible to calculate joint probability models. A joint probability model is calculated as the probabilities of a set of conditions relative to the variables of interest. For example,

$$P\{\phi(\mathbf{u}_\alpha) > 5\% \text{ and } k(\mathbf{u}_\alpha) > 800\,\text{mD}\}, \quad \forall\alpha \in V$$

is the joint probability at each location, $\mathbf{u}_\alpha$, in the model that porosity and permeability values will exceed a specific threshold. This is calculated by applying the respective indicator transform to all locations.

$$i^\ell(\mathbf{u}_\alpha;\phi_t,k_t) = \begin{cases} 1, & \text{if } \phi^\ell(\mathbf{u}_\alpha) > \phi_t \text{ and } \\ & \quad k^\ell(\mathbf{u}_\alpha) > k_t \\ 0, & \text{otherwise} \end{cases} \tag{5.1}$$

Then the resulting indicator values are averaged over all realizations to calculate the associated probability of occurrence:

$$P\{\phi(\mathbf{u}_\alpha) > \phi_t \text{ and } k(\mathbf{u}_\alpha) > k_t\} = \frac{1}{L}\cdot\sum_{\ell=1}^{L} i^\ell(\mathbf{u}_\alpha;\phi_t,k_t), \qquad \forall\alpha \in V$$

This may be a useful way to visualize and summarize uncertainty relative to important thresholds. For example, a joint probability model may state the probability a specific minimum porosity,

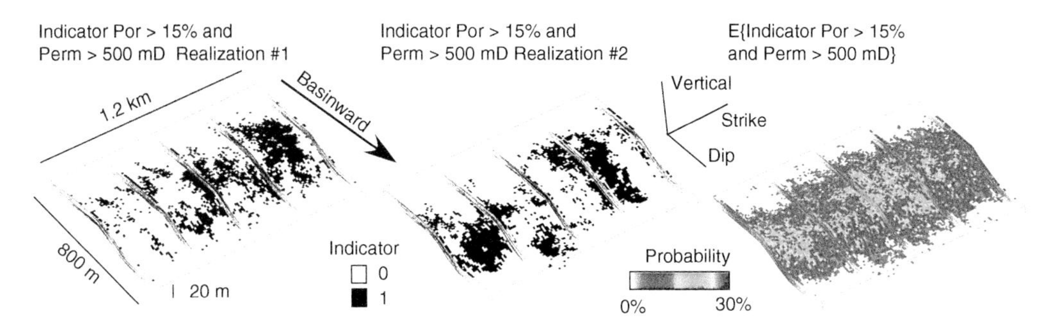

FIGURE 5.31: Oblique View of Indicator Results Derived from the Joint 10 Realizations of Porosity and Permeability (See Three Realizations in Figure 5.28).

The indicator models are for two realizations, and the expected model is based on 10 realizations.

permeability, and oil saturation deemed necessary for reservoir (i.e., potential sweet spots). An example of joint indicators is shown in Figure 5.31. The indicator is set to 1 if $\phi(\mathbf{u}_\alpha) > 15\%$ and $k(\mathbf{u}_\alpha) > 500$ mD, and the expectation model provides the joint probability of porosity and permeability models exceeding their respective thresholds.

More complexity may be added by combining a local moving window summarization, as discussed previously. Instead of a cell scale measure, the probabilities may be calculated from the indicators over a moving window to provide joint probabilities at a larger scale or with a simple connectivity calculation to add the concept of connected volume that is discussed next.

Geobodies and Connectivity

Connectivity measures and connected geobodies may be calculated from realizations to summarize model heterogeneity. There are various methods available to calculate 3-D connectivity (Deutsch, 1998a; Mehlhorn, 1984; Preparata and Shamos, 1988). The Deutsch (1998a) method applies the multivariate indicator transform [see Eq. (5.1)] to assign reservoir and nonreservoir and then applies a fast method to assign unique indices to each geobody and calculate geobody size.

This method is a potential candidate for ranking realizations (see next section), but may also be applied to summarize and compare model realizations. In addition, the results may be (a) summarized locally with connected volumes associated with each location in the model and (b) averaged over multiple realizations to provide expected connected volume and probabilities of exceeding a specific connected volume.

Spatial Features

In addition to connectivity, other advanced measures may be good model summaries. The introduction of Section 4.7, Optimization for Model Construction, mentioned various advanced statistical measures for describing heterogeneity. These include point-based and raster-based statistics, such as the Ripley's K function and lacunarity, respectively. While these spatial statistics are not currently imposed on geostatistical realizations, they provide a unique opportunity to post-process and summarize the geological model. The raster-based methods may be directly applied to realizations. Lacunarity provides a measure of the fractal space-filling behavior of a realization across all relevant scales. Point-based statistics are more difficult, because they require the additional step of assigning points to the model to represent heterogeneity. This is not likely practical by hand for 3-D models, and methods to automate point assignment will likely require supervision and/or checking.

5.2.6 Work Flow

This section has covered the use of post-processing to scale, modify, or summarize models. Figure 5.32 provides a general work flow for model post-processing. Multiple realizations are compiled. They are checked for data, trend, and distribution reproduction (see Section 5.1, Model Checking) and corrected as needed. Then the scale of the models are adjusted as required. Pointwise and joint summary statistics are calculated and stored as needed. If issues are identified from these model summaries, then the simulation setup may be revisited and the realizations may be recalculated.

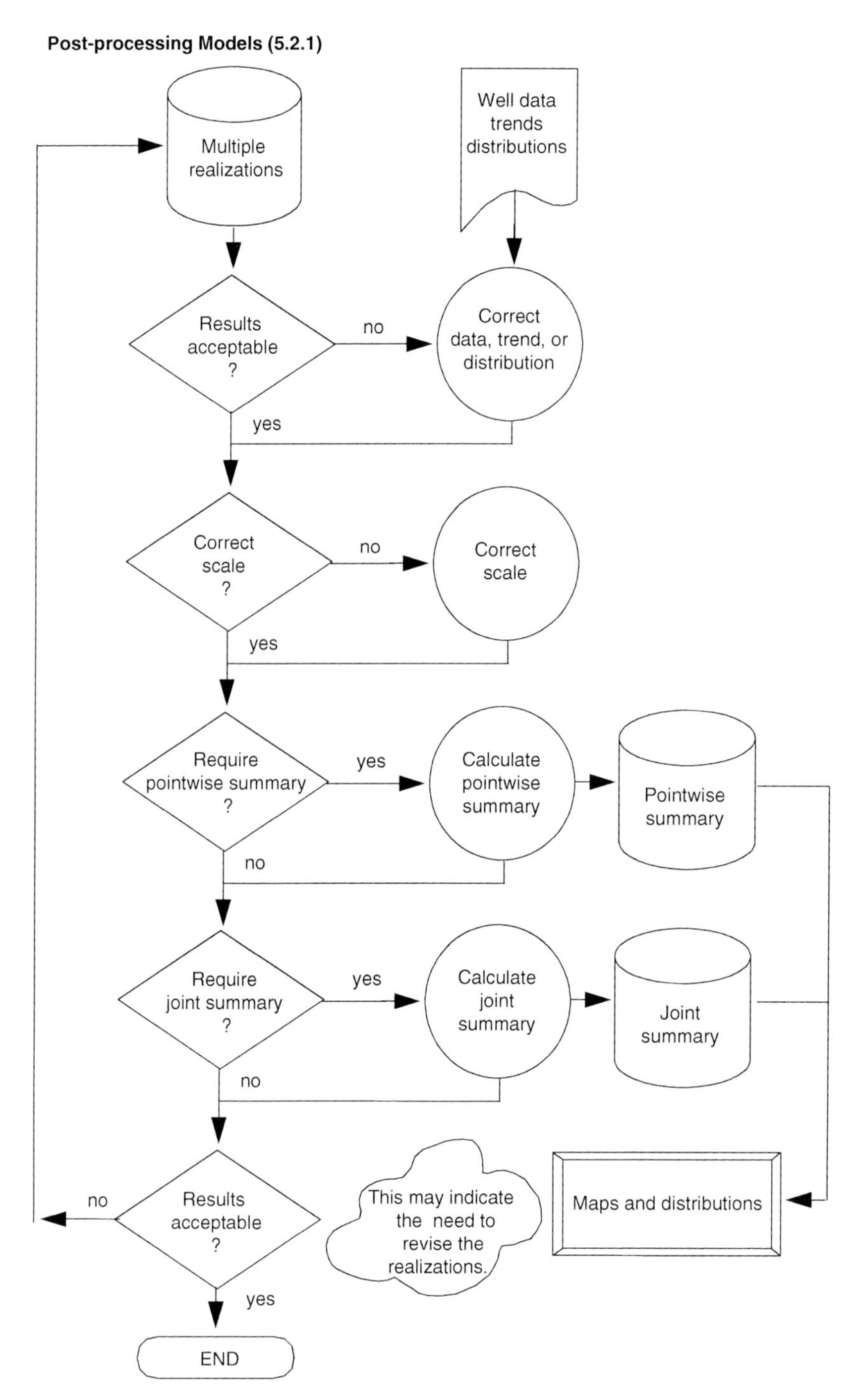

FIGURE 5.32: Work Flow for Post-processing Models.

5.2.7 Section Summary

Model post-processing is a suite of tools that may be applied to the geological model, including all realizations of all properties, or to a single realization to modify or fix the models, to scale the models, or to summarize the models. In some work flows, these operators are an integral part of the work flow to correct data conditioning and input statistics. With other work flows, these operators are important tools for better understanding and learning from the realizations and/or the entire geological model represented by the set of realizations.

The next section covers the topic of uncertainty management for the geological model.

5.3 UNCERTAINTY MANAGEMENT

The geological model is a set of multiple geostatistical realizations that reproduce the input data and geological concepts equally well. This chapter addresses the inevitable questions related to the space of uncertainty, summarizing uncertainty, validating geostatistical models, ranking realizations, the number of realizations required, the value of ranking realizations, and decision making in the presence of uncertainty.

The *Background* subsection addresses the space of uncertainty represented by geostatistical reservoir models. Setting the conceptual geological model and the input statistics constant for all realizations may be unrealistic. An increasingly popular alternative is to vary input statistics and the conceptual model within alternative scenarios to more completely represent our state of incomplete knowledge.

The *Summarizing Uncertainty* subsection covers various methods and formats to summarize uncertainty. This includes discussion on work flows to determine the impact of well density on the level of uncertainty and arguments for purely geometric methods to determine the level of uncertainty. While the latter may seem contrarian to geostatistical modeling, these arguments provide insightful considerations on issues related to probabilistic uncertainty models.

The number of realizations required for a particular reservoir modeling study depends on the aspect of uncertainty being characterized and the precision with which the uncertainty must be known. The *How Many Realizations?* subsection presents guidelines to determine the number of required realizations.

Often, flow simulation can only be performed on a limited number of realizations due to CPU requirements and the need to consider uncertainty in fluid properties, well locations, and other aspects of the reservoir development plan. The *Ranking Realizations* subsection presents techniques for ranking and selecting realizations for flow simulation.

The *Decision Making with Uncertainty* subsection addresses decision making in the presence of uncertainty. Finally, the *Work Flow* subsection presents some work flow diagrams for uncertainty assessment and ranking.

> The methods presented in this section are accessible to novice and intermediate modelers and will provide useful guidance to uncertainty management. For expert modelers, the new uncertainty summarizations methods and arguments for geometrical constraints may be useful.

5.3.1 Background

A geostatistical reservoir model is a "set" of spatially distributed parameters including (1) the structural definition of each stratigraphic layer, (2) facies and/or other regions within each stratigraphic layer, and (3) petrophysical properties such as porosity, permeability, and residual saturations on a by-layer and by-facies basis. Each model consists of *one* realization of each parameter. For example, multiple porosity realizations are not built for the same facies realization; one porosity realization is associated with each facies realization (see Section 5.2, Model Post-processing). Recall that the complete set of realizations is the"geological model."

Multiple geostatistical realizations provide an assessment of the "model" or "space" of uncertainty. Each realization is a Monte Carlo sample from the space of uncertainty defined by all decisions implicit to the modeling approach. As discussed in Chapter 1, Introduction, there is no objective or correct space of uncertainty. The space of uncertainty created by multiple realizations is realistic when the conceptual geological framework and statistical parameters, such as the variograms and size distributions, are well known. These parameters are *not* well known early in the lifecycle of a reservoir; therefore, there is more uncertainty than measured by a set of geostatistical realizations generated with the same set of underlying parameters.

A more realistic space of uncertainty is determined by a combination of a scenario-based approach and conventional geostatistical modeling. Figure 5.33 gives a schematic illustration. In this example, the reservoir could fit any of three different conceptual geological models (M-I, M-II, or M-III) with probability 0.5, 0.3, and 0.2. There may be different sets of modeling parameters (for example, variograms or size distributions) for each conceptual geological model. There are 3, 2, and 3 sets of possible parameters for the conceptual geological

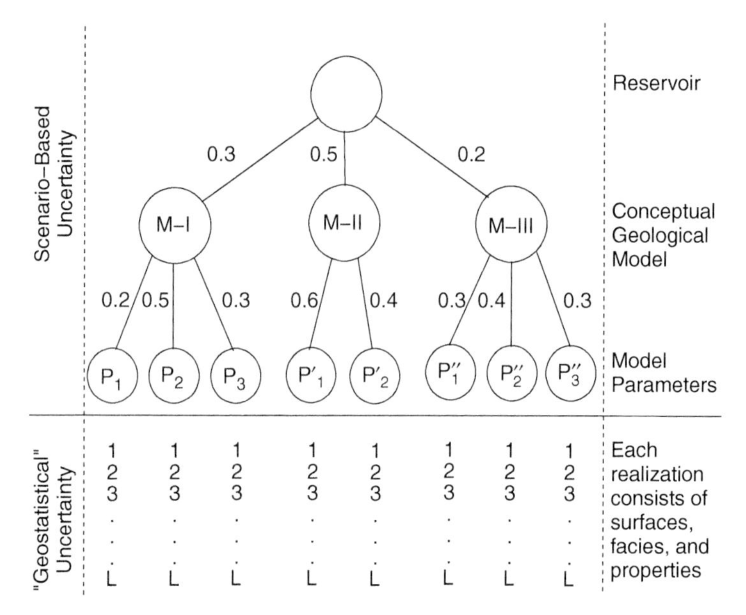

FIGURE 5.33: Schematic Illustration of Combining Scenario-Based and Geostatistics-Based Models of Uncertainty. A number of realizations, L, for each scenario is constructed. The probability of each realization is calculated by recursive application of Bayes relation.

models in Figure 5.33. Once again the probability of each set of parameters must be specified. Then, a set of geostatistical realizations is generated for each of the eight scenarios. These are combined for a more realistic space of uncertainty.

Conditional probabilities are needed at every step of the scenario "tree." For example, in Figure 5.33, given the scenario M-I, the probabilities for parameter sets P_1, P_2, and P_3 are 0.2, 0.5, and 0.3, reflecting the expert judgment that the parameter set P_2 is the most probable given this conceptual model. These probabilities are necessarily subjective and must be specified by the project team or relevant expert. The conditional probability values must sum to 1 at every level.

Recursive application of Bayes' law is used to combine uncertainty at all levels—that is, assign the probability to each geostatistical realization. Recall the definition of a conditional probability that is closely related to Bayes' law from Section 2.2, Preliminary Statistical Concepts:

$$\text{Prob}\{A|B\} = \frac{\text{Prob}\{A \text{ and } B\}}{\text{Prob}\{B\}}$$

where A and B are events. $\text{Prob}\{A|B\}$ is the conditional probability of event A given that B has occurred, $\text{Prob}\{A \text{ and } B\}$ is the joint probability of A and B happening together, and $\text{Prob}\{B\}$ is the marginal probability of event B occurring. Bayes' law tells us that the joint probability at different levels is established by multiplication; for example, the probability of the parameter set (P_1) and facies type one (I) would be

$$\begin{aligned}\text{Prob}\{P_1 \text{ and MI}\} &= \text{Prob}\{\text{MI}\} \cdot \text{Prob}\{P_1|\text{MI}\} \\ &= 0.3 \cdot 0.2 = 0.06 \\ &= 0.06\end{aligned}$$

This relation may be applied recursively for any number of scenarios and any level of complexity.

At the lowest level (Figure 5.33) there are L realizations. Each realization is equally probable given the set of parameters used to create it. Once again, by Bayes' law the probability of each realization is calculated by multiplying the probability of each realization $(1/L)$ by the probability of each scenario. The sum of all the probabilities is 1. The probabilities of the eight scenarios in Figure 5.33 are 0.06, 0.15, 0.09, 0.30, 0.20, 0.06, 0.08, and 0.06 from left to right, adding up to one. A combined distribution of uncertainty in any particular variable, such as oil in place or recovery, is assembled pooling all of the realizations with the correct probability. All summary statistics and probabilities are calculated using these probability weights.

A reasonable definition of scenarios and assignment of conditional probabilities is critical. The scenarios can reflect different aspects of uncertainty—for example, depositional style (estuarine channels versus tidal dominated), fault seal (sealing versus partial sealing), level of fracturing, and facies definition. The scenarios could also reflect uncertainty in critical statistical parameters such as the net-to-gross ratio (low, medium, and high), the trend model, or the horizontal variogram range. There can be many levels to the scenario tree and a different number of levels down each branch of the tree. There could also be a different number of geostatistical realizations for each scenario.

The large-scale discrete aspects of uncertainty are quantified with the scenario-based approach. Uncertainty due to incomplete data is quantified with multiple geostatistical realizations. Reservoir uncertainty addressed with geostatistical realizations *alone* is reasonable late in the reservoir life cycle where there is little large-scale uncertainty; small-scale models of heterogeneity are adequate to reflect uncertainty.

While there is no method for assessment of a completely objective uncertainty model, methods were discussed in Section 5.1, Model Checking, for validating geostatistical models of uncertainty. These methods can demonstrate if an uncertainty model is reasonable with respect to data constraints. Checking local uncertainty is possible, but it is difficult to check large scale resource/reserve uncertainty. There are alternative techniques for uncertainty modeling, but these methods are not discussed here. The interested reader is referred to Caers (2011), Cherpeau et al. (2012), Scheidt and Caers (2008), and Suzuki et al. (2008)

5.3.2 Uncertainty Considerations

The uncertainty model must be specified with an associated volume of interest, V. The level of uncertainty may vary significantly over the reservoir. Well density and seismic quality may vary over the reservoir, and some parts of the model may involve extrapolation with limited data support and the potential for major changes in local reservoir character. Care should be taken to document this and avoid summarizations that overly average out and lose this information.

An attempt must be made to identify and account for the various sources of uncertainty. This includes: uncertainty in the data measures, uncertainty in the interpretation of the data, uncertainty in the data locations, and uncertainty in the input statistics and concepts inferred from data and analogs. This may require randomizing and applying scenarios for the modeling inputs. Yet, we must realize that there are practical limits with this uncertainty model. Consider that there is uncertainty in this uncertainty model. It may be tempting to consider uncertainty in uncertainty, and so on, but this is not typically productive. We limit ourselves to building a defendable uncertainty model that accounts for direct sources of uncertainty.

It is important to consider the scale of the uncertainty model. Uncertainty may be represented at reservoir cell, region, reservoir, or larger scales. Volume variance relations (see Section 2.3, Quantifying Spatial Correlation) indicate that as the modeling scale increases, local variance and uncertainty decrease. If the scale is set as overly large, then uncertainty may appear exceptionally low. Often, uncertainty at large scales relative to the reservoir are related to large-scale trends and transition and not simple statistical fluctuations of the input statistics. This is perhaps the most subjective and challenging part of uncertainty characterization. In these cases, expert geological judgment is preferred over methods that attempt to calculate parameter uncertainty, given assumed statistics (see the discussion on parameter uncertainty in Section 2.2). Scale of support of uncertainty should be directly related to the project goals and should be explicitly stated with the uncertainty model and with any associated summarizations

The next subsection discusses the methods for determining the number of realizations required to formulate an uncertainty model. This is appropriate before presenting procedures for using the resulting space of uncertainty, such as summarization, ranking, and decision making.

5.3.3 How Many Realizations?

An important contribution of geostatistical reservoir modeling is a framework for uncertainty assessment through Monte Carlo sampling from the space of uncertainty in reservoir responses. Monte Carlo sampling proceeds by (1) drawing L realizations from a probabilistic model, (2) processing the L realizations through some performance calculation, and (3) assembling a histogram of the L responses to represent a distribution of uncertainty in the output(s).

Classical Monte Carlo simulation requires the L realizations to be drawn randomly; therefore, they each go into the distribution of uncertainty with equal probability. The question of how many realizations must be addressed.

Significant CPU time is required to process realizations through a flow simulator, which is the most common performance calculator. Often, a small number of realizations are considered for this reason alone. The following discussion on the required number of realizations is for those cases that we *can* consider more realizations for a better assessment of uncertainty.

The required number of realizations depends on (1) the aspect of uncertainty or the "statistic" being quantified and (2) the precision with which the uncertainty assessment is required. Few realizations may be required to assess an average statistic such as the average porosity. A large number of realizations are required, however, to assess the 1% and 99% percentiles of the oil-in-place distribution.

The reporting procedures of environmental standards and political opinion polls will help us define the basis for the results of geostatistical analysis. Consider a poll to answer the question "Have recent price increases for energy caused any financial hardship for you or your household?" The answer would be reported as percentages of yes and no responses—for example, 56% yes and 44% no. Responsible polling agencies would add a caveat to convey the uncertainty in this result. They may report the number of respondents; or, more likely, they would apply basic statistics to more completely communicate uncertainty in the response—for example:

> The true percentage is within 3% of this reported result (56% yes) 19 times out of 20.

These two additional numbers are used to summarize uncertainty in the reported result.

The reported statistics from geostatistical analysis are particular quantiles of some response variable (for example, oil in place)—say, the $F(p) = 0.1, 0.5,$ and 0.9 quantiles. As introduced above, two tolerance parameters are also required: (1) a tolerance for the quantiles, say $\Delta_F = \pm 0.01$ and (2) the minimum probability of being within probability Δ_F, say $t_F = 0.8$ or 80%. A typical case would be to require the P_{10}, P_{50}, and P_{90} of recovery factor within 2%, 80% of the time. A more stringent requirement would be the P_1 and P_{99} of net present value within 0.1%, 95% of the time ($\Delta_F 0.001$ and $t_F = 0.95$).

Thus, two parameters are used: (1) Δ_F, which is the deviation, in units of cumulative probability, from the reported quantile and (2) t_F, which is the fraction of times the true value will fall within $\pm$ of the reported statistic. In this context, we can derive the required number of realizations L to meet a certain specified standard—that is, a specified Δ_F and t_F. Alternatively, the parameters Δ_F and t_F can be established for a given L; that is, it is possible to calculate the uncertainty of our results due to a small number of realizations. This is illustrated in Figure 5.34.

The number of realizations L to meet specified Δ_F and t_F can be calculated. The sampling distributions for the cdf values tend to normal distributions. This is not surprising since the cdf value is the sum of a large number of values (the indicator transform at the correct threshold) that are independent (the

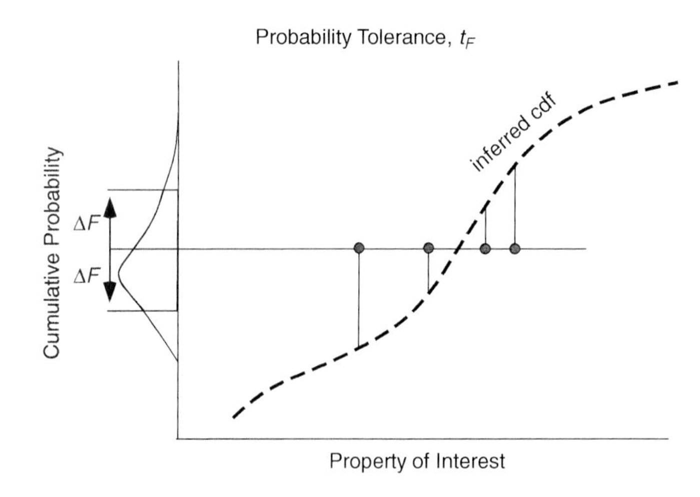

FIGURE 5.34: Illustration Uncertainty in a Specific Percentile Due to Limited Sampling.

With increased number of realizations, this uncertainty expressed as probability of the truth being within a tolerance, Δ_F, decreases; therefore, the precision of the uncertainty assessment increases as more realizations are considered.

		± Tolerance					
		5%	4%	3%	2%	1%	0.5%
	50%	15	25	45	100	410	1600
Probability to be in Tolerance	60%	30	40	70	160	640	2500
	70%	40	60	105	240	970	3900
	60%	60	90	160	370	1500	5900
	90%	100	150	270	600	2400	9700
	95%	140	220	380	860	3500	14000
	99%	240	370	660	1500	6600	24000

FIGURE 5.35: The Number of Realizations Required for a Set of Tolerances, Δ_F, and Probability to Be within Tolerance, t.

realizations are random) and identically distributed. The central limit theorem tells us that the sampling distribution in this case tends toward a normal distribution as the number increases. The number we are considering (L) is large for the central limit theorem; therefore, it is expected that the distribution will be normal. This has been verified numerically.

Given these assumptions, we can derive the following expression to calculate the required number of realizations:

$$L = \frac{F(1-F)}{\left(\frac{\Delta_F}{G^{-1}((t_F+1)/2)}\right)^2}$$

The number of realizations for a specified precision can be calculated directly.

Figure 5.35 shows a look-up table with a number of realizations required to reach the specified tolerance, Δ_F, of 5%, 4%, 3%, 2%, 1%, and 0.5% with probability to be within tolerance t_F of 50%, 60%, . . ., 90%, 95%, and 99%. These values were calculated numerically and tell us "how many realizations are needed for a specified precision" and "how good are the probability values for a given number of realizations."

Practical Results: How Many Realizations?

The analytical result presented above should be used if there is a requirement for quantitative measures of uncertainty. This may lead to an unrealistically large number of realizations. In general, a staged approach should be considered.

- Enumerate the geological scenarios to be considered and generate a single geostatistical realization of all variables for each scenario. Validate that each realization honors the input data within acceptable statistical fluctuation. Calculate the response variables (oil in place, flow performance, and so on) with each realization.
- Generate, say, five realizations for all important scenarios. Important scenarios have a large probability of occurrence of response variables that are unusually low or high. The expert probabilities assigned to each scenario and the response variable from the first realization must be used to make this judgment. Calculate the response variables with these realizations.
- Consider more realizations for those scenarios where the first five random realizations are quite different from each other. The response variables could be calculated on all of these additional realizations.

Calculating the response variables of interest can involve running a flow simulator at significant CPU and professional cost. The generation of geostatistical realizations is relatively quick. This leads to the idea of generating a large number of realizations, ranking them by some fast-to-calculate statistic, and then processing a limited number through the full flow simulator.

5.3.4 Summarizing Uncertainty

This subsection covers methods to summarize uncertainty and visualize uncertainty. While we provide practical advice on internal project reporting and documentation, no effort is made in this book to address formal reserves and resource reporting regulations (e.g., as set by the United States Securities and Exchange Commission). For information on this topic see the current published standards and regulations in the appropriate jurisdiction.

Wilde and Deutsch (2010a) proposed a set of uncertainty formats that are useful for expressing acceptable uncertainty. Any probabilistic uncertainty specification should include the following;

1. Identification of the population or sample being considered
2. Measure of ± uncertainty range
3. Probability of the inaccessible truth value being within the specified uncertainty range
4. Proportion of volumes, specified as $S_\alpha \in V$, within the volume of interest, V, required to meet the proceeding criteria

For example, *the porosity will be ±5% of the predicted value, 19 times out of 20 for at least 90% of the volumes* $S_\alpha \in V$. Often the final criterion is replaced with a definition of the volume of interest. For example, *the porosity will be ±5% of the estimated, 19 times out of 20 over the mapped fairway.* Uncertainty could also be expressed at any scale, even down to individual model grid cells $\mathbf{u}_\alpha$. Here the predicted value indicates an estimate or expected value of a distribution of uncertainty.

These formats include relative uncertainty, absolute uncertainty, and misclassification. The previously mentioned format is relative: a measure of the percentage of the predicted value. Absolute formats may also be applied. These formats specify uncertainty independent of the predicted value (see Figure 5.36 for a schematic of uncertainty that is dependent and independent of the predicted value). Absolute formats include ± a specified tolerance, Δ, ± a specified number of standard deviations, σ, and a range based on specified percentiles—for example, local P90–local P10. Any absolute format may be converted to a relative format by division with the predicted value. For example, the standard deviation is divided by the predicted results in a coefficient of variation, a useful uncertainty format.

A critical difference between relative and absolute formats is that the relative measures account, to some extent, for heteroscedasticity; the variability or uncertainty will scale with the predicted value. Relative measures will better summarize heteroscedastic uncertainty models. Homoscedasticity is assumed in kriging, since the kriging variance is not dependent on the estimate, but the forward and back transform to Gaussian space in sequential Gaussian simulation often imparts heteroscedasticity to some degree (see Section 2.4, Preliminary Mapping Concepts). Also, the decomposition of a variable into trend and residual will result in a homoscedastic uncertainty model, because the trend and residual are independent. Therefore, depending on the work flow, the models may express a mixture of homoscedastic and heteroscedastic uncertainty. The use of a relative format on an uncertainty model that does not scale with the predicted value tends to inflate uncertainty in low predicted values.

A final format recommended by Wilde and Deutsch (2010a) is misclassification. This is expressed as the probability of one category being misclassified as another. A common application is to utilize economic thresholds to define rock as reservoir and nonreservoir. This format is attractive because it may be related directly to risk, since the PType I is the risk of omitting a good reservoir and PType II is the risk of producing from a nonreservoir.

Three uncertainty formats are demonstrated to summarize uncertainty for the Carbonate Reservoir, from the porosity realizations discussed in Section 5.2, Model Post-processing (refer to Figure 5.28 and see Figure 5.37).

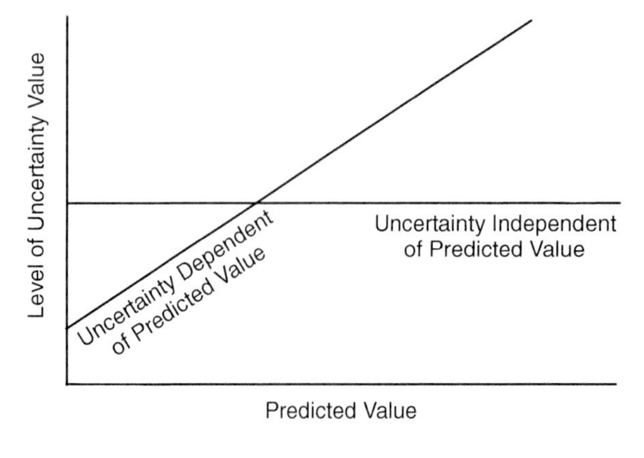

FIGURE 5.36: The Difference Between Uncertainty That Is Dependent on the Predicted Value and Uncertainty That Is Independent of the Predicted Value. For those dependent on the predicted value, use relative formats; otherwise use absolute formats.

Figure modified from Wilde and Deutsch (2010a).

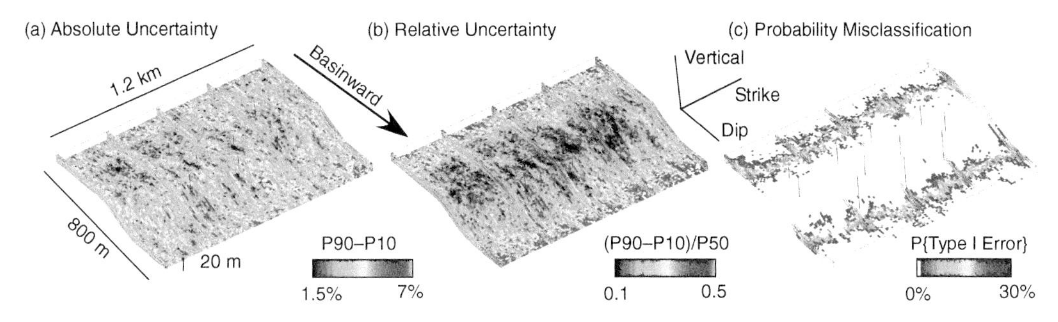

FIGURE 5.37: Uncertainty Summaries for the Carbonate Reservoir, Including: (a) Absolute Measure from the Local P90–Local P10, (b) Relative Measure from the (Local 90–Local P10)/Local P50 and Probability of Type I Misclassification. Note that zero probability of Type I is set as transparent.

5.3.5 Uncertainty Versus Well Spacing

Wilde and Deutsch (2010b) proposed a method to evaluate uncertainty as a function of data spacing with realizations of sequential Gaussian simulation. Their method proceeds as follows:

1. Simulate realizations of the true distribution.
2. Sample the simulated true distributions at regular spacing and add sampling error.
3. Generate realizations conditioned to the simulated data and scale-up.
4. Summarize uncertainty measures for each data density.

By averaging over multiple truth models, the results are more robust. Sampling error is added to mimic imperfect sampling and interpretation. Scale-up is required because the simulation is conducted at an assumed point support size and uncertainty is typically required at reservoir model cell size. The measure of uncertainty may be any of the formats discussed previously. Wilde and Deutsch (2010a) suggested that the measure be summarized over the entire model for a realization and then averaged over all realizations for the summary of uncertainty versus data density.

$$\overline{U}^{(\ell)} = \frac{1}{n}\sum_{\alpha=1}^{n} U^{(\ell)}(\mathbf{u}_\alpha)$$

$$\overline{U}_{(d)} = \frac{1}{L}\sum_{\alpha=1}^{L} \overline{U}^{(\ell)}$$

where there are $\ell = 1,\ldots,L$ realizations and $\alpha = 1,\ldots,n$ locations in the model.

This method requires a strong reliance on the stationary application of the Gaussian model through the sequential Gaussian simulation method. If more information is available, then this approach may be augmented with specified regions and nonstationary trends.

The relationship between data spacing from Wilde and Deutsch (2010b) is shown in Figure 5.38. Note the general increase in the uncertainty metrics as the data spacing increases. This is a useful communication tool to demonstrate the value of well data and its relationship with uncertainty and to aid in the design of well density. In addition, the diminishing incremental value of wells is apparent.

5.3.6 Case for Geometric Criteria

While probabilistic methods, the subject of this book, are useful to describe uncertainty, Deutsch et al. (2007) suggest that geometric criteria, such as drill hole spacing, backed up by uncertainty analysis should be the primary tool for resource allocation (see Figure 5.39). While we do not discuss resource reporting in this book, their argument provides some interesting warnings for the applications of probabilistic uncertainty methods.

Consider the following warnings concerning the application of probabilistic uncertainty:

- Uncertainty is model-dependent and stationarity-dependent. Uncertainty can be changed dramatically by minor changes to these decisions. For example, while trends are generally challenging to infer, changing the strength of a trend directly impacts the overall model of uncertainty as variability is balanced between certain trend and uncertain residual.
- Many parameters affect the distributions of uncertainty in a nonintuitive and nontransparent manner. For example, an

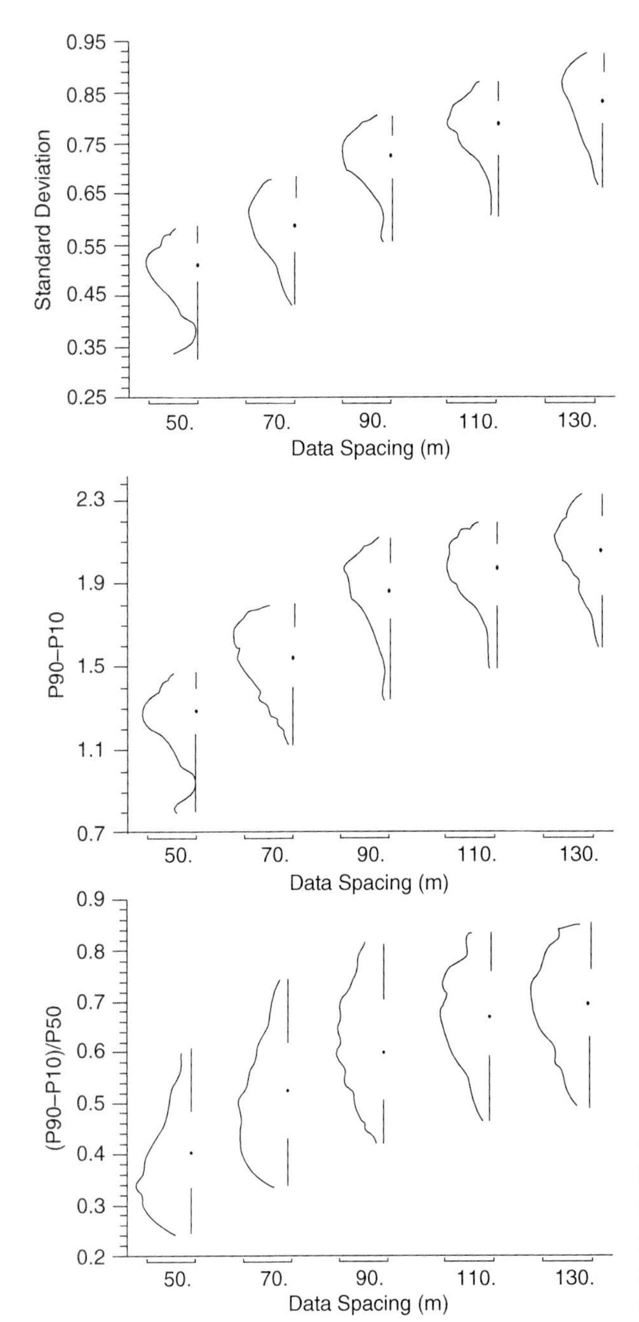

FIGURE 5.38: Relationships between Standard Deviation, P90–P10, (P90–P10)/P50, and Data Spacing. The average (shown as a point) and entire distribution of local uncertainties over the reservoir model are shown.

From Wilde and Deutsch (2010b).

increased nugget effect (i.e., more uncertainty away from wells) drastically reduces the uncertainty after upscaling as properties quickly average out.

- Inference of the histogram—and especially the histogram variance and tails—is often challenging. In addition, inference of uncertainty in the histogram may be very challenging. These all have a first-order control on the resulting uncertainty model.
- Choosing the uncertainty format for classification cannot be universal and is highly case-specific. For example, choosing relative uncertainty versus absolute uncertainty causes low-grade regions to have larger uncertainty.

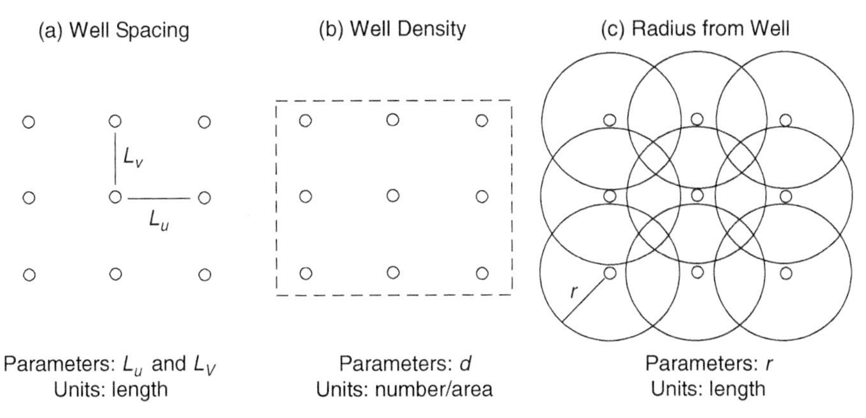

FIGURE 5.39: Schematic Illustration of the Different Geometric Measures of the Amount of Data: Drill Hole Spacing, Drill Hole Density, and Radius from the Drill Hole.

From Deutsch (2005c).

The limitations and challenges of model-dependent uncertainty should not stop us from quantifying uncertainty. A defendable, repeatable statement of uncertainty is more useful than no estimate whatsoever.

5.3.7 Ranking Realizations

Each geostatistical realization reproduces the input data equally well and is equally probable given the scenario or random function model under consideration. Nevertheless, the realizations may be ranked according to some criteria or measure not used as data. For example, the volume of hydrocarbon-filled pore space connected to the well locations can be used to rank a large number of geostatistical realizations.

There is no unique ranking index when there are multiple flow response variables, and no ranking measure is perfect. Nevertheless, the value of ranking realizations will be a reduction in the number of realizations that must be processed to arrive at the same level of uncertainty.

A good ranking statistic correctly identifies low and high realizations. Before describing a number of ranking measures, consider some cases where ranking is problematic or unnecessary:

1. When each realization leads to nearly the same answer.
2. When the aspect of uncertainty being assessed is easy to calculate; for example, the uncertainty in oil-in-place may be simply assessed by calculating the pore volume of all realizations.
3. When there are many independent reservoir responses of interest; that is, it is impossible to conceive of a ranking index that would lead to a unique reliable ranking.

There are times, however, when significant professional or CPU time is required to evaluate each realization and the number of realizations considered must be limited. In these situations, it is worthwhile to consider ranking the realizations to limit the number of fine-scale simulations and yet to obtain an idea of uncertainty in the flow response.

Most simple ranking measures require each cell in all geostatistical realizations to be assigned a *net* indicator. This indicator is one if the cell is reservoir quality and zero otherwise. Reservoir quality is achieved in certain facies by porosity and permeability simultaneously above specified thresholds. The net-to-gross ratio for each realization can be easily calculated and used as a simple ranking measure. Another simple ranking measure is the total oil volume—that is, the product of the net-to-gross indicator, the oil saturation, and the pore volume. This accounts for porosity and saturation variations. An example of this method is shown below.

$$CV = \sum_{\alpha=1}^{N} i(\mathbf{u}_\alpha) \cdot \phi(\mathbf{u}_\alpha) \cdot (1 - S_w(\mathbf{u}_\alpha))$$

where α are all locations over the model or within a drainage radius, N.

Flow performance depends on the connectivity of the reservoir-quality rock; there are many measures of connectivity that may be quickly calculated without running a full simulator (Ballin et al., 1992; Berteig et al., 1988; Deutsch and Srinivasan, 1996;

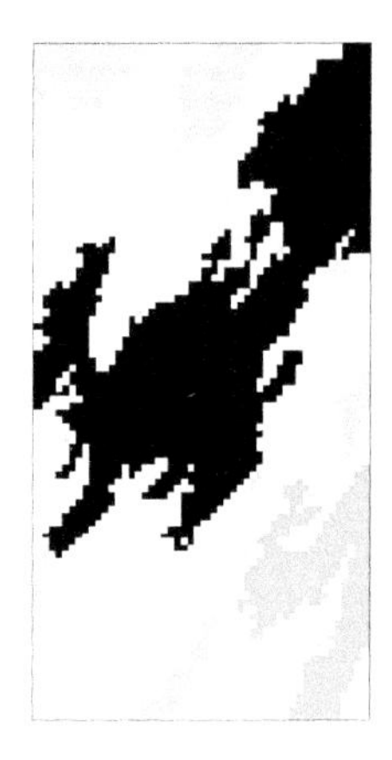

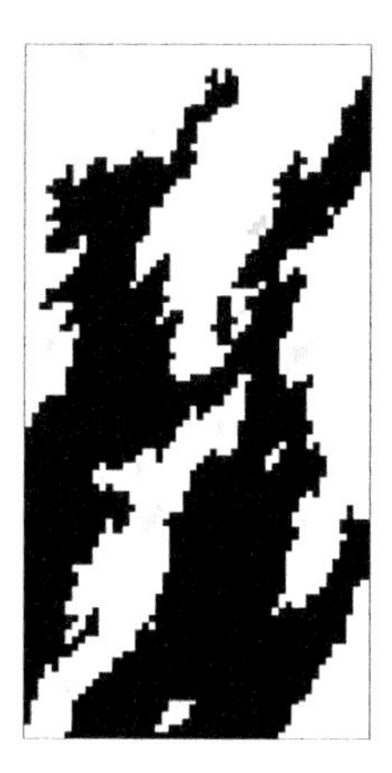

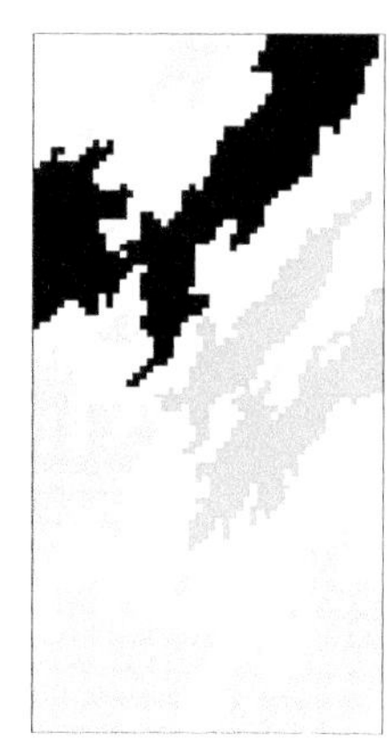

FIGURE 5.40: Slices through Three Facies Models. The black region in each realization is the largest geo-object. Smaller geo-objects are shaded as yellow or blue.

Fetel and Caumon, 2008; Li and Deutsch, 2008; McLennan and Deutsch, 2005). A first measure of connectivity is available by determining the sets of *net* geological modeling cells that are connected in 3-D space. There are fast algorithms to scan through 3-D binary net/non-net realizations aggregating those reservoir-quality cells that are connected net cells (Deutsch, 1998b). The result will be a 3-D specification of the number N_{geo} of geo-objects or connected objects each with an associated volume $V_{\text{geo},j}, j = 1, \ldots, N_{\text{geo}}$. Figure 5.40 shows the geo-objects for three SIS facies realizations.

The geo-objects of a set of realizations can be used in a number of ways to rank the realizations:

- The fraction of cells within the first, say, 5, geo-objects could be used as a ranking measure. A realization is more "connected" when this fraction is large.
- The connected hydrocarbon volume within some radius of the production wells could be considered if the well locations are known. This ranking measure accounts for local information about connectivity.
- The connected volume between "pairs" of injection and production well pairs could be used if the well pair locations are known.

An alternative use of geo-objects is for the selection of well locations. An optimization scheme could be devised to select the well locations that maximize the geo-object volume within some drainage radius.

Geo-objects measure static connectivity; that is, they do not account for tortuosity, permeability, attic oil, and the interaction between multiple producing well locations. A simple proxy could be formulated to account for static connectivity as follows (Li and Deutsch, 2008):

$$Q_s = \sum_{\text{iw}=1}^{N_w} \sum_{\alpha=1}^{N_{\{i\}(\text{iw})}} V(\mathbf{u}_\alpha) \cdot \phi(\mathbf{u}_\alpha) \cdot (1 - S_w(\mathbf{u}_\alpha)) \cdot \left(\frac{d_{\max}}{d_{\text{iw},\mathbf{u}_\alpha}}\right)^{\text{dw}} \cdot \left(\frac{k_{\text{iw},\mathbf{u}_\alpha}}{k_{\max}}\right)^{\text{kw}}$$

where Q_s is a static quality score for the realization given the wells, N_w is the number of wells, $N_{\{i\}(iw)}$ is the number of cells within the drainage distance of a specific well, iw, $V(u_\alpha)$ is the cell volume, $\phi(\mathbf{u}_\alpha)$ is the cell scale porosity, $S_w(\mathbf{u}_\alpha)$ is the water saturation, and the additional two terms represent the tortuous distance and permeability (geometric mean) along the shortest path to the well. d_w and k_w are weights to scale the influence of each factor, and $d_{\max}$ and $k_{\max}$ are maximum values that convert the distance and permeability along flow path, $d_{\text{iw},\mathbf{u}_\alpha}$ and $k_{\text{iw},\mathbf{u}_\alpha}$ respectively, into relative scores (Da Cruz, 2000; Li and Deutsch, 2008).

There are some more sophisticated alternatives that measure dynamic connectivity and, yet, are still simpler than running the full flow simulator. There are random walk algorithms that measure "dynamic" continuity between injecting and producing locations. These methods often call for a solution to the pressure field (single phase flow) given assumed well rates. Particles are then tracked through the media, and the distribution of "times" or "lengths" between injecting and producing wells provides a measure of connectivity that could be used for ranking (Batyeky et al. 1997).

There are other relatively simple and fast flow models including (1) tracer simulation, (2) simulation based on a network of 1-D streamlines (Izgec et al., 2011; Shook and Mitchell, 2009), and (3) a water flood simulation in lieu of a more complex miscible or compositional-type simulation. The time of 5% watercut is likely a good measure for the breakthrough of other miscible components in a more complex process.

Another ranking approach is to use the correct physics or flow equations, but with the geological models scaled to such a coarse resolution that the computer time is acceptable. The coarseness of the underlying grid will compromise the direct usefulness of the results. Nevertheless, the relative ranking of the results may be used to rank the underlying geological realizations.

Practical Result: Ranking

Ranking for quantitative statements of probability should be avoided. Published applications of ranking consider the ranking measure to be "quantile preserving" for example, the P_{10} of the full performance assessment is determined by running the full assessment on the P_{10} of the ranked realizations. There is no statement of how close the resulting P_{10} is to the true value. This procedure will work when the correlation between the "rank of the ranking measure" and "rank of the full assessment" is large; however, that is not usually the case.

Ranking should be applied more qualitatively to select realizations to get a rough idea of "low," "expected," and "high" performance. The following are some additional cautionary notes on ranking.

The rank assignment must include clear indication on the method and assumptions and resulting limits in applicability of the rank measure. The rank is only valid for the region of investigation, specific scale, and ranking measure. For example, models ranked by a global measure of oil in place will possibly rank very differently for a local measure of connectivity, and a model that is high for global connectivity may not be high for a specified region of the model. The ranking criteria (region, scale, and measure) must be retained and documented with the rank assignments, because rank assignments are often very "sticky" and frequently misapplied. Often the labels of low or high model outlasts the original application and get applied for new problems. For any new question, the modeler must once again analyze the full geological model and reapply a matched ranking scheme.

Also, retaining too few realizations from the geological model from ranking may be risky. Given the imperfection of the ranking method, a small set of realizations may not span the uncertainty model after the transfer function. Deutsch and Begg (2001) and Deutsch (2005a) suggested a methodology to relate the number of ranked realizations equivalent to 100 Monte Carlo realizations given the correlation between true and estimated rank (goodness of the ranking measure) to determine the number of required ranked models.

5.3.8 Decision Making with Uncertainty

Making the best decision in face of geological uncertainty is an important topic. Some general ideas are given below; the reader will be required to go to the literature for more information (Bickel and Bratvold, 2008; Bratvold and Begg, 2006).

Petroleum exploration and production are inherently risky activities. Decisions regarding those activities depend on forecasts of future hydrocarbon production revenue. Such forecasts are uncertain because of (1) uncertainty about the reservoir geometry and the spatial distribution of petrophysical properties, (2) uncertainty about the fluid properties, (3) uncertainty about the actual behavior of the rock and fluid when subjected to external stimuli, (4) model limitations, and (5) uncertainty about costs and future prices. This book addresses the uncertainty of the geological model due to sparse sampling of the reservoir.

Exploration and production are sequential. First, exploration finds a promising geologic structure using knowledge of the sedimentary basin and seismic data. Then, a well is drilled to prove the existence of a hydrocarbon reservoir. Other exploratory wells may be drilled to delimit the reservoir, depending on the field size. Next, a development plan is generated to provide the necessary data for the production cash flow analysis. If the organization decides to invest in that project, the development plan is implemented and hydrocarbons are produced.

There are three main types of decisions involved in exploration and production: (1) the decision of whether or not to drill an exploratory well, (2) the technical decision of selecting the best development plan to optimize the profitability of the reservoir production, and (3) the business decision to invest in a project or not. Decision making in exploration typically accounts for uncertainty through decision tables which relate alternative actions to various outcomes (Harbaugh, 1995).

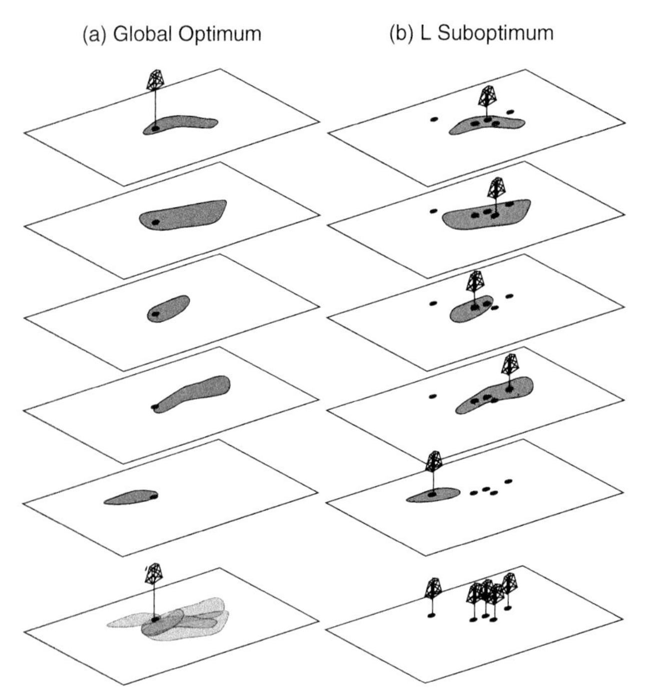

FIGURE 5.41: Illustration of (a) Optimal Well Site Selection jointly over All Realizations and (b) Suboptimal Well Site Selection for Each Realization. Five realizations are indicated by the five stacked planes. Reservoir area for each realization is indicated by the closed shape projected onto the planes.

After exploratory reservoir delimitation, a reservoir development plan is devised that determines the number, type, and location of additional wells and presents the rig work schedule and the curves for injection and production of fluids. Once the development plan is defined, it is possible to transfer some aspects of data uncertainty to the production forecasts. The probability distribution of flow responses can be used to evaluate the expected monetary value of the project and to guide the business decision of whether or not to invest in a particular project.

Accounting for geological uncertainty in the selection of the best development plan will be presented briefly. This is by far the most important decision in reservoir management.

The "conventional" approach to define the development plan ignores geological uncertainty. The conventional procedure consists of (1) building a single deterministic geological model of the reservoir, (2) defining some different possible production scenarios (numbers of wells, configuration for each number of wells, vertical or horizontal, producer or injector, fluid to inject, and so on), (3) running a flow simulator for each scenario to generate the respective production and injection curves, (4) performing a cash flow analysis for each scenario, and (5) selecting the scenario that provides the maximum profit.

Well Site Selection

A number of approaches have been developed for well site selection: (1) integer programming using geo-objects (Vasantharajan and Cullick, 1997), (2) optimization combined with the results of simulating with one well at a time (da Cruz et al., 1999), and (3) experimental design and response surface methodology (Aanonsen et al., 1995; Dejean and Blanc, 1999; Vincent et al., 1990). This optimization problem is particularly difficult due to the large combination of well locations.

Regardless of the approach, the well site selection should be optimized over the entire geological model, not over a single realization or estimated model. Optimization of well placement on a single realization will not likely be optimal over all realizations, while optimizing over all realizations may appear to be suboptimal for any individual realization (see Figure 5.41).

Development Plans

Optimization of the reservoir development plan is also particularly difficult due to:

- The large combination of parameters to select.
- Difficult to capture the effect of timing

- Limitations in the physics accounted for in flow simulation
- Location of injectors/displacement
- Specifics of well completion

The flow simulator has been combined with classical optimization procedures for a single deterministic geological model (Bittencourt and Horne, 1997). Other optimization procedures have been used to account for multiple realizations (da Cruz, 2000). Experimental design and response surface methodology have been applied (Damsleth et al., 1992a; Egeland et al., 1992; Jones et al., 1997) to obtain the distribution of flow responses for each scenario and to retain the best scenario in the presence of uncertainty; however, the consideration of a reduced number of realizations required by those methodologies yields an incomplete assessment of the uncertainty in the flow responses (Stripe et al., 1993).

A comprehensive study was performed by da Cruz (2000). A general procedure for decision making in the presence of geological uncertainty was presented:

1. Generate L geostatistical realizations of the geological model $l = 1, \ldots, L$. The notation for the geological model "l" is intentionally simple, but actually l is a spatially distributed vector of numerical models representing structural surfaces, facies, porosity, permeability, and fluid saturations.
2. Define the possible reservoir management scenarios: $s = 1, \ldots, S$. Each scenario is a complete specification of one possible solution for the problem. For example, one scenario would define the number of wells, their locations, the completion intervals, the surface facilities. The number of scenarios could be very large, but an inspection of the L realizations and prior sensitivity flow analysis based on just one realization should reduce this number substantially. Note that these scenarios are reservoir management scenarios and *not* the geological scenarios discussed at the beginning of this section.
3. Establish a quantitative measure of profit P to be maximized. The measure of profit would increase with increased hydrocarbon production and would decrease as more wells and facilities are required. The profit depends on the related costs, hydrocarbon prices, and taxes. A good unit to measure the profit is the present value of the discounted cash flow.
4. Calculate the profit for each reservoir management scenario and each realization: $P_{s,l}$, $s = 1, \ldots, S$, $l = 1, \ldots, L$. The fluid production and injection curves are obtained by running a flow simulator, and the profit measure is calculated from the scenario specifications and curves for each case (s and l).
5. Calculate the expected profit $\overline{P}_s$ for each reservoir management scenario, based on the average profit over all realizations.
6. Define the optimal reservoir management scenario s^* as the scenario that has the maximum optimal estimate of profit $\overline{P}_s$.

Implementation of this procedure requires care in the specification of the scenarios and the execution of many flow simulations.

Loss Functions

In the preceding discussion, the optimal scenario was defined as the one that maximized the expected value of the profit over the geostatistical realizations. Taking the expected value or average profit implicitly assumes that the consequences of under- and overestimation are the same, more specifically, a least squares criterion is used. In practice, however, the consequences of under- and overestimation are rarely the same and rarely follow a least squares relationship.

The idea of a "loss function" (Journel, 1989; Srivastava, 1987a, 1990a) quantifies the notion of different "loss" or consequences for under- and overestimation. The unknown true profit P will be estimated by a single value p^* with a given error $e = p^* - P$. The function Loss(e) must be specified by the organization or person in charge of the economic decisions in the company. The error e is not known, however, because the true value P is not known; therefore, an expected loss value for each scenario s can be determined using the distribution of P:

$$E\{\text{Loss}\}_s = \frac{1}{L}\sum_{l=1}^{L} \text{Loss}(p_s^* - P_{l,s})$$

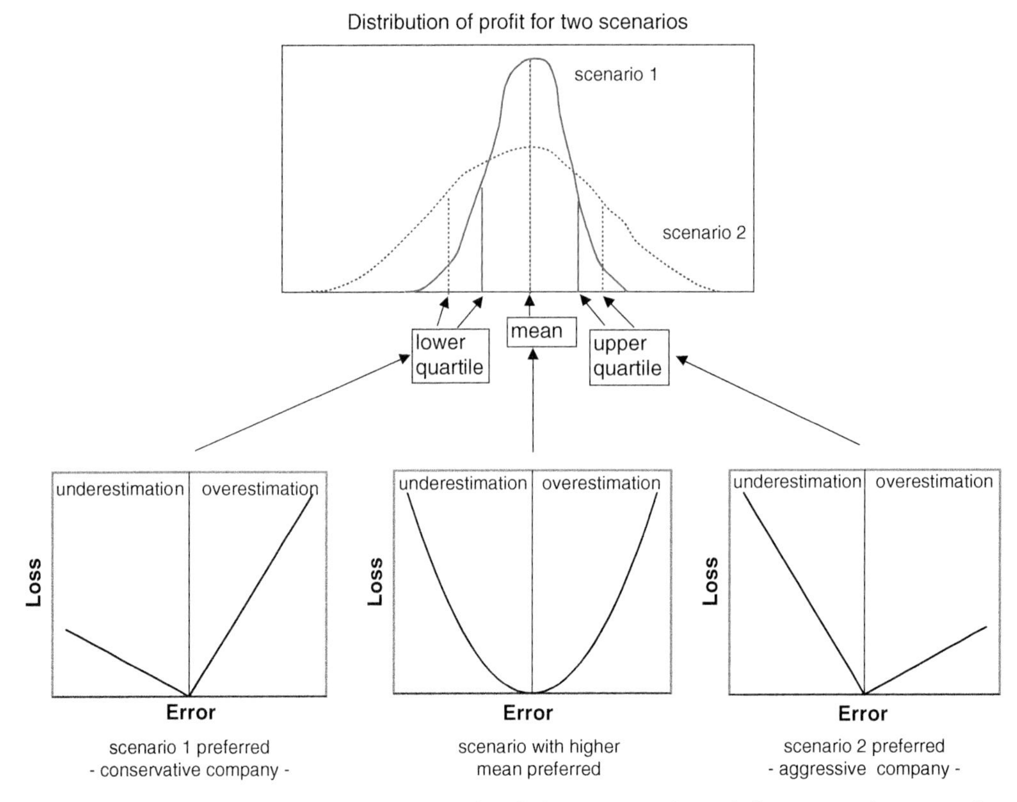

FIGURE 5.42: Example of Probability Distribution of Profit for Two Scenarios and Three Types of Loss Function That Yield Different Values of the Retained Profit Value for Each Scenario and Different Decisions about the Best Scenario.

Modified from Paulo Sérgio da Cruz (da Cruz, 2000).

recall that s is a particular reservoir management scenario, L is the number of geostatistical realizations, and $P_{l,s}$ is the profit for scenario s using realization l. The best estimate of profit for the scenario s is $\hat{P}_s$ such that the expected loss is minimum when taking $p_s^* = \hat{P}_s$.

Figure 5.42 shows a small example. The distributions of profits for two scenarios are shown at the top. Both scenarios have the same average profit, but scenario 1 has less uncertainty that scenario 2. Some remarks:

- The least squares loss function, shown in the center, would not distinguish these scenarios because the mean values are the same.
- The loss function to the left shows a greater penalty for overestimation than for underestimation. This corresponds to a "conservative" company that does not want to overestimate the value of the reservoir. In this case, the profit value retained would be below the mean; for example, the lower quartile of the distribution is retained if overestimation is three times as consequential. Between two scenarios with the same mean profit, a company using this type of loss function would prefer the one with smaller uncertainty (scenario 1).
- The loss function to the right penalizes underestimation more than overestimation. This corresponds to an "aggressive" company that is seeking profit and would take more risk to achieve that profit. In this case, the profit value retained would be above the mean—for example, the upper quartile of the distribution. Between two scenarios with the same mean profit, a company using this type of loss function would prefer the one with greater uncertainty (scenario 2).

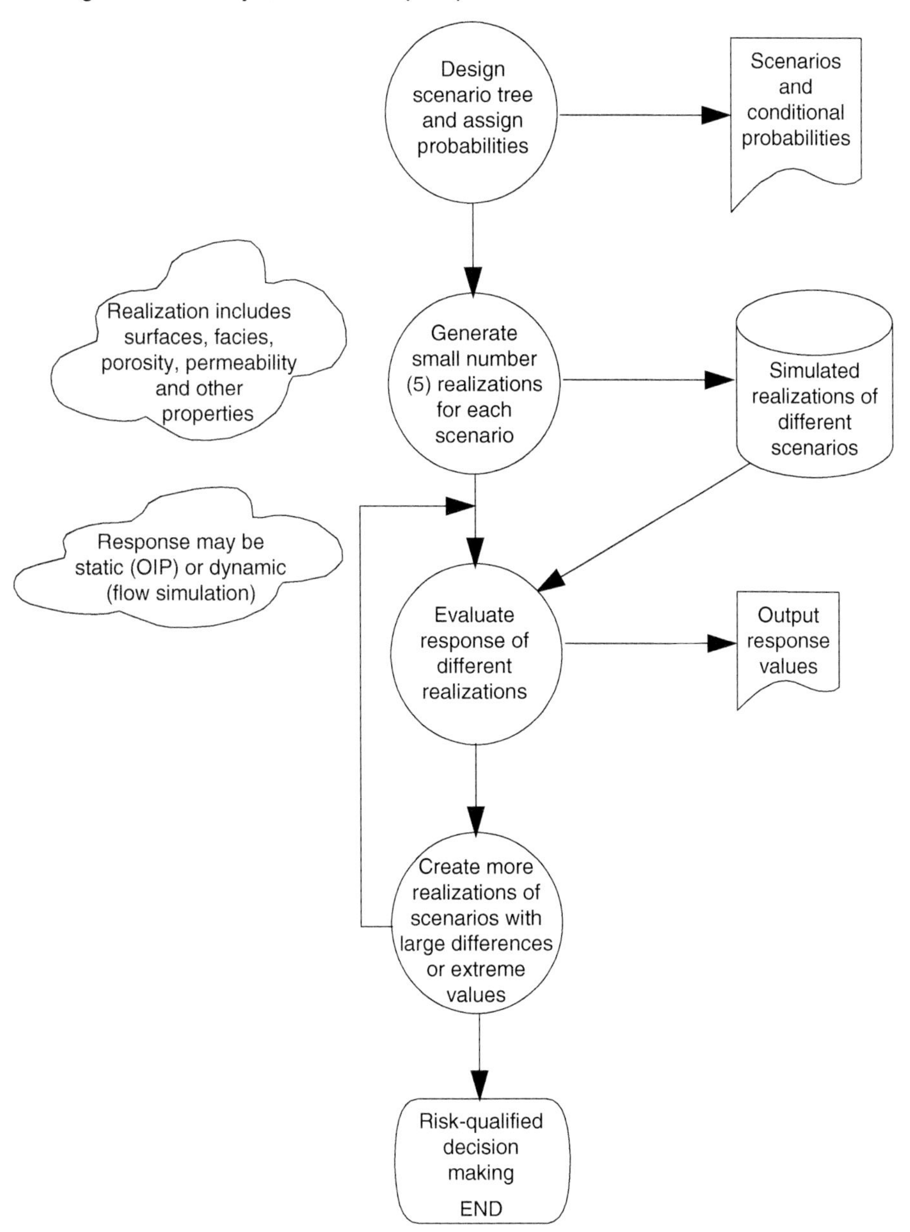

FIGURE 5.43: Work Flow with the Design of Uncertainty Quantification by Scenarios and Geostatistical Realizations. There are many steps within the "create realizations" bubble.

Defining the correct loss function is critical. The optimal decision can be easily determined with a loss function $L(e)$, a distribution of uncertainty ($l = 1, \ldots, L$ realizations), and a set of possible reservoir management scenarios ($s = 1, \ldots, S$). The challenge is to quantify the most appropriate attitude toward risk—that is, the loss function.

The use of loss functions—that is, the minimization of expected loss or the maximization of expected monetary value—is relevant for repeated decisions such as the placement of many wells. One-off or relatively infrequent decisions should be considered on a case-by-case basis, and risk mitigation strategies should be explored with the possibility of gathering new information or staging decisions to permit greater flexibility.

5.3.9 Work Flow

Figure 5.43 illustrates the work flow for quantification of uncertainty using a combination of scenario-based approach and geostatistical realizations.

5.3.10 Section Summary

This section has dealt with fundamental questions related to the space of uncertainty, the number of realizations required, summarizing uncertainty, the value of ranking realizations, and decision making in the presence of uncertainty.

This concludes Chapter 5 and the discussion on Model Applications. In Chapter 6 some special topics are covered briefly with comments on some other emerging techniques and final thoughts.

6

Special Topics

Geostatistical reservoir modeling is the subject of ongoing research and development. This chapter presents some frontier areas that are expected to become important in the future.

The *Unstructured Grids* section presents the problems and methods related to conforming geostatistical property models to modern flexible gridding schemes. There have been various research efforts to formulate simulation algorithms that are able to simulate directly to unstructured grids. In general, the problems have not been worked out, and data or small-scale support simulation and scale-up are now preferred. Methods that utilize two grids,—one large-scale unstructured grid and another high-resolution more uniform cell size—are commonly adopted. Much effort has been placed in improving the variety of spatial heterogeneities that may be reproduced with categorical, facies simulation (see Sections 4.2–4.5). While it is understood that facies heterogeneity is often the most important constraint, work has been conducted on improving *continuous variable heterogeneity* (Section 6.2). This includes efforts to seamlessly conform continuous variable heterogeneity to categorical simulated realizations and efforts to impose heterogeneities in continuous simulation beyond the maximum entropy variogram-based Gaussian model.

Estimation is an important step in the general modeling work flow. Yet, there are various simulation methods that do not have estimation embodiment available. The *More Estimation Methods* section presents a recently developed method for estimation of truncated Gaussian techniques as an example to address this issue.

The *Spectral Methods* section provides a brief overview of simulation in the spectral domain. These methods are very fast, and in spectral domain the variogram has greater flexibility, yet there are limitations that constrain broad applications. Data conditioning and reproduction of the histogram and variogram remain as challenges.

The *Surface-Based Modeling* section reviews the basic approach of stochastic modeling of surfaces. While this had been developed several years ago, utilized in process-mimicking methods to track topography and support rules, broad application as a stand-alone method has not been realized. Given the flexibility to integrate data conditioning, geometries, and trends, it is still anticipated that this method will have future importance.

Ensemble Kalman filtering (Section 6.6) and other ensemble methods have received a lot of attention over the past several years as a method to integrate production data. The method is based on an efficient recursive filter of a suite of model realizations to impose constraints. With a good prior model, these methods have been effective.

Furthermore, we recognize the efforts among earth scientists to develop *advanced geological characterization* (Section 6.7) to efficiently characterize reservoir heterogeneity (mentioned in Section 4.7). These statistics have demonstrated the ability to quantify reservoir architectures and provide objective measures for description, classification, and comparison. The challenge is for geostatistics to integrate these statistics when they are impactful on reservoir heterogeneity.

The *Other Emerging Techniques* section is a miscellaneous collection of methods and opportunities that may be considered in the future. These include new opportunities with increasing computational power and visualization and data sources. In addition, highly customized modeling methods may be an opportunity to deal with these unique data sources at the loss of general applicability.

The *Final Thoughts* section gives some final thoughts for this introductory book on geostatistical reservoir modeling. Geostatistical methods have many limitations. Nevertheless, they are appropriate for modeling heterogeneity and assessing uncertainty in the presence of sparse data.

6.1 UNSTRUCTURED GRIDS

Unstructured grids have been under consideration for reservoir flow analysis for some time. This has presented a unique challenge to geostatistics due to the assumption of the same grid element size across the model grid. The initial research into algorithms for simulation to unstructured grids focused on direct sequential simulation (Manchuk et al., 2005; Tran et al., 2001; Xu and Journel, 1994). This approach did not come to fruition due to several outstanding issues, including difficulties in reproducing distribution models and specifically in accounting for heteroscedasticity—that is, the relationship between the mean and variance that is commonly observed in natural data sets (Leuangthong, 2005).

$$\sigma^2(\mathbf{u}) = f(m(\mathbf{u}))$$

Direct sequential simulation offers the opportunity to model without a distribution assumption and the associated forward and back transforms and at first glance offers new opportunities. Yet, while local distributions may be modeled in a flexible nonparametric manner, the pervasiveness of the central limit theorem results in multi-Gaussianity in the global distributions. It follows that without consistency between the local and global distributions or without a transform, the global histogram is not reproduced. It may appear that direct sequential simulation provides the opportunity to model the proportional effect directly. Yet, the underlying building block remains as kriging with its homoscedastic assumption (mean and variance are independent of each other) that cannot reproduce heteroscedastic distributions.

Given these challenges, current methods for geostatistics on unstructured grids implement an underlying structured grid such that existing geostatistical theory and algorithms are applicable (Caumon et al., 2005a). The underlying structured grid is typically at a fine enough scale to accurately represent the unstructured grid geometries and to provide reasonable upscaled results.

Manchuk and Deutsch (2010) proposed an approach based on the premise that when the coarse grid is unstructured, regular grids are not the optimal choice. Rather, a grid that eliminates the issues and maintains the same set of advantages as with a fine-scale conforming structured grids is proposed. That is, the fine-scale grid is a simplex grid that is able to exactly discretize practically any coarse grid specification including structured, tetrahedral, and perpendicular bisector (PEBI) grids. The simplex grid is subsequently scaled up to the unstructured grid.

The concern with using geostatistics to populate simplex elements is scale. Element volumes are not equal. This is handled by modeling properties at a pseudo-point scale on the discretization (see Section 2.4). It is shown that the error between the pseudo-point scale and simplex element scale variance is mitigated through the upscaling processes.

Some advantages include:

- The discretization fills unstructured elements exactly so that coarse and fine-element interfaces coincide, which simplifies upscaling processes.
- Flow simulation methods are available for simplex grids, and flow-based upscaling methods are applicable.
- Simplex grid generation is flexible, and the resolution can vary locally similar to the coarse unstructured grid—for example, to achieve high resolution near flowing wells.
- The discretization can be constrained to reproduce geological features that are not captured by the coarse grid.

A clear disadvantage of this work flow is that geological modeling is carried out twice; once during preliminary mapping, since the coarse grid design depends on geological heterogeneity, and again after grid refinement. However, geostatistical modeling work flows are easily automated once they are parameterized. This involves assessing statistical properties of sample data including probability distribution functions, first- and second-order moments, multivariate relationships, and spatial covariance. All of these statistics are dependent on the scale of the sample data from which they are computed. Once all data and parameters are defined for a specific case, geostatistical models are constructed automatically with a computer.

6.2 CONTINUOUS VARIABLE HETEROGENEITY

Continuous variable heterogeneity is often considered a secondary issue, because facies heterogeneity typically provides first-order constraint on reservoir spatial continuity. Nevertheless, it may be important to capture specific continuous reservoir property heterogeneities within the facies model or to improve continuous reservoir property heterogeneity in reservoir models without facies constraint.

Cavelius et al. (2012) suggested various methods to better constrain continuous reservoir properties within MPS (or any other cell-based) facies models. These include post-processing the facies model with a distance transform to calculate a within-facies coordinate scheme that can be used to calculate a continuous within-facies trend model. Clearly, this method will encounter difficulties if the facies have short-range noise and a high degree of amalgamation. Image cleaning will be a useful step to improve performance (see Section 4.2).

Another option is to produce non-Gaussian continuous simulations. For example, Cavelius et al. (2012) proposed continuous MPS simulation where the categories represent the truncation of the continuous property distribution. Then the continuous distribution is simulated with the categorical variable as collocated secondary data and a high correlation coefficient. Admittedly this method is ad hoc. Also see work by Daly (2005).

Methods such as FilterSIM (Wu et al., 2008; Zhang et al., 2005) provide more robust methods to reproduce continuous variable heterogeneity. As with MPS, FilterSIM relies on a training image, but first proceeds by applying various filter scores to the training image. These scores provide essential dimensional reduction of the problem (i.e., consider the immense combinatorial of possible patterns over a template) through the classification of local training image patterns into pattern groups. Simulation proceeds by selecting patterns with the closest match group to local conditioning and placing them within the model. This method is able to reproduce very complicated continuous property heterogeneities at the cost of increased complexity, including the need to specify meaningful filters.

With the MPS approach the spatial continuity model is a discrete model at the size support of the model cells. This is contrasted by the variogram/covariance model applied in Gaussian and indicator simulation that provide a model of spatial continuity in all possible directions and distances and may be scaled to account for change in model cell support size. To retain this flexibility and the advantages of characterization beyond two point descriptions, an attempt has been made to extend the concept of the variogram to higher order spatial statistics, spatial cumulants. As opposed to MPS this method attempts to fully characterize the high order spatial behavior using maps of higher-order statistics.

Dimitrakopoulos and others (2010, 2011) proposed an associated high-order simulation algorithm. This provides methods for continuous model construction and checking based on higher order statistics. Goodfellow and others (2012) note that these statistics may not be possible to infer from the available data, but may be more practical through decomposition; models that formulate the higher order statistics from combinations of lower order statistics that may be calculated from the available data.

6.3 MORE ESTIMATION METHODS

As discussed in the generalized work flow introduced in Chapter 1, estimation is still an important tool in reservoir modeling. Specifically, estimation provides a valuable display of the integration of available data without the overprint of stochastically simulated heterogeneity that may hide important features. There is need for estimation methods concomitant to each simulation method. For example, Biver et al. (2012) introduced an estimation method for truncated Gaussian techniques. This method proceeds by:

- Transformation of the secondary variable (e.g., related geophysical property) to its associated cumulative probability, a uniform distribution between 0 and 1.
- Performing simple indicator kriging with any locally variable proportions as the local mean and the conditioning data at each location, **u**, within the model. This represents the local probability density function for facies.
- Sampling from the local probability density function with the secondary variable uniform score value.

This is a simple and convenient method to build an estimation model with constraint on ordering relations typical of truncated Gaussian simulation. Similar methods are conceivable to build estimation models based on other simulation methodologies.

6.4 SPECTRAL METHODS

Spectral methods have interesting opportunities. First we describe the theory briefly and then discuss opportunities. The following is based on the comprehensive literature review and summary of spectral methods provided by Wilde (2010). The fundamental concept in spectral simulation is to transform the covariance function from the spatial to the frequency domain, calculate a model in the frequency domain, and then back transform the model to the spatial domain.

It has been shown by classical spectral representation theorem that a set of values, $z(\mathbf{u}_\alpha), \alpha = 1, \ldots, N$, with a defined covariance $C_z(\mathbf{h})$ can be expressed as a discrete inverse Fourier transform of a finite series of Fourier coefficients, $A(\omega)$ (Yao, 2004).

$$Z(\mathbf{u}) = \sum_{\omega=0}^{N-1} A(\omega) e^{i2\pi\omega\mathbf{u}/N}, \qquad \omega = 0, \ldots, N-1$$

where the Fourier coefficients are composed of the amplitude and phase spectra.

$$A(\omega) = |A(\omega)| e^{-i\phi(\omega)}, \qquad \omega = 0, \ldots, N-1$$

where $|A(\omega)|$ is the amplitude spectrum and $\phi(\omega)$ is the phase spectrum. The amplitude spectrum is related to the spectral density $S\omega$.

$$|A(\omega)|^2 = S(\omega), \qquad \omega = 0, \ldots, N-1$$

The spectral density is the Fourier transform of the covariance function (Pardo-Iguzquiza and Chica-Olmo, 1993).

$$S(\omega) = \frac{1}{N} \sum_{h=0}^{N-1} C_Z(\mathbf{h}) e^{-i2\pi\omega\mathbf{h}/N}, \qquad \omega = 0, \ldots, N-1$$

The amplitude spectrum is related to the covariance model through the spectral density. The phase spectrum is not related to the covariance model. Given these relationships, we can simulate an unconditional stationary Gaussian field with defined covariance with the following steps:

1. Define the covariance function.
2. Calculate the spectral density by Fourier transform of the covariance function (move to the spectral domain).
3. Compute the amplitude spectrum as the square root of the spectral density.
4. Draw a random phase spectrum.
5. Combine the amplitude and phase spectra into Fourier coefficients.
6. Inverse Fourier transform (move back to spatial domain) the coefficients to calculate an unconditional realization in standard normal space.

There are some issues with this method. For example, Yao (2004) showed that over multiple realizations the distribution variance is underestimated. Efforts to correct the variance result in the covariance not being honored. Also, Fourier simulation methods can only generate unconditional realizations. This requires a second step to condition, such as an expensive data mismatch kriging step. While Yao (1998) proposed methods for conditioning, the results are only approximative and remove the speed advantage of spectral methods.

Spectral methods allow for greater flexibility in modeling variograms and cross variograms (Emery and Lantuéjoul, 2008). It is straightforward to correct a covariance function to be positive definite in the spectral domain by setting all real components to be greater than zero and constraining them to sum to the variance (Pyrcz and Deutsch, 2006a; Yao and Journel, 1998). Also, there is greater flexibility than with the typical linear model of coregionalization for modeling variograms for multiple variables (Pardo-Iguzquiza and Chica-Olmo, 1993). Finally, spectral models provide a continuous, in space, representation of the random function model, $Z(\mathbf{u})$, that can be sampled at any location, without local resimulation required with other simulation method. This allows for efficient model refinement.

While there are challenges and advantages to simulation in the spectral domain, the speed advantage of simulation by Fourier transform as opposed to sequential solution of the kriging system at each simulated location has made this a method of choice for unconditional models. Further work is required to realize its potential in conditional settings.

6.5 SURFACE-BASED MODELING

Reservoirs were formed by a succession of depositional and erosional events (refer back to Section 2.1, Preliminary Geological Modeling Concepts). Most sedimentary sequences of interest to the petroleum industry were formed along continental margins, where the sedimentation is a result of the interaction of tectonic activity and eustasy (Emery and Mayers, 1996; Galloway, 1998; Mulholland, 1998). Over long time periods associated with large-scale geological events, major changes in accumulation and erosion result in large changes to the sediment distribution. Within these large-scale changes, there are more frequent events such as the rising and falling of sea level that superimposes cyclical features on the large-scale architecture of the reservoir. Each large-scale sediment package consists of a number of sediment packages at smaller scales, which reflect more frequent depositional events.

Surfaces are boundaries of distinct sediment packages and these boundaries represent the start and end of different geological events. Generally, there are significant changes in the sediment properties at the boundaries due to the change of geological conditions, whereas within the sediment package the petrophysical properties are more homogeneous or change with an identifiable trend. Therefore, surfaces provide large-scale connectivity and continuity control of facies and petrophysical properties that is critical for reservoir performance prediction. At smaller scale, surfaces may represent continuous heterogeneities, such as mud drapes, that may represent important barriers, baffles, or conduits to flow. Considering surfaces in the modeling process often provides better constraints for the modeling of facies and petrophysical properties.

Some large-scale time surfaces are visible with seismic data. They are typically modeled deterministically with the aid of well data. These are the stratigraphic layers discussed in Section 4.1. Smaller-scale time surfaces, while not visible with seismic data, can be observed from core or well logs. In general, with few wells, small-scale surfaces may not be modeled deterministically. The purpose of surface modeling is not to get surfaces per se, but to provide constraints for modeling facies and petrophysical properties. Specific facies can be present at surface boundaries. Surfaces may also control grain size and facies trends such as fining or coarsening upward.

Surfaces may be stochastically positioned with parameterized surface templates. Local well data can be explicitly honored at the correct depth. Figure 6.1 gives example realizations that reproduce surface intersections at three well locations. While surface-based methods have been applied in process-mimicking facies models (see Section 4.5) to track the aggradation of the facies model and evolving topography to impose constraints on subsequent events, simple stochastic surface-based methods are a practical variant of object-based modeling (see Section 4.4).

Some early work on surface-based modeling includes work by Deutsch et al. (2001) along with various subsequent efforts by Leiva (2009), Pyrcz et al. (2005a), Sech et al. (2009), and Zhang et al. (2009). Yet, the place of stand-alone surface-based modeling in reservoir modeling has not been established. The concepts are important, but relatively new; time will tell if the methodology is adopted in practice.

6.6 ENSEMBLE KALMAN FILTERING

Data integration as in the case of static data integration in the form of local measures, trends, secondary variables, and massive multivariate settings have been discussed throughout this book. Admittedly, dynamic data integration is still a major challenge

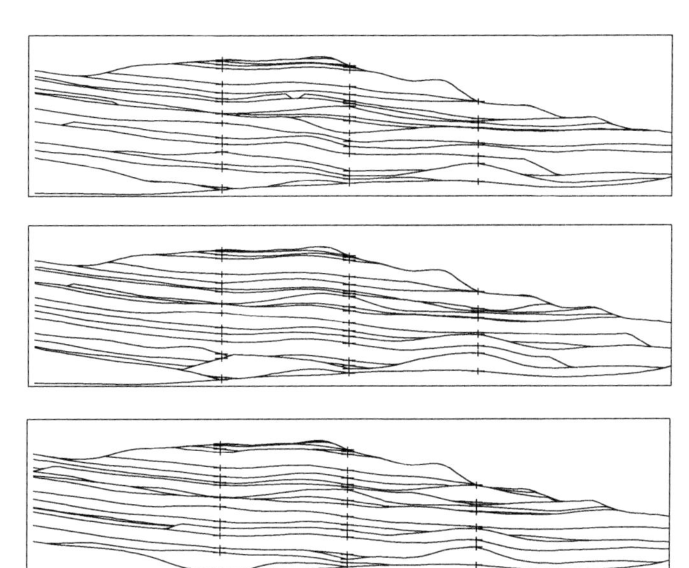

FIGURE 6.1: Three Realizations of Stratigraphic Surfaces within a Larger Layer. The realizations are 3-D. These slices are through three wells. Note the reproduction of the well intersections and the different geometry of the surfaces away from the wells.

in geostatistical reservoir modeling. These data include production rates, composition of fluids, and even down-hole temperature and pressure data.

Ensemble Kalman filtering (EnKF) has recently gained popularity for production data integration. An overview of the technique and development is provided in the note by Almendral-Vazquez and Syversveen (2006) and the monograph by Zagayevskiy et al. (2010). EnFK is based on the Kalman filter by Kalman (1960), an efficient recursive filter that estimates the state of a linear dynamic system from a series of noisy measurements. EnFK, developed by Evensen (1994), is a Monte Carlo implementation of the Kalman Filter that replaces the state vector with a state matrix that contains a set of possible values of the state variables (known as the ensemble). While EnKF assumes Gaussianity, it has been applied to weakly nonlinear systems with normal scores of lognormal transformation of the data. Also, EnKF relies on the Bayesian framework with the necessity to develop a reasonable prior model (i.e., set of realizations conditioned to static properties) and updating this model to a posterior that integrates dynamic data through a likelihood function.

A complete iteration includes forecast and analysis steps that are run sequentially. In the forecast step the process is applied to each ensemble to update the dynamic variables (the static variables are retained from the previous time step).

$$Y_n^f = F(Y_{n-1}^a) + W_{n-1}$$

where $F(*)$ is the model operator (such as flow simulation), Y_{n-1}^a represents updated static and dynamic properties from the previous time step, W_{n-1} is the measurement error covariance matrix for the measured dynamic observations, and Y_n^f is the matrix with updated dynamic variables and static variables from the previous time step.

The Kalman gain matrix K_n is calculated from the covariance of the updated dynamic variables and static variables from the previous times step [see Zagayevskiy et al. (2010)].

In the analysis step the static properties are updated with dynamic variable observations and the relationships between the static and dynamic properties.

$$Y_n^a = Y_n^f + K_n(D_n - H_n Y_n^f)$$

where D_n is the matrix of observations, H_n is simply a binary matrix to indicate what data observations are available at time step n. The analysis step is repeated until the required precision is obtained.

The following are some considerations in applying EnKF to dynamic data integration. If model resolution is not sufficient, then a bias may result. Lødøen and Omre (2008) have documented this and developed a method for dealing with this potential bias. It is important to start with a good initial prior model. Choice of a poor prior may result in more iterations to reach convergence and/or a poor estimate (Oliver and Chen, 2009).

6.7 ADVANCED GEOLOGICAL CHARACTERIZATION

There has been significant effort recently within geology to improve quantification of observations from outcrops, seismic data, numerical process models, and flumes. This work is motivated by the need to provide more objective, repeatable classification and description of architectural heterogeneity. This quantification is not limited to variograms and multiple-point statistics, and it includes metrics that relate directly to concepts of stacking and space filling. As mentioned in Section 4.7, Optimization for Model Construction, a good summary of this effort and various statistics are provided with the *STRATISTICS* software package from Professor Roy Plotnick. This work includes various point- and raster-based statistics that describe spatial heterogeneities (Middleton et al., 1995; Perlmutter and Plotnick, 2002; Plotnick, 1986). For example, lacunarity is a raster-based statistic that describes the space-filling nature of spatial heterogeneity across scales (Plotnick, 1986), and the Ripley's K function describes the clustering, randomness, or repulsion of points in space (Ripley, 2004). These statistical descriptions are now being actively applied to outcrop to improve quantification and classification of these natural examples and flume experiments to allow for comparison of results and sensitivity analysis of input initial and boundary conditions (Hajek et al., 2010; Wang et al., 2011).

There is no current method to impose such statistical descriptions in stochastic simulation within a direct framework. As these statistical descriptions become more accepted and more commonly available from reservoir analogs, there will be a growing need to develop reservoir modeling methodologies that integrate these statistics. While optimization methods provide the opportunity to immediately integrate them into reservoir models through the formulation of specific objective functions, work is required to provide efficient methodologies. It may be useful to develop new direct modeling methods that may honor these statistics. The result will be a greater information bandwidth from geology to geostatistical reservoir models and improved geological realism.

6.8 OTHER EMERGING TECHNIQUES

In this section we have the opportunity to attempt to forecast radical new developments, now on the horizon, that may impact geostatistics in the future. This type of "prediction" is likely somewhat faulty: Ideas that seem good are often not; conversely, crazy ideas often turn out to be good. With that acknowledgment, we cast a wide net and suggest a broad set of other emerging techniques.

Much of the current research is aimed at new ways to solve old problems related to data processing, geological modeling, and post-processing. The advancement in computational hardware and software has mostly helped provide access to new ideas from machine learning and numerical analysis. It is possible to visualize and summarize our data sources and models in new and efficient ways. In general, it is easier to check models and to create and update them quickly. Of course, this is limited by a natural tendency to build larger models given these processing resources.

New data sources are resulting in integration opportunities and challenges. These include emerging work on microseismic methods that utilizes seismic monitoring during reservoir production and production monitoring that track injection and production rates, composition, and pressures in real time. Both of these data sources may provide dense temporal data that may be inverted to indicate reservoir heterogeneity and connectivity. 4-D seismic surveillance provides a sequence of 3-D seismic snapshots of the reservoir to infer fluid flow at a larger temporal scale, but potentially with local accuracy. All of these data sources require methods for integration into geostatistical models accounting for data accuracy, support size, and redundancy between data sources. The new data sources may necessitate massive multivariate work flow.

Massively multivariate modeling is an emerging area where a large number, perhaps hundreds, of data variables are used simultaneously for the prediction of important reservoir variables. The data variables are derived from geophysical, structure, and geological sources. Complex relationships arising from compositional data or highly non-Gaussian features are being understood and will be accounted for in future reservoir models. While some methods were discussed in this book, further advancement in this area is needed.

Flexible methods to account for nonstationarity,—that is, local variations in statistical parameters—are emerging. The use of locally varying anisotropy where distances are computed in a non-Euclidean manner to track through geological features promises to provide greater local precision in estimation and uncertainty quantification. The regionalization of other parameters such as training images, histograms, and correlation coefficients also promises to improve local precision. Some developments have been discussed and referenced throughout his book. More work is provided to provide practical and robust work flows.

Special tools are being developed for the extreme data-rich applications of mature reservoirs and the data-poor applications of deep offshore development. The tools required in these two applications are very different; the decisions required to be made are also very different. Thus, the decision support information required from geostatistics is also different. Customized tools for these different environments are being developed. For example, methods to condition models may vary with data density. In sparse data sets, object-based model conditioning may be practical with minor model geometric adjustments, while in denser data sets a broader optimization with greater flexibility may be required. Also, in sparse data settings, tools to further integrate conceptual information may be required; while in dense data sets, tools to integrate and reconcile data sources may be needed.

Along with tools customized to data setting are tools customized to geological settings. In the case of process-mimicking models, the geological processes relevant to the setting are coded directly into the simulation algorithms. While there is a loss of general applicability, these methods provide opportunities to further improve the integration of conceptual information into the reservoir models. Care must be taken to understand this conceptual information and to ensure that uncertainty in concepts is integrated and the resulting uncertainty models are fair.

The limitations of process-based methods that prevent them from being applied as reservoir models were discussed previously, including CPU intensity and conditioning difficulties. With advancements in computational power, numerical methods, and improved process knowledge and more accurate physical concepts and constraints, process-based models may become more practical in reservoir modeling. This may be in the form of fast process approximations.

Geomechanical process may be applied to model at a larger scale than the sedimentation and stratigraphic process models considered above. Geomechanical models consider the response of a basin to sediment loading and unloading, with the resulting compaction, faulting, fracturing, and changes in rock properties. There may be opportunities to further integrate and even couple geomechanical models with geostatistical reservoir models.

When dealing with model downscaling, it is natural to consider the opportunity to describe the reservoir within a framework that is infinitely resolvable. For example, fractal and spectral representations allow for simultaneous characterization across scales and limitless interpolation of the property of interest. Currently, these methods have their associated limitations that restrict their heterogeneity models and prevent practical conditioning.

There are many opportunities related to the integration of concepts and methods from other scientific disciplines. For example, the nature-inspired optimization methods have been useful for building models that integrate various data sources. As discussed previously, genetic algorithms have optimized model parameters and simulated annealing has effectively constructed models that honor challenging conditioning such as production data. Also, consider the advancements in graphics processing and visualization in the gaming community. Such developments should be useful for building and visualizing large, complicated heterogeneity models. The related field of mathematical morphology has useful synergies. More could be done.

Finally, there are opportunities with respect to building uncertainty models. Methods to account for data and concepts are typically static. Yet, new data are collected and concepts are changed during the development of a reservoir. Simulated learning methods provide the opportunity to model uncertainty more dynamically and account for the impact of data collection strategy and the resulting temporal sequence of information additions. These methods may provide fairer uncertainty models and assessment of the value of information.

6.9 FINAL THOUGHTS

High-resolution 3-D models of reservoir properties are required for resource evaluation, flow studies, and reservoir management. This book has presented geostatistical techniques to build such 3-D models. These geostatistical techniques have limitations. A list of the most important limitations would include the following: (1) Relative to conventional methods such as contouring, splines, or distance weighting, geostatistical methods take more time to apply, require more expertise and training, and have the potential for a greater number of mistakes; (2) there is a need for complex and complete input statistics, and yet there is often no reliable source for most of these input parameters; and (3) there is an implicit assumption that geostatistical models reflect facies and property variations at all smaller scales—for example, from the core scale through the geological modeling scale. These limitations are being addressed through professional training, improved software, and research.

Even with these significant limitations, geostatistical techniques provide the best tools available for constructing reliable numerical reservoir models because they (1) provide means to introduce a controllable degree of heterogeneity in facies and reservoir properties, (2) permit geological trends and data of different types to be accounted for by construction rather than laborious "hand" processing, (3) make it easier to perform sensitivity studies, and (4) provide a means to assess uncertainty.

The objective of this book was to collect the important aspects of geostatistical theory and practice applied to petroleum reservoir modeling. That objective has been partially met. There is clearly a need for more specialized books and monographs.

APPENDIX A

Glossary and Notation

A.1 GLOSSARY

The more complete glossary edited by Ricardo Olea (1991) is a useful supplement to the terse presentation below.

absolute permeability: the permeability (see below) of a rock saturated completely with one fluid.

accommodation: a term used in sequence stratigraphy that refers to the volume available for the sediment to accumulate.

accuracy plot: a method and graphical representation of the goodness of an uncertainty model based on cross validation between withheld data and the local uncertainty model.

aeolian: or eolian sediments are those transported and deposited by wind energy—for example, sand dunes.

affine variance correction: an approach to modify Z values to have lesser variability—that is, to account for a larger-volume support. The affine correction leaves the shape of the histogram unchanged and simply shifts data values closer to the mean, depending on the magnitude of variance reduction—for example, $z' = \sqrt{f} \cdot (z - m) + m$.

allogenic processes: geological controls on the reservoir genesis that are external to the basin, including eustasy, climate, and tectonics. These controls typically result in large-scale quality trends and controls on reservoir presence and extent.

analog data: information from outcrops, more densely drilled reservoirs, and high-resolution shallow seismic data of a similar depositional environment or numerical process and experimental stratigraphy of similar depositional processes. Analog data are used to determine statistics (e.g., horizontal variograms) and to support concepts for a field of study.

anisotropy: directional dependence of a parameter—for example, the variogram range. In the context of the variogram, geometric anisotropy is anisotropy that may be corrected by an affine correction of the coordinate system. Zonal anisotropy may not be corrected by an affine transformation and indicates additional variability specific to a particular direction.

annealing: see simulated annealing.

autogenic processes: geological controls on the reservoir genesis that are internal to the basin, including process related to sediment transport and deposition, such as channel formation and avulsion. Autogenic controls typically result in smaller scale/noisier features than allogenic controls, but have important controls on reservoir heterogeneity and connectivity.

Bayesian statistics: a branch of statistics with a fundamentally different frame of reference than the frequentist perspective. Bayesian statistics utilizes belief and probability logic to build useful probability relationships.

bivariate: related to the joint probability distribution of two variables—for example, porosity and permeability.

bombing model: a particular indicator variogram model based on the result of randomly dropping (or bombing) ellipsoids or any other fixed shapes, coded as "0s" into a background matrix, coded as "1s."

bootstrap: a statistical resampling procedure whereby uncertainty in a calculated statistic is derived from the data itself. Monte Carlo simulation is used (with replacement) to resample from the data distribution. Strong assumptions of data independence and representativity are required.

boundary conditions: in the context of determining the effective properties of a grid block, boundary conditions are the pressure and flow rate conditions surrounding the block of interest.

bounding surface: a stratigraphic surface that separates two reservoir units. A bounding surface may correspond to some fixed geological time or represent a composite surface formed over a period of time.

calibration: in the context of geostatistical modeling, calibration is the exercise where the relationship of soft secondary data to the "hard" data being modeled is established. The calibration relationship is almost always probabilistic in the sense that there is no one-to-one relationship.

carbonate: a geological reservoir type consisting of calcareous deposits deposited as the shells of reef-dwelling organisms.

cell-based: a technique to build a facies model where facies codes are assigned to cells such that the statistical relations between the cells reproduce some realistic variations. Compare with "object-based" and "surface-based" techniques.

cell declustering: a method for declustering that weights data such that each subarea (or cell) receives the same weight, see index.

cleaning: a (geo)statistical procedure whereby a categorical variable realization is modified to remove short-scale variations that are deemed unreliable.

clinoform: a geological structure formed during deposition. A stratigraphic layer may be made up of clinoforms blurring or complicating the identification of the principal directions of continuity.

cloud transform: a probability field (see *p*-field) technique of simultaneously drawing from conditional distributions. The term "cloud transform" may be used interchangeably with "*p*-field;" however, there is the connotation with cloud transform that the conditional distributions are coming from a bivariate cross plot.

cokriging: Kriging applied with more than one data type. The same basic set of equations are called for; however, the covariance between the different data types is also required.

collocated: at the same location.

conditional: made to reproduce certain local data at their locations. Used in the context of geostatistical simulation.

conditional finite domain: a parameter uncertainty method based on recursive simulation that accounts for the domain relative to the data locations and results in decreasing parameter uncertainty with increasing spatial continuity.

conditional variance model: the model of the variance of the local realizations across the area of study.

connectivity: the "persistence" or linkage between reservoir units that are good quality and poor quality. Connectivity is a straightforward concept, but more difficult to quantify numerically. The variogram is one measure of connectivity, but it does not capture multipoint connectivity that would be measured by a flow simulator.

coregionalization: the noun used to refer to a random function of two or more variables. A "regionalization" is the probability law of one particular variable.

cosimulation: simulation of one variable using data of multiple types or, more commonly, realizations from a previous variable for conditioning.

covariance: measure of correlation in units of the variance. The covariance is the expected value of the product of the two variables Y and Z, that is, $E\{[Y - \overline{Y}][Z - \overline{Z}]$. Often, the covariance is standardized by the standard deviations of Z and Y, it is then called the correlation coefficient.

cross validation: the procedure of leaving each data (or the entire well) out of the data set and estimating the reservoir property at that well location. This is repeated for all data, and the true values and their estimates can be compared to judge the goodness of the estimation technique.

cross variogram: the variogram (see below) between two different variables. The variogram is defined as $2\gamma(\mathbf{h}) = E\left\{\left[Z(\mathbf{u}) - Z(\mathbf{u}+\mathbf{h})\right]^2\right\}$ and the cross variogram is defined as

$$2\gamma(\mathbf{h}) = E\left\{\left[Z(\mathbf{u}) - Z(\mathbf{u}+\mathbf{h})\right]\left[Y(\mathbf{u}) - Y(\mathbf{u}+\mathbf{h})\right]\right\}$$

where Z and Y are two different variables.

CV: acronym for coefficient of variation, which is the standard deviation of a distribution divided by the mean. This measure of variation is sometimes preferred since it is dimensionless.

cyclicity: a particular variogram signature, called the "hole effect," that is induced by geological periodicity that causes repeated patterns.

declustering: assignment of weights whereby redundant data are given less weight in the construction of a representative probability distribution or the calculation of summary statistics.

depositional setting: the geological environment (fluvial river channels, deltaic deposits, deepwater turbidite deposits, carbonate reefs, and so on) in which the sediments were formed. The geostatistical methods we employ are different within different depositional settings.

despiking: a technique to break ties in a data set; for example, certain techniques would require *spikes* of constant values such as 10% zero's to be ranked from smallest to largest for the purposes of transformation and back transformation. The ties would be broken by local averages.

diagenetic: an alteration to the character of reservoir sediments after deposition. Diagenetic alteration could take the form of silica or carbonate cements or remineralization.

direct: adjective used before geostatistical simulation to mean that the values have not been transformed by Gaussian, indicator, or any other transformation method.

dispersion variance: the expected value of the spatial variance of values of support a in larger volume support b, commonly denoted $\sigma^2(a, b)$.

distribution: a word used loosely for many different purposes. A probability distribution is the histogram of cumulative probability values of a random variable. A spatial distribution is the set of values in space that describe a reservoir.

dynamic data: data that are inherently time-varying such as a well test or historical production data. The challenge is to relate the time variations to spatial variations.

e-type model: a model of the expected value of each local conditional distribution for a specific continuous reservoir property across the area of study.

equally likely to be drawn: the expression applied to simulated realizations that are drawn by (pseudo) random numbers. The result of each realization will enter a histogram of uncertainty with an equal increment.

equiprobable: a term applied to simulated realizations that are considered equally probable given the modeled probability distribution. This is distinct from "equally likely to be drawn." Geostatistical realizations should rightly be referred to as "equally likely to be drawn."

ergodic: a study area is considered "ergodic" when it is large with respect to the range of correlation.

erosion: removal of sediment, which affects the geological correlation style for geostatistical modeling.

experimental stratigraphy: the use of "scaled" experiments to study geological processes.

exponential variogram: a common variogram model.

eustasy: large-scale change in structure related to changes in sea level.

evolutionary algorithms: nature-inspired optimization methods that utilize a genetic representation of the solution set and mechanisms analogous to the evolution of species. Evolutionary algorithms have been applied to optimize model parameters for production data integration.

facies: categorical classification of reservoir rock into different types based on distinct petrophysical properties. Facies are based on lithology and other considerations. There are many different bases for facies specifications.

faults: structural deformation of the reservoir layers that could lead to a displacement and disconnection between the layers. Faults could also be conduits for fluid flow.

flow unit: a particular portion of the reservoir that is believed (through empirical observation or flow simulation) to be in communication. Historical practice of mapping flow units forces a priori expectations of reservoir continuity into the model, which is good if that knowledge is based on actual measurements and dangerous if not.

fluvial: river or stream-derived sediments. Fluvial reservoirs may be modeled by object-based procedures where sand-filled channels and other facies are modeled as "clean" objects.

forward modeling: the "forward" problem is the processing of static geostatistical models to calculate flow responses. The "inverse" problem is the determination of the static geostatistical models that lead to known flow responses.

frequentist statistics: a branch of statistics that focuses on the concept of pooling sample replicates to construct and model probability distributions.

gamma bar: more correctly denoted as $\overline{\gamma}(V, v)$, this is the average variogram between volumes V and v—that is, the average variogram value where one end of the lag vector independently defines V and the other independently defines *v*. *V* and *v* could be the same.

Gaussian variable: a variable that follows a Gaussian or normal distribution. Often, data are transformed to follow a Gaussian distribution for ease of calculation.

Gaussian variogram: a common variogram model. This variogram model implies smooth variability and is normally observed with geological layer thickness or topographic elevations.

GeoEAS: a geostatistical package put together by the EPA for environmental site characterization. The ASCII file format adopted by the GeoEAS software is widely used by other geostatistical software.

geological model: the suite of all realizations of the area of interest, including all properties with all multivariate and spatial relationships, that jointly represent the reservoir uncertainty model.

geometric mean: exponentiation of the average of logarithms of a data set. The geometric mean can be written in a number of forms; see Index.

geomorphology: a type of analog data. The study of the evolution of landforms, focused on form and the responsible processes.

geostatistics: study of phenomena that vary in space and/or time.

Gibbs sampler: a special case of the Metropolis algorithm that draws from complicated joint distributions by resampling individual variables conditional to the others. This avoids the challenging or impossible task of sampling directly from the complete joint distribution. Gibbs sampler has been applied in MPS.

global kriging: a kriging-based method to calculate parameter uncertainty in the mean with the kriging variance by kriging an estimate of the entire reservoir, accounting for volume support with averaged variogram values (gamma-bar).

grain size: the size of the rock particles that make up a sedimentary rock. In general, coarser grain size leads to greater permeability. Poorly sorted grains—that is, of many different sizes—often lead to reduced porosity and permeability.

grid-free modeling: representation of heterogeneity as continuous functions and procrastination of rasterization to a grid. This allows for ease in moving between scales and design of irregular grids conforming to reservoir architecture.

hard data: direct measurements on the petrophysical property of interest. Hard data are considered the reference data for all other potential sources of information. See also *soft data*.

heterogeneities: variations in reservoir properties. These variations often appear as random or unpredictable because of the relatively large data spacing.

hierarchical modeling: modeling in a manner that accounts for heterogeneity across scales. This may be based on a specific architectural hierarchy scheme and may include nested models with various modeling methods, coordinate transformations, trend models, and spatial continuity models at each scale. Examples discussed in this book include hierarchical fluvial object-based models and hierarchical trend models.

higher-order checks: a set of model checks (beyond those discussed in minimum acceptance criteria) to check multiple-point simulation models.

horizon markers: the depth or elevation locations of important geological bounding surfaces.

hydrocarbon volume: the volume, expressed in cubic meters or barrels, of a reservoir that contains hydrocarbon.

indicator: a data transformation method whereby a continuous or categorical data variable is transformed to a probability value. A continuous variable is transformed to the *probability to be less than a certain threshold,* and a categorical variable is transformed to the *probability of that category prevailing*. The indicator transform is always between zero and one.

indicator map or model: the map or model of the indicator transform of the data or a model realization.

input statistics: the set of statistics (including univariate, bivariate, and spatial) that are inferred from the local data, concepts, and analog data with associated uncertainty that are applied to build a geostatistical reservoir model.

inverse modeling: see *forward modeling*

isochore: thickness of a stratigraphic formation measured in the vertical direction.

isopach: thickness of a stratigraphic formation measured perpendicular to the dip of the formation.

isotropic: showing no directional preference or *anisotropy*.

jackknife: a portion of the data is withheld from the geostatistical analysis until the very end; then, the reservoir properties are estimated at these data locations. The closeness of the withheld true values to the estimates is a measure of the goodness of the estimation procedure.

kriging: spatial linear regression named in honor of Professor Krige in South Africa. The kriging equations are derived for the best linear estimator, given a stationary covariance model and a particular model for the trend. The "normal equations," developed at about the same time, are the same as kriging.

KT: kriging with a trend model (see also *universal kriging*).

lag deposits: sediment deposited at the base of a depositional unit; for example, shale clasts, may be found at the base of certain fluvial deposits.

layercake: geological model made up of deterministic layers of constant properties. Although these models are not necessarily realistic, they provide a measure of stratification that typically exists in a reservoir.

lithofacies: reservoir rock type.

local percentile model: a model of the local p-value based on the local distribution of uncertainty over an area of study. For example, a P10 local percentile model indicates the value at each location for which there is a 90% probability of exceedance.

lognormal: refers to a distribution where the logarithm of the variable follows a normal or Gaussian probability distribution. This adjective is often wrongly used to refer to any histogram that is lower-bounded by zero and positively skewed.

low-pass filter: a term from geophysical data processing, which amounts to filter out high-frequency (short-scale or periodic) variations. Kriging is sometimes called a low-pass filter because of its smoothing.

Markov: a model whereby a limited number of nearest neighbors screen the information of all further away data.

minimum acceptance criteria: a set of basic models checks to ensure that data and input statistical constraints are reproduced. These include checking data, histogram, variogram, and correlation coefficient reproduction.

missing scale: the volume scale difference between the data and the grid cells being modeled. This is termed the "missing" scale when no explicit attempt is made to capture the change in statistical properties between the data scale and model scale.

model: a much-abused word in science and engineering typically referring to a numerical replacement for some physical or mathematical quantity. The context is typically required to sort out the exact meaning. A variogram *model* is an equation that gives the variogram for all distances and directions. A reservoir *model* is a gridded set of facies, porosity, and permeability values that represent the entire volume of interest.

modeling cell: a regular volume in transformed stratigraphic coordinate space that is assigned facies, porosity, and permeability by geostatistical means. Modeling cells are typically 1–3 feet thick and 50–500 feet in areal dimension.

Monotonic: consistently increasing or decreasing, for example, a cdf is a nondecreasing or monotonic function.

Monte Carlo: the procedure of drawing or sampling from a specified probability distribution. Monte Carlo simulation typically proceeds in two steps: (i) generating a pseudo-random number that follows a uniform distribution between zero and one and (ii) reading the quantile, associated with that random number, from the target distribution.

multiphase: normally used to refer to the flow regime in the reservoir; for example, multiphase flow is the flow of multiple phases such as oil, water, or gas.

multiple-point density function: the multiple-point statistic represented by the frequency of all possible facies combination over a multiple point template. There are k^{n+1} possible combinations for n points in the data template and the unknown location, and there is no natural ordering. As a result, it is not typically practical to visualize a multiple-point density function. Efficient calculation, storage, and retrieval are a critical challenge in MPS.

multiple-point simulation: (MPS) a simulation algorithm that primarily utilizes multiple-point statistics.

multiple-point statistics: statistics related to multiple points taken simultaneously. Variograms are two-point statistics. Three, four, and more points could be simultaneously considered for multiple-point or multipoint statistics.

nested: structures refer to multiple variogram structures that are combined to model the spatial variation in complex settings.

net: a "net" rock type is one that has hydrocarbons that could be produced unless it is bypassed. For example, high-porosity sandstone is a net rock type and shale would be a *non-net* rock type.

net-to-gross: the fraction of net rock type in a reservoir.

nonstationarity model: a model of the change in a model input statistic over the area of interest. These statistics may include: mean or proportion (commonly called a trend model), variogram, orientation and anisotropy, correlation coefficient, and geometry (in the case of object-based models).

normal variable: the word "normal" is interchangeable with "Gaussian." A normally distributed variable is one that has a Gaussian probability distribution. The use of one word or the other is largely dictated by convention; for example, we often write *normal score transform* and *sequential Gaussian simulation,* but rarely do we write *Gaussian score transform* or *sequential normal simulation.*

numerical analog model: see *training image.*

numerical process model: numerical models for geological processes including three distinct types: landscape evolution models, sediment transport models, and morphodynamic models. While these models integrate the geological processes, they are typically not practical for reservoir modeling because they cannot be efficiently conditioned, are complicated to apply, require unavailable initial and boundary conditions, and are CPU-intensive. They are utilized to provide analog data or to tune process-mimicking models.

object-based modeling: a technique to build a facies model where whole geological objects such as fluvial channels, remnant shales, or sand dunes are embedded within a matrix facies. Although the final model may be cellular, it is still called "object-based" because of how it was constructed in the first place.

onlap: a geological correlation style where the "horizontal" lines of continuity truncate against a lower surface that may have represented the present-day topography at time of deposition.

order relations: the required properties of a probability distribution; that is, no negative probabilities and the sum of all probabilities must equal one. The context of this expression is typically indicator kriging that does not enforce order relations by construction; they must be enforced by a posteriori correction.

***p*-field:** the technique of separating the construction of local conditional probability distributions and the Monte Carlo simulation from these distributions. Correlated probability values, or a *p-field,* are used to perform the Monte Carlo simulation.

paleoslope: the slope (and direction of slope) that existed at the time of deposition. Knowledge of this direction may help in determining directions of continuity for variogram calculation.

parametric: refers to a probability distribution that is characterized by a limited set of numbers or parameters such as the mean and variance. The most common parametric probability distribution would be the normal or Gaussian distribution.

particle swarm optimization: nature-inspired optimization method that utilizes a population (swarm) of solutions (particles) and rules that constrain random and attractive behavior of the particles over iterations. This method has been applied to update reservoir models to honor seismic data.

pdf: acronym for probability density function.

periodicity: see *cyclicity.*

permeability: defined as a constant relating fluid flux to pressure gradient in a porous substance.

petroleum: hydrocarbons of (often) commercial value to be extracted from a reservoir.

petrophysical property: some attribute of a given volume of rock—for example, porosity or density.

polygonal declustering: a method for declustering that uses polygonal areas of influence, see index.

pool: a particular reservoir; for example, there are twenty pools in this basin. "Pool" is also used as a verb to mean combining data together; for example, data from all sand types may be pooled together for geostatistical calculations.

population: a statistical or geological group of data that relates to a particular reservoir zone or facies. The interbedded sand facies within a particular reservoir zone would be an example.

porosity: void space in rock. Effective porosity is the void space that could contribute to fluid flow. Total porosity is the total void space including those pores that could never be connected and drained. We are typically interested in effective porosity.

positive definite: the mathematical property of a covariance model that ensures that the variance of all linear combinations is strictly greater than or equal to zero. The same property of the corresponding variogram is referred to as "non-negative."

post-processing: a suite of operations that are applied to the geological model to fix, modify, scale, or summarize.

power law scale-up: expressing the average or effective property of a region as a ω or power law average where $\omega = 1$ for arithmetic and $\omega = -1$ for harmonic averaging. The ω is calibrated from flow simulation.

power law variogram: a particular type of variogram model without sill that may be interpreted as fractal.

primary variable: the variable currently being modeled. All other data are considered secondary.

process-mimicking modeling: a method for reservoir modeling that integrates geological information through geological rules. These methods are typically forward models (built from the base up), use objects (see *object-based modeling*), track surfaces (see *surface-based modeling*), and produce models with a high degree of complexity, but are challenging to condition to dense data.

proportional correlation: a particular type of stratigraphic correlation style where the "horizontal" continuity is parallel between two grid surfaces. There is no top erosion or bottom onlap.

Q–Q plot: a Q–Q or quantile-quantile plot compares two different histograms or univariate distributions: a scatterplot of matching (according to cumulative probability) quantiles from two distributions.

quantile: any Z-variable value that separates a histogram into two parts. A Z-quantile value is qualified by the cumulative probability associated with it, for example, the 0.5 quantile is the median.

quartile: the first, second, and third quartiles are the 0.25, 0.5, and 0.75 quantiles, respectively. They divide a distribution into four equal parts.

random function: a statistical model of a variable distributed in space (say, porosity) characterized by a multivariate probability distribution. Continuous variable random functions (RFs) are often assumed to be multivariate Gaussian because geostatisticians can handle them easily.

ranking: order of a set of realizations from low to high based on some criterion—for example, net-to-gross or average porosity. Selected realizations can then be carried through flow simulation.

realization: specific output of lithofacies, porosity, and permeability drawn by Monte Carlo simulation.

regionalized variable: a variable (such as lithofacies, porosity, or permeability) that may take different values at different spatial locations.

regridding: the process of representing a reservoir at a different grid system, for example, a different stratigraphic and flow simulation grid system are often used and there is a need to go between the two grid systems.

relative permeability: saturation-dependent factor that specifies what fraction of the total or absolute permeability that is available for a particular fluid. There is less relative permeability for fluids with low saturation since they are likely disconnected.

reserves: amount of a resource that can be economically extracted.

reservoir: some interesting finite volume of rock and fluid in the subsurface. Enhancing the economic extraction of important fluids from a reservoir is often the ultimate goal of reservoir characterization.

resource: amount of hydrocarbon that could potentially be extracted.

saturation function: a variable such as relative permeability or capillary pressure that is a function of fluid saturation.

secondary variable: variable related to the current primary variable being modeled.

sequential: in geostatistics, the sequential adjective typically refers to the order of simulation—that is, one grid node after another using previously simulated grid nodes as conditioning for the current and future grid nodes.

shallow seismic: a form of analog data. Due to the higher frequencies available at shallow depths, may be available as part of a specific study or available from imaging deeper objectives.

siliciclastic: silica-type sand/shale transported by either water or air, deposited, and lithified to sandstone/shale.

simulation: drawing realizations from a probability distribution that may be analytically defined or only observable after the drawing of many realizations. The realizations typically involve multiple variables.

simulated annealing: an optimization technique based on an analogy with the physical process of annealing. Simulated annealing does not require calculating derivatives, but may be difficult to code efficiently.

smoothing: this is used in a number of contexts. Histogram or distribution smoothing refers to a procedure to fit or "fill-in" a sparse distribution with one that removes spikes and adds values that should be there, but are not sampled. Kriging is also said to smooth the data because the kriged values have less variability than the data from which they were calculated.

soft data: indirect measurements of the petrophysical property of interest. These data must be calibrated to available *hard* data, or direct measurements, before being used in geostatistical modeling. See also *hard data.*

space of uncertainty: the range of possible outcomes from a particular probabilistic model. The space of uncertainty is typically sampled by multiple realizations and is inaccessible in practice.

spherical variogram: a common variogram model that takes its name from the geometric interpretation of its mathematical expression; the standardized spherical variogram function for lag distance **h** is 1.0 minus the intersection volume of two unit spheres separated by lag distance **h**.

spread: the deviation of data from the center of the data values. The variance and interquartile range are two common measures of spread.

static: an intrinsic property of the porous media such as porosity and permeability that does not change over the practical time scale of the reservoir. Compare with a *dynamic* measurement that is time-dependent.

stationarity: decision that statistics do not depend on location within the study area; for example, the variogram is often assumed "stationary" in the sense that it is applicable over the entire area over which it has been calculated.

statistical fluctuations: in the realizations' reproduction of input statistics. These are minimum when the study area is ergodic (see ergodic). Previously, these fluctuations alone were considered in the parameter uncertainty model, while the modern approach is to explicitly model parameter uncertainty and pass it through the reservoir realizations.

stochastic: a numerical procedure is called "stochastic" when it involves the use of a random variable.

stratigraphy: geological study of the large-scale geometry and continuity of reservoir facies and properties. One typically talks about stratigraphic layers (large-scale zones of reservoir rock that belong together), stratigraphic coordinates (conversion of the subsea depth or elevation coordinates to one convenient to model), or sequence stratigraphy (relates to correlating layers according to the chronography or time of deposition).

structural analysis: typically the analysis and modeling of geometric architecture of the reservoir— that is, faults, folds, and deformation. Sometimes used in place of *variography*.

support: the measurement volume of a particular type of data. For example, the support of a core plug is a few cubic centimeters, which is less than a well log measurement, which is less than a well test measurement.

surface-based modeling: the procedure of constructing a reservoir model where the reservoir layers are built up by a stochastic succession of surfaces that define depositional events at a scale smaller than the well spacing. Commonly applied within process-mimicking facies modeling methods to assist in geological rule implementation by tracking evolution of topography.

systems tract: a particular geological setting for deposition leading to particular lithofacies and trends in porosity and permeability. Some systems tracts include fluvial, deltaic, shoreface systems, deepwater depositional systems, platform carbonates, and so on.

training image: an exhaustive model at the correct scale and without locality that provides a spatial continuity model for multiple-point simulation. Training images may be constructed to export other statistics such as the variogram or to assess parameter uncertainty as in the case of spatial bootstrap (in these cases they are further denoted as numerical analog models).

transfer function: any operation applied to a geostatistical reservoir model or jointly to a set of geosatistical reservoir models to support a reservoir exploitation decision. This may include volumetric calculations or flow simulation.

transformation: there are different types of transformation: *data* transformation that converts the data to a different histogram, and *coordinate* transformation used to convert real reservoir depth or elevation coordinates to a system that is more convenient for modeling.

trend: vertical or areal variability that is predictable. For example, there is often a fining-upwards trend in alluvial sand bodies since coarse grains settle quicker than fine grains. The variability of most natural phenomena is both deterministic or predictable (the trend) and stochastic. The stochastic part is modeled by geostatistics.

truncated Gaussian: a technique to model a categorical lithofacies variable. A continuous

"Gaussian" variable is simulated first and then divided (or truncated) into classes.

uncertainty: lack of complete information about some quantity. Uncertainty is often represented by a probability distribution, whose variability or "spread" tells us the level of uncertainty.

universal kriging: kriging where the mean is specified to follow a particular form of trend. The coefficients of the trend are estimated implicitly by the kriging procedure. More commonly referred to as "kriging with a trend."

variogram: measure of spatial variability for a separation lag distance $\mathbf{h}$. Defined as the average squared difference in the variable value for that distance, that is, $2\gamma(\mathbf{h}) = E\{[Z(\mathbf{u}) - Z(\mathbf{u}+\mathbf{h})]^2\}$. The variogram is $2\gamma(\mathbf{h})$ and the *semi*variogram is $\gamma(\mathbf{h})$.

variography: the iterative procedure of calculating, interpreting, and modeling variograms.

volumetrics: a reporting of the volume of hydrocarbon in a particular reservoir interval (over some limited areal extent). Volumetrics typically refer to a static resource before consideration of a particular recovery scheme or development plan.

Walther's Law: a geological *law* that (very roughly) states that facies associations encountered in the horizontal direction are also encountered in the vertical direction. This is often brought up as an argument to use the same type of horizontal variogram as vertical variogram, just with a different range of correlation.

water cut: fraction of water to total flow rate. This is an important response variable in reservoir management.

zonal anisotropy: directional variability induced by stratigraphic or areal zonation. The variogram in one direction would not encounter the full variability because of this stratification (see Index).

A.2 NOTATION

$\forall$: whatever

a: range parameter

$B(z)$: Markov–Bayes calibration parameter

$C(\mathbf{0})$: covariance value at separation vector $\mathbf{h} = 0$. It is also the stationary variance of random variable $Z(\mathbf{u})$

$C(\mathbf{h})$: stationary covariance between any two random variables $Z(\mathbf{u}), Z(\mathbf{u}+\mathbf{h})$ separated by vector $\mathbf{h}$

$C(\mathbf{u}, \mathbf{u}')$: nonstationary covariance of two random variables $Z(\mathbf{u}), Z(\mathbf{u}')$

$C_l(\mathbf{h})$: nested covariance structure in the linear covariance model $C(\mathbf{h}) = \sum_{l=1}^{L} C_l(\mathbf{h})$

$C_I(\mathbf{h}; z_k)$: stationary indicator covariance for cutoff z_k

$C_I(\mathbf{h}; z_k, z_{k'})$: stationary indicator cross covariance for cutoffs z_k, $z_{k'}$; it is the cross covariance between the two indicator random variables $I(\mathbf{u}; z_k)$ and $I(\mathbf{u}+\mathbf{h}; z_{k'})$

$C_I(\mathbf{h}; s_k)$: stationary indicator covariance for category s_k

$C_{ZY}(\mathbf{h})$: stationary cross covariance between the two random variables $Z(\mathbf{u})$ and $Y(\mathbf{u}+\mathbf{h})$ separated by lag vector $\mathbf{h}$

$D^2(a, b)$: dispersion variance is the variance of values of volume support a in a larger volume b.

$d\mathbf{u}$: denotes an infinitesimal area (volume) centered at location $\mathbf{u}$

$E\{\cdot\}$: expected value

$E\{Z(\mathbf{u})|(n)\}$: conditional expectation of the random variable $Z(\mathbf{u})$ given the realizations of n other neighboring random variables (called data)

$\mathrm{Exp}(\mathbf{h})$: exponential semivariogram model, a function of vector $\mathbf{h}$

$F(\mathbf{u}; k|(n))$: nonstationary conditional probability distribution function of the categorical variable $Z(\mathbf{u})$

$F(\mathbf{u}; z)$: nonstationary cumulative distribution function of random variable $Z(\mathbf{u})$

$F(\mathbf{u}; z|(n))$: nonstationary conditional cumulative distribution function of the continuous random variable $Z(\mathbf{u})$ conditioned by the realizations of n other neighboring random variables (called data)

$F(\mathbf{u}_1, \ldots, \mathbf{u}_K; z_1, \ldots, z_K)$: K variate cumulative distribution function of the K random variables $Z(\mathbf{u}_1), \ldots, Z(\mathbf{u}_K)$

$F(z)$: cumulative distribution function of a random variable Z, or stationary cumulative distribution function of a random function $Z(\mathbf{u})$

$F^{-1}(p)$: inverse cumulative distribution function or quantile function for the probability value $p \in [0, 1]$

$f_k(\cdot)$: function of the coordinates used in a trend model

$\gamma(\mathbf{h})$: stationary semivariogram between any two random variables $Z(\mathbf{u})$, $Z(\mathbf{u}+\mathbf{h})$ separated by lag vector $\mathbf{h}$

$\gamma_I(\mathbf{h}; z)$: stationary indicator semivariogram for lag vector $\mathbf{h}$ and cutoff z: it is the semivariogram of the binary indicator random function $I(\mathbf{u}; z)$

$\gamma_I(\mathbf{h}; p)$: same as above, but the cutoff z is expressed in terms of p quantile with $p = F(z)$

$\gamma_{ZY}(\mathbf{h})$: stationary cross semivariogram between the two random variables $Z(\mathbf{u})$ and $Y(\mathbf{u}+\mathbf{h})$ separated by lag vector $\mathbf{h}$

$\overline{\gamma}\,(v(\mathbf{u}), V(\mathbf{u}'))$: the gamma bar or volume integrated variogram between volumes v and V

$G(y)$: standard normal cumulative distribution function

$G^{-1}(p)$: standard normal quantile function: $G(G^{-1}(p)) = p \in [0,1]$

$\mathbf{h}$: separation vector

$i(\mathbf{u}; s_k)$: binary indicator value at location $\mathbf{u}$ and for category s_k

$I(\mathbf{u}; s_k)$: binary indicator random function at location $\mathbf{u}$ for category s_k

$I(\mathbf{u}; z)$: binary indicator random function at location $\mathbf{u}$ and for cutoff z

$i(\mathbf{u}; z)$: binary indicator value at location $\mathbf{u}$ and for cutoff z

$j(\mathbf{u}; z)$: binary indicator transform arising from constraint interval

λ_α, $\lambda_\alpha(\mathbf{u})$: kriging weight associated to datum α for estimation at location $\mathbf{u}$. The superscripts (OK), (KT), and (m) are used when necessary to differentiate between the various types of kriging

M: stationary median of the distribution function $F(z)$

m: stationary mean of the random variable $Z(\mathbf{u})$

$m(\mathbf{u})$: mean at location $\mathbf{u}$, expected value of random variable $Z(\mathbf{u})$; trend component model in the decomposition $Z(\mathbf{u}) = m(\mathbf{u}) + R(\mathbf{u})$, where $R(\mathbf{u})$ represents the residual component model

$m^*_{KT}(\mathbf{u})$: estimate of the trend component at location $\mathbf{u}$

$\mu, \mu(\mathbf{u})$: Lagrange parameter for kriging at location $\mathbf{u}$

$N(\mathbf{h})$: number of pairs of data values available at lag vector $\mathbf{h}$

$P(A)$: marginal probability of event A

$P(A|B)$: conditional probability of event A given event B

$P(A \text{ and } B)$: joint probability of event A and event B

$\prod_{i=1}^{n} y_i = y_1 \cdot y_2 \cdot y_3 \cdots y_n$: product

$p_k = E\{I(\mathbf{u}; s_k)\}$: stationary mean of the indicator of category k

$q(p) = F^{-1}(p)$: quantile function, that is, inverse cumulative distribution function for the probability value, $p \in [0,1]$

$R(\mathbf{u})$: residual random function model in the decomposition $Z(\mathbf{u}) = m(\mathbf{u}) + R(\mathbf{u})$, where $m(\mathbf{u})$ represents the trend component model

ρ: correlation coefficient $\in [-1,+1]$

$\rho(\mathbf{h})$: stationary correlogram function $\in [-1,+1]$

$\sum_{i=1}^{n} y_i = y_1 + y_2 + \cdots + y_n$: summation

$\Sigma(\mathbf{h})$: matrix of stationary covariances and cross covariances

σ^2: variance

$\sigma^2_{OK}(\mathbf{u})$: ordinary kriging variance of $Z(\mathbf{u})$

$\sigma^2_{SK}(\mathbf{u})$: simple kriging variance of $Z(\mathbf{u})$

$Sph(\mathbf{h})$: spherical semivariogram function of separation vector $\mathbf{h}$

s_k: kth category

$\mathbf{u}$: coordinates vector

$Var\{\cdot\}$: variance

$Y = \varphi(Z)$: transform function $\varphi(\cdot)$ relating two random variables Y and Z

$Z = \varphi^{-1}(Y)$: inverse transform function $\varphi(\cdot)$ relating random variables Z and Y

Z: generic random variable

$Z(\mathbf{u})$: generic random variable at location $\mathbf{u}$, or a generic random function of location $\mathbf{u}$

$Z^*_{COK}(\mathbf{u})$: cokriging estimator of $Z(\mathbf{u})$

$Z^*_{KT}(\mathbf{u})$: estimator of $Z(\mathbf{u})$ using some form of prior trend model

$Z^*_{OK}(\mathbf{u})$: ordinary kriging estimator of $Z(\mathbf{u})$

$Z^*_{SK}(\mathbf{u})$: simple kriging estimator of $Z(\mathbf{u})$

$\{Z(\mathbf{u}), \mathbf{u} \in A\}$: set of random variables $Z(\mathbf{u})$ defined at each location $\mathbf{u}$ of a zone A

$z(\mathbf{u})$: generic variable function of location $\mathbf{u}$

$z(\mathbf{u}_\alpha)$: z-datum value at location $\mathbf{u}$

z_k: kth cutoff value

$z^{(l)}(\mathbf{u})$: lth realization of the random function $Z(\mathbf{u})$ at location $\mathbf{u}$

$z_c^{(l)}(\mathbf{u})$: lth realization conditional to some neighboring data

$z^*(\mathbf{u})$: an estimate of value $z(\mathbf{u})$

$[z(\mathbf{u})]^*_E$: E-type estimate of value $z(\mathbf{u})$, obtained as an arithmetic average of multiple simulated realizations $z^{(l)}(\mathbf{u})$ of the random function $Z(\mathbf{u})$

BIBLIOGRAPHY

Aanonsen, S. I., Eide, A. L., and Holden, L. Optimizing reservoir performance under uncertainty with application to well location. In *1995 SPE Annual Technical Conference and Exhibition*. Society of Petroleum Engineers, October 1995. SPE Paper # 30710.

Aarts, E., and Korst, J. *Simulated Annealing and Boltzmann Machines*. John Wiley & Sons, New York, 1989.

Abrahamsen, P., Fjellvoll, B., Hauge, R., Howell, J., and Aas, T. Process based on stochastic modeling of deep marine reservoirs. In *EAGE Petroleum Geostatistics 2007*, 2008.

Alabert, F. G. Stochastic imaging of spatial distributions using hard and soft information. Master's thesis, Stanford University, Stanford, CA, 1987.

Alabert, F. G. Constraining description of randomly heterogeneous reservoirs to pressure test data: a Monte Carlo study. In *SPE Annual Conference and Exhibition, San Antonio*. Society of Petroleum Engineers, October 1989.

Alabert, F. G. and Corre, B. Heterogeneity in a complex turbiditic reservoir: impact on field development. In *66th Annual Technical Conference and Exhibition*, pages 971–984, Dallas, TX, October 1991. Society of Petroleum Engineers. SPE Paper # 22902.

Alabert, F. G. and Massonnat, G. J. Heterogeneity in a complex turbiditic reservoir: stochastic modelling of facies and petrophysical variability. In *65th Annual Technical Conference and Exhibition*, pages 775–790. Society of Petroleum Engineers, September 1990. SPE Paper # 20604.

Alahaidib, T. and Deutsch, C. V. A Gaussian framework for multivariate multiscale data integration. Technical Report 116, CCG Annual Report 12, Edmonton, AB, 2010.

Alapetite, A., Leflon, B., Gringarten, E., and Mallet, J.-L. Stochastic modeling of fluvial reservoirs: the YACS approach. In *SPE Annual Technical Conference and Exhibition*, number SPE Paper 97271, Dallas, USA, 2005. Society of Petroleum Engineers.

Allen, J. R. L. A review of the origin and characteristics of recent alluvial sediments. *Sedimentology*, 5:89–191, 1965.

Allen, J. R. L. Studies in fluviatile sedimentation: an exploratory quantitative model for the architecture of avulsion-controlled alluviate suites. *Sedimentary Geology*, 21:129–147, 1978.

Almeida, A. S. and Journel, A. G. Joint simulation of multiple variables with a Markov-type coregionalization model. *Mathematical Geology*, 26:565–588, 1994.

Almeida, A. S. D. *Joint Simulation of Multiple Variables with a Markov-Type Coregionalization Model*. PhD thesis, Stanford University, Stanford, CA, 1993.

Almendral-Vazquez, A. and Syversveen, A. R. The ensemble Kalman filter-theory and applications in oil industry. *Norwegian Computing Center*, 2006.

Alshehri, N. *Quantification of reservoir uncertainty for optimal decision making*. PhD thesis, University of Alberta, Edmonton, AB, 2009.

Anderson, M. P. Comment on universal scaling of hydraulic conductivities and dispersivities. *Water Resources Research*, 27(6):1381–1382, 1991.

Anderson, T. *An Introduction to Multivariate Statistical Analysis*. John Wiley & Sons, New York, 1958.

Andrews, H. C. and Hunt, B. R. *Digital Image Restoration*. Prentice Hall, Englewood Cliffs, NJ, 1989.

Armitage, D. A., Romans, B. W., Covault, J. A., and Graham, S. A. Tres Pasos mass transport deposit topography; the sierra contreras. In Fildani, A., Hubbard, S. M., and Romans, B. W., editors, *Stratigraphic evolution of deep-water architecture; Examples on controls and depositional styles from the Magallanes Basin*, Chile, 2009. SEPM Field Guide.

Armstrong, M. Improving the estimation and modeling of the variogram. In Verly, G., editor, *Geostatistics for Natural Resources Characterization*, pages 1–20. Reidel, Dordrecht, 1984.

Armstrong, M., Galli, A. G., Beucher, H., Le Loc'h, G., Renard, D., Doligez, B., Eschard, R., and Geffroy, F.

Plurigaussian Simulations in Geosciences. Springer, Berlin, 1st edition, 2003.

Arpat, B. and Caers, J. A multi-scale, pattern-based approach to sequential simulation. In Leuangthong, O. and Deutsch, C. V., editors, *Geostatistics Banff 2004, Quantitative Geology and Geostatistics*, pages 255–264. Springer Netherlands, Dordrecht, 2005.

Aziz, K. and Settari, A. *Petroleum Reservoir Simulation*. Elsevier Applied Science, New York, 1979.

Aziz, K., Arbabi, S., and Deutsch, C. V. Why it is so difficult to predict the performance of nonconventional wells? SPE Paper # 37048, 1996.

Babak, O. and Deutsch, C. V. Accounting for parameter uncertainty in reservoir uncertainty assessment: the conditional finite-domain approach. *Natural Resources Research*, November 2007a.

Babak, O. and Deutsch, C. V. An intrinsic model of coregionalization that solves variance inflation in collocated cokriging. *Computers & Geosciences*, 2007b.

Ballin, P. R., Journel, A. G., and Aziz, K. A. Prediction of uncertainty in reservoir performance forecasting. *JCPT*, 31(4), April 1992.

Bashore, W. M. and Araktingi, U. G. Importance of a geological framework and seismic data integration. In Yarus, J. M. and Chambers, R. L., editors, *Stochastic Modeling and Geostatistics: Principles, Methods, and Case Studies*, pages 159–176. AAPG Computer Applications in Geology, No. 3, 1995.

Batycky, R. P., Blunt, M. J., and Thiele, M. R. A 3D field-scale streamline-based reservoir simulator. *SPE Reservoir Engineering*, pages 246–254, November 1997.

Beaubouef, R. T., Rossen, C., Zelt, F., Sullivan, M. D., Mohrig, D., and Jennette, D. C. Deep-water sandstones, Brushy Canyon Formation. In *Field Guide for AAPG Hedberg Field Research Conference*, number 40, page 48, West Texas, April 1999. AAPG Continuing Education Course.

Behrens, R. A., Macleod, M. K., Tran, T. T., and Alimi, A. O. Incorporating seismic attribute maps in 3D reservoir models. *SPE Reservoir Evaluation & Engineering*, 1(2):122–126, 1998.

Benediktsson, J. A. and Swain, P. H. Consensus theoretic classification methods. *IEEE Transactions on Systems, Man, and Cybernetics*, 22(4), 1992.

Benkendorfer, J. P., Deutsch, C. V., LaCroix, P. D., Landis, L. H., Al-Askar, Y. A., Al-AbdulKarim, A. A., and Cole, J. Integrated reservoir modeling of a major arabian carbonate reservoir. In *SPE Middle East Oil Show*, Bahrain, March 1995. Society of Petroleum Engineers. SPE Paper # 29869.

Berteig, V., Halvorsen, K. B., Omre, H., Hoff, A. K., Jorde, K., and Steinlein, O. A. Prediction of hydrocarbon pore volume with uncertainties. In *63rd Annual Technical Conference and Exhibition*, pages 633–643. Society of Petroleum Engineers, October 1988. SPE Paper # 18325.

Bertram, G. T. and Milton, N. J. Seismic stratigraphy. In Emery, D. and Myers, K., editors, *Sequence Stratigraphy*, pages 45–60. Blackwell Publishing Ltd., Oxford, 1996.

Besag, J. On the statistical analysis of dirty pictures. *J. R. Statistical Society B*, 48(3):259–302, 1986.

Beucher, H., Galli, A., Le Loc'h, G., and Ravenne, C. Including a regional trend in reservoir modelling using the truncated Gaussian method. In Soares, A., editor, *Geostatistics—Troia*, volume 1, pages 555–566. Kluwer, Dordrecht, 1993.

Bickel, E. J. and Bratvold, R. B. From uncertainty quantification to decision making in the oil and gas industry. *Energy, Exploration & Exploitation*, 26(5):311–325, 2008. doi: 10.1260/014459808787945344.

Bittencourt, A. C. and Horne, R. N. Reservoir development and design optmization. In *1997 SPE Annual Technical Conference and Exhibition Formation Evaluation and Reservoir Geology*. Society of Petroleum Engineers, October 1997. SPE Paper # 38895.

Biver, P., Pivot, F., and Henrion, V. Estimation of most likely lithology map in context of truncated Gaussian techniques. *Journal of the American Statistical Association*, 2012.

Boggs, S. *Principles of Sedimentology and Stratigraphy*. Prentice Hall, Upper Saddle River, NJ, 3rd edition, 2001.

Boisvert, J. B. *Geostatistics with Locally Variable Anisotropy*. PhD thesis, University of Alberta, Edmonton, AB, 2010.

Boisvert, J. B. Generating locally varying azimuth fields. Technical Report 103, CCG Annual Report 13, Edmonton, AB, 2011.

Boisvert, J. B. and Deutsch, C. V. Programs for kriging and sequential Gaussian simulation with locally varying anisotropy using non-euclidean distances. *Computers & Geosciences*, 37(4):495–510, 2011.

Boisvert, J. B., Pyrcz, M. J., and Deutsch, C. V. Multiple-point statistics for training image selection. *Natural Resources Research*, 16(4):313–321, 2007a.

Boisvert, J. B., Pyrcz, M. J., and Deutsch, C. V. Multiple point statistics for training image selection. *Natural Resources Research*, 16(4):313–321, 2007b.

Boisvert, J. B., Pyrcz, M. J., and Deutsch, C. V. Multiple point metrics to assess categorical variable models. *Natural Resources Research*, 19(3):165–174, 2010.

Bortoli, L. J., Alabert, F., Haas, A., and Journel, A. G. Constraining stochastic images to seismic data. In Soares, A., editor, *Geostatistics Troia 1992*, volume 1, pages 325–334. Kluwer, Dordrecht, 1993.

Bosch, M., Mukerji, T., and Gonzalez, E. F. Seismic inversion for reservoir properties combining statistical rock physics and geostatistics: a review. *Geophysics*, 75(5):75A165–75A176, 2010.

Boucher, A. Considering complex training images with search tree partitioning. *Computers & Geosciences*, 35(6):1151–1158, 2009.

Boucher, A. Strategies for modeling with multiple-point simulation algorithms. In *2011 Gussow Geoscience Conference*, Banff, Alberta, 2011.

Boucher, A., Gupta, R., Caers, J., and Satija, A. Tetris: a training image generator for SGeMS. Technical report, Stanford Center for Reservoir Forecasting, 2010.

Bras, R. L. and Rodrıguez-Iturbe, I. *Random Functions and Hyrdrology*. Addison-Wesley, Reading, MA, 1982.

Bratvold, R. B. and Begg, S. H. Education for the real world: equipping petroleum engineers to manage uncertainty. In *SPE Annual Technical Conference and Exhibition*, 2006. doi: 10.2118/103339-MS.

Bratvold, R. B., Holden, L., Svanes, T., and Tyler, K. STORM: integrated 3D stochastic reservoir modeling tool for geologists and reservoir engineers. SPE Paper # 27563, 1994.

Bridge, J. S. A FORTRAN IV program to simulate alluvial stratigraphy. *Computers & Geosciences*, 1979.

Bridge, J. S. and Leeder, M. R. A simulation model of alluvial stratigraphy. *Sedimentology*, 26:617–644, 1979.

Buland, A., Kolbjørnsen, O., and Omre, H. Rapid spatially coupled AVO inversion in the fourier domain. *Geophysics*, 68(3):824–836, May 2003. doi: 10.1190/1.1581035.

Caers, J. *Petroleum Geostatistics*. Society of Petroleum Engineers, 2005.

Caers, J. *Modeling Uncertainty in the Earth Sciences*. John Wiley & Sons, Hoboken, NJ, 2011.

Caers, J., Srinivasan, S., and Journel, A. G. Stochastic reservoir simulation using neural networks trained on outcrop data. *SPE Paper # 49026*, October 1999.

Caers, J., Strebelle, S., and Payrazyan, K. Stochastic integration of seismic data and geologic scenarios a west africa submarine channel saga. *The Leading Edge*, 22(3):192–196, 2003.

Caers, J., Hoffman, T., Strebelle, S., and Wen, X. H. Probabilistic integration of geologic scenarios, seismic, and production data - a west africa turbidite reservoir case study. *The Leading Edge*, 25(3):240–244, 2006.

Campion, K. M., Sprague, A. R., and Sullivan, M. D. Outcrop expression of confined channel complexes. *SEPM, Pacific Section*, 2005.

Cardwell, W. T. and Parsons, R. L. Average permeabilities of heterogeneous oil sands. *Transactions of the AIME*, 160:34–42, 1945.

Carr, J. and Myers, D. E. COSIM: a Fortran IV program for co-conditional simulation. *Computers & Geosciences*, 11(6):675–705, 1985.

Carrera, J. and Neuman, S. P. Estimation of aquifer parameters under transient and steady state conditions: 1. maximum likelihood method incorporating prior information. *Water Resources Research*, 22(2):199–210, 1986.

Casella, G. and George, E. I. Explaining the Gibbs sampler. *The American Statistician*, 46(3):167–174, August 1992.

Castro, S., Caers, J., Otterlei, C., Andersen, T., Hoye, T., and Gomel, P. A probabilistic integration of well log, geological information, 3D/4D seismic, and production data: application to the Oseberg field. In *SPE Annual Technical Conference and Exhibition*, 2006. SPE Paper # 103152.

Catuneanu, O. *Principles of Sequence Stratigraphy*. Elsevier, Boston, 3rd edition, 2006.

Catuneanu, O., Abreu, V., Bhattacharya, J. P., Blum, M. D., Dalrymple, R. W., Eriksson, P. G., Fielding, C. R., Fisher, W. L., Galloway, W. E., Gibling, M. R., Giles, K. A., Holbrook, J. M., Jordan, R., Kendall, C. G. S. C., Macurda, B., Martinsen, O. J., Miall, A. D., Neal, J. E., Nummedal, D., Pomar, L., Posamentier, H. W., Pratt, B. R., Sarg, J. F., Shanley, K. W., Steel, R. J., Strasser, A., Tucker, M. E., and Winker, C. Toward the standardization of sequence stratigraphy. *Earth-Science Reviews*, 92(1–2):1–33, 2009.

Caumon, G., Grosse, O., and Mallet, J.-L. High resolution geostatistics on coarse unstructured flow grids. In Leuangthong, O. and Deutsch, C. V., editors, *Geostatistics Banff 2004*, volume 1 of *Quantitative Geology and Geostatistics*, pages 703–712. Springer Netherlands, Dordrecht, 2005a.

Caumon, G., Lévy, B., Castanié, L., and Paul, J. C. Visualization of grids conforming to geological structures: a topological approach. *Computers & Geosciences*, 31(6):671–680, 2005b.

Caumon, G., Collon-Drouaillet, P., de Veslud, L. C., C., Viseur, S., and Sausse, J. Surface-based 3D modeling of geological structures. *Mathematical Geosciences*, 41(8):927–945, 2009.

Cavelius, C., Pyrcz, M. J., and Stebelle, S. MPS improvements. In Abrahamsen, P., Hauge, R., and Kolbjørnsen, O., editors, *Geostatistics Oslo 2012*, volume 17 of *Quantitative Geology and Geostatistics*, Oslo, Norway, 2012.

Černý, V. Thermodynamical approach to the travelling salesman problem: an efficient simulation algorithm. *Journal of Optimization Theory and Applications*, 45:41–51, 1985.

Cherpeau, N., Caumon, G., Caers, J., and Lévy, B. Method for stochastic inverse modeling of fault geometry and connectivity using flow data. *Mathematical Geosciences*, pages 1–22, 2012.

Chilès, J. P. and Delfiner, P. *Geostatistics: Modeling Spatial Uncertainty*. Wiley Series in Probability and Statistics, John Wiley & Sons, Hoboken, NJ, 2nd edition, April 2012.

Chu, J. and Journel, A. G. Conditional fBm simulation with dual kriging. In Dimitrakopoulos, R., editor, *Geostatistics for the Next Century*, pages 407–421. Kluwer, Dordrecht, 1994.

Chugunova, T. L. and Hu, L. Y. Multiple-point simulations constrained by continuous auxiliary data. *Mathematical Geosciences*, 40:133–146, 2008.

Clemensten, R., Hurst, A. R., Knarud, R., and Omre, H. A computer program for evaluation of fluvial reservoirs. In Buller, A. T., editor, *North Sea Oil and Gas Reservoirs II*. Graham and Trotman, London, 1990.

Cojan, I., Fouche, O., and Lopez, S. Process-based reservoir modelling in the example meandering channel. In Leuangthong, O. and Deutsch, C. V., editors, *Geostatistics Banff 2004*, volume 14 of *Quantitative Geology and Geostatistics*, pages 611–620. Springer Netherlands, Dordrecht, 2005.

Collins, J. F., Kenter, J. A. M., Harris, P. M., Kuanysheva, G., Fischer, D. J., and Steffen, K. L. Facies and reservoir quality variations in the late visean to bashkirian outer platform, rim, and flank of the Tengiz Buildup, Precaspian Basin, Kazakhstan. *Giant Hydrocarbon Reservoirs of the World: from Rocks to Reservoir Characterization and Modeling*, pages 55–95, 2006.

Cox, D. L., Lindquist, S. J., Bargas, C. L., Havholm, K. G., and Srivastava, R. M. Integrated modeling for optimum management of a giant gas condensate reservoir, Jurassic eolian nugget sandstone, Anschutz Ranch East Field, Utah overthrust (USA). In Yarus, J. M. and Chambers, R. L., editors, *Stochastic Modeling and Geostatistics: Principles, Methods, and Case Studies*, pages 287–321. AAPG Computer Applications in Geology, No. 3, 1995.

Cressie, N. *Statistics for Spatial Data*. John Wiley & Sons, New York, 1991.

Cressie, N. and Hawkins, D. Robust estimation of the variogram. *Mathematical Geology*, 12(2):115–126, 1980.

da Cruz, P. S., Horne, R. N., and Deutsch, C. V. The quality map: a tool for reservoir uncertainty quantification and decision making. In *1999 SPE Annual Technical Conference and Exhibition*. Society of Petroleum Engineers, October 1999.

da Cruz, P. S. *Reservoir Management Decision-Making in the Presence of Geological Uncertainty*. PhD thesis, Stanford University, Stanford, CA, 2000.

Dagbert, M., David, M., Crozel, D., and Desbarats, A. Computing variograms in folded strata-controlled deposits. In Verly, G., editor, *Geostatistics for Natural Resources Characterization*, pages 71–89. Reidel, Dordrecht, 1984.

Daly, C. Higher order models using entropy, Markov random fields and sequential simulation. In Leuangthong, O. and Deutsch, C. V., editors, *Geostatistics Banff 2004*, volume 14 of *Quantitative Geology and Geostatistics*, pages 215–224. Springer Netherlands, Dordrecht, 2005.

Damsleth, E., Hauge, A., and Volden, R. Maximum information at minimum cost: a North Sea Field development study with an experimental design. *Journal of Petroleum Technology*, pages 1350–1356, December 1992a.

Damsleth, E. and Tjølsen, C. B. Scale consistency from cores to geologic description. *SPE Formation Evaluation*, 9(4):295–299, 1994.

Damsleth, E., Tjølsen, C. B., Omre, H., and Haldorsen, H. H. A two-stage stochastic model applied to a north sea reservoir. *Journal of Petroleum Technology*, pages 402–408, April 1992b.

Datta-Gupta, A., Lake, L. W., and Pope, G. A. Characterizing heterogeneous permeability media with spatial statistics and tracer data using sequential simulation annealing. *Mathematical Geology*, 27(6):763–787, 1995.

David, M. *Geostatistical Ore Reserve Estimation*. Elsevier, Amsterdam, 1977.

Davis, B. M. Uses and abuses of cross-validation in geostatistics. *Mathematical Geology*, 19(3):241–248, 1987.

Davis, J. M., Phillips, F. M., Wilson, J. L., Lohman, R. C., and Love, D. W. A sedimentological–geostatistical model of aquifer heterogeneity based on outcrop studies. *EOS Trans*, 73(14), 1992.

Dejean, J. P. and Blanc, G. Managing uncertainties on production predictions using integrated statistical methods. In *1999 SPE Annual Technical Conference and Exhibition*. Society of Petroleum Engineers, October 1999. SPE Paper # 56696.

Delfiner, P. and Haas, A. Over thirty years of petroleum geostatistics. In Bilodeau, M., Meyer, F., and Schmitt, M., editors, *Space, Structure and Randomness*, volume 183 of *Lecture Notes in Statistics*, pages 89–104. Springer, New York, 2005.

Deutsch, C. V. Calculating effective absolute permeability in sandstone/shale sequences. *SPE Formation Evaluation*, pages 343–348, September 1989a.

Deutsch, C. V. DECLUS: a Fortran 77 program for determining optimum spatial declustering weights. *Computers & Geosciences*, 15(3):325–332, 1989b.

Deutsch, C. V. *Annealing Techniques Applied to Reservoir Modeling and the Integration of Geological and Engineering (Well Test) Data*. PhD thesis, Stanford University, Stanford, CA, 1992a.

Deutsch, C. V. *Annealing Techniques Applied to Reservoir Modeling and the Integration of Geological and Engineering (Well Test) Data*. PhD thesis, Stanford University, Stanford, CA, 1992b.

Deutsch, C. V. Algorithmically defined random function models. In Dimitrakopoulos, editor, *Geostatistics for the Next Century*, pages 422–435. Kluwer, Dordrecht, 1994.

Deutsch, C. V. Constrained modeling of histograms and cross plots with simulated annealing. *Technometrics*, 38(3):266–274, August 1996a.

Deutsch, C. V. Direct assessment of local accuracy and precision. In Baafi, E. Y. and Schofield, N. A., editors, *Fifth International Geostatistics Congress*, pages 115–125, Wollongong, Australia, September 1996b.

Deutsch, C. V. Cleaning categorical variable (lithofacies) realizations with maximum a-posteriori selection. *Computers & Geosciences*, 24(6):551–562, 1998a.

Deutsch, C. V. Fortran programs for calculating connectivity of three-dimensional numerical models and for ranking multiple realizations. *Computers & Geosciences*, 24(1):69–76, 1998b.

Deutsch, C. V. Notes on ranking. Technical Report 113, CCG Annual Report 7, Edmonton, AB, 2005a.

Deutsch, C. V. A new trans programs for histogram and trend reproduction. Technical Report 306, CCG Annual Report 7, Edmonton, AB, 2005b.

Deutsch, C. V. Post processing object based models for data reproduce well data: MAPSpp. Technical Report 204, CCG Annual Report 7, Edmonton, AB, 2005c.

Deutsch, C. V. A sequential indicator simulation program for categorical variables with point and block data: BlockSIS. *Computers & Geosciences*, 32(10):1669–1681, 2006.

Deutsch, C. V. Multiple scale geological models for heavy oil reservoir characterization. *AAPG Memoir on Oil Sands*, page 32, October 2010a.

Deutsch, C. V. Display of cross validation/jackknife results. Technical Report 406, CCG Annual Report 12, University of Alberta, Edmonton, AB, 2010b.

Deutsch, C. V. Guide to best practice in geostatistics. Technical report, Centre for Computational Geostatistics, Edmonton, AB, 2011.

Deutsch, C. V. and Begg, S. H. The use of ranking to reduce the required number of realizations. Technical Report 12, CCG Annual Report 3, Edmonton, AB, 2001.

Deutsch, C. V. and Cockerham, P. W. Practical considerations in the application of simulated annealing to stochastic simulation. *Mathematical Geology*, 26(1):67–82, 1994b.

Deutsch, C. V. and Journel, A. G. *GSLIB: geostatistical Software Library and User's Guide*. Oxford University Press, New York, 1992.

Deutsch, C. V. and Journel, A. G. Integrating well test-derived absolute permeabilities. In Yarus, J. M. and Chambers, R. L., editors, *Stochastic Modeling and Geostatistics: principles, Methods, and Case Studies*, pages 131–142. AAPG Computer Applications in Geology, No. 3, 1995.

Deutsch, C. V. and Journel, A. G. *GSLIB: geostatistical Software Library and User's Guide*. Oxford University Press, New York, 2nd edition, 1998.

Deutsch, C. V. and Kupfersberger, H. Geostatistical simulation with large-scale soft data. In Pawlowsky-Glahn, V., editor, *Proceedings of IAMG'97*, volume 1, pages 73–87. CIMNE, 1993.

Deutsch, C. V. and Lewis, R. W. Advances in the practical implementation of indicator geostatistics. In *Proceedings of the 23rd International APCOM Symposium*, pages 133–148, Tucson, AZ, April 1992. Society of Mining Engineers.

Deutsch, C. V. and Srinivasan, S. Improved reservoir management through ranking stochastic reservoir models. In Baafi, E. Y. and Schofield, N. A., editors, *SPE/DOE Tenth Symposium on Improved Oil Recovery, Tulsa, OK*, pages 105–113, Washington, DC, April 1996. Society of Petroleum Engineers. SPE Paper # 35411.

Deutsch, C. V. and Tran, T. T. FLUVSIM: a program for object-based stochastic modeling of fluvial depositional systems. *Computers & Geosciences*, 28(3):525–535, May 2002.

Deutsch, C. V. and Wang, L. Hierarchical object-based stochastic modeling of fluvial reservoirs. *Mathematical Geology*, 28(7):857–880, 1996.

Deutsch, C. V. and Zanon, S. D. Direct prediction of reservoir performance with Bayesian updating. *JCPT*, February 2007.

Deutsch, C. V., Xie, Y., and Cullick, A. S. Surface geometry and trend modeling for integration of stratigraphic data in reservoir models. In *2001 Society of Petroleum Engineers Western Regional Meeting*, number SPE Paper 6881, Bakersfield, California, March 2001. British Society of Reservoir Geologists.

Deutsch, C. V., Ren, W., and Leuangthong, O. Joint uncertainty assessment with a combined Bayesian updating/LU/P-Field approach. *Proceedings of IAMG 2005: GIS and Spatial Analysis*, 1:639–644, 2005.

Deutsch, C. V., Leuangthong, O., and Ortiz, J. M. A case for geometric criteria in resources and reserves classification. *SME Transactions*, 322:11, December 2007.

Deutsch, J. L. and Deutsch, C. V. Checking and correcting categorical variable trend models. Technical Report 132, CCG Annual Report 11, Edmonton, AB, 2009.

Deutsch, J. L. and Deutsch, C. V. Some geostatistical software implementation details. Technical Report 412, CCG Annual Report 12, Edmonton, AB, 2010.

Deutsch, J. L. and Deutsch, C. V. Accuracy plots for categorical variables. Technical Report 404, CCG Annual Report 14, Edmonton, AB, 2012.

de Vries, L. M., Carrera, J., Falivene, O., Gratacs, O., and Slooten, L. J. Application of multiple point geostatistics to non-stationary images. *Mathematical Geosciences*, 41(1):29–42, 2008.

Dietrich, W. E., Bellugi, D. G., Sklar, L. S., Stock, J. D., Heimsath, A. M., and Roering, J. J. Geomorphic transport laws for predicting landscape form and dynamics. *Prediction in Geomorphology, Geophysical Monograph Series*, 135:103–132, 2003. doi: 10.1029/135GM09.

Dimitrakopoulos, R. HOSIM: a high-order stochastic simulation algorithm for generating three-dimensional complex geological patterns. *Computers & Geosciences*, 2011. doi: 10.1016/j.cageo.2010.09.007.

Dimitrakopoulos, R., Mustapha, H., and Gloaguen, E. High-order statistics of spatial random fields: exploring spatial cumulants for modeling complex non-Gaussian and non-linear phenomena. *Mathematical Geosciences*, 42(1):65–99, 2010. doi: 10.1007/s110004-009-9258-9.

Doyen, P. M. Porosity from seismic data: a geostatistical approach. *Geophysics*, 53(10):1263–1275, 1988.

Doyen, P. M. and Guidish, T. M. Seismic discrimination of lithology in sand/shale reservoirs: a Bayesian approach. Expanded Abstract, SEG 59th Annual Meeting, 1989, Dallas, TX., 1989.

Doyen, P. M., Guidish, T. M., and de Buyl, M. Monte Carlo simulation of lithology from seismic data in a channel sand reservoir. SPE Paper # 19588, 1989.

Doyen, P. M., Psaila, D. E., and Strandenes, S. Bayesian sequential indicator simulation of channel sands from 3-D seismic data in the oseberg field, Norwegian North Sea. In *69th Annual Technical Conference and Exhibition*, pages 197–211, New Orleans, LA, September 1994. Society of Petroleum Engineers. SPE Paper # 28382.

Doyen, P. M., den Boer, L. D., and Pillet, W. R. Seismic porosity mapping in the Ekofisk field using a new form of collocated cokriging. In *1996 SPE Annual Technical Conference and Exhibition Formation Evaluation and Reservoir Geology*, pages 21–30, Denver, CO, October 1996. Society of Petroleum Engineers. SPE Paper # 36498.

Doyen, P. M., Psaila, D. E., and den Boer, L. D. Reconciling data at seismic and well log scales in 3-D earth modelling. In *1997 SPE Annual Technical Conference and Exhibition Formation Evaluation and Reservoir Geology*, pages 465–474, San Antonio, TX, October 1997. Society of Petroleum Engineers. SPE Paper # 38698.

Dubrule, O. A review of stochastic models for petroleum reservoirs. In Armstrong, M., editor, *Geostatistics*, volume 2, pages 493–506. Kluwer, Dordrecht, 1989.

Dubrule, O. Introducing more geology in stochastic reservoir modeling. In Soares, A., editor, *Geostatistics—Troia*, volume 1, pages 351–370. Kluwer, 1993.

Dubrule, O. *Geostatistics for Seismic Data Integration in Earth Models: 2003 Distinguished Instructor Short Course*, volume 6. Society of Exploration Geophysicists, 2003.

Dubrule, O., Thibaut, M., Lamy, P., and Haas, A. Geostatistical reservoir characterization constrained by seismic data. *Petroleum Geoscience*, 2(2), 1998.

Dueck, G. and Scheuer, T. Threshold accepting: a general purpose optimization algorithm appearing superior to simulated annealing. *Journal of Computational Physics*, 90:161–175, 1990.

Earlougher, R. C. *Advances in Well Test Analysis*. Society of Petroleum Engineers, New York, 1977.

Efron, B. *The Jackknife, the Bootstrap, and Other Resampling Plans*. Society for Industrial and Applied Math, Philadelphia, 1982.

Efron, B. and Tibshirani, R. J. *An Introduction to the Bootstrap*. Chapman & Hall, New York, 1993.

Egeland, T., Hatlebakk, E., Holden, L., and Larsen, E. A. Designing better decisions. In *European Petroleum Conference*, Stavanger, Norway, 1992. SPE Paper # 24275.

Eillis, D. V. and Singer, J. M. *Well Logging for Earth Scientists*. Prentice Hall, Springer, Upper saddle River, NJ, 2010.

Einsele, G. *Sedimentary Basins: evolution, facies and sediment budget*. Springer, New York, 3rd edition, 2000.

Emery, D. and Mayers, K. J. *Sequence Stratigraphy*. Blackwell Science, London, 1996.

Emery, X. Simulation of geological domains using the plurigaussian model: new developments and computer programs. *Computers & geosciences*, 33(9):1189–1201, 2007.

Emery, X. and Lantuéjoul, C. TBSIM: a computer program for conditional simulation of three-dimensional Gaussian random fields via the turning bands method. *Computers & Geosciences*, 32(10):1615–1628, 2006.

Emery, X. and Lantuéjoul, C. A spectral approach to simulating intrinsic random fields with power and spline generalized covariances. *Computational Geosciences*, 12(1):121–132, 2008.

Evensen, G. Sequential data assimilation with a nonlinear quasi-geostrophic model using Monte Carlo methods to forecast error statistics. *Journal of Geophysical Research*, 99:10–10, 1994. doi: 10.1029/94JC00572.

Fælt, L. M., Henriquez, A., Holden, L., and Tjelmeland, H. MOHERES, a program system for simulation of reservoir architecture and properties. In *European Symposium on Improved Oil Recovery*, pages 27–39, 1991.

Farmer, C. L. The generation of stochastic fields of reservoir parameters with specified geostatistical distributions. In Edwards, S. and King, P. R., editors, *Mathematics in Oil Production*, pages 235–252. Clarendon Press, Oxford, 1988.

Farmer, C. L. Numerical rocks. In King, P. R., editor, *The Mathematical Generation of Reservoir Geology*, Oxford, 1992. Clarendon Press. Proceedings of a conference held at Robinson College, Cambridge, 1989.

Feitosa, G. S., Chu, L., Thompson, L. G., and Reynolds, A. C. Determination of reservoir permeability distributions from well test pressure data. In *1993 SPE Western Regional Meeting*, pages 189–204, Anchorage, Alaska, May 1993a. Society of Petroleum Engineers. SPE Paper # 26047.

Feitosa, G. S., Chu, L., Thompson, L. G., and Reynolds, A. C. Determination of reservoir permeability distributions from pressure buildup data. In *1993 SPE Annual Technical Conference and Exhibition Formation Evaluation and Reservoir Geology*, pages 417–429, Houston, TX, October 1993b. Society of Petroleum Engineers. SPE Paper # 26457.

Fetel, E. and Caumon, G. Reservoir flow uncertainty assessment using response surface constrained by secondary information. *Journal of Petroleum Science and Engineering*, 60(3):170–182, 2008.

Feyen, L. and Caers, J. Quantifying geological uncertainty for flow and transport modeling in multi-modal heterogeneous formations. *Adv. Water Resour.*, 29(6):912–929, 2006.

Fildani, A., Fosdick, J. C., Romans, B. W., and Hubbard, S. M. Stratigraphic and structural evolution of the Magallanes Basin, Southern Chile. In Fildani, A., Hubbard, S. M., and Romans, B. W., editors, *Stratigraphic Evolution of Deep-Water Architecture; Examples on Controls and Depositional Styles from the Magallanes Basin, Chile*. SEPM Field Guide, 2009.

Fitzgerald, J. J. *Black gold with grit*. Evergreen Press Limited, Vancouver, B.C., 1978.

Folkestad, A., Veselovsky, Z., and Roberts, P. Utilising borehole image logs to interpret delta to estuarine system: a case study of the subsurface Lower Jurassic Cook Formation in the Norwegian northern North Sea. *Marine and Petroleum Geology*, 29:255–275, January 2012.

Fournier, F. and Derain, J.-F. A statistical methodology for deriving reservoir properties from seismic data. *Geophysics*, pages 1437–1450, September–October 1995.

Francis, A. Limitations of deterministic and advantages of stochastic seismic inversion. *CSEG Recorder*, pages 5–11, February 2005.

Froidevaux, R. Probability field simulation. In Soares, A., editor, *Geostatistics Troia 1992*, volume 1, pages 73–84. Kluwer, Dordrecht, 1993.

Frykman, P. and Deutsch, C. V. Geostatistical scaling laws applied to core and log data. In *1999 SPE Annual Technical Conference and Exhibition*. Society of Petroleum Engineers, October 1999. SPE Paper 56822.

Gadallah, M. R. and Fishe, R. *Exploration Geophysics*. Prentice Hall, Springer, Upper Saddle River, NJ, 2010.

Galli, A., Le Loc'h, G., Geffroy, F., and Eschard, R. An application of the truncated pluri-Gaussian method for modeling geology. In Coburn, T. C., Yarus, J. M., and Chambers, R. L., editors, *Stochastic modeling and geostatistics*, volume II, pages 109–122. AAPG Computer Applications in Geology 5, 2006.

Galloway, W. E. Clastic depositional systems and sequences: applications to reservoir prediction, delineation, and characterization. *The Leading Edge*, pages 173–180, 1998.

Galloway, W. E. and Hobday, D. K. *Terrigenouus Clastic Depositional Systems: Applications to Fossil Fuel and Groundwater Resources*. Springer, New York, 1996.

Gandin, L. S. *Objective Analysis of Meteorological Fields*. Gidrometeorologicheskoe Izdatel'stvo (GIMEZ), Leningrad, 1963. Reprinted by Israel Program for Scientific Translations, Jerusalem, 1965.

Gavalas, G. R., Shah, P. C., and Seinfeld, J. H. Reservoir history matching by Bayesian estimation. *SPE Journal*, pages 337–349, December 1976.

Gawith, D. E., Gutteridge, P. A., and Tang, Z. Integrating geoscience and engineering for improved field management and appraisal. In *1995 SPE Annual Technical Conference and Exhibition Formation Evaluation and Reservoir Geology*, pages 11–23, Dallas, TX, October 1995. Society of Petroleum Engineers. SPE Paper # 29928.

Gelfand, A. E. and Smith, A. F. M. Sampling-based approaches to calculating marginal densities. *Journal of the American statistical association*, pages 398–409, 1990.

Gelhar, L. W., Welty, C., and Rehfeldt, K. R. A critical review on field-scale dispersion in aquifers. *Water Resources Research*, 28(7):1955–1974, 1992.

Geman, S. and Geman, D. Stochastic relaxation, Gibbs distributions, and the Bayesian restoration of images. *IEEE Transactions on Pattern Analysis and Machine Intelligence*, PAMI-6(6):721–741, November 1984.

Georgsen, F. and Omre, H. Combining fibre processes and Gaussian random functions for modeling fluvial reservoirs. In Soares, A., editor, *Geostatistics Troia 1992*, volume 2, pages 425–440. Kluwer, Dordrecht, 1992.

Goldberger, A. Best linear unbiased prediction in the generalized linear regression model. *JASA*, 57:369–375, 1962.

Gómez-Hernández, J. J. and Srivastava, R. M. ISIM3D: an ANSI-C three dimensional multiple indicator conditional simulation program. *Computers & Geosciences*, 16(4):395–410, 1990.

Gómez-Hernández, J. J., Sahuquillo, A., and Capilla, J. E. Stochastic simulation of transmissivity fields conditional to both transmissivity and piezometric data, 1. the theory. *Journal of Hydrology*, 203(1–4):162–174, 1998.

Gonzalez, E., McLennan, J., and Deutsch, C. V. Non-stationary Gaussian transformation: a new approach to sgs in presence of a trend. *APCOM 2007*, 1:12, 2007. doi: 10.1029/2008WR007408.

Goodfellow, R., Mustapha, H., and Dimitrakopolous, R. Approximations of high-order spatial statistics through decomposition. *Geostat 2012*, pages 91–102, 2012.

Goovaerts, P. Comparative performance of indicator algorithms for modeling conditional probability distribution functions. *Mathematical Geology*, 26(3):385–410, 1994a.

Goovaerts, P. Comparison of CoIK, IK and mIK performances for modeling conditional probabilities of categorical variables. In Dimitrakopoulos, R., editor, *Geostatistics for the Next Century*, pages 18–29. Kluwer, Dordrecht, 1994b.

Goovaerts, P. Comparative performance of indicator algorithms for modeling conditional probability distribution functions. *Mathematical Geology*, 26:389–411, 1994c.

Goovaerts, P. *Geostatistics for Natural Resources Evaluation*. Oxford University Press, New York, 1997.

Goovaerts, P. Impact of simulation algorithm, magnitude of ergodic fluctuations and number of realizations on the spaces of uncertainty of flow properties. *Stochastic Environmental Research and Risk Assessment*, 13(2):161–182, 1999.

Granjeon, D. 3D stratigraphic modeling of sedimentary basins. In *AAPG Annual Convention and Exhibition*,

Denver, Colorado, June 2009. AAPG Search and Discovery Article # 90090 ©2009.

Gringarten, E. and Deutsch, C. V. Methodology for variogram interpretation and modeling for improved petroleum reservoir characterization. In *1999 SPE Annual Technical Conference and Exhibition Formation Evaluation and Reservoir Geology*, Houston, TX, October 1999. Society of Petroleum Engineers. SPE Paper # 56654.

Guardiano, F. and Srivastava, R. M. Multivariate geostatistics: beyond bivariate moments. In Soares, A., editor, *Geostatistics Troia 1992*, volume 1, pages 133–144. Kluwer, Dordrecht, 1993.

Gull, S. F. and Skilling, J. The entropy of an image. In Smith, C. R. and Gandy Jr., W. T., editors, *Maximum Entropy and Bayesian Methods in Inverse Problems*, pages 287–301. Reidel, Dordrecht, 1985.

Gundesø, R. and Egeland, O. SESIMIRA—a new geologic tool for 3-D modeling of heterogeneous reservoirs. In Buller, A. T., editor, *North Sea Oil and Gas Reservoirs II*. Graham and Trotman, London, 1990.

Gutjahr, A. L. Fast Fourier transforms for random fields. Technical Report No. 4-R58-2690R, Los Alamos, NM, 1989.

Haas, A. and Dubrule, O. Geostatistical inversion—a sequential method of stochastic reservoir modeling constrained by seismic data. *First Break*, 12(11):561–569, 1994.

Hajek, L., Heller, P., and Sheets, B. A. Significance of channel-belt clustering in alluvial basins. *Geology*, 38(6):535–538, 2010.

Haldorsen, H. H. and Chang, D. M. Notes on stochastic shales: from outcrop to simulation model. In Lake, L. W. and Caroll, H. B., editors, *Reservoir Characterization*, pages 445–485. Academic Press, New York, 1986.

Haldorsen, H. H. and Lake, L. W. A new approach to shale management in field-scale models. *SPE Journal*, pages 447–457, April 1984.

Halliburton, 2012. *Landmark Graphics Suite of Software*. www.halliburton.com.

Hamilton, D. E. and Jones, T. A. *Computer Modeling of Geologic Surfaces and Volumes*. 1992. American Association of Petroleum Geologists.

Hammersley, J. M. and Handscomb, D. C. *Monte Carlo Methods*. John Wiley & Sons, New York, 1964.

Harbaugh, J. W. *Computing Risk for Oil Prospects: Principles and Programs*. Pergamon Press, New York, 1995.

Harding, A., Strebelle, S., Levy, M., Thorne, J., Xie, D., Leigh, S., and Preece, R. Reservoir facies modeling: new advances in mps. In Leuangthong, O. and Deutsch, C. V., editors, *Geostatistics Banff 2004*, volume 14 of *Quantitative Geology and Geostatistics*, pages 559–568. Springer Netherlands, Dordrecht, 2005.

Hassanpour, M. and Deutsch, C. An introduction to grid-free object-based facies modeling. Technical Report 107, CCG Annual Report 12, Edmonton, AB, 2010.

Hassanpour, M., Pyrcz, M. J., and Deutsch, C. V. Improved geostatistical models of inclined heterolithic strata for McMurray Formation. AAPG Bulletin, 2013.

Hatløy, A. S. Numerical facies modeling combining deterministic and stochastic method. In Yarus, J. M. and Chambers, R. L., editors, *Stochastic Modeling and Geostatistics: principles, Methods, and Case Studies*, pages 109–120. AAPG Computer Applications in Geology, No. 3, 1995.

Hauge, R. and Syversveen, L. Well conditioning in object models. *Mathematical Geology*, 39:383–398, 2007.

Hein, J. F. and Cotterill, D. K. The Athabasca oil sands—a regional geological perspective. *Natural Resources Research*, **15**(2), June 2006. doi: 10.1007/s11053-006-9015-4.

Hektoen, A. and Holden, L. Bayesian modelling of sequence stratigraphic bounding surfaces. In Baafi, E. Y. and Schofield, N. A., editors, *Geostatistics Wollongong 1996*, pages 339–349. Kluwer, Dordrecht, 1997.

Helgesen, J., Magnus, I., Prosser, S., Saigal, G., Aamodt, G., Dolberg, D., and Busman, S. Comparison of constrained sparse spike and stochastic inversion for porosity prediction at Kristin Field. *The Leading Edge*, 40, April 2000.

Henriquez, A., Tyler, K., and Hurst, A. Characterization of fluvial sedimentology for reservoir simulation modeling. *SPEFEJ*, pages 211–216, September 1990.

Hewett, T. A. Fractal distributions of reservoir heterogeneity and their influence on fluid transport. SPE Paper # 15386, 1986.

Hewett, T. A. Modelling reservoir heterogeneity with fractals. In Soares, A., editor, *Geostatistics—Troia*, volume 1, pages 455–466. Kluwer, Dordrecht, 1993.

Hewett, T. A. Fractal methods for fracture characterization. In Yarus, J. M. and Chambers, R. L., editors, *Stochastic Modeling and Geostatistics: Principles, Methods, and Case Studies*, pages 249–260. AAPG Computer Applications in Geology, No. 3, 1995.

Honarkhah, M. and Caers, J. Stochastic simulation of patterns using distance-based pattern modeling. *Mathematical Geosciences*, 42:487–517, 2010. doi: 10.1007/s11004-010-9276-7.

Honeycutt, C. E. and Plotnick, R. Image analysis techniques and gray-level co-occurrence matrices (glcm) for calculating bioturbation indices and characterizing biogenic sedimentary structures. *Computers & Geosciences*, 34(11):1461–1472, 2008.

Hong, G. and Deutsch, C. V. Fluvial channel size determination with indicator variograms. *Petroleum Geosciences*, 16:161–169, 2010.

Hong, S. *Multivariate Analysis of Diverse Data for Improved Geostatistical Reservoir Modeling*. PhD thesis, University of Alberta, Edmonton, AB, 2009.

Hong, S. and Deutsch, C. V. Methods for integrating conditional probabilities for geostatistical modeling. Technical Report 105, CCG Annual Report 9, Edmonton, AB, 2007.

Hong, S. and Deutsch, C. V. On secondary data integration. Technical Report 101, CCG Annual Report 11, Edmonton, AB, 2009a.

Hong, S. and Deutsch, C. V. 3D trend modeling by combining lower order trends. Technical Report 130, CCG Annual Report 11, Edmonton, AB, 2009b.

Hong, S. and Deutsch, C. V. Evaluation of probabilistic models for categorical variables. Technical Report 131, CCG Annual Report 11, Edmonton, AB, 2009c.

Horne, R. N. *Modern Well Test Analysis. A Computer-Aided Approach*. Petroway Inc, 926 Bautista Court, Palo Alto, CA, 94303, 2nd edition, 1995.

Horta, A., Caeiro, M., Nunes, R., and Soares, A. Simulation of continuous variables at meander structures: application to contaminated sediments of a lagoon. In Atkinson, P. M. and Lloyd, C. D., editors, *GeoENV VII Ű-Geostatistics for Environmental Applications*, volume 16 of *Quantitative Geology and Geostatistics*, pages 161–172. Springer Netherlands, Dordrecht, 2010. doi: 10.1007/978-90-481-2322-3_15.

Hovadik, J. M. and Larue, D. K. Stratigraphic and structural connectivity. *Geological Society*, 347:219–242, 2010. doi: 10.1144/SP347.13.

Hovadik, J. M. and Larue, D. K. Predicting waterflood behavior by simulating earth models with no or limited dynamic data: from model ranking to simulating a billion-cell model. In Ma, Y. Z. and Pointe, P. R. L., editors, *Uncertainty analysis and reservoir modeling*, volume 96, pages 29–55. AAPG Memoir, 2011.

Hove, K., Olsen, G., Nilsson, S., Tonnesen, M., and Hatløy, A. From stochastic geological description to production forecasting in heterogeneous layered reservoirs. In *SPE Annual Conference and Exhibition, Washington, DC*, Washington, DC, October 1992. Society of Petroleum Engineers. SPE Paper # 24890.

Howard, A. D. Modeling channel migration and floodplain sedimentation in meandering streams. In Carling, P. A. and Petts, G. E., editors, *Lowland Floodplain Rivers*. John Wiley & Sons, New York, 1992.

Howell, J. A. and Flint, S. S. Siliciclastics case study: the book cliffs. *The Sedimentary Record of Sea-level Change*, pages 135–208, 2003.

Hoyal, D. C. J. D. and Sheets, B. A. Intrinsic controls on the range of volumes, morphologies, and dimensions of submarine lobes. *Journal of Geophysical Research*, 114, 2009. doi: 10.1029/2007JF000882.

Hu, L. Y. and Ravalec-Dupin, M. L. On some controversial issues of geostatistical simulation. In Leuangthong, O. and Deutsch, C. V., editors, *Geostatistics Banff 2004*, volume 14 of *Quantitative Geology and Geostatistics*, pages 175–184. Springer Netherlands, Dordrecht, 2005.

Hubbard, S. M., Smith, D. G., Nielsen, H., Leckie, D. A., Fustic, M., Spencer, R. L., and Bloom, L. Seismic geomorphology and sedimentology of a tidally influenced river deposit, lower cretaceous athabasca oil sands, alberta, canada. *AAPG Bulletin*, 95(7):1123–1145, 2011.

Isaaks, E. H. *The Application of Monte Carlo Methods to the Analysis of Spatially Correlated Data*. PhD thesis, Stanford University, Stanford, CA, 1990.

Isaaks, E. H. and Srivastava, R. M. *An Introduction to Applied Geostatistics*. Oxford University Press, New York, 1989.

Izgec, O., Sayarpour, M., and Shook, G. M. Maximizing volumetric sweep efficiency in waterfloods with hydrocarbon f-phi curves. *Journal of Petroleum Science and Engineering*, 78:54–64, 2011.

Jacquard, P. and Jain, C. Permeability distribution from field pressure data. *SPE Journal*, pages 281–294, December 1965.

Jensen, J. L., Corbett, P. W. M., Pickup, G. E., and Ringrose, P. S. Permeability semivariograms, geological structure, and flow performance. *Mathematical Geology*, 28(4):419–435, 1996.

Jerolmack, D. J. and Paola, C. Complexity in a cellular model of river avulsion. *Geomorphology*, 91:259–270, 2007.

Johnson, N. L. and Kotz, S. *Continuous Univariate Distributions—1*. John Wiley & Sons, New York, 1970.

Jones, A. D. W., Al-Qabandi, S., Reddick, C. E., and Anderson, S. A. Rapid assessment of pattern waterflooding uncertainty in a giant oil reservoir. In *1997 SPE Annual Technical Conference and Exhibition Formation Evaluation and Reservoir Geology*. Society of Petroleum Engineers, October 1997. SPE Paper # 38890.

Jones, T. A., Hamilton, D. E., and Johnson, C. R. *Contouring Geologic Surfaces with the Computer*. Van Nostrand Reinhold, New York, 1986.

Journel, A. G. Non-parametric estimation of spatial distributions. *Mathematical Geology*, 15(3):445–468, 1983.

Journel, A. G. Constrained interpolation and qualitative information. *Mathematical Geology*, 18(3):269–286, 1986a.

Journel, A. G. Geostatistics: models and tools for the earth sciences. *Mathematical Geology*, 18(1):119–140, 1986b.

Journel, A. G. *Fundamentals of Geostatistics in Five Lessons*. Volume 8 Short Course in Geology. American Geophysical Union, Washington, DC, 1989.

Journel, A. G. Resampling from stochastic simulations. *Environmental and Ecological Statistics*, 1:63–84, 1994.

Journel, A. G. The abuse of principles in model building and the quest for objectivity: Opening keynote address. In Baafi, E. Y. and Schofield, N. A., editors, *Fifth International Geostatistics Congress*, Wollongong, Australia, September 1996.

Journel, A. G. Markov models for cross-covariances. *Mathematical Geology*, 31(8):955–964, 1999.

Journel, A. G. Combining knowledge from diverse sources: an alternative to traditional data independence hypotheses. *Mathematical Geology*, 34(5), 2002.

Journel, A. G. and Alabert, F. G. Focusing on spatial connectivity of extreme valued attributes: stochastic indicator models of reservoir heterogeneities. SPE Paper # 18324, 1988.

Journel, A. G. and Alabert, F. G. New method for reservoir mapping. *Journal of Petroleum Technology*, pages 212–218, February 1990.

Journel, A. G. and Bitanov, A. Uncertainty in n/g ratio in early reservoir development. *Journal of Petroleum Science and Engineering*, 44(1–2):115–130, 2004.

Journel, A. G. and Deutsch, C. V. Entropy and spatial disorder. *Mathematical Geology*, 25(3):329–355, April 1993.

Journel, A. G. and Gómez-Hernández, J. J. Stochastic imaging of the Wilmington clastic sequence. *SPEFE*, pages 33–40, March 1993. SPE Paper # 19857.

Journel, A. G. and Huijbregts, C. J. *Mining Geostatistics*. Academic Press, New York, 1978.

Journel, A. G. and Isaaks, E. H. Conditional indicator simulation: application to a Saskatchewan uranium deposit. *Mathematical Geology*, 16(7):685–718, 1984.

Journel, A. G. and Kyriakidis, P. C. *Evaluation of Mineral Reserves: A Simulation Approach*. Oxford University Press, New York, 1st edition, 2004.

Journel, A. G. and Xu, W. Posterior identification of histograms conditional to local data. *Mathematical Geology*, 26:323–359, 1994.

Journel, A. G., Deutsch, C. V., and Desbarats, A. J. Power averaging for block effective permeability. In *56th California Regional Meeting*, pages 329–334. Society of Petroleum Engineers, April 1986. SPE Paper # 15128.

Kalla, S., White, C., Gunning, J., and Glinsky, M. Consistent downscaling of seismic inversion thicknesses to cornerpoint flow models. *SPE Journal*, 13(4):412–422, 2008.

Kalman, R. E. A new approach to linear filtering and prediction problems. *Journal of Basic Engineering 82*, (1):35–45, 1960.

Kedzierski, P., G. Caumon, J.-L. M., Royer, J.-J., and Durand-Riard, P. 3D marine sedimentary reservoir stochastic simulation accounting for high resolution sequence stratigraphy and sedimentological rules. In Ortiz, J. and Emery, X., editors, *Eighth International Geostatistics Congress*, pages 657–666, Gecamin Ltd., September 2008.

Kennedy, J. and Eberhart, R. Particle swarm optimization. *Proceedings Annual Conference of the International Association of Mathematical Geologists*, IV:1942–1948, 1995.

Kennedy Jr., W. J. and Gentle, J. E. *Statistical Computing*. Marcel Dekker, Inc, New York, 1980.

Khan, A., Horowitz, D., Liesch, A., and Schepel, K. Semi-amalgamated thinly-bedded deepwater GOM turbidite reservoir performance modeling using object-based technology and Bouma lithofacies. In *1996 SPE Annual Technical Conference and Exhibition Formation Evaluation and Reservoir Geology*, pages 443–455, Denver, CO, October 1996. Society of Petroleum Engineers. SPE Paper # 36724.

Kirkpatrick, S., Gelatt Jr., C. D., and Vecchi, M. P. Optimization by simulated annealing. *Science*, 220 (4598):671–680, May 1983.

Koltermann, C. E. and Gorelick, S. M. Heterogeneity in sedimentary deposits: a review of structure-imitating, process-imitating, and descriptive approaches. *Water Resources Research*, 32(9):2617–2658, September 1996.

Korvin, G. Axiomatic characterization of the general mixture rule. *Geoexploration*, 19:267–276, 1981.

Krige, D. G. A statistical approach to some mine valuations and allied problems at the Witwatersrand. Master's thesis, University of Witwatersrand, South Africa, 1951.

Krishnana, S. *Combining diverse and partially redundant information in the earth sciences*. PhD thesis, Stanford University, Stanford, CA, 2004.

Krygowski, D., Asquith, A., and Gibson, C. Basic well log analysis. *Geology*, page 244, 2004.

Kupfersberger, H. and Deutsch, C. V. Methodology for integrating analogue geologic data in 3-D variogram modeling. *AAPG Bulletin*, 83(8):1262–1278, 1999.

Kyriakidis, P. C., Deutsch, C. V., and Grant, M. L. Calculation of the normal scores variogram used for truncated Gaussian lithofacies simulation: theory and FORTRAN code. *Computers & Geosciences*, 25(2):161–169, 1999.

Langlais, V. and Doyle, J. Comparison of several methods of lithofacies simulation on the fluvial Gypsy Sandstone of Oklahoma. In Soares, A., editor, *Geostatistics—Troia*, volume 1, pages 299–310. Kluwer, Dordrecht, 1993.

Lantuéjoul, C. *Geostatistical Simulation: Models and Algorithms*. Springer, Berlin, 1st edition, 2001.

Lantuéjoul, C. Ergodicity and integral range. *Journal of Microscopy*, 161(3):387–403, 2011.

Lantuéjoul, C., Beucher, H., Chilès, J.-P., Lajaunie, C., Wackernagel, H., and Elion, P. Estimating the trace length distribution of fractures from line sampling data. In Leuangthong, O. and Deutsch, C. V., editors, *Geostatistics Banff 2004*, volume 14 of *Quantitative Geology and Geostatistics*, pages 165–174. Springer Netherlands, Dordrecht, 2005.

Larrondo, P. F. Accounting for geological boundaries in geostatistical modeling of multiple rock types. Master's thesis, University of Alberta, Edmonton, AB, 2004.

Larrondo, P. F. and Deutsch, C. V. Application of local non-stationary lmc for gradational boundaries. Technical Report 131, CCG Annual Report 6, Edmonton, AB, 2004.

LeBlanc, R. J. Distribution and continuity of sandstone reservoirs—part 1. *JPT*, pages 776–792, July 1977a.

LeBlanc, R. J. Distribution and continuity of sandstone reservoirs—part 2. *JPT*, pages 793–804, July 1977b.

Lee, T., Richards, J. A., and Swain, P. H. Probabilistic end evidential approaches for multisource data analysis. *IEEE Transactions on Geoscience and Remote Sensing*, PAMI-6(6):721–741, November 1987.

Leeder, M. R. A quantitative stratigraphic model for alluvium with special reference to channel deposit density and interconnectedness. In Miall, A. D., editor, *Fluvial Sedimentology*, pages 587–596. Mem. Canadian Society of Petroleum Geologists, 1978.

Leiva, A. Construction of hybrid geostatistical models combining surface based methods with object-based simulation: use of flow direction and drainage area. Master's thesis, Stanford University, 2009.

Leuangthong, O. *Stepwise conditional transform for geostatistical simulation.* PhD thesis, University of Alberta, Edmonton, AB, 2003.

Leuangthong, O. The promises and pitfalls of direct simulation. In Leuangthong, O. and Deutsch, C. V., editors, *Geostatistics Banff 2004*, pages 305–314. Springer Netherlands, 2005.

Leuangthong, O. and Deutsch, C. V. Stepwise conditional transformation for simulation of multiple variables. *Mathematical Geology*, 35(2):155–173, 2003.

Leuangthong, O., McLennan, J. A., and Deutsch, C. V. Minimum acceptance criteria for geostatistical realizations. *Natural Resources Research*, 13(3):131–141, 2004.

Li, H. and White, C. D. Geostatistical models for shales in distributary channel point bars (Ferron Sandstone, Utah): from ground-penetrating radar data to three-dimensional flow modeling. *AAPG Bulletin December*, 87(12):1851–1868, December 2003. doi: 10.2118/103268-PA.

Li, S. and Deutsch, C. V. A petrel plugin for ranking realizations. Technical Report 407, CCG Annual Report 10, Edmonton, AB, 2008.

Liu, G. R. *Mesh Free Methods: Moving Beyond the Finite Element Method.* CRC, Boca Raton, FL, 2002.

Liu, Y., Harding, A., Abriel, W., and Strebelle, S. Multiplepoint statistics simulation integrating wells, seismic data and geology. *AAPG Bulletin*, 88(7):905–921, 2004.

Lo, T. and Bashore, W. M. Seismic constrained facies modeling using stochastic seismic inversion and indicator simulation, a north-sea example. Expanded Abstract, SEG 69th Annual Meeting, 1999.

Loc'h, G. L. and Galli, A. Truncated plurigaussian method: theoretical and practical points of view. In Baafi, E. and Schofield, N., editors, *Fifth International Geostatistics Congress*, volume 1, pages 211–222, Wollongong, Australia, Kluwer, Dordrecht, 1996.

Lødøen, O. and Omre, H. Scale-corrected ensemble kalman filtering applied to production-history conditioning in reservoir evaluation. *SPE Journal*, 13(2):177–194, 2008.

Lopez, S., Galli, A., and Cojan, I. Fluvial meandering channelized reservoirs: a stochastic and process-based approach. In *Proceedings Annual Conference of the International Association of Mathematical Geologists*, Cancun, Mexico, CD-ROM, 2001. International Association of Mathematical Geologists.

Luenberger, D. G. *Optimization by Vector Space Methods.* John Wiley & Sons, New York, 1969.

Lyster, S. Theoretical justification for iterative simulation methods such as the Gibbs sampler. Technical Report 114, CCG Annual Report 9, University of Alberta, Edmonton, AB, 2007.

Lyster, S. Reproducing local proportions in MPS simulation. Technical Report 129, CCG Annual Report 10, University of Alberta, Edmonton, AB, 2008.

Lyster, S. and Deutsch, C. An entropy-based approach to establish MPS templates centre for computational geostatistics. Technical Report 114, CCG Annual Report 8, University of Alberta, Edmonton, AB, 2006.

MacDonald, A. C. and Aasen, J. O. A prototype procedure for stochastic modeling of facies tract distribution in shoreface reservoirs. In Yarus, J. M. and Chambers, R. L., editors, *Stochastic modeling and geostatistics; principles, methods, and case studies*, pages 91–108. American Association of Petroleum Geologists Computer Applications in Geology, 1994.

Mackey, S. D. and Bridge, J. S. A revised FORTRAN program to simulate alluvial stratigraphy. *Computers & Geosciences*, 18(2):119–181, 1992.

Maharaja, A. *Global net-to-gross uncertainty assessment at reservoir appraisal stage.* PhD thesis, Stanford University, Stanford, CA, 2007.

Maharaja, A. TiGenerator: object-based training image generator. *Computers & Geosciences*, 34(7):1753–1761, December 2008.

Mallet, J.-L. Structural unfolding. http:www.ensg. u-nancy. frGOCAD, 1999.

Mallet, J.-L. *Geomodelling.* Oxford Uuniversity Press, New York, 2002.

Manchuk, J. G. and Deutsch, C. V. Sensitivity analysis and the value of information in Gaussian multivariate prediction. Technical Report 115, CCG Annual Report 7, Edmonton, AB, 2005.

Manchuk, J. G. and Deutsch, C. V. Geostatistical assignment of reservoir properties to unstructured grids. Technical report, 30th Gocad Meeting, April 2010.

Manchuk, J. G. and Deutsch, C. V. A short note on trend modeling using moving windows. Technical Report 403, CCG Annual Report 13, Edmonton, AB, 2011.

Manchuk, J. G., Leuangthong, O., and Deutsch, C. V. Direct geostatistical simulation of unstructured grids. In Leuangthong, O. and Deutsch, C. V., editors, *Geostatistics Banff 2004*, volume 14 of *Quantitative Geology and Geostatistics*, pages 85–94. Springer Netherlands, Dordrecht, 2005.

Manchuk, J. G., Mlacnik, M. J., and Deutsch, C. V. Upscaling permeability to unstructured grids using the multipoint flux approximation. *Petroleum Geosciences*, 18:239–248, 2012.

Mariethoz, G. and Kelly, B. F. J. Modeling complex geological structures with elementary training images and transform-invariant distances. *Water Resources Research*, 47(7):W07527, 2011.

Mariethoz, G., Renard, P., Cornaton, F., and Jaquet, O. Truncated plurigaussian simulations to characterize aquifer heterogeneity. *Ground Water*, 47(1):13–24, 2009. doi: 10.1111/j.1745-6584.2008.00489.x.

Mariethoz, G., Renard, P., and Straubhaar, J. The direct sampling method to perform multiple-point simulations. *Water resources Research*, 46(11), 2010. doi: 10.1029/2008WR007621.

Marsaglia, G. The structure of linear congruential sequences. In Zaremba, S. K., editor, *Applications of Number Theory to Numerical Analysis*, pages 249–285. Academic Press, London, 1972.

Martínez, J. L. F., Gonzalo, E. G., Muñiz, Z. F., and Mukerji, T. How to design a powerful family of particle swarm optimizers for inverse modelling. *Transactions of the Institute of Measurement and Control*, 34(6):705–719, 2012.

Massart, B. Y. G., Jackson, M. D., Hampson, G. J., Legler, B., Johnson, H. D., Jackson, C. A. L., Ravnas, R., and Sarginson, M. Three-dimensional characterization and surface-based modeling of tide-dominated heterolithic sandstones. In *EAGE Petroleum Geostatistics 2011*, 2011.

Matern, B. *Spatial Variation*, volume 36 of *Lecture Notes in Statistics*. Springer-Verlag, New York, second edition, 1980. First edition published by Meddelanden fran Statens Skogsforskningsinstitut, Band 49, No. 5, 1960.

Matheron, G. Traité de géostatistique appliquée. Vol. 1 (1962), Vol. 2 (1963), ed. Technip, Paris, 1962.

Matheron, G. Structure et composition des perméabilités. *Revue de l'IFP Rueuil*, 21(4):564–580, 1966.

Matheron, G. La théorie des variables régionalisées et ses applications. Fasc. 5, École National Supériure des Mines, Paris, 1971.

Matheron, G., Beucher, H., de Fouquet, H., Galli, A., Guerillot, D., and Ravenne, C. Conditional simulation of the geometry of fluvio-deltaic reservoirs. SPE Paper # 16753, 1987.

Maxwell, S. C. and Urbancic, T. I. The role of passive microseismic monitoring in the instrumented oil field. *The Leading Edge, Society of Exploration Geophysicists*, 20(6):636–639, December 2001.

McConway, K. J. Marginalization and linear opinion pools. *Journal of American Statistical Association*, 76(374), 1981.

McHargue, T., Pyrcz, M. J., Sullivan, M. D., Clark, J. D., Fildani, A., Romans, B. W., Covault, J. A., Levy, J. A., Posamentier, H. W., and Drinkwater, N. J. Architecture of turbidite channel systems on the continental slope: patterns and predictions. *Marine and Petroleum Geology*, 28(3):728–743, 2010.

McHargue, T., Pyrcz, M. J., Sullivan, M., Clark, J., A., F., Levy, M., Drinkwater, N., Posamentier, H., Romans, B., and Covault, J. Event-based modeling of tubidite channel fill, channel stacking pattern and net sand volume. In Martinsen, O. J., Pulham, A. J., Haughton, P. D., and Sullivan, M. D., editors, *Outcrops Revitalized: tools, Techniques and Applications*, number 10, pages 163–174. SEPM Concepts in Sedimentology and Paleontology, 2011.

McLennan, J. and Deutsch, C. V. BOUNDSIM: implicit boundary modeling. *APCOM 2007*, 1:9, 2007.

McLennan, J. A. *The decision of stationarity*. PhD thesis, University of Alberta, Edmonton, AB, 2008.

McLennan, J. A. and Deutsch, C. V. Local ranking of geostatistical realizations for flow simulation. In *SPE International Thermal Operations and Heavy Oil Symposium*, Alberta, Canada, 2005. SPE Paper # 98168.

McLennan, J. A., Allwardt, P. F., Hennings, P. H., and Farrell, H. E. Multivariate fracture intensity prediction: application to oil mountain anticline, wyoming. *AAPG Bulletin*, 93(11):1585–1595, November 2009.

Mehlhorn, K. *Multi-Dimensional Searching and Computational Geometry*. Springer-Verlag, New York, 1984.

Metropolis, N., Rosenbluth, A. W., Rosenbluth, M. N., Teller, A. H., and Teller, E. Equations of state calculations by fast computing machines. *Journal of Chemical Physics*, 21(6):1087–1091, 1953.

Miall, A. D. *The Geology of Fluvial Deposits*. Springer-Verlag, New York, 1996.

Michael, H. A., Li, H., Boucher, A., Sun, T., Caers, J., and Gorelick, S. M. Combining geologic-process models and geostatistics for conditional simulation of 3-D subsurface heterogeneity. *Water Resources Research*, 46(5):W05527, 2010. doi: 10.1029/2009WR008414.

Middleton, G., Plotnick, R., and Rubens, D. Fractals and non-linear dynamics: New numerical techniques for sedimentary data. *SEPM Short Course*, 36:174, 1995.

Miller, J., Sun, T., Li, H., Stewart, J., Genty, C., Li, D., and Lyttle, C. Direct modeling of reservoirs through forward process-based models: can we get there. *International Petroleum Technology Conference*, pages 259–270, December 2008.

Montgomery, D. C. and Runger, G. C. *Applied statistics and probability for engineers*. John Wiley & Sons, Hoboken, NJ, 4th edition, 2007.

Mulholland, J. W. Sequential stratigraphy: basic elements, concepts, and terminology. *The Leading Edge*, pages 37–40, 1998.

Murray, C. J. Identification and 3-D modeling of petrophysical rock types. In Yarus, J. M. and Chambers, R. L., editors, *Stochastic Modeling and Geostatistics: Principles, Methods, and Case Studies*, pages 323–338. AAPG Computer Applications in Geology, No. 3, 1995.

Myers, D. E. Matrix formulation of co-kriging. *Mathematical Geology*, 14(3):249–257, 1982.

Myers, D. E. Cokriging-new developments. In Verly, G., editor, *Geostatistics For Natural Resources Characterization*, pages 295–305. Reidel, Dordrecht, Holland, 1984.

Myers, D. E. Pseudo-cross variograms, positive-definiteness, and cokriging. *Mathematical Geology*, 23(6):805–816, 1991.

Myers, K. J. and Milton, N. J. Concepts and principles. In Emery, D. and Myers, K. J., editors, *Sequence Stratigraphy*. Blackwell Publishing Ltd., 1996.

Neufeld, C. and Deutsch, C. V. Incorporating secondary data in the prediction of reservoir properties using Bayesian updating. Technical Report 114, CCG Annual Report 6, Edmonton, AB, 2004.

Nordlund, U. FUZZIM: forward stratigraphic modeling made simple. *Computers & Geosciences*, 25(4):449–456, 1999.

Norrena, K. and Deutsch, C. V. Automatic determination of well placement subject to geostatistical and economic constraints. In *SPE International Thermal Operations and Heavy Oil Symposium and International Horizontal Well Technology Conference*, 2002.

Novakovic, D., White, C. D., Corbeanu, R. M., Hammon, W. S., Bhattacharya, J. P., and McMechan, G. A. Hydraulic effects of shales in fluvial-deltaic deposits: ground-penetrating radar, outcrop observations, geostatistics, and three dimensional flow modeling for the Ferron Sandstone, Utah. *Mathematical Geology*, 34(7):857–893, 2002.

Olea, R. A. *Geostatistical Glossary and Multilingual Dictionary*. Oxford University Press, New York, 1991.

Olea, R. A. Fundamentals of semivariogram estimation, modeling, and usage. In Yarus, J. M. and Chambers, R. L., editors, *Stochastic Modeling and Geostatistics: principles, Methods, and Case Studies*, pages 27–36. AAPG Computer Applications in Geology, No. 3, 1995.

Oliver, D. S. The averaging process in permeability estimation from well test data. *SPE Formation Evaluation*, pages 319–324, September 1990a.

Oliver, D. S. Estimation of radial permeability distribution from well test data. In *SPE Annual Conference and Exhibition, New Orleans, LA*, pages 243–250, New Orleans, LA, September 1990b. Society of Petroleum Engineers. SPE Paper # 20555.

Oliver, D. S. Incorporation of transient pressure data into reservoir characterization. *In Situ*, 18(3):243–275, 1994.

Oliver, D. S. Conditioning channel meanders to well observations. *Mathematical Geology*, 34:185–201, 2002.

Oliver, D. S. and Chen, Y. Improved initial sampling for the ensemble Kalman filter. *Computational Geosciences 13*, (1):13–26, 2009.

Omre, H. *Alternative Variogram Estimators in Geostatistics*. PhD thesis, Stanford University, Stanford, CA, 1985.

Omre, H. Heterogeneity models. In *SPOR Monograph: Recent Advances in Improved Oil Recovery Methods for North Sea Sandstone Reservoirs*, Norway, 1992. Norwegian Petroleum Directorate.

Oreskes, N., Shrader-Frechette, K., and Belitz, K. Verification, validation, and confirmation of numerical models in the earth sciences. *Science*, 263:641–646, February 1994.

Ortiz, J. M. and Deutsch, C. V. Calculation of uncertainty in the variogram. *Mathematical Geology*, 34(2):169–183, 2002.

Ortiz, J. M. and Deutsch, C. V. A practical approach to validate the variogram reproduction from geostatistical simulation. Technical Report 125, CCG Annual Report 9, Edmonton, AB, 2007.

Paola, C. Quantitative models of sedimentary basin filling. *Sedimentology*, 47:121–178, 2000.

Paola, C., Mullin, J., Ellis, C., Mohrig, D. C., Swenson, J., Parker, G., Hickson, T., Heller, P., Pratson, L., Syvitski, J., Sheets, B., and Strong, N. Experimental stratigraphy. *GSA Today*, 11(7):4–9, 2001.

Paola, C., Straub, K., Mohrig, D., and Reinhardt, L. The "unreasonable effectiveness" of stratigraphic and geomorphic experiments. *Earth-Science Reviews*, 97(1): 1–43, 2009.

Paradigm Suite of Software. Paradigm, 2012. www.pdgm.com.

Pardo-Iguzquiza, E. and Chica-Olmo, M. The fourier integral method: an efficient spectral method for simulation of random fields. *Mathematical Geology*, 25(2):177–217, 1993.

Parker, G. 1-D sediment transport morphodynamics with applications to rivers and turbidity currents, e-book. 2012. URL http://hydrolab.illinois.edu/people/parkerg/morphodynamics_e-book.htm.

Pedersen, M. E. H. and Chipperfield, A. J. Simplifying particle swarm optimization. *Applied Soft Computing*, 10(2):618–628, March 2010. doi: 10.1016/j.asoc.2009.08.029.

Perlmutter, M. A. and Plotnick, R. E. Predictable variations in the marine stratigraphic record of the northern and southern hemispheres and reservoir potential. In *Sequence Stratigraphic Models for Exploration and Production: evolving Methodology, Emerging Models and Application Histories*, pages 231–256, 2002. doi: 10.5724/gcs.02.22. GCSSEPM 22nd Bob F. Perkins Research Conference.

Plotnick, R. E. A fractal model for the distribution of stratigraphic hiatuses. *The Journal of Geology*, pages 885–890, 1986.

Plotnick, R. E., Gardner, R. H., Hargrove, W. W., Prestegaard, K., and Perlmutter, M. Lacunarity analysis: a general technique for the analysis of spatial patterns. *Physical Review E*, 53(5):5461–5468, 1996.

Posamentier, H. Depositional elements associated with a basin floor channel-levee system: case study from the Gulf of Mexico. *Marine and Petroleum Geology*, 20:677–690, 2003.

Posamentier, H. W., Davies, R. J., Cartwright, J. A., and Wood, L. Seismic geomorphology—an overview. *Special Publication-Geological Society of London*, 277, 2007.

Pranter, M. J. and Sommer, N. K. Static connectivity of fluvial sandstones in a lower coastal-plain setting: an example from the Upper Cretaceous Lower Williams Fork Formation, Piceance Basin, Colorado. *AAPG Bulletin*, 95(1-4):899–923, June 2001. doi: 10.1306/12091010008.

Prélat, A., Covault, J. A., Hodgson, D. M., Fildani, A., and Flint, S. S. Intrinsic controls on the range of volumes, morphologies, and dimensions of submarine lobes. *Sedimentary Geology*, 232(1–4):66–76, 2010. doi: 10.1016/j.sedgeo.2010.09.010.

Preparata, F. P. and Shamos, M. I. *Computational Geometry: An Introduction*. Springer-Verlag, New York, 1988.

Press, W. H., Flannery, B. P., Teukolsky, S. A., and Vetterling, W. T. *Numerical Recipes*. Cambridge University Press, New York, 1986.

Pringle, J. K., Brunt, R. L., Hodgson, D. M., and Flint, S. S. Capturing stratigraphic and sedimentological complexity from submarine channel complexes outcrops to 3D digital models, Karoo basin, South Africa. *Petroleum Geoscience*, 16(1-4):307–330, 2010. doi: 10.1144/1354-079309-028.

Pyles, D. R. and Jennette, D. Process and facies associations in basin-margin strata of structurally confined submarine fans: example from the carboniferous ross sandstone (ireland). *Marine and Petroleum Geology*, 29:1974–1996, 2009.

Pyrcz, M. J. *Integration of geologic information into geostatistical models*. PhD thesis, University of Alberta, Edmonton, AB, 2004.

Pyrcz, M. J. and Deutsch, C. V. Two artifacts of probability field simulation. *Mathematical Geology*, 33(7):775–799, 2001.

Pyrcz, M. J. and Deutsch, C. V. Debiasing for improved inference of the one-point statistic. In *30th International Symposium on Computer Applications in the Mineral Industries (APCOM)*, Phoenix, Arizona, February 2002.

Pyrcz, M. J. and Deutsch, C. V. Conditional event-based simulation. In Leuangthong, O. and Deutsch, C. V., editors, *Geostatistics Banff 2004*, Quantitative Geology and Geostatistics, pages 135–144. Springer Netherlands, 2005.

Pyrcz, M. J. and Deutsch, C. V. Spectrally corrected semivariogram models. *Mathematical Geology*, 38(7):277–299, 2006a. doi: 10.1007/s11004-006-9053-9.

Pyrcz, M. J. and Deutsch, C. V. Semivariogram models based on geometric offsets. *Mathematical Geology*, 38(4), 2006b.

Pyrcz, M. J. Gringarten, E., Frykman, P., and Deutsch, C. V. Representative input parameters for geostatistical simulation. In Coburn, T. C., Yarus, R. J., and Chambers, R. L., editors, *Stochastic Modeling and Geostatistics: Principles, Methods and Case Studies, Volume II: AAPG Computer Applications in Geology 5*, pages 123–137, 2006.

Pyrcz, M. J. and Strebelle, S. Event-based geostatistical modeling of deepwater systems. *GCSSEPM 26th Bob F. Perkins Research Conference*, pages 893–922, 2006.

Pyrcz, M. J. and Strebelle, S. Event-based geostatistical modeling. In Ortiz, J. M. and Emery, X., editors, *Geostatistics Santiago 2008*, pages 135–144. Springer, Netherlands, 2008.

Pyrcz, M. J., Catuneanu, O., and Deutsch, C. V. Stochastic surface-based modeling of turbidite lobes. *AAPG Bulletin*, 89:177–191, December 2005a.

Pyrcz, M. J., Leuangthong, O., and Deutsch, C. V. Hierarchical trend modeling for improved reservoir characterization. *International Association of Mathematical Geology*, August 2005b.

Pyrcz, M. J., Sullivan, M., Drinkwater, N., Clark, J., Fildani, A., and Sullivan, M. Event-based models as a numerical laboratory for testing sedimentological rules associated with deepwater sheets. *GCSSEPM 26th Bob F. Perkins Research Conference*, pages 923–950, 2006.

Pyrcz, M. J., Boisvert, J., and Deutsch, C. V. A library of training images for fluvial and deepwater reservoirs and associated code. *Computers & Geosciences*, 2007. doi: 10.1016/j.cageo.2007.05.015.

Pyrcz, M. J., Boisvert, J., and Deutsch, C. V. Alluvsim: a conditional event-based fluvial model. *Computers & Geosciences*, 2009. doi: 10.1016/j.cageo.2008.09.012.

Pyrcz, M. J., Sullivan, M. D., McHargue, T. R., Fildani, A., Drinkwater, N. J., Clark, J., and Posamentier, H. W. Numerical modeling of channel stacking from outcrop. In Martinsen, O., Pulham, A., Haughton, P., and Sullivan, M., editors, *Outcrops Revitalized: tools, Techniques and Applications*. SEPM special publication, 2011.

Pyrcz, M. J., McHargue, T., Clark, J., Sullivan, M., and Strebelle, S. Event-based geostatistical modeling: description and applications. In Abrahamsen, P., Hauge, R., and Kolbjøÿrnsen, O., editors, *Geostatistics Oslo 2012*, volume 17 of *Quantitative Geology and Geostatistics*, pages 27–38. Springer Netherlands, Dordrecht, 2012.

Raghavan, R. *Well Test Analysis*. PTR Prentice-Hall, Englewood Cliffs, NJ, 1993.

Rahman, A., Tsai, F. T. C., White, C. D., and Willson, C. S. Coupled semivariogram uncertainty of hydrogeological and geophysical data on capture zone uncertainty analysis. *Journal of Hydrologic Engineering*, 13(10):915–925, 2008. doi: 10.1061/(ASCE)1084-0699(2008)13:10(915).

Rasheva, S. and Bratvold, R. B. A new and improved approach for geological dependency evaluation for multiple-prospect exploration. In *SPE Annual Technical Conference and Exhibition*, 2011. doi: 10.2118/147062-MS.

Razavi, S. F. and Deutsch, C. V. Scaling up of effective absolute permeability. Technical Report 405, CCG Annual Report 14, Edmonton, AB, 2012.

Reading, H. G. *Sedimentary Environments: processes, facies and stratigraphy*. Blackwell Science, Oxford, 3rd edition, 1996.

Ren, W. Large scale modeling by Bayesian updating techniques. Technical Report 129, CCG Annual Report 9, Edmonton, AB, 2007.

Ren, W. and Deutsch, C. V. Exact downscaling in geostatistical modeling. Technical Report 101, CCG Annual Report 7, Edmonton, AB, 2005.

Ren, W., Cunha, L., and Deutsch, C. V. Preservation of multiple point structure when conditioning by kriging. Technical Report 108, CCG Annual Report 6, University of Alberta, Edmonton, AB, 2004.

Ren, W., McLennan, J. A., Leuangthong, O., and Deutsch, C. V. Reservoir characterization of McMurray Formation by 2-D geostatistical modeling. *Natural Resources Research*, 13:7, March 2006.

Ren, W., Deutsch, C. V., Garner, D., Wheeler, T. J., Richy, J. F., and Mus, E. Quantifying resources for the Surmont lease with 2D mapping and multivariate statistics. *SPE Reservoir Evaluation & Engineering*, page 9, February 2008.

Renard, P. and de Marsily, G. Calculating equivalent permeability: a review. *Advances in Water Resources*, 20:253–278, 1997.

Ripley, B. D. *Spatial Statistics*. John Wiley & Sons, New York, 1981.

Ripley, B. D. *Spatial Statistics*. John Wiley and Sons, New Jersey, 2004.

Ritzi, R. W. and Dominic, D. F. Evaluating uncertainty in flow and transport in heterogeneous buried-valley aquifers. In *Groundwater Modeling Conference*, Golden, Colorado, 1993.

Robert, C. P. and Casella, G. *Monte Carlo Statistical Methods*. Springer Science and Business Media LLC, New York, 2nd edition, 2004.

Rock, N. M. S. Numerical geology. In Bhattacharji, S. et al., editors, *Lecture Notes in Earth Sciences*, volume 18. Springer Verlag, New York, 1988.

Romero, C. E. and Carter, J. N. Using genetic algorithms for reservoir characterization. In M. Nikravesh, L. A. Z. F. Aminzadeh, editor, *Soft computing and intelligent data analysis in oil exploration*, volume 228, Dallas, 2003.

Romero, C. E., Carter, J. N., Zimmerman, R. W., and Gringarten, A. C. Improved reservoir characterization through evolutionary computation. In *SPE Annual Technical Conference and Exhibition*, number SPE 62942, Dallas, 2000.

Rossini, C., Brega, F., Piro, L., Rovellini, M., and Spotti, G. Combined geostatistical and dynamic simulations for developing a reservoir management strategy: a case history. *Journal of Petroleum Technology*, pages 979–985, November 1994.

Rothman, D. H. Nonlinear inversion, statistical mechanics, and residual statics estimation. *Geophysics*, 50:2784–2796, 1985.

Sabet, M. A. *Well Test Analysis*, volume 8 of *Contributions in Petroleum Geology and Engineering*. Gulf Publishing Company, Houston, 1991.

Salles, T., Mulder, T., Gaudin, M., Cacas, M. C., Lopez, S., and Cirac, P. Simulating the 1999 turbidity current occurred in Capbreton Canyon through a cellular automata model. *Geomorphology*, 97(3–4):516–537, 2008.

Sams, M., Atkins, D., Siad, N., Parwito, E., and van Riel, P. Stochastic inversion for high resolution reservoir characterization in the central sumatra basin. In *1999 SPE Asia Pacific Improved Oil Recovery Conference*, Kuala Lumpur, Malaysia, October 1999. Society of Petroleum Engineers. SPE Paper # 57260.

Saussus, D. Model building. In *MCMC for geostatistical inversion*, October 2009.

Scheidt, C. and Caers, J. A new method for uncertainty quantification using distances and kernel methods: application to a deepwater turbidite reservoir. *SPE Journal*, 14(4):680–692, 2008. doi: 10.2118/118740-PA.

Schlumberger Software, 2012, http://www.software.slb.com

Schnetzler, E. Visualization and cleaning of pixel-based images. Master's thesis, Stanford University, Stanford, CA, 1994.

Scholle, P. A. and Spearing, D., editors. *Sandstone Depositional Environments*. The American Association of Petroleum Geologists, Tulsa, Oklahoma, 1982.

Schulze-Riegert, R. W. and Ghedan, S. Modern techniques for history matching. *9th International Forum on Reservoir Simulation*, pages 9–13, 2007.

Schulze-Riegert, R. W., Axmann, J. K., Haase, O., Rian, D. T., and Scandpower, Y. Evolutionary algorithms applied to history matching of complex reservoirs. *SPE Reservoir Evaluation & Engineering*, 5(2):163–173, 2002.

Scott, D. W. *Multivariate Density Estimation: Theory, Practice, and Visualization*. John Wiley & Sons, New York, 1992.

Sech, R. P., Jackson, M. D., and Hampson, G. J. Three-dimensional modeling of a shoreface-shelf parasequence reservoir analog: part 1. surface-based modeling to capture high-resolution facies architecture. *AAPG Bulletin*, 93:1155–1181, 2009.

Senger, R. K., Lucia, F. J., Kerans, C., Ferris, C., and Fogg, G. E. Geostatistical/geological permeability characterization of carbonate ramp deposits in San Andres Outcrop, Algerita Escarpment, New Mexico. SPE Paper # 23967, 1992.

Sharma, P. V. Geophysical methods in geology. *Elsevier Science Pub. Co., Inc., New York, NY*, page 700, 1986.

Shmaryan, L. E. and Deutsch, C. V. Object-based modeling of fluvial/deepwater reservoirs with fast data conditioning: methodology and case studies. *SPE Annual Technical Conference and Exhibition, Society of Petroleum Engineers*, 1999.

Shmaryan, L. E. and Journel, A. G. Two Markov models and their application. *Mathematical Geology*, 31(8):965–988, 1999.

Shook, G. and Mitchell, K. A robust measure of heterogeneity for ranking earth models: the F-PHI curve and dynamic lorenz coefficient. In *SPE Annual Technical Conference and Exhibition*, New Orleans, LA, 2009.

Silverman, B. W. *Density Estimation for Statistics and Data Analysis*. Chapman and Hall, New York, 1986.

Sivia, D. S. *Data Analysis: A Bayesian Tutorial*. Oxford University Press, Oxford, 1996.

Slingerland, R. and Kump, L. *Mathematical Modeling of Earth's Dynamical Systems*. Princeton University Press, Princeton, 2011.

Smith, L. Spatial variability of flow parameters in a stratified sand. *Mathematical Geology*, 13(1):1–21, 1981.

Soleng, H. H. Oil reservoir production forecasting with uncertainty estimation using genetic algorithms. *Proceedings of the 1999 Congress of Evolutionary Computing*, 1999.

Sprague, A. R., Patterson, P. E., Hill, R. E., Jones, C. R., Campion, K. M., Wagoner, J. C. V., Sullivan, M. D., Larue, D. K., Feldman, H. R., Demko, T. M., Wellner, R. W., and Geslin, J. K. The physical stratigraphy of fluvial strata: a hierarchical approach to the analysis of genetically related stratigraphic elements for improved reservoir prediction. In *(Abstract) AAPG Annual Meeting*, volume 87, page 10. AAPG Bulletin, 2002.

Srivastava, R. M. Minimum variance or maximum profitability? *CIM Bulletin*, 80(901):63–68, 1987a.

Srivastava, R. M. A non-ergodic framework for variogram and covariance functions. Master's thesis, Stanford University, Stanford, CA, 1987b.

Srivastava, R. M. An application of geostatistical methods for risk analysis in reservoir management. SPE Paper # 20608, 1990a.

Srivastava, R. M. *INDSIM2D: An FSS International Training Tool*. FSS International, Vancouver, Canada, 1990b.

Srivastava, R. M. Iterative methods for spatial simulation: stanford center for reservoir forecasting. Stanford Center for Reservoir Forecasting: Report Number 5, 1992.

Srivastava, R. M. Matheronian geostatistics: where is it going? In Baafi, E. Y. and Schofield, N. A., editors, *Fifth International Geostatistics Congress*, Wollongong, Australia, September 1996.

Srivastava, R. M. and Froidevaux, R. Probability field simulation: a retrospective. In *SPE Annual Technical Conference and Exhibition*, pages 55–64. Springer, 2005.

Srivastava, R. M. and Parker, H. M. Robust measures of spatial continuity. In Armstrong, M., editor, *Geostatistics*, pages 295–308. Reidel, Dordrecht, 1989.

Stalkup, F. I. Permeability variations observed at the faces of crossbedded sandstone outcrops. In Lake, L. W. and Carroll, H. B., editors, *Reservoir Characterization*, pages 141–175. Academic Press, New York, 1986.

Stanley, K. O., Jorde, K., Raestad, N., and Stockbridge, C. P. Stochastic modeling of reservoir sand bodies for input to reservoir simulation, Snorre Field, Northern North Sea. In Buller, A. T., editor, *North Sea Oil and Gas Reservoirs II*. Graham and Trotman, London, 1990.

Stoyan, D., Kendall, W. S., and Mecke, J. *Stochastic Geometry and its Applications*. John Wiley & Sons, New York, 1987.

Straub, K. M., Mohrig, D., McElroy, B., Buttles, J., and Pirmez, C. Interactions between turbidity currents and topography in aggrading sinuous submarine channels: a laboratory study. *Geologic Society of America Bulletin*, 120(3–4):368–385, 2008.

Straubhaar, J., Renard, P., Mariethoz, G., Froidevaux, R., and Besson, O. An improved parallel multiple-point algorithm using a list approach. *Mathematical Geosciences*, 43(3):305–328, 2010.

Strebelle, S. *Sequential Simulation: Drawing Structures from Training Images*. PhD thesis, Stanford University, Stanford, CA, 2000.

Strebelle, S. Conditional simulation of complex geological structures using multiple-point statistics. *Mathematical Geology*, 34(1):1–21, 2002.

Strebelle, S. Sequential simulation for modeling geological structures from training images. In M., Y. J. and L., C. R., editors, *Stochastic Modelling and Geostatistics: Principles, Methods and Case Studies*, volume 2, pages 139–149, 2006.

Strebelle, S. Multiple-point geostatistics: from theory to practice. *9th International Geostatistics Congress*, 2012.

Strebelle, S. and Remy, N. Post-processing of multiple-point geostatistical models to improve reproduction of training patterns. In Leuangthong, O. and Deutsch, C. V., editors, *Geostatistics Banff 2004*, volume 14 of *Quantitative Geology and Geostatistics*, pages 979–988. Springer Netherlands, Dordrecht, 2005.

Stright, L. Modeling, upscaling and history matching thin, irregularly-shaped flow barriers; a comprehensive approach for predicting reservoir connectivity. In *SPE Annual Technical Conference and Exhibition*, number SPE Paper # 106528, San Antonio, USA, 2006. Society of Petroleum Engineers.

Stripe, J. A., Kazuyoshi, A., and Durandeau, M. Integrated field development planning using risk and and decision analysis to minimize the impact of reservoir and other uncertainties: a case study. In *Middle East Oil Technical Conference and Exhibition*, pages 155–167. Society of Petroleum Engineers, April 1993. SPE Paper # 25529.

Sullivan, J. A. *Non-parametric Estimation of Spatial Distributions*. PhD thesis, Stanford University, Stanford, CA, 1985.

Sullivan, M. D., Foreman, J. L., Jennette, D. C., Stern, D., Jensen, G. N., and Goulding, F. J. An integrated approach to characterization and modeling of deep-water reservoirs, Diana Field. In P. Harris, G. E. and Grammer, M., editors, *AAPG Memoir 80: integration of Outcrop and Modern Analogs in Reservoir Modeling*, pages 215–234, Western Gulf of Mexico, 2004. British Society of Reservoir Geologists.

Sun, T., Meakin, P., and Josang, T. A simulation model for meandering rivers. *Water Resources Research*, 32(9), 1996.

Suzuki, S. and Strebelle, S. Real-time post-processing method to enhance multiple-point statistics simulation. *Petroleum Geostatistics 2007*, 2007.

Suzuki, S., Caumon, G., and Caers, J. Dynamic data integration for structural modeling: model screening approach using a distance-based model parameterization. *Computational Geosciences*, 12(1):105–119, 2008.

Sylvester, Z., Pirmez, C., and Cantelli, A. A model of submarine channel-levee evolution based on channel trajectories: implications for stratigraphic architecture. *Marine and Petroleum Geology*, 2010. doi: 10.1016/j.marpetgeo.2010.05.012.

Syvitski, J. Earth-surface dynamics modeling & model coupling course. Community Surface Modeling Dynamics Systems, 2012. URL http://csdms.colorado.edu/wiki/Earth-Surface_Dynamics_Modeling.

Tetzlaff, D. M. Limits to the predictive ability of dynamic models the simulation clastic sedimentation. In Cross, T., editor, *Quantitative Dynamic Stratigraphy*, pages 55–65. Prentice-Hall, Netherlands, 1990.

Thomas, R. G., Smith, D. G., Wood, J. M., Visser, J., Calverley-Range, E. A., and Koster, E. H. Inclined heterolithic stratification-terminology, description, interpretation and significance. *Sedimentary Geology*, 53(1–2):123–179, 1987.

Tjelmeland, H. and Omre, H. Semi-Markov random fields. In Soares, A., editor, *Geostatistics Troia 1992*, volume 2, pages 493–504. Kluwer, Dordrecht, 1993.

Torres-Verdin, C., Victoria, M., Merletti, G., and Pendrel, J. Trace-based and geostatistical inversion of 3-D seismic data for thin sand delineation: an application to san Jorge Basin, Argentina. *The Leading Edge*, 40, September 1999.

Traer, M. M., Hilley, G. E., Fildani, A., and McHargue, T. The sensitivity of turbidity currents to mass and momentum exchanges between these underflows and their surroundings. *Journal of Geophysical Research*, 117, 2012.

Tran, T., Deutsch, C. V., and Yulong, X. Direct geostatistical simulation with multiscale well, seismic, and production data. In *SPE Annual Technical Conference and Exhibition*, New Orleans, 2001.

Tran, T. T. Improving variogram reproduction on dense simulation grids. *Computers & Geosciences*, 20(7):1161–1168, 1994.

Tran, T. T. *Stochastic Simulation of Permeability Fields and Their Scale-Up for Flow Modeling*. PhD thesis, Stanford University, Stanford, CA, 1995.

Tran, T. T. The missing scale and direct simulation of block effective properties. *Journal of Hydrology*, 182:37–56, 1996.

Tversky, A. and Kahneman, D. Judgment under uncertainty: heuristics and biases. *Science*, 185(4157):1124–1131, 1974.

Tyler, K., Henriquez, A., Georgsen, F., Holden, L., and Tjelmeland, H. A program for 3D modeling of heterogeneities in a fluvial reservoir. In *3rd European Conference on the Mathematics of Oil Recovery*, pages 31–40, Delft, June 1992a.

Tyler, K., Henriquez, A., MacDonald, A., Svanes, T., and Hektoen, A. L. MOHERES—a collection of stochastic models for describing heterogeneities in clastic reservoirs. In *3rd International Conference on North Sea Oil and Gas Reservoirs III*, pages 213–221. 1992b.

Tyler, K., Svanes, T., and Henriquez, A. Heterogeneity modelling used for production simulation of fluvial reservoir. *SPE Formation Evaluation*, pages 85–92, June 1992c.

Tyler, K., Henriquez, A., and Svanes, T. Modeling heterogeneities in fluvial domains: a review on the influence on production profile. In Yarus, J. M. and Chambers, R. L., editors, *Stochastic Modeling and Geostatistics: Principles, Methods, and Case Studies*, pages 77–89. AAPG Computer Applications in Geology, No. 3, 1995.

Vasantharajan, S. and Cullick, A. S. Well site selection using integer programming optimization. In *Third Annual Conference*, volume 1, pages 421–426, International Association for Mathematical Geology. Vera Pawlowsky Glahn (ed.), CIMNE Press, Barcelona, Spain, September 1997.

Verly, G. *Estimation of Spatial Point and Block Distributions: The MultiGaussian Model*. PhD thesis, Stanford University, Stanford, CA, 1984.

Villalba, M. E. and Deutsch, C. V. Computing uncertainty in the mean with a stochastic trend approach. Technical Report 109, CCG Annual Report 12, Edmonton, AB, 2010.

Vincent, G., Corre, B., and Thore, P. Managing structural uncertainty in a mature field for optimal well placement. In *1990 SPE Annual Technical Conference and Exhibition*. Society of Petroleum Engineers, October 1990. SPE Paper # 48953.

Viseur, S., Shtuka, A., and Mallet, J.-L. New fast, stochastic, boolean simulation of fluvial deposits. In *SPE Annual Technical Conference and Exhibition*, New Orleans, LA, 1998. Society of Petroleum Engineers.

Wackernagel, H. Geostatistical techniques for interpreting multivariate spatial information. In Chung, C., editor,

Quantitative Analysis of Mineral and Energy Resources, pages 393–409. Reidel, Dordrecht, 1988.

Wagoner, J. C. V., Mitchum, R. M., Campion, K. M., and Rahmanian, V. D. *Siliclastic Sequence Stratigraphy in Well Logs, Cores, and Outcrops: Concepts for High-Resolution Correlation of Time Facies*. The American Association of Petroleum Geologists, Tulsa, Oklahoma, 1990.

Walker, R. G. and James, N. P. *Facies Models: Response to Sea Level Changes*. Geologic Association of Canada, pages 454, 1992.

Wang, F. J. and Wall, M. A. Incorporating parameter uncertainty into prediction intervals for spatial data modeled via a parametric variogram. *Journal of Agricultural, Biological, and Environmental Statistics*, 8(3):296–309, 2003.

Wang, Y., Straub, K. M., and Hajek, E. A. Scale dependant compensational stacking: an estimate of autogenic timescales in channelized sedimentary deposits. *Geology*, 39(9):811–814, 2011.

Wasson, R. J. Last-glacial alluvial fan sedimentation in the Lower Derwent Valley, Tasmania. *Sedimentology*, 24(6):781–799, 1977.

Watson, A. T., Seinfelf, J. H., Gavalas, G. R., and Woo, P. T. History matching in two-phase petroleum reservoirs. *SPE Journal*, pages 521–532, December 1980.

Webb, E. K. *Simulating the Spatial heterogeneity of Sedimentological and Hydrogeological Characteristics for Braided Stream Deposits*. PhD thesis, University of Wisconsin, Madison, Madison, WI, 1992.

Weber, K. J. Influence of common sedimentary structures on fluid flow in reservoir models. *JPT*, pages 665–672, March 1982.

Weber, K. J. and dL. C. Van Geuns. Framework for constructing clastic reservoir simulation models. *JPT*, pages 1248–1253, 1296–1297, October 1990.

Weber, L. J., Francis, B. P., Harris, P. M., and Clark, M. Stratigraphy, lithofacies and reservoir distribution, Tengiz Field, Kazakhstan. *Permo-Carboniferous Carbonate Platform and Reefs: SEPM special publication*, 78:351–394, 2003.

Wen, R. SBED studio: an integrated workflow solution for multi-scale geo modelling. In *European Association of Geoscientists and Engineers 67th Conference*, Madrid, 2005.

Wen, X. H. and Gómez-Hernández, J. J. Upscaling of hydraulic conductivity in heterogeneous media: an overview. *Journal of Hydrology*, 183:ix–xxxii, 1996.

White, C. D. and Willis, B. J. A method to estimate length distributions from outcrop data. *Mathematical Geology*, 32(4):389–419, 2000.

Wietzerbin, L. J. and Mallet, J.-L. Parameterization of complex 3D heterogeneities: a new CAD approach. *SPE Annual Technical Conference and Exhibition, Society of Petroleum Engineers*, 1993.

Wikramaratna, R. S. ACORN—a new method for generating sequences of uniformly distributed pseudorandom numbers. *Journal of Computational Physics*, 83:16–31, 1989.

Wilde, B. Application of spectral techniques to geostatistical modeling. Technical Report 108, Centre for Computational Geostatistics (CCG) Guidebook Series, University of Alberta, 2010.

Wilde, B. Programs to aid the decision of stationarity. volume 14, Edmonton, AB, February 2011. University of Alberta, Centre for Computational Geostatistics Guidebook Series.

Wilde, B. and Deutsch, C. V. Formats for expressing acceptable uncertainty. Technical Report 298, CCG Annual Report 12, Edmonton, AB, 2010a.

Wilde, B. and Deutsch, C. V. Methodology for calculating uncertainty versus data spacing. Technical Report 108, CCG Annual Report 12, Edmonton, AB, 2010b.

Wilde, B. and Deutsch, C. V. Simulating boundary realizations. Technical Report 403, CCG Annual Report 13, Edmonton, AB, 2011.

Williams, G. J. J., Mansfield, M., MacDonald, D., and Bush, M. D. Top-down reservoir modelling. *SPE Paper 89974 presented at the ATCE 2004*, 39(9):26–29, 2004.

Willis, B. J. and Tang, H. Three-dimensional connectivity of point-bar deposits. *Journal of Sediment Research*, 80:440–454, 2010.

Willis, B. J. and White, C. D. Quantitative outcrop data for flow simulation. *Journal of Sediment Research*, 70:788–802, 2000.

Winkler, R. L. Combining probability distributions from dependent information sources. *Management Science*, 27(4), 1981.

Wu, J., Zhang, T., and Journel, A. Fast FILTERSIM simulation with score-based distance. *Mathematical Geosciences*, 40(7):773–788, 2008.

Xie, Y., Deutsch, C. V., and Cullick, A. S. Surface-geometry and trend modeling for integration of stratigraphic data in reservoir models. In G., K. W. J. . K. D., editor, *GEOSTATS 2000: Cape Town, Proceedings of the 6th International Geostatistics Congress*, Cape Town, South Africa, April 2000.

Xu, W. *Stochastic Modeling of Reservoir Lithofacies and Petrophysical Properties*. PhD thesis, Stanford University, Stanford, CA, 1995.

Xu, W. and Journel, A. G. GTSIM: Gaussian truncated simulations of reservoir units in a West Texas carbonate field. SPE Paper # 27412, 1993.

Xu, W. and Journel, A. G. DSSIM: a general sequential simulation algorithm. In *Report 7, Stanford Center for Reservoir Forecasting*, Stanford, CA, May 1994.

Xu, W. and Journel, A. G. Histogram and scattergram smoothing using convex quadratic programming. *Mathematical Geology*, 27:83–103, 1995.

Xu, W., Tran, T. T., Srivastava, R. M., and Journel, A. G. Integrating seismic data in reservoir modeling: the collocated cokriging alternative. In *67th Annual Technical Conference and Exhibition*, pages 833–842, Washington, DC, October 1992. Society of Petroleum Engineers. SPE Paper # 24742.

Yang, C. T., Chopra, A. K., and Chu, J. Integrated geostatistical reservoir description using petrophysical, geological, and seismic data for Yacheng 13-1 gas field. In *1995 SPE Annual Technical Conference and Exhibition Formation Evaluation and Reservoir Geology*, pages 357–372, Dallas, TX, October 1995. Society of Petroleum Engineers. SPE Paper # 30566.

Yao, T. Conditional spectral simulation with phase identification. *Mathematical Geology*, 30(3):285–308, 1998.

Yao, T. Reproduction of the mean, variance, and variogram model in spectral simulation. *Mathematical Geology*, 36(4):487–506, 2004.

Yao, T. and Journel, A. G. Automatic modeling of (cross) covariance tables using fast fourier transform. *Mathematical Geology*, 30(6):589–615, 1998.

Yarus, J. and Chambers, R. Practical geostatistics—an armchair overview for petroleum reservoir engineers. *Journal of Petroleum Technology*, 58(11):78–86, 2006.

Yeh, W. W.-G. Review of parameter identification procedures in groundwater hydrology: the inverse problem. *Water Resources Research*, 22(2):95–108, 1986.

Zagayevskiy, Y., Hosseini, A., and Deutsch, C., editors. *Ensemble Kalman Filtering for Geostatistical Applications*, volume 10, Edmonton, AB, 2010. University of Alberta, Centre for Computational Geostatistics Guidebook Series.

Zarra, L. Chronostratigraphic framework for the Wilcox Formation (Upper Paleocene–Lower Eocene) in the deep-water Gulf of Mexico: biostratigraphy, sequences, and depositional systems. In *The Paleogene of the Gulf of Mexico and Caribbean basins: processes, events, and petroleum systems: Gulf Coast Section SEPM 27th Annual GCSSEPM Foundation Bob F. Perkins Research Conference Proceedings*, pages 81–145, Houston, TX, 2007. SEPM.

Zhang, K., Pyrcz, M. J., and Deutsch, C. V. Stochastic surface-based modeling for integration of geological information in turbidite reservoir model. *Petroleum Geoscience and Engineering*, 2009. doi: j.petrol.2009.06.019.

Zhang, T., Switzer, P., and Journel, A. G. Sequential conditional simulaiton using classification of local training pattern. In Leuangthong, O. and Deutsch, C. V., editors, *Geostatistics Banff 2004*, volume 1, pages 265–273. Springer Netherlands, Dordrecht, 2005.

Zhang, T., Switzer, P., and Journel, A. G. Filter-based classification of training image patterns for spatial simulation. *Mathematical Geology*, 38(1):63–80, 2006. doi: 10.1007/s11004-005-9004-x.

Zhu, H. *Modeling Mixture of Spatial Distributions with Integration of Soft Data*. PhD thesis, Stanford University, Stanford, CA, 1991.

Zhu, H. and Journel, A. G. Formatting and integrating soft data: stochastic imaging via the Markov–Bayes algorithm. In Soares, A., editor, *Geostatistics Troia 1992*, volume 1, pages 1–12. Kluwer, Dordrecht, 1993.

INDEX